Laser Processing and Chemistry

Springer
*Berlin
Heidelberg
New York
Barcelona
Budapest
Hong Kong
London
Milan
Paris
Santa Clara
Singapore
Tokyo*

Dieter Bäuerle

Laser Processing and Chemistry

Second Edition

With 262 Figures

 Springer

Professor Dr. Dieter Bäuerle
Institut für Angewandte Physik
Johannes-Kepler-Universität Linz
A-4040 Linz, Austria

Title of the first edition: *Chemical Processing with Lasers,*
Springer Series in Material Sciences, Vol. 1

ISBN 3-540-60541-X 2nd Edition
Springer-Verlag Berlin Heidelberg New York

ISBN 3-540-17147-9 1st Edition
Springer-Verlag Berlin Heidelberg New York

Library of Congress Cataloging-in-Publication Data. Bäuerle, D. (Dieter), 1940– Laser processing and chemistry/Dieter Bäuerle. p. cm. Includes bibliographical references. ISBN 3-540-60541-X (hardcover:alk. paper) 1. Laser-Industrial applications. 2. Surfaces–Effect of radiation on. 3. Materials–Effect of radiation on. 4. Surface chemistry–effect of radiation on. I. Title. TA1677.B39 1996 660'.293'028-dc20 96-95

This work is subject to copyright. All rights are reserved, whether the whole or part of the material is concerned, specifically the rights of translation, reprinting, reuse of illustrations, recitation, broadcasting, reproduction on microfilm or in any other ways, and storage in data banks. Duplication of this publication or parts thereof is permitted only under the provisions of the German Copyright Law of September 9, 1965, in its current version, and permission for use must always be obtained from Springer-Verlag. Violations are liable for prosecution under the German Copyright Law.

© Springer-Verlag Berlin Heidelberg 1986, 1996
Printed in Germany

The use of general descriptive names, registered names, trade marks, etc. in this publication, does not imply, even in the absence of a specific statement, that such names are exempt from the relevant protective laws and regulations and therefore free for general use.

Cover design: Springer-Verlag, Design & Production
Typesetting: Thomson Press (India) Ltd., Madras
SPIN: 10518209   54/3144/SPS – 5 4 3 2 1 0 – Printed on acid-free paper

For Anne, Christoph, and Horst

Preface

Materials processing with lasers is an expanding field which is captivating the attention of scientists, engineers, and manufacturers alike. The aspect of most interest to scientists is the basic interaction mechanisms between the intense light of a laser and materials exposed to a chemically reactive or nonreactive surrounding medium. Engineers and manufacturers see in the laser a tool which will not only make manufacturing cheaper, faster, cleaner, and more accurate but also open up entirely new technologies and manufacturing methods that are simply not available using standard techniques. The most established applications are laser machining (cutting, drilling, shaping) and laser welding. Increasingly, however, lasers are also being used for surface hardening, annealing, and glazing. Laser *chemical* processing (micro-patterning and extended-area processing by laser-induced etching, material deposition, chemical transformation, etc.) has actual and potential applications in micromechanics, metallurgy, integrated optics, semiconductor manufacture, and chemical engineering.

This book concentrates on various aspects of laser–matter interactions, in particular with regard to laser material processing. Special attention is given to laser-induced chemical reactions at gas–, liquid–, and solid–solid interfaces. The intention is to give scientists, engineers, and manufacturers an overview of the extent to which new developments in laser processing are understood at present, of the various new possibilities, and of the limitations of laser techniques. Students may prefer to read the book selectively, not troubling themselves unduly with detailed calculations or descriptions of single processes.

The book is divided into six parts, each of which consists in turn of several chapters. The main symbols, conversion factors, abbreviations, acronyms, and mathematical functions and relations used throughout the text are listed in Appendices A.1–A.3. The different materials investigated are listed in Appendices B.1–B.10. These give readers a quick and comprehensive overview of the "state of the art" and direct them to the original literature of a particular area of interest. Tables I–V are intended to encourage the reader to use the formulas presented for rapid estimation of various quantities. An extensive subject index can be found at the end of the book.

I wish to thank my students and all my staff for valuable discussions and critical reading of various parts of the manuscript. I am deeply indebted to Dr. N. Arnold and Prof. B. Luk'yanchuk for many suggestions for improvements to the manuscript. Last but not least, I wish to express my deep gratitude to my secretary, Dipl.-Ing. Irmengard Haslinger, for her tireless assistance in writing this book.

Linz, July 1995   *Dieter Bäuerle*

Contents

Part I: Overview and Fundamentals 1

1 Introduction 3
   1.1 Conventional Laser Processing 4
   1.2 Laser Chemical Processing 7

2 Thermal, Photophysical, and Photochemical Processes . . . 13
   2.1 Excitation Mechanisms, Relaxation Times 13
   2.2 The Heat Equation 19
      2.2.1 The Source Term 19
      2.2.2 Dimensionality of Heat Flow 21
      2.2.3 Kirchhoff and Crank Transforms 21
      2.2.4 Phase Changes 22
      2.2.5 Limits of Validity 23
   2.3 Selective Excitations of Molecules 25
      2.3.1 Electronic Excitations 26
      2.3.2 Infrared Vibrational Excitations 29
   2.4 Surface Excitations 34
      2.4.1 External Photoeffect 34
      2.4.2 Internal Photoeffect 35
      2.4.3 Electromagnetic Field Enhancement,
           Catalytic Effects 37
      2.4.4 Adsorbed Molecules 37

3 Reaction Kinetics and Transport of Species 39
   3.1 Photothermal Reactions 41
   3.2 Photochemical Reactions 43
   3.3 The Concentration of Species 44
      3.3.1 Basic Equations 45
      3.3.2 Dependence of Coefficients on Temperature
           and Concentration 49
   3.4 Heterogeneous Reactions 51
      3.4.1 Stationary Equations 52

3.4.2 Transport Limitations	53
3.4.3 Dynamic Solutions	57
3.4.4 Heterogeneous Versus Homogeneous Activation	58
3.5 Combined Heterogeneous and Homogeneous Reactions	59
3.5.1 The Boundary-Value Problem	59
3.5.2 Approximate Solutions	60
3.6 Homogeneous Photochemical Activation	61

4 Nucleation and Cluster Formation ... 62

4.1 Homogeneous Processes	62
4.1.1 Classical Kinetics	63
4.1.2 Droplets within a Laser Beam	65
4.1.3 Transport of Clusters, Thermophoresis, Chemophoresis	68
4.2 Heterogeneous Processes	69
4.2.1 Nucleation in Laser-CVD	69
4.2.2 Coalescence	71
4.2.3 Liquid-Solid and Solid-Solid Interfaces	72

5 Lasers, Experimental Aspects, Spatial Confinement ... 73

5.1 Lasers	73
5.1.1 CW Lasers, Gaussian Beams	73
5.1.2 Pulsed and High-Power CW Lasers	75
5.2 Experimental Aspects	76
5.2.1 Microprocessing	77
5.2.2 The Reaction Chamber; Typical Setup	79
5.2.3 Large-Area Processing	80
5.2.4 Substrates	81
5.3 Confinement of the Excitation	82
5.3.1 The Thermal Field	83
5.3.2 Non-thermal Substrate Excitations	83
5.3.3 Gas-, Liquid- and Adsorbed-Phase Excitations	83
5.3.4 Plasma Formation	84
5.3.5 Material Damages	84
5.3.6 Nonlinearities	84

Part II: Temperature Distributions and Surface Melting ... 89

6 General Solutions of the Heat Equation ... 91

6.1 The Boundary-Value Problem	91
6.1.1 The Attenuation Function $f(z)$	92

6.1.2 Boundary and Initial Conditions 93
6.2 Analytical Solutions 96
6.3 Pulse Shapes 98
6.4 Beam Shapes 101
6.5 Characteristics of Temperature Distributions 102
6.6 Numerical Techniques 104

7 Semi-infinite Substrates 106

7.1 The Center Temperature Rise 106
7.2 Stationary Solutions
for Temperature-Independent Parameters 108
  7.2.1 Surface Absorption 109
  7.2.2 Finite Absorption 110
7.3 Stationary Solutions
for Temperature-Dependent Parameters 113
7.4 Scanned CW Laser 115
7.5 Pulsed-Laser Irradiation 117
  7.5.1 Gaussian Intensity Profile 117
  7.5.2 Uniform Irradiation 119
7.6 Dynamic Solutions
for Temperature-Dependent Parameters 120

8 Infinite Slabs 124

8.1 Strong Absorption 124
  8.1.1 Thermally Thin Film 124
  8.1.2 Scanned CW Laser 125
8.2 The Influence of Interferences 128
8.3 Coupling of Optical and Thermal Properties 130
8.4 Average Temperature Distributions 131

9 Non-uniform Media 132

9.1 Continuous Changes in Optical Properties 132
9.2 Absorption of Light in Multilayer Structures 133
  9.2.1 Thin Films 134
  9.2.2 Two-Layer Structures 136
  9.2.3 Three-Layer Systems 137
9.3 Temperature Distributions for Large-Area Irradiation 137
  9.3.1 Stationary Solutions for Thin Films 137
  9.3.2 Dynamic Solutions 139
9.4 Temperature Distributions for Focused Irradiation . 140
  9.4.1 Strong Film Absorption 141
  9.4.2 Finite Film Absorption 142
9.5 The Ambient Medium 144

9.5.1	Influence on Substrate Temperature	144
9.5.2	Indirect Heating	146
9.5.3	Free Convection	147
9.5.4	Temperature Jump	150

10 Surface Melting 152

10.1	Temperature Distributions, Interface Velocities	152
	10.1.1 Boundary Conditions	156
	10.1.2 Temperature Dependence of Parameters	160
10.2	Solidification	160
10.3	Process Optimization	164
10.4	Convection	164
10.5	Surface Deformations	166
10.6	Welding	167
10.7	Liquid-Phase Expulsion	169

Part III: Material Removal 171

11 Vaporization, Plasma Formation 173

11.1	Vaporization	173
	11.1.1 Energy Balance	175
	11.1.2 One-Dimensional Model	175
	11.1.3 Minimum Intensity	177
	11.1.4 The Recoil Pressure	178
	11.1.5 Influence of a Liquid Layer	179
	11.1.6 Limitations of Model Calculations	180
11.2	Plasma Formation	181
	11.2.1 Optical Properties of Plasmas, Energy Coupling	182
	11.2.2 Laser-Supported Combustion Waves (LSCW): $I_p \leq I \leq I_d$	184
	11.2.3 Laser-Supported Detonation Waves (LSDW): $I \geq I_d$	185
	11.2.4 Superdetonation	186
11.3	Abrasive Laser Machining	187
	11.3.1 Cutting, Drilling, Shaping	187
	11.3.2 Comparison of Techniques	189
	11.3.3 Non-metals	190

12 Pulsed-Laser Ablation 191

12.1	Surface Patterning	192
12.2	Interactions Below Threshold	195
12.3	The Threshold Fluence ϕ_{th}	196

12.4	Ablation Rates	198
	12.4.1 Dependence on Pulse Number	199
	12.4.2 Dependence on Fluence	200
	12.4.3 Influence of Spot Size	202
	12.4.4 Time-Resolved Dynamics	203
12.5	Material Damage, Localization of Excitation Energy	204
	12.5.1 Strong Absorption	204
	12.5.2 Finite Absorption	206
12.6	Influence of an Ambient Atmosphere	207
	12.6.1 Debris	207

13 Modelling of Pulsed-Laser Ablation ... 209

13.1	Model	210
13.2	Photothermal Ablation	213
	13.2.1 Stationary Conditions	214
	13.2.2 The Regime $t < t_v$	215
	13.2.3 Average Ablation Velocity	215
13.3	Photophysical Ablation	217
	13.3.1 Stationary Solutions	219
	13.3.2 Non-Stationary Ablation	221
13.4	Photochemical Ablation	223
	13.4.1 Dissociation of Polymer Chains	224
	13.4.2 Defect-Related Processes, Incubation	224
13.5	Thermo- and Photomechanical Ablation	226

14 Etching of Metals and Insulators ... 229

14.1	Photochemistry of Precursor Molecules	230
	14.1.1 Halides	231
	14.1.2 Halogen Compounds	232
14.2	Concentration of Reactive Species	234
	14.2.1 Ballistic Approximation	235
	14.2.2 Diffusion	236
	14.2.3 Influence of Reaction Chamber	237
	14.2.4 Gas-Phase Recombination	239
	14.2.5 Gas-Phase Heating	241
14.3	Dry-Etching of Metals	241
	14.3.1 Spontaneous Etching Systems	241
	14.3.2 Diffusive Etching Systems	242
	14.3.3 Passivating Reaction Systems	243
14.4	Dry-Etching of Inorganic Insulators	244
	14.4.1 SiO_2 Glasses	245
	14.4.2 Oxides	247
14.5	Wet-Etching	249

15 Etching of Semiconductors ... 251

- 15.1 Dark Etching ... 251
- 15.2 Laser-Induced Etching of Si in Cl_2 ... 254
  - 15.2.1 Surface Patterning ... 254
  - 15.2.2 Photochemical and Thermal Etching ... 255
  - 15.2.3 Chlorine Radicals ... 257
  - 15.2.4 Electron-Hole Pairs ... 258
  - 15.2.5 Crystal Orientation and Doping ... 260
- 15.3 Si in Halogen Compounds ... 261
  - 15.3.1 Si in XeF_2 ... 261
  - 15.3.2 Si in SF_6 ... 262
- 15.4 Microscopic Mechanisms ... 264
- 15.5 Dry-Etching of Compound Semiconductors ... 265
  - 15.5.1 III-V Compounds ... 266
  - 15.5.2 Laser Etching of Atomic Layers ... 267
  - 15.5.3 Dopants, Impurities, and Defects ... 267
- 15.6 Wet-Etching ... 268
  - 15.6.1 Silicon ... 269
  - 15.6.2 Compound Semiconductors ... 269
  - 15.6.3 Interpretation of Results ... 271
  - 15.6.4 Spatial Resolution, Waveguiding ... 274

Part IV: Material Deposition ... 279

16 Laser-CVD of Microstructures ... 281

- 16.1 Precursor Molecules ... 281
- 16.2 Pyrolytic LCVD of Spots ... 282
  - 16.2.1 Deposition from Halides ... 282
  - 16.2.2 Deposition from Carbonyls ... 286
- 16.3 Modelling of Pyrolytic LCVD ... 287
  - 16.3.1 Gas-Phase Processes ... 288
  - 16.3.2 The Coupling Between $T(x)$ and $h(x)$ ... 291
- 16.4 Temperature Distributions on Circular Deposits ... 293
- 16.5 Simulation of Pyrolytic Growth ... 296
- 16.6 Photolytic LCVD ... 298
  - 16.6.1 Metals ... 300
  - 16.6.2 Other Materials ... 302
  - 16.6.3 Process Limitations ... 303

17 Growth of Fibers ... 304

- 17.1 In Situ Temperature Measurements ... 305
- 17.2 Microstructure and Physical Properties ... 306
- 17.3 Kinetic Studies ... 307

17.4 Gas-Phase Transport 309
    17.4.1 The Coupling of Fluxes 309
    17.4.2 Thermal Diffusion (Soret Effect) 312
17.5 Simulation of Growth 315

18 Direct Writing 317

18.1 Characteristics of Pyrolytic Direct Writing 317
    18.1.1 Dependence on Laser Parameters
           and Substrate Material 318
    18.1.2 Electrical Properties 320
18.2 Temperature Distributions in Direct Writing . . . 321
    18.2.1 Center Temperature Rise 322
    18.2.2 One-Dimensional Approach, $\kappa^* \gg 1$ 323
    18.2.3 Numerical Solutions 324
18.3 Simulation of Direct Writing 325
    18.3.1 One-Dimensional Model 326
    18.3.2 Comparison with Experimental Data 327
    18.3.3 Two-Dimensional Model 331
18.4 Photophysical LCVD 332
18.5 Applications of LCVD in Microfabrication 334
    18.5.1 Planar Substrates 334
    18.5.2 Non-planar Substrates, 3-D Objects 336

19 Thin-Film Formation by Laser-CVD 338

19.1 Direct Heating 339
    19.1.1 Stationary Solutions 339
    19.1.2 Non-stationary Solutions 342
19.2 Pyrolytic Processing Rates 343
    19.2.1 Diffusion 344
    19.2.2 Recombination 346
19.3 Photolytic Processing Rates 347
19.4 Metals 348
    19.4.1 Deposition from Metal Halides 348
    19.4.2 Deposition from Alkyls and Carbonyls . . . 350
19.5 Semiconductors 352
    19.5.1 Photodecomposition of Silanes 352
    19.5.2 Crystalline Ge and Si 354
    19.5.3 Amorphous Hydrogenated Silicon (a-Si:H) . 354
    19.5.4 Compound Semiconductors 359
    19.5.5 Carbon 361
19.6 Insulators 362
    19.6.1 Oxides 362
    19.6.2 Nitrides 365
19.7 Heterostructures 366

19.8 Comparison of Laser-CVD
and Standard Techniques 368

20 Adsorbed Layers, Laser-MBE 370

20.1 Fundamental Aspects 371
20.2 Deposition from Adsorbed Layers 375
    20.2.1 Vacuum 376
    20.2.2 Gaseous Ambient 378
20.3 Combined Laser and Molecular/Atomic Beams .. 383
    20.3.1 Laser-MBE 383
    20.3.2 Laser-ALE 385

21 Liquid-Phase Deposition, Electroplating 387

21.1 Liquid-Phase Processing Without External EMF . 387
    21.1.1 Thermal Decomposition 387
    21.1.2 Electroless Plating 389
    21.1.3 Metal-Liquid Interfaces 390
    21.1.4 Semiconductor-Liquid Interfaces 392
    21.1.5 Further Experimental Examples 393
21.2 Electrochemical Plating 393

22 Thin-Film Formation by Pulsed-Laser Deposition and Laser-Induced Evaporation 397

22.1 Experimental Requirements 398
    22.1.1 Congruent and Incongruent Ablation 399
    22.1.2 Targets 401
22.2 Volume and Surface Processes, Film Growth ... 402
    22.2.1 Plasma and Gas-Phase Reactions 403
    22.2.2 Substrate Temperature,
            Laser-Pulse Repetition Rate 404
    22.2.3 Energetic Species 405
    22.2.4 Particulates 405
    22.2.5 Chemical Composition of Films 410
22.3 Overview of Materials and Film Properties 410
    22.3.1 Metal Targets 411
    22.3.2 Semiconductors 411
    22.3.3 Diamond-Like Carbon 412
    22.3.4 Insulators 413
22.4 High-Temperature Superconductors 414
    22.4.1 Non-reactive Deposition 415
    22.4.2 Reactive Deposition 416
    22.4.3 Buffer Layers, Technological Aspects 418
    22.4.4 Heterostructures 419

22.5 Metastable Compounds, Mixed Systems 419
22.6 Laser-Induced Forward Transfer 420

Part V: Surface Transformations, Synthesis and Structure Formation 423

23 Structural Transformations 425

23.1 Transformation Hardening 425
23.2 Laser Annealing, Recrystallization 427
    23.2.1 Ion-Implanted Semiconductors 427
    23.2.2 Thin Films 429
23.3 Glazing 430
23.4 Shock Hardening 430

24 Doping 432

24.1 Solid-Phase Diffusion 432
24.2 Liquid-Phase Transport 435
24.3 Sheet Doping 436
    24.3.1 Silicon 436
    24.3.2 Compound Semiconductors 440
24.4 Local Doping 440

25 Cladding, Alloying, and Synthesis 442

25.1 Laser-Assisted Cladding 442
25.2 Alloying 443
    25.2.1 Laser-Surface Alloying 443
    25.2.2 Formation of Metastable Materials 444
    25.2.3 Silicides 444
25.3 Synthesis 445
    25.3.1 Thin Films 446
    25.3.2 Fibers 447
    25.3.3 Polymerization, Waveguides 449

26 Oxidation, Nitridation 450

26.1 Basic Mechanisms 451
26.2 Metals 456
    26.2.1 Photothermal Oxidation 456
    26.2.2 Photochemical Contributions 458
    26.2.3 Oxidation by Pulsed-Laser Plasma Chemistry 459
    26.2.4 Nitridation 459
26.3 Elemental Semiconductors 460
    26.3.1 Photothermal Oxidation of Si 460

	26.3.2 Photochemically Enhanced Oxidation of Si	462
	26.3.3 Nitridation of Silicon	463
26.4	Compound Semiconductors	464
26.5	Oxide Transformation, Reoxidation	465

27 Depletion and Exchange of Species ... 467

27.1	Reduction and Metallization of Oxides	467
	27.1.1 Oxidic Perovskites and Related Materials	467
	27.1.2 Superconductors	469
	27.1.3 Qualitative Description	470
27.2	Surface Modification of Polymers	472
	27.2.1 Laser-Enhanced Adhesion	472
	27.2.2 Changes in Crystallinity	473
	27.2.3 Chemical Degradation	474
	27.2.4 Photochemical Exchange of Species	476
27.3	Chemical Transformation of Solid Films	476

28 Instabilities and Structure Formation ... 479

28.1	Coherent and Non-coherent Structures	479
28.2	Ripple Formation	482
	28.2.1 Interference Pattern	483
	28.2.2 Distribution of Energy	485
	28.2.3 Feedback	486
	28.2.4 Comparison of Experimental and Theoretical Results	488
28.3	Spatio-temporal Oscillations	490
	28.3.1 Zero Isoclines	492
	28.3.2 Instabilities in Laser-Induced Oxidation	493
	28.3.3 Explosive Crystallization	494
	28.3.4 Exothermal Reactions	495
	28.3.5 Instabilities in Direct Writing	496
	28.3.6 Discontinuous Deposition and Bistabilities	500
28.4	Instabilities in Laser Ablation	502
28.5	Hydrodynamic Instabilities	507
28.6	Stress-Related Instabilities	507
28.7	Technological Aspects	511

Part VI: Measurement Techniques, Diagnostics ... 513

29 Measurement Techniques ... 515

29.1	Characterization of Laser-Beam Profiles	515
29.2	Homogenization of Laser Beams	515
29.3	Deposition, Etch, and Ablation Rates	517
29.4	Temperature Measurements	521

	29.4.1 Photoelectric Pyrometry	521
	29.4.2 Other Optical Techniques	524
	29.4.3 Other Techniques	525

30 Analysis of Species, Plasmas, and Surfaces 526

30.1 Precursor and Product Species 526
    30.1.1 Optical Spectroscopy 526
    30.1.2 Mass Spectrometry 527
30.2 Species in Vapor and Plasma Plumes 529
    30.2.1 Species at Subthreshold Fluences 530
    30.2.2 Atomic and Molecular Neutral 530
    30.2.3 Ions 531
    30.2.4 Electrons, X-rays 532
    30.2.5 Fragments and Clusters
           in Polymer Ablation 532
30.3 Shock Waves, Plume Expansion 533
    30.3.1 Propagation in Gases 533
    30.3.2 Propagation in Liquids 537
30.4 Processed Surfaces and Thin Films 539
    30.4.1 Optical Techniques 540
    30.4.2 Other Techniques 540
    30.4.3 Transport Measurements 540

Appendix A: Definitions and Formulas 542

A.1: Symbols and Conversion Factors 542
A.2: Abbreviations, Acronyms 547
A.3: Mathematical Functions and Relations 550
A.4: The Density of Dissociated Species 553
A.5: The $\mathscr{F}$-Function 554

**Appendix B: Tabular Presentation
of the Materials Investigated** 558

B.1: Ablation of Inorganic Materials 558
B.2: Ablation of Organic Polymers
     and Biological Materials 559
B.3: Materials Etching 559
B.4: LCVD of Microstructures 561
B.5: Thin-Film Formation by LCVD 563
B.6: Deposition from Adsorbed Layers, Laser-MBE,
     Laser-ALE 565
B.7: Deposition from Liquids 565
B.8: Formation of Thin Films and Heterostructures
     by PLD 566

B.9: Surface Oxidation and Nitridation 570
B.10: Surface Modifications, Transformation
of Solid Films 571

Tables 572

References 589

Subject Index 641

Part I: Overview and Fundamentals

1 Introduction

The current interest in the use of lasers, be it for scientific investigations or for industrial applications, is directly linked to the unique properties of laser light. The high spatial coherence achieved with lasers permits extreme focusing and directional irradiation at high energy densities. The monochromaticity of laser light, together with its tunability, opens up the possibility of highly selective narrow-band excitation. Controlled pulsed excitation offers high temporal resolution and often makes it possible to overcome competing dissipative mechanisms within the particular system under investigation. The combination of all of these properties offers a wide and versatile range of quite different applications.

Materials processing with lasers takes advantage of virtually all of the characteristics of laser light. The high energy density and directionality achieved with lasers permits strongly localized heat- or photo-treatment of materials with a spatial resolution of better than 1 µm. Pulsed lasers or scanned cw lasers allow time controlled processing between about 10^{-14} s and continuous operation. The monochromaticity of laser light allows for control of the depth of heat treatment or selective, nonthermal excitation – either within the surface of the material or within the molecules of the surrounding medium – simply by changing the laser wavelength. Because laser light is an essentially massless tool, there is no need for mechanical holders with all the attendant problems these pose in the case of either brittle or soft materials (workpieces). Furthermore, laser beams can be moved at speeds which can never be obtained using mechanical tools or conventional heat sources. Contrary to mechanical tools, laser light is not subject to wear and tear. This avoids any contamination of the material being processed and, if the beam is properly controlled, also guarantees constant processing characteristics. With medical and biological applications it is also important that laser beams are absolutely sterile tools. Laser technology is completely compatible with present-day electronic control techniques. Naturally, a particular processing application will require only one or a few of these properties.

Laser processing can be classified into two groups: conventional laser processing and laser chemical processing. *Conventional* laser processing can be performed, at least in principle, in an inert atmosphere and can take place without any changes in the overall chemical composition of the material being processed. This is the most important difference to laser *chemical* processing which is characterized by an overall change in the chemical composition of the material or the activation of a real chemical reaction. In many situations, a unique classification into chemical and non-chemical laser processing is difficult or impossible.

1.1 Conventional Laser Processing

The interaction mechanisms between laser light and matter depend on the parameters of the laser beam and the physical and chemical properties of the material. Laser parameters are the wavelength, intensity, spatial and temporal coherence, polarization, angle of incidence, and the dwell time (illumination time at a particular site). The material is characterized by its chemical composition and microstructure (the arrangement of atoms or molecules within a solid) which determine the type of elementary excitations and the interactions between them.

Conventional laser processing is mainly performed with *infrared* (IR) laser light. This can excite free electrons within a metal, or vibrations within an insulator. In semiconductors both types of excitations are possible in the IR. In general, the excitation energy is dissipated into heat within a time which is short compared to any other time involved in the process. As a consequence, with low to medium intensities, the laser beam can just be considered as a heat source which induces a temperature rise on the surface and within the bulk of the material. This is schematically shown in Fig. 1.1.1a. The temperature distribution is determined by the optical and thermal properties of the material

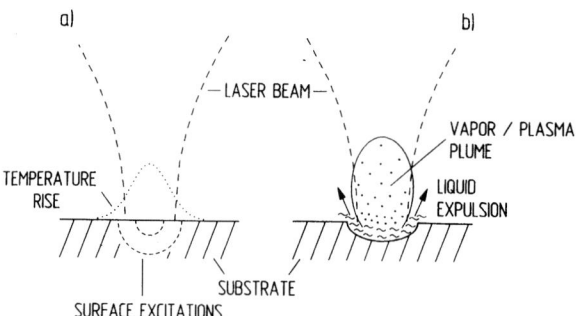

Fig. 1.1.1a, b. The picture illustrates two regimes of laser–material interactions employed in conventional laser processing. (a) Heat treatment of surface with laser-light intensity below the vaporization threshold, $I < I_v$. The absorbed laser light causes a temperature rise with or without surface melting. This intensity regime is mainly employed with laser-induced surface modifications. If surface melting takes place, abrasive processing by liquid-phase expulsion using a gas jet is possible. (b) The intensity regime $I > I_v$ is characterized by liquid-phase expulsion, vaporization, and plasma formation. This regime is employed in most cases of abrasive laser processing

1.1 Conventional Laser Processing

and, near phase transitions, by transformation energies for crystallization, melting, boiling, etc.

When the laser-light intensity, I, reaches a certain intensity, I_v, which causes significant material vaporization, a vapor plume above the substrate surface is formed (Fig. 1.1.1b). With further increasing intensity, the number of species within the plume increases and interactions between the laser light and the vapor become important. These result in an ionization of species. Above a certain intensity, I_p, the vapor is more appropriately denoted as a plasma. The plasma strongly absorbs the laser light which now couples to the substrate, mainly via the plasma plume. This is the dominating interaction mechanism in most types of abrasive metal processing by means of CO_2 lasers. Because of the strong nonlinearity in this interaction, small changes in laser parameters may cause strong changes in processing results. For this reason, proper control of the various parameters is a prior condition for reproducible processing.

Figure 1.1.2 gives an overview of the various applications and parameter regimes employed in laser processing. The intensities and interaction times shown in the figure refer to different types of lasers.

The best-established application is abrasive laser machining such as drilling, scribing, cutting, trimming, and shaping. Here, the material is removed from the workpiece as liquid, vapor, or plasma (Fig. 1.1.1b). The liquid is expelled by the recoil pressure of the vapor or by an additional gas jet. With conduction-limited laser welding and with bonding, the material is only melted. Typical light intensities employed in these applications range from about 10^5 to 10^{10} W/cm^2 with irradiation times between 10^{-7} s $< \tau_l < 10^{-1}$ s. Most commonly used are high power CO_2 lasers ($\lambda \approx 10.6 \,\mu m$) and Nd:YAG lasers ($\lambda \approx 1.064 \,\mu m$; see Table I). Here, in general, the laser beam is focused onto the workpiece to a spot size of, typically, between some ten micrometers and several millimeters.

Besides these well-established applications, laser-induced morphological, structural and compositional transformations of material surfaces and thin films are attracting increasing interest. These applications are performed, in general, with laser-light intensities $I < I_v$ (Fig. 1.1.1a). Among the most common processing applications of this type are laser annealing, transformation hardening, glazing, recrystallization, many types of alloying, and shock hardening. Except of the latter, these applications are based on the short heating and cooling cycles achieved with lasers. Here, the laser-induced surface temperature is either kept just below the melting temperature or slightly above it. Shock hardening is performed with high intensity short pulses that cause a strong mechanical compression.

Processing applications that involve an overall change in the chemical composition of the surface – without the necessity of a chemical reaction – are cladding and those cases of alloying where new material is added to the surface. These processes, in general, require surface melting.

The somewhat lower laser-light intensities employed in many of these applications, typically $10^3 - 10^8$ W/cm^2, make it possible to process larger

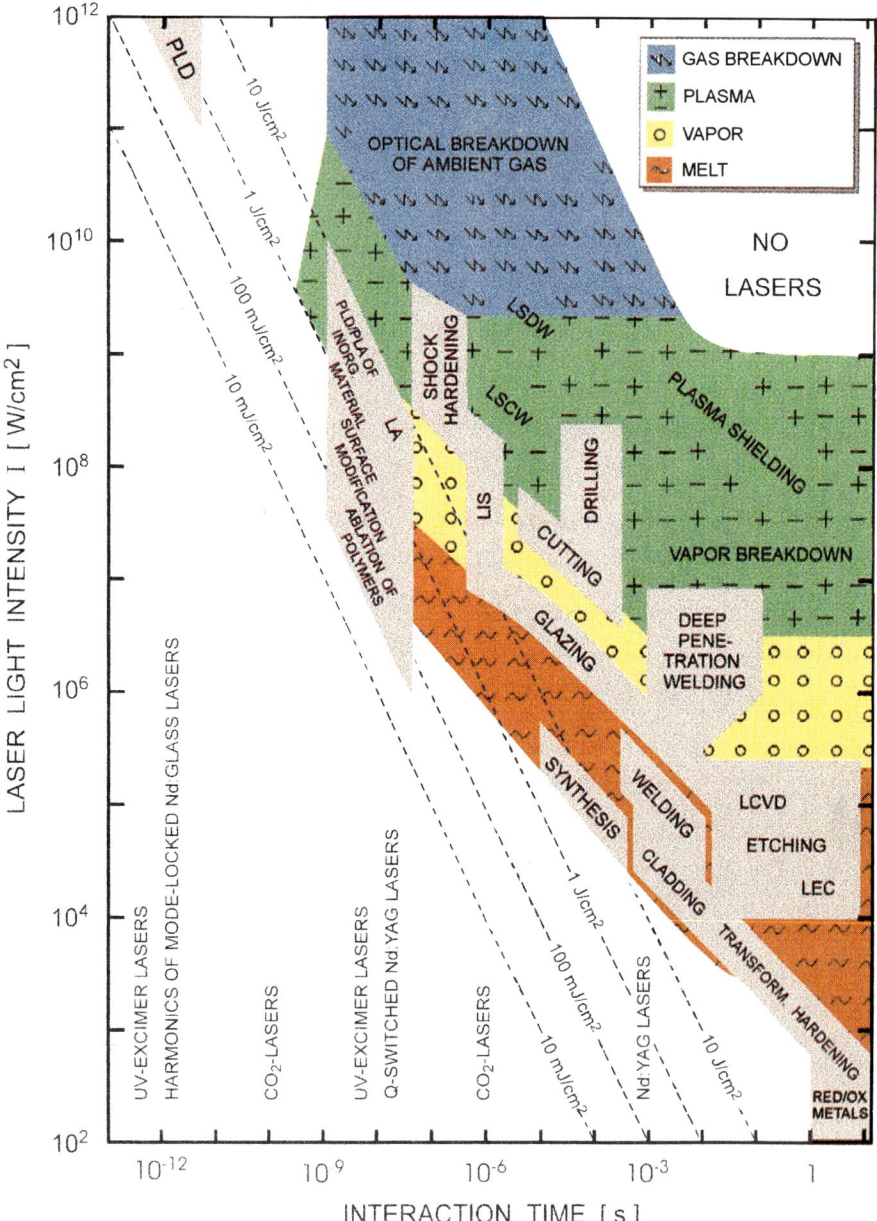

Fig. 1.1.2. Applications of lasers in materials processing. PLD: pulsed-laser deposition, and ablation of inorganic materials. Ablation of organic materials takes place at fluences that are, typically, one order of magnitude lower. Surface modifications include excimer- and Nd:YAG-laser-induced oxidation/nitridation of metals, surface doping, etc. LA: laser annealing. LIS: laser-induced isotope separation/IR-laser photochemistry. LSDW, LSCW: laser-supported detonation/combustion waves. LCVD: laser-induced CVD. LEC: laser-induced electrochemical plating/etching. RED/OX: long pulse or cw CO_2-laser-induced reduction/oxidation

areas by using unfocused or defocused laser beams. The lateral dimension of the area that can be processed in a single scan is therefore much wider, typically up to several centimeters. The processed depth is between some ten Ångstroms and several centimeters, depending on the material being processed and the type of laser employed. With some of these large-area processing applications, and in particular with the annealing of semiconductor surfaces, high-intensity lamps instead of lasers are used on real production lines.

1.2 Laser Chemical Processing

The object of laser-induced chemical processing (LCP) of materials is the patterning, coating and physicochemical modification of solid surfaces by activation of real chemical reactions. An overview of the various possibilities is presented in Fig. 1.2.1. The figure includes reactions that result in material deposition, etching, ablation, synthesis, surface modification (doping, oxidation, nitridation, reduction, metallization), and polymerization. Laser-induced activation or enhancement of a reaction can take place heterogeneously or homogeneously or via a combination of both. A *heterogeneous* reaction is induced in an adsorbate–adsorbent system, at a gas–solid or liquid–solid interface, or within the solid surface itself (Figs. 1.2.1a, b, d, e). A *homogeneous* reaction is activated within the ambient medium (Fig. 1.2.1c) or within the bulk of the material. Symbolically, the first step in a laser-induced reaction can be described, in many cases, by

$$AB + M + \text{Photons} \rightarrow A(\downarrow) + B(\uparrow) + M, \tag{1.2.1}$$

including the case $B \equiv A$. 'A' shall be the relevant species for surface processing. If $B \neq A$, the interaction of species B (atoms or molecules) with the substrate surface shall be weak or negligible. 'M' can be a gas, a liquid solvent, or a solid.

Both heterogeneous and homogeneous laser-induced reactions may be activated *thermally* (photothermally, pyrolytically), or *photochemically* (photolytically).

Thermal activation

We shall denote a reaction as thermally activated if the thermalization of the (laser) excitation energy is fast compared to the reaction. In this case, the laser can be considered simply as a heat source. This is schematically shown in Fig. 1.2.2. In case a) the laser light shall be absorbed exclusively by the substrate. If we ignore any heat flow into the ambient medium, molecules AB are decomposed only within the laser-heated area. While species A shall stick on or subsequently further react with the substrate, species B shall desorb.

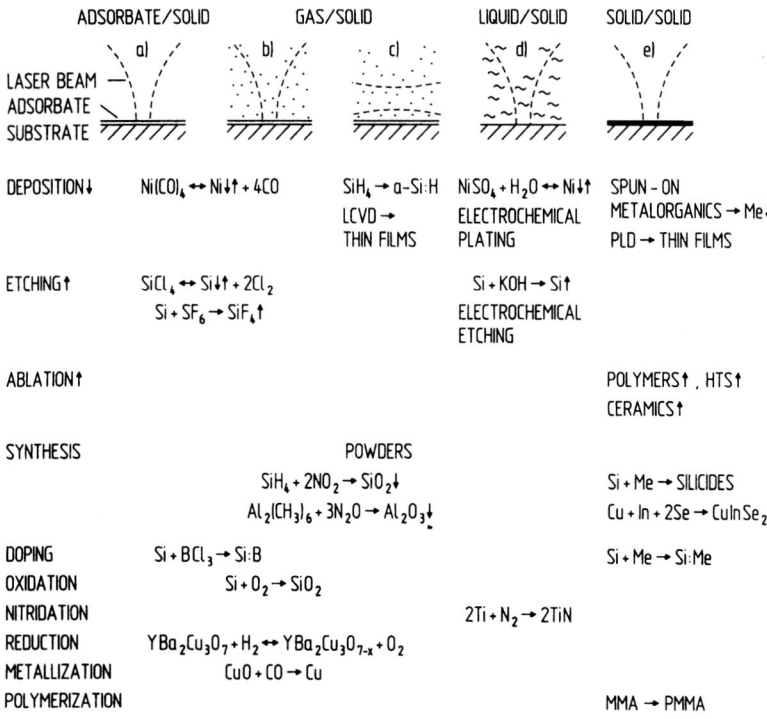

Fig. 1.2.1. Laser chemical processing at or near solid surfaces. The laser beam is always shown at perpendicular incidence to the substrate (workpiece) except in case c) where it propagates parallel to the surface. For simplicity, not all reaction products are included in the formulas. Arrows (↓) refer to deposition or condensation of products and arrows (↑) to desorption of species, surface etching, or ablation. ↔ denotes reactions that can be reversed by shifting the chemical equilibrium to the other side. Additional abbreviations are: Me = metal; HTS = high-temperature superconductors; PLD = pulsed-laser deposition; MMA = methyl-methacrylate; PMMA = polymethyl-methacrylate

Heating of the ambient medium via the substrate may result in homogeneous activation of the reaction.

In Fig. 1.2.2b the laser beam propagates parallel to the surface and can only be absorbed by molecules AB. The excitation energy shall be rapidly transformed into heat and the molecules are thermally decomposed. On their way to the substrate, some of the fragments A and B recombine. Those species A (or products AC formed in a subsequent reaction $A + CD \rightarrow AC + D$) that reach the substrate can either just stick on the surface or react with it. The situation is similar in cases where gas-phase molecules are not directly decomposed but thermally excited into higher vibrational states, AB*. Some of the molecules AB* reach the substrate, while others thermalize via collisions with molecules AB that are in the vibrational ground state or in a lower excited state. Molecules

1.2 Laser Chemical Processing

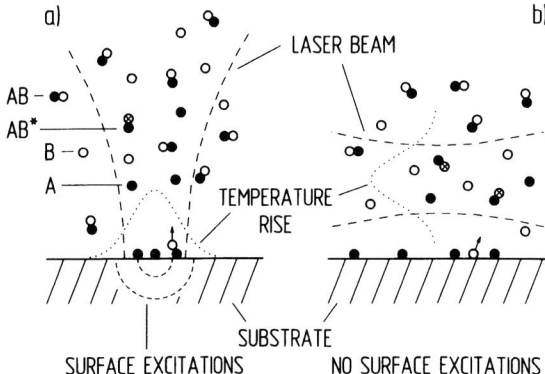

Fig. 1.2.2a, b. Illustration of laser-induced chemical processing (LCP) at perpendicular (case a) and parallel laser-beam incidence (case b). Thermal (photothermal, pyrolytic) LCP at perpendicular incidence is, in general, performed with precursors AB that do not absorb the laser radiation. Non-thermal (photochemical, photolytic) LCP is based on selective excitation/dissociation of precursor molecules and/or the substrate. In a purely photochemical process, the laser-induced temperature rise can be ignored

AB* may also decompose due to collisions with each other or with the substrate. The sticking probabilities of AB* and AB on the substrate surface may differ significantly.

Non-thermal activation

In photochemical laser processing, the first reaction step is faster than the thermalization of the excitation energy. Non-thermal excitations are included in Fig. 1.2.2. With both perpendicular and parallel irradiation, precursor molecules AB are dissociated as a consequence of *selective* electronic or vibrational excitation within the volume of the laser beam. Species A will diffuse and, if they do not recombine on their way, may hit the surface. The situation is similar if AB is not decomposed but only selectively excited. With perpendicular incidence the laser light may also excite the solid surface, thermally or non-thermally. Non-thermal surface excitations are photoelectrons generated on metal surfaces, electron–hole pairs in semiconductors, selective excitations of surface polaritons, etc. Surface excitations may significantly influence reactions between species A or AB* and the substrate, and they often permit one to localize the reaction in space. For purely photochemical processes the temperature rise within the ambient medium and on the substrate surface can be ignored.

In many cases, different excitation mechanisms contribute *simultaneously* to the reaction, but often one of them dominates. Frequently, a reaction is initiated photochemically and proceeds thermally, and vice versa.

Finally, changes in the optical, thermal, mechanical, and chemical properties of the processed surface or the ambient medium may cause various *feedbacks* and different kinds of *instabilities*.

Local and large-area processing

As in conventional laser processing, laser-chemical processing can be performed locally or on a more extended scale.

Laser-induced *microchemical* processing allows for single-step direct substrate patterning with lateral dimensions down into the submicrometer range. This can be performed by scanning a focused laser beam across the substrate surface (direct writing), by projecting the laser light via a mechanical mask, or by interference of laser beams.

Large-area chemical processing can be performed either with the laser beam propagating perpendicular (normal incidence) or parallel to the surface. The latter irradiation geometry permits thin-film fabrication with or without uniform substrate heating.

Comparison of techniques

Local and large-area material deposition, oxidation, nitridation, reduction, metallization, doping, compound formation, and etching are needed in many areas of technology, as in mechanics, electronics, integrated optics, and chemical technology. In virtually all of these fields light-assisted processing, and in particular laser processing, offers new and unique processing possibilities that are impossible with currently available technologies. On the other hand, there are many applications where laser processing has to compete with standard and well-established techniques. Among those are: conventional chemical vapor deposition (CVD); plasma-CVD (PCVD) or plasma etching (PE); electron-beam (EB) processing; ion-beam processing, e.g., reactive ion-beam etching (RIE), etc.

With the exception of electron- and ion-beams, these conventional techniques are all large-area processing techniques. Thus, surface patterning can be achieved only in combination with mechanical masking or lithographic methods. Here, typically, 10 to 20 different dry- and wet-processing steps are required to produce a particular pattern or single feature. The repeated physical and chemical treatment influences the *whole* substrate or device in each cycle. The fabrication of a microstructure, for example by CVD, requires the substrate to be uniformly heated up to several hundred degrees. In conventional liquid-phase etching, the exposure of the whole substrate to an aggressive etching solution may result in serious damage. Problems associated with photoresist masks are sometimes also difficult to overcome. In other words, the conventional techniques may become problematic or even inadequate whenever sensitive materials or prefabricated devices are to be processed.

1.2 Laser Chemical Processing

With lasers, thermal and chemical treatment can be strongly localized, thereby leaving the material otherwise unaffected. Consequently, the laser technique allows one to process heat-sensitive materials such as compound semiconductors, high-temperature superconductors (HTS), piezoelectric ceramics, polymers, etc. For instance, in laser-induced CVD (LCVD) heating takes place only locally or not at all. Furthermore, laser processing avoids material damage from ion or electron bombardment, or from overall vacuum-ultraviolet (VUV) radiation, which is inherent in plasma processing. Laser processing is *not* limited to planar substrates but also allows three-dimensional fabrication. The nonlinearity of laser-induced chemical reactions makes it possible to increase the process resolution over that achieved in standard photolithography. Unlike ion- or electron-beams, laser radiation can propagate through a great variety of media, or it can be made strongly absorbable, e.g., by changing its wavelength. These and other properties of laser processing may become important in micromechanics and semiconductor device technology, and a prior condition for the fabrication of new multicomponent microdevices. Laser processing for surface profiling by doping, etching, etc., often yields superior results with respect to those achieved with conventional techniques.

It is evident that in all cases where standard techniques can be applied equally well or where the quality of a particular processing step can be tolerated, economical arguments will be decisive. Here, the most serious limitations of laser processing are, at present, the total process rates and throughputs. This severe problem refers not only to laser-direct writing, but also to many cases of large-area laser processing. The local processing rates for deposition, etching, etc., can be extremely high, $100\,\mu m/s$ and more. Nevertheless, the processed surface area per unit time is quite small. Many of the conventional large-area techniques permit fabrication of a large number of devices simultaneously. Thus, despite being multiple step processes, the total throughput is very high. Therefore, in the foreseeable future, laser direct writing of complete complex structures may be interesting for the design of prototypes, but not for mass production.

To resolve the throughput problem one needs more powerful laser systems than those currently available, and further developments in optical projection, interference, and fiber techniques. For these reasons, laser micro-processing should be considered for the time being as a *complementary* technique that can be used when standard techniques become inadequate. In such cases fabrication of tools, devices, wafers, etc., on a piece-by-piece basis becomes quite conceivable. Even today laser micro-processing substitutes conventional techniques in cases where small area complementations and modifications (for example for customization), or repair of prefabricated devices or tools are necessary. In such cases multiple-step conventional techniques become very inefficient or cannot be applied at all. Here, laser micro-processing can significantly improve the total production yield. Furthermore, localized deposition may be advantageous when recovery of precious materials such as rare metals from liquid or solid admixtures, for example from photoresists, is expensive or altogether uneconomic.

Planar and non-planar processing

Laser processing is frequently classified into two-dimensional planar processing and three-dimensional nonplanar processing:
- If the substrate is planar *and* if the lateral dimensions of the processed feature are larger than or comparable to the axial dimensions, we define processing to be planar (two-dimensional). This includes large-area etching, deposition, alloying, and compound formation, but also most cases of laser direct writing on planar substrates.
- If the substrate is nonplanar, or if the lateral dimensions of the laser processed feature are small compared to its axial dimensions, we define processing to be nonplanar (three-dimensional). This includes many cases of material cutting and drilling, the etching of deep grooves and via holes, the growth of fibers, etc. Clearly, this classification is sometimes somewhat arbitrary.

2 Thermal, Photophysical, and Photochemical Processes

A proper definition of *thermal* (photothermal) and *non-thermal* (photochemical) laser processing would require a detailed knowledge of the fundamental interactions between laser light and matter, and of the various relaxation times involved. This information is available only for a few special systems. For this reason, the definitions usually employed are not very strict. We shall consider a laser-induced process as thermally activated if the thermalization of the excitation energy is fast compared to both the excitation rate and the initial processing step. The term photochemical is used if the laser-induced process proceeds *mainly* non-thermally. If both thermal and non-thermal mechanisms are significant, we denote the process as *photophysical*. Particularly in connection with photo-decomposition processes, we frequently use, instead of photo-thermal and photochemical, the terms pyrolytic and photolytic, respectively.

If laser processing is thermally activated, the state of the system is described by the temperature and the total enthalpy. The latter is relevant only if phase changes or chemical reactions take place. For a quantitative analysis and optimization of a particular process, the laser-induced temperature distribution must be known. In laser-microchemical processing, direct temperature measurements have been performed with a reliable degree of accuracy in only a very few cases. Frequently, laser-induced temperatures can only be calculated. In fact, many features in thermal processing can be qualitatively, and in some cases even quantitatively, analyzed on this basis.

Photochemical laser processing is determined by the *selectivity* of the excitation. In a gas or liquid, the selectivity is characterized by the number density of (selectively) excited, ionized, or dissociated species. In a solid, the degree of selectivity is determined by the number density of non-equilibrium photoelectrons, electron–hole pairs, photodissociated bonds, etc.

2.1 Excitation Mechanisms, Relaxation Times

The *primary* interactions between light and matter are always non-thermal. In laser processing, the relevant excitations can be classified into those of the solid substrate to be processed, those of the ambient medium, and those of the adsorbate–adsorbent system.

In solids, light can interact with elementary excitations that are optically active. Among those are different types of electronic excitations (inter- and intraband excitations, excitons, plasmons, etc.), excitations of phonons, polaritons, magnons, etc. Additionally, there may be localized or non-localized electronic or vibrational states that are related to defects, impurities, or to the solid surface itself. Some of these transitions are schematically shown in Fig. 2.1.1. The energy E_g describes the distance between the highest valence band and the lowest conduction band.

In liquids and gases, light can induce electronic, vibrational, and rotational transitions within single molecules.

If molecules or atoms become adsorbed on a solid surface, their electronic and vibrational properties change. As a consequence, one observes changes in absorption cross sections and selection rules for optical absorption, additional vibrational transitions, etc. Light can increase or decrease the density of adsorbed species, e.g., via excitations of the solid, the (free) gas- or liquid-phase molecules, or the adsorbate–adsorbent complex.

In all systems, different elementary excitations are coupled via anharmonic or higher order dipolar (multipolar) interactions.

High laser-light intensities allow high excitation densities to be generated, thermally or non-thermally. The density of excited molecules, atoms, ions,

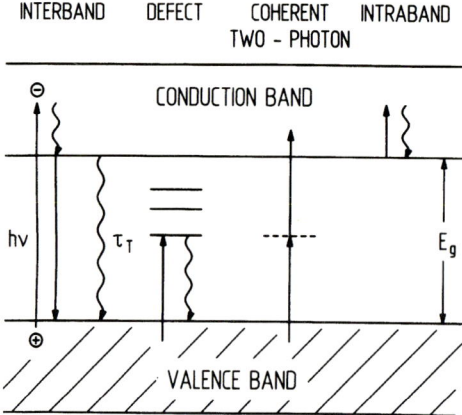

Fig. 2.1.1. Schematic picture of different types of electronic excitations in a solid. Only the highest valence band (VB) and the lowest conduction band (CB) are shown. Straight lines indicate absorption or emission of photons with different energies, $h\nu$. Oscillating lines indicate non-radiative processes. Interband transitions VB → CB take place if $h\nu \geq E_g$. In this process, electron–hole pairs are generated. Band-gap excitations are located in the near infrared (NIR) and visible (VIS) for semiconductors and in the ultraviolet (UV) for insulators. Defect, impurity and surface states often permit sub-band-gap excitations with $h\nu < E_g$. At high laser-light intensities, sequential multiphoton excitations via defect states or coherent multiphoton excitations become important. Intraband electronic transitions are typical for CO_2-laser excitations in metals, and in semiconductors at elevated temperatures

radicals, electrons, etc., can exceed 10^{22} species/cm^3. The coupling of elementary excitations among each other and with the intense laser radiation can cause a number of *new* phenomena. Prominent examples are changes in absorption cross sections, thermal runaway in metals and semiconductors, thermal self-focusing in transparent media – including the ambient medium, high densities of free carriers generated by interband excitation or impact ionization in semiconductors and insulators. With even higher laser-light intensities, *nonlinear* optical phenomena such as self-focusing, multiphoton processes, etc., become important. With very high intensities, the formation of plasmas, shock waves, detonation waves, etc., is observed.

The time for the thermalization of the excitation energy depends on the type of material and the laser parameters.

In metals, light is absorbed by electron transitions from lower to higher energy states within the conduction band. The time between electron-electron collisions, τ_{e-e}, is of the order of 10^{-14} to 10^{-13} s. Electron–phonon relaxation times, τ_{e-ph}, are typically one to two orders of magnitude longer.

In non-metals, interband electronic excitations can last much longer, ranging from, typically, 10^{-12} to 10^{-6} s. Excitations of localized electronic states associated with defects, impurities or surfaces, may have much longer lifetimes. Phonons in non-metals can be directly excited by infrared (IR) light.

In single, isolated molecules, electronic excitations decay within 10^{-14} to 10^{-6} s. The lifetime of *lowly* excited vibrational levels is, typically, 10^{-3} s. With the (high) molecular densities employed in laser-chemical processing, energy randomization between molecules with vibrational mismatches $\leq k_B T$, occurs via collisions within, typically, 10^{-10} to 10^{-4} s.

It should be emphasized that with the *high* light intensities achieved with lasers, excitation and energy relaxation mechanisms can significantly be altered with respect to those relevant at low to medium intensities. For example, in polyatomic molecules highly excited vibrational states may have lifetimes of 10^{-13} to 10^{-11} s only.

Thermal processes

The thermalization of the excitation energy shall be described by the relaxation time τ_T. For a thermal process $\tau_T \ll \tau_R$ where τ_R characterizes the initial processing step or the inverse excitation rate, depending on which is smaller. For some systems, however, this condition must be modified. τ_R can be the time for desorption of species from the surface, or for structural rearrangements of atoms or molecules within the surface, the time which characterizes the *initial* step in a chemical reaction, etc.

Let us consider a simple example: Assume a reaction that is mediated by collisions of gas-phase molecules with a Si surface. The reaction probabilities for thermal and non-thermal processes shall be equal. The Si shall be irradiated by a picosecond laser pulse with $hv > E_g(\text{Si})$ and an intensity that generates a carrier density $N_c \approx 10^{22}/\text{cm}^3$ near the surface. This carrier density decreases via Auger recombination to 1% within about 10^{-10} s (Sect. 2.4). Due to electron-phonon coupling, the recombination energy is dissipated into heat within, typically, 10^{-12} to 10^{-11} s. With a gas pressure of 100 mbar and a temperature of 300 K, the impingement rate onto the surface is some 10^{22} molecules/cm^2 s. The number of surface atoms is about 10^{14} atoms/cm^2. The time for the initial reaction step between a gas-phase molecule and a surface atom is then $\tau_R \geqslant 10^{-8}$ s which is very long compared to $\tau_T \leqslant 10^{-10}$ s. Thus, the effect of laser radiation is purely thermal. In other words, if $\tau_T \ll \tau_R$ the detailed excitation mechanisms become irrelevant and the laser can be considered simply as a heat source.

In spite of their thermal character, laser-driven thermophysical and thermochemical processes may be quite different from those initiated by a conventional heat source. There are various reasons for this: The laser-induced temperature rise can be localized in space and time. Temperatures of more than 10^4 K can be induced in a *small* volume which is defined by the focused laser beam. With short high intensity laser pulses, heating rates up to more than 10^{15} K/s can be achieved. With such heating rates, the chemical relaxation time may be slow in comparison, and the chemical reaction takes place far from equilibrium. Another possibility is the selective excitation of a particular species, e.g., in a gas mixture. Furthermore, laser heating may change the optical properties of the medium and thereby introduce nonlinearities in the interaction process. As a consequence of these various mechanisms, novel chemical reaction pathways and reaction products, novel material microstructures and phases, novel surface morphologies, and novel evaporation characteristics may occur.

Photochemical processes

The term photochemical (photolytic) laser processing is used if the overall thermalization of the excitation energy is slow, i.e., if $\tau_T \geqslant \tau_R$. This condition frequently holds for chemical reactions of excited molecules among themselves or with the substrate surface, for photoelectron transfer and subsequent chemisorption of species on solid surfaces, for photochemical desorption of species from surfaces, etc. If we consider the example discussed above, but assume that the molecules are already adsorbed on the Si surface, photocarriers can *directly* interact with the adsorbate and thereby initiate a reaction. In this case τ_R is the time for charge transfer (Sect. 15.1). With purely photochemical processes, the temperature of the system remains (almost) unchanged under laser-light irradiation.

Due to the high excitation densities, laser photochemistry can be quite different from standard photochemistry using lamps.

2.1 Excitation Mechanisms, Relaxation Times

Photophysical processes

Thermal and photochemical processes can be considered as limiting cases of photophysical processes. We denote a process as photophysical if both thermal and non-thermal mechanisms *directly* contribute to the overall processing rate. The degree of thermal and non-thermal contributions depends on the relative yield of the respective reaction channels.

A simple model

The classification into thermal, photophysical, and photochemical processes is often quite complex. Consider the situation shown in Fig. 2.1.2. A and A* shall characterize the system in the ground state and in the excited state, respectively. If we ignore spontaneous emission, non-radiative transitions $A^* \to A$ are described by the thermal relaxation time, τ_T. The characteristic times for the reaction of A and A* with C are $\tau_A(T)$ and $\tau_{A^*}(T)$. It is often convenient to use instead of relaxation times τ_i, rate constants $k_i \equiv k_i(T) = \tau_i^{-1}$. Let us consider *low* excitation rates where $\tau_T \leqslant h\nu/\sigma I$ (σ is the excitation cross section).

If $\tau_T \ll \tau_A, \tau_{A^*}$ and $\tau_A \ll \tau_{A^*}$ the excitation energy is immediately dissipated into heat and the reaction is thermally activated. The reaction rate is determined by k_A.

If $\tau_T > \tau_{A^*}$ and $\tau_{A^*} \ll \tau_A$ the process is mainly photochemically activated. The reaction takes place via excited species A*.

If $\tau_T \ll \tau_A, \tau_{A^*}$ but $\tau_{A^*} \ll \tau_A$, or if all these times are comparable, both the "thermal channel" and the "photochemical channel" are important. We denote this process as photophysical. Let us study this situation in further detail and consider the kinetic equations

$$\frac{dN_A}{dt} = \frac{\sigma I}{h\nu}\left(N_{A^*} - N_A\right) + \frac{N_{A^*}}{\tau_T},$$

$$\frac{dN_{A^*}}{dt} = \frac{\sigma I}{h\nu}\left(N_A - N_{A^*}\right) - \frac{N_{A^*}}{\tau_T}. \tag{2.1.1}$$

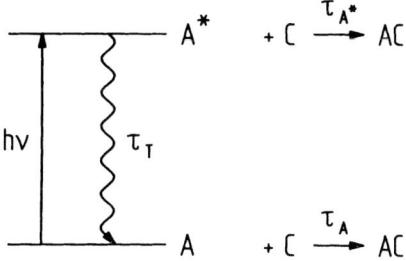

Fig. 2.1.2. A simple model for the competition between a thermal and a photochemical reaction. Stimulated emission is not indicated

Here, the reaction of A and A* with C has been assumed to be so slow that the related changes in N_A and N_{A^*} can be ignored. The total density of species is then $N = N_A + N_{A^*}$ and quasi-stationary conditions are achieved after a time $t \approx \tau_T$. With $dN_A/dt = dN_{A^*}/dt = 0$ we obtain from (2.1.1) the quasi-stationary densities $\tilde{N}_A$ and $\tilde{N}_{A^*}$. The sum of reaction rates via the thermal channel $W_T \approx k_A \tilde{N}_A$ and the photochemical channel $W_{PC} \approx k_{A^*} \tilde{N}_{A^*}$ is then

$$W = W_T + W_{PC} = k_A \tilde{N}_A \left[1 + \frac{k_{A^*}}{k_A} \left(1 + \frac{h\nu}{\sigma I \tau_T} \right)^{-1} \right]. \tag{2.1.2}$$

The relative importance of the photochemical and the thermal channel is determined by the second term in the parenthesis. When the laser beam is switched off, the contribution of the photochemical channel decays within the characteristic time τ_T. Afterwards, the process is determined by k_A only.

This example shows, in a transparent way, how the competition between thermal and photochemical processes is determined by different time constants.

Chemical relaxation

Let us consider a thermochemical process which is characterized by a chemical relaxation time, τ_{ch}, given by

$$\frac{1}{\tau_{ch}} \equiv k_{ch} = -\frac{1}{\Delta N_i} \frac{dN_i}{dt}, \tag{2.1.3}$$

where N_i is the number density of species i, and ΔN_i its deviation from the equilibrium value. If the heating rate

$$\frac{1}{\tau_t} \equiv k_t = \frac{1}{T} \frac{dT}{dt} \tag{2.1.4}$$

is small compared to k_{ch}, the system is in chemical equilibrium. If, on the other hand, $k_t \gg k_{ch}$, the system is far from equilibrium. This is the regime of non-equilibrium thermochemistry where novel reaction pathways and reaction products may be observed. The situation is analogous in the case of fast cooling (quenching) of systems. In this case, a non-equilibrium chemical state can be frozen.

Non-equilibrium chemistry is important in laser-induced surface modification, alloying, synthesis, etc. Quenching of non-equilibrium states is important also with structural transformations where no chemical reactions take place (Chap. 23).

2.2 The Heat Equation

Temperature distributions induced by the absorption of laser radiation within gases, liquids, and solids have been calculated on the basis of the heat equation. In the most general case, the temperature $T \equiv T(\mathbf{x},t) = T(x_\alpha, t)$ is a function of both the spatial coordinates, x_α, and the time, t. With fixed laser parameters the temperature distribution depends on the optical absorption within the irradiated zone, on the transport of heat out of this zone and, if relevant, on transformation enthalpies for crystallization, melting, vaporization, and on chemical reaction enthalpies (exothermal or endothermal), etc. In the absence of heat transport by convection and thermal radiation, the heat equation can be written, in a coordinate system that is fixed with the laser beam, as

$$\rho(T)c_p(T)\frac{\partial T(\mathbf{x},t)}{\partial t} - \nabla[\kappa(T)\nabla T(\mathbf{x},t)]$$
$$+ \rho(T)c_p(T)\mathbf{v}_s \nabla T(\mathbf{x},t) = Q(\mathbf{x},t), \qquad (2.2.1)$$

where $\rho(T)$ is the mass density and $c_p(T)$ the specific heat at constant pressure. $\mathbf{v}_s$ is the velocity of the substrate (medium) relative to the heat source, $Q\,[\mathrm{W/cm^3}]$ (in a coordinate system that is fixed with the laser beam, $\mathbf{v}_s$ is in the opposite direction to the scanning velocity shown in Fig. 6.1.1. If the substrate (workpiece) is uniform and isotropic, its thermal properties are characterized by a single thermal conductivity, κ, and a single heat diffusivity, D, which are related by

$$D = \frac{\kappa}{\rho c_p} \qquad (2.2.2)$$

If all temperature dependences in material parameters are ignored, the heat equation becomes linear. Values of ρ, c_p, κ, and D are listed for various materials in Table II.

2.2.1 The Source Term

Subsequently, we assume that the light energy absorbed within the medium is totally transformed into heat. The source term can then be written as

$$Q(\mathbf{x},t) = -\nabla\langle \mathbf{S}\rangle + U(\mathbf{x},t). \qquad (2.2.3)$$

The function $U(\mathbf{x},t)$ shall describe the additional energy per volume and time that is required or provided, if phase changes or chemical reactions take place. $\langle \mathbf{S}\rangle = c\langle \mathbf{E}\times\mathbf{H}\rangle/4\pi = I\hat{\mathbf{k}}_l$ is the time average of the Poynting vector. $\hat{\mathbf{k}}_l$ is a unit vector in the direction of the propagating light. This general expression for $\langle S \rangle$ should be used if interference phenomena, optical inhomogeneities in the material, etc., must be considered. The propagation of light is characterized by the dielectric and magnetic permittivity of the medium, $\varepsilon = \varepsilon' + i\varepsilon''$ and $\mu = \mu' + i\mu''$. Henceforth, we set $\mu' = 1$ and $\mu'' = 0$. The complex index of refraction can then be written as $\tilde{n} = \varepsilon^{1/2} = n + i\kappa_a \equiv n(1 + i\kappa_0)$. Frequently, only the real part, n, is termed refractive index. $\kappa_a = n\kappa_0$ is the absorption index, and κ_0 the attenuation index. In the case of weak absorption and $\varepsilon'' \ll \varepsilon'$ we have $n \approx \varepsilon'^{1/2}$ and $\kappa_a \approx \varepsilon''/2\varepsilon'^{1/2}$. In the low-density approximation which holds, for example, for dilute gases $n - 1 \approx [\varepsilon' - 1]/2$ and $\kappa_a \approx \varepsilon''/2$.

For monochromatic light and an isotropic medium, the first term in (2.2.3) can be written as

$$-\nabla \langle S \rangle = \frac{\omega}{8\pi} \varepsilon'' |E_0|^2,$$

where $E(x,t) = [E_0(x) \exp(-i\omega t) + \text{c.c.}]/2$ is the electric field. In the approximation of a (single) plane-wave and *low* absorption we obtain

$$\langle S \rangle = \frac{c}{8\pi} n |E_0|^2 \hat{k}_l = I \hat{k}_l.$$

If we assume that the laser beam propagates in z-direction, these two expressions yield the Bouguer–Lambert–Beer law[1]

$$\frac{dI(z)}{dz} = -\alpha I(z), \tag{2.2.4}$$

where

$$\alpha = \frac{4\pi \kappa_a}{\lambda} = \frac{2\omega \kappa_a}{c} = \frac{4\pi n \kappa_0}{\lambda} \tag{2.2.5}$$

is the (linear) absorption coefficient. λ is the wavelength in vacuum. Values of α are listed in Table III for different materials. Instead of α, we often introduce the optical penetration depth, $l_\alpha = \alpha^{-1}$. The absorbed energy per volume and time is αI.

If the attenuation of the laser radiation is not solely caused by absorption but also by scattering, described by α_s, we have to replace α in (2.2.4) by the extinction coefficient, $\beta = \alpha + \alpha_s$.

In a strict sense, Beer's law can be applied only if ε is uniform in space. Nevertheless, (2.2.4) is often also employed in photothermal processing where the refractive index is *not* constant. Then, α is simply replaced by $\alpha = \alpha(T)$. Furthermore, for a tightly focused laser beam, the plane-wave approximation holds only within the Rayleigh length, z_R (Chap. 5).

---

[1] Throughout the literature, (2.2.4) is often termed the Lambert–Beer law or simply Beer's law; it was first derived experimentally in its integral form by P. Bouguer (1729), and theoretically by I. G. Lambert (1760). A mesoscopic interpretation of α was first given by A. Beer (1852).

2.2 The Heat Equation

For a divergent laser beam and $U(\mathbf{x}, t) = 0$, the source term can be written as
$$Q(\mathbf{x}, t) = I(x, y, z, t) f(z), \tag{2.2.6}$$
where $f(z)$ describes the absorption of the laser light along the z-direction.

2.2.2 Dimensionality of Heat Flow

An important quantity in thermal processing is the heat diffusion length, l_T. In the literature, various definitions are employed. Henceforth, we approximate l_T by
$$l_T \approx 2(D\tau_l)^{1/2} \tag{2.2.7}$$
in *all* cases of transient heating that are characterized by a laser-beam dwell time, τ_l. The definition (2.2.7) describes the $1/e$ (spatial) decay in the temperature distribution
$$T(\mathbf{x}, t) \approx \frac{Q}{\rho c_p (4\pi D t)^{m/2}} \exp\left(-\frac{|\mathbf{x}|^2}{4Dt}\right),$$
which is a (fundamental) solution of the linear heat equation for a *point source* in the infinite space
$$\frac{\partial T}{\partial t} = D \nabla^2 T + \frac{Q}{\rho c_p} \delta(\mathbf{x}, t).$$
The number m characterizes the (spatial) dimensionality of the problem, i.e., $m = 1, 2, 3$. Q is the total energy release and ρ the density in [g/cm^m].

It must be emphasized, however, that in all other cases the characteristic length, l_T', which defines $T(l_T', t)/T(0, t) = 1/e$, depends on the particular boundary–value problem under consideration and may significantly differ from (2.2.7).

The *dimensionality* of the heat flow is characterized by the relative size of l_T and other characteristic quantities such as the radius of the laser beam, w, the optical penetration depth, l_α, the thickness of the substrate, h_s, etc. If, for example, $l_T \gg w, l_\alpha$ one has to consider the propagation of heat in three dimensions. On the other hand, for moderate absorption with $l_\alpha \leq l_T$ and $l_T \ll w$ lateral heat flow can be ignored and the temperature distribution in z-direction is obtained from the one-dimensional heat equation.

2.2.3 Kirchhoff and Crank Transforms

If the irradiated medium is isotropic, the temperature dependence of the thermal conductivity, $\kappa(T)$, can be eliminated from the heat equation by performing the Kirchhoff transform
$$\theta(T) = \int_{T(\infty)}^{T} \frac{\kappa(T')}{\kappa(T(\infty))} dT', \tag{2.2.8}$$

where θ is a linearized temperature, and $T(\infty)$ the temperature at infinity, i.e., far away from the processed region. If κ is independent of temperature, the linearized temperature is equal to the temperature rise, i.e., $\theta = \Delta T$. In terms of θ, the heat equation (2.2.1) has the form

$$\frac{1}{D(T(\theta))}\frac{\partial \theta}{\partial t} - \nabla^2 \theta + \frac{v_s}{D(T(\theta))}\nabla \theta = \frac{Q}{\kappa(T(\infty))} \ . \tag{2.2.9}$$

An analytical solution of (2.2.9) is possible only in special cases. For an arbitrary geometry and temperature-dependent parameters D, α, and the reflectivity R, only numerical solutions can be found.

With similar types of equations where D depends on time only, i.e. $D = D(t)$, and where $v_s = 0$ (see, e.g., Sect. 24.1) the Crank transform

$$\tau = \int_0^t D(t')dt' \tag{2.2.10}$$

yields the linear equation

$$\frac{\partial \theta}{\partial \tau} - \nabla^2 \theta = \frac{Q}{\kappa(T(\infty))} \ . \tag{2.2.11}$$

which can be solved analytically in many cases.

2.2.4 Phase Changes

If the laser power exceeds the threshold power for surface melting and evaporation, the temperature distribution can be calculated from the heat equation only when the latent heat of melting, ΔH_m, and evaporation, ΔH_v, is taken into account (because the experiments are, in general, performed at constant pressure, we use the enthalpy). Henceforth, the enthalpy is used with different dimensions as convenient. The conversion of the enthalpy per *atom* is $\Delta H^a \equiv \Delta H \text{ [J/atom]} = \Delta H \text{ [J/mol]}/L = \Delta H \text{ [J/cm}^3] \times M/[\rho L] = \Delta H \text{ [J/g]} \times M/L$ where L is the Avogadro number and M the atomic weight per mol. Values of ΔH_m and ΔH_v are listed in Table IV. The table shows that the latent heat of evaporation is, typically, 2–4 eV/atom (50 to 100 kcal/mol) while the latent heat of melting is, typically, 0.1–0.5 eV/atom (2 to 10 kcal/mol).

If phase changes take place, $U(x,t)$ in (2.2.3) is non-zero and given by

$$U(x,t)dV \approx v_{ls}\Delta H_m dF_{ls} + v_{vl}\Delta H_v dF_{vl} \tag{2.2.12}$$

where $v_{ls}(x,t), v_{vl}(x,t)$ and dF_{ls}, dF_{vl} are the respective velocities and surface elements of the liquid–solid and vapor–liquid interface within the volume element dV. It is often convenient to introduce the *total* enthalpy which can be approximated by

$$\Delta H(T) \approx \int_{T(\infty)}^{T} \rho(T')c_p(T')dT' + \mathscr{H}(T-T_m)\Delta H_m + \mathscr{H}(T-T_v)\Delta H_v \ . \tag{2.2.13}$$

2.2 The Heat Equation

Here, the kinetic energy of the vapor is ignored. The first term describes the enthalpy density [J/cm^3; with solids and liquids this is equal to the energy density] required to heat the material from the temperature $T(\infty)$ to T. Note that in the case of vaporization this term includes the enthalpy change within the solid, liquid, and gas phase. The latter can be approximated, for an ideal gas, by $\Delta H_G = \gamma R_G \rho_V \Delta T / M$ [$\gamma - 1$]. $\gamma = c_p/c_v$ is the adiabatic index.

The second term describes the additional energy density necessary for melting. $\mathscr{H}$ is the Heaviside function which is zero if $T < T_m$ and unity if $T > T_m$. The third term describes the latent heat of evaporation.

If we consider the liquid–solid system only and ignore density changes and also convective fluxes, the heat equation can be written together with (2.2.13) in the form

$$\frac{\partial \Delta H(\mathbf{x},t)}{\partial t} - \nabla[\kappa(T)\nabla T(\mathbf{x},t)] + \mathbf{v}_s \nabla \Delta H(\mathbf{x},t) = -\hat{\mathbf{k}}_l \nabla I(\mathbf{x},t), \qquad (2.2.14)$$

where $\mathbf{v}_s$ is the substrate velocity with respect to the laser beam. Equation (2.2.14) is most conveniently used with problems where latent heat effects play an important role. The situation is analogous with exothermal or endothermal chemical reactions when the energy of formation, ΔH, cannot be ignored with respect to the absorbed laser-light energy. In general, (2.2.14) can be solved only numerically (for a more general discussion see *Landau* and *Lifshitz*, Vol. VI, Fluid Mechanics).

2.2.5 Limits of Validity

The heat equation describes temperature distributions in many cases of thermal laser processing quite well. Nevertheless, one should be aware of the restrictions and uncertainties of calculated temperature distributions:

– The heat equation implies a macroscopic description of the medium averaged over a volume where thermal fluctuations are small. To estimate the length scale where such a description is appropriate, we consider a cube of side length l with N atoms (molecules) per unit volume. Then, the relative temperature fluctuation is $\delta T/T \approx (Nl^3)^{-1/2}$ and thus $l \approx (\delta T/T)^{-2/3} N^{-1/3}$. For $\delta T/T = 10^{-3}$ this yields $l \approx 0.02$ μm for solids ($N \approx 10^{23}$ atoms/cm^3) and $l \approx 1$ μm for gases with $N = 10^{18}$ atoms/cm^3. Thus, with submicrometer structures and with gases at low pressures the application of the heat equation becomes inappropriate (Sect. 9.5.4).
– The values of α, R, κ, D, etc., are usually derived from static or quasi-static measurements where only small temperature gradients are involved. In laser processing, however, the temperature gradients may be very strong and the interaction times very short. The temperature gradients are, typically, of the order $\nabla T \approx \Delta T/l$. Here, l is a characteristic length; for example, the radius of

the laser focus w, the heat diffusion length l_T, the optical penetration depth l_α, etc., depending on the particular problem. In any case, ∇T may be $10^5 - 10^{10}$ K/cm. As a consequence, the parameter values relevant in photothermal processing may significantly differ from those in conventional heat conduction problems.
- The optical properties of a specific medium depend on the laser parameters which, in turn, affect the thermophysical properties via their temperature dependences. These dependences are often known in small temperature intervals only.
- The optical and thermal properties of a solid depend also on its surface morphology and crystallinity (amorphous, ceramic, poly- or single-crystalline), on surface contaminations (adsorbates, oxide layers, etc.), on defects (both physical defects such as dislocations, cracks, etc., and chemical defects such as isolated impurities, aggregate centers, etc.).
- The optical and thermal properties of liquids and gases depend on admixtures; with gases they depend also on pressure.
- Changes in parameter values originating from laser-induced changes in material properties introduce additional complications: Let us consider laser-CVD (Fig. 1.2.2a). Before nucleation takes place, α, R, D, κ, and the total emissivity, ε_t, are determined by the physical properties of the substrate material. When deposition commences, these quantities will rapidly change with the density and size of nuclei and therefore with time. When a compact film is formed, e.g., a metal film, and when the penetration depth of the laser light is small compared to the film thickness, α, R, and ε_t will refer only to this deposited film. Similarly, D and κ will be quite different for such a combined structure compared to a uniform plane substrate. The situation is very similar in laser-induced surface modification and compound formation. Further complications arise if changes in surface geometry become significant. This is sometimes the case in materials deposition, etching, and ablation.
- The coupling between different degrees of freedom (e.g., the temperature, the density of species N_i, etc.) cause feedbacks in the laser–matter interactions. Thus, from a theoretical point of view, a proper description of laser processing would require, in many cases, consideration of coupled non-linear equations.
- Whenever Knudsen effects are important, the kinetic Boltzmann equation instead of the heat equation should be solved. In reality, it is often possible to solve the problem for the Knudsen layer separately and derive modified boundary conditions for the heat equation.
- With ultrashort pulses where $\tau_l \leqslant D/v_0^2$ (v_0 is the sound velocity) the finite velocity of the heat front must be taken into account. Thus, a term $v_0^{-2} \partial^2 T/\partial t^2$ should be added to $D^{-1} \partial T/\partial t$ in the heat equation. This will result in a significant increase in temperature because the energy cannot be removed at a sufficient rate (see, e.g., *Vedavarz* et al. 1994).

In spite of these difficulties and restrictions, we will demonstrate in later chapters that essential features observed in laser processing can be understood from model calculations. In inhomogeneous media the propagation of light and heat must be calculated in a different way (Chap. 9).

In any case, the knowledge of laser-induced temperature distributions is a prerequisite for the modelling of processing rates, the clarification of the chemical kinetics, and the enlightenment of the basic microscopic interaction mechanisms. It is evident that it is desirable to measure as many of the relevant quantities as possible in situ, i.e., during laser processing.

2.3 Selective Excitations of Molecules

Laser photochemistry near or at molecule–solid interfaces can be based on selective electronic excitations of both the molecules and the solid surface, on selective vibrational excitations of the molecules, or on a combination of those. The excitation energy can also be transferred indirectly via an intermediate species as in photosensitization.

Electronic transitions of molecules are located mainly in the UV and VIS. Vibrational transitions are located in the IR. Both electronic transitions and vibrational transitions can be excited by single-photon (linear) processes or by multiphoton (MP) nonlinear processes.

Laser-photochemical (non-thermal) processing is frequently based on *electronically* excited molecules and photofragments of those. There are, however, very few examples where selective single- or multiphoton *vibrational* excitations are of importance.

Different fundamental mechanisms involved in selective optical excitations of molecules and, to some extent, of solids have been studied for model systems. However, apart from a very few exceptions, only little is known about the photochemistry of systems relevant to LCP. Here, the physical conditions are very complex compared to the conditions in model systems.

The degree of selectivity achieved in a particular photoexcitation process is determined by the ratio of the excitation rate, W_{exc}, and the relaxation rate, W_{relax}. The selectivity is more pronounced the better the condition

$$W_{exc} > W_{relax} \tag{2.3.1}$$

is fulfilled. We shall term an excitation as selective if the system is *not* in local equilibrium. This shall include cases where the laser light induces a local temperature rise but without complete thermalization, e.g., between vibrational and translational degrees of freedom. In a mixture, selectivity can also denote the excitation of a particular kind of species. The term non-selective or thermal excitation is used if the absorbed light energy is, at least locally, thermalized between the different degrees of freedom *and* the different kinds of species.

Subsequently, we shall summarize some basic aspects on selective electronic excitations and IR vibrational excitations.

2.3.1 Electronic Excitations

Electronic excitations of molecules can be based on single-photon or multi-photon processes. They are, in general, accompanied by simultaneous changes in the vibrational and rotational energy of the molecule.

Single-photon excitations

Let us consider some characteristic cases of single-photon (linear) excitations. Figure 2.3.1 shows potential energy curves for the electronic ground state and excited states of different molecules. According to the Franck–Condon principle, transitions occur vertically between maxima in the density $|\psi_1|^2$ and $|\psi_2|^2$ where ψ_i are vibrational wave functions for the lower and upper electronic states. If the excited electronic state is unstable (Fig. 2.3.1a), excitation results in direct dissociation within times of, typically, 10^{-14} to 10^{-13} s. Clearly, relaxation and energy transfer between gas-phase molecules is unlikely within such short times. If the excited electronic state is stable, dissociation only occurs for photon energies $h\nu \geqslant E'_D$ (Fig. 2.3.1b). However, in many cases

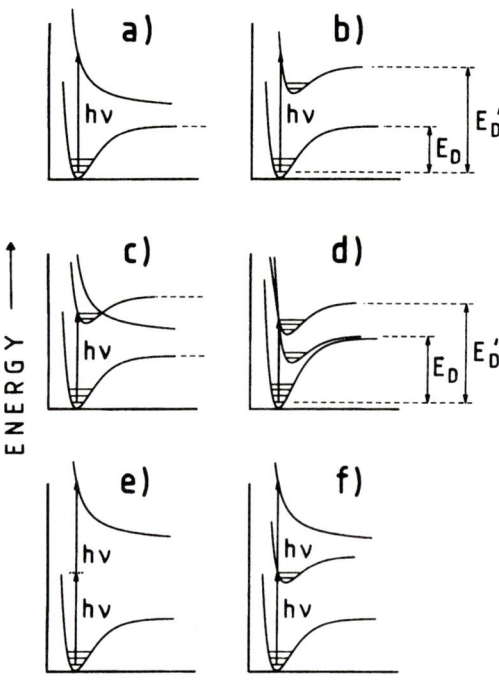

Fig. 2.3.1 a–f. Potential energy curves for the electronic ground state and excited states of molecules showing different cases of optical excitation and dissociation. E_D and E'_D are dissociation energies. Vibrational energy levels are only indicated. Rotational levels are not shown at all. Cases (**a**) to (**d**) show *single*-photon excitations. (**e**) Coherent two-photon excitation. (**f**) Sequential two-photon excitation. The energies of photons in cases (**e**) and (**f**) are not necessarily equal

dissociation of isolated molecules is even observed for $hv \leq E'_D$ (Figs. 2.3.1c,d). This phenomenon is termed spontaneous predissociation. It is related to transitions from the initially excited electronic state to an unstable state (Fig. 2.3.1c) or to a stable electronic state whose dissociation energy is below the originally excited state (Fig. 2.3.1d). The final state can also be the electronic ground state itself; then, the molecule dissociates if $hv \geq E_D$. Such intramolecular radiationless transitions result from the mixing of states near crossings of potential curves. They are therefore more common in polyatomic molecules than in diatomic molecules. The typical times for predissociation are between 10^{-12} and 10^{-6} s. Radiationless transitions are also termed internal conversion and intersystem crossing [*Avouris* et al. 1977; *Bixon* and *Jortner* 1968] or as Landau–Zener transitions [*Akulin* and *Karlov* 1992].

The main limitation of single-photon excitation/dissociation processes relevant to laser-chemical processing, is the lack of flexibility of available lasers to match the maxima of dissociative transitions in the medium to far UV.

Densities of excited and dissociated species

In photochemical laser processing, reaction rates are directly related to the average number of excited or dissociated molecules. Let us consider the problem for the simple photochemical process

$$AB_\mu + hv \underset{\tau_{em}}{\overset{\sigma}{\rightleftarrows}} AB^*_\mu \underset{\tau'_{rec}}{\overset{\tau_d}{\rightleftarrows}} A + \mu B \ . \tag{2.3.2}$$

$$\underset{\tau_{rec}}{\xleftarrow{\hspace{2cm}}}$$

The excitation of molecules AB_μ is characterized by the effective cross section, σ, at the particular laser wavelength. σ depends on the type of reactant, the gas pressure, etc. The effect of pressure broadening, line shifts, etc., also depends on the band width of the laser light. The situation is similar for species AB_μ dissolved in a liquid. The effective cross section can significantly differ from the excitation (absorption) cross section for a single *isolated* molecule, σ_a. The latter is measured under collisionless conditions, and it has large values only if the photon energy matches the distance between respective energy levels of the molecule and if the transition is allowed by symmetry (selection rules), i.e., if it is optically active. For $hv \geq E'_D$ and negligible fluorescence, the absorption cross section is equal to the dissociation cross section, σ_d.

The relaxation time for deactivation of AB^*_μ is denoted by τ_{em}. τ_d describes the time for dissociation of AB^*_μ in a first order decomposition process (Chap. 3). τ_{rec} and τ'_{rec} characterize the recombination of A and B to AB_μ and AB^*_μ, respectively. The relaxation times depend on gas pressure.

Electronic absorption and dissociation cross sections of molecules that are of particular relevance in laser processing are summarized in Table V for different laser wavelengths. Most of the values σ found in the literature refer to effective cross sections.

For an estimation of photochemical processing rates, the concentrations of species A and AB_μ^*, x_A and x_{AB^*}, must be known. Stationary values of those are given in Appendix A.4.

Multiphoton excitations

Multiphoton (MP) processes open up additional excitation/dissociation channels and thereby permit one to use the laser light at a particular wavelength more efficiently or to use a much wider variety of precursor molecules.

The number of molecules excited in a MP process depends nonlinearly on photon flux. Figures 2.3.1e and f show two different kinds of multiphoton excitations. If the photon energy is smaller than the energy difference between the first optically active excited state and the ground state, excitation is possible only via *coherent* two-photon absorption (case e). The absorption cross section for coherent n-photon excitations is henceforth denoted by $^{(n)}\sigma$. The situation is different in case f. Here, the molecule is transferred to the first excited state by absorption of a single photon. The absorption of an additional photon results in dissociation. This process is denoted as *sequential* two-photon absorption. In the simplest case, the cross section of a sequential n-photon excitation is proportional to $\Pi_i^n {}^{(1)}\sigma_i$ where $^{(1)}\sigma_i$ is the single-photon absorption cross section; Π denotes the product. The energy of the single photons involved in a multiphoton process are not necessarily equal.

Efficient MP processing can only be performed with high-power pulsed lasers. Because such laser pulses may cause substrate damage, most applications of MP processing are performed with an irradiation geometry where the laser beam propagates parallel to the substrate surface (Fig. 1.2.2b). The relevant processing step can be based on multiphoton ionization (MPI), multiphoton dissociation (MPD), or multiphoton excitation of the precursor molecules.

Photosensitization

Photosensitization denotes a process where photons are absorbed by intermediate species which transfer their excitation energy to acceptor molecules via collisions [*Calvert* and *Pitts* 1966]. For example, direct photolysis of CH_4 is only possible below 144 nm

$$CH_4 + h\nu(\lambda < 144 \text{ nm}) \rightarrow CH_2 + H_2 , \qquad (2.3.3)$$

while the Hg-photosensitized reaction can take place at a longer wavelength

$$Hg(^1S_0) + h\nu(\lambda = 253.7 \text{ nm}) \rightarrow Hg(^3P_1) \qquad (2.3.4)$$

$$Hg(^3P_1) + CH_4 \qquad \rightarrow Hg(^1S_0) + CH_3 + H . \qquad (2.3.5)$$

As can be seen from this example, the photo-products are not necessarily the same in both cases.

Photosensitized reactions are very common in photochemical studies, but cannot be employed in laser microchemistry due to the delocation of the reaction in this process. However, the technique has been applied for large-area low temperature growth of epitaxial layers of HgTe [*Irvine* et al. 1984], the deposition of hydrogenated amorphous silicon (a–Si: H) [*Kamimura* and *Hirose* 1986], and etching reactions [*Loper* and *Tabat* 1984].

2.3.2 Infrared Vibrational Excitations

In this subsection we shall discuss some fundamentals on vibrational excitations of free molecules in the electronic ground state. Special emphasis is put on aspects that are relevant to LCP.

Excitation of isolated diatomic molecules

Vibrational excitation of single *isolated* molecules can be realized within the collisionless environment of a molecular beam. Figure 2.3.2 shows an anharmonic potential which shall represent the electronic ground state. Because of anharmonicity, the vibrational levels are not equally spaced. Rotational levels are ignored in the figure although they are very essential in excitation processes. For simplicity, we always use the term vibrational transition, even when the rotational state of the molecule is changed simultaneously.

The simplest absorption process is a one-photon (linear) excitation of the vibrational state $v = 1$, as shown by arrow 'a'. The excitation energy $hv = E_{v=1} - E_{v=0}$ is, typically, between 100 cm^{-1} and some 1000 cm^{-1} (about 0.01 eV to some 0.1 eV). Arrows 'b' and 'c' indicate excitations of the third and fourth vibrational level by *coherent* two- and three-photon processes, respectively.

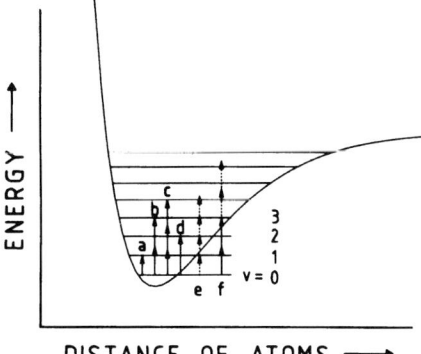

Fig. 2.3.2. Various types of IR vibrational excitations of a single isolated diatomic molecule. For simplicity, rotational levels have been ignored. 'a' one-photon excitation; 'b', 'c' coherent two- and three-photon excitation; 'd' one-photon overtone excitation; 'e' sequential four-photon excitation; 'f' two-photon coherent excitation (full arrows) followed by sequential excitation (dotted arrows). The energies of photons employed in multiphoton excitation processes are not necessarily equal

Such MP processes become quite unlikely for excitations $v > 4$, because of the rapid decrease in σ with such highly nonlinear processes. Vibrational levels $v = 2, 3, \ldots$ can also be excited, though with low probability, in a one-photon overtone absorption process using a photon energy $hv' = E_v - E_{v=0}$. This is indicated for $v = 2$ by arrow 'd'. MP absorption by *sequential* excitation (case 'e') becomes quite unlikely for high vibrational levels as well, simply because of the (increasing) mismatch between the photon energy and the vibrational energy levels. A combined excitation process is shown in case 'f'. Here, two-photon coherent absorption (full arrows) is followed by two-photon sequential absorption (dotted arrows). In this way, higher vibrational levels can be excited.

Excitation of isolated polyatomic molecules

In contrast to diatomic molecules, polyatomic molecules can absorb a great number of monochromatic photons even under collisionless conditions. This can be seen from Fig. 2.3.3. The 'superposition' of different vibrational level systems, corresponding to different normal modes of the molecule, results in different regions of vibrational level densities. At low energies the vibrational levels are discrete. With increasing energy, their density increases rapidly. The region above a certain energy, E_s, is denoted as *quasi-continuum*. E_s corresponds to, typically, 3 to 10 vibrational quanta for simple polyatomic molecules and to only one vibrational quantum for molecules consisting of many atoms, or such with heavy atoms.

Selective excitation of the particular mode that is in resonance with the infrared laser frequency, takes place as discussed with diatomic molecules. If

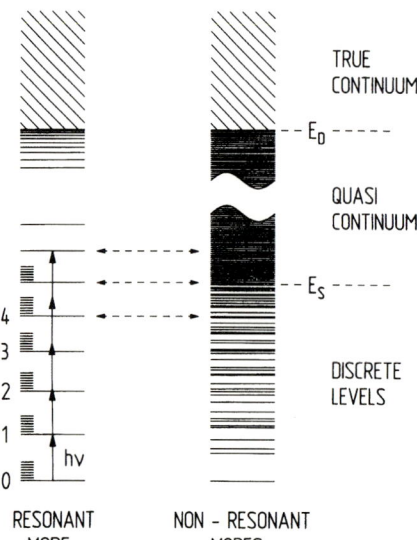

Fig. 2.3.3. Multiphoton vibrational excitation and dissociation of a polyatomic molecule by intense IR radiation. Left: Vibrational-rotational levels for the mode that is selectively excited by the IR radiation. Right: Three regimes of vibrational level densities: Discrete levels of non-resonant modes, vibrational quasi-continuum, and true continuum

this resonant mode is excited up to the quasi-continuum, an even weak intermode anharmonicity is sufficient to cause stochastization of the vibrational energy. In other words, when the vibrational energy stored in the selectively driven mode approaches E_s, it will spread over all the different modes. This mechanism diminishes the number of vibrational quanta in the resonant mode and thereby permits further laser-light absorption. This process takes place again and again. Thus, the vibrational degrees of freedom are subjected to strong heating. The true continuum is reached at the dissociation energy, E_D. Dissociation of the vibrationally excited molecule will take place, in general, via the *lowest* dissociation channel. With the high excitation rates that can be achieved with intense IR lasers, many-photon *superexcitation* of polyatomic molecules far above the dissociation energy has been observed [*Steinfeld* 1981; *Bagratashvili* et al. 1983].

The rate of vibrational excitation of a single polyatomic molecule is often written as

$$W_{ex} = \sigma \frac{I}{h\nu}, \qquad (2.3.6)$$

where σ is the *average* absorption cross section which depends on laser fluence and pulse length. For polyatomic molecules, σ has values of, typically, 10^{-20} to 10^{-18} cm^2. The average number of IR photons absorbed per pulse by a *single* molecule is $\langle n \rangle = \sigma\phi/h\nu$. With polyatomic molecules, 10 to 100 IR photons can be absorbed with fluences $\phi \approx 1 - 10 \, \text{J/cm}^2$.

Collisionless IR-MP excitation and dissociation of many molecules which are used as precursors in laser-chemical processing, such as SF_6, BCl_3, $CO(CF_3)_2$, CF_3I, and CDF_3, is consistent with the model in Fig. 2.3.3. Highly vibrationally excited molecules and radicals produced by IR-MP excitation/dissociation, interact with solid surfaces quite differently than molecules in the vibrational ground state. Examples will be given in various chapters.

The role of collisions

Collisions will not only change the lifetime of a particular excitation, but permit high level vibrational excitation and dissociation of even diatomic molecules via near-resonant energy transfer. For *pure* gaseous CO, this process can be described by

$$\begin{aligned} CO(v=0) + h\nu &\leftrightarrow CO(v=1) \\ CO(v=1) + CO(v=1) &\leftrightarrow CO(v=2) + CO(v=0) \\ CO(v=2) + CO(v=1) &\leftrightarrow CO(v=3) + CO(v=0) \end{aligned} \qquad (2.3.7)$$

etc. The transition $v = 0 \to 1$ which is in resonance with the photon energy $h\nu$ is excited in a one-photon process (Fig. 2.3.2). Subsequently, the energy is transferred to other excited molecules. Thereby, high vibrational states, up to the dissociation limit, can be reached. The same process can take place with lower efficiency (unless there is a coincidence in vibrational energies), between *different* molecules, A and B, including isotopes.

Selectivity

The selectivity of a particular photo-excitation process is determined by (2.3.1). The redistribution of vibrational energy is determined by different relaxation times:

- The time for spontaneous radiative (dipolar) transitions between low lying well-separated vibrational levels which determines the natural linewidth. This is, typically, of the order 10^{-3} s.
- The time for *intra*-molecular transfer of vibrational energy between different vibrational modes being excited, τ_{v-v}^{A}. This time decreases with increasing vibrational anharmonicity and increasing density of vibrational levels. Within the quasi-continuum it is, typically, of the order of 10^{-13} to 10^{-11} s.
- The time for *inter*-molecular transfer of vibrational energy via collisions between molecules of either the *same* kind, τ_{v-v}^{A-A}, or of *different* kind, τ_{v-v}^{A-B}. For low level excitations, energy exchange between molecules of different kind is less efficient because of the mismatch of vibrational energy levels. For highly excited states the type of colliding molecules becomes almost unimportant.
- The time for molecular vibrational energy to be transferred to translational degrees of freedom, τ_{v-T}. This is the time to reach thermal equilibrium within the molecular mixture.

Clearly, τ_{v-v}^{A-A}, τ_{v-v}^{A-B}, and τ_{v-T} vary with experimental conditions such as the molecular density, temperature, and the type of admixtures or solvents. Figure 2.3.4 shows, schematically, various energy transfer processes and the corresponding relaxation times that are typical for *binary collisions* of gas-phase molecules at 300 K and 1000 mbar. With these conditions, the time between successive collisions is $\tau_c \approx 10^{-10}$ s ($\tau_c^{-1} \approx N \langle \sigma_c v \rangle$; σ_c is the cross section for collisions and v the velocity of molecules).

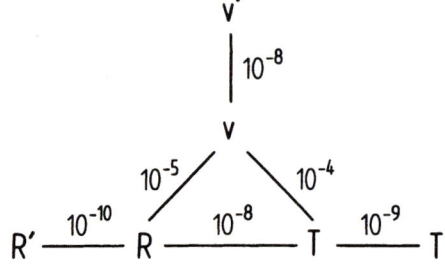

Fig. 2.3.4. Different relaxation channels for energy-transfer during binary collisions of molecules. v'—v stands for vibrational-vibrational, v—T for vibrational-translational, R—T for rotational-translational processes. The numbers represent typical relaxation times for pure gases at 300 K in units of s bar. [*Eyring* et al. 1980]

2.3 Selective Excitations of Molecules

Classification of IR multiphoton photochemistry

Infrared multiphoton photochemistry based on pulsed-laser excitation of high vibrational states can be classified into four different cases:

Mode- or *bond-selective* photochemistry requires an excitation rate that is large compared to the rate of intra-molecular vibrational energy transfer. This would need both a mode which is fairly isolated from other vibrational modes, and high intensity picosecond or femtosecond resonant excitation. Mode isolation is well fulfilled for diatomic molecules, because they have only one vibrational degree of freedom. In fact, vibrationally enhanced photochemical reactions based on low level excitations of diatomic molecules have been reported. Collisionless multiphoton, high-level excitation/dissociation by monochromatic infrared radiation is, however, unlikely/impossible (Fig. 2.3.2). For polyatomic molecules the situation is somewhat complementary. They permit high level vibrational excitation but no bond selectivity (Fig. 2.3.3). Laser processing based on bond-selective vibrational excitation/dissociation of precursor molecules has not been demonstrated.

The (three) remaining cases of selective vibrational excitations have been employed in laser processing:

Molecule-selective excitation requires

$$W_{ex} \gg \frac{1}{\tau_{v-v}^{A-B}} \,. \tag{2.3.8}$$

The vibrational energy within molecules A which interact with the infrared light is in equilibrium. Other molecules within the mixture, B, that are *not* directly excited are in lower vibrational states. Thus, molecules in resonance with the laser frequency acquire a higher vibrational temperature than all other molecules. High level molecule-selective vibrational excitation can take place as discussed together with Fig. 2.3.3, and via collisions analogous to (2.3.7). The time for energy transfer from A to B or from A to surface atoms, τ_{v-v}^{A-B}, must be long compared to the time of resonant energy transfer, A to A.

Molecule-selective excitation and dissociation is of practical interest, e.g., in laser *isotope separation* [*Letokhov* 1983]. With isotopes whose mismatch between vibrational levels is of the order of $k_B T$, collisional excitations of the type (2.3.7) must be avoided.

Non-equilibrium excitation is achieved if

$$W_{ex} \gg \frac{1}{\tau_{v-T}} \,. \tag{2.3.9}$$

In this case, there may be vibrational equilibrium among all molecules in the mixture, but no equilibrium between vibrational and translational degrees of freedom. Condition (2.3.9) can only be fulfilled if the gas mixture does not contain any component with fast v–T relaxation. For example, with pure SF_6 one finds that with gas pressures $p(SF_6) \approx 0.1$ mbar and low cw CO_2-laser-light intensities, about 50% of the molecules can be in a non-equilibrium state. Because of the difference in the vibrational and translational temperature, nonselective vibrational photochemistry is possible when the time constant for the *fastest* reaction channel is shorter than τ_{v-T}. The most important application is IR laser-induced radical synthesis [*Letokhov* 1988].

Photothermal excitation is characterized by

$$W_{ex} \ll \frac{1}{\tau_{v-T}}. \qquad (2.3.10)$$

All molecules within the reaction volume defined by the laser beam are in thermal equilibrium. The vibrational energy is immediately thermalized.

Since σ in (2.3.6) and τ_{v-T} depend on temperature, (2.3.9) transforms to

$$W_{ex} \tau_{v-T} = \sigma(T) \frac{I}{hv} \tau_{v-T}(T) \gg 1. \qquad (2.3.11)$$

The decrease in relaxation time with temperature can be described by the Landau–Teller relation $\tau_{v-T}(T) = \tau_{v-T}(0) \exp(\mu/T^{1/3})$ where $\mu > 0$. The cross section $\sigma(T)$ can increase or decrease with temperature. Because $T = T(I)$ relation (2.3.11) is a non-monotonic function of intensity with regions that correspond to either thermal or non-thermal gas-phase excitations.

2.4 Surface Excitations

In this section we give an overview on non-thermal or not purely thermal excitations of solid surfaces and adsorbate–adsorbent systems.

2.4.1 External Photoeffect

The external photoeffect denotes the ejection of electrons from a solid surface that is irradiated with photons $hv \geq hv_G$ where v_G denotes a threshold frequency. With metals, v_G is located within the VIS and UV [λ_G(Cs, Cu, Pt) $\approx$ 639, 277, 231 nm]. If the solid is immersed in a reactive ambient, molecules that capture an electron can become unstable. Spontaneous decay or partial fragmentation of the molecule may be the consequence [*Schröder* et al. 1987]. Further fragmentation can take place via collisions with other molecules or with the substrate surface (Fig. 2.4.1).

2.4.2 Internal Photoeffect

The internal photoeffect is the generation of electron-hole pairs in semiconductors or insulators that are irradiated with photons $hv > E_g$ (Fig. 2.1.1; we will not bother with direct and indirect processes as this is outlined in standard textbooks). Electrons and holes change the optical properties of the material and thereby its interaction with laser light (Sect. 7.6). Moreover, photocarriers play a fundamental role in many types of molecule–surface interactions relevant in laser-CVD, surface modification, and etching (Fig. 15.1.1). The analysis of such processes requires detailed information on the carrier distribution.

Carrier densities

For an intrinsic semiconductor, the carrier density can be described, in a simple approximation, by

$$\frac{\partial N_c(x,t)}{\partial t} = \alpha(v)\frac{I(x,t)}{hv} - k^{rec}[N_c(x,t) - \bar{N}_c(T)] + \nabla[D_c(x,t)\nabla N_c(x,t)]. \quad (2.4.1)$$

The first term describes the generation of photocarriers by interband absorption. The second term stands for the loss of carriers by recombination, where $k^{rec} = \tau_{rec}^{-1}$ is the rate constant for electron–hole pair recombination and $\bar{N}_c(T)$ the carrier density in (thermal) equilibrium. The last term describes the diffusion of carriers.

The recombination time, τ_{rec}, depends on the material and the concentration of photocarriers. It is determined by direct or indirect band-to-band recombination, multicarrier (Auger) recombination, and by defects and impurities. Thus, values of τ_{rec} near the surface differ somewhat from those within the bulk. τ_{rec} is, typically, between a few picoseconds and several seconds.

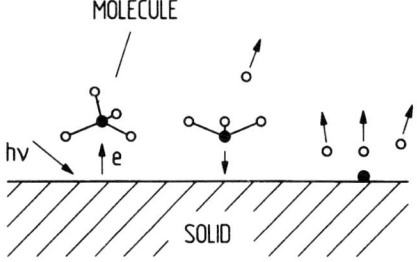

Fig. 2.4.1. Decomposition of a molecule by photoelectron capture

The diffusion coefficient can be written as

$$D_c = \frac{\sigma_e D_e + \sigma_h D_h}{\sigma_e + \sigma_h}. \tag{2.4.2}$$

D_e and D_h are the actual diffusion coefficients of electrons and holes and σ_e and σ_h the corresponding conductivities.

Equation (2.4.1) ignores laser-induced heating, and collective (plasma) phenomena which are observed at very high carrier densities [*Yoffa* 1980].

Let us consider electron–hole pair generation in some more detail for silicon. Because E_g (Si; 300 K) $\approx$ 1.1 eV band-gap excitations become possible with $\lambda < 1\,\mu$m. With increasing temperature, E_g decreases, and thereby the minimum photon energy for excitation. With carrier densities $N_c > 10^{18}$/cm^3, Auger recombination becomes dominating. Then, the carrier lifetime decreases with increasing concentration as

$$\tau_{rec} = \frac{1}{k^{rec}} \propto \frac{1}{N_c^2}. \tag{2.4.3}$$

For room temperature, the second term in (2.4.1) can be substituted by ζN_c^3 with $\zeta \approx 4 \cdot 10^{-31}$ cm^6/s. For example, an initial carrier density of $N_c = 10^{22}$/cm^3 decreases via Auger recombination within about 10^{-10} s to 1%.

Avalanche ionization

With high laser-light intensities, the rate of electron excitation may overtake the rate of energy loss via generation of phonons. Then, electrons become highly excited and, eventually, attain sufficient energy to generate secondary electron–hole pairs by impact ionization of lattice atoms/molecules (the contribution of holes can often be ignored because of their low mobility, in particular in insulators and large band-gap semiconductors). Because of the positive feedback involved in this process, very high electron densities can be generated. This effect is often termed avalanche ionization. With such conditions, even originally highly transparent materials can become strongly absorbing and, as a consequence, *optical breakdown* and plasma formation is often observed.

With very high laser-light intensities, electrons may even be generated when $h\nu < E_g$. This is mediated via highly nonlinear processes such as defect enhanced or coherent MP absorption (Fig. 2.1.1), light-induced defect formation (Sect. 13.4), thermal ionization, and MP ionization (MPI).

2.4.3 Electromagnetic Field Enhancement, Catalytic Effects

Various types of electromagnetic field enhancements observed on solid surfaces can significantly alter both surface morphologies and reaction rates in laser-chemical processing [*Akhmanov* et al. 1985; *Chen* and *Osgood* 1983; *van Driel* et al. 1985; *Nitzan* and *Brus* 1981; *Monreal* and *Apell* 1990]. Such field enhancements may be related to surface roughnesses, nucleation centers, clusters, the excitation of surface polaritons, interference phenomena, etc.

The physical and chemical properties of surfaces change during laser ablation, etching, deposition, doping, and surface modification. Changes in surface properties may cause autocatalytic effects, as observed during laser-induced metal deposition, catalyze electroless plating, as observed after polymer ablation, etc.

Some of these different effects are discussed in detail in other chapters.

2.4.4 Adsorbed Molecules

Adsorption of molecules on solid surfaces changes their electronic and vibrational properties and thereby the absorption cross section for the interaction with light. Additionally, the number of vibrational degrees of freedom can increase. With a diatomic molecule such as CO, adsorption changes the number of vibrational degrees of freedom from one to six (Fig. 2.4.2). *Selective* elec-

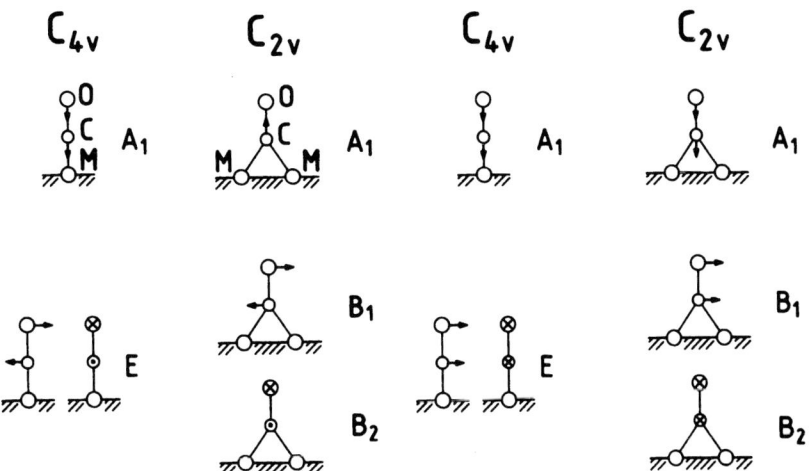

Fig. 2.4.2. Normal modes of CO molecules adsorbed on a metal surface in top site (C_{4v} symmetry) and bridging site (C_{2v} symmetry) positions. $\otimes$ and $\odot$ indicate elongations perpendicular to the plane of the drawing. Vibrations with A and B symmetry are non-degenerate, vibrations with E symmetry are two-fold degenerate

tronic or vibrational excitation of adsorbate–adsorbent systems may result in selective desorption or photolysis of adsorbed species, in changes in the catalytic properties of the surface, etc. [*Aussenegg* et al. 1983].

It should be emphasized that the consideration of *single* effects oversimplifies the situation. A real understanding of chemical reactions at interfaces requires simultaneous treatment of the different interactions in the gas phase, adsorbed phase, and solid phase and, in addition, the often subtle couplings between them.

3 Reaction Kinetics and Transport of Species

An estimation of processing times in laser-chemical processing (LCP) requires the knowledge of reaction rates. In this chapter we shall outline some fundamentals on the kinetics and mass transport of laser-induced chemical reactions, with special emphasis on heterogeneous reactions at gas–solid interfaces (Fig. 1.2.1).

Heterogeneous activation of a chemical reaction by direct laser-light irradiation of the substrate can take place within adsorbed layers, or at gas–solid, liquid–solid, or solid–solid interfaces (Figs. 1.2.1a, b, d, e) The reaction zone, i.e., the area on the substrate surface where the reaction takes place, is not necessarily the same size as the laser spot but it can be smaller or larger (Sect. 5.3). If laser light at parallel incidence is used to excite or dissociate species that do *not* react within the ambient medium (except recombination) but only on the substrate surface, we still term the reaction heterogeneous.

With *homogeneous* activation, the first step of excitation *and* subsequent reaction takes place within a certain volume of the gas, the liquid, or the solid. In LCP, such reactions are activated mainly by using parallel incidence of the laser beam (Fig. 1.2.1c).

Let us consider a chemical reaction of the type

$$\zeta_{AB}AB + \zeta_{CD}CD + \cdots \rightleftarrows \zeta_{AD}AD + \zeta_{BC}BC + \cdots \qquad (3.0.1)$$

where AB, CD, etc., are reactants which may include the substrate as, for example, in laser-induced chemical etching of Si in Cl_2. AD, BC, etc., are reaction products which may desorb from the substrate surface or may stick on it. The ζ_i are stoichiometric coefficients which characterize the particular reaction path and which may differ for heterogeneous and homogeneous reactions. In general, reactions of type (3.0.1) will consist of a number of consecutive steps:

- Transport of reactants into the reaction volume.
- Transport of product atoms or molecules generated, for example in the gas phase, to the surface with possible recombination or secondary reactions on the way.
- Adsorption of one or more reactants on the surface.
- Pyrolytic or photolytic activation of molecules at or near the substrate surface.

- Pyrolytic or photolytic activation of the substrate surface.
- Transport of electrons, atoms, molecules, etc., within the solid surface.
- Condensation or further reactions of excited molecules on the surface.
- The chemical reaction itself.
- Desorption of reaction products from the surface.
- Transport of reaction products out of the reaction volume.

Clearly, in different types of LCP, one or more of these steps will either not occur at all or else will differ significantly.
The *net* reaction rate in (3.0.1) is given by

$$W = W_\rightarrow - W_\leftarrow , \qquad (3.0.2)$$

where $W_\rightarrow$ and $W_\leftarrow$ denote the reaction rates in forward and backward direction. The (dynamic) equilibrium is characterized by $W_\rightarrow = W_\leftarrow$. Away from equilibrium, the concentration of reactants and reaction products changes with time. If the concentration is uniform and if mass transport limitations are ignored, the temporal change in reaction rate for a closed system is described by

$$W = -\frac{1}{\zeta_{AB}}\frac{dN_{AB}}{dt} = -\frac{1}{\zeta_{CD}}\frac{dN_{CD}}{dt} = \cdots = \frac{1}{\zeta_{AD}}\frac{N_{AD}}{dt} = \frac{1}{\zeta_{BC}}\frac{N_{BC}}{dt} = \cdots . \qquad (3.0.3)$$

The total number density of species is given by

$$N = \sum_i N_i . \qquad (3.0.4)$$

In homogeneous gas-, liquid-, or solid-phase reactions, N_i is the number of atoms (molecules) per unit volume. Here, N_i includes both reactants and reaction products. In adsorbed-phase processing, N_i is the number of species per unit area. Instead of N_i we often introduce molar ratios $x_i = N_i/N$ with $\Sigma x_i = 1$.

In many cases of LCP, the concentration of reaction products is kept small so that $x_{AD}, x_{BC} \ll 1$. Thus, $W_\rightarrow \gg W_\leftarrow$ and the net reaction rate can be approximated by $W \approx W_\rightarrow$.

Irrespective of the type of reaction or the detailed activation mechanism, the reaction rate is either limited by the chemical kinetics or by mass transport.

Within the *kinetically* controlled regime, the reaction rate depends on the detailed activation mechanisms, the density of reactants, the physical and chemical properties of the substrate, and the laser parameters.

Within the *mass transport* limited regime, the reaction rate depends on the maximum flux of species into the reaction zone, but not on the detailed activation mechanisms. Here, strong gradients in the concentration of reactants occur near the reaction zone.

3.1 Photothermal Reactions

For thermally activated reactions of type (3.0.1) and $W_\rightarrow \gg W_\leftarrow$, the (net) reaction rate can be described by

$$W(x,t) = k(T) N_{AB}{}^{\gamma_{AB}} N_{CD}{}^{\gamma_{CD}} \cdots$$
$$= W_0(x,t) \exp\left(-\frac{\Delta E}{k_B T(x,t)}\right), \tag{3.1.1}$$

where the (phenomenological) rate constant is given by the Arrhenius law

$$k(T) = k_0 \exp\left(-\frac{\Delta E}{k_B T(x,t)}\right), \tag{3.1.2}$$

where ΔE is the *apparent* chemical activation energy. In gas-phase reactions, typical values of ΔE are between 0.5 to 5 eV (about 10 to 100 kcal/mol). k_0 is a preexponential factor whose dimension depends on the total reaction order. It also depends on temperature, the spatial orientation of colliding species (steric factor), the distribution of rotational and vibrational energy of molecules, etc., [*Hirschfelder* et al. 1964; *Hänggi* et al. 1990]. k_B is the Boltzmann constant (if ΔE is given in kcal/mol, k_B is replaced by the gas constant, R_G). The temperature distribution can be written as $T(x,t) = T(\infty) + \Delta T(x,t)$ where $\Delta T(x,t)$ is the laser-induced temperature rise and $N_i \equiv N_i(x,t)$. γ_i are partial reaction orders. The total reaction order is

$$\gamma = \gamma_{AB} + \gamma_{CD} + \cdots = \sum_i \gamma_i. \tag{3.1.3}$$

Here, the summation includes the reactants only. γ is, typically, within the range $0 \leq \gamma \leq 3$ and includes fractional numbers. In thermodynamic equilibrium, or close to it, the partial reaction orders γ_i coincide, for elementary reactions, with the stoichiometric coefficients, ζ_i, as introduced in (3.0.1).

Sometimes, it is more convenient to use instead of (3.1.2) the Frank–Kamenetsky expansion

$$k(T) \approx k_0 \exp\left(-\frac{\mathscr{E}}{T_c}\right) \exp\left(-\frac{\mathscr{E}}{T_c} \frac{T_c - T}{T_c}\right), \tag{3.1.4}$$

where T_c is the maximum temperature and $\mathscr{E} \equiv \Delta E / k_B$.

The reaction enthalpy (reaction heat at constant pressure) is given by $\Delta H = \Delta H_\rightarrow - \Delta H_\leftarrow$ where $\Delta H_\rightarrow \equiv \Delta E$ (Fig. 3.1.1). We term a reaction exothermal if $\Delta H < 0$ and endothermal if $\Delta H > 0$. The reaction coordinate represents the energetically most favorable path in the (multidimensional) phase space in which the molecules participating in the reaction are described.

If the reaction proceeds within a closed system, the *net* reaction rate decreases continuously with time and, finally, becomes zero when $W_\rightarrow = W_\leftarrow$. From the law of mass action we can calculate the equilibrium constant for a

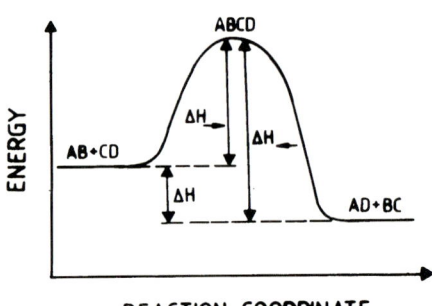

Fig. 3.1.1. Illustration of the reaction enthalpy $\Delta H = \Delta H_\rightarrow - \Delta H_\leftarrow$. The activation energy for the reaction $AB + CD \rightarrow AD + BC$ is $\Delta E \equiv \Delta H_\rightarrow$. The reaction is exothermal if $\Delta H < 0$ and endothermal if $\Delta H > 0$

homogeneous reaction

$$K_{ce} \equiv K_{ce}(T) = \frac{k_\rightarrow}{k_\leftarrow} = \frac{N_{AD}^{\gamma_{AD}} N_{BC}^{\gamma_{BC}} \ldots}{N_{AB}^{\gamma_{AB}} N_{CD}^{\gamma_{CD}} \ldots}. \qquad (3.1.5)$$

The index ce denotes chemical equilibrium.

Equation (3.1.5) can also be applied to heterogeneous systems in which one of the reactants or reaction products is solid while the other constituents are gaseous or liquid. Consider, for example, etching of Si in Cl_2 atmosphere

$$Si + 2Cl_2 \rightleftarrows SiCl_4. \qquad (3.1.6)$$

The law of mass action can be written in the form

$$K'_{ce} = \frac{k_\rightarrow}{k_\leftarrow} = \frac{N_{SiCl_4}}{N_{Cl_2}^2 N_{Si}}, \qquad (3.1.7)$$

where N_{Si} is the *gas-phase* concentration of Si atoms which is determined by the saturation pressure p_{Si} above the (solid) silicon surface. Because p_{Si} depends only on temperature, it is commonly included into K_{ce}, i.e.,

$$K_{ce} = \frac{N_{SiCl_4}}{N_{Cl_2}^2}, \qquad (3.1.8)$$

where $K_{ce} = K'_{ce} N_{Si}$. In other words, in reactions of type (3.1.6) the solid phase is, in general, not explicitly considered in the law of mass action. The situation is similar in homogeneously induced reactions if one of the reaction products condenses. Consider (3.1.6) from right to left. Assume that gaseous $SiCl_4$ is decomposed within a laser beam at parallel incidence to the substrate (Fig. 1.2.1c). If the partial pressure of Si atoms exceeds the saturation pressure, Si condenses and forms clusters and a solid film.

The Arrhenius law implies several assumptions. In particular, it is valid when the translational, vibrational and rotational temperatures are equal, i.e., if $T_{trans} = T_{vib} = T_{rot}$ and if the system is close to equilibrium. In many cases, however, the Arrhenius law can even be applied to systems that are far from equilibrium.

A further comment seems to be appropriate. Equation (3.0.1) describes a *net* reaction which may include many intermediate steps and parallel reaction pathways. Each of these is characterized by a (different) rate constant k_j. The *overall* reaction rate observed experimentally, is dominated by the slowest step within the fastest reaction channel, and this channel can change with experimental parameters. This is the reason why k_0 and ΔE in (3.1.2) can be considered as constants only within certain ranges of temperatures, molecular densities, etc. It also explains, why ΔE in (3.1.2) is denoted as an *apparent* activation energy. For the same reasons, the reaction orders γ_i do not coincide, in general, with the stoichiometric coefficients of the corresponding net reaction.

As a final point we note that the reaction rate W is proportional to but not identical with the processing rate. This becomes evident from the (heterogeneous) reaction

$$C_2H_2 \to 2C(\downarrow) + H_2, \qquad (3.1.9)$$

where the *deposition* rate is given by

$$W_D[\text{cm/s}] = \frac{Zm}{\rho} W[\text{species/cm}^2\text{s}]. \qquad (3.1.10)$$

Here Z is the number of atoms (molecules) deposited per formula unit, and m and ρ their mass and mass density, respectively.

3.2 Photochemical Reactions

For photochemically activated reactions the rate can be described by

$$W(\mathbf{x}, t) = W_0' \prod_i \left[{}^{(n)}\sigma_i \left(\frac{I(\mathbf{x}, t)}{h\nu} \right)^n \right]^{\gamma_i} = k'(I) N_{AB}{}^{\gamma_{AB}} N_{CD}{}^{\gamma_{CD}} \dots, \qquad (3.2.1)$$

where

$$W_0' = k_0' N_{AB}{}^{\gamma_{AB}} N_{CD}{}^{\gamma_{CD}} \dots, \qquad (3.2.2)$$

and

$$k'(I) = k_0' \prod_i \left[{}^{(n)}\sigma_i \left(\frac{I(\mathbf{x}, t)}{h\nu} \right)^n \right]^{\gamma_i}. \qquad (3.2.3)$$

Π denotes the product over all species i to be excited. k_0', and thereby W_0', depends on the relaxation times relevant for selective excitation of species (Chap. 2). To illustrate (3.2.1) we consider a number of different cases:

- Single-photon excitation of a single type of reactant AB that reacts in a first order reaction which is characterized by $i = AB$, $n = 1$; $\gamma_{AB} = 1$, and $\gamma_{CD} = 0$.

Thus, the rate constant is

$$k'(I) = k'_0 \sigma_{AB} \frac{I}{h\nu}. \tag{3.2.4a}$$

$\sigma_{AB} \equiv {}^{(1)}\sigma_{AB}$ is the single-photon excitation cross section.
- Coherent n-photon excitation of species AB that react in a first order reaction where $i = AB$, $n \neq 1$, $\gamma_{AB} = 1$, and $\gamma_{CD} = 0$

$$k'(I) = k'_0 \, {}^{(n)}\sigma_{AB} \left(\frac{I}{h\nu}\right)^n. \tag{3.2.4b}$$

- Single-photon excitation with two excited species AB and CD that react with partial reaction orders γ_{AB} and γ_{CD}. Thus, $i = AB$, CD and $n_{AB} = n_{CD} = 1$ so that

$$k'(I) = k'_0 \left(\frac{\sigma_{AB} I}{h\nu}\right)^{\gamma_{AB}} \left(\frac{\sigma_{CD} I}{h\nu}\right)^{\gamma_{CD}}. \tag{3.2.4c}$$

- Sequential n-photon excitation of a single reactant AB that reacts with partial reaction order γ_{AB}

$$k'(I) = k'_0 \left({}^{(1)}\sigma_{AB} \frac{I}{h\nu}\right)^{n\gamma_{AB}}. \tag{3.2.4d}$$

Here, the cross sections for sequential excitation steps have been assumed to be equal, and any deactivation of species during these steps has been ignored.

In the case of a mixture of AB and CD where, however, only molecules AB are excited in a coherent n-photon process, the reaction rate is described by

$$W = k'_0 \left[{}^{(n_{AB})}\sigma_{AB} \left(\frac{I}{h\nu}\right)^{n_{AB}} \right]^{\gamma_{AB}} N_{AB}^{\gamma_{AB}} N_{CD}^{\gamma_{CD}} \tag{3.2.5}$$

3.3 The Concentration of Species

The reaction rates in laser-chemical processing depend on the density of relevant species within the reaction zone. This density can be calculated from the following basic equations:

- The equations of continuity for the individual species (diffusion equations).
- The equation of overall continuity.
- The heat transport equation.
- The Navier–Stokes equation. If the viscosity is ignored, this is also called the Euler equation.

3.3 The Concentration of Species

Besides the transport equations, we have the equation of state which gives a relation between the pressure, the temperature, and the concentration of species.

The formulas presented subsequently apply to both heterogeneous and homogeneous laser-induced reactions. However, the meaning of the various quantities, the reaction rates, and the reaction pathways may be quite different in both cases. With heterogeneous reactions, W describes the number of species that react per unit time per unit area, while with homogenous reactions W is the number of species that react per unit time per unit volume. Furthermore, with a heterogeneous reaction the relevant temperature distribution $T(x,t)$ is that induced on the solid surface, while with a homogeneous reaction it is that induced within the volume.

3.3.1 Basic Equations

Let us consider a non-equimolecular first order reaction of the type

$$AB_\mu + M \underset{k_2, k_4}{\overset{k_1, k_3}{\rightleftarrows}} A + \mu B + M . \tag{3.3.1}$$

The decomposition of molecules AB_μ shall take place heterogeneously at the solid surface, i.e., at the interface between the solid and the ambient medium or/and homogeneously within the volume just above the substrate surface. Surface and volume forward reactions shall be characterized by rate constants k_1 and k_3, respectively. M shall be a carrier gas, a liquid solvent, or a solid. In gas-phase reactions a carrier gas (diluent) is often added simply because it is more convenient to work at normal total pressure instead of reduced pressure. The carrier gas M can be inert or it can participate in the reaction as, e.g., H_2 during deposition of Si from $SiH_4 + H_2$. In any case, in the presence of a carrier gas, $N_{AB} \ll N_M$. A and B are products of the forward reaction. A shall be the relevant species for surface processing. If A is generated in the volume, it must first diffuse to reach the substrate surface to be processed. Species A can simply stick on the surface and form a deposit or else react further. In the backward reaction, species A react with B to form the original constituent. The backward reaction between condensed species A and molecules B is characterized by k_2 while the recombination between A and B within the volume is described by k_4. Henceforth we make the following assumptions:

- The reaction shall be so slow that pressure gradients can be ignored (isobaric conditions).
- External forces are ignored.
- The contribution of the Dufour effect (transport of energy originating from a concentration gradient) to the thermal flux is omitted.

- Any generation of heat due to the finite viscosity of the medium is ignored.
- The substrate shall not be melted.

The basic equations will now be listed for the example of the reaction (3.3.1).

Diffusion equations

The equations of continuity for the *individual* species can be written in the form

$$\frac{\partial N_i}{\partial t} + \nabla \mathbf{J}_i = Q_{v,i}, \tag{3.3.2}$$

where i stands for all species within the volume of the ambient medium, i.e., $i = 1, 2, \ldots, j$. For the example of (3.3.1) we have $i \equiv AB_\mu$, A, B, and M. $\mathbf{J}_i \equiv \mathbf{J}_i(\mathbf{x}, t)$ is the flux, and $Q_{v,i}$ the source term, which is the *net* rate of species i generated within the volume

$$Q_{v,AB} = -W_v = -k_3(T)N_{AB} + k_4(T)N_A N_B^\mu. \tag{3.3.3}$$

The flux of species i with respect to stationary coordinates is

$$\mathbf{J}_i = N_i \mathbf{v}_i, \tag{3.3.4}$$

where $\mathbf{v}_i \equiv \mathbf{v}_i(\mathbf{x}, t)$ is the mean velocity of species i, i.e., the sum of the velocities of molecules i within a small volume element divided by its number. With the approximations made, each flux $\mathbf{J}_i$ can be written as a sum of the diffusion flux and the hydrodynamic flux, for example

$$\mathbf{J}_{AB} = -ND_{AB}\left(\nabla \frac{N_{AB}}{N} + k_T^{AB} \nabla \ln T\right) + N_{AB} \mathbf{v}, \tag{3.3.5}$$

where $N = \Sigma_i N_i$. The first term in (3.3.5) describes ordinary diffusion, where $D_{AB} \equiv D_{AB}(T(\mathbf{x}, t))$ is the molecular diffusion coefficient. The second term describes thermal diffusion of species AB_μ where $k_T^{AB} \equiv k_T^{AB}(x_i(\mathbf{x}, t))$ is the thermal diffusion ratio and $x_i = N_i/N$. Diffusion is superimposed on the hydrodynamic flow (third term). $\mathbf{v}$ is the average velocity of *all* species and it is given by

$$\mathbf{v} = \frac{1}{N}\sum_i N_i \mathbf{v}_i = \frac{1}{N}\sum_i \mathbf{J}_i = \frac{1}{N}\mathbf{J}, \tag{3.3.6}$$

where $\mathbf{J}(\mathbf{x}, t)$ is the total flux. $\mathbf{v}$ can be determined from the equation of (overall) continuity.

Equation (3.3.5) is exact only for binary diffusion. For a gas mixture, in principle, the multicomponent Stefan–Maxwell equations for diffusion should be employed [Bird et al. 1960]. However, it is a common approximation to use (3.3.5) even in such cases, but consider D_{AB} as an effective molecular diffusion coefficient. If a diluent is present, binary diffusion is certainly a good

3.3 The Concentration of Species

approximation as long as $N_{AB} \ll N_M$. The flux of the other species can be written in analogy to (3.3.5).

The equation of continuity

The sum of equations (3.3.2) yields the equation of (overall) continuity

$$\frac{\partial N}{\partial t} + \nabla(N v) = \sum_i Q_{v,i} . \qquad (3.3.7)$$

Heat transport equation

For the present problem, the heat equation is most conveniently written in the form

$$\frac{\partial}{\partial t}\left(\sum_i N_i E_i^a\right) + \nabla S_T = Q , \qquad (3.3.8)$$

where S_T is the total energy flux which is given by

$$S_T = \sum_i J_i H_i^a - \kappa \nabla T . \qquad (3.3.9)$$

The (internal) energy per molecule is

$$E_i^a(T) = E_{i0}^a + \int_{T(\infty)}^T c_{v,i}(T')\,dT' .$$

The corresponding enthalpy is

$$H_i^a(T) = H_{i0}^a + \int_{T(\infty)}^T c_{p,i}(T')\,dT' .$$

Here $c_{v,i}$ and $c_{p,i}$ are the heat capacities per molecule at constant volume and constant pressure, respectively. Equation (3.3.8) means that the change in energy per unit volume is determined by the flux of enthalpy and the heat flux. The source term Q [W/cm^3] includes the absorbed laser power density only. Any heats of reaction are included in the first term. $\partial N_i/\partial t$ can be eliminated by means of (3.3.2). With an isobaric system E_i^a can be replaced by H_i^a. For gas mixtures with $N_M \gg N_{AB}, N_A, N_B$ the energy is determined by the carrier gas only. The same approximation holds for liquids where the precursor molecules are diluted within a solvent. Additionally, for liquids $c_p \approx c_v$.

If we assume the molecular heat capacities to be equal for all species and ignore their temperature dependences we obtain

$$c_v \frac{\partial}{\partial t}(NT) + \nabla(c_p N T v - \kappa \nabla T) = Q . \qquad (3.3.10)$$

Equation of state

Because the diffusion equations for the individual species and the equation of continuity are not independent of each other, we need one additional equation which is the equation of state

$$N(x,t) = \frac{N(\infty)}{T^{*q}(x,t)}. \tag{3.3.11}$$

$N(\infty)$ is the total number density far away from the reaction zone. $T^*(x,t) = T(x,t)/T(\infty)$ is the normalized temperature. For an ideal gas, the exponent is $q = 1$, while for a liquid the approximation $q \approx 0$ can be employed.

The boundary-value problem for N_i, N, v, and T is then described by the $j + 3$ equations (3.3.2, 7, 8 and 11). One of the variables N_i, and therefore one of the equations (3.3.2), can be excluded because $N = \Sigma N_i$.

Boundary conditions

The temperature at the interface is assumed continuous

$$T(x_s, t) = T_s. \tag{3.3.12}$$

Thus, any temperature jump at the surface x_s is ignored. T_s must be calculated by taking into account possible changes in the shape of the reaction zone. At sufficiently large distances from the reaction zone

$$T(x \to \infty, t) = T(\infty). \tag{3.3.13}$$

At the surface x_s the net flux (normal component; the surface normal shall be directed from the solid surface into the ambient medium) of species AB_μ shall be equal to the net reaction rate $W_s \equiv W(x_s, t)$

$$-J_{AB}(x_s, t) = W_s, \tag{3.3.14}$$

where $W_s = k_1(T_s)N_{AB} - k_2(T_s)N_B^\mu$ is the net amount of molecules AB decomposed at the surface per unit area and time. Correspondingly we have

$$J_B(x_s, t) = \mu W_s. \tag{3.3.15}$$

From (3.3.6, 14, and 15) we obtain for the normal component of the average velocity at the surface

$$v(x_s, t) = \frac{b W_s}{N}, \tag{3.3.16}$$

where $b = \mu - 1$ decribes the net increase in particle number density per formula unit. Far away from the reaction zone we assume

$$N_{AB}(x \to \infty, t) = N_{AB}(\infty) \text{ and } N_A(x \to \infty, t) = N_B(x \to \infty, t) = 0, \tag{3.3.17}$$

and at the surface

$$N_A(x_s, t) = 0. \qquad (3.3.18)$$

The latter condition implies that all species A impinging onto the surface x_s stick on it.

Subsequently, we will investigate different solutions of this boundary-value problem. In many laser processing situations, one can assume steady state conditions.

The assumptions made clearly oversimplify the real situation of laser-activated chemical reactions at or near interfaces. In particular, the omission of ordinary (free) convection is certainly a crude approximation if we are dealing with high-density ambient media, i.e., with high gas pressures or with liquids. It should also be emphasized that the equations discussed in this section are coupled. Thus, the single contributions to the reaction rate are not independent of each other. The *coupling* of fluxes results in the appearance of *new* phenomena (Sect. 17.4.1).

3.3.2 Dependence of Coefficients on Temperature and Concentration

For *gases*, the temperature and pressure dependence of the *molecular* diffusion coefficient, for example of species AB_μ, can often be described by

$$D_{AB} \equiv D_{AB}(T_G, p) \approx D_{AB}(T(\infty), p_s) \frac{T_G^{*n}}{p^*}, \qquad (3.3.19)$$

where $T_G^* \equiv T_G(x,t)/T(\infty)$. $D_{AB}(T(\infty), p_s)$ is the diffusion coefficient at a standard pressure, p_s, far away from the reaction zone. Within the elementary kinetic theory of gases the exponent is $n = 1.5$. Experimental values are within ranges $1.5 \leq n \leq 2$ [*Hirschfelder* et al. 1964]. At low to medium gas pressures, $D_{AB} \propto 1/p^*$ where $p^* = p/p_s$. The total pressure is $p = \Sigma p_i$ if $N_M = 0$, and $p \approx p_M$ if $N_M \neq 0 (N_M \gg N_i)$.

The dependence of D_{AB} on the relative concentrations of gaseous constituents can be ignored, in good approximation [*Ferziger* and *Kaper* 1972].

In the phenomenological theory of *liquids*, the temperature dependence of the molecular diffusion coefficient is described by

$$D_i(T_l) = D_0 T_l^* \exp\left(-\frac{\mathscr{E}^*}{T_l^*}\right) \approx D_0 T_l^{*n}. \qquad (3.3.20)$$

The latter approximation can be employed for small temperature intervals where n is within ranges $1 \leq n \leq 10$. Typical values of diffusion coefficients in liquids are $D_i(300 \text{ K}) \approx 10^{-5} \text{ cm}^2/\text{s}$. Subsequently, we often use the abbreviation $D_i(\infty) \equiv D_i(T(\infty))$.

Thermal conductivity

The thermal (heat) conductivity of a *gas* mixture depends on temperature and concentration

$$\kappa \equiv \kappa(T_G, x_i) = \sum \kappa_i(T_G) x_i \approx \kappa_M ,\qquad(3.3.21)$$

where κ_i are the thermal conductivities and $x_i \equiv x_i(\mathbf{x}, t)$ the molar ratios of (pure) gases i. The latter approximation refers to $N_M \neq 0$. The temperature dependence can be described by

$$\kappa_i(T_G) = \kappa_i(T(\infty)) \, T_G^{*m_i}.\qquad(3.3.22)$$

The exponents m_i are, in the general case, different for different gases. In the elementary kinetic theory of gases, all m_i are equal with $m_i \equiv m = 1/2$. Experimental values are within the range $0.5 \leqslant m \leqslant 1.5$. If not otherwise indicated, the exact concentration dependence of κ is ignored and an average value is used instead.

The thermal conductivity of *liquid solutions* can also be described, in good approximation, by (3.3.21) and (3.3.22) [*Hirschfelder* et al. 1964].

Thermal diffusion ratio

The thermal diffusion ratio k_T for species AB_μ and $N_M = 0$ can be described by

$$k_T^{AB} = \alpha_T x_{AB}(1 - x_{AB}) ,\qquad(3.3.23a)$$

and for $N_M \neq 0$ by

$$k_T^{AB} = \alpha_T x_{AB} x_M \approx \alpha_T x_{AB} .\qquad(3.3.23b)$$

α_T is the thermal diffusion coefficient whose sign depends on the relative size of masses and the interaction between different types of molecules. For rigid spheres and $N_M = 0$

$$\alpha_T = \alpha_0 \frac{m_{AB} - m_B}{m_{AB} + m_B} ,\qquad(3.3.24a)$$

while with $N_M \neq 0$ we have, correspondingly

$$\alpha_T = \alpha_0' \frac{m_{AB} - m_M}{m_{AB} + m_M} .\qquad(3.3.24b)$$

Henceforth, we assume $\alpha_T = $ const.; in reality, α_T depends weakly on concentration and temperature and, with very high temperatures ($\geqslant 10^3$ K), it may even change sign. If $m_{AB} \ll m_M$ we can use the approximation $\alpha_T \approx 1/2$.

3.4 Heterogeneous Reactions

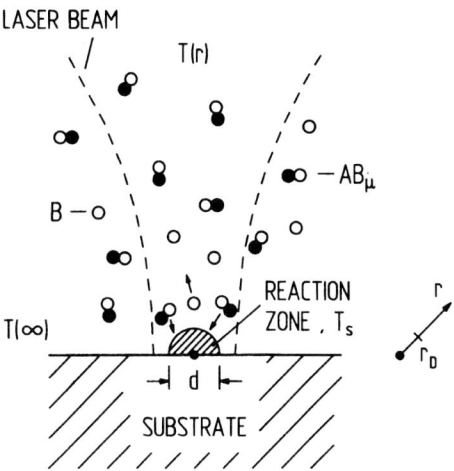

Fig. 3.4.1. Spherical reaction zone with radius $r_D = d/2$. The surface temperature, T_s, shall be uniform. The origin of the radius vector, r, is in the center of the hemisphere. Laser radiation is exclusively absorbed on the surface $r = r_D$. The model applies to laser-induced gas- and liquid-phase processing with ambient temperature $T(r)$. Carrier gas or solvent molecules possibly present, are not indicated

3.4 Heterogeneous Reactions

This section deals mainly with the modelling of reaction rates in laser-induced gas- and liquid-phase processing. In particular, we shall investigate static and dynamic solutions of the boundary-value problem formulated in the preceding section. The model employed is schematically shown in Fig. 3.4.1. The reaction zone is described by a hemisphere of radius $r_D = d/2$ which is placed on a semiinfinite substrate. The reaction shall take place exclusively on the surface of the hemisphere whose temperature shall be uniform and given by

$$T_s = T(\infty) + \Delta T_s , \qquad (3.4.1)$$

where $\Delta T_s \equiv \Delta T_s(r_D)$. We assume $T_s = T(r_D)$, where $T(r_D)$ is the temperature of the gas or liquid at the surface r_D. This approximation holds as long as r_D is much larger than the mean free path of molecules, λ_m. The heat and particle (mass) flux calculated within this model shall have *spherical* symmetry with respect to the center of the hemisphere. Thus, we ignore ordinary convection.

In order to permit a direct comparison between the various results, we often introduce normalized quantities which are indicated by asterisks.

If we assume the temperature of the ambient medium to be *uniform*, i.e., $T = T(\infty)$, the situation applies to those cases of thermal processing, where gas-phase heating via the laser-heated surface can be ignored. It applies also, however, to photochemical processing from an adsorbed phase that is in dynamic equilibrium with the gas phase. Here, the *whole* system, including the surface $r = r_D$ can be isothermal.

Let us consider again a first order non-equimolecular heterogeneous reaction

$$AB_\mu + M \rightarrow A(\downarrow) + \mu B(\uparrow) + M . \qquad (3.4.2)$$

In contrast to (3.3.1) decomposition of precursor (reactant) molecules takes place exclusively at the interface $r = r_D$ and no backward reaction is considered. Any heat of reaction is ignored.

3.4.1 Stationary Equations

It is convenient to solve the equations given in Sect. 3.3.1 in dimensionless variables: $r^* = r/r_D$, $v^* = v/k_0$, $T^* = T/T(\infty)$, $\mathscr{E}^* = \Delta E/k_B T(\infty)$, $J_i^* = J_i r_D/N(\infty)D_i(\infty)$, $S_T^* = S_T/c_p k_0 N(\infty) T(\infty)$, and $\kappa^* = \kappa/c_p k_0 N(\infty) r_D$. We assume $N_M = 0$, and c_p to be constant and equal for all components. From the equation of continuity and the equation of state we obtain

$$v^*(r^*) = \left(\frac{T^*}{T_s^*}\right)^q \frac{v^*(1)}{r^{*2}}, \qquad (3.4.3)$$

where $v^*(1) \equiv v^*(r^* = 1)$. With (3.4.3), (3.3.5 and 11) the flux can be written as

$$J_{AB}^*(r^*) = \frac{J_{AB}^*(1)}{r^{*2}} = -\frac{D_{AB}}{D_{AB}(\infty)T^{*q}}\left(\frac{\partial x_{AB}}{\partial r^*} + k_T^{AB}\frac{\partial}{\partial r^*}\ln T^*\right)$$

$$+ k_0^* \frac{v^*(1)}{T_s^{*q}} \frac{x_{AB}}{r^{*2}}. \qquad (3.4.4)$$

$k_0^* = k_0 r_D/D_{AB}(\infty)$ is the preexponential factor in the rate constant $k \equiv k_\rightarrow$ in (3.4.2). The energy flux (3.3.9) can be written as

$$S_T^*(r^*) = \frac{S_T^*(1)}{r^{*2}} = \frac{T^*}{T_s^{*q}}\frac{v^*(1)}{r^{*2}} - \kappa^*(T^*)\frac{\partial T^*}{\partial r^*}. \qquad (3.4.5)$$

In analogy to (3.3.14 and 16) we obtain

$$-J_{AB}^*(1) = k_0^* \frac{x_{AB}(1)}{T_s^{*q}} \exp\left(-\frac{\mathscr{E}^*}{T_s^*}\right), \qquad (3.4.6)$$

and

$$v^*(1) = b x_{AB}(1) \exp\left(-\frac{\mathscr{E}^*}{T_s^*}\right). \qquad (3.4.7)$$

The *coupled* (ordinary) differential equations (3.4.4 and 5) can be solved together with the conditions

$$T^*(1) = T_s^*; \quad T^*(\infty) = 1; \quad x_{AB}(r^* \to \infty) = x_{AB}(\infty). \qquad (3.4.8)$$

From these equations we obtain $x_{AB}(r^*)$, $T(r^*)$, and $S_T(1)$. $x_{AB}(1)$ can then be determined self-consistently.

The application of this boundary-value problem to heterogeneous pyrolytic *gas-phase* reactions, and the phenomena that arise from the coupling of fluxes are discussed in Sect. 17.4.1.

3.4.2 Transport Limitations

An adequate description of laser-induced reactions requires one to solve the transport equations simultaneously. In many cases, however, it is quite illuminating to investigate the effect of *single* contributions to the reaction rate. Let us consider the problem discussed in the preceding subsection, but ignore any temperature gradients within the ambient medium. Then, we obtain from (3.3.6) for the total flux $J = J_{AB} + J_B = -bJ_{AB} = Nv$. Together with (3.3.5) this yields

$$J_{AB}(r) = -\frac{ND_{AB}\nabla x_{AB}}{1 + bx_{AB}(r)} \, . \tag{3.4.9}$$

The factor $[1 + bx_{AB}(r)]$ is often denoted as (dimensionless) drift velocity. This means that diffusion is superimposed on an overall flow caused by the change in particle number density (chemical convection).

Together with (3.4.4) and boundary conditions analogous to those in (3.4.6) and (3.4.8) we obtain

$$\frac{bx_{AB}(r^*) + 1}{bx_{AB}(\infty) + 1} = \exp\left(-\frac{bk^*}{r^*} x_{AB}(1)\right), \tag{3.4.10}$$

with $k^* = kr_D/D_{AB}(\infty) = k_0^* \exp(-\mathscr{E}^*/T_s^*)$. If the reaction order $\gamma_{AB} \neq 1$ the product $k^* x_{AB}(1)$ in the exponent has to be replaced by $k'^* N^{\gamma-1} x_{AB}^{\gamma}(1)$. Equation (3.4.10) permits one to calculate the molar ratio at the surface and thereby the reaction rate

$$W = kNx_{AB}(1). \tag{3.4.11}$$

With isothermal conditions, N is independent of r, i.e., $N(r) = N(\infty)$. The solution is unique with all parameter values. The average velocity of the gas in radial direction is

$$v(r^*) = \frac{bkx_{AB}(1)}{r^{*2}} \equiv \frac{v(1)}{r^{*2}} \, . \tag{3.4.12}$$

Equimolecular reactions: $b = 0$

With equimolecular reactions, the total number of species remains constant, i.e., with each molecule AB decomposed, a single atom/molecule B is generated. The solution (3.4.10) yields

$$x_{AB}(r^*) = x_{AB}(\infty)\left(1 - \frac{k^*}{r^*(1+k^*)}\right). \tag{3.4.13}$$

The reaction rate is then

$$W = \frac{kN_{AB}(\infty)}{1+k^*}. \quad (3.4.14)$$

This relation is often termed the Smoluchowski equation. If diffusion is fast compared to the reaction, i.e., if $D_{AB}/k \gg r_D$ and thus $k^* \ll 1$, the reaction is *kinetically* controlled. In this approximation, the concentration of species AB is almost uniform and given by

$$x_{AB}(r^*) \approx x_{AB}(\infty)\left(1 - \frac{k^*}{r^*}\right) \approx x_{AB}(\infty). \quad (3.4.15)$$

The reaction rate within the kinetically-controlled regime becomes

$$W^{\text{kin}} \approx kN_{AB}(\infty). \quad (3.4.16)$$

In this regime the reaction rate is proportional to the rate constant and does not depend on any geometrical factor. If, on the other hand, $D_{AB}/k \ll r_D$ and thus $k^* \gg 1$, the reaction rate is limited by the *transport* of species

$$W^{\text{tr}} \approx \frac{D_{AB} N_{AB}(\infty)}{r_D}. \quad (3.4.17)$$

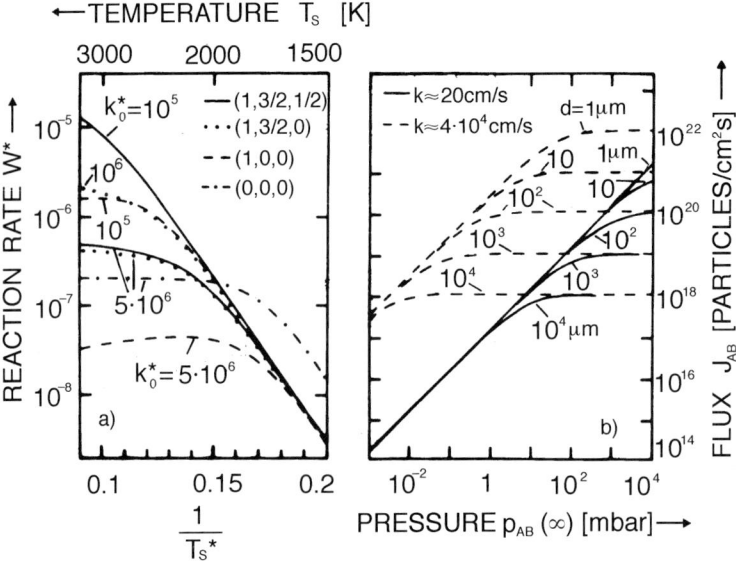

Fig. 3.4.2. (a) Arrhenius plot of normalized reaction rate W^* for various values of k_0^* and parameter sets (q, n, m). $\mathscr{E}^* = 90$ and $T(\infty) = 300$ K, adapted from [Bäuerle et al. 1990a]. (b) Dependence of reaction flux on pressure $p_{AB}(\infty)$ in a reaction of type (3.4.2) with $N_M = 0$, $b = 0$, $p_{AB}(\infty) \times D_{AB} \approx 80$ mbar cm^2/s, and $r_D = d/2$ (Fig. 3.4.1). Full curves: $k = 20$ cm/s ($\eta = 10^{-3}$). Dashed curves: $k \approx 4 \times 10^4$ cm/s ($\eta \to 1$)

3.4 Heterogeneous Reactions

Within the approximations made, this is the *highest* rate that can be achieved in a specific chemical reaction. Within the transport limited regime, the detailed activation mechanisms are unimportant. The transition from the kinetically controlled regime to the mass transport limited regime takes place when $k^* \approx 1$. An Arrhenius plot of the normalized reaction rate $W^* = W/k_0 N_{AB}(\infty)$ for $b = 0$ is included in Fig. 3.4.2a by the dash-dotted curve. The kinetically controlled regime and the transport limited regime can clearly be visualized.

From (3.4.16) it becomes evident that within the kinetically controlled regime of a first order reaction the particle flux, and thus the reaction rate, increases linearly with partial pressure $p_{AB}(\infty)$. This is shown in Fig. 3.4.2b. Full and dashed curves refer to rate constants $k \approx 20$ cm/s ($\eta = 10^{-3}$) and $k \approx 4 \times 10^4$ cm/s ($\eta \to 1$), respectively. Sometimes, we use instead of k, reaction probabilities, η. If $\eta \ll 1$ we can use the approximation $k \approx \eta \langle v_{AB} \rangle / 4$. This relation can also be applied for arbitrary η, including $\eta \to 1$ as long as $r_D \leqslant \lambda_m$.

If $N_{AB}(\infty) \gg N_B(\infty)$ the total pressure is $p \approx p_{AB}(\infty)$. With the approximation $D_{AB} \propto 1/p \approx 1/p_{AB}(\infty)$ the particle flux within the transport-limited regime becomes independent of pressure.

The influence of b

Chemical convection affects only the transport-limited regime. This is quite understandable. Within the kinetically controlled regime $k^* \ll 1$ and the

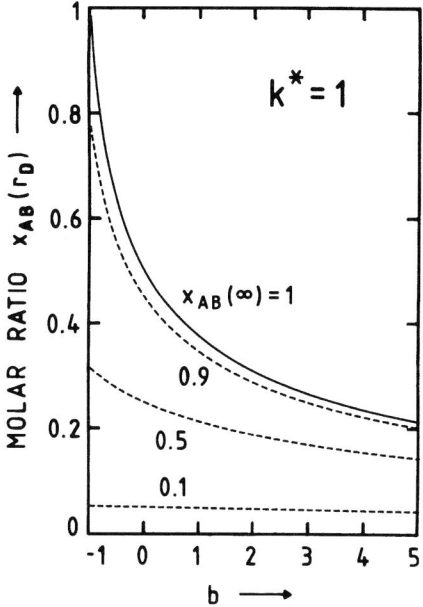

Fig. 3.4.3. Molar ratio $x_{AB}(r_D)$ as a function of b for $k^* = 1$ and various values of $x_{AB}(\infty)$. The dashed curves describe the influence of finite concentrations $x_B(\infty)$. With the restriction $x_{AB} \ll x_M$ the results can be adapted to the case $x_M \neq 0$ [*Kirichenko* et al. 1990]

consumption/generation of species is so small that the total number density remains almost unchanged; consequently, the influence of a hydrodynamic flow can be ignored.

The molar ratio at the surface of the reaction zone, $x_{AB}(r_D)$, calculated from (3.4.10) as a function of b is shown in Fig. 3.4.3. The full curve refers to $x_{AB}(\infty) = 1$; the dashed curves show the influence of finite concentrations $x_B(\infty)$. The reaction rates with $b > 0$ are significantly smaller than those with $b = 0$. The physical reason is that species B generated within the reaction zone hinder the transport of species AB_μ into this zone.

In thermal processing, the coupling of fluxes may significantly diminish the effect of b and, with certain parameters, may even cause an inverse behavior (Sect. 17.4.1). In the limit of full absorption of species AB_μ at the surface r_D which is described by $b = -1$, and with $x_B(\infty) = 0$, the transport of AB_μ is purely convective and the concentration $x_{AB}(r_D)$ equals unity.

The velocity of the convective flow can be calculated from (3.4.12). With $r^* = 1$, $x_{AB}(\infty) = 1.0$, $k^* = 1$, $D_{AB} = 0.1\,\text{cm}^2/\text{s}$ and $r_D = 1\,\mu\text{m}$ we obtain $v(r_D) \approx 4\,\text{m/s}$ and $v(r_D) \approx 10\,\text{m/s}$ for $b = +1$ and $b = -1$, respectively (the velocity vectors are oriented in opposite directions).

The influence of scanning and convection

Scanning of the laser beam or ordinary (free) convection provides an additional supply of reactant molecules to the reaction zone. This can be taken into account, in a crude approximation, by substituting k^* in (3.4.14) by

$$k^* = \frac{k}{v + D_{AB}/r_D}, \qquad (3.4.18)$$

where v is the scanning velocity of the laser beam, $v \equiv v_s$, or the velocity of the convective flow, $v \equiv v_c$ (Sect. 9.5.3). The influence of convection on mass transport can be ignored as long as $v_c r_D/D_{AB} \ll 1$.

The influence of gas-phase heating

Let us consider the influence of gas-phase heating on the reaction rate for $b = 0$ and *arbitrary* exponents q, n, m which describe the temperature dependences in the number density of species (3.3.11), the diffusion coefficient (3.3.19), and the thermal conductivity (3.3.22). Different cases are included in Fig. 3.4.2a. The temperature dependences of N and D influence the reaction rate considerably, while the temperature dependence of κ has only little influence.

3.4.3 Dynamic Solutions

For spherical symmetry (Fig. 3.4.1) with $T(r > r_D) = T(\infty)$, $\mu = 1$, and $W_v = 0$ we obtain from the diffusion equation (3.3.2), the boundary conditions (3.3.14) to (3.3.18), and the initial conditions $N_{AB}(t=0) = N_{AB}(\infty)$ the time-dependent density of species AB

$$N_{AB}(r^*, t^*) = N_{AB}(\infty) \left\{ 1 - \frac{k^*}{r^*[1+k^*]} \left(\text{erfc}\left(\frac{r^*-1}{2t^{*1/2}}\right) \right.\right.$$

$$- \exp\left[(1+k^*)(r^*-1) + (1+k^*)^2 t^*\right]$$

$$\left.\left. \times \text{erfc}\left[\frac{r^*-1}{2t^{*1/2}} + (1+k^*)t^{*1/2}\right] \right) \right\}, \qquad (3.4.19)$$

where $t^* = D_{AB}t/r_D^2$. In the limit $t \to \infty$ this yields the steady-state profile (3.4.13). The (normalized) particle flux at the surface r_D is given by

$$J_{AB}^*(r_D, t) = \frac{J_{AB}(r_D, t)}{J_{AB}(r_D, 0)} = -\frac{J_{AB}(r_D, t)}{kN_{AB}(\infty)} = \frac{1 + k^*\Psi}{1 + k^*}, \qquad (3.4.20)$$

with

$$\Psi = \exp[(1+k^*)^2 t^*] \text{ erfc }[(1+k^*)t^{*1/2}].$$

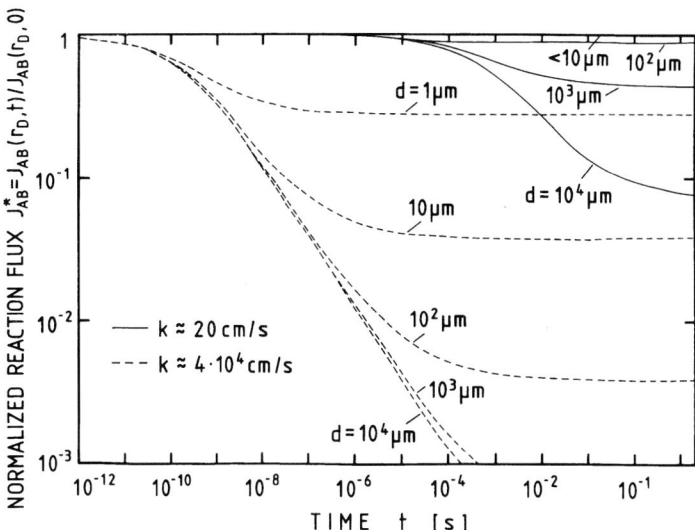

Fig. 3.4.4. Temporal dependence of the normalized flux (3.4.20) for various diameters of the reaction zone, $d = 2r_D$. The parameters employed were $p_{AB}(\infty)D_{AB} = 80$ mbar cm^2/s with $p \approx p_{AB}(\infty) = 100$ mbar

Figure 3.4.4 shows the temporal dependence of J_{AB}^* for various values of $d = 2r_D$. Full curves have been calculated for $k \approx 20$ cm/s and dashed curves for $k \approx 4 \times 10^4$ cm/s. Within the kinetically controlled regime ($k^* \ll 1$) the flux J_{AB}^* remains almost constant, while in the diffusion limited regime ($k^* \gg 1$) it strongly decreases with increasing time.

The characteristic time to reach the *stationary* flux is given by

$$\tau(\delta) \approx \frac{r_D^2 k^{*2}}{D_{AB}\pi[1+k^*]^2}\frac{1}{\delta^2}, \tag{3.4.21}$$

where $\delta \equiv [J_{AB}(t) - J_{AB}(\infty)]/J_{AB}(\infty)$. The time $\tau(\delta)$ increases with increasing diameter of the reaction zone. With $\delta = 0.1$ and $k^* \ll 1$, we obtain $\tau(0.1) \approx 2d^4 k^2/D_{AB}^3$ while with $k^* \gg 1$ this time becomes $\tau(0.1) \approx 8d^2/D_{AB}$. If both k and d are small, the reaction flux is almost unity and independent of time.

3.4.4 Heterogeneous Versus Homogeneous Activation

Up to now we have assumed "purely" heterogeneous activation of the chemical reaction. There are, however, many situations where a clear separation of heterogeneous and homogeneous contributions to thermally activated reactions becomes difficult.

Let us again consider the model in Fig. 3.4.1. Even if the ambient medium does *not* absorb the laser light, it will be indirectly heated via the laser-induced temperature rise on the surface $r = r_D$. As a consequence, a *homogeneous* reaction within a hemispherical shell above this surface can be activated. The thickness of this shell, estimated from (3.1.4) and (9.5.6), is

$$\Delta r = r - r_D < \frac{k_B T_s}{\Delta E} r_D. \tag{3.4.22}$$

For $T_s = 2000$ K, $\Delta E = 0.4$ eV (10 kcal/mol), and $r_D = 1$ μm, we obtain $\Delta r \leqslant 0.4$ μm. Within this model, a distinction between homogeneous and heterogeneous reactions loses sense when $\Delta r \ll r_D$.

In reality, the situation is more complicated. First, a proper estimation of the width of the heated zone must include consideration of convection, the temperature jump at the interface, if relevant, and the influence of temperature dependences in parameters. Second, the densities of species relevant to surface processing can differ significantly with heterogeneous and homogeneous activation. For example, in adsorbates the densities of species, N_i, as well as their cross sections for dissociation, σ_i, may differ significantly from the corresponding gas-phase values. There are also many other light–molecule–surface interactions which influence heterogeneous but not homogeneous reactions. The situation is similar with photochemically activated reactions.

3.5 Combined Heterogeneous and Homogeneous Reactions

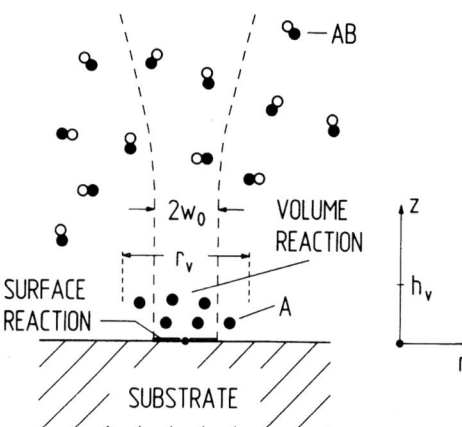

Fig. 3.5.1. Cylindrical model for the description of pyrolytic laser-induced chemical processing. The origin of the coordinate system is on the substrate surface in the center of the laser beam. The laser light is exclusively absorbed on the surface $z = 0$. The temperature distribution on the surface is $T_s(r)$; in the gas phase it is $T(r, z)$ [Kirichenko and Bäuerle 1992]

3.5 Combined Heterogeneous and Homogeneous Reactions

A model particularly useful for the description of many types of laser-induced surface modifications and the deposition and etching of flat patterns is depicted in Fig. 3.5.1. Here, we assume *cylindrical* symmetry. The laser light which shall exclusively be absorbed by the substrate causes a surface temperature rise, and thereby indirect heating of the adjacent medium. We consider a thermally activated reaction similar to (3.3.1)

$$\zeta_{AB} AB + \zeta_C C + M \underset{k_2}{\overset{k_1, k_3}{\rightleftarrows}} \zeta_A A(\downarrow) + \zeta_{BC} BC + M \ . \tag{3.5.1}$$

In contrast to the spherical model we assume a Gaussian intensity distribution of the laser beam and *inhomogeneous* substrate heating. The maximum temperature, T_c, occurs in the center of the beam at $r = z = 0$. Due to inhomogeneous heating, the rate constants and transport properties of species will vary from point to point. For simplicity, we assume that the temperature distribution remains unchanged during laser processing. This is a good approximation as long as we consider only flat structures and ignore changes in the physical properties within the irradiated area. With stationary conditions, the temperature distribution on the substrate surface is given by

$$T_s(r) \equiv T(r, z = 0) = T(\infty) + \Delta T_s(r) \ . \tag{3.5.2}$$

3.5.1 The Boundary-Value Problem

Approximate solutions of the boundary-value problem for a reaction of type (3.5.1) and gas-phase precursors have been derived by *Kirichenko* and *Bäuerle*

(1992). Chemical convection and thermal diffusion have been ignored. With stationary conditions the equations in Sect. 3.3.1 yield for species AB

$$-\nabla(ND_{AB}\nabla x_{AB}) = Q_{v,AB} = -W_v,$$

$$x_{AB}(r, z \to \infty) = x_{AB}(\infty),\qquad(3.5.3)$$

$$-J_{AB} = ND_{AB}\frac{\partial x_{AB}}{\partial z}\bigg|_{z=0} = W_s.$$

W_v and W_s are the net reaction rates in the volume and on the surface, respectively. These rates are given by

$$W_v = k_3(T) N_{AB}^{\gamma_{AB}^v} N_c^{\gamma_c^v},$$

and

$$W_s = k_1(T_s) N_{AB}^{\gamma_{AB}} N_c^{\gamma_c} - k_2(T_s) N_{BC}^{\gamma_{BC}}.$$

γ_i^v and γ_i are partial reaction orders for volume and surface reactions, respectively (Sect. 3.1). Analogous equations hold for species BC and A.

The temperature dependence of D_i is described by (3.3.19); with an exponent $n = 1$ we obtain with

$$ND_i = N(\infty) D_i(T(\infty))\qquad(3.5.4)$$

and $N \equiv N(T) = N(\infty)/T^*$ the conservation law

$$N(T)[D_{AB}(T)x_{AB}(r,z) + D_{BC}(T)x_{BC}(r,z)]$$
$$= N(\infty) D_{AB}(T(\infty)) x_{AB}(\infty),\qquad(3.5.5)$$

where we have assumed $x_{BC}(r, z \to \infty) = 0$.

3.5.2 Approximate Solutions

An analytical solution of the boundary-value problem (3.5.3) to (3.5.5) is not possible. Solutions can be obtained by employing an iteration approach where the concentrations x_{AB} and x_{BC} are calculated self-consistently.

The rate constants k_j contain exponential factors for which, typically, $T \ll \mathscr{E}_j$. This permits further simplifications. Near the origin $r = z = 0$ we can use the expansion

$$T(r,z) \approx T_c - \Delta T_c\left(\frac{r^{*2}}{2} + \frac{2z}{\sqrt{\pi}w_0}\right).\qquad(3.5.6)$$

For the *surface* reactions, the rate constants (3.1.4) can then be written as

$$k_j(T_s) = k_j(T_c) \exp\left(-\frac{r^2}{r_j^2}\right), \qquad (3.5.7\text{a})$$

with $j = 1, 2$. For the *volume* reaction, $j = 3$, we obtain

$$k_3(T) \approx k_3(T_c) \exp\left(-\frac{r^2}{r_3^2} - \frac{z}{h_v}\right), \qquad (3.5.7\text{b})$$

where

$$h_v = \frac{\sqrt{\pi}}{2} \frac{w_0 T_c^2}{\mathscr{E}_3 \Delta T_c} \quad \text{and} \quad r_j = w_0 \left(\frac{2 T_c^2}{\mathscr{E}_j \Delta T_c}\right)^{1/2}. \qquad (3.5.8)$$

h_v describes the extension of the volume reaction in z direction. Explicit expressions for the molar ratios x_i and further details on the calculations can be found in Kirichenko and Bäuerle (1992). An illustration of the results is given in Sect. 16.2.

Reaction rates achieved by homogeneous gas-phase heating using *parallel* laser-beam incidence are discussed in Chap. 19.

3.6 Homogeneous Photochemical Activation

The analysis of photochemical processing requires the modelling of the spatial and temporal distribution of photogenerated species and of reaction rates. Consider, for example, a photolytic reaction of the type

$$A_{\mu_A} B_{\mu_B} + M + h\nu \to \mu_A A(\downarrow) + \mu_B B + M. \qquad (3.6.1)$$

Such reactions are employed in laser-CVD of metals from alkyls or carbonyls ($B \neq A$), and in many cases of dry etching ($B \neq A$ or $B \equiv A$). Equation (3.6.1) can be applied also to situations where dissociation of $[A_{\mu_A} B_{\mu_B}]^*$ takes place via subsequent molecular collisions or other interactions.

Molecules AB are excited/dissociated within the total *volume* of the incident laser beam (Fig. 1.2.2). The photoproducts diffuse into the surrounding medium and recombine, or they react at the substrate surface and, to some extent, at the walls of the reaction chamber. In the case of localized irradiation (Fig. 1.2.2a), species A will also condense *outside* the reaction zone and thereby decrease the spatial confinement. Non-thermal excitations of the substrate surface (Sect. 2.4) may also play an important or even decisive role in the molecule–surface interactions.

Because of the interrelation between the different quantities determining single effects, the concentration of photoproducts and the reaction rates depend on the size of the reaction zone, the reaction probabilities at the different material surfaces, the size of the reaction chamber, etc. This will be outlined in Sect. 14.1.

4 Nucleation and Cluster Formation

The formation of nuclei and clusters from single atomic or molecular species plays an important or even decisive role in laser-induced (gas-phase) synthesis of powders, laser-CVD, etc. Of similar importance are (heterogeneous) nucleation processes in laser-annealing, recrystallization, thin-film formation, etc. Depending on the particular application, nucleation and cluster formation can be desirable or undesirable. An example of the latter case is pulsed-laser deposition (PLD) where clusters formed within the vapor plume may condense on the substrate and thereby deteriorate the film quality.

Laser radiation can influence both homogeneous and heterogeneous nucleation and cluster formation in various ways:

- Laser-induced decomposition of precursor molecules generates atomic or molecular radicals, ions, etc., which may serve as nucleation centers.
- Selective laser-light absorption results in heating or ionization of nuclei/clusters and thereby favors either their further growth or decomposition, depending on the particular system under consideration.
- The degree of non-equilibrium that can be achieved in laser-induced processes is significantly higher than in any classical system. For example, within the initial phase of the vapor plume generated in pulsed-laser ablation, the degree of oversaturation exceeds that of classical systems by many orders of magnitude.
- The strong gradients in temperature, concentration of species, etc., influence both the kinetics of nucleation and cluster formation and the transport of particles. For example, in (heterogeneous) pyrolytic LCVD the temperature gradients near the deposit are, typically, 10^7 K/cm. With such temperature gradients, thermophoretic and chemophoretic forces, thermal diffusion, etc., become important.

4.1 Homogeneous Processes

In this section we consider homogeneous nucleation and cluster formation within a metastable phase, mainly an oversaturated vapor.

4.1.1 Classical Kinetics

Nuclei are homogeneously formed within an oversaturated vapor due to fluctuations. The growth of nuclei is favored by the free energy of condensation and suppressed by the interface energy. For spherical nuclei (droplets) and isothermal conditions, the critical radius can be described by

$$r_c = \frac{2\sigma V_n}{\mu(p) - \mu_n(p)} = \frac{2\sigma V_n}{k_B T \ln S}, \quad (4.1.1)$$

where σ is the surface tension coefficient, and V_n the volume available per atom/molecule within the nucleus. The chemical potentials for the metastable vapor and the nucleus, μ and μ_n, depend on partial pressure, $p \equiv p_i$. $S = p/p_s(T) = N(\infty)/N_s$ characterizes the degree of oversaturation, and $p_s(T) \equiv p_s(T, r \to \infty)$, the saturated vapor pressure over a plane surface. Figure 4.1.1 shows the Gibbs free energy, ΔG, as a function of particle radius. The dashed curve corresponds to the situation observed without laser light. If a nucleus with radius $r > r_c$ is formed, it continues to grow. If, however, $r < r_c$ the nucleus decomposes.

The probability for the formation of a nucleus with radius r_c is given by [*Landau* and *Lifshitz*; Statistical Physics, Pt. 1]

$$w \propto \exp\left(-\frac{16\pi\sigma^3 V_n^2 p^2}{3T^3 \Delta p^2}\right), \quad (4.1.2)$$

with $\Delta p = p - p_s \ll p$. This equation can be rewritten by using the Clausius–

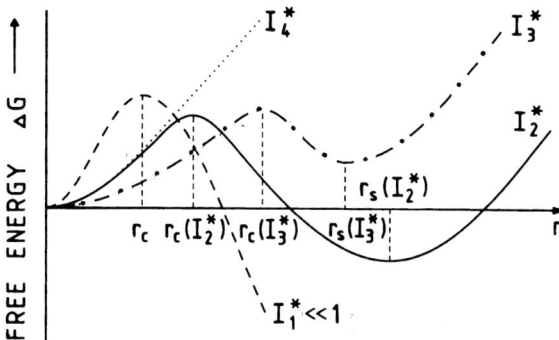

Fig. 4.1.1. Free energy of droplets as a function of their radius. The dashed curve corresponds to the situation without laser light or with very low intensities, $I_1^* \ll 1$ [for normalization see (4.1.16)]. For medium laser light intensities (full curve) there exists a critical radius, $r_c(I_2^*) > r_c$ and a stable size of the particle, $r_s(I_2^*)$. For very high intensities (dotted curve) formation of a stable nucleus becomes impossible. The dash-dotted curve shows an intermediate situation

Clapeyron relation

$$\Delta p = \frac{\Delta H^a}{T_0 V} \Delta T. \qquad (4.1.3)$$

ΔH^a is the heat of condensation per molecule. $\Delta T = T - T_0$ where T_0 characterizes the phase equilibrium with $p = p_s(T_0)$. V is the volume available per molecule/atom within the gas phase.

The temporal behavior of the droplet radius is schematically shown in Fig. 4.1.2. The ambient atmosphere of droplets shall consist of the vapor and a buffer gas.

- The formation of critical nuclei takes place within a latent time $t_n \propto \omega^{-1}$.
- In the kinetically controlled region, the nuclei are small compared to their mean free path within the gas, i.e., their radius is within the range $r_c < r \ll \lambda_m$. Their growth in volume is proportional to the number of collisions which, in turn, is proportional to the cross section, so that $d/dt(r^3) \propto r^2$. Thus, the radius of nuclei increases linearly with time, $r \propto t$.
- In the diffusion-limited region, r is comparable to λ_m and much larger than r_c. The nuclei are in quasi-equilibrium with the surrounding vapor. The diffusion equation yields $d/dt\,(r^3) \propto J \propto r^2 D_i [N(\infty) - N_s(r)]/r$ where J is the total flux onto the surface of the droplet. $N_s(r)$ is the density of saturated vapor over the curved surface of the droplet with radius r. It is related to the saturated vapor over a plane surface, N_s, via $N_s(r) = N_s(1 + 2\sigma V_n/k_B T r)$ (Laplace formula). Together with (4.1.1) this yields for low oversaturation $N_s(r) = N_s + [N(\infty) - N_s] r_c/r$. Thus, $d/dt(r^3) \propto r^2 [N(\infty) - N_s](r - r_c)/r^2 \propto (r - r_c) \propto r$ and thereby $r \propto t^{1/2}$.
- Within the transition region, IV, the radius of droplets remains almost constant. This can be due to the competition in growth between different droplets of similar size [*Pflügl* and *Titulaer* 1993], or due to changes in local

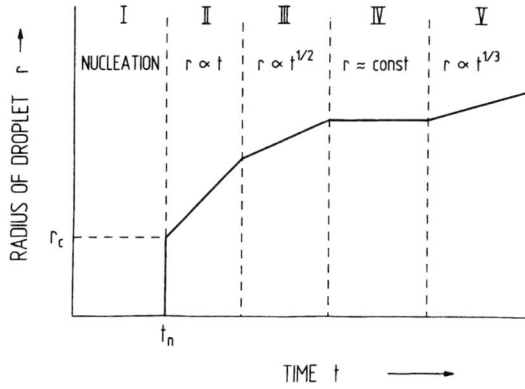

Fig. 4.1.2. Evolution of droplet formation: I nucleation; II kinetically controlled growth; III diffusion limited growth; IV transition region; V Ostwald ripening region

vapor pressure related to fluctuations in the density of nuclei [*Tokuyama* and *Enomoto* 1992].
- In the Ostwald ripening region, V, growth can be described by $r \propto t^{1/3}$ [*Gunton* et al. 1983; *Lifshitz* and *Slyozov* 1961].

Due to the growth of nuclei, the degree of oversaturation, S, decreases and r_c increases. As a consequence, all nuclei are approximately of critical size, i.e., $r \approx r_c$. We can apply the diffusion equation as in region III, but replace $N(\infty)$ by N_s so that $N(\infty) - N_s(r) \approx N_s - N_s(r) \propto N_s/r$ where the Laplace formula was again employed. Thus, $d/dt \, (r^3) \propto r^2[N(\infty) - N_s(r)]/r \propto N_s \propto$ const. and thereby $r \propto t^{1/3}$.

This mean-field type description of the growth process ignores *direct* interactions between droplets. This holds for many experimental situations in laser-chemical processing. It certainly cannot be applied to pulsed-laser ablation where the plasma plume contains high densities of electrons, ions, and charged clusters.

A detailed introduction to nucleation and cluster formation in standard systems can be found, e.g., in [*Abraham* 1978; *Maissel* and *Glang* 1970].

4.1.2 Droplets within a Laser Beam

Let us consider an oversaturated vapor with droplets formed by vapor condensation in the presence of laser light. The laser radiation shall be absorbed by the droplets but not by the vapor. Thus, the temperature of droplets, T_n, increases with respect to the temperature of the ambient vapor, T, and their growth behavior will be changed.

For large droplets (typically $r > 10$ nm) which can be described by macroscopic variables, the absorptivity can be calculated on the basis of the Mie theory [*Born* and *Wolf* 1980]. The energy balance for such droplets can be written as

$$c_p \frac{d}{dt}(m_n T_n) = \pi r^2 \, AI - 4\pi r^2 \, I_{loss} + 4\pi r^2 \rho \Delta H_v \frac{dr}{dt}, \qquad (4.1.4)$$

where $m_n = 4\pi r^3 \rho/3$ is the mass, ρ the density, c_p the specific heat, and A the effective absorptivity of the droplet. I_{loss} describes the heat exchange with the ambient vapor. The last term describes the heat of condensation.

For droplets whose radius is small compared to both the laser wavelength, λ, and the penetration depth, l_α, the effective absorptivity can be approximated by

$$A = \frac{\sigma_a}{\pi r^2} = \frac{32}{3}\pi^2 \frac{r}{\lambda} \text{Im}\{\alpha_e\} = 8\pi \frac{r}{\lambda} \text{Im}\left\{\frac{\varepsilon - 1}{\varepsilon + 2}\right\} = \frac{r}{\lambda} f(\varepsilon). \qquad (4.1.5)$$

Here, terms of higher orders in r/λ have been ignored. σ_a is the absorption cross section. The electric polarizability of the particle, α_e, has been approximated by $\alpha_e = 3(\varepsilon-1)/4\pi(\varepsilon+2)$. With $\varepsilon = \tilde{n}^2 = [n+i\kappa_a]^2$ the function $f(\varepsilon)$ becomes

$$f(\varepsilon) = \frac{48\pi n \kappa_a}{(n^2 - \kappa_a^2 + 2)^2 + 4n^2 \kappa_a^2} \tag{4.1.6}$$

In the simplest case, we can approximate the loss term by

$$I_{\text{loss}} = \eta'(T_n - T). \tag{4.1.7}$$

If $r \gg \lambda_m$ the heat exchange coefficient can be approximated by $\eta' \approx \kappa_G/r$ where κ_G is the thermal conductivity of the vapor. If $r \ll \lambda_m$, we can use $\eta' \approx c'_v J$ where c'_v is the specific heat per (gas) molecule and J the flux of molecules onto the droplet surface. In the elementary kinetic theory, J is described by

$$J = \frac{1}{4} N \langle v \rangle = \frac{1}{4} N \left(\frac{8 k_B T}{\pi m} \right)^{1/2}, \tag{4.1.8}$$

where $\langle v \rangle$ is the mean velocity of gas molecules of mass m. More detailed calculations yield

$$\eta' = \zeta(l+1) \frac{p_t}{T} \left(\frac{k_B T}{8\pi m} \right)^{1/2}, \tag{4.1.9}$$

where $l = 6Z - 6$. Z is the number of atoms per molecule. For a linear molecule $l = 6Z - 5$, and for a monatomic gas $l = 3$. p_t is the total gas pressure, and $\zeta \leq 1$ the accommodation coefficient which characterizes the degree of inelasticity of collisions.

For *stationary* conditions, (4.1.4) yields

$$T_n = T + \frac{If(\varepsilon)}{4\eta'\lambda} r. \tag{4.1.10}$$

Due to the difference in temperatures, $T_n \neq T$, the critical radius of droplets in the presence of laser light, $r_c(I)$, will differ from (4.1.1). In addition, droplets with finite (stable) radii, $r_s(I)$, will appear. The equilibrium values $r_e(I) = r_c(I)$, $r_s(I)$ can be calculated from the condition that the flux of condensing molecules, $J = J(T)$, must be equal to the flux of evaporating molecules, $J_v = J_v(T_n)$. This is analogous to the condition that the saturated vapor pressure near the surface of the droplet, $p_s(T_n, r_e(I))$, is equal to the (oversaturated) partial pressure of the vapor, p, within the ambient gas, i.e.,

$$p_s(T_n, r_e(I)) = p. \tag{4.1.11}$$

The pressure p_s near a surface of radius r and temperature T is given by

$$p_s(T_n, r) = p_o \exp\left(-\frac{\Delta E_v^a(r)}{k_B T_n} \right), \tag{4.1.12}$$

where

$$\Delta E_v^a(r) = \Delta E_v^a - \frac{2\sigma V_n}{r}. \tag{4.1.13}$$

p_o is a constant. $\Delta E_v^a \approx \Delta H_v^a$ is the enthalpy of vaporization per atom (molecule) from a plane surface. With $S = p/p_s(T)$ we obtain from (4.1.11) to (4.1.13)

$$\frac{2\sigma V_n}{k_B T_n r_e(I)} = \ln S + \frac{\Delta H_v^a}{k_B T_n} - \frac{\Delta H_v^a}{k_B T}. \tag{4.1.14}$$

In the absence of laser radiation $T = T_n$ and we obtain (4.1.1). Equations (4.1.14) and (4.1.1) yield

$$r_e^* = \left[1 + \left(1 - \frac{\mathcal{H}_v^*}{\ln S}\right)\Delta T_n^*\right]^{-1}. \tag{4.1.15}$$

Here, we have introduced dimensionless quantities $r_e^* = r_e(I)/r_c$, $\mathcal{H}_v^* = \Delta H_v^a/k_B T$, and $\Delta T_n^* = T_n/T - 1$. Note that $\mathcal{H}_v^*/\ln S > 1$. Equation (4.1.10) can then be rewritten as

$$\Delta T_n^* = \left(\frac{If(\varepsilon)r_c}{4\eta'\lambda T}\right) r_e^* = I^* r_e^*. \tag{4.1.16}$$

The two functions $r^*(\Delta T_n^*)$ are schematically drawn in Fig. 4.1.3. The dashed lines represent (4.1.16). The behavior of (4.1.15) is shown by the combined full and dash-dotted curve. The number of solutions depends on the

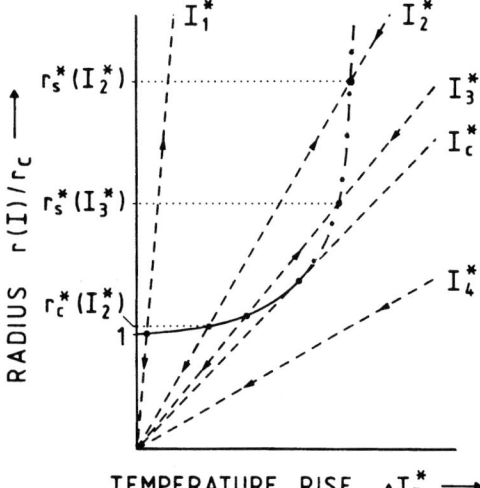

Fig. 4.1.3. Critical and stable radii of droplets normalized to the radius $r_c \equiv r_c(I=0)$ in the presence of laser light as a function of the temperature rise $\Delta T_n^* = (T_n/T) - 1$. The (normalized) laser-beam intensities are $I_1^* < I_2^* < I_3^* < I_c^* < I_4^*$. The branch of critical solutions is drawn by the full curve and the branch of stable solutions by the dash-dotted curve

parameter values. Two solutions exist for laser-beam intensities

$$I^* < \frac{1}{4}\left(\frac{\mathcal{H}_v^*}{\ln S} - 1\right)^{-1}. \tag{4.1.17}$$

The branch of *critical* solutions corresponds to the full curve and the branch of stable solutions to the dash-dotted curve. For very *low* laser-light intensities, I_1^*, the situation is essentially the same as in the absence of laser light. The (lower) intersection point yields a critical radius, $r_c(I_1^*) \approx r_c$, as given by (4.1.1). The second intersection point between curves is beyond the range of applicability of the model. For *medium* laser-light intensities, I_2^*, there are two intersection points. The lower corresponds to the critical radius, $r_c(I_2^*)$. If $r < r_c(I_2^*)$, the nucleus decomposes. If $r > r_c(I_2^*)$ the droplet grows until it reaches a stable size $r_s(I_2^*)$. This behavior is included in Fig. 4.1.1 (full curve). If the laser-light intensity is increased to I_3^* the critial radius increases. The radius of stable droplets decreases and the minimum becomes more shallow (see dash-dotted curve in Fig. 4.1.1). For intensities I_c^* the critical points collapse and *no* stable solution exists. For $I_4^* > I_c^*$ no solution exists (dotted curve in Fig. 4.1.1). The formation of droplets is suppressed by laser-induced evaporation. Thus, if we choose an appropriate laser-light intensity, for example I_2^*, droplets of almost equal size, $r_s(I_2^*)$, can be grown.

In the present model we have employed macroscopic quantities to describe droplets of spherical shape. In cases of strong oversaturation which are, in fact, relevant in laser processing, critical nuclei may consist of a small number of molecules only. For such small aggregates a macroscopic description is inadequate.

In spite of its simplicity, the model shows that droplet formation in the presence of laser radiation differs significantly from the standard behavior. The possibility to grow droplets of almost equal size is used for the synthesis of fine-grain powders with narrow distributions in particle sizes [*Steinfeld* 1981]. In most of these investigations CO_2-laser radiation has been employed. Depending on the particular experimental conditions, the grain sizes are, typically, between 10 nm and some μm. Among the materials that have been synthesized are ultrafine powders of Al_2O_3 [*Borsella* et al. 1993a], MgO [*Vorobyev* et al. 1991], SiC [*Scholz* et al. 1993], and Si, Si_3N_4, ternary Si-C-N, etc., [*Borsella* et al. 1993b and references therein].

4.1.3 Transport of Clusters, Thermophoresis, Chemophoresis

With laser-light intensities that produce high concentrations of product species by homogeneous decomposition of precursor molecules within gases or liquids, clusters may significantly contribute to the processing rate. With the condensation of clusters, the deposition rate strongly increases and the morphology of the deposit changes, in general, to large-grain low-density material.

Laser light not only changes the kinetics of the nucleation process, but also influences the transport of nuclei and clusters. Besides their Brownian motion, the particles will be accelerated by convective forces related to laser-induced temperature gradients. Due to these temperature gradients, additional driving forces based on thermophoresis and chemophoresis may also become important [*Karlov* et al. 1993]. Thermophoretic forces are directed from the hotter to the colder regions, because the momentum transfer from molecules that hit the cluster on the hotter side exceeds that from molecules on the colder side. The situation is similar to thermal diffusion of heavy molecules. If a chemical reaction takes place, differences in momentum transfer from precursor and product molecules on the hotter and the colder side of the cluster give rise to chemophoretic forces. While both thermophoretic and chemophoretic forces are directed from the hotter to the colder regions they can contribute to the transport of clusters towards the substrate surface. The reason is simple: With the formation of clusters, the temperature distribution within the medium may strongly be changed. Both the energy of condensation and direct absorption of laser light by the clusters can produce a maximum in temperature distribution above the surface.

4.2 Heterogeneous Processes

This section deals with nucleation and cluster formation at interfaces, mainly such between gases and solids. Nucleation processes at interfaces are extremely important in standard and laser-induced thin-film growth, annealing, surface melting, solidification, crystallization, material evaporation, etc.

4.2.1 Nucleation in Laser-CVD

In the presence of laser light, nucleation is strongly modified with respect to standard thin-film growth and additional mechanisms must be considered. Laser light can change the sticking probability of species and their migration on and desorption from the substrate surface, and it can photodecompose molecules within the gaseous atmosphere or within adsorbed layers, etc. In any case, for nucleation to take place, a certain concentration of species, and thereby a threshold intensity for deposition is required. Subsequently, we consider two cases: Absorbing substrates, and transparent substrates that do not or only very slightly absorb the laser light.

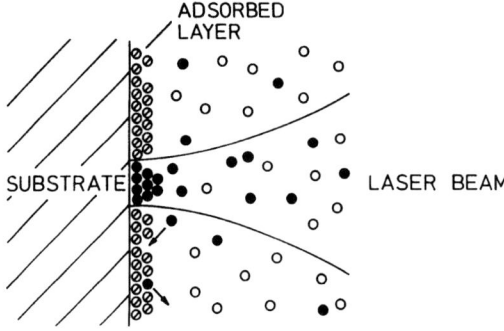

Fig. 4.2.1. Influence of adsorbed-layer photolysis on nucleation and condensation of gas-phase photofragments. Free gas-phase precursors and photofragments are indicated by empty and full circles, respectively

Absorbing substrates

With (strongly) absorbing substrates, precursor molecules are thermally dissociated at or near the laser-heated area on the substrate surface. The free atoms form clusters, which provide nucleation centers for further film growth. With most processing conditions, the time of nucleation, t_n, is short compared to the laser-beam illumination time, τ_l. The main differences to nucleation in standard CVD are related to temporal changes in the laser-induced temperature distribution, e.g., due to changes in the reflectivity and thermal conductivity provided by the nuclei. In laser micropatterning, nucleation processes will also be influenced by the confinement of the temperature distribution and the strong (lateral) temperature gradients.

Transparent substrates

For transparent substrates, nucleation may depend on the quality of the substrate surface. Surface defects such as pinholes and scratches, but also dust particles, etc., absorb the laser light and thereby may allow nucleation to be initiated at such "hot spots".

Nucleation may also be initiated by atoms that result from selective excitation/dissociation of *adsorbed* molecules (Fig. 4.2.1). Because of the high density of adsorbed molecules, the free atoms may form clusters which, even when of subcritical size, may strongly absorb the laser radiation. Such heated clusters can provide nucleation sites and film growth can proceed mainly thermally. Both the degree of photodecomposition of adsorbed species and the light intensity absorbed by the nuclei, depend sensitively on the laser wavelength [adsorption may significantly change the effective absorption/dissociation cross section with respect to free gas-phase molecules (Chap. 20)]. This is an important difference to nucleation on strongly absorbing substrates where the temperature rise is determined by the optical and thermal properties of the substrate. With transparent substrates, latent times for nucleation of several

4.2 Heterogeneous Processes

seconds or even minutes have been observed. Examples are Ar^+- or Kr^+- laser-induced metal deposition from alkyls or carbonyls on quartz or (standard) glass substrates.

In photolytic CVD based on (non-thermal) dissociation of gas- or adsorbed-phase molecules, multiatom clusters originally formed on the substrate surface by thermal or non-thermal processes, may serve as nucleation centers for film growth. For example, metal microstructures have been grown on transparent substrates by UV-laser photolysis of gas-phase molecules. While the metal atoms are produced within the *total* volume of the laser beam, they condense preferentially on the nuclei generated within the irradiated area (Fig. 4.2.1). For Cd, for example, the critical number of atoms necessary for the formation of a stable nucleus in the gas phase is about 10 at 300 K. For nuclei adsorbed on solid surfaces, this number depends on the physical properties of the material. Atoms that neither form stable nuclei nor attach themselves to nucleation centers formed within the area of the laser focus, will diffuse across the surface and then evaporate, with high probability. The surface diffusion length is, approximately, $l \approx (2D_i t_v)^{1/2}$ where D_i is the coefficient for surface diffusion of atoms and t_v the average residence time before reevaporation. If an atom impinges onto the substrate within an area defined by the radius $r \leqslant r_n + l$, where r_n is the radius of a stable nucleus, it will, on average, be captured by this nucleus. Atoms impinging onto the substrate at distances outside of this zone will reevaporate prior to capture. In other words, the sticking probability for free gas- or liquid-phase atoms, or small clusters of atoms, impinging on or near the nuclei formed within the irradiated substrate area is much bigger than anywhere else on the substrate surface. For Cd atoms, the sticking coefficient is about unity on a Cd film, but $< 10^{-3}$ on SiO_2 glass. Thus, nucleation thresholds suppress isotropic deposition and thereby enhance the contrast in photolytic micropatterning.

4.2.2 Coalescence

After the initial stage of formation, nuclei grow on the substrate surface and finally coalesce. In the presence of laser light, this process takes place in a different way than in the case of uniform substrate heating. This has been demonstrated for surface oxidation of metals by means of cw CO_2-laser radiation and standard oven heating [*Alimov* et al. 1984]. For isothermal (equilibrium) oxidation, the distribution function for the grain size shows a single maximum which shifts to larger values during oxidation. For laser-induced (non-equilibrium) oxidation, the distribution function shows *several* peaks which correspond to grains of different sizes. The coalescence of grains will then give rise to additional maxima.

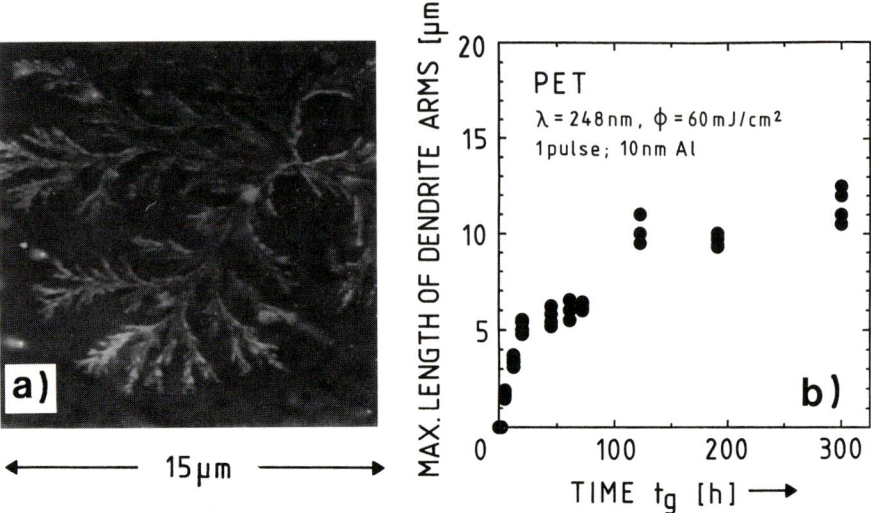

Fig. 4.2.2a,b. Dendritic growth on PET after single-pulse KrF-laser irradiation in vacuum. (**a**) Atomic force microscope (AFM) picture of dendrite ($\phi = 41$ mJ/cm^2). Such dendrites are elevated above the surface by, typically, 5 to 30 nm and extended into the surface by about the same amount. (**b**) Length of dendrite arms versus time, t_g. [*Heitz* et al. 1994]

4.2.3 Liquid–Solid and Solid–Solid Interfaces

In the case of liquid-solid and solid-solid phase transitions, some peculiarities in the growth of individual clusters occur. Among those is the growth of dendritic structures as shown for PET (polyethylene-terephthalate) in Fig. 4.2.2a. UV laser light amorphizes the PET surface within a depth $l_\alpha \approx 0.1$ μm. This layer recrystallizes in part. Recrystallization can be suppressed by coating the surface with a thin metal layer. The growth of dendrites starts at different nucleation centers within the irradiated area. The temporal evolution of a dendrite is shown in Fig. 4.2.2b.

5 Lasers, Experimental Aspects, Spatial Confinement

5.1 Lasers

The lasers most commonly used in materials processing are listed in Table I. Subsequently, we will discuss only some characteristic properties of various lasers and their particular areas of application in materials processing.

5.1.1 CW Lasers, Gaussian Beams

Microfabrication by laser direct writing is mainly performed with continuous wave (cw) Ar^+- or Kr^+-lasers, including frequency-doubled lines. The good spatial coherence of such lasers permits tight focusing; together with their good stability in beam profile (mode), output power, frequency, and beam pointing, these lasers allow accurate microfabrication with constant and well-defined morphology. The main disadvantages of ion lasers for applications are their low efficiencies and high costs.

The TEM_{00} mode, which has been used in most of the experiments, is of Gaussian shape (Fig. 5.1.1). The laser-beam intensity within the focal plane has the form

$$I(r) = I_0 \exp\left(-\frac{r^2}{w_0^2}\right), \tag{5.1.1}$$

where w_0 is the radius of the laser focus defined by $I(w_0) = I_0/e$ (frequently, one uses instead the definition $I(w_e) = I_0/e^2$ so that $w_e = \sqrt{2}\, w_0$). The *total laser power* is

$$P = 2\pi \int_0^\infty r I(r)\, dr = \pi w_0^2 I_0. \tag{5.1.2}$$

Throughout this book, the laser power always refers to the *effective* power incident onto the substrate surface, i.e., it is corrected for losses at the entrance window of the reaction chamber, etc.

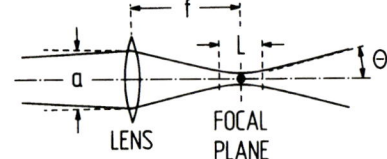
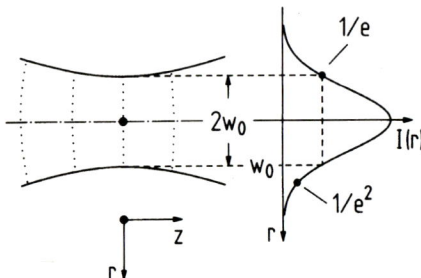

Fig. 5.1.1. Intensity distribution and shape of a Gaussian laser beam near the focal plane. $2w_0$ is the beam waist, $L = 2z_R$ the length (depth) of the focus, and Θ the beam divergence angle. The dotted curves indicate the shape of the wave front

In order to obtain a tight focus, the laser beam is first expanded and then focused by a lens (Fig. 5.1.1). For small divergence of the incident beam and $a \gg \lambda$, the beam waist occurs approximately at the focal distance (length), f, from the lens. The radius of the *laser focus* within the focal plane is approximately given by

$$w_0 \approx \zeta \frac{f\lambda}{\pi a}. \tag{5.1.3}$$

ζ depends on the definition of a, which is the diameter of the beam in the middle of the focusing lens. If we define this diameter by the $1/e$ and $1/e^2$ intensity level, $a(1/e)$ and $a(1/e^2)$, the respective values are $\zeta = 1$ and $\zeta = \sqrt{2}$. The intensity profile within the focal plane is Gaussian only if the diameter of the lens significantly exceeds the diameter of the laser beam.

If we confine the diameter of the laser beam by a real aperture of diameter d, the intensity profile within the focal plane becomes non-Gaussian. With $d = a(1/e)$ and $d = a(1/e^2)$, the transmitted power is 63% and 86%, respectively. With $d = \pi a(1/e)/\sqrt{2}$ about 99% of the total power is transmitted. Within the focal plane, the variation in intensity due to diffraction is then $\pm 17\%$. With $d \approx 3.3\, a(1/e)$ this variation is diminished to below $\pm 1\%$ [*Siegman* 1986].

It is evident that the diffraction-limited diameter of the laser focus, $2w_0$, decreases with decreasing wavelength. Clearly, w_0 is one of the essential parameters determining the lateral resolution of patterns. If not otherwise noted, we henceforth use w_0 and a defined by the $1/e$ intensity level.

The *radius of the laser beam* at distance z from the focal plane is

$$w(z) = w_0 \left[1 + \left(\frac{z}{z_R} \right)^2 \right]^{2/2}, \tag{5.1.4}$$

where z_R is the Rayleigh length, i.e., the distance over which the diameter of the focused beam changes by a factor of $\sqrt{2}$. The region $|z| \leq z_R$ is often denoted as the *Rayleigh range*. The length (depth) of the laser focus is defined by

$$L = 2z_R = \frac{4\pi w_0^2}{\lambda} = \frac{4f^2 \lambda}{\pi a^2}. \qquad (5.1.5)$$

Sometimes, L is also denoted as the confocal parameter. With decreasing w_0, accurate positioning of the substrate into the focal plane becomes more and more difficult. For $z \gg z_R$, the beam radius increases linearly with z and the *divergence angle* (Fig. 5.1.1) is

$$\Theta = \frac{w_0}{z_R} = \frac{a}{2f}. \qquad (5.1.6)$$

5.1.2 Pulsed and High-Power CW Lasers

Pulsed lasers are the preferred sources in all types of laser processing where the light energy shall be deposited onto the substrate (workpiece) surface within a short time. High-intensity short pulses cause high heating rates and well-defined localization of the energy input within the substrate. The optimal laser fluence for processing and the heat diffusion length are both controlled via the laser-pulse length, τ_l. The intensity profile of pulsed lasers is, in general, non-Gaussian. The beam width is then frequently defined by the full width at half maximum (FWHM) which is henceforth denoted by $2w$.

Excimer lasers

For laser micropatterning and some types of surface modifications and thin film deposition, excimer lasers are almost ideal sources. This is related to their high photon energies (short wavelengths), their short pulse lengths (typically 10 to 40 ns), and their relatively poor coherence (an excimer laser's output is highly multimode and contains as many as 10^5 transverse modes). The high photon energy of excimer lasers allows for direct photodissociation of many molecules and short optical penetration depths in many solids. The short pulse length is a prerequisite for spatially well-defined and chemically stoichiometric ablation with low damage of the surrounding material, in particular for heat sensitive and multicomponent substrates. The poor spatial coherence of excimer-laser light diminishes interference effects. Interferences cause severe problems in imaging applications with lasers of high spatial coherence. High photon energies, rapid heating and cooling rates, and good energy localization within a thin surface layer are prior conditions in laser lithography, in sheet doping of semiconductor surfaces, and in *congruent* laser ablation for surface patterning and thin film formation by pulsed-laser deposition (PLD). Among the disadvantages of excimer lasers are their low efficiency (the laser output power is only about 1–3% of the electrical input power), their relatively low repetition rates, problems in the pulse-to-pulse stability, and the high operating costs [*Laude* 1994].

With some types of applications, excimer lasers may be substituted by excimer lamps [*Kogelschatz* and *Esrom* 1990].

Nd:YAG lasers

High power pulsed and cw Nd:YAG lasers are employed in various types of laser machining (scribing, trimming, cutting, etc.) and surface modifications. Its frequency-multipled (higher harmonic) lines are also used in different areas of laser microprocessing.

CO_2 lasers

CO_2 lasers with high average power in cw or pulsed operation are the most commonly used sources in laser machining (cutting, drilling, shaping, etc.), laser welding, and many types of large-area surface modifications (hardening, glazing, etc.). CO_2 lasers combine a high efficiency with great reliability and low-cost operation. They are presently the most important laser sources for technical applications.

The specific requirements on CO_2 and Nd:YAG lasers in conventional laser processing are widely discussed in the literature [*Steen* 1991; *Duley* 1976].

At high laser-light intensities ($> 10^8$ W/cm^2) the optical properties of the material to be processed (workpiece) become less important. In this regime, all materials become absorbing at any wavelength due to surface breakdown and plasma formation (Chap. 11).

Among the problems that arise with the presently commercial pulsed lasers are: Difficulties in the precise control of the output power, at least on a shot-to-shot basis, and changes in the beam profile and pointing. These problems may result in surface damage due to uncontrolled melting or ablation, and in a poor uniformity and morphology of generated patterns in microprocessing. Additionally, with UV lasers, in particular with excimer lasers and frequency tripled or quadrupled Nd:YAG lasers, only low pulse energies (Table I) are available.

5.2 Experimental Aspects

The main components to be considered in laser processing are the laser, the imaging optics, the substrate and, in many cases of laser–chemical processing, the reaction chamber. Subsequently, we shall discuss different kinds of

5.2 Experimental Aspects

irradiation geometries and experimental setups employed in different kinds of localized and large-area laser processing.

5.2.1 Microprocessing

Laser microprocessing allows for *single-step* direct substrate patterning with lateral dimensions down into the submicrometer range. Laser microprocessing can be performed by direct writing, by projection of the laser light via a mechanical mask, by employing a direct-contact mask, or by the interference of laser beams (Fig. 5.2.1).

Direct writing

In direct writing, the laser beam, in general a cw laser, is expanded and then focused at normal incidence onto the substrate surface (Fig. 5.2.1a). In most

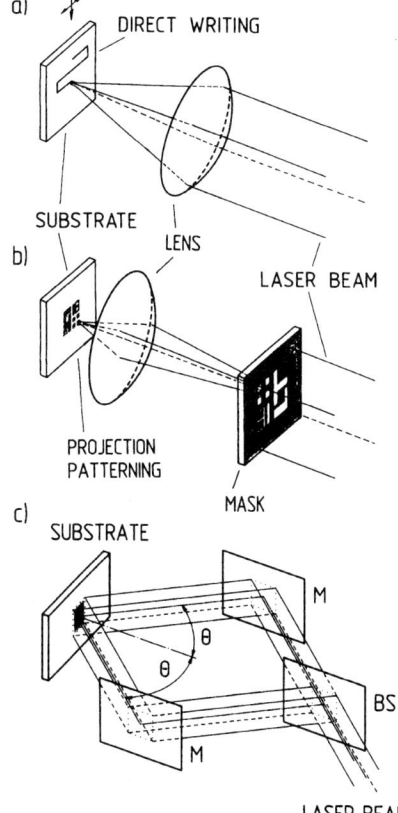

Fig. 5.2.1a–c. Optical configurations employed in laser microprocessing. (a) Direct writing. (b) Projection patterning (the optical path is indicated for one feature only). (c) Patterning by interference of laser beams (M: mirror, BS: beam splitter, Θ: angle of incidence)

cases, patterning is accomplished by translating the substrate with respect to the fixed laser beam.

The diffraction-limited diameter of a Gaussian laser beam focus is given by (5.1.3). Thus, for $f/a \approx 1$, the minimum laser spot size becomes $2w_{0\,\text{min}} \approx \lambda$. With Ar^+- and Kr^+-lasers this is, typically, around 0.5 µm. In practical applications, however, values $f/a > 1$ are frequently employed. There are several reasons for this: In order to minimize diffraction effects, the diameter of the laser beam at the focusing lens, $a(1/e)$, should be considerably smaller than the diameter of the lens, d, and thus $f/a > f/d \geqslant 1$. To minimize image aberrations, special lenses or elaborate lens combinations must be used, and this increases f/d even further. Moreover, all types of laser processing require a certain working distance between the lens and the substrate to be processed, e.g., to suppress contaminations of the lens by species desorbing/evaporating from the substrate, or due to a window of a reaction chamber, etc. Furthermore, some applications require a certain depth of focus, L, e.g., for processing of non-planar substrates. This also limits the size of the focus that can be employed, since $L \propto f^2/a^2$. In summary, in practical applications one has to compromise between the size of the focus, the working distance, the focal depth, and the price of the imaging optics.

Projection patterning

Laser-light projection (Fig. 5.2.1b) allows one to generate whole patterns with a single or a few laser shots. The resolution achieved is somewhat lower than in direct writing.

The smallest resolvable feature size is often defined by the smallest distance between two points that can be resolved according to the Rayleigh criterion. It is given by $d = \xi \lambda / N_A$, where $N_A = n \sin \Theta$ (2Θ is the angle of the focused beam near the image) is the numerical aperture of the imaging system, and ξ a factor which depends on the spatial and temporal coherence of the light and which is, typically, within the range $0.6 \leqslant \xi \leqslant 0.8$.

Besides the simple imaging shown in the figure, telecentric optical schemes, Schwarzschild telescopes, etc., have been employed.

Surface patterning using a contact mask can be performed with laser light that is either focused to a line or unfocused with perpendicular or/and parallel incidence (Fig. 5.2.3). Here, the resolution is determined by the mask.

Interference

Laser-beam interference (Fig. 5.2.1c) allows one to generate patterns with periods

$$\Lambda = \frac{\lambda}{2 \sin \Theta} \tag{5.2.1}$$

5.2 Experimental Aspects

over several square centimeters. The technique has been successfully demonstrated for material deposition, surface modifications, chemical etching, and (well-defined) surface roughening. It permits one to fabricate diffraction gratings, holograms, etc.

5.2.2 The Reaction Chamber; Typical Setup

In laser–chemical processing, the reaction chamber is often operated with a constant flow of the gaseous or liquid reactant with or without a carrier. In microchemical processing, the reaction chamber can be sealed off, in many cases, because of the small amount of species consumed in most of the reactions. The type of material used for the fabrication of the reaction chamber is often of great importance. Spontaneous reactions of precursor molecules, intermediate species, reaction products, and carrier gases or solvents with the chamber material can result in significant changes in the number densities of species, reaction pathways and, more importantly, in additional reaction products that contaminate the substrate surface, the deposited film, the microstructure, etc. Thus, proper materials selection for the reactor, including its windows, o-rings, etc., is a prerequisite for well-defined processing.

A typical setup employed in direct writing is schematically shown in Fig. 5.2.2. Laser-beam illumination times are electronically controlled via a mechanical shutter or via an electro-optical modulator. In micropatterning, the laser beam is first expanded and then focused onto the substrate, for example by employing an objective. An eyepiece, or a CCD camera in combination with a

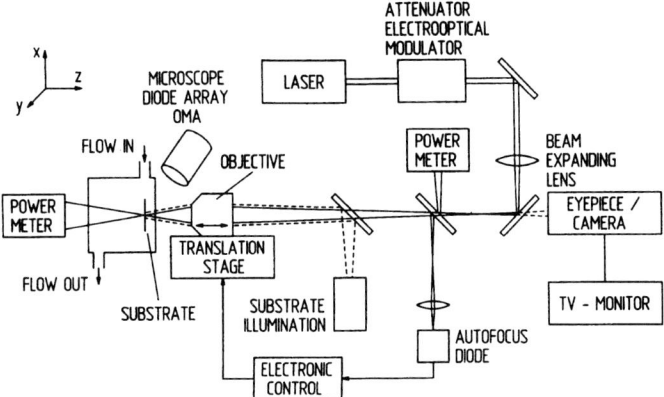

Fig. 5.2.2. Typical experimental setup employed in microchemical processing. The substrate is mounted on an xyz stage. The position of the objective is optically and electronically controlled (autofocus). The eyepiece, or a CCD camera together with a monitor, is used for direct observation of patterns

monitor, is used for direct observation of patterns. The position of the objective is optically and electronically controlled (autofocus). Additionally, a polarizer and a $\lambda/4$ plate are introduced, in general, in order to avoid optical feedback from the substrate or the processed area. Optical coupling can generate relaxation oscillations, mode instabilities, or chaotic temporal behavior, which result in spatial and temporal intensity fluctuations. The lasers most commonly used in direct writing are cw lasers, mainly Ar^+- and Kr^+-lasers. Typical (effective) power densities employed, are between 10^3 and some 10^7 W/cm^2. The main features of setups used in micropatterning by laser-light projection and interference are similar.

5.2.3 Large-Area Processing

Extended thin film formation by laser-CVD and different kinds of large-area surface transformations, modifications, etching, and ablation are performed with optical arrangements shown in Fig. 5.2.3.

For *perpendicular* (normal) laser-beam incidence, the substrate is scanned with respect to the laser beam which is either unfocused or defocused, or

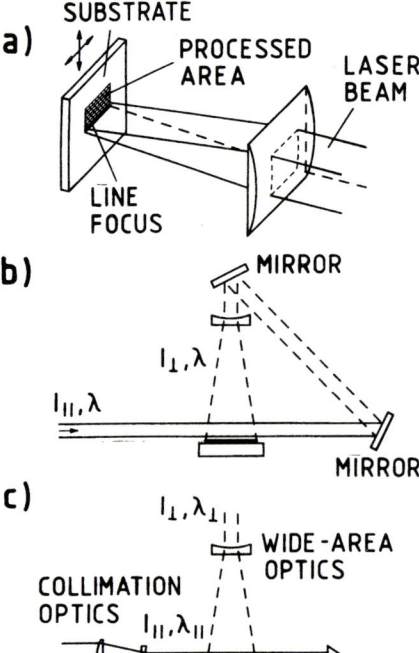

Fig. 5.2.3a–c. Large-area surface processing by employing: (a) A line focus. (b) Parallel laser-beam incidence, eventually combined with perpendicular incidence of the transmitted light. (c) Combined parallel and perpendicular incidence of different laser beams

focused to a line by means of a cylindrical lens. The latter arrangement is shown in Fig. 5.2.3a. The experimental setup and laser–molecule–substrate interactions are very similar to those described for microprocessing. Direct substrate irradiation can clean the surface, assist nucleation processes, enhance surface diffusion of species, promote surface catalyzed reactions, etc. With large-area gas- or liquid-phase processing, transport limitations become important at lower (thickness) deposition rates than in microchemical processing. This is a consequence of the dimensionality of transport. Transport will be determined by two-dimensional diffusion if a tight line focus is used, and by one-dimensional diffusion if the laser beam is unfocused or defocused. Besides irradiating the substrate through the ambient medium, the entrance window can itself be used as a substrate (Fig. 9.5.1b). This irradiation geometry can be employed with strongly absorbing media, or with gases of moderate absorption at high pressures.

Films deposited at perpendicular laser-beam incidence are often of better morphology and higher purity than those deposited at parallel incidence. This is mainly a consequence of laser-induced surface heating.

Parallel laser-beam incidence opens up the ability to investigate the influence of species excited within the gas phase only. With both pyrolytic and photolytic processes, excitation of species will take place within the total volume of the laser beam. The absence of direct substrate or film irradiation, permits to employ high power cw- or pulsed-lasers. In order to increase the laser-light intensity above the surface, the mirror on the right side in Fig. 5.2.3b can be positioned perpendicular to the laser beam.

For the production of high-quality films, the substrate is often preheated to a certain temperature, T_s. This allows for proper control of the surface temperature and film thickness. The substrate temperatures typically employed are much lower than those used in standard CVD. Laser-CVD permits one to control film thicknesses within, typically, 0.01 Å/pulse. This allows accurate growth of multilayer structures.

Combined parallel and perpendicular irradiation can be achieved by either directing the emerging beam onto the surface as shown in Fig. 5.2.3b or by using two lasers, for example at different wavelengths (Fig. 5.2.3c). In the latter case, one can separately optimize homogeneous and heterogeneous pyrolytic or photolytic reactions by proper selection of the intensities and wavelengths of beams at parallel $(I_\|, \lambda_\|)$ and perpendicular $(I_\perp, \lambda_\perp)$ incidence. For perpendicular irradiation the laser can often be substituted by a lamp.

5.2.4 Substrates

Proper pretreatment, material selection, and temperature control of the substrate is of great importance in many types of laser processing.

The surface quality (cleanliness, roughness, morphology, surface oxides, etc.) and, if relevant, the crystallographic orientation of a particular substrate, influence the laser-induced surface temperature, the sticking of impinging species, solid-phase in/out diffusion of species, the adherence of deposited films, etc.

Cleaning of the substrate surface is performed by employing the standard techniques that are widely described in the literature on gas-phase epitaxy, CVD, plasma deposition, etc. Efficient surface cleaning can sometimes be performed by irradiating the substrate (mainly in vacuum or an inert atmosphere) prior to surface processing with a single or a few laser pulses. Here, excimer lasers have proved to be of particular suitability.

During LCVD or PLD, the substrate can be kept at either ambient or elevated temperature. This permits deposition onto heat sensitive materials such as organic polymers, compound semiconductors, piezoelectric ceramics, etc. The as-deposited films are amorphous, polycrystalline, or monocrystalline, depending on the specific film and substrate material, surface temperature, and ambient atmosphere. For the growth of high-quality films, the substrate must fulfil a number of conditions:

– The mismatch in thermal expansion between substrate and film should be as small as possible. Otherwise, strains and even microcracks build up during cooling or post-deposition annealing.
– High chemical and thermal stability of the substrate material are required in order to avoid interface reactions. Interdiffusion of substrate/film elements causes contaminations and changes in film stoichiometry.
– The lattice constants of the substrate surface should closely match the lattice constants of the deposited material. This is a prior condition for oriented large-grain polycrystalline or epitaxial film growth.

The mismatch in lattice parameters and thermal expansion, as well as material interdiffusion can often be diminished by means of an intermediate (buffer) layer.

5.3 Confinement of the Excitation

The resolution achieved in laser micropatterning is determined by the width of the laser focus, the spatial confinement of the laser-induced excitation, by material damages, and by different types of nonlinearities. The quantity which determines the confinement of the excitation depends on the particular system under consideration. In some types of conventional laser processing and most cases of pyrolytic microchemical processing, the important parameters are the width and depth of the thermal field induced on the substrate (workpiece) to be processed. In photochemical laser processing, the confinement of the interaction process is determined by (non-thermal) excitations of the ambient gas or liquid, of adsorbed layers, *and* of the substrate. With high laser powers, the spatial extension of the plasma plume becomes the relevant quantity.

5.3 Confinement of the Excitation

5.3.1 The Thermal Field

The width and depth of the thermal field induced within the irradiated substrate is discussed in various chapters, and in particular in Sect. 6.5. It is essentially determined by the width of the laser focus, the heat diffusion length, and the optical penetration depth, depending on the particular system.

5.3.2 Non-thermal Substrate Excitations

The spatial resolution achieved in a particular processing application can depend on non-thermal excitations of the substrate (Sect. 2.4). For example, the diffusion of photo-generated carriers in crystalline semiconductors may decrease the resolution in laser-induced dry-etching (Chap. 15).

5.3.3 Gas-, Liquid- and Adsorbed-Phase Excitations

Apart from substrate excitations, adsorbed-phase and homogeneous gas- or liquid-phase excitations may decrease the spatial confinement in LCP. Species that are photoexcited/dissociated within the gaseous or liquid ambient medium will randomly diffuse towards the solid surface. Consequently, deposition on, or etching of areas beyond the laser spot will occur. The loss in resolution can be minimized by a careful selection of the processing parameters:

In *pyrolytic* gas- or liquid-phase processing, thermal or photochemical excitations within the volume of the laser beam can be avoided, to a large extent, by a proper choice of the intensity and wavelength of the laser radiation.

In *photochemical* gas- or liquid-phase processing, a confinement of the reaction to the illuminated area is possible only if the dissociative continuum of adsorbed molecules is considerably shifted (in general, towards longer wavelengths). In such cases, proper selection of the laser-light frequency allows mainly adsorbed-phase instead of gas- or liquid-phase photolysis (Chap. 20). If the interaction between adsorbed species and the substrate is strong enough, adsorbed-layer photolysis can also be performed by removing the ambient medium prior to laser-light irradiation. In all other cases, photoexcitation/dissociation will take place within the *total* volume of the laser beam. Here, tight focusing not only limits the area of excitation on the substrate, but also confines the relevant volume of excitation within the ambient medium. Species that are generated at a distance larger than their mean free path for collision-induced deactivation, recombination, or reactions with parent or other molecules, will not reach the substrate. In other words, the flux of such species onto the substrate is limited to a small region around the laser spot. This depends on the collision rate and the incident light flux. In gas-phase processing, the reaction volume can sometimes be further diminished by proper selection of the type of buffer gas and the respective partial pressures of gaseous constituents.

The spatial confinement is also decreased by surface diffusion of adsorbed precursors, or of photofragments impinging onto the substrate.

For the broad range of experimental conditions used in LCP, the relevant diffusion lengths may be smaller, comparable, or larger than the diffraction limited diameter of the laser beam.

5.3.4 Plasma Formation

With the formation of a plasma, the confinement of the laser–solid interaction decreases, in both conventional and chemical laser processing. With dense plasmas, laser–solid interactions are mediated only via the plasma. The processing width is then determined by the size of the plasma plume.

5.3.5 Material Damages

Laser-induced material damages and disturbances around the laser-processed zone may substantially decrease the spatial resolution in laser micropatterning: Strong temperature gradients may cause material cracking, the depletion of a certain component, material segregation, etc. Changes in the optical index of refraction, convection, turbulence, bubbling, etc., play an important role, in particular in liquid-phase processing and in all cases where the surface is melted. The desorption of species or fragments from the surface, the formation of clusters, the coating of entrance windows, etc., may attenuate and scatter the incident light.

5.3.6 Nonlinearities

It has been demonstrated in different processing applications that, under certain conditions, it is possible to produce patterns with lateral dimensions that are significantly *smaller* than the diffraction-limited diameter of the laser focus. This observation is related to nonlinearities in the laser–matter interactions.

Thermal activation

To illustrate the influence of nonlinearities in thermally activated laser processing, we have plotted in Fig. 5.3.1 the intensity profile of a Gaussian laser beam

5.3 Confinement of the Excitation

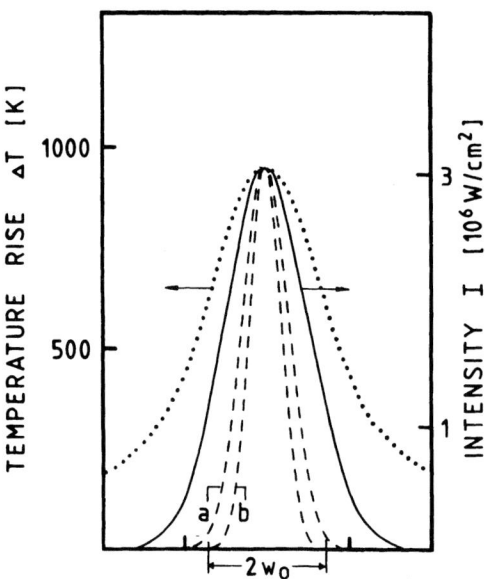

Fig. 5.3.1. Spatial confinement in pyrolytic laser processing. Full curve: Intensity profile of the laser beam ($2w_0$ is the 1/e diameter). Dotted curve: Calculated temperature rise ($\kappa = 0.50$ W/cmK, $P_{abs} = 0.3$ W, $\alpha \to \infty$, $w_0 \approx 1.8$ μm). Dashed curves: Normalized processing rates for $\Delta E = 22$ kcal/mol (curve a) and $\Delta E = 46.6$ kcal/mol (curve b)

(full curve), the temperature distribution induced on a semi-infinite substrate (dotted) and (normalized) processing rates $W^*(r) \propto \exp[-\Delta E/k_B T(r)]$ (dashed). Activation energies of $\Delta E = 22$ kcal/mol (curve a) and $\Delta E = 46.6$ kcal/mol (curve b) are characteristic, for example, for laser-induced deposition of Ni from $Ni(CO)_4$ and of Si from SiH_4. The nonlinear dependence of the processing rate on temperature causes the lateral variation in W to be substantially narrower than in temperature, $T(r)$, and the laser-beam intensity, $I(r)$. The spatial confinement of the reaction increases with increasing activation energy ΔE.

Let us now study, in more detail, the influence of ΔE and the center temperature, T_c, on the spatial confinement. The increase in spatial confinement shall be described by the ratio w_0/r_e where r_e is given by the 1/e point in processing rate, $W(r)$. For a circular laser beam, the temperature distribution can be written as

$$T(r) = T_0 + \Delta T(r) = T_0 + \Delta T_c f(r) . \tag{5.3.1}$$

If we assume a Gaussian beam, surface absorption, and temperature-independent parameters, $f(r)$ is given by (7.2.4). The definition of r_e yields for an *arbitrary* function $f(r)$

$$r_e = f^{-1}\left[\frac{1 - T_0(1 + T_0/\Delta T_c)/\mathscr{E}}{1 + T_0(1 + \Delta T_c/T_0)/\mathscr{E}}\right] \equiv f^{-1}(\gamma) , \tag{5.3.2}$$

where $f^{-1}(\gamma)$ is the inverse function of $f(r)$ and $\mathscr{E} \equiv \Delta E/k_B$. In Fig. 5.3.2 we have plotted the ratio w_0/r_e versus $\Delta T_c^* \equiv \Delta T_c/T_0$ for $T_0 = T(\infty) = 300$ K. For

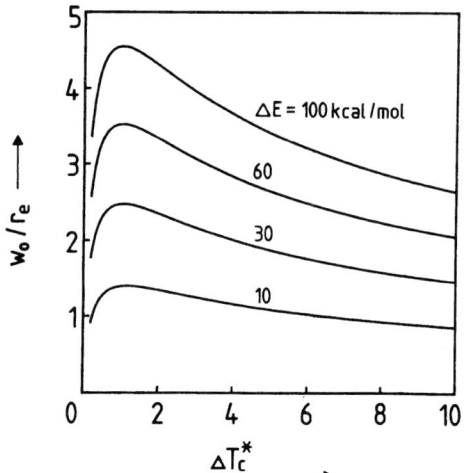

Fig. 5.3.2. Ratio w_0/r_e $[W(r_e) = W(0)/e]$ versus (normalized) laser-induced center temperature rise $\Delta T_c^* = \Delta T_c/T(\infty)$ for various activation energies, ΔE

a certain value of ΔE, there exists a maximal value of w_0/r_e at a certain optimal center temperature $T_{c\,opt}$. Because the function $f^{-1}(\gamma)$ is monotonic, the temperature $T_{c\,opt}$ can easily be calculated from

$$\frac{\partial r_e}{\partial \Delta T_c} = \frac{\partial f^{-1}}{\partial \gamma} \frac{\partial \gamma}{\partial \Delta T_c} = 0,$$

which yields

$$\Delta T_{c\,opt} = T_0 \left(\frac{\mathscr{E} + T_0}{\mathscr{E} - T_0} \right) \approx T_0.$$

Thus, with systems where $\mathscr{E} \gg T_0$ we find $T_{c\,opt} \approx 2T_0$. Clearly, efficient processing requires the practical center temperature to be well above the threshold temperature for the particular processing application, i.e., $T_{c\,prac} > T_{c\,opt}$. Thus, the confinement will be decreased with respect to the optimal value.

If $\Delta T_c \approx T_c \gg T_0$ the processing rate can be approximated by

$$W = k_0 \exp\left(-\frac{\mathscr{E}}{T(r)}\right) \approx k_0 \exp\left(-\frac{\mathscr{E}}{T_c}\right) \exp\left(-\frac{\mathscr{E}}{T_c} \frac{r^2}{w_T^2}\right). \tag{5.3.3}$$

Here, we have used the expansion $T(r) \approx T_c(1 - r^2/w_T^2)$. Equation (5.3.3) then yields

$$r_e = w_T \left(\frac{T_c}{\mathscr{E}} \right)^{1/2}.$$

For the case of surface absorption we obtain $w_T \approx 1.5\, w_0$ (Sect. 6.5).

Other nonlinearities, for example in the absorptivity $A = A(T)$, $A(h)$, etc., may also influence the confinement of the reaction.

5.3 Confinement of the Excitation

Photophysical and photochemical processing

Non-thermal excitations depend on the particular interaction process. For example, in laser photolysis based on *multiphoton* dissociation of adsorbed molecules, or surface bonds, the reaction rate is given by $W \propto I^n$, where n is the number of coherently absorbed photons. This nonlinearity increases the confinement in photolytic processing.

Nucleation

Whenever nucleation processes are involved in a particular processing step, such as in material deposition, compound formation, recrystallization, etc., they may be of importance for the spatial confinement, simply because of their strong nonlinearity (Chap. 4).

Instabilities, periodic structures

While the nonlinear interactions mentioned above will, in general, improve the confinement of the laser process, the occurrence of instabilities and periodic structures will, within certain parameter regions, deteriorate it. This will be discussed in detail in Chap. 28.

Part II : Temperature Distributions and Surface Melting

One of the basic quantities in laser processing is the temperature rise induced by the absorbed laser radiation on a material surface or within its bulk. A knowledge of the temperature distribution is a prerequisite for both fundamental investigations and technical applications. In fact, before starting laser processing at all, the temperature distribution, or at least the maximum temperature, should first be estimated. Only on this basis one can select the type of laser and the parameters adequate for the particular processing application.

Laser-induced temperature distributions can be estimated by solving the heat equation for the particular irradiation geometry and the substrate under consideration. This is the main content of this part of the book. The limits of such calculations have already been discussed in Chap. 2.

6 General Solutions of the Heat Equation

6.1 The Boundary-Value Problem

The substrate shall be an infinite slab of uniform thickness, h_s, that is irradiated by a cw- or pulsed-laser beam which is either focused or extended over a wider area (Fig. 6.1.1). For *localized* irradiation, the absorbed laser light generates a local temperature rise, $\Delta T(x, t)$, which can be calculated by solving the *three-dimensional* heat equation (2.2.1). For *large-area* (uniform) irradiation, the temperature is uniform within planes $z = $ const., and the temperature rise, $\Delta T(z, t)$, can be calculated by solving the *one-dimensional* heat equation.

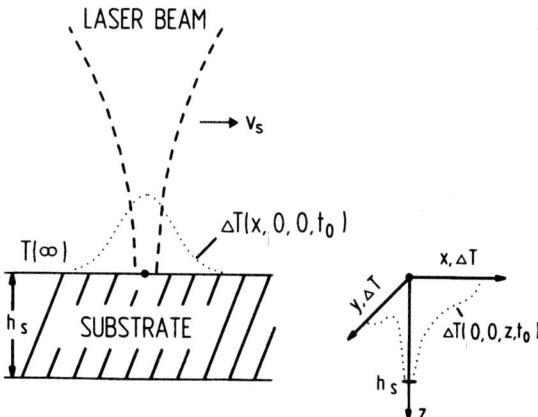

Fig. 6.1.1. Infinite slab of uniform thickness, h_s, irradiated by a laser beam at perpendicular incidence. In the absence of scanning, the origin of the coordinate system is on the surface in the center of the laser beam. If $v_s \neq 0$, the origin of the coordinate system is either fixed with the substrate or with the laser beam, depending on the particular problem under consideration. The laser-induced temperature rise on the surface $z = 0$ and along the z-direction is indicated by dotted curves. The temperature far away from the irradiated area is $T(\infty)$. With increasing focal width, $2w$, the temperature distribution becomes wider. In the limit $w \to \infty$ (large-area irradiation) the temperature rise becomes uniform within planes $z = $ const.

The source term

In most cases of thermal laser processing, the Rayleigh length is long compared to the optical penetration depth. Then, in a coordinate system that is fixed with the laser beam, the source term in the heat equation can be written in the form

$$Q(x_\alpha, t) = I(x, y, t)(1 - R)f(z) = I_a g(x, y)f(z)q(t), \tag{6.1.1}$$

where $I_a = I_0(1 - R)$ is the (maximum) laser-light intensity that is not reflected from the surface $z = 0$. $g(x, y)$ describes the (arbitrary) intensity distribution (beam shape) within the xy-plane. $f(z)$ describes the attenuation of the laser light in z-direction, and $q(t)$ its temporal dependence (pulse shape). Thus, we have $\max[g(x, y)] = \max[q(t)] = 1$. Frequently, we introduce cylindrical coordinates, so that $I = I(r, \varphi, t)$ where φ describes the angle between the radius vector $\mathbf{r}$ and the x-axis.

$R \equiv R(T, \lambda)$ denotes the temperature- and wavelength-dependent normal-incidence reflectivity within the processed area. In the general case, the reflectivity depends on the angle of incidence, the polarization of the laser beam, and the thickness of the slab. The latter dependence can be ignored if the optical penetration depth $l_\alpha \ll h_s$. If, on the other hand, $l_\alpha > h_s$, interference phenomena due to multiple reflections of the laser light may become important. The reflectivity, and also the absorptivity, will then depend on h_s. In this chapter we ignore multiple reflections within the slab.

Values of R and α are listed in Table T3 for different materials and various wavelengths. For metals, reflectivity values in the near ultraviolet (UV) and visible (VIS) spectral range are, typically, between 0.4 and 0.95. In the infrared (IR), typical values are between 0.9 and 0.99. It should be noted, however, that these values can be applied only if the wavelength, λ, is small compared to the radius of the laser spot, w. Additionally, in the dependence $R = R(T, \lambda)$ we can employ the surface temperature $T = T(x, y, 0, t)$ only if the variation of $T \equiv T(\mathbf{x}, t)$ in z-direction is slow over the distance l_α, i.e., if $l_\alpha(dT/dz) \ll T$. Finally, with very high laser-light intensities, optical nonlinearities become important.

6.1.1 The Attenuation Function f(z)

The function $f(z)$ in (6.1.1) describes the attenuation of the laser beam in z-direction. For a *uniform* material, it can be written as

$$f(z) = \alpha(T(z)) \exp\left[-\int_0^t \alpha(T(z'))\, dz'\right], \tag{6.1.2}$$

where $\alpha(T)$ is the temperature-dependent linear absorption coefficient at the laser wavelength under consideration. Because of this temperature dependence, the optical properties of the material become inhomogeneous upon laser-light

6.1 The Boundary-Value Problem

irradiation. This case is discussed in Chap. 9. In the present chapter we mainly consider either finite absorption with α being a constant, or surface absorption.

If α is *finite* but independent of temperature, we obtain

$$f(z) = \alpha \exp(-\alpha z). \tag{6.1.3a}$$

For *surface absorption*, $l_\alpha = \alpha^{-1}$ is small compared either to the heat diffusion length, $l_T \approx 2(D\tau_l)^{1/2}$, or to another characteristic length, $\mathscr{L}$, depending on which is smaller. $\mathscr{L}$ can be the radius of the laser beam, w, the thickness of the slab, h_s, a length which characterizes the heat exchange, κ/η, the surface curvature or roughness, etc. The criterion for surface absorption can be written as $l_\alpha \ll \min(l_T, \mathscr{L})$. Henceforth, surface absorption is often denoted, symbolically, by $\alpha \to \infty$ (for metals, one often introduces instead of l_α the skin depth $l_s = c/(2\pi\sigma\omega)^{1/2}$ where σ is the frequency-dependent ac conductivity). In many practical cases, the assumption of surface absorption is satisfied with coefficients $\alpha > 10^4 \text{cm}^{-1}$. In this approximation, the light intensity that penetrates into the material can be ignored. Thus, the source term Q vanishes, except at the irradiated surface, where it is given by the absorbed laser power, so that

$$f(z) = \delta(z), \tag{6.1.3b}$$

where $\delta(z) [\text{cm}^{-1}]$ is the delta function. This approximation holds for metals within the near UV to near IR spectral region where the absorption coefficients are, typically, between 10^5 and 10^7cm^{-1}. It also holds for many semiconductors at elevated temperatures and laser wavelengths $\lambda \ll hc/E_g$ (E_g denotes the band-gap energy). For crystalline Si, the approximation $\alpha \to \infty$ is reasonably well fulfilled for laser wavelengths $\lambda < 0.5\,\mu\text{m}$.

6.1.2 Boundary and Initial Conditions

The solution of the heat equation (2.2.1) requires the consideration of boundary conditions. We shall assume that the slab is immersed in a gaseous or liquid medium M and that the temperature is continuous at the interfaces, i.e.,

$$T_s(x_\alpha, t) = T_M(x_\alpha, t) \quad \text{at} \quad z = 0, h_s. \tag{6.1.4}$$

Another boundary condition is obtained from the balance of energy fluxes at the surface $z = 0$

$$-\kappa_s \frac{\partial T_s(x_\alpha, t)}{\partial z}\bigg|_{z=0} - J_{ch}(x, y, 0, t) = -\kappa_M \frac{\partial T_M(x_\alpha, t)}{\partial z}\bigg|_{z=0}. \tag{6.1.5}$$

The same boundary condition applies at the surface $z = h_s$, except the change in sign of J_{ch}. The heat conductivities κ_s and κ_M refer to the substrate and the ambient medium, respectively. J_{ch} is the heat flux originating from phase changes or/and chemical reactions at the surface $z = 0$. Clearly, the heat flux at $z = 0$ is continuous only if $J_{ch} = 0$. In the case of surface absorption, the source term, Q, is included in (6.1.5). The boundary conditions (6.1.4) and (6.1.5) are good approximations as long as any temperature jump at the interfaces can be ignored. This approximation holds if the mean free path of molecules, λ_m, is small compared to the size of the heated zone. For the case of localized irradiation (Fig. 6.1.1) this is fulfilled if $\lambda_m \ll w$ where w is the radius of the laser focus. For a liquid ambient medium, this condition always holds. For laser microchemical processing in gases at medium to low pressures it may not be fulfilled and $\lambda_m \geqslant w$. In this case, *no* complete thermal equilibrium between the laser-heated area and the adjacent gas is reached (Sect. 9.5.4).

In many practical cases one can use instead of (6.1.5) the *effective* boundary condition

$$-\kappa_s \left.\frac{\partial T_s(x_\alpha, t)}{\partial z}\right|_{z=0} \approx J_{ch}(x, y, 0, t) - J_{loss}(x, y, 0, t), \qquad (6.1.6)$$

where

$$J_{loss} = J_c + J_r$$
$$\approx \eta[T_s(x, y, 0, t) - T_M(\infty)] + \sigma_r \varepsilon_t [T_s^4(x, y, 0, t) - T_M^4(\infty)]. \qquad (6.1.7)$$

J_c is the energy flux into the ambient medium. Here, η is the surface conductance (coefficient of surface heat transfer). It depends on the thermal conductivities κ_s and κ_M, on the geometry of the substrate, the flow velocity of the ambient – if relevant, on the surface morphology, and the thickness of surface coverages – for example, native oxide layers, adsorbates, water films, etc. In the absence of convection one can often use the approximation $\eta \approx \kappa_M/l$ where l is some characteristic length.

In the presence of *free* convection, η is temperature dependent and can be described by

$$\eta(T_s) = \eta_0' \left(\frac{T_s - T_M(\infty)}{T_M(\infty)}\right)^{1/4}. \qquad (6.1.8)$$

For substrates with areas of several square centimeters that are immersed in air at standard conditions, typical values of η_0' are around 10^{-4} W/cm²K. For small areas, η_0' may exceed this value by a factor of 10 or more. For a liquid, η_0' is, typically, within the range 0.1 to 0.3 W/cm²K.

With *forced* convection one can often employ the approximation

$$\eta_f = \eta(T) + \eta_1(T)\left(\frac{v_c}{v_1}\right)^{1/2}, \qquad (6.1.9)$$

as long as $v \ll v_0$ where v_0 is the sound velocity and v_c the velocity of the ambient with respect to the substrate. For air at standard conditions $v_1 \approx 1$ m/s

6.1 The Boundary-Value Problem

and $\eta_1 \approx 0.2\,\eta$. All coefficients η_i depend on the geometry of the substrate and its orientation with respect to the gas flux.

The consideration of the heat conductance in the boundary conditions is quite useful, because it permits to estimate the influence of an ambient medium on the substrate temperature by solving the (single) heat equation for the substrate only. A more physical description of the problem is given in Sect. 9.5.

J_r describes the energy flux by thermal radiation. The Stefan–Boltzmann constant is $\sigma_r \approx 5.7 \cdot 10^{-12}\,\text{W/cm}^2\text{K}^4$. $\varepsilon_t \equiv \varepsilon_t(T_s)$ is the total emissivity. For polished metals, $\varepsilon_t \approx 0.02$ to 0.05. For thermally oxidized metals, $\varepsilon_T \approx 0.6$ to 0.7. For standard glass and fused quartz, $\varepsilon_t \approx 0.93$. For (carbon) soot, $\varepsilon_t \approx 0.98$.

The comparison of fluxes J_c and J_r shows that with substrate materials immersed in gases at standard conditions and with $\varepsilon_t \approx 0.4$, $J_c > J_r$ for temperatures $T_s < 1000\,\text{K}$ and $J_c < J_r$ if $T_s > 1000\,\text{K}$. At $T_s = 1000\,\text{K}$ we have $J_c \approx J_r \approx 1$ to $3\,\text{W/cm}^2$. In most processing cases, this energy loss can be ignored in comparison to absorbed laser-light intensities commonly employed.

If there are no phase transformations or/and chemical reactions, or if the related energies are small, we can also ignore J_{ch}.

Summary

The boundary-value problem described by the heat equation, the source term (6.1.1), and the above boundary and initial conditions, shall now be summarized. For convenience, we introduce the linearized temperature, θ, see (2.2.8). The heat equation can then be written as

$$\frac{\kappa}{D}\frac{\partial \theta}{\partial t} - \kappa \nabla^2 \theta = Q, \tag{6.1.10a}$$

where $D \equiv D(T(\theta))$ and $\kappa \equiv \kappa(T(\infty))$ refer to the substrate (we drop indices if the meaning of quantities is evident). Henceforth, we mainly assume D to be independent of temperature. In a coordinate system that is fixed with the slab (for more general representations of Q see Sect. 2.2) the source term (6.1.1) can be written as

$$Q = I_a(x, y, t) f(z) = I_a g(x - v_s t, y) f(z)\, q(t). \tag{6.1.10b}$$

If $v_s \neq 0$, the laser beam is scanned with constant velocity in x-direction. If losses by thermal radiation are ignored, the boundary conditions have the form

$$-\kappa \frac{\partial \theta}{\partial z}\bigg|_{z=0} = -\eta_0 \left[T(\theta(z=0)) - T(\infty) \right] \tag{6.1.10c}$$

$$-\kappa \frac{\partial \theta}{\partial z}\bigg|_{z=h_s} = \eta_h \left[T(\theta(z=h_s)) - T(\infty) \right] \tag{6.1.10d}$$

$$\theta(r \to \infty) = 0. \tag{6.1.10e}$$

Here, we have introduced different surface conductances η_0 and η_h for planes $z = 0$ and $z = h_s$, respectively. The initial condition is

$$\theta(t \leq 0) = 0 . \tag{6.1.10f}$$

For surface absorption, the source, Q, is more conveniently considered in the boundary condition. The right side of (6.1.10c) is then replaced by $I_a(z=0) - \eta_0[T(\theta(z=0)) - T(\infty)]$.

6.2 Analytical Solutions

The most frequently employed analytical techniques to solve the boundary-value problem described in the preceding section are the method of integral transformations and the Green's function technique [*Carslaw* and *Jaeger* 1988]. If analytical solutions are found, they can frequently be presented in integral form only. These integrals can be calculated either numerically or by employing additional assumptions. There are also other analytical techniques which are less well known but very useful, in particular with nonlinear problems: Among those are methods that are based on the theory of functions with complex variables (method of conformal mapping) and methods which are related to the symmetry properties of the heat equation. The latter techniques are based on the theory of Lie groups. In addition, there are a number of approximation methods such as variational techniques, the method of averaging, the Galerkin method, etc.

Subsequently, we first present quite general solutions of the boundary-value problem (6.1.10) and then consider special cases. For convenience, we introduce dimensionless variables. The normalized coordinates

$$x^* = \frac{x}{l} ; \quad y^* = \frac{y}{l} ; \quad z^* = \frac{z}{l} , \tag{6.2.1a}$$

where l is some characteristic length. In many problems, the radius or width of the laser beam is an appropriate normalization factor, i.e., $l \equiv w_0, w, w_x$ or w_y (which refers to a Gaussian beam, a circular beam with constant intensity, or a rectangular beam). For thin infinite slabs of thickness h_s we sometimes choose $l \equiv h_s$. The normalized time and scanning velocity are defined as

$$t^* = \frac{D}{l^2} t \quad \text{and} \quad v_s^* = \frac{v_s l}{D} . \tag{6.2.1b}$$

We also introduce the dimensionless absorption coefficient and thickness

$$\alpha^* = \alpha l \quad \text{and} \quad h_s^* = \frac{h_s}{l} . \tag{6.2.1c}$$

6.2 Analytical Solutions

The dimensionless coefficients of heat transfer are

$$\eta_0^* = \frac{\eta_0 l}{\kappa} \quad \text{and} \quad \eta_h^* = \frac{\eta_h l}{\kappa}. \tag{6.2.1d}$$

We also introduce the dimensionless quantities $\kappa_0^* = \kappa_0/\kappa$ and $\kappa_h^* = \kappa_h/\kappa$ which are equal to unity, except if one of the surfaces is kept at a constant temperature, e.g., if $\theta(z = h_s) = 0$, we have $\kappa_h^* = 0$.

The most general solution of the boundary-value problem (6.1.10) obtained with the Green's function technique can be written as

$$\theta(\mathbf{x}^*, t^*) = \frac{l^2}{\kappa} \int_0^{t^*} \int_0^{h_s^*} \int_{-\infty}^{+\infty} \int_{-\infty}^{+\infty} \mathbb{G}(\mathbf{x}^*, t^*, \mathbf{x}_1^*, t_1^*) Q(\mathbf{x}_1^*, t_1^*) d\mathbf{x}_1^* dt_1^*,$$

where $Q = Q(\mathbf{x}_1^*, t_1^*)$ is an arbitrary source term. Here, we have assumed linear boundary conditions. The Green's function $\mathbb{G}$ can be obtained by the method of integral transformations. For a coordinate system that is fixed with the laser beam, $\mathbb{G}$ is given by

$$\mathbb{G} = \frac{1}{4\pi(t^* - t_1^*)} \exp\left(-\frac{(x^* - x_1^*)^2 + (y^* - y_1^*)^2}{4(t^* - t_1^*)}\right) \mathbb{F}(z^*, z_1^*, t^*, t_1^*)$$

and

$$\mathbb{F}(z^*, z_1^*, t^*, t_1^*) = \sum_{n=-\infty}^{\infty} B_n Z_n(z^*) Z_n(z_1^*) \exp[-v_n^2(t^* - t_1^*)],$$

while

$$Z_n(z^*) = \cos v_n z^* + \frac{\eta_0^*}{\kappa_0^* v_n} \sin v_n z^*$$

$$B_n = \frac{\zeta \kappa_0^{*2} v_n^2}{(\kappa_0^{*2} v_n^2 + \eta_0^{*2})(\zeta h_s^* + \kappa_h^* \eta_h^*) + \zeta \kappa_0^* \eta_0^*},$$

where $\zeta = \kappa_h^{*2} v_n^2 + \eta_h^{*2}$. Here, v_n are roots of the equation

$$\tan(h_s^* v_n) = \frac{v_n(\kappa_0^* \eta_0^* + \kappa_h^* \eta_h^*)}{\kappa_0^* \kappa_h^* v_n^2 - \eta_0^* \eta_h^*}.$$

Subsequently, we shall consider this solution in further detail by employing the source term (6.1.10b) with $f(z)$ given by (6.1.3a). Thus, we assume a finite but temperature-independent absorption coefficient and ignore interference phenomena within the slab. With this simplification we can perform the integration over z_1^*. We shall also assume equal surface conductances at both sides of the slab, i.e., $\kappa_0^* = \kappa_h^* = 1$ and $\eta_0^* = \eta_h^* = \eta^*$. The laser beam shall be switched on at $t = 0$ in the position $x = y = 0$ and it shall be scanned with constant velocity, v_s, along the x-direction. For a coordinate system that is fixed with the slab, the solution can be written as

$$\theta(x^*, t^*) = \frac{l}{\kappa} I_a \int_0^{t^*} q(t^* - t_1^*) \mathscr{G}(x^*, y^*, t^*, t_1^*) \mathscr{F}(z^*, t_1^*) dt_1^*. \tag{6.2.2}$$

The function $\mathscr{G}$ is given by

$$\mathscr{G}(x^*, y^*, t^*, t_1^*) = \frac{1}{4\pi t_1^*} \int_{-\infty}^{\infty} dx_1^* \int_{-\infty}^{+\infty} dy_1^* g(x_1^*, y_1^*)$$

$$\times \exp\left(-\frac{[(x_1^* - x^*) - v_s^*(t_1^* - t^*)]^2 + (y_1^* - y^*)^2}{4t_1^*}\right). \tag{6.2.3}$$

The solution (6.2.2) can be applied to many practical cases. It contains three *independent* functions. $\mathscr{G}$ and $\mathscr{F}$ describe the temperature field within the xy-plane and along the z-direction, respectively. $\mathscr{F}$ depends on the absorption coefficient, the heat exchange coefficient, and the thickness of the substrate (Appendix A.5).

6.3 Pulse Shapes

We now consider various *temporal* dependences of laser-beam intensities that are most commonly used in laser processing. With a static cw-laser beam, the intensity is *constant* with respect to time, i.e.,

$$I(t) \equiv I_0 q(t) = I_0 \mathscr{H}(t), \tag{6.3.1}$$

where $\mathscr{H}$ is the Heaviside function and $I_0 = $ const. This is a good approximation also for long laser pulses where stationary solutions ($t^* \to \infty$) of (6.2.2) can be employed.

Single rectangular pulse

For a single rectangular pulse of duration τ_l (full curve in Fig. 6.3.1a) the intensity can be described by

$$I(t) = I_0 \mathscr{H}(\tau_l - t) \mathscr{H}(t). \tag{6.3.2}$$

The laser fluence (energy density) is

$$\phi = I_0 \tau_l. \tag{6.3.3}$$

The temperature rise calculated from (6.2.2) is then

$$\Delta T \equiv \theta(x^*, y^*, z^*, t^*) = \frac{I_a l}{\kappa} \int_{t_0^*}^{t^*} \mathscr{G}(x^*, y^*, t^*, t_1^*) \mathscr{F}(z^*, t_1^*) dt_1^*, \tag{6.3.4}$$

6.3 Pulse Shapes

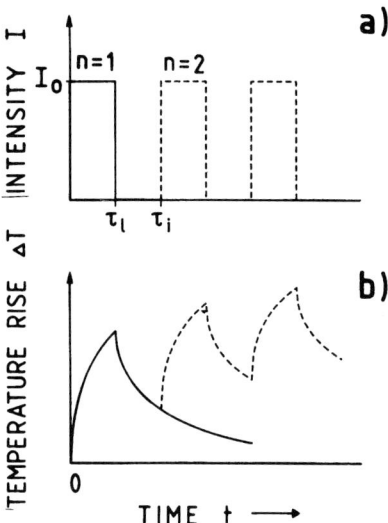

Fig. 6.3.1. Temporal variation of the temperature rise for single-pulse (full curve) and multiple-pulse (dashed curve) irradiation

where $t_0^* = 0$ with $t^* < \tau_l^* \equiv D\tau_l/l^2$ and $t_0^* = t^* - \tau_l^*$ with $t^* > \tau_l^*$. The temporal dependence of ΔT is schematically shown in Fig. 6.3.1b by the full curve.

Triangular pulse

For a triangular pulse (Fig. 6.3.2a) the intensity can be written as $I(t) = I_0 q(t)$ with

$$q(t) = \begin{cases} \dfrac{t}{\tau_0} \mathscr{H}\left(\dfrac{t}{\tau_0}\right) & \text{for } t < \tau_0 \\ \dfrac{\tau_1 - t}{\tau_1 - \tau_0} & \text{for } \tau_0 \leqslant t \leqslant \tau_1 , \\ 0 & \text{for } t > \tau_1 \end{cases} \qquad (6.3.5)$$

where I_0 is the maximum intensity that is reached at τ_0. The duration of the pulse defined by the full width at half maximum (FWHM) is $\tau_l = (1/2)\tau_1$.

Smooth pulse

The intensity of the pulse shown in Fig. 6.3.2b can be approximated by

$$I(t) = I_0 \left(\dfrac{t}{\tau_0}\right)^\beta \exp\left[\beta\left(1 - \dfrac{t}{\tau_0}\right)\right], \qquad (6.3.6)$$

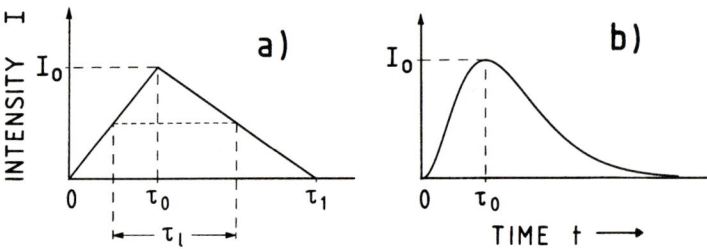

Fig. 6.3.2. Temporal variation of the intensity for a triangular and a smooth pulse

where β describes the temporal shape of the pulse. The laser fluence is

$$\phi = I_0 \tau_0 \exp(\beta) \frac{\Gamma(\beta+1)}{\beta^{\beta+1}},$$

where we have introduced the Γ function (Appendix A.3).

Multiple-pulse irradiation

The temperature rise for *multiple*-pulse irradiation can also be calculated from (6.2.2). With laser pulses that are rectangular in time (Fig. 6.3.1a) the intensity can be described by

$$I(t) = I_0 \sum_{n=1}^{m} [\mathcal{H}(t - \tau_i^n) - \mathcal{H}(t - \tau_f^n)], \qquad (6.3.7)$$

where $\tau_i^n = (n-1)\,\tau_i$ and $\tau_f^n = \tau_i^n + \tau_l$. The time dependence of ΔT is shown, qualitatively, by the dashed curve in Fig. 6.3.1b. The temperature oscillates between a maximum value at the end of the pulse and a minimum value before of the next pulse. The *average* temperature rise can often be calculated by replacing $I(t)$ by its average value $\langle I(t) \rangle$. The figure shows that the average temperature rise $\langle \Delta T(t) \rangle$ initially increases and levels off when the absorbed laser-light energy is balanced by heat losses.

With some types of pulsed lasers or modulated cw lasers, the temporal behavior of the intensity can be described by

$$I(t) = I_0 \left[1 - \cos\left(\frac{\pi t}{\tau_l}\right) \right]. \qquad (6.3.8)$$

6.4 Beam Shapes

Among the most important beam shapes employed in laser processing is the *Gaussian beam* whose (spatial) intensity distribution is given by (5.1.1)

$$I(r^*) = I_0 \exp(-r^{*2}). \tag{6.4.1}$$

The characteristic length for normalization of variables is $l = w_0$. Thus, $r^* = r/w_0$ with $r^2 = x^2 + y^2$. For a moving laser beam the distribution (6.4.1) remains unchanged in a system fixed with the beam. The $\mathscr{G}$–function that enters (6.2.2) is obtained by integration of (6.2.3) with

$$g(x^*, y^*) = \exp(-x^{*2} - y^{*2}).$$

This yields

$$\mathscr{G}(x^*, y^*, t^*, t_1^*) = \frac{1}{1 + 4t_1^*} \exp\left(-\frac{[x^* - v_s^*(t^* - t_1^*)]^2 + y^{*2}}{1 + 4t_1^*}\right). \tag{6.4.2}$$

Circular beam

For a circular beam with constant intensity over its cross section $F = \pi w^2$ the intensity can be written as

$$I(r^*) = I_0 \mathscr{H}(1 - r^*), \tag{6.4.3}$$

where $r^* = r/l = r/w$. Integration of (6.2.3) yields

$$\mathscr{G}(x^*, y^*, t^*, t_1^*) = \frac{1}{2t_1^*} \exp\left(-\frac{[x^* - v_s^*(t^* - t_1^*)]^2 + y^{*2}}{4t_1^*}\right)$$

$$\times \int_0^1 r^* I_0(\xi r^*) \exp\left(-\frac{r^{*2}}{4t_1^*}\right) dr^*, \tag{6.4.4}$$

where I_0 is the modified Bessel function and

$$\xi = \frac{1}{2t_1^*} \left\{ \left[x^* - v_s^*(t^* - t_1^*)\right]^2 + y^{*2} \right\}^{1/2}.$$

Rectangular beam

For a rectangular beam with constant intensity over its cross section $F = 2w_x \, 2w_y$ we can write

$$I(x, y) = I_0 \mathscr{H}(w_x^2 - x^2) \mathscr{H}(w_y^2 - y^2). \tag{6.4.5}$$

Integration of (6.2.3) yields

$$\mathcal{G}(x^*, y^*, t^*, t_1^*) = \frac{1}{4}\sum_{\pm}\mathrm{erf}\left(\frac{w_x^* \pm x^* \mp v_s^*(t^* - t_1^*)}{2\sqrt{t_1^*}}\right)$$

$$\times \sum_{\pm}\mathrm{erf}\left(\frac{w_y^* \pm y^*}{2\sqrt{t_1^*}}\right), \qquad (6.4.6)$$

where $w_x^* = w_x/l$, $w_y^* = w_y/l$. It is convenient to use either $l = w_x$ or $l = w_y$. The $\Sigma_\pm$ implies the sum of terms taken once with the upper and once with the lower sign, i.e., $\Sigma_\pm \exp(a \pm b \mp c) = \exp(a + b - c) + \exp(a - b + c)$.

Uniform illumination

For large-area uniform illumination of the substrate surface the intensity is constant within the plane $z = 0$, i.e.,

$$I(x, y) = I_0 = \mathrm{const.}$$

and we obtain $\mathcal{G} = 1$.

6.5 Characteristics of Temperature Distributions

In many applications it is essential to estimate the characteristic features of the laser-induced temperature distribution. These are the maximum temperature, and the width of the temperature distribution within the xy-plane and along the z-direction.

Center temperature rise

With the beam shapes given in Sect. 6.4 and the assumptions made throughout this chapter, the maximum temperature rise on the substrate surface occurs in the *center* of the laser beam at $x^* = 0$. This center temperature rise, $\Delta T_c(0, 0, 0) \equiv \theta_c$, is estimated in Chaps. 7 and 8 for different cases.

Width of distribution

For *transient* laser-beam irradiation and $\eta = 0$, the width of the thermal field in axial direction, w_T^a, and in lateral direction within the xy-plane, w_T^l, is of the order

$$w_T^a(t) \approx l_\alpha + l_T \quad \text{and} \quad w_T^l(t) \approx w \qquad (6.5.1\mathrm{a})$$

6.5 Characteristics of Temperature Distributions

for $l_\alpha \ll w$ and times $t \ll w^2/D$. If $l_\alpha \gg w$ we obtain

$$w_T^a(t) \approx l_\alpha \quad \text{and} \quad w_T^l(t) \approx w + l_T, \tag{6.5.1b}$$

with $t \ll l_\alpha^2/D$. As before we use, in general, $l_T \approx 2(D\tau_l)^{1/2}$. Note, however, that in many cases of (thermal) laser processing the temperature distribution is non-exponential and described by a power-like dependence on coordinates. Thus, there is *no* universal characteristic length, l (see also Sect. 2.2.2). The *stationary* temperature distribution is characterized by

$$w_T^a(t \to \infty) \approx \max(\xi^a w, l_\alpha) \quad \text{and} \quad w_T^l(t \to \infty) \approx \xi^l w, \tag{6.5.1c}$$

where ξ^a and ξ^l are constants. The temporal dependence of w_T^a and w_T^l is shown by the isotherms in Fig. 6.5.1. Here, a semi-infinite substrate, a Gaussian laser beam, and $l_\alpha = 0$ have been assumed. The isotherms are defined by $T(r^*, z^*, t^*)/T_{max}(0, 0, t^*) = 1/e$ with $r^* = r/w_0$, $z^* = z/w_0$, and $t^* = Dt/w_0^2$. The stationary values are $w_T^a(t \to \infty) \approx 1.25 \, w_0$ and $w_T^l \approx 1.73 \, w_0$. The temporal development of the temperature distribution for surface absorption and finite absorption is also shown in Fig. 7.2.3.

For focused *cw-laser* irradiation, the width w_T^l of the stationary temperature field is often described by the relation

$$\theta(x = w_T^l) = \zeta \theta_c, \tag{6.5.2}$$

where ζ has values of 0.5, $1/e$ or $1/e^2$. With a Gaussian beam and a semi-infinite substrate with $\alpha^* \to \infty$ and $\eta^* = 0$, the respective values yield $w_T^l \approx 1.38 \, w_0$, $1.50 \, w_0$, and $1.79 \, w_0$.

Another definition of w_T^l is given by the so-called *parabolic approximation* (Fig. 6.5.2). Here, the real temperature distribution,

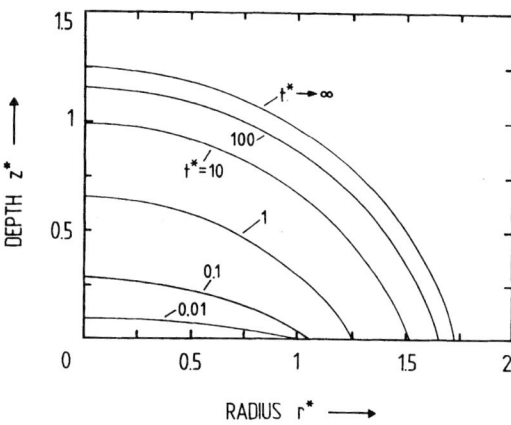

Fig. 6.5.1. Temporal dependence of isotherms calculated for a semi-infinite substrate, $\alpha \to \infty$, and a Gaussian laser beam

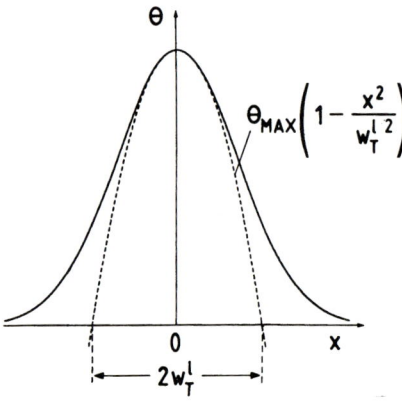

Fig. 6.5.2. Width of temperature distribution, $2w_T^l$, defined by the parabolic approximation

$\theta(x)$, is approximated by $\theta_c(1 - x^2/w_T^{l2})$ with

$$w_T^l = \left(-\frac{1}{2\theta}\frac{d^2\theta}{dx^2}\right)^{-1/2}_{x=0} \tag{6.5.3}$$

With a Gaussian beam and a semi-infinite substrate this yields $w_T^l(t \to \infty) = \sqrt{2}\,w_0$. Thus, the different definitions yield similar results. This is true also for other definitions used throughout the literature. The reason is that all these definitions are related to the size of the laser beam.

With $\eta \neq 0$, the temperature distribution becomes more localized, i.e., the widths w_T^a and w_T^l are diminished.

6.6 Numerical Techniques

Numerical techniques for solving differential equations such as the heat equation, permit one to easily include in the calculations arbitrary variations in material parameters. Furthermore, such calculations can be performed for almost any geometry of the substrate (work piece). The main disadvantage is that such solutions do not provide general relationships and they are always limited to finite regions. In order to satisfy boundary conditions at infinity, calculations should be performed over a wide range of variables. Here, one has to compromise between the accuracy in calculations and the total computer time required. Additionally, inappropriate discretizations of the problem can cause artefacts like oscillations, instabilities, and even chaos in nonlinear equations.

Depending on the geometry of the physical problem, it is advantageous to employ either *finite difference* or *finite element* techniques. These techniques have been widely described in the literature [*Marsal* 1976; *Davies* 1980; *Press* et al. 1992], and they will not be discussed any further in this book.

In the next two chapters we often compare analytical solutions, where they are possible, with numerical solutions based on finite difference or finite element techniques.

7 Semi-infinite Substrates

In this chapter we discuss various specific solutions of the heat equation for semi-infinite substrates. The irradiation geometry and the definition of the coordinate system shall be the same as in Fig. 6.1.1 but with $h_s \to \infty$. Any phase changes and chemical reaction energies are ignored.

7.1 The Center Temperature Rise

For many applications, the knowledge of the *maximum* laser-induced temperature rise is of great importance. With the beam shapes discussed in Sect. 6.4, the maximum surface temperature occurs always in the center of the laser beam at $x^* = 0$. Subsequently, we present analytic expressions for semi-infinite substrates, stationary conditions, and temperature-independent material parameters. Thus, the linearized temperature is equal to the laser-induced temperature rise, see (2.2.8), so that $\theta_c(\alpha^*, \eta^*) \equiv \Delta T(x^* = 0, \alpha^*, \eta^*)$. With low to moderate values of η^* or with $\alpha^* \to \infty$, the center-temperature rise at the substrate surface is equal to the maximum temperature rise.

Gaussian beam

From the equations presented in Sects. 6.2 and 6.4 we obtain for a Gaussian beam, finite absorption, and finite heat losses

$$\theta_c^G(\alpha^*, \eta^*) = \frac{I_a w_0}{\kappa} \left\{ \int_0^\infty \frac{dt^*}{1 + 4t^*} \alpha^* \exp(\alpha^{*2} t^*) \operatorname{erfc}(\alpha^* t^{*1/2}) \right.$$

$$- \frac{\sqrt{\pi}}{2} \frac{\alpha^* \eta^*}{\eta^* - \alpha^*} \int_0^\infty \frac{dt^*}{(1 + 4t^*)^{1/2}} [\eta^* \exp(\eta^{*2} t^*)$$

$$\left. \times \operatorname{erfc}(\eta^* t^{*1/2}) - \alpha^* \exp(\alpha^{*2} t^*) \operatorname{erfc}(\alpha^* t^{*1/2})] \right\}. \qquad (7.1.1)$$

7.1 The Center Temperature Rise

In the absence of heat losses, i.e., with $\eta^* = 0$, we obtain

$$\theta_c^G(\alpha^*, \eta^* = 0) = \frac{\alpha^* I_a w_0}{\kappa} \int_0^\infty \frac{dt^*}{2t^* + \alpha^*} \exp(-t^{*2}). \tag{7.1.2}$$

For surface absorption, $\alpha^* \to \infty$, equation (7.1.1) yields

$$\theta_c^G(\eta^*) = \frac{\sqrt{\pi}}{2} \frac{I_a w_0}{\kappa} \left\{ 1 - \eta^* \int_0^\infty \frac{dt^*}{(1 + 4t^*)^{1/2}} \right.$$

$$\left. \times \left[\frac{1}{(\pi t^*)^{1/2}} - \eta^* \exp(\eta^{*2} t^*) \operatorname{erfc}(\eta^* t^{*1/2}) \right] \right\}. \tag{7.1.3}$$

For surface absorption and no heat losses we obtain

$$\theta_c \equiv \theta_c^G(\alpha^* \to \infty, \eta^* = 0) = \frac{\sqrt{\pi}}{2} \frac{I_a w_0}{\kappa} = \frac{P(1-R)}{2\sqrt{\pi}\kappa w_0} \approx 0.89 \frac{I_a w_0}{\kappa}. \tag{7.1.4}$$

Subsequently, θ_c is often used for normalization.

Circular laser beam

For a circular beam we obtain with $\alpha^* \to \infty$ and $\eta^* = 0$

$$\theta_c^C(\alpha^* \to \infty, \eta^* = 0) = \frac{I_a w}{\kappa}. \tag{7.1.5}$$

Rectangular beam

For a rectangular beam of area $F = 2w_x 2w_y$ and with $\alpha^* \to \infty$ and $\eta^* = 0$ the center temperature is

$$\theta_c^R(\alpha^* \to \infty, \eta^* = 0) = \frac{2I_a(w_x w_y)^{1/2}}{\pi\kappa} \left[\left(\frac{w_x}{w_y}\right)^{1/2} \ln\left(\frac{(w_x^2 + w_y^2)^{1/2} + w_y}{(w_x^2 + w_y^2)^{1/2} - w_y}\right) \right.$$

$$\left. + \left(\frac{w_y}{w_x}\right)^{1/2} \ln\left(\frac{(w_x^2 + w_y^2)^{1/2} + w_x}{(w_x^2 + w_y^2)^{1/2} - w_x}\right) \right]. \tag{7.1.6}$$

For a square beam with $w_x = w_y = w_s$ this yields

$$\theta_c^S(\alpha^* \to \infty, \eta^* = 0) = \frac{2I_a w_s}{\pi\kappa} \ln\left(\frac{\sqrt{2}+1}{\sqrt{2}-1}\right) \approx 1.12 \frac{I_a w_s}{\kappa}. \tag{7.1.7}$$

7.2 Stationary Solutions for Temperature-Independent Parameters

In this section the values of the absorption coefficient, $\alpha = \alpha(T_0)$, the thermal conductivity, $\kappa = \kappa(T_0)$, and the reflectivity, $R = R(T_0)$, are taken at a fixed temperature, T_0. For this (linear) problem stationary solutions of the heat equation can be obtained from (6.2.2) with $v_s^* = 0$, $h_s \to \infty$ and an upper integration variable $t^* \to \infty$. This yields

$$\theta(x) = \frac{I_a l}{\kappa} \int_0^\infty \mathscr{G}(x^*, y^*, t_1^*) \, \mathscr{F}(z^*, t_1^*) \, dt_1^*, \qquad (7.2.1)$$

where

$$\mathscr{G}(x^*, y^*, t_1^*) = \frac{1}{4\pi t_1^*} \int_{-\infty}^\infty dx_1^* \int_{-\infty}^\infty dy_1^* \, g(x_1^*, y_1^*)$$

$$\times \exp\left(-\frac{(x_1^* - x^*)^2 + (y_1^* - y^*)^2}{4 t_1^*}\right), \qquad (7.2.2a)$$

or, in cylindrical coordinates

$$\mathscr{G}(r^*, \varphi, t_1^*) = \frac{1}{4\pi t_1^*} \int_0^\infty dr_1^* \, r_1^* \int_0^{2\pi} d\varphi_1 \, g(r_1^*, \varphi_1)$$

$$\times \exp\left(-\frac{r^{*2} + r_1^{*2} - 2 r^* r_1^* \cos(\varphi - \varphi_1)}{4 t_1^*}\right), \qquad (7.2.2b)$$

with $x^* = r^* \cos \varphi$ and $y^* = r^* \sin \varphi$. The $\mathscr{F}$-function for semi-infinite substrates is listed for different cases in Appendix A.5. The characteristic length, l, for the normalization of variables is the laser beam radius w_0 or w, or, in the case of a rectangular beam, w_x or w_y. Solutions for uniform large-area illumination are obtained with $l \to \infty$.

With a *Gaussian* beam, we obtain from (7.2.1) in the absence of heat losses to the surrounding medium

$$\theta(r^*, z^*) = \frac{\alpha^*}{\sqrt{\pi}} \theta_c \int_0^\infty \frac{d\xi}{\alpha^{*2} - \xi^2} J_0(\xi r^*)$$

$$\times \left[\alpha^* \exp(-\xi z^*) - \xi \exp(-\alpha^* z^*)\right] \exp\left(-\frac{\xi^2}{4}\right). \qquad (7.2.3)$$

$\theta_c = \pi^{1/2} I_a w_0 / 2\kappa$ is the center-temperature rise given by (7.1.4). J_0 is the Bessel function of order zero, and ξ an integration variable. We shall now discuss (7.2.3) for surface absorption and for finite absorption.

7.2 Stationary Solutions for Temperature-Independent Parameters

7.2.1 Surface Absorption

With $\alpha^* \to \infty$, the temperature rise within the plane $z = 0$ becomes

$$\Delta T(r^*, 0) = \theta_c I_0\left(\frac{r^{*2}}{2}\right) \exp\left(-\frac{r^{*2}}{2}\right), \tag{7.2.4}$$

where I_0 denotes the modified Bessel function of order zero with $I_0(0) = 1$. The result (7.2.4) is included in Fig. 7.2.1 by the dashed curve. In order to permit a comparison with temperature distributions that apply to other cases (see below), we have chosen parameter values that correspond to crystalline Si.

The temperature distribution around the center of the irradiated area can be approximated by the interpolation formula

$$\Delta T(r^*, 0) \approx \frac{\theta_c}{(1 + r^{*2})^{1/2}} \tag{7.2.5}$$

(dotted curve in Fig. 7.2.1). For $1 < r^* \leq 3$ the temperature rise is $\Delta T(r^*) \approx \theta_c / r^*$. The temperature rise along the z-axis is

$$\Delta T(0, z^*) = \theta_c \, \text{erfc}(z^*) \exp(z^{*2}) \tag{7.2.6}$$

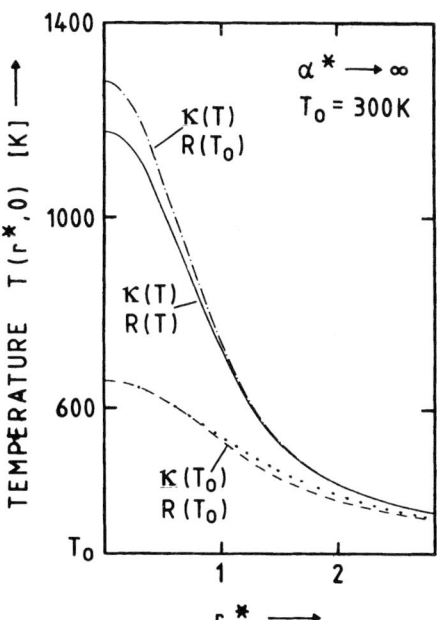

Fig. 7.2.1. Stationary temperature distribution induced by a Gaussian laser beam at normal incidence ($P = 0.5$ W, $w_0 \approx 1.8\,\mu$m, $T_0 = T(\infty) = 300$ K). The radius is measured from the center of the laser beam. All cases calculated apply to surface absorption, i.e., $\alpha^* \to \infty$. The other parameters employed were: Dashed curve: $\kappa = \kappa(\text{Si}, T_0) \approx 1.50$ W/cmK, $R(\text{Si}, T_0) = 0.324$. Dotted curve: Interpolation formula (7.2.5) with otherwise same parameters as for dashed curve. Dash-dotted curve: $\kappa = \kappa(\text{Si}, T)$, see (7.3.2). Full curve: Numerical solution for $\kappa = \kappa(\text{Si}, T)$, $R = R(\text{Si}, T)$, see (7.3.5)

With large values of either r or z, equation (7.2.3) approaches

$$\Delta T(r^* \gg 1 \text{ and/or } z^* \gg 1) = \frac{1}{\sqrt{\pi}} \frac{\theta_c}{(r^{*2} + z^{*2})^{1/2}}, \qquad (7.2.7)$$

which holds for values, typically, $r^*, z^* \geqslant 10$; it holds also for finite absorption if $r^*, z^* \gg \alpha^{*-1}$.

The assumption of surface absorption and temperature-independent parameters is a fairly good approximation with many metals. Here, the thermal conductivity is dominated by conduction electrons and can be described, in a first approximation, by the Wiedemann–Franz law $\kappa = \pi^2 k_B^2 T \sigma / 3 e^2$ where σ is the electrical conductivity which, with temperatures greater than the Debye temperature, θ_D, can be described by $\sigma \propto T^{-1}$ [*Ashcroft* and *Mermin* 1976]. Inspection of Table II reveals, however, that with some metals the thermal conductivity can significantly decrease, in particular for high temperatures. This is mainly a consequence of the increase in electron–phonon interactions.

7.2.2 Finite Absorption

We shall characterize the influence of finite absorption by the normalized temperature rise

$$\Delta T^*(0, 0; \alpha^*) = \frac{\Delta T(0, 0; \alpha^*)}{\Delta T(0, 0; \alpha^* \to \infty)} \equiv \frac{\theta_c(\alpha^*)}{\theta_c}, \qquad (7.2.8)$$

where θ_c is given by (7.1.4). Figure 7.2.2 shows $\theta_c(\alpha^*)/\theta_c$ versus α^* (full curve). The figure demonstrates that the approximation $\Delta T(0, 0) \approx \theta_c$ holds very well as long as $\alpha^* > 10^2$. With decreasing values of α^*, however, the center temperature

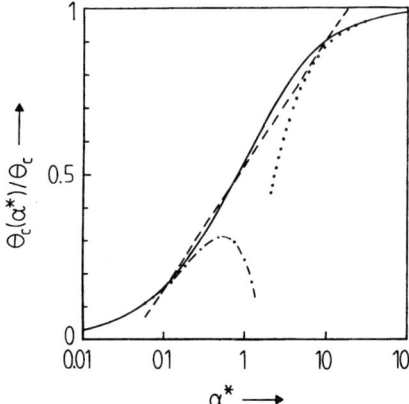

Fig. 7.2.2. Normalized center temperature rise induced on a semi-infinite substrate of finite absorption as a function of $\alpha^* = \alpha w_0$ (full curve). Dotted, dash-dotted, and dashed curves refer to approximations (7.2.9a–c)

7.2 Stationary Solutions for Temperature-Independent Parameters

strongly decreases. With $\alpha^* > 10$, ΔT^* can be approximated by

$$\Delta T^*(0,0;\alpha^* > 10) \approx 1 - \frac{2}{\sqrt{\pi}} \frac{1}{\alpha^*} \tag{7.2.9a}$$

(dotted curve in Fig. 7.2.2). With $\alpha^* < 0.1$ we have

$$\Delta T^*(0,0;\alpha^* \ll 1) \approx \frac{\alpha^*}{\sqrt{\pi}} \left(\ln\frac{2}{\alpha^*} - \frac{C}{2} \right) \tag{7.2.9b}$$

(dash-dotted curve). $C = 0.577$ is Euler's constant. The intermediate region $0.1 \leqslant \alpha^* \leqslant 10$ can be approximated by the linear dependence

$$\Delta T^*(0,0;0.1 \leqslant \alpha^* \leqslant 10) \approx 0.53 + 0.165 \ln \alpha^* \tag{7.2.9c}$$

(dashed curve). The steady-state temperature distribution, in both r- and z-directions, calculated from (7.2.3) with $\alpha = 3 \times 10^3 \text{cm}^{-1}$ is included in the lower part of Fig. 7.2.3 by the full curves. The comparison with curves for $\alpha^* \to \infty$ (upper part of figure) shows that with a penetration depth $l_\alpha \approx 3 \, \mu\text{m}$ ($l_\alpha^* = l_\alpha/w_0 \approx 1.67$), the steady-state center-temperature rise on the surface, $\Delta T(0,0;\alpha^*)$, decreases by more than a factor of two. On the other hand, in both r- and z-directions, the temperature distributions are only little affected with distances $r, z > l_\alpha$.

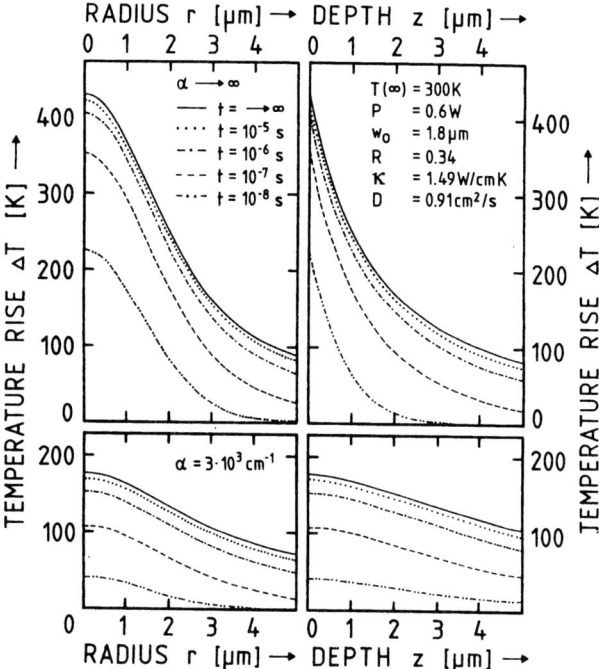

Fig. 7.2.3. Temperature distribution in radial (left-hand side) and axial (right-hand side) directions for various durations of laser light irradiation. Upper part: Strong absorption with $\alpha \to \infty$. Lower part: Finite absorption with $\alpha = 3 \times 10^3 \text{ cm}^{-1}$. All other parameters employed were equal for both cases with values as indicated in the figure [$\kappa = \kappa$ (Si, 300 K); $R = R$ (Si, 300 K); $D = D$ (Si, 300 K)]

With $z \ll w_0$ and $l_\alpha \ll w_0$, the temperature rise can be described by

$$\Delta T(r^*, z^*) = \theta_c \left\{ I_0\left(\frac{r^{*2}}{2}\right) \exp\left(-\frac{r^{*2}}{2}\right) \right.$$
$$\left. - \frac{2}{\sqrt{\pi\alpha^*}} [\alpha^* z^* + \exp(-\alpha^* z^*)] \exp(-r^{*2}) \right\}, \quad (7.2.10)$$

where I_0 is the modified Bessel function of zero order.

The consideration of a finite penetration depth is important in all cases where the absorption coefficient of the material at the laser wavelength employed, $\alpha(\lambda)$, is smaller than, typically, 10^4 cm^{-1}. This is the case with insulators and semiconductors for photon energies within the range $h\nu_{ph} < h\nu < E_g$ where $h\nu_{ph}$ is the phonon (vibrational) energy of the highest IR-active dispersion oscillator (Reststrahl oscillator). Figure 7.2.4 shows the absorption coefficient and the reflectivity as a function of wavelength for two metals, a semiconductor, and an insulator.

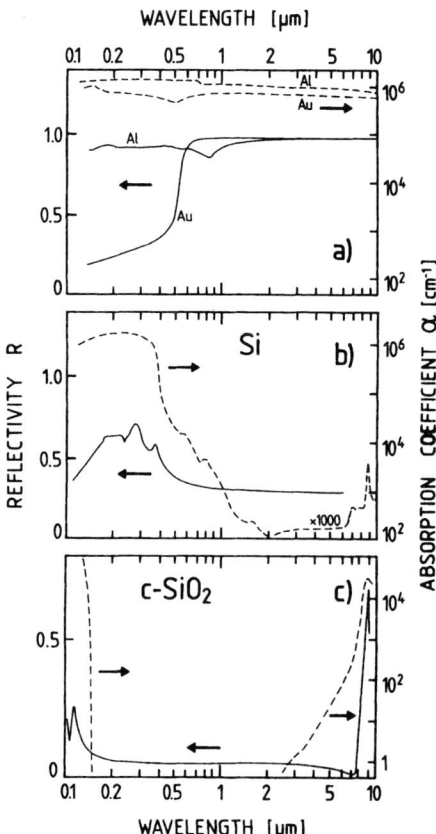

Fig. 7.2.4a–c. Reflectivity R (full curves) and absorption coefficient α (dashed curves) as a function of wavelength for two typical metals, a semiconductor, and an insulator. (a) Evaporated films of Al and Au. (b) Crystalline silicon (Si). (c) Crystalline quartz (c-SiO$_2$). Adapted from [*Von Allmen* and *Blatter* 1995]

7.3 Stationary Solutions for Temperature-Dependent Parameters

It is important to note, however, that at higher temperatures and high laser-light intensities, the absorption coefficient may change considerably due to free carrier generation via highly nonlinear mechanisms such as thermal ionization, multiphoton ionization, and impact ionization. These mechanisms become important in particular with pulsed-laser irradiation.

7.3 Stationary Solutions for Temperature-Dependent Parameters

In this section we investigate changes in temperature distribution caused by temperature-dependent parameters $\kappa = \kappa(T)$ and $R = R(T)$.

From a mathematical point of view, consideration of a temperature dependence in the *thermal conductivity* yields no new aspects. We can use all of the solutions given in the previous section and calculate the temperature from the Kirchhoff transform (2.2.8). Subsequently, we assume surface absorption and ignore heat losses to the ambient medium, i.e., $\eta^* = 0$.

Case 1: $\alpha \to \infty$, $\kappa(T)$, $R(T_0)$.

For crystalline insulators and semiconductors, the thermal conductivity is dominated by phonon–phonon interactions and decreases with increasing temperature. With temperatures much higher than the Debye temperature, the thermal conductivity can be described by

$$\kappa(T) \approx \frac{\kappa(T(\infty))}{T^{*m}}, \tag{7.3.1}$$

where $T^* = T/T(\infty)$. The exponent has values within the range $1 \leq m \leq 2$. By employing the Kirchhoff tranform we obtain with (7.3.1) and arbitrary $m \neq 1$

$$T(\theta) = T(\infty)\left[1 + (1-m)\frac{\theta}{T(\infty)}\right]^{1/(1-m)}, \tag{7.3.2a}$$

and with $m = 1$

$$T(\theta) = T(\infty)\exp\left(\frac{\theta}{T(\infty)}\right). \tag{7.3.2b}$$

For Si, the temperature dependence of κ can be described, in good approximation, by $\kappa(T(\infty)) = 1.54$ [W/cmK] and $m = 1.22$. The experimental data within the range $300\ K < T < 1400\ K$ can be fitted even more properly by [Ho et al. 1974]

$$\kappa(T) \approx \frac{k}{T - T_k} \quad \text{where } k = 299\,\text{W/cm and } T_k = 99\,\text{K}. \tag{7.3.3}$$

With (7.3.3) we obtain

$$T(\theta) = T_k + [T(\infty) - T_k] \exp\left(\frac{\theta}{T(\infty) - T_k}\right). \tag{7.3.4}$$

The dash-dotted curve in Fig. 7.2.1 is calculated for this case. Clearly, this curve can be obtained directly from $\Delta T(r, 0) = \theta(r, 0)$ as given by (7.2.4), the Kirchhoff transform (2.2.8), and the temperature dependence of $\kappa(T)$. From the comparison of curves it becomes evident that consideration of a temperature-dependent κ with $d\kappa/dT < 0$ strongly increases the temperature near the center of the laser beam. The region $r \gg w_0$ is only slightly affected. In other words, with $d\kappa/dT < 0$, the width of the temperature distribution decreases significantly. This *sharpening* of the temperature field plays an important role in many cases of thermal laser processing.

Case 2: $\alpha \to \infty$, $\kappa(T)$, $R(T)$.

The reflectivity of most insulators and semiconductors shows a complex dependence on temperature. For the example of Si and photon energies $h\nu < 3$eV we can use for temperatures $T \leqslant 1000$K the approximation [*Jellison* and *Modine* 1983]

$$R(\lambda, T) \approx R(\lambda, T_0) + 5 \times 10^{-5} T \quad (T < 1000\text{K}), \tag{7.3.5a}$$

where $T_0 = 300$K. With higher temperatures and 694 nm ruby-laser radiation the experimental data can be fitted by [*Toulemonde* et al. 1985]

$$R(694 \text{ nm}, T) \approx 0.584 - 4.8 \times 10^{-4} T + 2.6 \times 10^{-7} T^2$$

$$(1000\text{K} < T < T_m). \tag{7.3.5b}$$

The result of numerical calculations that employ (7.3.5) with $R(\lambda = 694$ nm, $T_0) = 0.324$ is included in Fig. 7.2.1 by the full curve. The increase in reflectivity diminishes the laser-induced temperature at the center of the irradiated zone by more than 10%. The figure clearly demonstrates that the temperature dependences in the material parameters κ and R strongly influence the laser-induced temperature distribution.

The increase in the reflectivity of *semiconductors* observed in the low to medium temperature range originates from thermally activated electrons.

The reflectivity of *metals*, in general, decreases with increasing temperature. This is mainly a consequence of the increase in electron–phonon interactions. With many metals, this behavior can be described, in good approximation, by the linear relationship [*Ujihara* 1972]

$$R(T) = R(T_0) - \delta(T - T_0), \tag{7.3.6}$$

where δ is a constant. For Ag, Al, Au, and Cu, typical values of δ within the near IR are around 2×10^{-5} K^{-1}. With laser-beam intensities $I \gg 10^2$ W/cm^2 one has $I\delta \gg \eta$ for typical values of η (Sect. 6.1). For this reason, consideration of a temperature-dependent reflectivity is more important than the consideration of heat losses to a gaseous ambient medium.

One should be aware, however, that at high temperatures with almost all materials, changes are observed in the reflectivity due to surface contaminations (oxidation, nitridation, etc.) and, near melting, due to surface deformations (corrugations, ripples, etc.). These effects are often of greater importance than those related to (inherent) reflectivity changes.

7.4 Scanned CW Laser

We now consider temperature distributions induced by scanned cw-laser irradiation of a substrate of finite absorption and temperature-independent material parameters, $\eta = 0$ and $l \equiv w_0$. Scanning shall be performed in x-direction with constant velocity, v_s (Fig. 6.1.1). The laser beam shall be of Gaussian shape, switched on at $t = -\infty$ and it shall pass the position $x = y = 0$ at $t = 0$. In a reference frame that is fixed with the substrate, the temperature rise at the origin, $\Delta T(0, t^*; v_s^*) \equiv \Delta T(0, 0, 0, t^*; v_s^*)$, becomes

$$\Delta T(0, t^*; v_s^*) = \frac{2}{\sqrt{\pi}} \alpha^* \theta_c \int_0^\infty \frac{dt_1^*}{1 + 4t_1^*} \exp\left(-\frac{v_s^{*2}(t^* - t_1^*)^2}{1 + 4t_1^*}\right)$$
$$\times \operatorname{erfc}(\alpha^* t_1^{*1/2}) \exp(\alpha^{*2} t_1^*), \qquad (7.4.1)$$

where θ_c is given by (7.1.4). With $\alpha^* \to \infty$ this solution becomes

$$\Delta T(0, t^*; v_s^*) = \frac{2}{\pi} \theta_c \int_0^\infty \frac{dt_1^*}{t_1^{*1/2}(1 + 4t_1^*)} \exp\left(-\frac{v_s^{*2}(t^* - t_1^*)^2}{1 + 4t_1^*}\right). \qquad (7.4.2)$$

In order to obtain further insight, we show in Fig. 7.4.1a the time evolution of the normalized temperature rise $\Delta T^*(0, t^*; v_s^*) \equiv \Delta T(0, t^*; v_s^*)/\Delta T(0, t^*; 0)$ as a function of $v_s^* t^*$ for $\alpha^* = 1$ and for different scanning velocities. Note that $v_s^* t^* = v_s t/w_0 = 2t/\tau_l$ where $\tau_l = 2w_0/v_s$ is the dwell time of the laser beam. The figure shows that the maximum temperature rise occurs at times $t > 0$, i.e., *after* the laser beam has passed the position $x_\alpha = 0$. This demonstrates that for high scanning velocities heat diffusion lags behind the beam center. For velocities $v_s^* \gg 1$ the heat diffusion length becomes smaller than the laser focus. If $l_T = 2(D\tau_l)^{1/2} \ll w_0$ the energy absorbed during the dwell time cannot diffuse out of the focus region.

Figure 7.4.1b shows for $t^* = 0$ the normalized temperature rise at the origin, $\Delta T^*(0, 0; v_s^*) \equiv \Delta T(0, 0; v_s^*)/\Delta T(0, 0; v_s^* = 0)$, and the maximum temperature rise as a function of v_s^*. The full curves have been calculated for the case of surface absorption, while the dashed curves refer to finite absorption with $\alpha^* = 1$. As v_s^* increases, ΔT^* decreases, since the total energy delivered to any point on the surface decreases with decreasing laser beam dwell time. Let us consider the result for $\alpha^* \to \infty$, $w_0 \approx 1.8\,\mu\text{m}$, and $D = D(\text{Si}, T_0 = 1680\,\text{K}) = 0.084\,\text{cm}^2/\text{s}$. From the figure we reveal that with these parameters and with scanning velocities $v_s < 10^2\,\text{cm/s}$, the decrease in center temperature is less than 5%. This

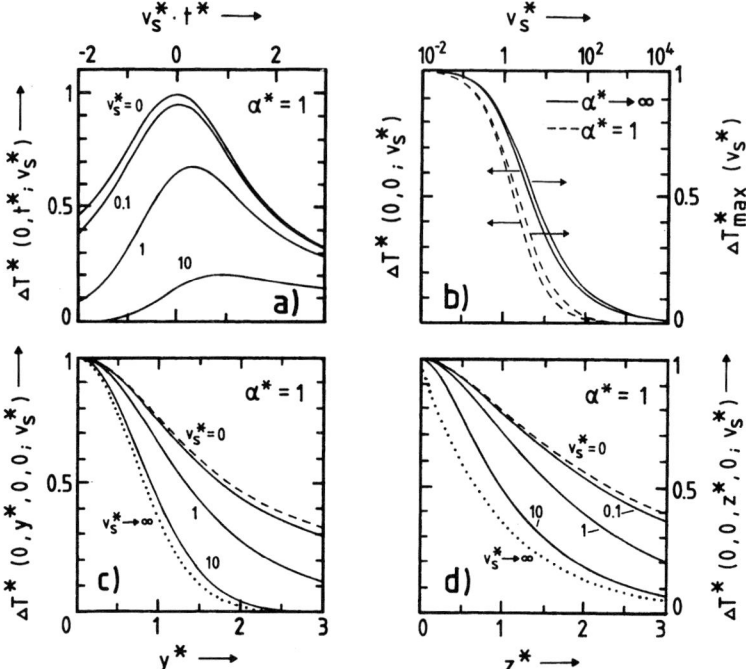

Fig. 7.4.1a–d. Normalized surface temperature rise induced by a Gaussian laser beam on a semi-infinite substrate (reference frame fixed with substrate). (**a**) As a function of the dimensionless product $v_s^* t^* = v_s t/w_0$ where $v_s^* \equiv v_s w_0 / D$. (**b**) As a function of normalized scanning velocity. Full curves refer to surface absorption, dashed curves to finite absorption, $\alpha^* = \alpha w_0$. For comparison, we have also included the maximum temperature rise. (**c**) Profiles perpendicular to scanning direction for various scanning velocities, v_s^*, $\alpha^* = 1$, and $t^* = 0$. (**d**) Same as (**c**) but along z-direction

approximation also holds, for example, in direct writing of patterns by LCVD, if the thermal conductivities of the deposit and the substrate are about equal. With scanning velocities $v_s > 10^2$ cm/s, which are common in laser annealing and laser synthesis, the surface temperature decreases significantly with increasing v_s.

The (normalized) surface temperature rise perpendicular to the scanning direction, $\Delta T^*(0, y, 0, 0; v_s^*) \equiv \Delta T(0, y, 0, 0; v_s^*)/\Delta T(0, 0, 0, 0; v_s^*)$, is illustrated in Fig. 7.4.1c. If $v_s^* \gg 1$, the profile can be described by

$$\Delta T^*(0, y^*, 0, 0; v_s^*) \approx \exp(-y^{*2}). \tag{7.4.3}$$

In this approximation, the temperature profile has the same shape as the (Gaussian) profile of the laser beam. No significant amount of heat can diffuse out of the laser-irradiated area. With $v_s^* \ll 1$ the temperature profile approaches the steady-state limit, $v_s^* = 0$. The limits $v_s^* \to \infty$ and $v_s^* \to 0$ are included in the figure by dotted and dashed curves, respectively.

Figure 7.4.1d shows the corresponding results for the z-direction. With $v_s^* \gg 1$ the profile approaches the form

$$\Delta T^*(0,0,z,0;v_s^*) \approx \exp(-\alpha^* z^*). \tag{7.4.4}$$

This is equal to the shape of the beam attenuation in z-direction. With decreasing v_s, the profile is broadened by heat transport.

A further point concerns the temperature-dependence of the material parameters. In the dynamic case one has to consider, in addition, the temperature dependence of the thermal diffusivity, D. For semiconductors, D decreases, in general, with increasing temperature. For the example of Si, we can use the approximation [*Ho* et al. 1974]

$$D(T) = \frac{C}{T-T_d}, \tag{7.4.5}$$

with $C = 128 \text{ cm}^2\text{K/s}$ and $T_d = 159 \text{ K}$. Consideration of a temperature-dependent diffusivity, requires one to solve the heat equation numerically. However, a rough estimation of the influence of $D(T)$ is obtained from Fig. 7.4.1b, if for v_s^* the value for the diffusivity is taken once at $T_0 = 300$ K and once at a temperature near the melting point, $T_0 \approx T_m$. The temperature dependence of D, which is mainly related to $\kappa = \kappa(T)$ in (7.3.3), causes a strongly nonlinear increase in the laser-induced temperature rise (Fig. 7.6.1).

7.5 Pulsed-Laser Irradiation

In this section we discuss pulsed-laser irradiation of a semi-infinite substrate with finite absorption, temperature-independent material parameters, and $\eta = 0$. The laser-beam intensity shall be Gaussian or uniform. Solutions for other beam shapes are obtained from (6.2.2) and the intensity distributions given in Sect. 6.4.

7.5.1 Gaussian Intensity Profile

For a Gaussian laser beam we obtain from (6.3.4) with (6.4.2) and (A.5.10) for a *single* rectangular pulse the temperature rise at the origin $r = 0$, $z = 0$

$$\Delta T(0,0,t) = \frac{2\alpha^*}{\sqrt{\pi}} \theta_c \int_{t_0^*}^{t^*} \frac{dt_1^*}{1+4t_1^*} \text{erfc}(\alpha^* t_1^{*1/2}) \exp(\alpha^{*2} t_1^*). \tag{7.5.1}$$

For the heating cycle ($0 \leq t \leq \tau_l$) we set $t_0^* = 0$ while for the cooling cycle ($t > \tau_l$) we set $t_0^* = t^* - \tau_l^*$. The equation can also be applied, in good approximation, to multiple-pulse irradiation at low repetition rates.

Surface absorption

Let us first consider the limiting case $\alpha^* \to \infty$. Then, (7.5.1) yields for the heating cycle

$$\Delta T(0,0,t^*) = \frac{2}{\pi} \theta_c \arctan(2t^{*1/2}), \quad (7.5.2)$$

with $\theta_c = \sqrt{\pi} I_a w_0 / 2\kappa$ from (7.1.4). Thus, for $t^* = Dt/w_0^2 \ll 1$ we find

$$\Delta T(0,0,t) = \frac{I_a l_T}{\sqrt{\pi}\kappa} \propto t^{1/2}, \quad (7.5.3)$$

where $l_T = 2(Dt)^{1/2}$. Clearly, the maximum temperature is reached for $t = \tau_l$. The average temperature rise can be estimated from the energy balance. The energy absorbed, $I_a t$, heats a layer of thickness l_T. Thus, $c_p \rho \Delta T = I_a t / l_T$.

If the heat equation is linear, the *cooling* cycle can be described by subtracting the solution for the heating cycle starting at $t = \tau_l$ from the heating cycle starting at $t = 0$; i.e., for $t > \tau_l$

$$\Delta T(0,0,t^*) = \frac{2}{\pi} \theta_c \{\arctan(2t^{*1/2}) - \arctan[2(t^* - \tau_l^*)^{1/2}]\}$$

$$= \frac{2}{\pi} \theta_c \arctan\left(\frac{2t^{*1/2} - 2(t^* - \tau_l^*)^{1/2}}{1 + 4t^{*1/2}(t^* - \tau_l^*)^{1/2}}\right). \quad (7.5.4)$$

For times $t \gg \tau_l$, equation (7.5.4) approaches

$$\Delta T(0,0,t^* \gg \tau_l^*) = \frac{\theta_c \tau_l^*}{2\pi t^{*3/2}} = \frac{I_a w_0^2 \tau_l}{4\sqrt{\pi} \kappa t (Dt)^{1/2}}. \quad (7.5.5)$$

The full curves in Fig. 7.5.1 show the normalized center temperature rise, $\Delta T_c^* = \Delta T(0,0,t^*; \alpha)/\Delta T(0,0,\infty;\alpha)$, as a function of normalized time for the heating and the cooling cycle, and for laser-pulse lengths between $\tau_{l1}^* = 10^{-5}$ and $\tau_{l4}^* = 10$. With increasing pulse length, ΔT_c^* first increases rapidly and then levels off when $\tau_l \geq w_0^2/D$. For pulse lengths, $\tau_l^* > 1$, the temperature rise can well be approximated by the *steady* state solution $\Delta T^* = 1$.

The temporal behavior of temperature distributions calculated from (6.2.2) for radial (xy-plane) and axial (z-axis) directions is included in the upper part of Fig. 7.2.3.

Finite absorption

The radial and axial temperature rise calculated from (7.5.1) with $\alpha = 3 \times 10^3 \text{ cm}^{-1}$ is included in the lower part of Fig. 7.2.3. Finite penetration of the laser light affects the temperature distributions around the center more strongly than for larger distances.

7.5 Pulsed-Laser Irradiation

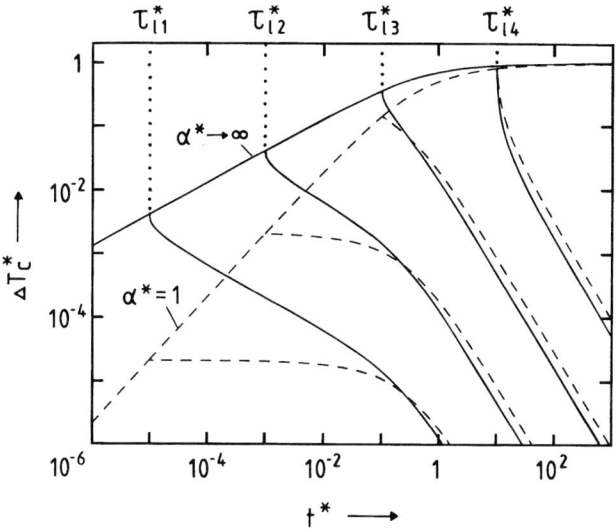

Fig. 7.5.1. Normalized center temperature rise as a function of $t^* = Dt/w_0^2$ for laser pulse lengths (dwell times) between $\tau_{l1}^* = 10^{-5}$ and $\tau_{l4}^* = 10$

In the limit $l_T \ll l_\alpha$, equation (7.5.1) yields

$$\Delta T(0,0,t) \approx \frac{\alpha I_a t}{\rho c_p}. \tag{7.5.6}$$

The solution (7.5.6) is termed energy density solution or calorimetric solution because it directly follows from the heat balance. The energy density absorbed, $I_a t$, heats a layer of thickness l_α. Thus, $\rho c_p \Delta T l_\alpha = I_a t$.

For the beginning of the cooling cycle, the energy balance yields, in analogy,

$$\frac{dT}{dt} \approx -\frac{\alpha^3 I_a D \tau_l}{\rho c_p}. \tag{7.5.7}$$

Thus, the cooling rate strongly increases with α.

7.5.2 Uniform Irradiation

For uniform (large-area) irradiation by a *single* rectangular laser pulse the temperature rise in z-direction can be described for the *heating cycle* by

$$\Delta T(z,t) = \frac{I_a}{\kappa} \left\{ l_T \operatorname{ierfc}\left(\frac{z}{l_T}\right) - \frac{1}{\alpha}\exp(-\alpha z) + \frac{1}{2\alpha}\exp\left(\frac{\alpha l_T}{2}\right)^2 \right. \\ \left. \times \sum_{\pm} \left[\exp(\pm \alpha z) \operatorname{erfc}\left(\frac{\alpha l_T}{2} \pm \frac{z}{l_T}\right) \right] \right\}, \tag{7.5.8}$$

where $l_T = 2(Dt)^{1/2}$. With $\alpha \to \infty$ only the first term in the parenthesis remains; with $z = 0$ the solution coincides with (7.5.3).

In analogy to (7.5.4), the *cooling cycle* after single-pulse irradiation ($t > \tau_l$) can be described by

$$\Delta T(z,t) = \frac{2 I_a D^{1/2}}{\kappa}\left[t^{1/2}\,\mathrm{ierfc}\left(\frac{z}{2(Dt)^{1/2}}\right) - (t-\tau_l)^{1/2}\right.$$
$$\left.\times\,\mathrm{ierfc}\left(\frac{z}{2\,D^{1/2}(t-\tau_l)^{1/2}}\right)\right]. \tag{7.5.9}$$

Note that $\mathrm{ierfc}(z=0) = 1/\sqrt{\pi}$.

For *multiple-pulse* irradiation and an intensity described by (6.3.8) with $I(r) = I_a = \mathrm{const}$, the temperature rise in z-direction is given by

$$\Delta T(z,t) = \frac{2 I_a (Dt)^{1/2}}{\kappa}\,\mathrm{ierfc}\left(\frac{z}{2(Dt)^{1/2}}\right) - \frac{I_a D^{1/2}}{\kappa\sqrt{\pi}}$$
$$\times \int_0^t \cos\left(\frac{\pi}{\tau_l}(t-t_1)\right)\exp\left(-\frac{z^2}{4D\,t_1}\right)\frac{dt_1}{t_1^{1/2}}. \tag{7.5.10}$$

7.6 Dynamic Solutions for Temperature-Dependent Parameters

We now discuss the influence of temperature-dependent parameters on temperature distributions induced by uniform (large-area) pulsed-laser irradiation.

Case 1: $0 \leq t \leq \tau_l$, $\alpha \to \infty$, $\kappa(T_0)$, $D(T_0)$, $R(T)$.

With many metals, the reflectivity decreases with increasing temperature, and it can be described by the ansatz (7.3.6).

For large-area pulsed-laser irradiation with uniform intensity, the boundary-value problem (6.1.10) can directly be solved by taking into account $I_a = I_0[1 - R(T(\infty)) + \delta\Delta T]$. For the heating cycle $0 \leq t \leq \tau_l$ we obtain

$$\Delta T(z,t) = \frac{[1-R(T(\infty))]\,I_0}{I_0\delta - \eta}\left[\mathrm{erf}\left(\frac{z}{l_T}\right) - 1\right.$$
$$\left.+\,\mathrm{erfc}\left(\frac{z}{l_T} - \Lambda\right)\exp\left(\Lambda^2 - \frac{2\Lambda}{l_T}z\right)\right], \tag{7.6.1}$$

where $T(\infty) = 300$ K, and

$$\Lambda = \frac{1}{2}l_T\left(\frac{I_0\delta}{\kappa} - \frac{\eta}{\kappa}\right). \tag{7.6.2}$$

The decrease in reflectivity causes the temperature to increase more rapidly. It can easily be proved that with $\delta, \eta \to 0$ equation (7.6.1) becomes identical to (7.5.8) with $\alpha^* \to \infty$.

7.6 Dynamic Solutions for Temperature-Dependent Parameters

Case 2: $0 \leq t \leq \tau_l$, $\alpha = \alpha(T)$, $\kappa = \kappa(T)$, $D = D(T)$, $R = R(T)$.

If the material parameters are temperature dependent, the laser-induced temperature rise can be calculated only numerically. The influence of temperature-dependent parameters on the temperature distribution is most pronounced for nonmetals, and in particular for semiconductors at temperatures below melting.

In semiconductors, the total absorption coefficient can be described by

$$\alpha = \alpha_f + \alpha_c = \alpha_f + \sigma_a N_c . \qquad (7.6.3)$$

α_f is the lattice or the interband absorption coefficient, depending on the laser wavelength under consideration (see Fig. 7.2.4). α_f depends only slightly on temperature, except for photon energies that are close to either the energy of a Reststrahl oscillator, $h\nu_{ph}$, or to the band gap energy, E_g. α_c describes the absorption by free carriers which depends on the concentration of electron–hole pairs, N_c, and on their absorption cross section, σ_a. In semiconductors, both electrons and holes are mobile and both contribute to the electrical conductivity. σ_a can be written in the form

$$\sigma_a = \frac{e^2}{\varepsilon_0 n c \omega^2} \left(\frac{1}{m_e^* \tau_e} + \frac{1}{m_h^* \tau_h} \right), \qquad (7.6.4)$$

where e is the electron charge and ε_0 the vacuum dielectric constant. n is the refractive index (real part), c the velocity of light, and $\omega = 2\pi\nu$. The quantities m_e^*, m_h^* are the effective masses and τ_e, τ_h the collision times for electrons and holes, respectively. Equation (7.6.4) reveals that the absorption cross section of the carriers increases with decreasing photon energy.

For Si, the absorption cross section of Nd:YAG laser radiation is $\sigma_a(300\,\text{K}, 1.06\,\mu\text{m}) \approx 5 \times 10^{-18}\,\text{cm}^2$ [*Svantesson* and *Nilsson* 1978] and for CO_2 laser radiation $\sigma_a(300\,\text{K}, 10.6\,\mu\text{m}) \approx 10^{-16}\,\text{cm}^2$ [*Bhattacharyya* and *Streetman* 1980].

The temperature dependence of the absorption coefficient (7.6.3) originates *mainly* from the temperature dependence in electron–hole pair concentration, $N_c(T) = N_e(T) = N_h(T)$. In thermal equilibrium, the carrier concentration in intrinsic (undoped) crystalline semiconductors is given by [*Ziman* 1972]

$$\bar{N}_c(T) = 2 \left(\frac{k_B T}{2\pi \hbar^2} \right)^{3/2} (m_e^* m_h^*)^{3/4} \exp\left(-\frac{E_g(T)}{2k_B T} \right), \qquad (7.6.5)$$

where, in general, E_g decreases with increasing temperature. From (7.6.3 and 5) it becomes evident that the absorption coefficient strongly increases with temperature, even faster than exponentially. This can cause *thermal runaway*: As the material heats up due to its absorption by initially present free carriers, lattice defects, impurities, etc., the concentration N_c increases. The increase in N_c produces an increase in absorption and thereby an increase in heating rate, etc.

Let us consider that in further detail for the example of crystalline Si. Here, the experimental data on the temperature dependence of the absorption coefficient can well be fitted, within the range $300\,\text{K} \leqslant T \leqslant 1000\,\text{K}$ and for photon energies below 3 ev ($\lambda > 410$ nm) by [*Jellison* and *Modine* 1983]

$$\alpha(T) \approx \alpha_0 \exp\left(\frac{T}{T_R}\right). \tag{7.6.6}$$

The parameters α_0 and T_R are listed in Table 7.6.1. For wavelengths $\lambda < 410$ nm, the fit (7.6.6) becomes less satisfactory. The range where the approximation $\alpha \to \infty$ becomes adequate can directly be derived from (7.6.6) and the parameters listed in the table.

Let us first consider thermal runaway for photon energies $h\nu < E_g$. This is the case with CO_2-laser radiation which is only weakly absorbed in pure Si [$E_g(300\,\text{K}) \approx 1.1$ eV; $\lambda = 10.6\,\mu\text{m} \, \hat{=} \, 0.12$ eV]. The absorption coefficient in pure (undoped) Si is $\alpha(300\,\text{K}; 10.6\,\mu\text{m}) \leqslant 0.3$ cm^{-1} [in heavily doped Si, $\alpha(300\,\text{K}; 10.6\,\mu\text{m}) \leqslant 10^3$ cm^{-1}]. Absorption is mainly related to the photoexcitation of free (thermally activated) carriers within the conduction band. These carriers transfer their energy rapidly to the lattice via electron–phonon scattering within, typically, 10^{-12} to 10^{-13} s. As a result, the lattice is locally heated and the absorption coefficient increases exponentially, see (7.6.6). Simultaneously, $l_\alpha = \alpha^{-1}$ shrinks and thus causes absorption to take place in a smaller volume. The heating rate is further enhanced by the decrease in $\kappa(T)$ and $D(T)$, see (7.3.3) and (7.4.5). This dynamic feedback increases the overall heating rate extremely rapidly.

If the photon energy exceeds the band gap energy, i.e., if $h\nu > E_g$, the laser radiation *directly* generates electron–hole pairs. With Si, the condition is fulfilled for visible light. The carrier concentration generated is shown in Fig. 15.2.5. Because the quantum yield for interband absorption is near unity, very high

Table 7.6.1. Temperature dependence of the absorption coefficient $\alpha(T) = \alpha_0 \exp(T/T_R)$ for c-Si

λ(nm)	α_0 (10^3 cm^{-1})	T_R(K)
10 μm	2×10^{-5}	110
694	1.34	427
633	2.08	447
532	5.02	430
515	6.28	433
488	9.07	438
485	9.31	434
458	14.5	429
405	55.1	420
308	1400 ($T \leqslant 1100$ K) 1800 ($T > 1100$ K)	4545

7.6 Dynamic Solutions for Temperature-Dependent Parameters

carrier densities can be produced. For $N_c > 10^{18}/\text{cm}^3$ carrier recombination is dominated by Auger processes (Sect. 2.4.2). The time of energy transfer to the lattice is then of the order of some picoseconds.

A theoretical treatment of transient heating requires the solution of the heat equation by taking into account the temperature dependences of material parameters. Figure 7.6.1 shows the results of one-dimensional calculations for Si and pulsed ruby-laser radiation. The dramatic rise in heating rate with increasing laser-pulse intensity becomes evident.

In *insulators* and, with certain conditions, also in semiconductors, the laser light may itself change the effective absorption coefficient via other non-linear mechanisms. For example, with very intense short pulses, substrate heating related to laser-induced impact ionization can become very important or even dominating.

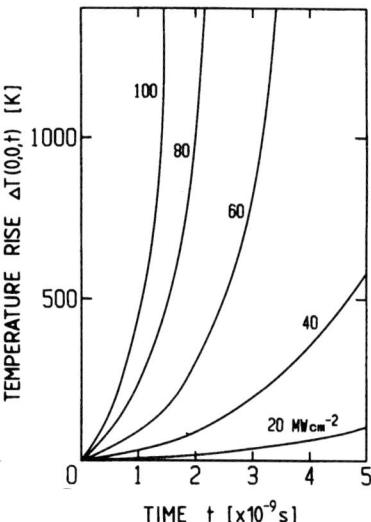

Fig. 7.6.1. Center temperature rise $\Delta T(0,0,t)$ for Si versus time $(t < \tau_l)$ for pulsed ruby-laser ($\lambda = 694$ nm) irradiation at various pulse intensities [*Kwong* and *Kim* 1983]

8 Infinite Slabs

In this chapter we consider the absorption of light and laser-induced temperature distributions in isotropic slabs (substrates) of finite uniform thickness, h_s, and infinite extension in the xy-plane (Fig. 6.1.1).

8.1 Strong Absorption

The laser-induced temperature distribution within a slab is similar to that within a semi-infinite substrate, as long as l_α, $l_T \ll h_s$. In such cases, the solutions presented in Chap. 7 can directly be applied.

Subsequently, we will consider solutions of (6.2.2) for several limiting cases and different types of laser-beam irradiation. The material parameters are assumed to be independent of temperature. The laser beam intensity shall be *Gaussian*. The temperature rise in a coordinate system that is fixed with the (scanned) laser beam can be written as

$$\Delta T(r^*, \varphi, z^*, t^*) = \frac{I_a w_0}{\kappa} \int_0^{t^*} \frac{dt_1^*}{1 + 4t_1^*} q(t^* - t_1^*)$$
$$\times \exp\left(-\frac{r^{*2} + v_s^* t_1^* (v_s^* t_1^* + 2r^* \cos\varphi)}{1 + 4t_1^*}\right) \mathscr{F}(z^*, t_1^*), \quad (8.1.1)$$

where $q(t)$ describes the (arbitrary) pulse shape of the Gaussian laser beam (Sect. 6.3). The characteristic length introduced for normalization is $l \equiv w_0$. The $\mathscr{F}$-function is given in Appendix A.5 for different cases characterized by the heat exchange coefficient, η, and the absorption coefficient, α.

8.1.1 Thermally Thin Film

The laser beam and the substrate shall be both fixed ($v_s^* = 0$) and illumination, pulsed or cw, shall start at $t^* = 0$. For a thermally thin film ($\eta^* h_s^* = \eta h_s/\kappa \ll 1$) the heat conductivity is so large, or the film so thin, that the temperature variation within the film is small compared to the average temperature. The

$\mathscr{F}$-function (A.5.7) can then be approximated by

$$\mathscr{F}(t_1^*) \approx \frac{1}{h_s^*} \exp\left(-\frac{2\eta^* t_1^*}{h_s^*}\right).$$

The temperature rise becomes

$$\Delta T(r^*, t^*) = \frac{I_a w_0^2}{\kappa h_s} \int_0^{t^*} \frac{dt_1^*}{1+4t_1^*} q(t^* - t_1^*) \exp\left(-\frac{r^{*2}}{1+4t_1^*}\right)$$
$$\times \exp\left(-\frac{2\eta^* t_1^*}{h_s^*}\right). \qquad (8.1.2)$$

For $q(t) = \text{const}$, the center temperature rise is

$$\Delta T_c(t^*) = \frac{I_a w_0^2}{4\kappa h_s} \left[\text{Ei}\left(-\zeta(1+4t^*)\right) - \text{Ei}(-\zeta)\right] \exp(\zeta), \qquad (8.1.3)$$

where $\zeta = \eta^*/2h_s^*$. The function Ei is given in Appendix A.3. With $\zeta t^* \ll 1$ we obtain

$$\Delta T_c(t^*) \approx \frac{I_a w_0^2}{4\kappa h_s} \ln(1+4t^*). \qquad (8.1.4)$$

Stationary solutions exist only in the case of finite heat losses. $\eta^* \neq 0$. With $t^* \to \infty$ we obtain from (8.1.3)

$$\Delta T_c = -\frac{I_a w_0^2}{4\kappa h_s} \exp(\zeta) \, \text{Ei}(-\zeta). \qquad (8.1.5)$$

With the approximation $\zeta \ll 1$ this yields

$$\Delta T_c \approx \frac{I_a w_0^2}{4\kappa h_s} \left[\ln\left(\frac{2h_s^*}{\eta^*}\right) - C\right]. \qquad (8.1.6)$$

With $\zeta \gg 1$ where $\text{Ei}(-\zeta) \approx (1-\zeta)\exp(-\zeta)/\zeta^2$ we obtain $\Delta T_c \approx I_a/2\eta$.

8.1.2 Scanned CW Laser

With scanned cw-laser irradiation and surface absorption, the temperature rise for quasi-stationary conditions ($t^* \to \infty$) and $\eta^* = 0$ can be obtained from (8.1.1) and (A.5.8). In a coordinate system that is fixed with the laser beam, we obtain

$$\Delta T(r^*, z^*, \varphi) = \frac{I_a w_0^2}{\kappa h_s} \int_0^\infty \frac{dt_1^*}{1+4t_1^*} \theta_3^J\left[\frac{\pi z^*}{2h_s^*} \bigg| \exp\left(-\frac{\pi^2 t_1^*}{h_s^{*2}}\right)\right]$$
$$\times \exp\left(-\frac{r^{*2} + v_s^* t_1^*(v_s^* t_1^* + 2r^* \cos\varphi)}{1+4t^{*\prime}}\right), \qquad (8.1.7)$$

where θ_3^J is the Jacobian theta function (Appendix A.3).

Figure 8.1.1 shows isotherms calculated from (8.1.7) with $h_s^* \equiv h_s/w_0 = 1$, $v_s^* = 2$ at a depth $z = 0.4 h_s$. The figure shows that the maximum temperature rise along the x-axis, $\Delta T_{max} \equiv \theta_{max} = \theta(x_0^*, z^*, \pi)$, remains behind the center of the laser beam at $x^* = y^* = 0$.

Figure 8.1.2 shows the maximum temperature rise, $\theta_{max}^* = \theta_{max}/\theta_c$, as a function of scanning velocity at the depth $z = h_s$, i.e., on the rear (non-irradiated) side of the substrate. If we set $\theta_{max}(z = h_s) \approx T_m^{eff} - T(\infty)$ with $T_m^{eff} = T_m + \Delta H_m/c_p$ we can estimate the scanning velocity for which the slab is melted through the whole thickness (ΔH_m is the enthalpy of melting, Chap. 10). This permits a rough estimation of cutting and welding velocities in laser machining.

The following approximations can be employed: For scanning velocities $v_s^* \ll 1$ one obtains [*Bunkin* et al. 1978] on the surface $z = 0$

$$x_0^*(z^* = 0) \approx -\frac{v_s^*}{4} \ln\left(\frac{16}{\xi v_s^{*2}}\right), \tag{8.1.8}$$

where $\xi = \exp(C + 1)$ and $C = 0.577$. The maximum temperature is

$$\Delta T_{max}(z^* = 0) \equiv \theta_{max}(z^* = 0) \approx \frac{\theta_c}{2\sqrt{\pi h_s^*}} \ln\left(\frac{16}{\xi v_s^{*2}}\right). \tag{8.1.9}$$

For high scanning velocities $v_s^* \gg 1$ one obtains for $z^* = 0$

$$x_0^*(z^* = 0) \approx 0.54$$
$$\Delta T_{max}(z^* = 0) \approx 1.37 \frac{\theta_c}{v_s^{*1/2}}. \tag{8.1.10}$$

Correspondingly, one obtains for the depth z^*

$$x_0^*(z^*) \approx x_0^*(0) + \frac{1}{2} v_s^* z^{*2} \tag{8.1.11}$$

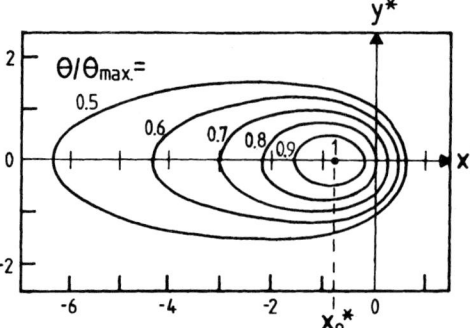

Fig. 8.1.1. Isotherms calculated at a depth $z = 0.4 h_s$ of a slab of thickness $h_s^* = 1$. The coordinate system x^*, y^* is fixed with the center of the Gaussian laser beam at $x^* = y^* = 0$. The scanning velocity is $v_s^* = 2$. The maximum temperature occurs at a distance $-x_0^*$ from the beam center [*Bunkin* et al. 1978]

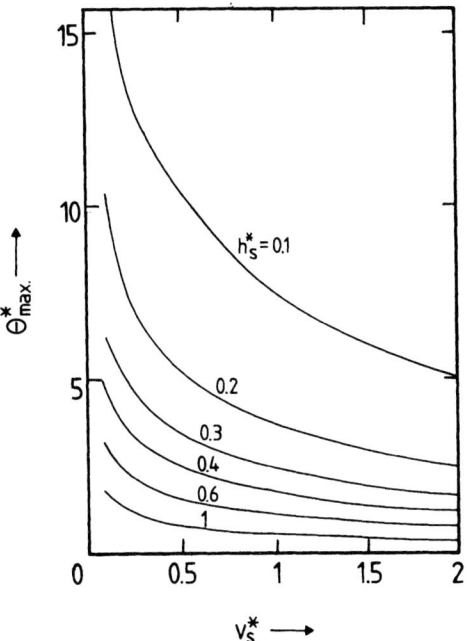

Fig. 8.1.2. Normalized maximum temperature rise at the rear side of a slab ($z = h_s$) as a function of normalized scanning velocity [Bunkin et al. 1978]

and

$$\Delta T_{\max}(z^*) \approx \frac{\theta_c}{0.74 v_s^{*1/2} + v_s^* z^*} . \tag{8.1.12}$$

The dependence (8.1.11) can also be derived from simple arguments. The time required for heat diffusion over a distance z is $t \approx z^2/4D$. Thus, $x_0(z) - x_0(0) \approx v_s t \approx \frac{1}{4} w_0 v_s^* z^{*2}$. Equation (8.1.12) may be rewritten as

$$z^* \approx \frac{\theta_c}{v_s^* \Delta T_{\max}(z^*)} - \frac{0.74}{v_s^{*1/2}} . \tag{8.1.13}$$

This equation can conveniently be used to estimate the processing depth for high velocities v_s^*.

The minimum energy deposit per unit length [J/cm], necessary to induce a temperature rise $\Delta T_{\max}$ at the depth z is

$$E \approx \frac{\pi w_0^2 I_a(0)}{v_s} , \tag{8.1.14}$$

where $v_s \equiv v_s(z, \theta_{\max}(z))$ must be determined from (8.1.13).

8.2 The Influence of Interferences

For weak to moderate absorption ($l_\alpha \gtrsim h_s$), multiple-reflections within the slab cannot be ignored and the absorptivity becomes dependent on the thickness of the slab, i.e., $A = A(h_s)$.

Let us consider the situation for *uniform* irradiation of a slab with permittivity ε_s in an ambient medium with $\varepsilon_M \approx 1$. Due to energy conservation, the absorptivity, A, is related to the reflectivity, $R = |r|^2$, and the transmittivity, $D = |d|^2$, by

$$A + R + D = 1. \tag{8.2.1}$$

The amplitude reflection and transmission coefficients are given by

$$r = \frac{r_{Ms}[1 - \exp(-i2\psi)]}{r_{Ms}^2 - \exp(-i2\psi)} \quad \text{and} \quad d = \frac{(r_{Ms}^2 - 1)\exp(-i\psi)}{r_{Ms}^2 - \exp(-i2\psi)},$$

where

$$r_{Ms} = \frac{1 - \sqrt{\varepsilon_s}}{1 + \sqrt{\varepsilon_s}}; \quad \psi = \frac{2\pi h_s}{\lambda}\sqrt{\varepsilon_s} = k_l h_s \sqrt{\varepsilon_s}$$

and $\sqrt{\varepsilon_s} = n + i\kappa_a$. Indices M and S refer to the ambient medium and the substrate (slab), respectively. Let us consider two limiting cases.

Case 1: Semi-infinite substrate, $h_s \to \infty$

For a semi-infinite substrate there is, of course, no interference and we obtain the well known Fresnel formulas

$$R = \frac{(1 - n)^2 + \kappa_a^2}{(1 + n)^2 + \kappa_a^2}; \quad A = \frac{4n}{(1 + n)^2 + \kappa_a^2}; \quad D = 0. \tag{8.2.2}$$

Case 2: Finite thickness, weak absorption, $l_\alpha \gg h_s$ and $\kappa_a \ll 1$.

If we consider only terms linear in κ_a and αh_s, we obtain

$$R = \frac{(1 - n^2)^2 \sin^2(\tfrac{1}{2}\beta h_s)}{\xi}(1 - A)$$

$$A = 2\kappa_a \frac{(1 + n)\beta h_s - (1 - n)\sin(\beta h_s)}{\xi} \tag{8.2.3}$$

$$D = \frac{4n^2}{\xi}(1 - A),$$

with

$$\xi = 4n^2 + (1 - n^2)^2 \sin^2\left(\frac{1}{2}\beta h_s\right),$$

8.2 The Influence of Interferences

and $\beta = 4\pi n/\lambda$. The heat source term for a slab of finite thickness is given by

$$Q(z) = I_0(1 - |r_{Ms}|^2)f(z), \qquad (8.2.4)$$

with $f(z) = \alpha f'(z)$ where

$$f'(z) = \left| \frac{r_{Ms} \exp(-i\psi z/h_s) - \exp(-i2\psi)\exp(i\psi z/h_s)}{r_{Ms}^2 - \exp(-i2\psi)} \right|^2. \qquad (8.2.5)$$

If $\kappa_a \ll n$ this equation can be approximated by

$$f'(z) = \frac{\exp[\alpha(h_s - z)] - 2\zeta\cos[\beta(h_s - z)] + \zeta^2\exp[-\alpha(h_s - z)]}{\exp(\alpha h_s) - 2\zeta^2\cos(\beta h_s) + \zeta^4\exp(-\alpha h_s)}, \qquad (8.2.6)$$

where $\zeta = (1-n)/(1+n)$. The full curves in Fig. 8.2.1 show $f'(z^*)$ as calculated from (8.2.5) with $z^* = z/h_s$. The dashed curves represent Beer's law, $f'(z) = \exp(-\alpha z)$. For thin films, considerable deviations between curves are observed. With $h_s = \lambda/4n$ the intensity at $z = 0$ becomes very small due to interference between the incident and the reflected beam. With larger thicknesses, $f'(z^*)$ shows oscillations. With $h_s > 2\lambda/n$ Beer's law becomes a good approximation.

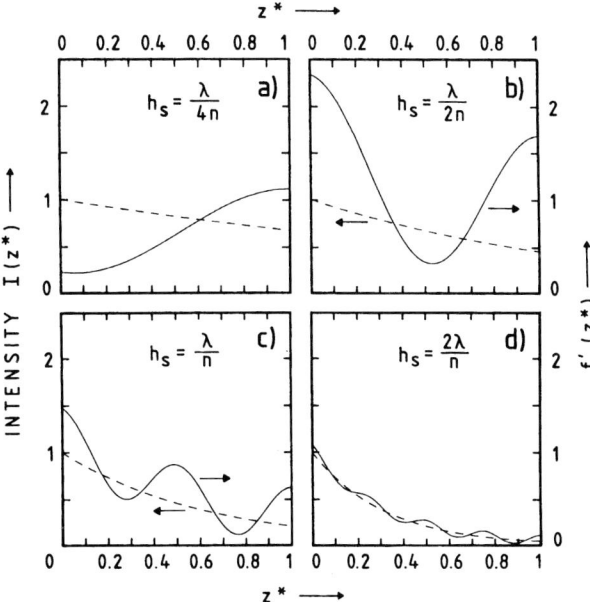

Fig. 8.2.1. Full curves show $f'(z^*)$ calculated from (8.2.5) for different thicknesses of an infinite slab, h_s. The parameters employed correspond to 10.6 μm CO_2-laser radiation and doped Ge ($n = 4$, $\kappa_a = 0.5$). The dashed curves have been calculated from Beer's law

8.3 Coupling of Optical and Thermal Properties

Some important features arising from the coupling between the optical and thermal properties of a material can be derived most simply from the heat balance. For a thermally thin slab and uniform irradiation this yields

$$h_s \rho c_p \frac{dT}{dt} = I_0 A - I_{\text{loss}}, \tag{8.3.1}$$

where $A \equiv A(T)$. The thermal losses can be described by

$$I_{\text{loss}}(T) = \eta [T - T(\infty)] . \tag{8.3.2}$$

$T(\infty)$ is the initial temperature. The absorptivity of the slab is given by (8.2.1). For *weak* absorption, A can be approximated by (8.2.3). This temperature dependence, $A = A(T)$, is related to the index of refraction, $n = n(T)$, the absorption index $\kappa_a = \kappa_a(T)$, and the thermal expansion of the slab, $h_s(T)$. If the temperature rise is small, we can use the expansion

$$n(T) = n(T(\infty)) + [T - T(\infty)] \frac{dn}{dT} \tag{8.3.3}$$

and

$$h_s(T) = h_s(T(\infty)) \{1 + \beta_T [T - T(\infty)]\} ,$$

where β_T is the linear thermal expansion coefficient. The temperature dependence of κ_a can be described in analogy to n; outside of resonances, it can be ignored.

For weak absorption, the laser-induced temperature can then be estimated from (8.3.1 to 3), together with (8.2.3). Due to the temperature dependence of its optical thickness, transient heating of the slab shows an oscillating behavior. Another phenomenon is the occurence of bistabilities or even multistabilities. This is schematically shown in Fig. 8.3.1. The stationary temperature rise $\Delta T \equiv \Delta T(t \to \infty)$ as a function of incident laser-beam intensity shows unstable branches which are indicated by dashed lines.

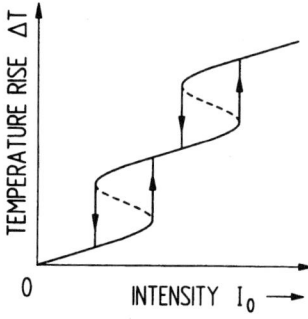

Fig. 8.3.1. Stationary temperature rise of an infinite slab as a function of laser-beam intensity for a temperature-dependent absorptivity $A(T)$

8.4 Average Temperature Distributions

If we ignore interference effects and assume temperature-independent material parameters, we can derive a large number of solutions from (6.2.2) with the $\mathscr{F}$-function given by (A.5.1). These solutions represent "average" temperature distributions. For example, for cw-laser irradiation by means of a Gaussian beam we obtain with $v_s^* = 0$ for the center temperature rise

$$\Delta T(0,0) \equiv \theta_c(h_s^*, \alpha^*, \eta^*)$$

$$= -\frac{\alpha^* I_a w_0}{4\kappa} \sum_{n=-\infty}^{+\infty} A_n \frac{v_n^2 \exp(v_n^2/4)}{(v_n^2 + \eta^{*2})h_s^* + 2\eta^*} \mathrm{Ei}(-v_n^2/4), \qquad (8.4.1)$$

where v_n and A_n are given by (A.5.4) and (A.5.5), respectively. Clearly, the stationary solution (8.4.1) exists only with $\eta^* \neq 0$.

9 Non-Uniform Media

Continuous or discontinuous changes in thermophysical and optical properties of materials significantly change laser-induced temperature distributions with respect to those estimated for plane uniform (homogeneous) substrates. *Continuous* changes in physical properties may be related to temperature dependences of material parameters or to slow changes in the material structure or composition. *Discontinuous* changes in physical properties occur in some types of composite materials, in multilayer structures, and in combined structures. Such structures may only be generated during laser processing as in laser-induced surface oxidation or deposition of thin films or microstructures.

9.1 Continuous Changes in Optical Properties

Laser-induced heating changes the optical properties of materials via their temperature dependence. The change in absorbed laser-light intensity, in turn, changes the temperature distribution. This feedback, if strong enough, may cause *new phenomena* in laser-matter interactions. Among the examples already discussed are thermal runaway (Sect. 7.6) and oscillations or multistabilities (Sect. 8.3). In more general cases, the reflectivity and absorptivity become complex functionals of $\tilde{n}(x)$. The distribution of dissipated energy must then be calculated from the Maxwell equations [see, e.g., *Born* and *Wolf* 1980]. The problem can be simplified considerably if we assume:

- Laser-beam irradiation at normal incidence.
- A temperature-dependent dielectric permittivity $\varepsilon = \varepsilon(T)$ only, and $\mu' = 1$, $\mu'' = 0$ (Sect. 2.2.1).
- A variation of ε only in z-direction, $\varepsilon = \varepsilon(z)$, which is a good approximation for large-area illumination.
- Weak absorption with $\kappa_a \ll n$ where $\kappa_a = \kappa_a(z)$.
- The validity of geometrical optics, i.e., the change in wavelength within the medium, $\lambda_M = \lambda/n$, shall be small compared to λ_M over the distance λ_M, so that $\lambda \, (d/dz) \, (1/n) \ll 1$.
- The beam shape to remain unchanged.

9.2 Absorption of Light in Multilayer Structures

With these assumptions, the source term in the heat equation can be written as

$$Q(z) = [1 - R(T(z=0))] \, I_0 f(z), \qquad (9.1.1)$$

where $f(z)$ is given by (6.1.2). If any of these assumptions becomes relaxed, the propagation of the laser light becomes more complicated.

Consider, for example, an optical inhomogeneity $n = n(z)$ with $dn/dz < 0$. This may originate from a temperature distribution $T(z)$, a chemical inhomogeneity $N_i(z)$, etc. Then, for oblique laser-beam incidence, the light may propagate as schematically shown in Fig. 9.1.1a. This is the situation in optical *waveguides* produced by laser-induced surface doping.

If n changes in the direction perpendicular to the wavevector of the laser beam, nonlinear refraction will result in self-focusing or defocusing phenomena (Fig. 9.1.1b). *Self-focusing* occurs if the refractive index of the medium changes with electric field so that $n(r) = n_0 + n_1 E^2(r)$ with $n_1 > 0$. This (non-inertial) nonlinearity of n depends directly on the amplitude of the field. In this case, the propagation of light can be described by the solution of the nonlinear Maxwell equations.

There are, however, other (inertial) nonlinearities where changes in optical properties are only indirectly related to the action of the laser light. An example is the laser-induced heating of a medium which causes an inhomogeneity $\varepsilon = \varepsilon(T(r, z))$. Then, a description of light propagation requires one to solve the Maxwell equations together with the heat equation. In media where the optical properties do not only change with temperature but also with chemical composition, i.e., $\varepsilon = \varepsilon(T, N_i)$, multiple nonlinearities arise [*Karlov* et al. 1992].

9.2 Absorption of Light in Multilayer Structures

Discontinuous changes in material properties occur in multilayer structures. Subsequently, we concentrate on the optical properties of multilayer structures

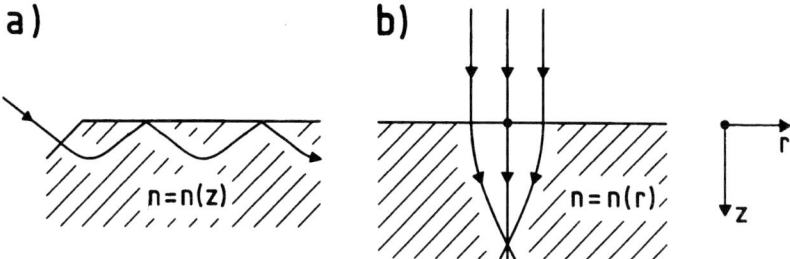

Fig. 9.1.1. (a) Optical waveguiding due to a refractive index $n = n(z)$ with $dn/dz < 0$. (b) Self-focusing related to (lateral) changes $n = n(r)$

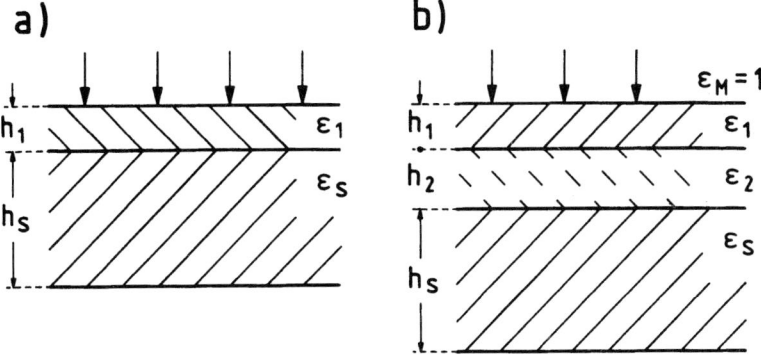

Fig. 9.2.1. Uniform irradiation of substrates covered with a single layer (**a**) and with two layers (**b**)

with special emphasis on systems that are relevant to laser processing. The situation considered is shown in Fig. 9.2.1. Different layers are characterized by their thickness $h_1, h_2, \ldots, h_s$ and their permittivity $\varepsilon_1, \varepsilon_2, \ldots, \varepsilon_s$ (the same indices are added to other related quantities). The ambient medium shall be air ($\varepsilon_M \approx 1$). The results obtained for uniform irradiation can also be employed with a collimated beam whose diameter is sufficiently large.

9.2.1 Thin Films

First, we consider a thin extended film placed on a substrate with $h_1 \ll h_s$ (Fig. 9.2.1a). If $h_s \gg l_{\alpha s}$ the substrate can be considered as semi-infinite with respect to its optical behavior. Because $D = 0$, we have

$$A + R = 1. \tag{9.2.1}$$

The amplitude reflection coefficient is

$$r_{M1s} = \frac{r_{1s} + r_{M1} \exp(-i2\psi_1)}{r_{M1} r_{1s} + \exp(-i2\psi_1)}, \tag{9.2.2}$$

with

$$r_{ij} = \frac{\sqrt{\varepsilon_i} - \sqrt{\varepsilon_j}}{\sqrt{\varepsilon_i} + \sqrt{\varepsilon_j}}; \quad \psi_i = \frac{2\pi h_i}{\lambda} \sqrt{\varepsilon_i} = k_i h_i \sqrt{\varepsilon_i},$$

where $\sqrt{\varepsilon_i} = n_i + i\kappa_{a_i}$. The indices i and j refer to the different media, in the present case to those characterized by $\varepsilon_M, \varepsilon_1$, and ε_s.

9.2 Absorption of Light in Multilayer Structures

Weak absorption

For weakly absorbing films, the absorptivity changes with film thickness, h_1, due to interferences. This situation applies to metal oxides on metal substrates irradiated with IR or VIS light, for which $\kappa_{a1} \ll n_1$. Additionally, for metals $A_s = 1 - |r_{Ms}|^2 \ll 1$. If we consider only terms linear in κ_{a1} and A_s, we obtain

$$A(h_1) \approx \frac{n_1^2 A_s + 2\kappa_{a1}[\beta h_1 - \sin(\beta h_1)]}{n_1^2 + (1 - n_1^2)\sin^2(\tfrac{1}{2}\beta h_1)}, \quad (9.2.3)$$

with

$$\beta = \frac{4\pi n_1}{\lambda} = 2k_l n_1. \quad (9.2.4)$$

The oscillations in the absorptivity (9.2.3) are due to interferences of the laser light within the thin layer. Their period is $\Delta h = 2\pi/\beta = \lambda/2n_1$. The number of oscillations is

$$Z = \frac{h_1}{\Delta h} \approx \frac{n_1}{2\pi \kappa_{a1}}. \quad (9.2.5)$$

The latter approximation estimates the number of *pronounced* oscillations which occur up to $h_1 \approx l_{\alpha 1} = \alpha_1^{-1}$. For an oxide layer of Cu_2O and 10.6 μm CO_2-laser radiation, the number of pronounced oscillations is $Z = 14$ (Fig. 9.2.2). If we consider, on the other hand, an SiO_2 layer on Si and UV-laser radiation, we find $Z \approx 10^6$. The number of oscillations observed experimentally depends also on the surface roughness, the spectral width of the laser light, etc. Via (9.2.5) the type of layer material can be analyzed.

For very thin films we can approximate (9.2.3) by

$$A(h_1) \approx A_s(1 + \xi h_1^2), \quad (9.2.6)$$

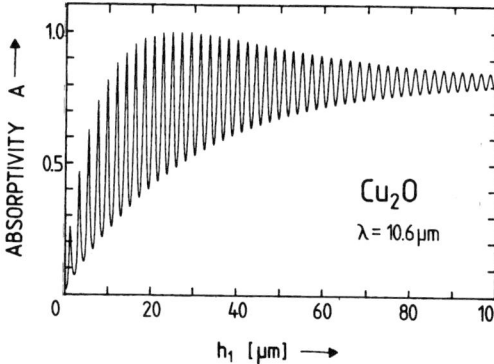

Fig. 9.2.2. Absorptivity calculated as a function of layer thickness. The parameters $A_s = 0.02$, $n_1 = 2.45$, and $\kappa_{a1} = 0.027$ correspond to a Cu_2O layer on a Cu substrate and CO_2-laser radiation

where $\xi = k_l^2(n_1^2 - 1)$ and $\xi h_1^2 \ll 1$. If we set $\kappa_{a1} = \kappa_{as} = 0$ we obtain from (9.2.2) the reflectivity

$$R(h_1) = |r_{M1s}|^2 = \frac{n_1^2(1-n_s)^2 \cos^2\varphi + (n_s - n_1^2)^2 \sin^2\varphi}{n_1^2(1+n_s)^2 \cos^2\varphi + (n_s + n_1^2)^2 \sin^2\varphi} \qquad (9.2.7)$$

where $\varphi = k_l n_1 h_1$.

Strong absorption

If $\kappa_{a1} > n_1$ the absorptivity changes smoothly from A_s to $A(h_1) \approx A_1 = 1 - |r_{M1}|^2$ within a film thickness of a few absorption lengths. The averaged absorptivity (with respect to the oscillations) is

$$\langle A(h_1) \rangle = \frac{1}{\Delta h_1} \int_0^{\Delta h_1} A(h_1 + x)\, dx \approx A_1 + (A_s - A_1) \exp(-2\alpha_1 h_1).$$

9.2.2 Two-Layer Structures

A two-layer structure (Fig. 9.2.1b) describes, for example, the situation in many cases of laser-induced oxidation of copper (Chap. 26). Here, h_1 corresponds to CuO, and h_2 to Cu_2O. The absorptivity is

$$A = 1 - |r_{M12s}|^2 \qquad (9.2.8)$$

where the amplitude reflection coefficient is

$$r_{M12s} = \frac{r_{12s} + r_{M1}\exp(-i2\psi_1)}{r_{M1}r_{12s} + \exp(-i2\psi_1)}. \qquad (9.2.9)$$

r_{M1} is defined as in (9.2.2). r_{12s} can be calculated from r_{M1s} by changing indices $M \to 1$ and $1 \to 2$. This procedure can be extended to an arbitrary number of layers.

For two-layer systems, the absorptivity may show double interference behavior. For example, high-frequency oscillations may be modulated by low-frequency oscillations. This has in fact been observed during cw Nd:YAG-laser oxidation of Cu.

The reflectivity of two free standing layers – without the semi-infinite substrate – is obtained from (9.2.9) by changing $S \to M$.

9.2.3 Three-Layer Systems

For a three-layer system, the amplitude reflection coefficient is given by

$$r_{M123s} = \frac{r_{123s} + r_{M1} \exp(-i2\psi_1)}{r_{M1} r_{123s} + \exp(-i2\psi_1)}, \quad (9.2.10)$$

where r_{123s} can be calculated from (9.2.9) by changing indices $M \to 1$, $1 \to 2$, $2 \to 3$.

This model can be applied, in a first approximation, to laser-induced oxidation of Fe. Here, h_1 corresponds to Fe_2O_3, h_2 to Fe_3O_4, and h_3 to FeO. While for Cu and Fe the boundaries between different oxide layers are relatively sharp, they are washed out for other oxides, e.g., those of Ti and V.

9.3 Temperature Distributions for Large-Area Irradiation

Consider uniform laser-light irradiation which directly or indirectly heats a thin film placed on a substrate (Fig. 9.2.1 a).

In the simplest case, the film changes only the absorptivity from A_s to A but has no influence on the overall *thermal* properties. A can be constant, or it can change with time, for example due to changes in film thickness. In the latter case, the (single) heat equation for the substrate must be solved together with the equation for $A = A(h)$. This approximation can be applied to laser-induced growth of films whose thermal properties are similar to those of the substrate.

If the film influences both the optical and thermal properties of the system, the heat equation must be solved for both the thin film and the substrate.

9.3.1 Stationary Solutions for Thin Films

The influence of the film on the surface temperature can be estimated from the boundary condition at the interface to the ambient medium

$$\eta \Delta T = \kappa_1 \frac{\partial T}{\partial z}\bigg|_{z=0} \approx \kappa_1 \frac{\delta T_1}{h_1}, \quad (9.3.1)$$

which yields

$$\frac{\delta T_1}{\Delta T} \approx \frac{\eta h_1}{\kappa_1}, \quad (9.3.2)$$

where δT_1 is the change in temperature within the film, and ΔT its *average* temperature rise. For gases $\eta h_1/\kappa_1 \ll 1$ and thus $\delta T_1 \approx 0$. The average temperature rise is then

$$\Delta T \approx \frac{I_0 A}{2\eta}, \qquad (9.3.3)$$

where $A = A_1 + A_s$ is the *total* absorptivity of the system. For surface absorption $A \equiv A_1$. For *finite* absorption, the distribution of the absorbed intensity must be considered in the source term for the film and the substrate so that

$$Q_1 = I_0 f_1(z) \quad \text{and} \quad Q_s = I_0 f_s(z). \qquad (9.3.4)$$

For simplification we assume the substrate as optically semi-infinite, i.e., $l_{\alpha s} \ll h_s$. This applies to metal oxidation and, for laser wavelengths $\lambda < 500$ nm, also to the oxidation of Si. In this approximation, the functions $f_1(z)$ and $f_s(z)$ can be written as

$$f_1(z) = \alpha_1 (1 - R_1) \left| \frac{r_{1s} \exp(-i\varphi) + \exp(-i2\psi) \exp(i\varphi)}{r_{M1} r_{1s} + \exp(-i2\psi)} \right|^2, \qquad (9.3.5)$$

and

$$f_s(z) = \alpha_s n_s \left| \frac{(1 + r_{M1})(1 + r_{1s})}{r_{M1} r_{1s} + \exp(-i2\psi)} \right|^2 \exp(\alpha_1 h_1) \exp[-\alpha_s (z - h_1)],$$

with

$$1 - R_1 = \frac{4n_1}{(1 + n_1)^2 + \kappa_{a1}^2}; \quad \varphi \equiv \varphi(z) = k_l z \sqrt{\varepsilon_s},$$

where r_{M1}, r_{1s}, r_{Ms} and ψ are given by (9.2.2). With the present assumptions, laser-beam interference takes place within the film, while the attenuation within the substrate in described by Beer's law. The total energy absorbed within the film and the substrate is proportional to

$$A_1 = \int_0^{h_1} f_1(z) \, dz \quad \text{and} \quad A_s = \int_{h_1}^{\infty} f_s(z) \, dz. \qquad (9.3.6)$$

Due to energy conservation

$$A = A_1 + A_s = 1 - |r|^2,$$

where the amplitude reflection coefficient is given by (9.2.2).

Figure 9.3.1 shows the absorptivities, reflectivities and transmissivities for a substrate of thickness h_s as a function of film thickness h_1. The parameters employed correspond to an SiO_2 layer on a 300 μm Si wafer. The results shown in Fig. 9.3.1a have been calculated from (9.3.6) and (9.2.2) for (uniform) XeCl-laser radiation. SiO_2 is transparent for 308 nm radiation while Si is strongly absorbing at this wavelength. Due to interferences within the SiO_2 layer, the reflectivity, R, oscillates and thereby the (surface) absorptivity of the silicon, A_s. The situation is different for CO_2-laser radiation (Fig. 9.3.1b). SiO_2 strongly absorbs at 10.6 μm while Si is almost transparent at this wavelength.

9.3 Temperature Distributions for Large-Area Irradiation

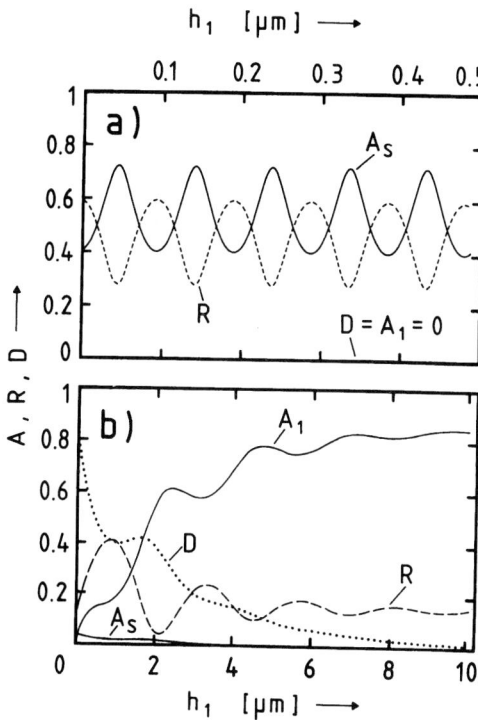

Fig. 9.3.1a,b. Absorptivities A, reflectivities R, and transmissivities D, calculated for uniform laser-light irradiation as a function of film thickness, h_1. The parameters employed correspond, approximately, to an SiO_2 layer on a Si wafer ($h_s = 300\,\mu m$). (a) 308 nm XeCl-laser radiation ($R_1 = 0.05$, $\alpha_1 = 0$, $R_s = 0.59$, $\alpha_s = 1.47 \times 10^6\,cm^{-1}$). (b) CO_2-laser radiation ($R_1 = 0.15$, $\alpha_1 = 4 \times 10^3\,cm^{-1}$, $R_s = 0.31$, $\alpha_s = 0.86\,cm^{-1}$)

For this reason, the finite thickness of the silicon wafer must be taken into account. Here, A_1 strongly increases with h_1 while the transmissivity D decreases. A_s can be described, in good approximation, by Beer's law. The oscillations in R, and thereby in A_1 and D, are due to interferences within the SiO_2 layer.

According to (9.3.3) the temperature rise behaves in the same way as the total absorptivity, $A = A_1 + A_s$.

9.3.2 Dynamic Solutions

For *large-area* uniform irradiation and $h_s \to \infty$ the temperature distribution within the film for $\alpha \to \infty$ and $\eta = 0$ is

$$\Delta T_1(z, t) = \frac{I_0(1 - R_1)l_T}{\kappa_1} \sum_{n=-\infty}^{+\infty} \Lambda^{|n|} \operatorname{ierfc}\left(\frac{|z - 2nh_1|}{l_T}\right), \tag{9.3.7}$$

where $l_T = 2(D_1 t)^{1/2}$ and

$$\Lambda = \frac{\kappa^* - D^{*1/2}}{\kappa^* + D^{*1/2}}, \quad \text{and} \quad D^* = D_1/D_s. \tag{9.3.8}$$

The effect of temperature-dependent thermal conductivities can be taken into account, if the functional form of the temperature dependences $\kappa_1(T)$ and $\kappa_s(T)$ is the same, so that $\kappa^* = \kappa_1(T)/\kappa_s(T)$ is independent of temperature. In this case, a Kirchhoff transform in analogy to (2.2.8) can be performed.

9.4 Temperature Distributions for Focused Irradiation

Subsequently, we shall investigate temperature distributions induced within a thin film and a substrate by a *focused* laser beam (Fig. 9.4.1). We assume $h_s \to \infty$ and cylindrical symmetry. Before presenting (approximate) solutions of the boundary–value problem, we shall estimate, in a crude way, the influence of the film on the surface temperature rise. Heat transport to the ambient medium is ignored, i.e., we set $\eta = 0$.

The temperature gradient induced by a focused laser beam absorbed at the surface is of the order $\nabla T \approx \Delta T/w$. The power transported from the illuminated area $F \approx \pi w^2$ into the substrate is $P_s \approx F \kappa_s \nabla T$. In the absence of the film and with stationary conditions, this must be equal to the laser power absorbed on the surface, $P_a \approx I_0 A \pi w^2$. This yields $\Delta T(z=0, h_1=0) \approx I_0 A w/\kappa_s$. In the presence of the film, we have to consider the change in lateral heat flux. If $\kappa_1 \gg \kappa_s$, the additional power transported in lateral direction is $P_1 \approx 2\pi w h_1 \times \kappa_1 \nabla T$. With $P_a = P_s + P_1$ the relative change in surface temperature rise caused by the film becomes

$$\Delta T^* \equiv \frac{\Delta T(z=0, h_1)}{\Delta T(z=0, h_1=0)} \approx \left(1 + \frac{2 h_1}{w} \frac{\kappa_1}{\kappa_s}\right)^{-1}. \tag{9.4.1}$$

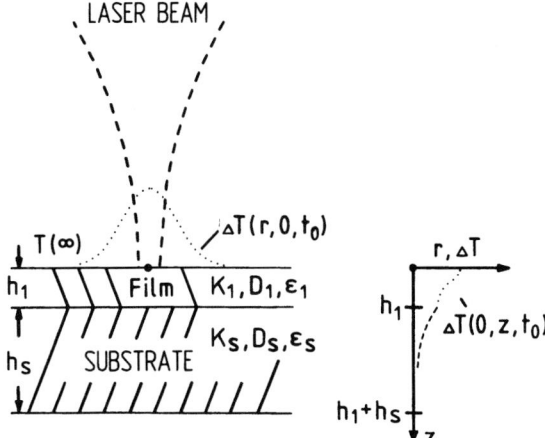

Fig. 9.4.1. Irradiation geometry and laser-induced temperature rise on a substrate covered with a thin film of thickness h_1. The origin of the coordinate system is on the film surface in the center of the laser beam

9.4 Temperature Distributions for Focused Irradiation

Thus, the increase in lateral heat flux decreases the surface temperature. The case $\kappa_1 \gg \kappa_s$ can be applied to thin films of metals or semiconductors on thermally insulating substrates. With $\kappa_1/\kappa_s \approx 50$, a laser focus $w = 10$ μm, and a film thickness of $h_1 = 100$ Å, the surface temperature will decrease by about 10%.

These simple arguments cannot be applied to thermally insulating films where $\kappa_1 \ll \kappa_s$. A more accurate estimation which also includes this case is given by (9.4.5).

9.4.1 Strong Film Absorption

For $\alpha_1 \to \infty$, temperature-independent material parameters, and $h_s \to \infty$, a Gaussian laser beam induces the steady-state temperature rise within the film [*Burgener* and *Reedy* 1982]

$$\Delta T_1(r^*, z^*) = \frac{P(1-R_1)}{2\pi w_0 \kappa_1} \int_0^\infty d\xi \, J_0(\xi r^*) \, \exp\left(-\frac{\xi^2}{4}\right)$$

$$\times \left[\exp(-\xi z^*) + \frac{(\kappa^* - 1)\cosh(\xi z^*)\exp(-\xi h_1^*)}{\kappa^* \sinh(\xi h_1^*) + \cosh(\xi h_1^*)} \right], \quad (9.4.2)$$

where ξ is an integration variable, J_0 the Bessel function of order zero, $r^* = r/w_0$, $z^* = z/w_0$, $h_1^* = h_1/w_0$, and $\kappa^* = \kappa_1/\kappa_s$. Figure 9.4.2 shows the center temperature rise $\Delta T_c^* = \Delta T(r = z = 0, h_1)/\Delta T(r = z = 0, h_1 = 0) \equiv \Delta T_c(h_1)/\Delta T_c(h_1 = 0)$ calculated from (9.4.2) for parameters h_1^* and κ^*. For $h_1^* > 1$, the temperature rise becomes independent of layer thickness.

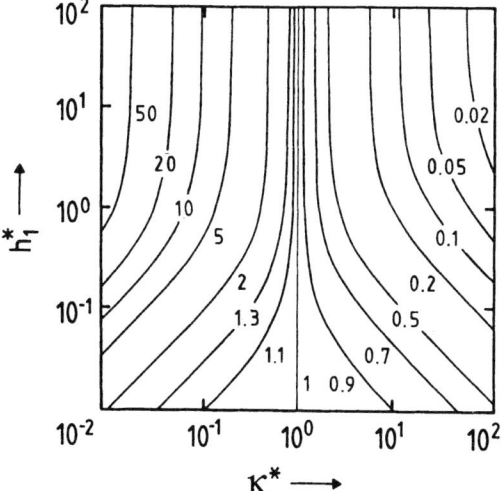

Fig. 9.4.2. Normalized center temperature rise $\Delta T_c^* = \Delta T_c(h_1)/\Delta T_c(h_1 = 0)$ (numbers on curves), as a function of film thickness h_1^* and heat conductivity κ^*

The temperature rise within the *substrate* is

$$\Delta T_s(r^*, z^*) = \frac{P(1-R_1)}{2\pi w_0 \kappa_1} \int_0^\infty d\xi \, J_0(\xi r^*) \exp\left(-\frac{\xi^2}{4}\right)$$

$$\times \left[\exp(-\xi z^*) + \frac{(\kappa^* - 1)\cosh(\xi h_1^*)\exp(-\xi z^*)}{\kappa^* \sinh(\xi h_1^*) + \cosh(\xi h_1^*)} \right]. \quad (9.4.3)$$

For large values of z and r this can be approximated by

$$\Delta T_s(r^* \gg 1, z^* \gg 1) = \frac{P(1-R_1)}{2\pi w_0 \kappa_s (r^{*2} + z^{*2})^{1/2}}. \quad (9.4.4)$$

This result is obvious. For large distances from the source, the temperature distribution must be equal to (7.2.7).

Another approximation seems to be interesting. If we assume $h_1^* \ll \min(\kappa^*, 1, \kappa^{*-1})$ we can expand (9.4.2) and obtain

$$\Delta T_c^* \equiv \frac{\Delta T_c(h_1)}{\Delta T_c(h_1 = 0)} = 1 + \frac{2}{\sqrt{\pi}} \frac{h_1}{w_0} \frac{(\kappa_s - \kappa_1)(\kappa_s + \kappa_1)}{\kappa_1 \kappa_s}. \quad (9.4.5)$$

For $\kappa_1 \gg \kappa_s$ this yields

$$\Delta T_c^* = 1 - \frac{2}{\sqrt{\pi}} \frac{h_1}{w_0} \frac{\kappa_1}{\kappa_s}.$$

Except for the factor $1/\sqrt{\pi}$, this result agrees with (9.4.1).

For $\kappa_1 \ll \kappa_s$ we obtain

$$\Delta T_c^* = 1 + \frac{2}{\sqrt{\pi}} \frac{h_1}{w_0} \frac{\kappa_s}{\kappa_1}. \quad (9.4.6)$$

Thus, the film increases the surface temperature. This case applies to surface oxidation of metals or semiconductors.

9.4.2 Finite Film Absorption

We now assume finite film absorption and all material parameters to be independent of temperature. Furthermore, if $h_1 \geq l_{\alpha 1}$ the real intensity distributions within the film can be approximated by an exponential (Fig. 8.2.1d). Thus, we can define an *effective* absorption coefficient

$$\alpha_1 = \frac{1}{h_1} \ln\left(\frac{P(z=0)}{P(z=h_1)}\right). \quad (9.4.7)$$

The reflection coefficient, R, depends on film thickness and oscillates with a period of, approximately, $\lambda/2n_1$.

9.4 Temperature Distributions for Focused Irradiation

Multilayer structures

The steady-state temperature distribution induced by cw-laser irradiation of multilayer structures has been calculated by *Calder* and *Sue* (1982).

For a semi-infinite substrate covered by a *single* film, the temperature rise induced within the film by a cylindrical beam of radius w can be described by

$$\Delta T_1(r^*, z^*) = \frac{P(1-R)\alpha_1^{*2}}{2\pi w \kappa_1 F(0)} \int_0^\infty d\xi \frac{F(\xi) J_0(\xi r^*)}{\alpha_1^{*2} - \xi^2}$$

$$\times \left\{ \left(\kappa^* \cosh[\xi(h_1^* - z^*)] + \sinh[\xi(h_1^* - z^*)] \right. \right.$$

$$+ \left[\frac{\xi}{\alpha_1^*} - \kappa^* + \left(1 - \frac{\xi}{\alpha_s^*}\right) \Lambda \exp(-\alpha_s^* h_1^*) \right]$$

$$\left. \times \exp(-\alpha_1^* h_1^*) \cosh(\xi z^*) \right)$$

$$\left. \times [\cosh(\xi h_1^*) + \kappa^* \sinh(\xi h_1^*)]^{-1} - \frac{\xi}{\alpha_1^*} \exp(-\alpha_1^* z^*) \right\}, \quad (9.4.8)$$

where $\Lambda \equiv \Lambda(\xi)$ is given by

$$\Lambda = \kappa^* \frac{\alpha_s^{*2} (\alpha_1^{*2} - \xi^2)}{\alpha_1^{*2} (\alpha_s^{*2} - \xi^2)},$$

and

$$F(\xi) = \int_0^\infty g(r^*) J_0(\xi r^*) r^* dr^*, \quad (9.4.9)$$

with $r^* = r/w$, $z^* = z/w$, and $\alpha_i^* = \alpha_i w$. The function $g(r^*)$ describes the radial intensity distribution of the (circular) laser beam.

With a Gaussian beam, $F(\xi)$ simplifies to

$$F(\xi) = \frac{1}{2} \exp\left(-\frac{\xi^2}{4}\right). \quad (9.4.10)$$

Figure 9.4.3 shows the (normalized) center temperature rise, $\Delta T_1(0,0)/P$, calculated from (9.4.8) and (9.4.10) as a function of the thickness of a p-Si film on a glass substrate. The overall change in surface temperature, and the interference pattern superimposed, originates from changes in the reflection coefficient (Sect. 9.2.1). If the film thickness exceeds the optical penetration depth, i.e., if $h_1 > l_{\alpha 1}$, the laser power absorbed remains almost constant, while the heat transport within the film ($\kappa_1 \gg \kappa_s$) further increases with h_1.

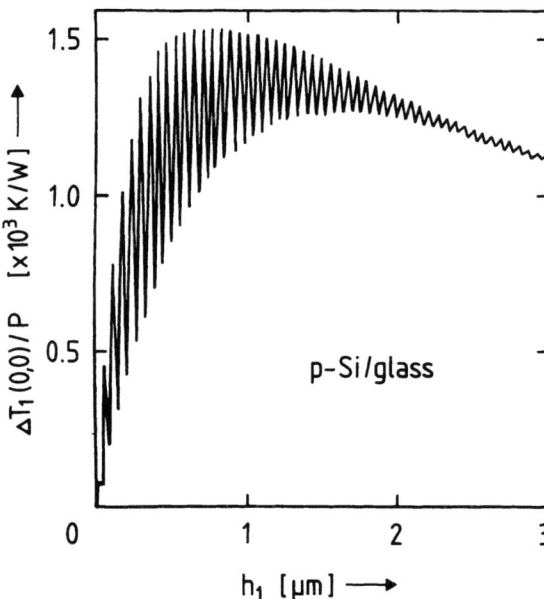

Fig. 9.4.3. Normalized temperature rise induced on the surface of a thin film on a semi-infinite substrate, calculated as a function of film thickness, h_1 ($P(\lambda = 501$ nm) is the incident laser power, $2w_0 = 40$ μm). The material parameters employed correspond to a polycrystalline silicon (p-Si) film and a glass substrate: $\kappa_s(T_0 = 300$ K$) = 0.02$ W/cmK, $\alpha_s(T_0) = 0$, $n_s(T_0, \lambda = 501$ nm$) \approx 1.35$, $\kappa_1(T_0 = 300$ K$) = 0.3$ W/cmK, $\alpha_1(T_0, \lambda = 501$ nm$) \approx 1.5 \times 10^4$ cm^{-1}, $n_1(T_0, \lambda = 501$ nm$) \approx 4.2$. Adapted from [*Calder* and *Sue* 1982]

9.5 The Ambient Medium

The analysis of laser processing rates requires, in many cases, the knowledge of the temperature distribution within the ambient medium, and the influence of this medium on the substrate temperature.

The ambient medium can be heated either *directly* if it absorbs the incident laser radiation (Sect. 19.1), or *indirectly* via the heated substrate area. Heat transport within a gas or a liquid takes place via heat conduction, convection, and thermal radiation. Convection may originate from density gradients related to temperature gradients (free convection), from changes in particle number density in non-equimolecular reactions (chemical convection), or from an external flow (forced convection).

9.5.1 Influence on Substrate Temperature

An ambient medium changes the substrate temperature. Let us consider Fig. 9.5.1.

Case a: The laser radiation shall exclusively be absorbed by the substrate. In the absence of a chemical reaction, the ambient medium will lower the surface temperature $T_s(z = 0)$. If an exothermal surface reaction related to the

9.5 The Ambient Medium

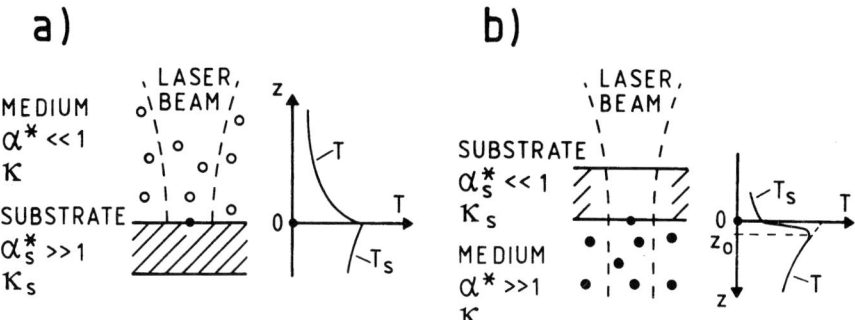

Fig 9.5.1a,b. Influence of an ambient medium on the substrate temperature, T_s. The quantity $\alpha^* = \alpha w_0$ is the normalized absorption coefficient, and κ the thermal conductivity. (a) Non-absorbing ambient medium, strongly absorbing substrate. (b) Non-absorbing substrate, strongly absorbing ambient

ambient medium takes place, T_s can increase or decrease, depending on the relative amount of the reaction enthalpy and the loss of energy by heat transport into the medium.

Case b: The laser radiation shall exclusively be absorbed by the medium. Then, the maximum temperature rise occurs within this medium at a distance $z_0 \propto \alpha^{-1}$. This irradiation scheme permits *localized* processing by means of strongly absorbing media. A practical example is CO_2 laser-induced etching of Si and Ge in an atmosphere of $CF_3I + SF_6$ [*Karlov* et al. 1985]. Here, SF_6 is used as a sensitizer that strongly absorbs CO_2-laser radiation.

Subsequently, we shall discuss the situation shown in Fig. 9.5.1a only. To estimate the change in laser-induced surface temperature caused by the ambient medium, a number of simplifying assumptions are made: The laser beam shall be fixed, of Gaussian shape, and exclusively absorbed on the substrate surface, i.e., $\alpha = 0$ and $\alpha_s \to \infty$. In the absence of convection, the boundary-value problem can be formulated as described in Sect. 6.1. The heat equation can be written in the form

$$\frac{1}{D}\frac{\partial T}{\partial t} - \nabla^2 T = 0. \tag{9.5.1}$$

The region $z \leqslant 0$ refers to the substrate with temperature T_s and thermal diffusivity $D \equiv D_s$ while $z > 0$ refers to the ambient medium with temperature T and diffusivity D. The boundary conditions are given by:

– The balance of energy fluxes at the interface $z = 0$

$$\kappa \frac{\partial T}{\partial z}\bigg|_{z=0} = \kappa_s \frac{\partial T_s}{\partial z}\bigg|_{z=0} - I_a(r). \tag{9.5.2}$$

For a Gaussian laser beam $I_a(r) = I(0)(1-R)\exp(-r^2/w_0^2)$.
- The continuity of temperatures at $z = 0$

$$T(r,0,t) = T_s(r,0,t). \tag{9.5.3}$$

- The temperature rise at infinity shall vanish

$$T(r \to \infty, z \to \infty, t) = T_s(r \to \infty, z \to -\infty, t) = T(\infty). \tag{9.5.4}$$

With steady-state conditions we obtain the same solution as in (7.2.4), except that the center temperature rise on the substrate surface, $\theta_c = \Delta T_c \equiv \Delta T_s(0,0)$, is now

$$\theta_c = \Delta T_c = \frac{P(1-R)}{2\sqrt{\pi} w_0 \kappa_s} \left(1 + \frac{\kappa}{\kappa_s}\right)^{-1}. \tag{9.5.5}$$

The physical reason for the additional term κ/κ_s can easily be understood. With stationary conditions, the heat flux from the surface is simply shared between the ambient medium and the substrate as $J:J_s = \kappa I_a/(\kappa + \kappa_s) : \kappa_s I_a/(\kappa + \kappa_s) = \kappa : \kappa_s$. Note that (9.5.5) coincides with (7.1.4) if $\kappa/\kappa_s \ll 1$. This is certainly a good approximation with gases at low to medium pressures. Thus, in the absence of convection, the temperature rise induced on a semi-infinite substrate changes only little by the presence of a gaseous atmosphere.

If convection becomes important, (9.5.5) cannot strictly be applied. This is the case with gases of medium to high pressures and with liquids and "long" laser pulses (Sect. 9.5.3). In general, the substrate temperature will be decreased by convection.

Some further comments seem to be appropriate: In Sect. 6.1 we have introduced the surface conductance η. This permits one to describe, phenomenologically, the influence of an ambient medium on the substrate temperature by solving the (single) heat equation for the substrate. If we consider energy transport by heat conduction only, a simple consideration of the energy flux yields $\eta \approx \kappa/w_0$. This can be proved by comparing (9.5.5) and (7.1.3). The second term in the parenthesis of (7.1.3) is of the order of $\eta^* = \eta w_0/\kappa_s$. Thus, if $\kappa/\kappa_s \ll 1$, (9.5.5) and (7.1.3) yield the same result if $\eta = \kappa/w_0$. It should be noted, however, that (7.1.3) also permits one to estimate the influence of convection, if the appropriate form of η is known (Sect. 6.1).

9.5.2 Indirect Heating

For *cylindrical* symmetry the stationary temperature field induced within a *non-absorbing* ambient medium due to heat transport from the laser heated solid surface, $T(r,z)$, can be approximated by (3.5.6) where ΔT_c is now given by (9.5.5).

9.5 The Ambient Medium

For *spherical* symmetry (Fig. 3.4.1) the temperature distribution is, in the simplest approximation,

$$T(r) = T(\infty) + \Delta T_s \frac{r_D}{r}. \tag{9.5.6}$$

If the temperature dependence of κ is taken into account [see (3.3.22)], we obtain instead

$$T^*(r^*) = \left[1 + (T_s^{*m+1} - 1) \frac{1}{r^*} \right]^{1/(m+1)}, \tag{9.5.7}$$

With $T_s^* = T_s/T(\infty)$ and $r^* = r/r_D$. Equation (9.5.7) can directly be derived from the heat equation by employing a Kirchhoff transform. $T^*(r^*)$ is shown in Fig. 9.5.2 for different values of m. Clearly, the temperature distribution depends on the surface temperature T_s and it becomes flatter with increasing exponent m. With the approximations made, (9.5.7) describes temperature distributions in gases *and* liquids.

9.5.3 Free Convection

Consideration of convection is certainly essential if we are dealing with gases at medium to high pressures or with liquids. Free convection is described by

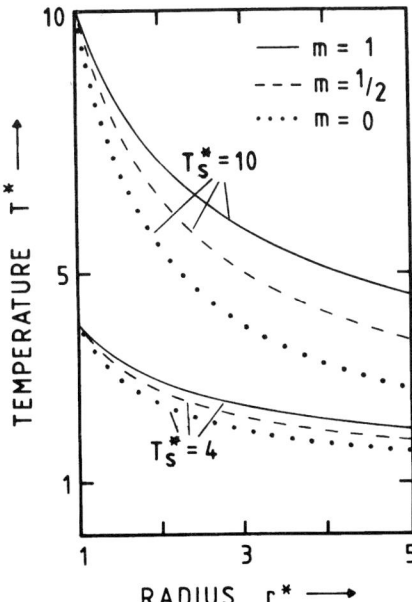

Fig. 9.5.2. Temperature distribution within a gaseous or liquid ambient medium, $T^*(r^*)$, for various temperature dependences of the thermal conductivity, $\kappa(T)$. Calculations have been performed for two surface temperatures $T_s^* = T^*(1)$

the set of (coupled) equations mentioned in the beginning of Sect. 3.3. It is evident that the overall problem can be solved only numerically and for each particular system (geometry, medium, etc.) only. For this reason, approximate solutions and even crude estimations are often quite useful. For example, the influence of free and forced convection on the substrate temperature can be estimated for many solutions given in Chaps. 7 to 9 if we approximate η by (6.1.8) and (6.1.9), respectively.

The temperature within the ambient medium can be described by the (simplified) heat equation

$$\rho c_p \frac{\partial T}{\partial t} - \nabla [\kappa(T) \nabla T] + \rho c_p v_c \nabla T = 0, \qquad (9.5.8)$$

where the third term describes heat transport by convection.

The velocity of the convective flow, v_c, can be described by [*Landau* and *Lifshitz*: Fluid Mechanics]

$$v_c \approx \frac{v_k}{l} f\left(\frac{x}{l}, \mathscr{P}\imath, \mathscr{G}\imath\right), \qquad (9.5.9)$$

where $v_k \,[\text{cm}^2/\text{s}]$ is the kinematic viscosity and l a characteristic length which has the dimension of the hot zone. Depending on the particular problem under consideration, l is given by the size of the laser focus, the size of the laser-processed structure, etc. For spherical symmetry $l \approx r_D T_c / T(\infty)$ (Fig. 3.4.1). The Prandtl number $\mathscr{P}\imath = v_k/D$ describes the properties of the medium where $D = \kappa / \rho c_p$.

Free convection is often characterized by the (dimensionless) Grashof number

$$\mathscr{G}\imath = g \beta_T l^3 \frac{\Delta T}{v_k^2}, \qquad (9.5.10)$$

where g is the acceleration due to gravity, β_T the coefficient of thermal expansion of the gas or liquid, and ΔT the temperature difference. For ideal gases with constant pressure

$$\beta_T = -\frac{1}{\rho} \frac{\partial \rho}{\partial T}\bigg|_p = \frac{1}{T} \propto p, \qquad (9.5.11)$$

where ρ is the (mass) density of the gas. If $\mathscr{G}\imath \to 0$ the function f in (9.5.9) tends to zero.

Instead of the Grashof number one often introduces the Rayleigh number

$$\mathscr{R}a = g \beta_T l^3 \frac{\Delta T}{D v_k} = \mathscr{G}\imath \times \mathscr{P}\imath. \qquad (9.5.12)$$

Note the strong dependence of $\mathscr{G}\imath$ and $\mathscr{R}a$ on the characteristic length, l. The viscosity of gases is about equal to the thermal diffusivity, $v_k \approx D$, and inversely proportional to the total pressure, $v_k \propto 1/p$, so that $\mathscr{R}a \approx \mathscr{G}\imath$.

9.5 The Ambient Medium

At $T \approx 300\,\mathrm{K}$, typical values of the kinematic viscosity are $v_\mathrm{k} \approx 0.15\,\mathrm{cm^2/s}$ for air, and $v_\mathrm{k} \approx 0.01\,\mathrm{cm^2/s}$ for water. The Prandtl number is for air $\mathscr{P}\mathnormal{r} \approx 0.73$, for water ≈ 6.75, glycerol ≈ 7250, and for Hg ≈ 0.044.

The heat flux into the medium is often described by

$$J_\mathrm{c} = \kappa \frac{\Delta T}{l} \mathscr{N}(\mathscr{G}\mathnormal{r}, \mathscr{P}\mathnormal{r}) = \eta \Delta T, \qquad (9.5.13)$$

where $\mathscr{N}$ is the Nusselt number. For $\mathscr{N} = 1$ heat transport takes place by conduction only. The function f in (9.5.9) and the Nusselt number $\mathscr{N}$ depend on the geometry of the problem. η is the surface conductance introduced in Sect. 6.1 and $\Delta T = T_\mathrm{s} - T(\infty)$. By comparing single terms in (9.5.8) the influence of heat transfer caused by convection and conduction can be estimated from $v_\mathrm{c}^* = \rho c_\mathrm{p} v_\mathrm{c} T l^{-1}/l^{-1} \kappa T l^{-1} = v_\mathrm{c} l / D$ where we have used $\nabla T \approx T/l$.

With *stationary* conditions and $-\nabla T \parallel \boldsymbol{g}$, heat transport by convection can be ignored as long as the Rayleigh (Grashof) number stays below a critical value, $\mathscr{R}a < \mathscr{R}a^\mathrm{cr} \approx 10^3$.

If, however, ∇T has a component $\perp \boldsymbol{g}$, the liquid is unstable and convection starts with all $\mathscr{R}a > 0$. For the stationary case, v_c can be estimated from the balance between buoyancy and (Stokes) friction forces within the liquid

$$\Delta \rho \, l^3 g \approx \rho l^2 v_\mathrm{k} \frac{v_\mathrm{c}}{l}.$$

With (9.5.11) and (9.5.12) this yields

$$v_\mathrm{c}^* = \frac{v_\mathrm{c} l}{D} = \frac{g \beta_\mathrm{T} l^3 \Delta T}{D v_\mathrm{k}} = \mathscr{R}a. \qquad (9.5.14)$$

These estimations show that the influence of convection becomes significant with Rayleigh numbers $\mathscr{R}a > 1$.

For *transient* laser heating, free convection becomes effective only with longer illumination times. The time constant involved can be estimated as follows: If we ignore viscosity, the acceleration in the initial phase of heating is

$$\frac{dv_\mathrm{c}}{dt} \approx \frac{\Delta \rho}{\rho} g = \beta_\mathrm{T} g \Delta T.$$

For a pulse length τ_l this yields

$$v_\mathrm{c} \approx \beta_\mathrm{T} g \Delta T \tau_l. \qquad (9.5.15)$$

The influence of free convection on the temperature distribution can be ignored if $v_\mathrm{c}^* \ll 1$. This if fulfilled for low laser pulse repetition rates and pulse lengths $\tau_l \ll D/\beta_\mathrm{T} g l \Delta T = l^2 / v_\mathrm{k} \mathscr{R}a$. Note that in liquids the influence of convection on mass transport (Sect. 3.4.2) may become effective earlier because $D_\mathrm{AB} \ll D$.

A convective flow can be laminar or turbulent. Turbulence is observed with Grashof numbers, typically, $\mathscr{G}\mathnormal{r} > 5 \times 10^4$. The influence of convection on the temperature distribution is sometimes estimated by substituting κ and D by other phenomenological parameters, κ' and D', that contain – besides κ and D – convective terms κ_c and D_c respectively [*Levich* 1962].

The effect of convection in laser-chemical processing can often be diminished by changes in the irradiation geometry or in the geometry of the reaction chamber. Estimations of temperature- and velocity-profiles near heated spots and lines [*Fujii* et al. 1973] and near heated cylinders [*Kuehn* and *Goldstein* 1980] have been performed. Temperature profiles near heated wires immersed in various gaseous atmospheres have been measured with high spatial resolution by employing Raman spectroscopy [*Leyendecker* et al. 1983a]. From these results we conclude that in laser microchemical gas-phase processing at low to medium pressures, the effect of free convection is small or negligible.

9.5.4 Temperature Jump

If the mean free path of molecules, λ_m, becomes comparable or larger than the size of the heated zone, r_D, a discontinuity (jump) in temperature at the interface between the laser-processed zone and the ambient medium occurs. Let us consider this in further detail for spherical symmetry (Fig. 3.4.1). If $\lambda_m \ll r_D$ we can ignore any temperature jump. If, however, $\lambda_m \gtrsim r_D$ the temperature at the laser-heated surface, $T_s(r_D)$, is *not* equal to the temperature of the gas at this surface, $T_G(r_D)$, but given by [*Smoluchowski* 1911]

$$T_s(r_D) = T_G(r_D) - g_T \frac{\partial T_G}{\partial r}\bigg|_{r_D}, \qquad (9.5.16)$$

where g_T is the temperature discontinuity coefficient which is inversely proportional to the gas pressure and thus directly proportional to the mean free path

$$g_T = \beta \lambda_m \propto \frac{1}{p}. \qquad (9.5.17)$$

Here, β is a factor of the order of unity which depends on the accomodation coefficient [*Landau* and *Lifshitz*: Physical Kinetics]. For rigid spheres, the mean free path can be described by

$$\lambda_m \approx \frac{1}{\sqrt{2}\sigma N} \propto \frac{1}{p}, \qquad (9.5.18)$$

where σ is the scattering cross section and N the number of gas molecules per volume, i.e., $N = p/k_B T$. For a two-component system consisting of molecules AB and M, the simplest approximation is $\sigma \approx \pi(r_{AB} + r_M)^2 \approx 4\pi r_{eff}^2$. Figure 9.5.3 shows λ_m as a function of temperature for various gas pressures and $r_{eff} = 2.5$Å. At low to medium pressures, λ_m becomes comparable or larger than typical

9.5 The Ambient Medium

dimensions of structures produced in laser microprocessing. In such cases, the temperature jump at the gas–solid interface cannot be ignored. For a temperature-independent thermal conductivity, the temperature distribution within the gas phase is then given by

$$T_G(r) = \frac{T_s(r_D) - T(\infty)}{1 + g_T/r_D} \frac{r_D}{r} + T(\infty). \tag{9.5.19}$$

Depending on the ratio g_T/r_D the temperature jump can significantly modify the gas-phase temperature. Its effect on the temperature distribution within the *substrate* can be ignored, in general. The heat flux from the solid surface to the gas is given by

$$J = \kappa_G \frac{T_s(r_D) - T(\infty)}{r_D + g_T}. \tag{9.5.20}$$

Thus, at a constant surface temperature T_s and radius r_D, the heat flux decreases with decreasing gas pressure.

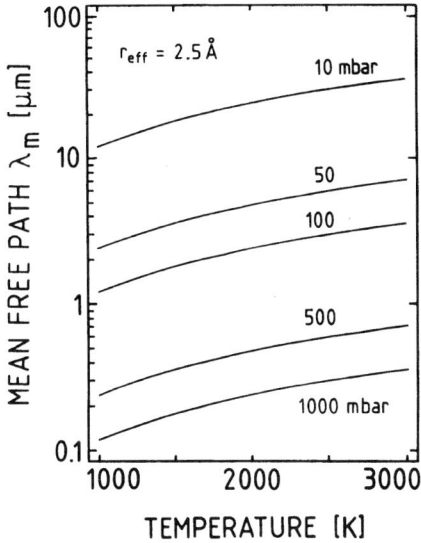

Fig 9.5.3. Mean free path of gas molecules as a function of gas-phase temperature for various pressures. The model of rigid spheres with an effective radius $r_{eff} = 2.5$ Å has been used

10 Surface Melting

Surface melting is involved in many types of *conventional* and *chemical* laser processing. Among the examples are surface homogenization, microstructure refinement, glass formation, sealing of porous materials, alloying, some types of surface hardening, selective laser sintering, surface cladding, many types of engraving and marking, laser welding, abrasive laser machining, some types of laser-chemical etching and deposition, most types of laser doping and synthesis.

With all kinds of laser-induced surface modifications that involve the diffusion of species or the mixing of material components, the processing rates are increased by several orders of magnitude when the surface melts. Sometimes, there are problems to achieve surface melting in a well-defined way. This originates mainly from temperature-dependent changes in the surface morphology and physical properties, and from various kinds of feedbacks. For many systems, the region of laser parameters that can be employed for controlled melting without significant surface damage is small. Surface warping and cracks resulting from mechanical stresses are frequently observed.

Clearly, material melting is a prerequisite in laser welding. The amount of energy required in abrasive laser machining such as drilling, cutting, and shaping is considerably smaller if the process is based on mainly liquid-phase expulsion instead of mainly vaporization.

Subsequently, we shall present simple models for melting and process optimization. Here, we consider laser-beam intensities that do not cause significant vaporization.

10.1 Temperature Distributions, Interface Velocities

Let us consider a semi-infinite substrate that is *uniformly* irradiated by a *single* rectangular laser pulse of duration τ_l (Fig. 10.1.1). Material evaporation shall be ignored.

The temporal behavior of the surface temperature, $T_s \equiv T(z=0)$, is schematically shown in Fig. 10.1.2. T_s reaches the (equilibrium) melting point, T_m, within a time $\tau_m(0)$. Subsequently, T_s increases only slightly. This is related to the enthalpy of melting, ΔH_m, which must be delivered by the absorbed laser light. After a time, $\tau_m(h_l)$, the surface is molten to a depth h_l. T_s increases

10.1 Temperature Distributions, Interface Velocities

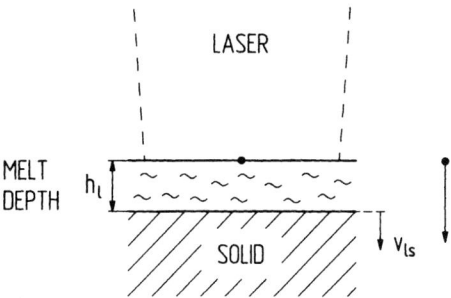

Fig. 10.1.1. Large-area laser-induced surface melting. The melt depth is denoted by h_l and the velocity of the liquid–solid interface by v_{ls}

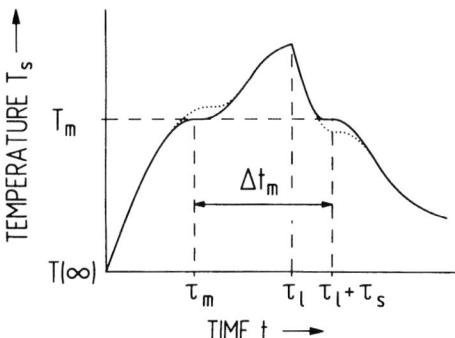

Fig. 10.1.2. Temporal dependence of the surface temperature induced by a single rectangular laser pulse of duration τ_l. τ_m is the time to reach the melting temperature, T_m. τ_s is the time of resolidification

further when the melt front matches the heat front. The maximum in T_s is reached at τ_l. For times $t > \tau_l$, the system cools and resolidifies within a time τ_s. In many types of laser material processing, the heating and cooling cycles are so short that, except with metals, overheating of the solid-phase and undercooling of the liquid-phase is significant (dotted curves). Clearly, the exact temporal behavior of T_s depends on the material under consideration, on the intensity, duration and shape of the laser pulse, on convective flows and, if relevant, on the type of the ambient medium. For example, with laser-induced melting of metals within an oxidizing atmosphere, the maximum in T_s is shifted to times $t > \tau_l$ due to the release in exothermal energy. In many applications the time during which the surface is molten, Δt_m, is of particular interest. This time depends on τ_m and thereby on the melt depth required in the process.

The solution of the general problem, i.e., the calculation of the total spatial and temporal behavior of the temperature distribution within the liquid and the solid material can be obtained only numerically. Here, one can solve the heat equation in the form (2.2.9) for the solid- and liquid-phase and include the heat of melting into the boundary condition for the moving interface. Alternatively, one can solve the (single) heat equation (2.2.14) for the whole temperature region and include the heat of melting in $c_p(T)$. For $T = T_m$ one can set $c_p = \Delta H_m \delta(T - T_m)$. In numerical calculations the δ-function is often approximated by a Gaussian profile.

Analytical solutions of the melting problem are possible only in a very few cases, as for example with uniform surface heating and special boundary conditions [*Carslaw* and *Jaeger* 1988; *Crank* 1988]. For this reason, crude estimations of the time, τ_m, the threshold intensity for surface melting, I_{th}, and the melt depth, h_l, are often quite useful. In a simple way, these quantities can be estimated from the solutions presented in Chaps. 7 and 8. For example, if we set for the laser-induced temperature rise

$$\Delta T(x) = T_m - T(\infty) , \qquad (10.1.1)$$

we obtain for $x = 0$ the intensity I_{th}. The time $\tau_m(h_l)$ can be estimated in a similar way.

The time $\tau_m(h_l)$

For large-area irradiation and finite absorption with $l_\alpha \gg l_T$ we find from (7.5.6)

$$\tau_m \equiv \tau_m(0) \approx \frac{\theta_m \kappa_s l_\alpha}{D I_a} , \qquad (10.1.2a)$$

with $\theta_m = T_m - T(\infty)$. The time $\tau_m(h_l)$ required to melt a layer of thickness $h_l = l_\alpha$ can be obtained by substituting T_m by $T_m^{eff} = T_m + \Delta H_m/c_p$.

For the case of surface absorption, i.e., with $l_\alpha \ll l_T$, we obtain from (7.5.3)

$$\tau_m \equiv \tau_m(0) \approx \frac{1}{D}\left(\frac{\theta_m \kappa_s}{I_a}\right)^2 . \qquad (10.1.2b)$$

For $h_l = l_T = 2(D\tau_m)^{1/2}$, the time $\tau_m(h_l)$ can again be estimated by substituting θ_m by θ_m^{eff}.

Maximum melt depth

For fluences around the melting threshold, i.e., with $\phi \geq \phi_{th} = I_{th}\tau_l \equiv \phi_m$, the maximum melt depth near T_m can be estimated in analogy. Here, the liquid–solid interface is considered as an isotherm which can be calculated from (10.1.1) by substituting T_m by T_m^{eff}.

Let us assume large-area irradiation (Fig. 10.1.1) and surface absorption. The solution (7.5.8) yields for fluences near the melting threshold $\phi \approx \phi_m$ the approximate relation

$$h_l^{max} \approx \frac{l_T}{\sqrt{\pi}} \frac{\phi - \phi_m}{\phi_m} , \qquad (10.1.3a)$$

where ϕ_m is the fluence necessary for surface melting

$$\phi_m = \frac{\sqrt{\pi}}{2} \kappa_s \theta_m \left(\frac{\tau_l}{D}\right)^{1/2} .$$

10.1 Temperature Distributions, Interface Velocities

For somewhat higher fluences, $\phi > \phi_m$, we can approximate (7.5.8) by

$$\Delta T(h_l^{max}, \tau_l) = \Delta T_{max} \exp\left(-\frac{h_l^{max2}}{l_T^2}\right) = T_m^{eff} - T(\infty),$$

where $\Delta T_{max} = I_a l_T / \sqrt{\pi} \kappa_s$ and $l_T = 2(D\tau_l)^{1/2}$. With $\Delta T \propto \phi$ this yields for $\phi \geqslant \phi_m$

$$h_l^{max} \approx l_T \left[\ln\left(\frac{\phi}{\phi_m}\right)\right]^{1/2} \approx l_T \left(\frac{\phi - \phi_m}{\phi_m}\right)^{1/2}. \tag{10.1.3b}$$

For $\phi < \phi_m$ we obtain $h_l^{max} = 0$. The latter approximation in (10.1.3b) is obtained for $\phi - \phi_m \ll \phi_m$.

It should be emphasized that such a description of the nonlinear melting problem by linear solutions is a crude approximation only.

For fluences well above the melting threshold, i.e., with $\phi > \phi_m$, the simplest estimation of the maximum melt depth is based on the *energy balance*

$$h_l^{max} \approx \frac{P(1-R) - P_L}{F\rho\Delta H} \tau_l \leqslant \frac{I_a}{\rho\Delta H} \tau_l \leqslant \frac{\phi_a}{\rho\Delta H}. \tag{10.1.3c}$$

R is the reflectivity and F the area of the molten surface. ΔH is the total enthalpy, i.e., $\Delta H = \Delta H_m + c_p[T_m - T(\infty)]$. P_L describes the energy loss by heat conduction and thermal radiation. The latter relation refers to very short high intensity laser pulses where P_L can often be ignored.

Both regimes, $h_l^{max} \propto (\phi - \phi_m)^n$ with $n \leqslant 1$ and $h_l^{max} \propto \phi$, have been verified in various experiments and with different materials. Figure 10.1.3 shows experimental data for Si. The linear regime (open circles) is clearly visible. The accuracy of data for fluences $\phi \approx \phi_m$ is not good enough to derive any functional dependence.

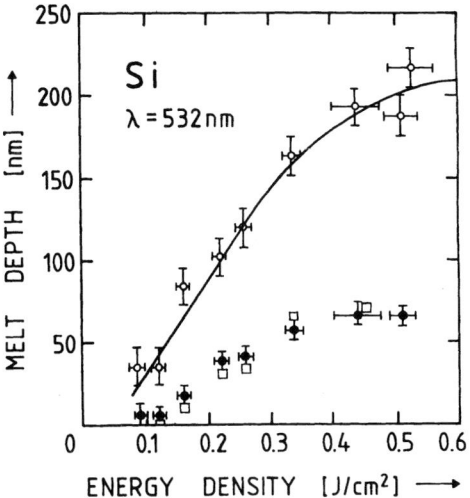

Fig. 10.1.3. Maximum melt depth in (100) Si as a function of Nd:YAG-laser fluence ($\lambda = 532$ nm, $\tau_l = 7.5$ ns). The data (open circles) have been derived from concentration profiles of Cu^+ ions implanted into the surface. The full curve has been calculated. Full circles and open squares indicate the position of the buried Cu peak derived from RBS and reflectivity measurements, respectively. This peak indicates the position where solidification fronts starting from the liquid–solid interface and from the surface collide (see also Sect. 10.2) [*Bruines* et al. 1986]

10.1.1 Boundary Conditions

Subsequently, we discuss different types of boundary conditions employed for solving the heat equation.

The Stefan problem

Within the frame of the Stefan problem, it is assumed that the temperature at the liquid–solid interface, $z = h_l(t)$, is continuous and equal to the melting temperature

$$T(z = h_l(t)) \equiv T_l(h_l) = T_s(h_l) = T_m \, . \tag{10.1.4}$$

The velocity of the liquid–solid interface, $v_{ls} = \dot{h}_l$, is described by

$$v_{ls}(t) \rho \Delta H_m - \kappa_s \frac{\partial T}{\partial z}\bigg|_{h_l} = J_{ls} \, . \tag{10.1.5}$$

Here, $\Delta H_m [\mathrm{J/g}]$ is the latent heat of melting. κ_s and κ_l are the thermal conductivities of the material in the solid and the liquid phase, respectively. J_{ls} is the energy flux from the liquid to the interface $z = h_l$. If convection within the liquid is ignored, J_{ls} is given by

$$J_{ls} = -\kappa_l \frac{\partial T}{\partial z}\bigg|_{h_l} .$$

Exact solutions of the Stefan problem can be found only in special cases; among those are the Neumann solutions.

In the presence of strong convective flows (Sect. 10.4) the temperature within the liquid is almost uniform, $T_l \approx T_m$ (note, however, that convection is only driven by temperature gradients). If we assume that the total energy absorbed at the liquid surface is (instantaneously) transported by convection to the interface, we can set $J_{ls} = I_a$. We then obtain

$$v_{ls} \rho \Delta H_m - \kappa_s \frac{\partial T}{\partial z}\bigg|_{h_l} = v_{ls} \rho \{ \Delta H_m + c_p [T_m - T(\infty)] \} = I_a \, ,$$

which yields

$$v_{ls} = \frac{I_a}{\rho \Delta H} \, .$$

This solution is equal to (10.1.3c). The temperature distribution for this case is schematically shown in Fig. 10.1.4 by the full curve.

If the laser light is absorbed at the liquid surface, and if we ignore convection, there is no analytical solution of the Stefan problem. In this case we have in addition to (10.1.4) and (10.1.5) the boundary condition

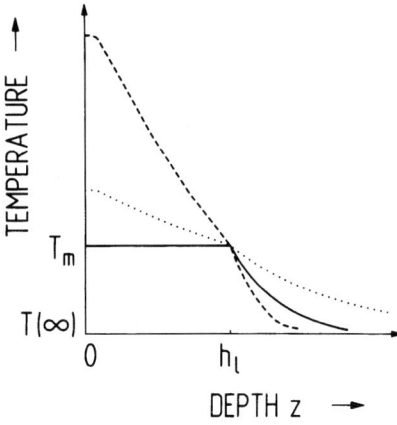

Fig. 10.1.4. Schematic picture of temperature distributions in laser surface melting. T_m is the melting temperature, z the distance from the surface, and h_l the melt depth. The full curve represents the temperature distribution in the presence of strong convection. The dotted curve refers to long laser pulses with $l_T \ll h_l$. The dashed curve refers to short high-intensity laser pulses which cause strong overheating

$$-\kappa_l \frac{\partial T}{\partial z}\bigg|_{z=0} = I_a \,.$$

Qualitatively, the temperature distribution will be similar to that shown by the dashed and dotted curves in Fig. 10.1.4.

The Stefan model is often employed to describe melting and solidification in cases where the interface velocities are very small compared to the sound velocity, so that overheating (undercooling) at the liquid–solid interface can be ignored. From a physical point of view, this model is wrong, because it assumes that the phase transition takes place instantaneously.

In reality, the temperature at the liquid–solid interface, T_i, should exceed the equilibrium melting temperature, i.e., $T_i > T_m$ (overheating), while in the case of solidification we should have $T_i < T_m$ (undercooling).

Kinetic model (Frenkel–Wilson law)

For the case of melting (solidification) with "moderate" overheating (undercooling) the velocity of the liquid–solid interface can be described by the difference in rate constants for melting, k_{sl}, and solidification, k_{ls}, i.e.,

$$v_{ls} = k_{sl} - k_{ls} \,. \tag{10.1.6}$$

Melting of a solid is a thermally activated process which can be described by

$$k_{sl} = k_{sl}^0 \exp\left(-\frac{\Delta E_{sl}^a}{k_B T_i}\right), \tag{10.1.7}$$

where k_{sl}^0 is a preexponential factor, ΔE_{sl}^a the activation energy per atom for melting. The situation is schematically shown in Fig. 10.1.5 for quasi-equilibrium conditions. In analogy, solidification can be described by

$$k_{ls} = k_{ls}^0 \exp\left(-\frac{\Delta E_{ls}^a}{k_B T_i}\right). \tag{10.1.8}$$

Clearly, the melting enthalpy per atom is given by $\Delta H_m^a = \Delta E_{sl}^a - \Delta E_{ls}^a$. Because $v_{ls} = 0$ at $T_i = T_m$, i.e., $k_{sl} = k_{ls}$, the velocity of the melt front can be described by

$$v_{ls} = v_0 \left[1 - \exp\left(-\frac{\Delta H_m^a}{k_B T_m} \frac{\Delta T_i}{T_i}\right)\right] \exp\left(-\frac{\Delta E_{sl}^a}{k_B T_i}\right). \tag{10.1.9}$$

Here, we have introduced the velocity $v_0 \equiv k_{sl}^0$ and $\Delta T_i = T_i - T_m$. Equation (10.1.9) is often termed the Frenkel–Wilson law and applies to homogeneous melting. During solidification $T_i < T_m$ and the velocity of the solid–liquid interface, v_{sl}, can be described by the same equation with $v_{sl} = -v_{ls}$ and the direction of the z-axis in Fig. 10.1.1 to be reversed. This law can well be applied to laser melting of semiconductors with interface velocities of 10^3 to 10^4 cm/s.

If the degree of overheating is small, we can expand (10.1.9) near $T_i \approx T_m$. This yields

$$v_{ls} \approx \tilde{v}_0 \frac{\Delta H_m^a}{k_B T_m^2} \Delta T_i \approx \tilde{v}_0 \frac{c_p \Delta T_i}{3 \Delta H_m [J/g]}. \tag{10.1.10}$$

This approximation can be employed with metals where, typically, $\tilde{v}_0 \equiv k_{sl}(T_m) \approx 10^4$ cm/s. The latter approximation is obtained with ΔH_m^a [J/atom] $= \Delta H_m [J/g] M/L \approx k_B T_m$ and $c_p \approx 3 L k_B / M$ (Dulong–Petit law), where M is the atomic weight per mol and L the Avogadro number. In analogy, (10.1.10) can also be applied to solidification.

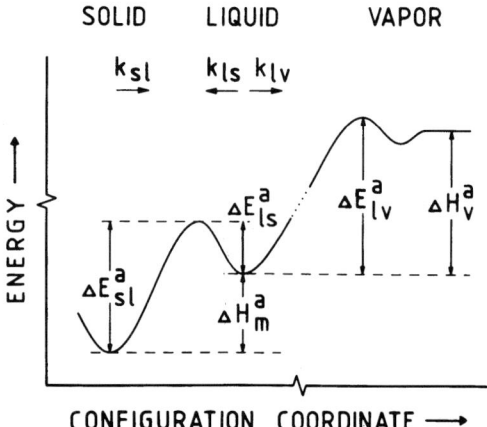

Fig. 10.1.5. Schematic picture to describe melting (solidification) and vaporization. ΔE_{sl}^a and ΔE_{ls}^a are the activation energies for melting and solidification; k_{sl} and k_{ls} are rate constants. The corresponding quantities for vaporization are included. ΔH_m^a and ΔH_v^a are the enthalpies (per atom/molecule) for melting and vaporization, respectively

Surface melting of Si

We now discuss numerical calculations for the example of a Si wafer. Here, the heat equation (2.2.14) has been solved together with (10.1.9) and

$$T_l(h_l) = T_s(h_l) = T_i$$

$$\left.\frac{\partial T}{\partial z}\right|_{z=0} = 0; \quad T(h_s, t) = T_0; \quad T(z, 0) = T_0. \tag{10.1.11}$$

Figure 10.1.6 shows temperature distributions calculated for different times during and after 532 nm Nd:YAG-laser irradiation. The arrows indicate the position of the liquid–solid interface. During the laser pulse, the solid surface is overheated (curve 1) and the heat front moves in z direction. The maximum melt depth is reached with $T_i = T_m$ where $v_{sl} = 0$. Subsequently, the liquid is undercooled due to the strong temperature gradient (curves 4 to 6). Due to solidification, the interface moves back towards the surface. The maximum in the temperature distribution that appears at the liquid–solid interface (curve 7) is related to the heat of crystallization. Note that this heat release takes place at $T_i \neq T_m$ where T_i depends on time. During solidification, the lowest temperature within the melt appears at the surface. Thus, nucleation in the undercooled

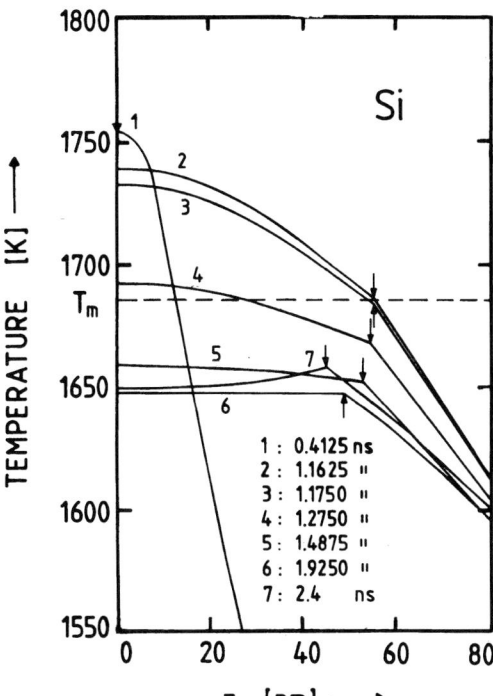

Fig. 10.1.6. Temporal development of temperature distribution during laser-induced surface melting and solidification of Si ($\lambda = 532$ nm Nd:YAG, $\phi = 0.3$ J/cm^2, $\tau_l = 0.5$ ns; $T_m = 1685$ K, $\Delta H_m = 0.495$ eV, $\Delta E_{sl} = 1.22$ eV, $v_0 = 6 \times 10^5$ m/s, and $T_0 = 150$ K). Only a *single* liquid–solid interface (arrows) was taken into account [*Stock* et al. 1985]

melt at the surface becomes quite likely. By this means, a second solidification front which starts from the surface may develop. This has been ignored in Fig. 10.1.6.

Strong overheating, localized irradiation

If the degree of overheating (undercooling) is very high, i.e., if $c_p \Delta T_i \geqslant \Delta H_m$, the Frenkel–Wilson law cannot be applied, mainly because of inhomogeneous melting (solidification) within the bulk of the solid (liquid) [*Motorin* 1983].

For high-intensity laser pulses and focused laser-beam irradiation, liquid-phase expulsion and vaporization become important. Here, the deformation of the surface (deep penetration melting, formation of droplets, etc.) must be taken into account.

10.1.2 Temperature Dependence of Parameters

At the melting temperature, some of the material parameters that enter the calculations, show discontinuities. Figures 10.1.7 show the behavior of $\kappa(T)$ and $c_p(T)$ for the example of Si and Cu.

Experimental data on $\kappa(T)$, $D(T)$, and $c_p(T)$ are listed for a large number of materials in Table II [see also, e.g., *Duley* 1976].

The reflectivity of liquid metals is, in general, smaller than for solid metals, $R_l < R_s$. For semiconductors, such as Ge and Si, the situation is opposite.

10.2 Solidification

For many applications it is a good approximation to consider materials processing only during the time Δt_m, where the surface is molten (Fig. 10.1.2). The estimation of this time is therefore very important. Precise calculations can be performed only numerically by solving the correct boundary-value problem. Figure 10.2.1 shows the results of such calculations for the example of Si. Here, the same boundary conditions and parameters as in Fig. 10.1.6 have been employed. The maximum melt depth is reached after about 0.8 ns. Without surface nucleation, solidification occurs exclusively at the liquid–solid interface (dashed curve). The interface velocity (derivative in Fig. 10.2.1) is first very high and then slows down. If surface nucleation is taken into account (full curves), a solidification front starts from the surface and both interfaces collide after about 10 ns at a depth of about 15 nm. While these details refer, of course, to the special system under consideration, the overall behavior shown in the figure is very typical for material solidification.

10.2 Solidification

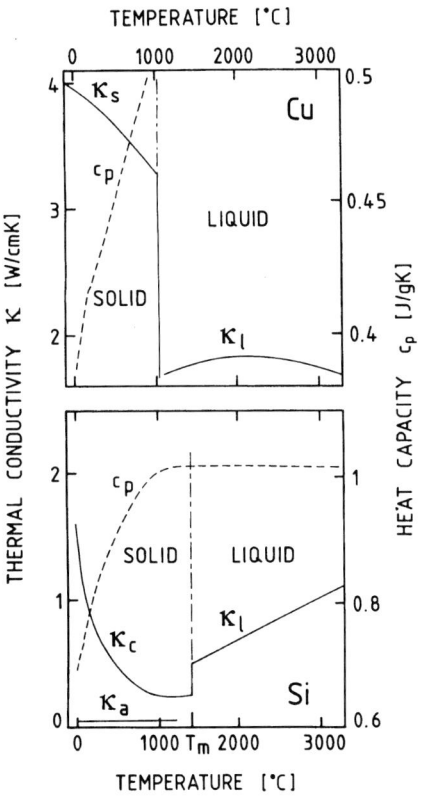

Fig. 10.1.7. Temperature dependence of the thermal conductivity, κ, and the specific heat, c_p. For Cu, κ_s and κ_l refer to the (crystalline) solid and the liquid phase. Adapted from [*Duley* 1976]. For Si, κ_c and κ_a denote thermal conductivities of crystalline and amorphous material. Adapted from [*Wood* and *Geist* 1986]

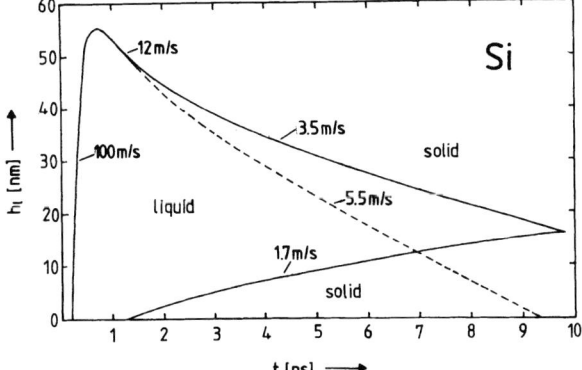

Fig. 10.2.1. Position of liquid-solid interfaces in Si versus time. Interface velocities (derivatives of curves) in different regions are indicated. For comparison, the path of one interface alone (dashed curve, no nucleation at the surface) is drawn. The parameters employed are the same as in Fig. 10.1.6 [*Stock* et al. 1985]

For Si, experimentally determined *solidification velocities* are of the order of 10 m/s; for metals they are of the order of 10^2 to 10^3 m/s.

Subsequently, we will present a simple estimation of Δt_m. Consider a laser pulse that induces large-area surface melting. With the definitions given in

Fig. 10.1.2 we find

$$\Delta t_m \approx \tau_l - \tau_m + \tau_s . \tag{10.2.1}$$

The time τ_m has already been estimated in Sect. 10.1. For *high-intensity* laser pulses that are long in comparison to τ_m, the approximation $\Delta t_m \approx \tau_l + \tau_s$ can often be employed. Then, we have to estimate only τ_s.

For *slow* solidification (in Fig. 10.2.1 for $t > 2$ ns) we can employ the Stefan model. Let us introduce a new time scale, $t' = t - \tau$, where τ is defined by the maximum melt depth, $h_l(\tau) = h_l^{max}$. For $t' > 0$ the thickness of the molten layer can then be described by

$$h_l(t') = h_l^{max} - h(t') . \tag{10.2.2}$$

Solidification shall start at a time τ where the liquid is at a (uniform) temperature, T_m. We assume that at this time the temperature distribution within the solid is characterized by the heat diffusion length $l_T \approx 2(D\tau)^{1/2}$ where $\tau \approx \tau_l$ and $T(z \to \infty) = T(\infty)$. The situation is described by the full curve in Fig. 10.1.4. Solidification ($t' \geq 0$) can then be described by a special Neumann solution of the Stefan problem where solidification starts at $t = 0$ at a depth $z' = h_l^{max} + \zeta l_T$ with $T_l(z \leq z') = T_m$ and $T_s(z \geq z') = T(\infty)$. For this case the solution is [*Carslaw* and *Jaeger* 1988]

$$h(t') = \zeta l_T \left[\left(1 + \frac{4Dt'}{l_T^2} \right)^{1/2} - 1 \right], \tag{10.2.3}$$

where ζ is given by

$$\zeta [1 + \mathrm{erf}(\zeta)] \exp(\zeta^2) \equiv f(\zeta) = \frac{c_p \theta_m}{\pi^{1/2} \Delta H_m} , \tag{10.2.4}$$

with $\theta_m = T_m - T(\infty)$. Physically reasonable values are within $0.25 < \zeta < 1$. From (10.2.3) the time of resolidification becomes

$$\tau_s = \frac{h_l^{max\,2}}{4\zeta^2 D} \left(1 + \frac{2\zeta l_T}{h_l^{max}} \right). \tag{10.2.5}$$

Because of the solidification front which starts from the surface, this time may be modified (Fig. 10.2.1). For fluences $\phi - \phi_m \ll \phi_m$ where h_l^{max} is very small, we obtain

$$\tau_s \approx \frac{l_T h_l^{max}}{2\zeta D} \approx \frac{l_T^2}{2\zeta D} \left(\frac{\phi - \phi_m}{\phi_m} \right)^{1/2} . \tag{10.2.6a}$$

The latter relation follows from (10.1.3b). Clearly, if we use instead (10.1.3a) we obtain $\tau_s \propto \phi - \phi_m$. For high fluences $\phi > \phi_m$ the melt depth h_l^{max} is large and the second term in (10.2.5) can be ignored, i.e.,

$$\tau_s \approx \frac{h_l^{max\,2}}{4\zeta^2 D} \approx \frac{(1-R)^2}{4D} \left(\frac{\phi - \phi_m}{\rho \Delta H \zeta} \right)^2 \approx \frac{1}{4\zeta^2} \frac{\phi^2}{\phi_m^2} \tau_l , \tag{10.2.6b}$$

10.2 Solidification

where we have substituted h_l^{max} from (10.1.3c) with $\phi_a = (1-R)(\phi - \phi_m)$. The latter approximation holds for $\phi_m \ll \phi$.

The functional dependence of τ_s expected from (10.2.6) was in fact observed experimentally. Figure 10.2.2 shows results for Q-switched Nd:Glass laser radiation and various semiconductor substrates. The inset shows the situation near the melting threshold where, in fact, τ_s can be approximated by $\tau_s \propto (\phi - \phi_m)^n$ with $0.5 \leq n \leq 1$. For large fluences the measured data can well be approximated by the parabolic law. If we consider the data for Si and 530 nm radiation with $\phi = 3 \text{ J/cm}^2$ we obtain from (10.2.6b) with $\zeta = 0.3$ for $\tau_s \approx 1450 \text{ ns}$. This value is in reasonable agreement with the experimental data.

The behavior of the solidification front velocity in the initial phase, i.e., immediately after the maximum melt depth has been reached, can qualitatively be understood from Fig. 10.1.4. For short laser pulses (dashed curve), the temperature gradient at $z = h_l$ is very large and the velocity v_{ls} will be very high. For long laser pulses (dotted curve) the temperature gradient is much smaller, and thus v_{ls}. By differentiation of $h(t')$ in (10.2.3) we obtain for times $t' \approx 0$ the (maximum) velocity

$$v_{ls}(t' \approx 0) \approx \frac{2\zeta D}{l_T} \approx \zeta \left(\frac{D}{\tau_l}\right)^{1/2}. \tag{10.2.7}$$

Note, however, that for very short laser pulses, temperature gradients within the liquid cannot be ignored.

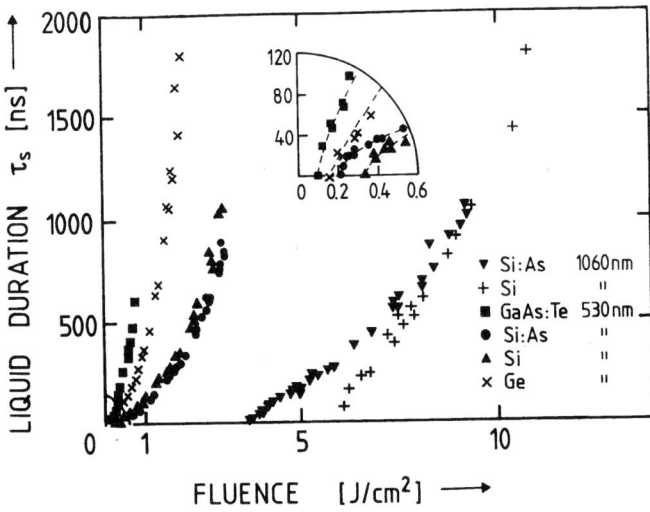

Fig. 10.2.2. Duration of the liquid phase at the surface of Si, Ge, and GaAs as a function of Nd:Glass-laser fluence ($\tau_l \approx 40$ ns; $\lambda = 1060$ nm and 530 nm) [*Auston* et al. 1979]

10.3 Process Optimization

For many technical applications the optimization of the surface melting process with respect to the energy required is of interest. For a certain melt depth, the smallest energy consumption can be found by minimizing the integral

$$\int_0^{\tau_l} P(t')\,dt' \to \text{minimum} . \qquad (10.3.1)$$

This is achieved, to a good approximation, if the maximum melt depth required, is of the order of the heat diffusion length. Thus, the condition to fulfill (10.3.1) is

$$h_l^{\max} \approx l_T \approx 2(D\tau_l)^{1/2}. \qquad (10.3.2)$$

Here, $h_l^{\max}$ is characterized by the effective melting temperature $T_m^{\text{eff}} = T_m + \Delta H_m/c_p$. Convection may strongly modify this condition.

The situation can most easily be understood from Fig. 10.1.4: From an energetical point of view, the full curve shows an almost optimal situation. The material is just molten up to the depth $h_l^{\max}$ and the heat front penetrates only slightly into the solid. Obviously, this idealized situation can never be achieved with realistic material parameters. For very long laser pulses with $l_T \gg h_l^{\max}$, there is a considerable waste of energy because the solid material is heated to a very large depth (dotted curve). For very short high-intensity laser pulses additional energy is consumed due to strong overheating at the surface (dashed curve). Thus, (10.3.2) is quite plausible. From a practical point of view, overheating of the surface or deep penetration heating may be advantageous or disadvantageous. A crude estimation of the optimized laser-beam intensities is obtained from (10.1.3c) and (10.3.2)

$$I_{\text{opt}} \approx \frac{2\rho\Delta H}{1-R}\left(\frac{D}{\tau_l}\right)^{1/2}, \qquad (10.3.3)$$

where $\Delta H \approx \Delta H_m + c_p T_m$.

For *finite* absorption and heat losses at the surface, the temperature distribution (in z-direction) is non-monotonic. Then, an estimation of the optimized laser-beam intensity becomes more complex.

10.4 Convection

Surface melting under the action of laser light may result in the excitation of convective fluxes within the liquid layer. Such convective fluxes play an important or even decisive role in material transport involved in many types of laser processing such as, e.g., in surface doping, surface alloying, etc.

With *uniform* laser-light irradiation, convection (mainly) originates from changes in material density related to temperature gradients in z-direction.

10.4 Convection

Free convection is characterized by the Rayleigh number $\mathcal{R}a$ (Sect. 9.5.3; here, l must be replaced by h_l).

With *focused* laser light, convective fluxes driven by changes in the surface tension of the material, σ, are dominating (Marangoni convection). If the changes in surface tension originate from gradients in the laser-induced temperature distribution along the surface, this phenomenon is also denoted as thermocapillary effect. The direction of the convective fluxes depends on the sign of $d\sigma/dT$. Marangoni convections frequently result in surface deformations, as schematically shown in Figs. 10.4.1 a, b.

The r-component of the velocity of the convective flow, v_c, related to the thermocapillary effect can be estimated from (Fig. 10.4.1)

$$\eta \frac{\partial v_c}{\partial z}\bigg|_{z=h_l(r)} = \frac{\partial \sigma}{\partial r} = \frac{d\sigma}{dT}\frac{\partial T}{\partial r}, \qquad (10.4.1)$$

and $v_c(z=0)=0$, which yields

$$v_c \approx \frac{h_l \Delta T}{\eta w_T}\frac{d\sigma}{dT}, \qquad (10.4.2)$$

where ΔT is the temperature rise, w_T the width of the temperature distribution, and $\eta = \rho v_k$ the dynamic viscosity. Typical values of v_c are between 1 cm/s and some 10 cm/s. For high laser-light intensities, v_c exceeds some critical value and the "laminar" convective flow loses its stability (Sect. 28.5).

The time for one convective cycle due to the thermocapillary effect can be estimated from $\tau_c \approx w_T/v_c$ and $h_l = 2(D\tau_c)^{1/2}$. With (10.4.2) this yields

$$\tau_c \approx \left(\frac{\eta w_T^2}{2D^{1/2}\Delta T}\frac{dT}{d\sigma}\right)^{2/3}. \qquad (10.4.3)$$

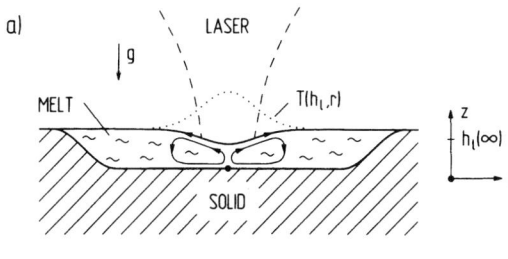

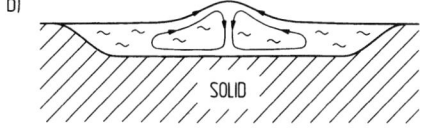

Fig. 10.4.1a,b. Convection due to localized melting. $h_l(\infty)$ is the melt depth away from the laser-irradiated zone and g the vector of gravity. The width of the molten layer is approximately given by the thermal width, w_T. (a) Surface tension decreases with increasing temperature, i.e., $d\sigma/dT<0$. (b) $d\sigma/dT>0$

Typical values of ΔT are 10^2 to 10^3 K. In a crude approximation we can set $d\sigma/dT \approx \sigma/T_{cr}$. For liquid metals, σ is of the order of 10^{-4} J/cm^2 and $\eta \approx 10^{-2}$ g/cm s. If we consider a situation similar to that shown in Fig. 10.4.1a with $w_T \approx 0.5$ cm we find that typical values of τ_c are between 10^{-3} and 10^{-5} s. It is evident that convective mixing is relevant only if $\tau_c < t_d = N\tau_l$ where t_d is the laser beam dwell time and N the number of laser pulses.

Another mechanism that excites convective flows is related to the recoil pressure due to laser-induced vaporization ($p_{rec} = p_{rec}(I)$; Sect. 11.1.4). Empirically, the time for one convective cycle can be estimated from (10.4.3) if we substitute $\Delta T \approx T_v$ and w_T by $w_s = w_s(I)$, which is a characteristic width of the surface perturbation that depends on the laser-light intensity. Typical values of τ_c are of the order of 10^{-6} s. The corresponding convection velocities are, typically, 10^2 to 10^3 cm/s. This mechanism is relevant to surface processing with µs CO_2 lasers.

Besides the instabilities already mentioned, other instabilities of the Rayleigh-Taylor or Kelvin-Helmholtz type (Sect. 28.5) can result in mixing times $\tau_c \leq 10^{-7}$ to 10^{-8} s. The typical velocities related to these instabilities are $v_c > 10^3$ cm/s.

10.5 Surface Deformations

In many cases, surface melting and convection results in surface deformations and in structural and chemical inhomogeneities of the resolidified material. The shape of the surface around the center of the laser beam is determined by the thermal expansion of the material and the surface tension of the liquid. If we assume the thickness of the molten layer to be small compared to the width of the laser focus, i.e., $h_l \ll 2w_0$, and ignore any temperature gradients in z-direction, we can estimate the shape of the surface from the solution of the corresponding hydrodynamical problem [*Landau* and *Lifshitz*: Fluid Mechanics]. In this approximation one obtains

$$h_l^2(r) \approx h_l^2(\infty) \left(\frac{\rho(\infty)}{\rho(r)}\right)^{3/4} + \frac{3}{g\rho(r)}[\sigma(r) - \sigma(\infty)], \qquad (10.5.1)$$

where $h_l(\infty)$ is the thickness of the molten layer and $\rho(\infty)$ the mass density for distances $r > w_0$. If we ignore thermal expansion, i.e., assume $\rho(r) \approx \rho(\infty)$, the thickness of the liquid layer within the beam center can be approximated by

$$h_l(0) \approx h_l(\infty) + \frac{3}{2\rho(\infty)gh_l(\infty)}[T(0) - T(\infty)]\frac{d\sigma}{dT}, \qquad (10.5.2)$$

where $T(0)$ is the center temperature (Fig. 10.4.1). For a rough estimation we can set $h_l(\infty) \approx l_T$, as before. The surface tension coefficient $\sigma(r) \equiv \sigma(T(r))$

decreases, in most cases, with increasing temperature as [*Hirschfelder* et al. 1964]

$$\sigma = \sigma_0 \left(1 - \frac{T}{T_{cr}}\right)^v, \qquad (10.5.3)$$

where σ_0 is a constant, T_{cr} the critical temperature, and $v \approx 1$. With $d\sigma/dT < 0$, equation (10.5.2) directly yields $h_l(0) < h_l(\infty)$. This is schematically shown in Fig. 10.4.1a.

It should be noted however, that with laser-light intensities generating significant material evaporation, the recoil of particles leaving the surface can produce a deformation similar to that depicted in the figure.

The situation may be even more complex. For example, in some cases of surface alloying, a surface deformation as shown in Fig. 10.4.1b is observed. This type of deformation can be obtained only if the direction of the flow changes sign due to the concentration dependence of σ. Consider the alloying of A and S. The concentration of A in S depends on temperature, i.e., $N_A = N_A(T)$. Thus, for certain types of species we may have

$$\frac{d\sigma}{dT} = \frac{\partial \sigma}{\partial T} + \frac{\partial \sigma}{\partial N_A} \frac{\partial N_A}{\partial T} > 0. \qquad (10.5.4)$$

The previous remarks apply also to cases where the laser beam is scanned with velocities $v_s^* = v_s w_0/D \ll 1$ perpendicular to the drawing plane in Fig. 10.4.1. Scanning of the laser beam changes the surface morphology mainly via the change in temperature distribution $\Delta T(v_s^*) = T(v_s^*) - T(v_s^* = 0)$ (Sect. 7.4). Changes in the hydrodynamics are of higher order in v_s^* and shall be ignored.

10.6 Welding

Lasers are used as a high-speed and high-quality welding tool. They permit autogeneous welding (without filler) at atmospheric pressure with little material distortion and contamination. The good localization of the temperature profile [small extention of heat affected zone (HAZ)] allows accurate and narrow welding even near temperature-sensitive material components. The lasers most commonly employed are cw CO_2 and Nd:YAG lasers. As with standard material welding, protection (shrouding) of the laser-processed zone, and of the laser optics, by an adequate inert gas is of importance. Laser welding can be classified according to two intensity ranges:

- With low to moderate laser-light intensities, the material is locally melted without significant vaporization. Heat transport is mediated via heat conduction and convection within the melt pool. This is schematically shown in Fig. 10.6.1a. This type of laser welding is denoted as *conduction-limited welding*. It is applied for low penetration (depth) welding.

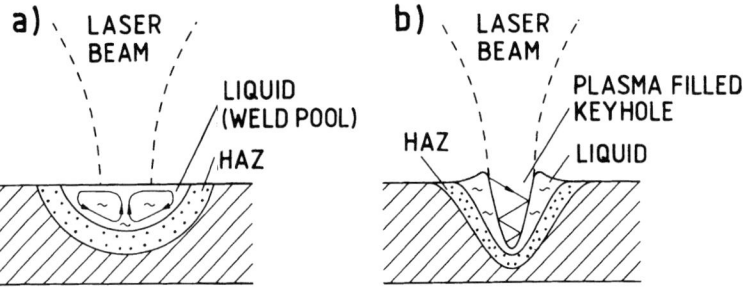

Fig. 10.6.1a,b. Models for the two intensity ranges of laser welding. HAZ stands for heat affected zone. (a) Conduction limited welding. (b) Deep penetration (keyhole) welding

– With laser-light intensities that cause sufficient material vaporization, a deep vapor cavity (keyhole) within the melt pool is formed (Fig. 10.6.1b). This is denoted as *deep penetration (keyhole) welding*. The incident laser light can penetrate deeply into the keyhole and is strongly absorbed due to multiple reflections and plasma-enhanced coupling (Chap. 11). Because the keyhole acts like a "light trap", this processing range is not very sensitive to the wavelength of the laser light. With a tight focus located below the material surface, very high aspect ratios, typically up to $\Gamma = h/d \approx 5$ to 20, can be achieved (h and d are the weld depth and width, respectively). The strong feedback between the amount of the absorbed laser light, the plasma, and the depth of the hole within the starting phase, may cause problems in process control. With optimized conditions, the material is welded through the whole thickness without any material dropout.

With a scanned laser beam, an egg-shaped melt region is formed (see, e.g., Fig. 8.1.1). The molten material ahead of the laser beam flows around the keyhole and solidifies behind it. The maximum welding speed can be estimated, in the simplest form, from the energy balance which yields

$$v_s^{max} \leqslant \frac{P(1-R) - P_L}{dh\rho\Delta H}, \qquad (10.6.1)$$

where $\Delta H \approx \Delta H_m + c_p T_m$. P_L corrects for heat losses by conduction and radiation. The penetration depth, h, increases with laser power and decreases with increasing scanning velocity. This is shown in Fig. 10.6.2 for stainless steel and CO_2-laser radiation.

For further details on laser welding the reader is referred to the extensive literature on this topic [*Steen* 1991; *Schuöcker* and *Kaplan* 1994].

10.7 Liquid-Phase Expulsion

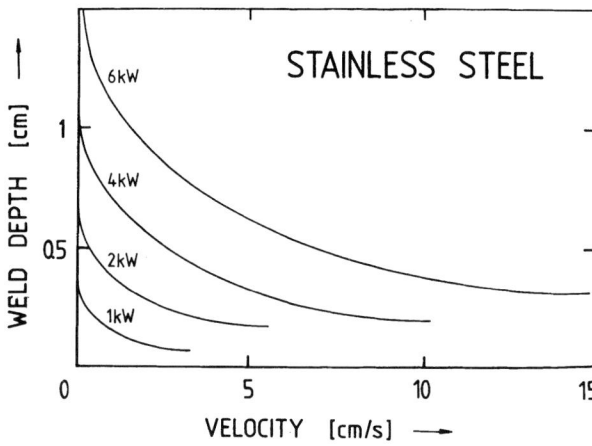

Fig. 10.6.2. Weld depth achieved in deep penetration (keyhole) welding of stainless steel (AISI 304) with a fast axial flow CO_2-laser as a function of scanning velocity. Adapted from [Industrial Laser Annual Handbook 1990]

10.7 Liquid-Phase Expulsion

Abrasive laser processing based on laser-induced melting requires *much* less energy than laser-induced vaporization. This can directly be verified from a simple estimation of the total enthalpy, ΔH, required in both cases (see Table IV). Thus, laser machining based mainly on laser-induced melting is to be preferred as long as the specifications of the particular application can be fulfilled. Efficient processing requires strong absorption of the laser radiation and, in the case of abrasive laser machining, fast removal of the liquefied material from the solid surface. The latter can be achieved in different ways:

- By the surface tension of the liquid (Sect. 10.5). This mechanism is of particular relevance for the fabrication of holes or direct writing of grooves in thin films or foils.
- By employing an ambient atmosphere which heavily reacts with the melt to a gaseous product.
- By using a high-pressure gas jet. This technique is denoted as liquid-phase expulsion. It is frequently employed in abrasive laser machining, in particular in metal processing (see also Sect. 11.3). The Mach numbers of the jet are, typically around 0.2. In many cases, the proper selection of a gas which exothermally reacts with the liquid, for example oxygen, permits one to not only compensate for gas-jet cooling but to even contribute to the overall energy input.

There are, of course, advantages and disadvantages of melt-phase processing (see also Sect. 10.3). In particular with metals, efficient coupling of CO_2-laser radiation to the substrate is mediated only via the plasma (Chap. 11). With the (high) laser-light intensities involved, the liquefied material is expelled by the recoil pressure of the vapor (Sect. 11.1).

Part III: Material Removal

Laser-induced material removal can be based on either *ablation* or *etching*. The basic mechanisms involved can be thermal, photophysical, or photochemical in nature.

The term laser-induced ablation, or simply laser ablation, is used if material removal can be performed, at least in principle, in vacuum or in an inert ambient medium. Thus, ablation can take place only if the laser light is directly absorbed by the material to be ablated.

Thermal ablation requires material melting and/or vaporization (or refers to materials that sublimate). At higher laser powers, interactions between the laser light and the plasma plume, and plasma–solid interactions become important. Laser-induced melting, vaporization, and plasma formation are the dominating mechanisms in conventional abrasive laser machining, in particular of metallic workpieces (Fig. 1.1.2). Here, the lasers most commonly employed are CO_2-lasers and Nd:YAG-lasers in either cw- or pulsed-mode operation. The dwell times and pulse widths used are, typically, between 10^{-7} and some 0.1 seconds (Chap. 11).

Photophysical and *photochemical* ablation mechanisms may become important if non-metals are subjected to short high-intensity UV-laser pulses obtained, e.g., from excimer lasers or frequency multiplied Q-switched Nd:YAG-lasers. Here, typical pulse lengths are between some 10^{-14} and 10^{-7} seconds. This range of pulsed-laser ablation is employed in micropatterning and thin film formation (Chaps. 12, 13).

Laser-induced chemical *etching* denotes thermal or non-thermal material removal in a *reactive* ambient medium (Chaps. 14, 15).

11 Vaporization, Plasma Formation

11.1 Vaporization

When the laser-light intensity is high enough to induce significant material vaporization, the density of species within the vapor plume strongly increases (Fig. 11.1.1). The vapor consists of clusters, molecules, atoms, ions, and electrons.

Species leaving the surface take along some kinetic and, eventually, internal energy. The binding energy required to remove an atom from a solid can be estimated from ΔH^a [J/atom] $= \Delta H$ [J/g]$/N_s \approx \Delta H_v$[J/g]$/N_s$ where ΔH_v is the enthalpy of evaporation; $N_s = L/M$ is the (atom) number density, L the Avogadro number, and M the atomic weight per mol. For fragments or clusters consisting of several atoms, the average energy per atom required for evaporation is smaller.

Thermalization of species leaving the surface is mediated via collisions within a few mean free paths, typically within some microns from the surface. This region is called the *Knudsen layer*. The strong temperature and pressure gradients in axial direction of the vapor plume – compared to the corresponding gradients in lateral direction – cause a strongly forward direction of the plume.

In the simplest model, the expansion of the vapor beyond the Knudsen layer is described by an adiabatically expanding gas [*Anisimov* et al. 1971]. Here, the temperature within the plume decreases with distance from the substrate surface. In vacuum, the expansion velocity increases with time up to some limiting value. In the presence of an ambient gas, the expansion velocity reaches some maximum and decreases afterwards. This hydrodynamic description does not consider any non-equilibrium effects which may become important with short high-intensity laser pulses. In this case, the species desorbed from the surface may have a *non*-Maxwellian velocity distribution which is "between" that for quasi-equilibrium thermal evaporation and that for a "monochromatic" molecular beam.

In any case, the species leaving the surface generate a recoil pressure onto the substrate. In the presence of a molten surface layer and with focused laser-beam irradiation, the recoil pressure expels, in part, the liquid (Fig. 11.1.1). The ablated material may also generate a shock wave (Sect. 30.3). The vapor plume will absorb and scatter the incident laser radiation.

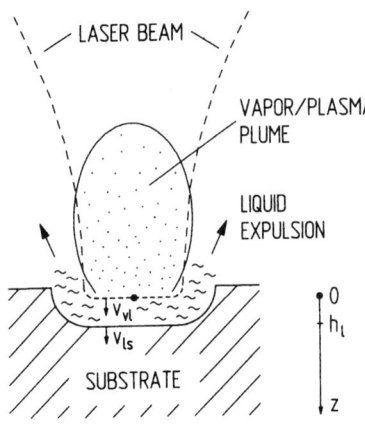

Fig. 11.1.1. Laser-induced surface melting and vaporization

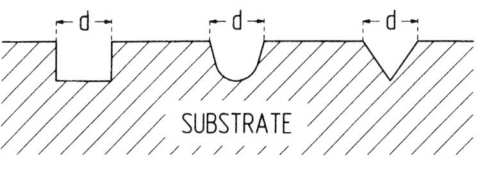

Fig. 11.1.2. Model structures for deep holes or grooves (oriented perpendicularly to the drawing). d is defined as the width at the substrate surface, while h is the maximum depth

In this section, we consider laser-light intensities which exceed the intensity for material vaporization, $I_v \equiv I_v(\lambda)$, but do not significantly ionize the vapor. For CO_2-laser radiation and metal substrates this intensity is within the range 10^4 to 10^6 W/cm^2. In this range, interactions of the laser light with the vapor can often be ignored. If this approximation is too crude, the laser-light intensity reaching the substrate can be described by an effective intensity which takes into account losses due to absorption and scattering within the plume. Local intensity changes due to distortions of the beam shape by the hot vapor plume can be taken into account in a similar way. With the intensities under consideration, the rate of material removal is determined by both evaporation *and* liquid-phase expulsion. In the presence of a reactive gas this rate can be enhanced by chemical reactions.

Subsequently, we will present simple models which permit one to estimate vaporization rates, mainly for flat structures whose depth is very small compared to their width, $h \ll d$. The problem becomes much more difficult for deep holes and grooves where $h \geqslant d$. Some typical cross sections which are observed in experiments, are shown in Fig. 11.1.2. The quantity $\Gamma \equiv h/d$ is often termed *aspect ratio*. With deep structures, calculations of the laser-induced temperature distribution must take into account the changes in geometry during the ablation/etching process. Such calculations can be performed only numerically.

11.1 Vaporization

11.1.1 Energy Balance

The simplest approximation to estimate processing rates, with both shallow and deep structures, is based on the energy balance.

Let us assume that material removal is governed by vaporization. The light energy absorbed can then be divided into bulk heating, melting, and vaporization of the material. For *surface absorption*, $l_\alpha \ll l_T$, and laser powers in excess of the threshold power for ablation, $P > P_{th}$, the depth, Δh, ablated during the dwell time of the laser beam, τ_l, can be estimated from

$$\Delta h \approx \frac{(P - P_s)(1 - R) - P_L}{F\rho\{\Delta H_v + \Delta H_m + c_p[T - T(\infty)]\}} \tau_l$$

$$\approx \frac{(\phi - \phi_s)(1 - R) - \phi_L}{\rho \Delta H_v}. \tag{11.1.1}$$

The power P_s corrects, in a heuristic way, for the shielding of laser light by the vapor plume. P_L includes all energy losses (heat conduction, radiation, convection, excess energy of the vapor, reaction enthalpies, etc.). $\phi\,[\text{J/cm}^2]$ are the corresponding fluences (energy densities). F is the area of the ablated structure. T is the temperature of the (stationary) interface. For the simplest estimation we can set $T = T_v$. Equation (11.1.1) ignores any temperature gradients within the heated material volume and temperature dependences in material parameters ρ and c_p. Within the present approximations, the total enthalpy, ΔH, can be replaced by the enthalpy of vaporization, $\Delta H_v\,[\text{J/g}]$ (Table IV). It is evident that (11.1.1) permits a simple estimation of Δh only for low to medium laser powers and *short* laser pulses for which, however, the condition $l_\alpha \ll l_T$ still holds. For laser powers that are around the threshold power for vaporization, the attenuation by the vapor plume is weak and we can set $P_s \approx 0$. Bulk heating is minimized if the heat diffusion length, l_T, is about equal to Δh. In this approximation we can set $P_L \approx 0$. Then, equation (11.1.1) suggests a linear increase in ablation rate with laser-light intensity, $\Delta h/\tau_l \approx I_a/\rho \Delta H_v$. With *long* irradiation times, however, the energy loss by conduction is significant, in particular for materials with a high thermal conductivity, such as metals. Then, Δh becomes nonlinearly dependent on intensity. Additional nonlinearities arise when laser–plasma–solid interactions become important.

11.1.2 One-Dimensional Model

For *shallow* structures, $h \ll d$, the surface temperature can be estimated from the one-dimensional heat equation. We ignore any liquid layer and use the simplified notation, $T_s \equiv T_{vs}$, for the temperature at the vapor–solid interface. In a reference frame that is attached to this interface and which moves with a velocity v_{vs} the heat equation has the form

$$\frac{\partial^2 T}{\partial z^2} + \frac{v_{vs}}{D}\frac{\partial T}{\partial z} = 0. \tag{11.1.2}$$

We employ the boundary conditions

$$-\kappa_s \frac{\partial T}{\partial z}\bigg|_{z=0} = I_a - \rho v_{vs}[\Delta H_t + c_p \Delta T(0)], \tag{11.1.3}$$

and

$$T(z \to \infty) = T(\infty), \tag{11.1.4}$$

where ρ is the mass density of the solid, and $\Delta H_t \approx \Delta H_m + \Delta H_v$ (Sect. 2.2). Any attenuation of the incident laser light by the vapor plume is ignored. Integration of (11.1.2) yields the temperature rise

$$\Delta T(z) = \Delta T(0) \exp\left(-\frac{v_{vs} z}{D}\right), \tag{11.1.5}$$

with

$$\Delta T(0) = T_s - T(\infty) = \frac{1}{2c_p}\left(\frac{I_a}{\rho v_{vs}} - \Delta H_t\right), \tag{11.1.6}$$

where $T_s \equiv T(0)$. In terms of the vaporization kinetics, the interphase velocity v_{vs} can be described by

$$v_{vs}(T_s) \approx v_0 \exp\left(-\frac{\mathscr{E}_v}{T_s}\right), \tag{11.1.7}$$

where v_0 is of the order of the sound velocity within the solid and $\mathscr{E}_v \equiv \Delta E_v/k_B$. The activation energy for vaporization can be approximated by the enthalpy of vaporization per atom/molecule, i.e., $\Delta E_v \approx \Delta H_v^a$ (see also Fig. 10.1.5). Equations (11.1.5) and (11.1.7) permit one to calculate the surface temperature, $T_s \equiv T(0)$, and the velocity v_{vs}. Figure 11.1.3 shows the normalized velocity $v_{vs}^* = v_{vs}/v_0$ as a function of the normalized intensity $I_a^* = I_a/(c_p \rho v_0 \mathscr{E}_v)$ for various values $C = [\Delta H_t/c_p - 2T(\infty)]/\mathscr{E}_v$. For conventional metal processing with CO_2- or Nd:YAG-lasers, $I_a^* \ll 1$. With short high-intensity pulses and materials with low activation temperature $\mathscr{E}_v$, the range $I_a^* > 1$ becomes relevant. With $[T_s - T(\infty)]c_p < \Delta H_t \approx \Delta H_v$, we obtain from (11.1.6) for low

11.1 Vaporization

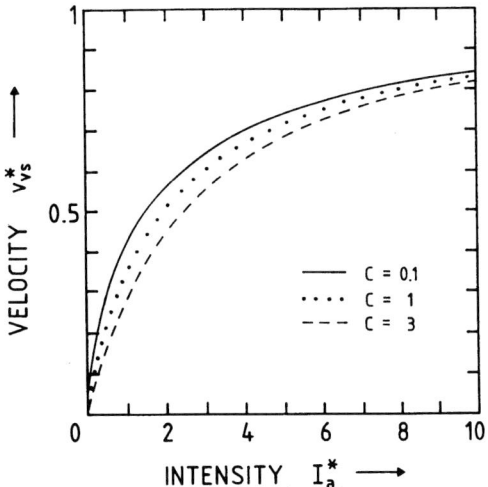

Fig. 11.1.3. Normalized interface velocity, v_{vs}^*, as a function of the normalized laser-beam intensity, I_a^*, for different values of C

laser-light intensities

$$v_{vs}^* \approx \frac{I_a^*}{C}. \tag{11.1.7a}$$

This is a good approximation for $I_a^* \leq 0.1\,C$. It is identical to the corresponding approximation obtained from the energy balance (11.1.1). For higher values of I_a^*, the velocity v_{vs} increases very slowly. However, due to plasma shielding, saturation may become effective at much lower values of I_a^*.

11.1.3 Minimum Intensity

In this subsection we will estimate the minimum absorbed laser-light intensity, $I_v^{\min}$, necessary to reach stationary material evaporation within the pulse length τ_l. Here, the velocity of the heat front, v_T, must be equal to the velocity of evaporation, v_{vs}, in (11.1.7), i.e.,

$$v_T = \frac{l_T}{\tau_l} = 2\left(\frac{D}{\tau_l}\right)^{1/2} \approx v_{vs}. \tag{11.1.8}$$

According to (7.5.3), the laser-induced surface temperature is

$$T_s(\tau_l) = T(\infty) + \frac{2I_a}{\sqrt{\pi}\kappa_s}\left(D\tau_l\right)^{1/2}. \tag{11.1.9}$$

From (11.1.7) to (11.1.9) we obtain

$$I_v^{\min} \approx \frac{\sqrt{\pi}}{2} \frac{\kappa_s \mathscr{E}_v}{D(1-R)} \left(\frac{D}{\tau_l}\right)^{1/2} \left(\zeta(\tau_l) - \frac{T(\infty)}{\mathscr{E}_v}\right), \qquad (11.1.10)$$

where

$$\zeta(\tau_l) = \frac{1}{\ln\left[\frac{v_0}{2}\left(\frac{\tau_l}{D}\right)^{1/2}\right]}.$$

For $\zeta \gg T(\infty)/\mathscr{E}_v$ we obtain with $\Delta H_v [\text{J/g}] \approx \mathscr{E}_v k_B L/M \approx c_p \mathscr{E}_v/3 \approx \kappa_s \mathscr{E}_v/3\rho D$

$$I_v^{\min} \approx \frac{\rho \Delta H_v}{1-R} \left(\frac{D}{\tau_l}\right)^{1/2} \zeta(\tau_l). \qquad (11.1.11)$$

This expression shows that $I_v^{\min}$ increases with decreasing laser-pulse length, while the corresponding fluence $\phi_v^{\min} = I_v^{\min} \tau_l$ decreases. For metals we obtain with $D(T_v) \approx 0.1$ cm^2/s, $\rho \Delta H_v \approx 40$ kJ/cm^3, $R(T_v) \approx 0.5$, $v_0 = 3 \times 10^5$ cm/s and a pulse length of $\tau_l = 10^{-2}$ s an intensity $I_v^{\min} \approx 2 \times 10^4$ W/cm^2; correspondingly, for $\tau_l = 10^{-8}$ s we obtain $I_v^{\min} \approx 6 \times 10^7$ W/cm^2. With this latter intensity, however, laser vapor/plasma interactions cannot be ignored, in general.

For *finite* absorption, $l_\alpha \gg l_T$, the surface temperature is

$$T_s(\tau_l) = T(\infty) + \alpha \frac{I_a}{\kappa_s} D\tau_l. \qquad (11.1.12)$$

With the same approximations as in (11.1.10), we obtain

$$I_v^{\min} = \frac{\kappa_s \mathscr{E}_v}{\alpha(1-R)D\tau_l} \left(\zeta(\tau_l) - \frac{T(\infty)}{\mathscr{E}_v}\right), \qquad (11.1.13)$$

or with $\zeta \gg T(\infty)/\mathscr{E}_v$

$$I_v^{\min} \approx \frac{\rho \Delta H_v}{\alpha(1-R)\tau_l} \zeta(\tau_l). \qquad (11.1.14)$$

Equations (11.1.9) and (11.1.12) overestimate the real surface temperature because the energy required for vaporization has been ignored. Together with the approximations made in (11.1.11) the real value of $I_v^{\min}$ can exceed the estimated values by one order of magnitude.

11.1.4 The Recoil Pressure

Because of momentum conservation, species evaporated from the surface cause a recoil pressure, p_{rec}. For surface absorption, this is of the order

11.1 Vaporization

$$p_{rec} \approx \frac{\zeta I_a \langle v^2 \rangle^{1/2}}{\Delta H_v + \xi \langle v^2 \rangle} \approx \zeta \rho \left(\frac{k_B T_s}{m} \right)^{1/2} v_0 \exp\left(-\frac{\mathscr{E}_v}{T_s} \right) \sim p_s(T_s). \quad (11.1.15)$$

ζ and ξ are functions of the adiabatic coefficient, $\gamma = c_p/c_v$; their values are of the order of unity. The average velocity of species leaving the surface is given by $m\langle v^2 \rangle/2 \approx k_B T_s/2$. The temperature at the solid–vapor interface, T_s, can be determined from (11.1.6) and (11.1.7). Note that $\langle v^2 \rangle$ increases with laser-light intensity. The second equality has been derived by substituting I_a by (11.1.7a). Thus, the recoil pressure is of the order of the saturated vapor pressure, $p_s(T_s)$, and increases nonlinearly with I_a.

For comparison, the radiation pressure is $p_{rad} \approx (1 + R)I/c \ll p_{rec}$ where c is the velocity of light.

11.1.5 Influence of a Liquid Layer

The next step of approximation takes into account the presence of the liquid layer. In this case one has to deal with *both* a vapor–liquid and a liquid–solid interface (Fig. 11.1.1). The recoil pressure will cause liquid-phase expulsion. For a focused laser-beam with radius w, the melt-ejection flux per unit (focus) area is

$$J_m \propto \frac{1}{\sqrt{w}} p_{rec}^{1/4}. \quad (11.1.16)$$

Numerical calculations which include both liquid-phase expulsion and vaporization, have been performed by many authors [*Chan* and *Mazumder* 1987; *Zweig* 1991]. The thickness of the liquid layer, calculated as a function of laser power is shown in Fig. 11.1.4 for different materials. The maximum thickness,

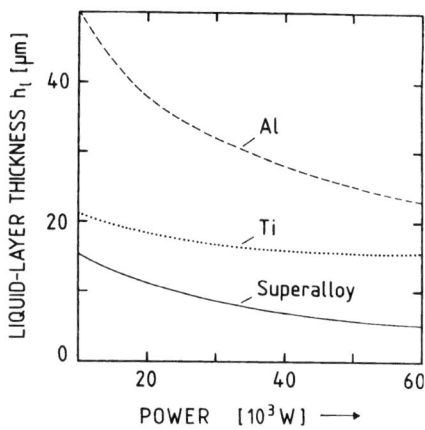

Fig. 11.1.4. Thickness of liquid layer, h_l, versus CO_2-laser power ($2w = 1$ mm) calculated for different materials [*Chan* and *Mazumder* 1987]

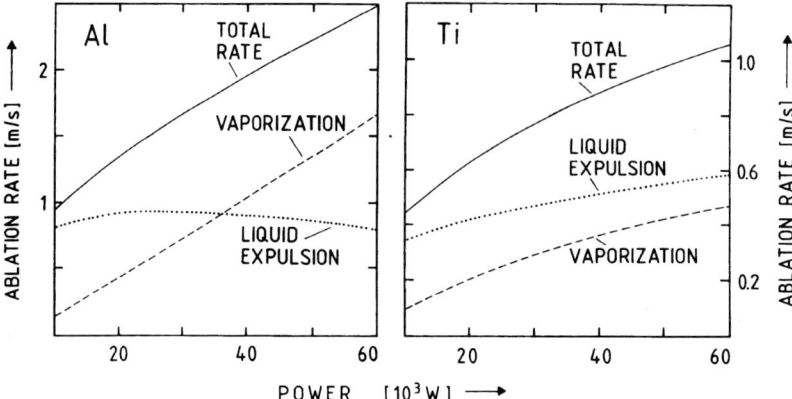

Fig. 11.1.5. Thermal ablation rate for Al and Ti calculated as a function of CO_2-laser power ($w = 0.5$ mm). The total rate is the sum of liquid-phase expulsion and vaporization [*Chan* and *Mazumder* 1987]

h_l^{max}, is some ten microns, depending on the physical properties of the material. The decrease in h_l with increasing power is due to the increase in vaporization rate, and thereby in recoil pressure. The total ablation rate is shown in Fig. 11.1.5 for Al and Ti. While the vaporization rate increases continuously with absorbed laser power, liquid-phase expulsion can increase or decrease, depending on the material properties and the laser parameters employed. This behavior is related to the interdependence between the recoil pressure and the thickness of the liquid layer. At low laser powers, material removal is governed by liquid-phase expulsion while at high laser powers vaporization dominates. It should be noted that the model ignores any direct coupling between the vapor plume and the laser beam. With the laser-light intensities considered, this is certainly a crude approximation.

11.1.6 Limitations of Model Calculations

Experimentally observed ablation rates often deviate considerably from theoretical predictions. There are many different reasons for this:

- The incident laser light will be attenuated by absorption and scattering within the vapor/plasma plume. Thus, the ablation rate will saturate at laser powers below those expected from the preceding model calculations.
- Part of the evaporated material recondenses within the area being processed. Besides its direct effect on ablation rates, recondensation diminishes surface cooling caused by vaporization and also changes the absorption behavior within the processed area.

11.2 Plasma Formation

- Surface melting changes the absorbed laser-light intensity, the dissipation of energy (including heat conduction, convection, liquid-phase expulsion, thermal radiation, etc.) and, in the presence of a reactive ambient medium, the surface chemistry and interaction with reactive species.
- Overheating and bubble formation can cause explosive melt ejection. This becomes effective, in particular, with a penetrating source, $l_\alpha > l_T$, and with long laser-pulse lengths. In such cases, the laser light heats a large material volume. Because evaporation cools the surface, the maximum temperature, and thus overheating, can occur *below* this surface (Fig. 13.2.1).
- With deep holes/grooves (Fig. 11.1.2), the parameters determining the absorption of laser radiation, the dissipation of heat, the shielding of the incoming laser light, etc., will change with the geometry of the structure, and thereby with time. Obviously, the various effects are mutually coupled.

These different contributions explain why calculated temperature distributions and processing rates, in particular for deep holes/grooves do not presently permit one a proper modeling of laser-induced ablation. Additional problems with quantitative analyses arise from the often significant inaccuracies in material parameters and their temperature dependences.

11.2 Plasma Formation

Up to now we have ignored any interactions between the incident laser beam and the vapor plume. With higher laser-light intensities, the vapor becomes substantially ionized and is more appropriately described as a plasma. With the onset of plasma formation, the coupling between the laser light and the substrate becomes strongly nonlinear.

To classify the scenario of laser–plasma–substrate interactions, it is convenient to introduce different laser-beam intensities that are characterized by their response onto the plasma and the solid surface.

We have already introduced the laser-light intensity which causes significant vaporization of the substrate surface, I_v. The onset of plasma formation shall be characterized by the intensity $I_p \equiv I_p(\lambda)$. Clearly, no sharp boundaries between these ranges exist. With these definitions, the intensities considered in the previous section are within $I_v \leqslant I < I_p$. With the formation of a plasma, the plume significantly absorbs the laser radiation and shields the substrate. When increasing the intensity above I_p, the plasma plume expands in volume and its forward direction becomes more pronounced (Fig. 11.1.1). If the laser-light intensity is even further increased, the plasma decouples from the substrate and propagates towards the incident beam. Such a plasma is often termed a laser-supported absorption wave (LSAW). The dynamic behavior of laser-supported absorption waves is determined by the laser-light energy absorbed within the plasma plume and the energy loss via heat conduction and via plasma and particle radiation. If the LSAW propagates towards the laser beam with *subsonic* velocity, it is also termed laser-supported combustion wave (LSCW). The propagation velocity of this wave increases with laser-light intensity and reaches the velocity of sound (with respect to the ambient medium) at the intensity I_d. Laser-supported absorption waves that propagate with *supersonic* velocity are also denoted as laser-supported detonation waves (LSDW).

Propagating LSAW may result in an *oscillating* behavior of the laser–plasma–substrate coupling. The different ranges of interactions have been extensively described in the literature.

11.2.1 Optical Properties of Plasmas, Energy Coupling

When approaching the intensity I_p, the laser light becomes increasingly absorbed by inverse Bremsstrahlung and by photo-excitation and photo-ionization of species within the plume. This results in *avalanche ionization* of the vapor and the formation of a plasma. This positive feedback is essential for both the hydrodynamic behavior of the plasma plume and many processing applications.

The characteristic time, t_p, for the development of a vapor/plasma, shall be defined by a decrease in density – with respect to the solid – by a factor of two, so that $\rho_v \approx \rho_s/2$. Thus, we can use the approximation $z_p \approx v_0 t_p$ where z_p can be estimated from the heated depth $z_p \approx D/v_{vs}$, see (11.1.5). This yields

$$t_p \approx \frac{D}{v_0^v v_{vs}}. \tag{11.2.1}$$

v_0^v is the sound velocity in the (high density) vapor which is of the same order of magnitude as the sound velocity within the solid, v_0^s, i.e., 10^5 to 10^6 cm/s. When assuming $v_{vs} = 10^3$ cm/s, which corresponds to an ablated layer thickness of 0.1 μm within a 10 ns laser pulse, and $D(T_v) = 0.1$ cm^2/s which is typical for a metal near the vaporization temperature, we obtain $t_p \approx 10^{-10}$ s.

The optical properties of the plasma are determined by the complex index of refraction, $\tilde{n} = \sqrt{\varepsilon}$. The dielectric constant of a fully ionized plasma is given by

$$\varepsilon = \varepsilon' + i\varepsilon'' = 1 - \frac{\omega_p^2}{\omega^2 + \omega_c^2} + i \frac{\omega_p^2}{\omega^2 + \omega_c^2} \frac{\omega_c}{\omega} \tag{11.2.2}$$

where ω_p is the plasma frequency and ω_c the frequency of collisions. ω_p is related to the density of electrons

$$\omega_p^2 = \frac{4\pi N_e e^2}{m_e}. \tag{11.2.3}$$

With *low* density plasmas, we can use for the absorption coefficient the approximation $\alpha = 2\omega \kappa_a/c \approx \omega \varepsilon''/c$ and thus obtain

$$\alpha \approx \frac{\omega_p^2 \omega_c}{c(\omega^2 + \omega_c^2)} \propto \frac{N_e N_v}{\omega^2}, \tag{11.2.4}$$

11.2 Plasma Formation

where c is the velocity of light. In the latter approximation we have assumed the frequency of collisions to be proportional to the density of species within the vapor, N_v, and $\omega_c \ll \omega$. Within this range, plasma absorption varies with $\alpha \propto \omega^{-2} \propto \lambda^2$. As a consequence, the threshold intensity for *optical breakdown*, I_p^{opt}, decreases with increasing λ. Except for very long wavelengths, the dependence can be approximated by $I_p^{opt} \propto \omega^2$. In clean air at atmospheric pressure, typical values of I_p^{opt} are about 3×10^{11} W/cm^2 for ruby-laser and about 2×10^9 W/cm^2 for CO_2-laser radiation [*Raizer* 1980]. Note that these intensities are considerably higher than the corresponding breakdown intensities, I_p, observed in front of solid or liquid targets.

For electron densities $N_e > 10^{-3} N_v$, plasma heating is dominated by electron–ion collisions. As long as the plasma plume stays near the substrate surface, both N_e and N_v increase with laser-beam intensity.

Because of the short times involved in the generation of the plasma and because of the avalanche-type increase in carrier concentration, the absorptivity rises almost instantaneously and, with certain experimental conditions, can reach a value near unity. This range of strong plasma absorption is employed in particular in metal processing (Sect. 11.3), and it has been investigated for a large number of different materials and a wide range of laser wavelengths [*Prokhorov* et al. 1990]. Figure 11.2.1 shows this effect for the example of steel. Here, the reflectivity is plotted versus the intensity of CO_2-laser radiation. At low intensities, the reflectivity corresponds to that of steel. With increasing surface temperature, this reflectivity decreases as with most metals (Sect. 7.3). At intensities $I \geqslant I_p$ a sharp drop in reflectivity occurs. The CO_2-laser radiation becomes now strongly absorbed within the laser-induced plasma. The figure shows that I_p is some 10^6 W/cm^2. For Nd:YAG-laser radiation plasma formation in front of metal targets is observed at laser-light intensities of some 10^8 W/cm^2. Due to electron and ion emission from the target, these values are lower than those obtained in clean air, roughly by two to three

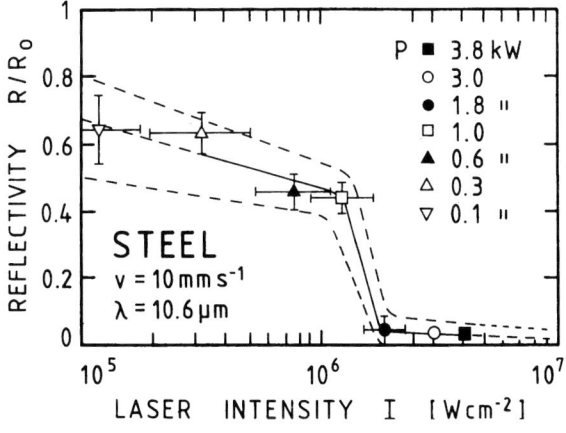

Fig. 11.2.1. Reflectivity as a function of CO_2-laser-beam intensity and different laser powers. The strong drop in reflectivity is due to plasma formation [*Herziger* and *Kreutz* 1986]

orders of magnitude. Additionally, the threshold intensity for optical breakdown near material surfaces is often described more properly by $I_p \propto \omega$.

With *high density* plasmas the situation changes considerably. This is the case if the electron density overcomes a critical density where the plasma frequency exceeds the laser frequency ω, i.e., $N_e > N_e^{cr}(\omega) = m\omega^2/4\pi e^2$. In this regime $\alpha \propto N_e N_v/\omega^2 \propto N_e^2/\omega^2 \propto \omega^2$. Thus, in contrast to the case considered above, plasma absorption increases with decreasing laser wavelength.

11.2.2 Laser-Supported Combustion Waves (LSCW): $I_p \leqslant I \leqslant I_d$

In this paragraph we consider the intermediate regime where the laser-beam intensity is high enough to cause optical breakdown within the gas in front of the substrate, but too low for generating a detonation wave. This is the regime of laser-supported combustion waves (LSCW). Here, the laser radiation is absorbed within a large volume of the plasma plume.

With laser-light intensities just above I_p, the plasma is confined to a region near the surface. This is shown in Fig. 11.2.2a for CO_2-laser radiation and steel. This intensity regime is employed in most types of laser machining. Here, the temperature of the plasma is, typically, of the order of

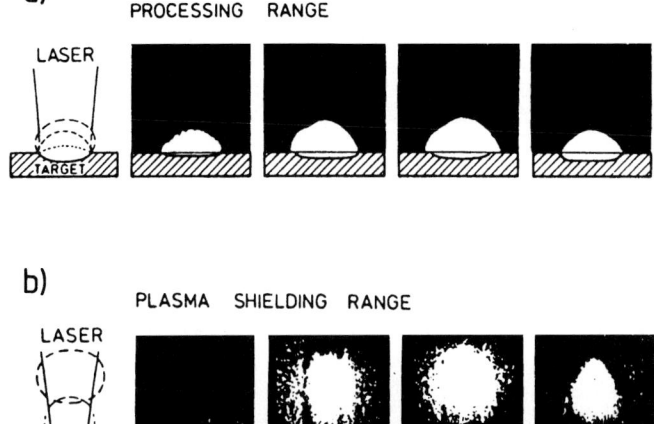

Fig. 11.2.2a,b. Development of a plasma in front of a steel target irradiated by 10.6 μm CO_2-laser light (TEM_{00}). The picture shows selected regimes (temporal distance 50 ns) of high-speed photography. (**a**) Intensity range employed in most types of laser machining. (**b**) Plasma shielding range [*Herziger* and *Kreutz* 1984]

11.2 Plasma Formation

10^4 K (≈ 1 eV). If the intensity is increased, the plasma plume expands towards the laser beam; nevertheless, it remains stationary. The spreading of the plasma increases the width of the temperature distribution on the substrate with respect to that which would be induced with the same laser focus in the absence of the plasma. Thus, the interaction of the laser radiation with the solid via the plasma diminishes the spatial resolution.

If the intensity reaches some critical value, typically between 10^7 W/cm$^2 < I^{cr} < 10^{10}$ W/cm^2, depending on wavelength, the laser light is essentially absorbed within the plasma and does not reach the substrate. This is the range of *plasma shielding*. Here, the coupling of the plasma to the substrate can become so weak that energy transfer is interrupted and, as a consequence, laser-induced material vaporization. Then the plasma decouples from the substrate surface (Fig. 11.2.2b). Due to the propagation and expansion of the plasma plume, the attenuation of the laser radiation is diminished. The laser-light intensity on the substrate surface then increases again, until plasma ignition is restarted.

The regime of plasma shielding can be shifted to higher intensities by appropriate changes of the ambient atmosphere, for example the admixture of He.

The propagation velocity of the LSCW towards the laser beam, v_{cw}, is dominated by heat conduction from the hot plasma to the cold gas of the medium ahead of the plume. A simple one-dimensional model, which strictly applies to a collimated laser beam only, yields [*von Allmen* and *Blatter* 1995]

$$v_{cw} \propto \left(\frac{\alpha I - J_L}{\kappa_{eff}[T_p - T(\infty)]} \right)^{1/2} \propto I^{1/2}, \qquad (11.2.5)$$

where J_L includes all energy losses. κ_{eff} is an effective transport coefficient which describes both heat conduction and thermal radiation. T_p is the temperature within the plasma plume. Any shock wave generated in this intensity range travels far ahead of the front of ablated material (Sect. 30.3). LSCW are very similar to combustion waves observed with (self-sustained) exothermal chemical reactions where $v_{cw} \propto \Delta H^{1/2}$ (ΔH is the reaction enthalpy).

If the absorbed laser-light energy just balances the energy losses, i.e., if $\alpha I = J_L$, the LSCW becomes stationary. Such stationary plasmas are frequently denoted as *plasmatrons*. They can be generated within focused laser beams even in the absence of any target. With cw CO_2 lasers, stationary temperatures within the plasmatron of, typically, 2×10^4 K can be achieved. This temperature is higher than that obtained, for example, with microwave (up to 6×10^3 K) or RF ($\approx 1 \times 10^4$ K) discharges.

11.2.3 Laser-Supported Detonation Waves (LSDW): $I \geqslant I_d$

At higher laser-light intensities, e.g., with CO_2 lasers and $I > 10^8$ W/cm^2, the plasma becomes strongly ionized and the frequency ω_p can exceed the laser-light frequency, ω. Then, the plasma becomes metal-like and absorbs the laser radiation within a thin layer (Sect. 11.2.1). The plasma temperature can reach more than 10^5 K (≈ 10 eV) and explosive propagation of the plasma with supersonic velocity is observed. Such a laser-supported detonation wave (LSDW), which propagates towards the laser beam, drives a shock wave into both the medium *and* the substrate material. In this regime, the velocity of the shock wave is about equal to the velocity of the ionization front. In a simple picture, the propagation of the LSDW within the gas can be described as follows: The strong ionization of the gas induces a shock wave. The shock wave heavily compresses the gas ahead of its front and thereby generates very high temperatures. Additional heating within this front is caused by the UV radiation of the plasma. With the temperatures generated in this way, the gas is ionized and, in turn, generates a shock wave.

In the case of large-area irradiation, the velocity of the detonation wave can be approximated by [*Zeldovich* and *Raizer* 1966]

$$v_{dw} \approx \left(2(\gamma^2 - 1)\frac{I}{\rho}\right)^{1/3} \propto I^{1/3}, \qquad (11.2.6)$$

where $I \gg I_p$. The adiabatic coefficient is $\gamma \approx 5/3$; ρ is the density of the ambient medium. With $I = 10^9$ W/cm^2 and air at standard conditions ($\rho \approx 1.3 \times 10^{-3}$ g/cm^3) we obtain $v_{dw} \approx 3 \times 10^6$ cm/s.

The pressure behind the detonation wave (11.2.6) can be approximated by

$$p \leqslant p_{dw} \approx \frac{\rho v_{dw}^2}{\gamma + 1}. \qquad (11.2.7)$$

With the above parameters, we obtain $p_{dw} \approx 4 \times 10^3$ atm. This pressure is almost equal to the pressure acting on the substrate surface.

The intensity regime under consideration is employed with special applications as, for example, in shock hardening (Chap. 23). Shock waves also play an important role in many cases of pulsed-laser ablation (Sect. 12.5) and material fragmentation.

11.2.4 Superdetonation

At very high laser-light intensities, typically, $\geqslant 10^9$ W/cm^2, new phenomena are observed: The ionization front propagates *ahead* of the shock wave. The reason is that species in front of the shock wave are first excited by the UV-plasma radiation and subsequently ionized by the laser light. This process is much faster than ionization via electron impact. The velocity of such fast ionization waves can be described by

$$v_{sd} \propto I^n,$$

with $n > 1$ [*Fisher* and *Kharash* 1982]. The properties of ionization waves depend on the laser parameters and on the type and pressure of the ambient gas. v_{sd} can reach values of some 10^9 cm/s. Due to the (non-thermal) excitation of species by the UV-plasma radiation, the temperature within the plume scales *inversely* with laser-light intensity, i.e., $T \propto I^\beta$ with $\beta < 0$.

The intensity regime considered in this paragraph is applied for the generation of pulsed X-rays and fast ions.

11.3 Abrasive Laser Machining

Applications of laser-induced vaporization include drilling, cutting, and shaping of materials, and also some types of trimming, engraving, marking, paint stripping, surface cleaning, etc. These applications have extensively been described in the literature [*Duley* 1983, 1976; *Ready* 1978, 1971]. The laser sources most commonly used in these applications are CO_2-lasers and Nd:YAG-lasers. The laser beam intensities employed depend on the laser wavelength. For CO_2-laser radiation they are, typically, between some 10^3 W/cm^2 and some 10^8 W/cm^2 (Fig. 1.1.2).

11.3.1 Cutting, Drilling, Shaping

Laser machining frequently requires laser-induced melting only. However, with many materials, and in particular with metals, efficient coupling of the laser-light energy to the substrate is mediated only via the generation of a plasma.

The energy flux onto the substrate surface is determined by both the laser-light intensity penetrating the plasma plume, and the net amount of energy transferred from the plasma. Important mechanisms of energy coupling between the plasma and the substrate are:

- Heat conduction. This is governed by the density and mean free path of electrons within the plasma plume.
- Plasma radiation. This contains a wide spectrum of frequencies, including UV radiation which is strongly absorbed by metals; it may even exceed the energy which would be directly absorbed from a CO_2-laser in the absence of the plasma. The situation can be different with an insulator when this has a strong dispersion oscillator whose frequency coincides with the CO_2-laser frequency (Fig. 7.2.4).
- Particle bombardment and condensation. Both contribute to the thermal energy available for substrate processing.

A proper estimation of the total energy absorbed by the substrate must consider all of the various contributions. From an experimental point of view, efficient plasma-enhanced coupling is observed as long as the plasma plume stays close to the surface (Fig. 11.2.2a).

Metals strongly reflect IR- and VIS-laser radiation (Table III). Thus, efficient processing becomes possible only via strong plasma absorption. This is the reason why laser machining of metals such as drilling, cutting, shaping, deep-penetration welding, etc., but also some laser-induced surface transformations, in particular surface alloying, are often performed in this regime.

For materials cutting, the maximum scanning velocity depends, for fixed laser parameters and a particular material, on the thickness of the workpiece, h_s. In the simplest approximation, the cutting speed can be estimated from the energy balance (11.1.1). If we assume, for example, a cw-laser beam of focus $2w$ and a scanning velocity v_s, the dwell time of the laser beam is $\tau_l \approx 2w/v_s$. With $\Delta h \equiv h_s$ equation (11.1.1) yields

$$v_s^{max} \leqslant \frac{P(1-R) - P_L}{d h_s \rho \Delta H}, \tag{11.3.1}$$

where $d \leqslant 2w$ is the *kerf width* which is, typically, some millimeters. ΔH is the total enthalpy. A better approximation is obtained from the equations given in Sect. 8.1. In most processing applications of this type, however, the laser beam is used in combination with a gas jet. The role of this jet can be twofold: It expels the liquefied material (Sect. 10.7), and it can also induce an exothermic reaction which can provide a significant amount of energy to the area being processed. An example is the cutting of steel by means of a CO_2-laser beam in

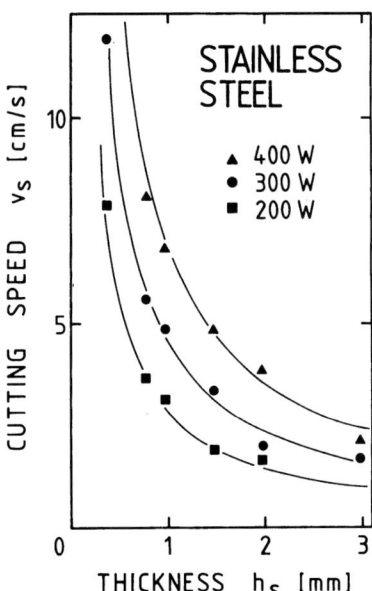

Fig. 11.3.1. Cutting speed versus thickness of stainless steel slabs for CO_2-laser radiation in combination with an oxygen gas jet (diameter of nozzle 1.2 mm; flow rate 20 normal liters per minute). Adapted from [*Sona* 1987]

11.3 Abrasive Laser Machining

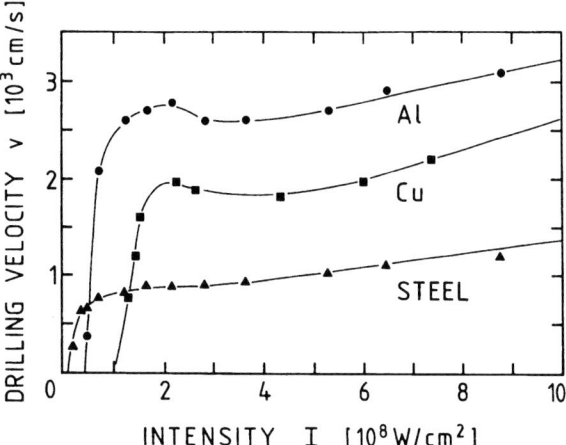

Fig. 11.3.2. Drilling velocity as a function of Nd:YAG-laser-light intensity ($\lambda = 1.06$ μm) for different metals [*Herziger* and *Kreutz* 1984]

combination with an oxygen gas jet. Figure 11.3.1 shows the cutting speed as a function of plate thickness for several laser powers. The dependence shown in the figure is qualitatively described by (11.3.1). It should be noted that with many applications reactive gases cannot be used because of their influence on material properties.

Figure 11.3.2 shows drilling velocities for Al, Cu, and steel as a function of Nd:YAG-laser-light intensity. The drilling velocity rises steeply just above the threshold for ablation and saturates at higher intensities. Here, the plasma determines the efficiency and quality of the process. Typical drilling velocities achieved in metal processing are between a few cm/s and some 10^3 cm/s.

11.3.2 Comparison of Techniques

The advantages of laser cutting, drilling, and shaping based mainly on material *evaporation* include smooth material edges (no or little solidified melt structure), and a relatively small extension of the transformed (e.g., oxidized) surface and the heat affected (damaged) zone (HAZ) [*Kar* and *Mazumder* 1990]. Among the disadvantages are the high laser-light intensities required, problems in process control, and the relatively low processing velocities.

The advantage of abrasive laser machining based mainly on material *melting* (Sect. 10.7) is the lower energy consumption which, for a certain laser beam intensity, permits higher processing velocities. Disadvantageous are the lower quality in surface morphology and, in general, the wider extension of the HAZ.

Abrasive laser processing based on material melting and *liquid-phase expulsion* by a reactive gas yields the highest processing velocities and permits cutting of thick metal slabs. Disadvantageous is the transformation of the material surface, e.g., by oxidation, the wide extension of the HAZ and, quite frequently, an even lower quality in surface morphology than in melt-phase processing using an inert gas atmosphere.

11.3.3 Non-metals

Laser machining has also been investigated for a large number of non-metals. Among those are semiconductors, inorganic insulators, organic materials, etc. Many of these materials, e.g., some types of ceramics and organic polymers, but also textile, paper, wood, etc., show *no* pronounced melt phase or even sublimate only. In such cases, abrasive processing with IR-laser radiation is based mainly on material decomposition and evaporation of fragments. Here, the laser-beam intensities employed are within 10^5 W/cm$^2 < I < 10^9$ W/cm^2.

The cutting speeds achieved with 0.5 kW cw CO_2-laser radiation and a kerf width of about $d \approx 0.2$ mm are, typically, between a few cm/s and some 10^3 cm/s (SiO_2, $h_s \approx 2$ mm, $v_s \approx 2$ cm/s; mylar, $h_s \approx 0.03$ mm, $v_s \approx 5 \times 10^2$ m/s; textile, 0.5 g/m^2, $v_s \approx 10^2$ m/s; newsprint paper 10^3 m/s) [*Spalding* 1987]. Depending on the material and the specific experimental conditions, the width of the damaged zone is some ten to several 10^3 µm wide.

12 Pulsed-Laser Ablation

Material removal caused by short high-intensity laser pulses is often termed pulsed-laser ablation. Throughout the literature, the terms laser-assisted evaporation and laser sputtering are also frequently used. Within the regime under consideration, material removal takes place far from equilibrium and may be based on thermal or non-thermal microscopic mechanisms. For this reason, we will prefer the term pulsed-laser ablation which is less suggestive with respect to the fundamental mechanisms involved in the process. Pulsed-laser ablation permits one to widely suppress the dissipation of the excitation energy beyond the volume that is ablated during the pulse. This is fulfilled if the thickness of the layer ablated per pulse, Δh, is of the order of the heat penetration depth, $l_T \approx 2(D\tau_l)^{1/2}$, or the optical penetration depth, $l_\alpha = \alpha^{-1}$, depending on which is the larger, i.e.,

$$\Delta h \approx \max(l_T, l_\alpha). \tag{12.0.1}$$

This (simplified) condition is, in fact, the basic requirement for applications of the technique. Pulsed-laser ablation can tentatively be classified into thermal, photophysical, and photochemical ablation.

Thermal (pyrolytic) ablation is based on laser-induced heating and (thermal) vaporization. Here, the dissipation of the excitation energy is so fast that the detailed excitation mechanisms become irrelevant.

In *photophysical* ablation, non-thermal excitations directly influence the ablation rate. Among those are electron–hole pairs, electronically excited species leaving the surface prior to energy transfer, etc.

Photochemical (photolytic) laser ablation is based on non-thermal bond breaking by either direct photodissociation or indirect energy transfer via defects, impurities, etc.

Laser ablation has been demonstrated to be a powerful tool in surface micropatterning of hard, brittle, and heat-sensitive materials, and in the fabrication of thin films with complex stoichiometry. The latter technique is termed pulsed-laser deposition (Chap. 22). It is evident that (12.0.1) is a crude estimation. Because of the fast heating and cooling rates achieved with pulsed lasers, material damage or material segregation in multicomponent systems can often be ignored even in cases where Δh is considerably smaller than the value obtained from (12.0.1). With many materials, (12.0.1) can be reasonably well fulfilled with UV-laser light and ns pulses. With VIS- and IR-laser radiation, this condition is often more difficult to fulfill because of the lower absorptivity observed with many materials at longer wavelength. Additionally, with increasing wavelength, laser plasma interactions become more pronounced; these result in plasma shielding, oscillations in the energy–substrate coupling, etc. (Chap. 11). With both longer wavelengths and enhanced laser-plasma interactions, the resolution achieved in micropatterning decreases.

Because pulsed-laser ablation permits one to preserve the stoichiometry during the ablation process (this is also known as congruent ablation) this technique, in combination with a mass spectrometer, can be used for chemical analysis of multicomponent materials.

Materials that are irradiated with short high-intensity laser pulses show a number of common features: Significant surface ablation is observed only if the laser fluence, ϕ, exceeds a certain threshold fluence, ϕ_{th}. Correspondingly, the experimental observations made with quite different materials can be classified into regimes $\phi < \phi_{th}$, $\phi \approx \phi_{th}$, and $\phi > \phi_{th}$. This classification has also been used in the organization of the present chapter.

12.1 Surface Patterning

Surface patterning by pulsed-laser ablation can be performed by direct focusing of the laser light onto the substrate, by direct masking, or by laser-light projection (Figs. 5.2.1, 5.2.3a). This has been demonstrated, in particular, for inorganic insulators, high-temperature superconductors (HTS), organic polymers, and biological materials. With many materials, reasonable suppression of material damage can be achieved with UV excimer-laser radiation. The ablation rates in surface micropatterning are, typically, between some 0.1 µm/pulse and several µm/pulse. The corresponding laser fluences are between 0.1 J/cm^2 and several J/cm^2. The physical properties required for estimating the thermal and optical penetration depth are listed in Tables I and III for various inorganic and organic materials.

Among the *inorganic* materials studied in most detail are oxidic perovskites, perovskite-related oxides including high-temperature superconductors, and some glasses. Some of the literature on laser ablation and its application to surface patterning is included in Appendix B.1.

Figures 12.1.1a,b show scanning electron microscope (SEM) pictures of grooves fabricated in ceramic PbTi$_{1-x}$Zr$_x$O$_3$ (PZT) by XeCl-laser radiation. The top surfaces next to the grooves show agitation due to radiation from the low fluence tail of the line focus. In the vicinity of groove walls, no changes in morphology or any material transformations have been detected. This result is quite different from that obtained with cw Ar$^+$-and Kr$^+$-laser radiation (Fig. 14.4.2). Both the short dwell time (pulse length) and the strong absorption of the UV-laser light are responsible for these differences. Figure 12.1.1c shows a SEM picture of a pattern produced by scanning a line focus over the

12.1 Surface Patterning

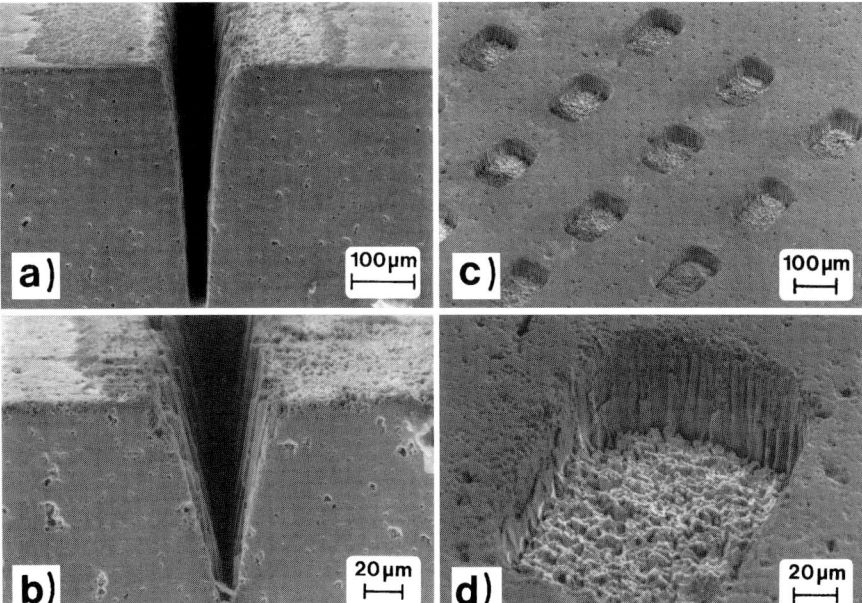

Fig. 12.1.1a–d. SEM pictures showing different patterns produced on ceramic PZT by means of 308 nm XeCl-laser radiation ($\tau_l \approx 15$ ns). The grooves in (**a**) and (**b**) have been obtained with a *stationary* line focus of $w = 50$ μm and a pulse repetition rate of 5 Hz. (**a**) $\phi = 10.8$ J/cm^2, $N = 4 \times 10^3$ shots. (**b**) $\phi = 3.2$ J/cm^2, $N = 10^4$ shots. (**c**) Line focus scanned perpendicularly to directly masked sample ($\phi = 15$ J/cm^2, $v_s = 0.84$ μm/s). (**d**) Magnification of (**c**) [*Eyett* et al. 1987]

directly masked sample surface. Direct masking permits one to avoid laser-induced surface damage.

Patterning by excimer-laser-light projection is demonstrated in Figs. 12.1.2. The bottom of the hole produced in LiNbO$_3$ is very smooth and almost no damage around the hole, apart from an approximately 1 μm-thick brittle layer at the rim, can be detected. Patterns of similar quality have also been produced in other materials. Figure 12.1.2b shows a HTS film which was patterned by KrF-laser-light projection. The deepening at the edge of the hole in a) and the fringes near the bar in b) originate from Fresnel diffraction.

Among the *organic polymers* studied in most detail are PET (polyethylene-terephthalate [MYLAR]), PI (polyimide), PMMA (polymethyl-methacrylate), and PTFE (polytetra-fluorethylene [Teflon]). The chemical structure of these polymers is shown in Fig. 12.1.3. An example for projection patterning of polymer films is shown in Fig. 12.1.4 for PMMA.

Figure 12.1.5 shows *biological* tissue ablated with 193 nm ArF- and 532 nm Nd:YAG laser radiation. The literature on experimental investigations on (mainly) pulsed-laser ablation of organic polymers and biological materials is summarized in Appendix B.2.

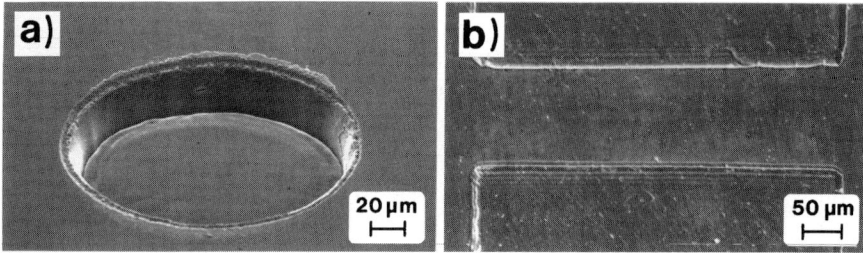

Fig. 12.1.2a,b. Projection patterning by excimer-laser ablation. (**a**) Single-crystalline LiNbO$_3$ ($\lambda = 308$ nm, $\phi = 2.7$ J/cm^2, $2w = 175$ μm, 500 pulses; vacuum) [*Eyett* and *Bäuerle* 1987]. (**b**) YBa$_2$Cu$_3$O$_7$ film on (100) SrTiO$_3$ substrate ($\lambda = 248$ nm, $\phi \approx 1.5$ J/cm^2, $\tau_l \approx 17$ ns; $h_1 \approx 0.1$ μm) [*Heitz* et al. 1990]

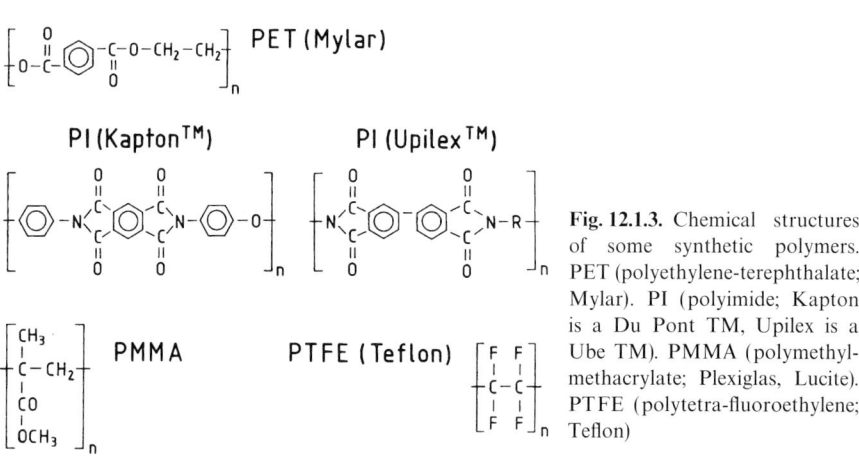

Fig. 12.1.3. Chemical structures of some synthetic polymers. PET (polyethylene-terephthalate; Mylar). PI (polyimide; Kapton is a Du Pont TM, Upilex is a Ube TM). PMMA (polymethyl-methacrylate; Plexiglas, Lucite). PTFE (polytetra-fluoroethylene; Teflon)

At present, pulsed-laser ablation is mainly applied in micropackaging [*Lankard* and *Wolbold* 1992; *Bachmann* 1989], wirestripping [*Brannon* and *Snyder* 1994], surface cleaning [*Tam* et al. 1995; *Lu* et al. 1994], in different types of trimming, and in link cutting, in particular in redundancy technology [*Richardson* and *Swenson* 1989]. In some applications, pulsed-laser ablation is employed for the fabrication of masters which are subsequently used for economic replication by standard techniques. Among the examples is the fabrication of masks, the combination of excimer-laser patterning with LIGA (LIGA combines deep lithography with galvano-forming and plastic moulding) [*J. Arnold* et al. 1995] etc. Furthermore, *medical* applications become increasingly important. Here, the most promising areas are angioplasty, ophthalmology, dermatology, and cellular microsurgery (Appendix B.2).

12.2 Interactions Below Threshold 195

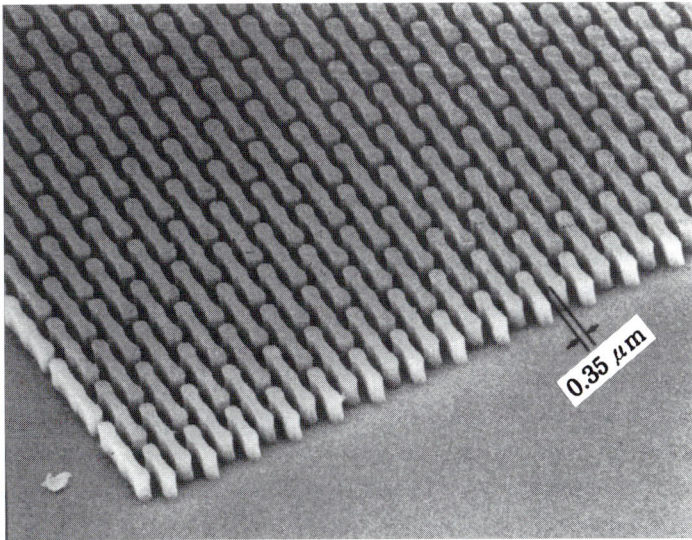

Fig. 12.1.4. Pattern produced in photo-resist by 10:1 refractive optics, using a KrF-laser as light source. The resist consists of multiple-layers of PMMA (1 μm)/SOG (0.5 μm)/OFPR 800 (1.2 μm) [*Horiike* et al. 1987]

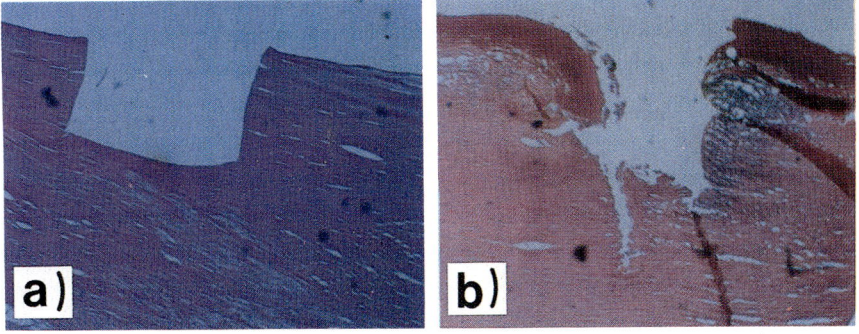

Fig. 12.1.5a,b. Cross section of luminal side of an aortic wall. (**a**) Trench (0.35 mm) produced by ArF-laser radiation ($\phi \approx 0.25$ J/cm^2, $\tau_l \approx 14$ ns). (**b**) Crater (0.4 mm) produced by 532 nm Nd:YAG laser radiation ($\phi \approx 1.0$ J/cm^2, $\tau_l \approx 5$ ns). The absorption coefficients of the material at the two wavelengths differ by about a factor of 10^3 [*Srinivasan* 1986]

12.2 Interactions Below Threshold

With laser fluences $\phi < \phi_{\text{th}}$, changes in surface morphology and microstructure, the generation of defects, and the depletion of one or several components of the material are frequently observed (see also Chaps. 27, 30).

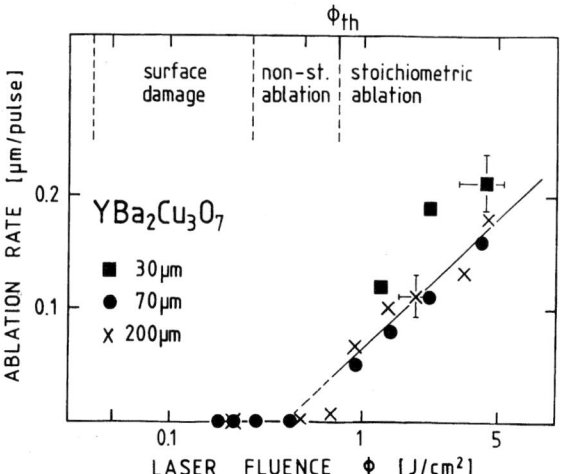

Fig. 12.2.1. Surface damage/ablation of $YBa_2Cu_3O_7$ films on (100) MgO substrates as a function of KrF-laser fluence. Different laser-beam spot sizes on the film surface, $2w$, are indicated by different symbols. Film thicknesses were between 0.5 and 1.5 µm [*Heitz* et al. 1990]

Figure 12.2.1 shows the behavior of $YBa_2Cu_3O_7$ (YBCO) films on (100) MgO substrates irradiated by KrF-laser light. Fluences of $\phi < 0.04$ J/cm² cause no detectable film damage. Within the range 0.04 J/cm² $\lesssim \phi \lesssim$ 0.27 J/cm², surface damage, together with the depletion of oxygen and small amounts of Cu, is observed. With $\phi \geqslant 0.27$ J/cm² non-stoichiometric ablation starts. Microprobe analyses reveal a depletion of Cu and, to a smaller extent, of Ba. This is expected from the vapor pressures of the metal oxides and metallic species formed during (thermal) decomposition of YBCO. The depletion of single material components increases with fluence. With $\phi > \phi_{th} \approx 0.75$ J/cm² ablation becomes stoichiometric. While the laser fluences corresponding to the different ranges of surface damage/ablation depend on the particular material and laser parameters, the overall behavior shown in the figure is characteristic of multicomponent materials subjected to short laser pulses.

12.3 The Threshold Fluence ϕ_{th}

Significant ablation is observed only above a certain threshold fluence, ϕ_{th}. With most inorganic insulators, ϕ_{th} is between 0.5 and 2 J/cm². With organic materials this range is, typically, $0.01 \leqslant \phi_{th} \leqslant 1$ J/cm².

For finite absorption, ϕ_{th} decreases with increasing absorption coefficient, irrespective of whether this is related to a decrease in laser wavelength, the addition of dopants, or to the generation of defects. A decrease of ϕ_{th} with increasing α is expected for *both* thermal and non-thermal ablation mechanisms, because the excitation energy will be localized within a smaller volume.

12.3 The Threshold Fluence ϕ_{th}

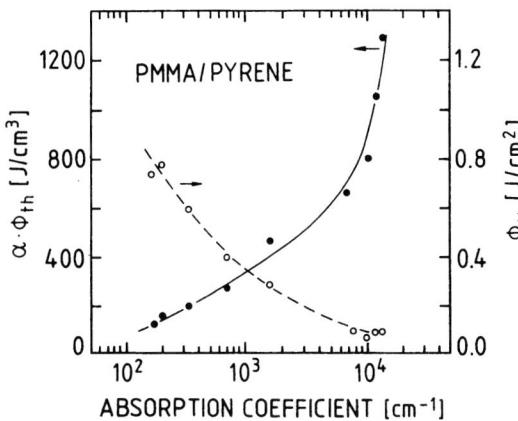

Fig. 12.3.1. Threshold fluence, ϕ_{th}, and $\alpha\phi_{th}$ for PMMA doped with pyrene. The absorption coefficient, α, refers to 308 nm XeCl-laser radiation [*Chuang* et al. 1988]

The *effective* absorption coefficient can be described by

$$\alpha = \alpha_0 + \sigma_D N_D + \alpha_i(N) + \alpha^{NL}. \quad (12.3.1)$$

Here, α_0 denotes the linear temperature-dependent absorption coefficient of the pure material (Fig. 7.2.4; note that α in Beer's law is the extinction coefficient which differs significantly for solids in crystalline and ceramic form). The second term describes the effect of light-absorbing dopants, where N_D is the number of dopant atoms/molecules per unit volume. The third term takes into account changes in absorption caused by radiation-induced defects (incubation centers; Sect. 13.4.2). α_i saturates after a certain number of laser pulses, N. With transient defects, α_i depends on the laser pulse repetition rate. The last term, α^{NL}, stands for multiphoton absorption processes. With very high laser powers when self-induced transparency, thermal runaway, avalanche ionization, etc., become important, the approximation (12.3.1) loses sense.

The effect of doping on ϕ_{th} is shown in Fig. 12.3.1 for PMMA. Here, the absorption coefficient for XeCl-laser radiation was tuned via the concentration of pyrene. By investigating additional host-dopant combinations, it has been found that the threshold fluence and the ablation rate are determined by the absorption strength of the dopant and not by its chemical nature or any charge transfer interactions between the excited dopant and the monomer units of the host polymer. Thus, the role played by the dopant is simply to absorb the light rather than to channel the electronic excitation energy to the host polymer. Similar results have been obtained with PTFE doped with PI [*Egitto* and *Davis* 1992]. However, with dopants that significantly change the thermal relaxation time, the interpretation of data may be different (Chap. 13).

Defects generated by the laser radiation itself are frequently denoted as incubation centers. Among those are color centers in ionic crystals, vacancies, broken bonds, molecular fragments, etc. Radiation-induced defects are of particular importance for the ablation behavior of wide band-gap materials and

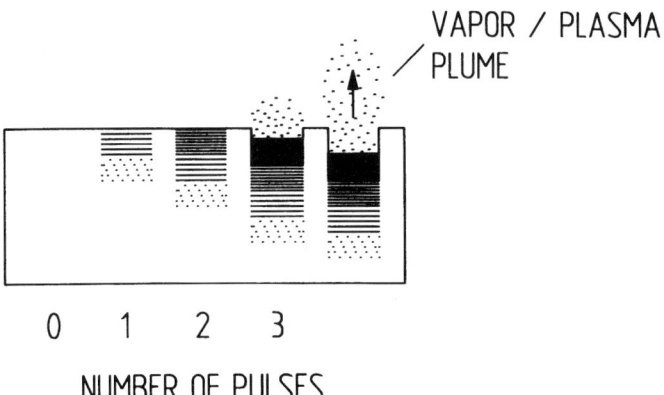

Fig. 12.3.2. Schematic picture showing the increase in absorption with successive laser pulses

photon energies $h\nu < E_g$. Here, successive laser pulses increase the number of defects and thereby the absorptivity within the irradiated volume. The increase in energy absorption causes a decrease in threshold fluence. Thus, ϕ_{th} for multiple-pulse ablation is lower than for single-pulse ablation. In other words, if ϕ is just below ϕ_{th} for single-pulse ablation, ablation starts after a certain number of pulses. With further pulses, stationary conditions are obtained (Fig. 12.3.2). ϕ_{th} can also be diminished via defects generated by electron- or ion-beam irradiation.

Another characteristic feature is the decrease in ϕ_{th} with pulse duration, τ_l (Fig. 12.4.5). With shorter pulses, the spatial dissipation of the excitation energy is diminished and ϕ_{th} is reached at lower fluences. This observation can be related to both, the decrease in heat penetration depth, and the increase in absorption coefficient due to nonlinear (multiphoton) excitation. The effect of nonlinear optical excitations on the threshold fluence may become important in particular for wide band-gap materials or/and optically strongly nonlinear materials.

An additional parameter becomes important with thin films on thermally insulating substrates. If the film thickness, h_1, is within the range $l_\alpha < h_1 < l_T$ the threshold fluence is diminished by a factor h_1/l_T. ϕ_{th} becomes independent of h_1 only when $h_1 > l_T$. This has been proved for metal films on fused quartz substrates [*Matthias* et al. 1994].

12.4 Ablation Rates

The ablation rate is defined by either the *total* layer thickness ablated per laser pulse, $\Delta h\,[\mu m/pulse]$, or by the average ablation velocity per pulse, $W_A \equiv \Delta h/\tau_l\,[\mu m/s]$. It depends on the photon energy, the laser fluence and

12.4 Ablation Rates

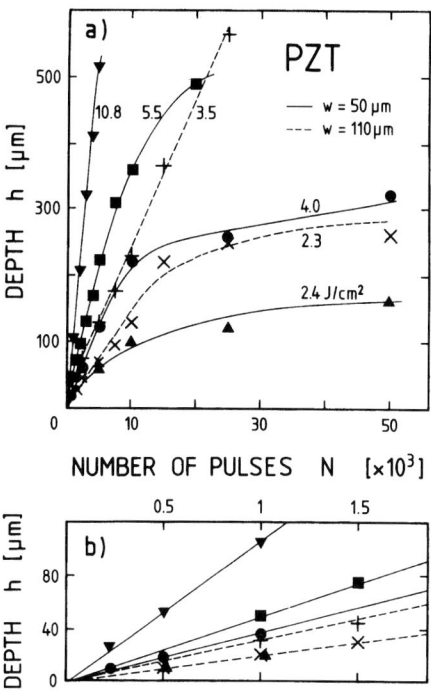

Fig. 12.4.1. (a) Depth of grooves as a function of the number of laser pulses for different fluences and widths of the line focus. Solid and dashed curves are guides for the eye. (b) Increase in depth for low numbers of pulses. Focus $w = 50\,\mu\text{m}$: ▲ $\phi = 2.4$ J/cm^2; ● $\phi = 4.0$ J/cm^2; ■ $\phi = 5.5$ J/cm^2; ▼ $\phi = 10.8$ J/cm^2. Focus $w = 110\,\mu\text{m}$: × $\phi = 2.3$ J/cm^2; + $\phi = 3.5$ J/cm^2 [*Eyett* et al. 1987]

width of focus, the heat or optical penetration depth, the enthalpy of vaporization, internal stresses, etc. With deep holes or grooves, W_A also becomes dependent on the number of laser pulses.

12.4.1 Dependence on Pulse Number

The typical dependence of the depth of grooves on the number of laser pulses, N, is shown in Fig. 12.4.1 for ceramic PZT. Initially, the depth increases linearly and then changes gradually to an approximately logarithmic dependence for high pulse numbers. A similar behavior has been found for many systems as, e.g., for excimer laser ablation of different glasses and polymers [*Srinivasan* 1994; *Braren* and *Srinivasan* 1988], etc.

The drop-off in rate observed with deep holes or grooves is related to different effects:

- The attenuation of the incident laser-light intensity by the ejected material. This can be made plausible from a simple estimation. The velocity of species ejected out of the hole is 10^5 to 10^6 cm/s. During a laser pulse of 20 ns, these species will travel about 20 to 200 μm. Thus, for groove depths under

consideration (Fig. 12.4.1) the attenuation of the incident light by scattering and secondary excitation of product species is important and becomes more efficient with increasing h. With higher fluences, secondary photolysis becomes more efficient and results in smaller fragments with smaller attenuation cross sections.
- With increasing depth, the transport of ablated species becomes less efficient and favors material recondensation within the groove. The depth at which the gradual decrease in slope is observed increases with the width of the line focus, $2w$. With wider holes, recondensation of species becomes less likely.
- With increasing hole depth, the loss of energy by heat conduction increases and thereby decreases the laser-induced temperature rise. This effect becomes significant when $h \approx w$.

With the change in slope in Fig. 12.4.1 a change in the cross section of grooves is observed.

12.4.2 Dependence on Fluence

The width of grooves is almost independent of the number of laser pulses but increases with increasing fluence. The latter dependence can be seen in

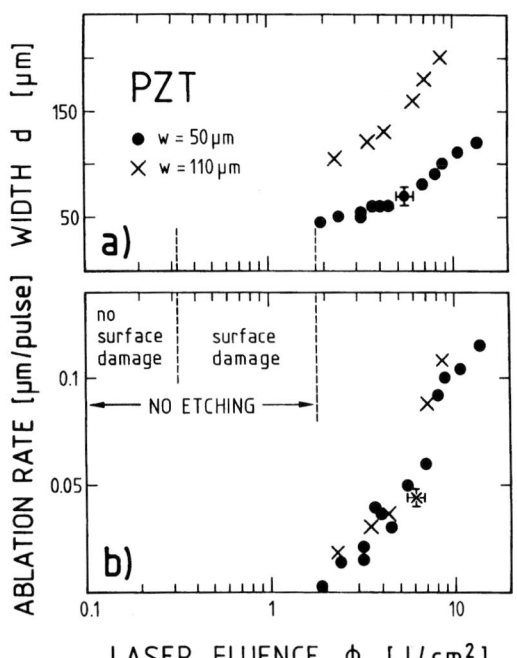

Fig. 12.4.2. Width of grooves and ablation rate as a function of incident laser fluence and two different widths of line focus, w (FWHM). ● $w = 50\,\mu m$; × $w = 110\,\mu m$ [*Eyett* et al. 1987]

12.4 Ablation Rates

Fig. 12.4.2a. It can tentatively be explained by the non-rectangular distribution of the laser fluence within the focus and the threshold behavior of the ablation process.

Figure 12.4.2b shows the ablation rate for PZT as a function of laser fluence, ϕ. The data have been derived from the initial linear increase in ablated depth with number of laser pulses (Fig. 12.4.1b). In the low fluence range, $\phi < \phi_{th}$, a similar behavior as in Fig. 12.2.1 is observed. For fluences $\phi > \phi_{th}$, the ablation rate increases initially with a slope that is approximately equal to the optical penetration depth, $l_\alpha = \alpha^{-1}$. Around 8 J/cm^2 a jump in the ablation rate seems to occur. In principle, this effect can be explained by the theoretical curve in Fig. 13.2.2b.

The present investigations do not permit an interpretation of the ablation mechanism. Although the surface morphology seems to indicate melting, non-thermal contributions due to direct band-gap excitation may be important.

Ablation by projection of excimer-laser-light has been demonstrated for various materials. Figure 12.4.3 shows ablation rates for LiNbO$_3$ as a function of laser fluence and for various laser-beam spot sizes produced by inserting apertures of various diameters into the beam path (see also Fig. 12.1.2a). Note the linear scale for the laser fluence. The ablation rates achieved are considerably higher than those obtained in ion-beam milling, which is the conventional method for patterning of LiNbO$_3$. No effect on the ablation rates was observed when working either in air or vacuum, although the latter avoids the condensation of debris and also yields somewhat better surface quality of patterns. Similar results have been achieved with single-crystalline BaTiO$_3$.

The ablation rate for PI (polyimide) is shown in Fig. 12.4.4 for different excimer-laser wavelengths. The decrease in threshold fluence with decreasing laser wavelength is related to the increase in intrinsic absorption of PI. For higher fluences the ablation rate achieved with longer wavelengths exceeds

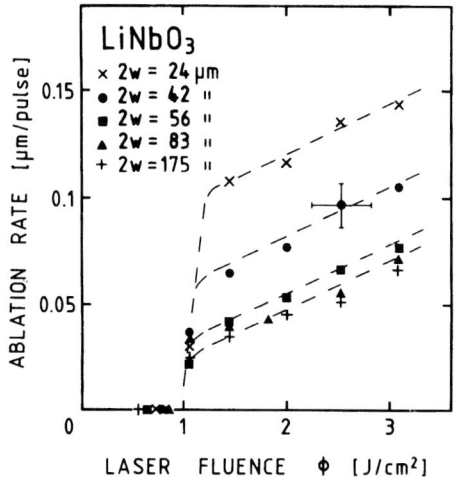

Fig. 12.4.3. Ablation rate of LiNbO$_3$ versus XeCl-laser fluence for various spot diameters ($\tau_l \approx 11$ ns). The dashed lines are to guide the eye [*Eyett* and *Bäuerle* 1987]

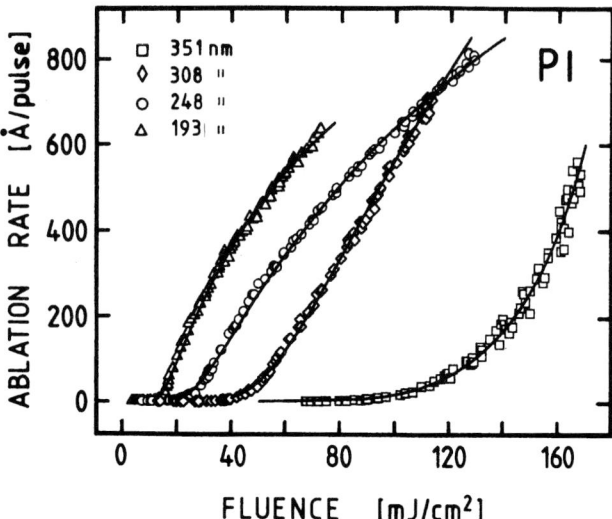

Fig. 12.4.4. Ablation rate of polyimide as a function of laser fluence for different excimer laser wavelength. Data points were derived from *Küper* et al. (1993). The full curves were calculated from the interpolation formula (13.3.13) using the parameters: △ 193 nm: $A = 883547$ Å/pulse, $B = 0.152$ J/cm^2, $\alpha_0 = 0.002$ Å^{-1}; ○ 248 nm: 29 716 Å/pulse, 0.176 J/cm^2, 0.0012 Å^{-1}; 308 nm: 87 096 Å/pulse, 0.37 J/cm^2, 0.00055 Å^{-1}; □ 351 nm: 32 561 Å/pulse, 0.76 J/cm^2, -0.0002 Å^{-1}

that observed for shorter ones. In this range, the absorbed laser-light intensity can more efficiently be used for ablation if the optical penetration depth is not too shallow. The full curves have been calculated from the interpolation formula (13.3.13).

12.4.3 Influence of Spot Size

The size of the illuminated spot on the substrate surface, $2w$, determines the width of the generated pattern and the expansion of the plasma plume. Both the transport of ablated species and the attenuation of the incident laser light are thereby related to w. With deep holes, both quantities depend also on the depth, h (Sect. 12.4.1). Subsequently, we consider the case $w \gg h$.

Figure 12.4.3 shows the dependence of the ablation rate on laser-beam spot size for LiNbO$_3$. The ablation rates are higher for smaller spot sizes. Above a "saturation" value, about 80 µm, the ablation rate becomes independent of w.

Similar observations have been made with various other materials, including organic polymers [*Wolff-Rottke* et al. 1995; *Heitz* et al. 1990].

The dependence of the ablation rate on w has been observed for many materials when using nanosecond or even longer laser pulses. This effect originates from the attenuation of the incident laser radiation by the expanding plasma plume. For semi-infinite surfaces and *shallow* patterns the attenuation decreases with decreasing spot size because three-dimensional transport of species becomes effective. This is consistent with time-dependent reflectivity measurements performed during such experiments. From a practical point of view, this effect has to be considered when employing projection masks with different feature sizes. In this case the depth of single features will differ from each other.

As already mentioned, the velocity of species ejected from the ablated surface is of the order of 1 to 10 µm/ns. Thus, even when we assume that ablation starts instantaneously, almost no plasma plume can develop during a picosecond or femtosecond pulse, and plasma shielding should be strongly diminished or even avoided. This has in fact been demonstrated. Figure 12.4.5 shows the ablation rate for $LiNbO_3$ as a function of fluence for XeCl-laser pulses of 1 ps duration. With such short pulses the ablation rate becomes independent of laser-beam spot size ($2w \approx d$ is the diameter of the hole). The ablation rate obtained in the regime of saturation is approximately equal to that obtained with the smallest spot size in Fig. 12.4.3.

These experiments show that the use of ultrashort pulses allows strong material excitation prior to the expansion of the plasma plume. This often simplifies the analysis of data.

12.4.4 Time-Resolved Dynamics

Time-resolved measurements yield fundamental information on the dynamics of the ablation process. In particular, they permit one to determine the importance

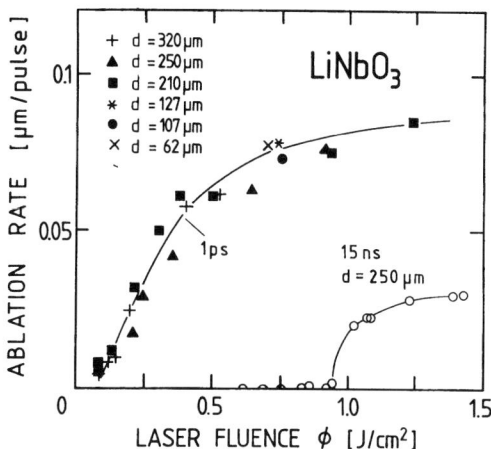

Fig. 12.4.5. Ablation rate of $LiNbO_3$ versus XeCl-laser fluence for two pulse durations and various spot diameters [*Beuermann* et al. 1990]

of multiphoton processes, the lifetimes of excited states, the "latent" time between the incident laser pulse and the ejection of species from the surface, the influence of plasma shielding, etc. Among the various techniques employed are acoustic methods, time-resolved reflectivity and beam deflection measurements, and various types of spectroscopic techniques (Chaps. 29, 30).

Measurements on latent times with high (temporal) resolution were performed by means of 0.5 ps KrF-laser *pulse pairs* of variable delay [*Preuss* et al. 1993]. With $LiNbO_3$ and ZrO_2, efficient plasma shielding of the second pulse occurs within delay times of a few ps only. This behavior may be characteristic for a large number of inorganic insulators. The situation is different with organic polymers. For example, with PMMA and PTFE plasma shielding can be ignored for delay times of up to at least 200 ps.

12.5 Material Damage, Localization of Excitation Energy

One of the most important questions for applications of pulsed-laser ablation in surface micro-patterning is the degree and extension of material damage beyond the volume ablated during the laser pulse. Among the different damages observed are defect formation, changes in morphology and chemical composition, material distortions, indications for melting, cracks, exfoliation, etc. The type and degree of damage caused by the ablation process depends on the laser parameters and the specific material, including its microstructure, pureness, internal stresses, etc. In many cases, material damages can be diminished by either increasing the absorption strength via the laser wavelength or material doping, or/and by decreasing the laser-pulse width (dwell time). This is understandable because the spread of the damaged zone is related to the degree of localization of the absorbed laser-light energy and thereby to the heat diffusion length and the optical penetration depth.

Another important point for applications is the smoothness of the ablated surface. With many materials, the ablated surface is relatively rough for fluences just above ϕ_{th} and becomes smoother with higher fluences, or shorter wavelengths and pulse lengths. This observation may be related to the suppression of surface instabilities, convective flows, or material segregation (Chaps. 10, 28). In some cases, the surface smoothness can also be improved by thermal annealing of the material prior to ablation (Sect. 28.6).

12.5.1 Strong Absorption

For strong absorption, (12.0.1) yields $\Delta h \approx l_T$. The condition $l_\alpha \ll l_T$ is well-fulfilled if the laser wavelength matches a strong elementary excitation of the material to be ablated. With inorganic insulators and semiconductors, strong absorption occurs when the photon energy exceeds the band-gap energy. On this basis, we can understand, for many systems, the degree of material damage observed in surface patterning. For oxidic perovskites, the band-gap energies are, typically, $E_g \approx 3$ eV. Thus, the requirement $h\nu > E_g$ can be achieved with laser wavelengths $\lambda \lesssim 410$ nm.

Let us consider the experiments performed with PZT (Fig. 12.1.1): With $D \approx 4 \times 10^{-3}$ cm^2/s and $\tau_l \approx 15$ ns we obtain $l_T(\text{XeCl}) \approx 0.1$ μm. This is comparable to the layer thickness ablated per pulse at the fluence employed (Fig. 12.4.2). Thus, the absence of any detectable damage on the side walls of grooves is consistent with this explanation. On the other hand, for typical dwell times employed with scanned cw Ar$^+$–or Kr$^+$–lasers, $\tau_l \approx 2$s (Fig. 14.4.2), we obtain $l_T(\text{cw Ar}^+, \text{Kr}^+) \approx 1000$ μm. Thus, thermal damage becomes widely extended.

The present argumentation can also be applied to some types of organic polymers, particularly those containing aromatic rings. With these polymers, the absorption cross section within the UV is very high. For PET at 193 nm we find $l_\alpha \approx 0.1$ μm which corresponds to an absorption cross-section of about 2×10^{-17} cm^2 per monomer. These materials can be patterned without detectable damage by means of ns UV laser pulses. With annealed material the ablated surface is very smooth (Sect. 28.6).

For materials with large values of D, ns pulses may be too long for (near) damage-free surface patterning. This is demonstrated in Figs. 12.5.1a and b for an YBa$_2$Cu$_3$O$_7$ film patterned by ns and fs KrF-laser light, respectively. In a) the ablated area is surrounded by a damaged zone whose width is, typically, 2 to 8 μm, depending on film thickness, laser fluence, pulse repetition rate, etc. The film damage is mainly ascribed to lateral heat transport. In b) almost *no* damage can be detected. Within the irradiated area the film is removed completely and no cracks are formed. The high quality of patterns has been proved by critical current density measurements (Chap. 22).

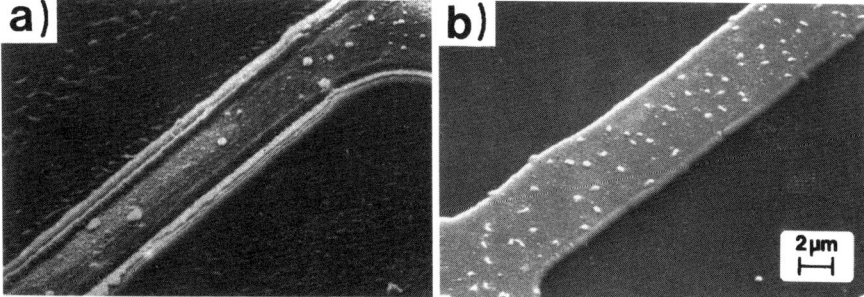

Fig. 12.5.1a,b. SEM pictures of YBa$_2$Cu$_3$O$_7$ bridges [$h_1 = 2500$ Å, (100) MgO substrate] fabricated by KrF-laser-light projection: (**a**) $\tau_l \approx 25$ ns ($\phi = 0.6$ J/cm^2, $N = 100$ pulses). (**b**) $\tau_l \approx 500$ fs ($\phi = 0.1$ J/cm^2, $N = 50$). The debris is due to recondensed material [*Proyer* et al. 1994]

12.5.2 Finite Absorption

Wide band-gap materials cannot be patterned in a well-defined way when employing standard excimer laser pulses of, typically, 10 to 50 ns duration. An exception are those materials in which, after a certain number of pulses, absorption is sufficiently increased due to defects (incubation centers) generated by the laser radiation itself (Sects. 12.3, 13.4). Many inorganic insulators are, however, quite insensitive to UV laser radiation. As a consequence, the number of pulses that would cause significant (defect) absorption is very high. For applications this is impractical. There are, however, a number of techniques to achieve good quality surface patterning also with such materials:

- One possibility is to generate surface or near-surface defects by either VUV radiation or an electron or ion beam, and subsequently ablate the surface by UV-laser radiation. This has been demonstrated for sodium trisilicate glass ($Na_2O \cdot 3SiO_2$), SiO_2, NaCl, and LiF [*Sugioka* et al. 1994; *Dickinson* et al. 1991].
- Another possibility is to employ ultrashort pulses. Figure 12.5.2 shows NaCl irradiated with 16 ns and 300 fs KrF-laser pulses. High quality patterning without detectable thermal damage is only achieved with fs pulses. This has been ascribed to coherent two-photon absorption which diminishes the optical penetration depth, and thus permits sufficient energy to be absorbed per volume and time to initiate ablation without damage. The ablation rate is shown in Fig. 12.5.3 as a function of fluence. The situation is similar with Teflon and PMMA.

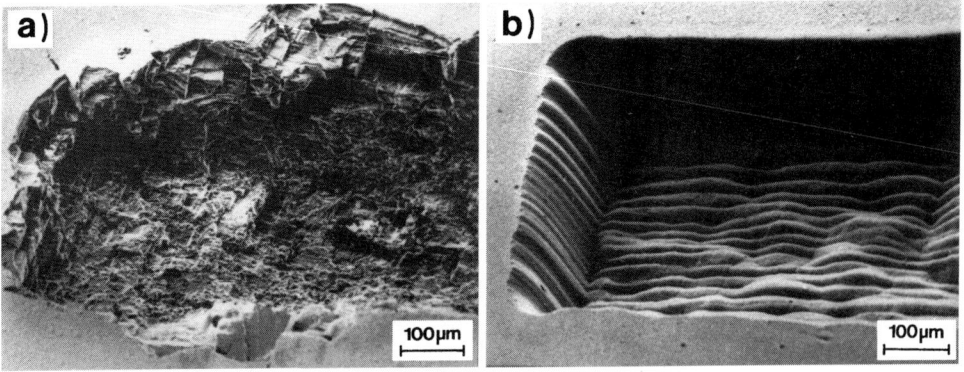

Fig. 12.5.2a,b. SEM pictures of NaCl surfaces ablated by 248 nm KrF-laser radiation. (a) Irradiation with ns excimer laser pulses ($\tau_l \approx 16$ ns, 15 pulses, $\phi = 4.2$ J/cm^2). An undefined crater is observed with cracks reaching deep into the surrounding material. (b) Irradiation with fs pulses ($\tau_l = 300$ fs, 500 pulses, $\phi = 500$ mJ/cm^2). The surface is relatively smooth and no cracks are observed in the surrounding material [*Küper* and S*tuke* 1989]

12.6 Influence of an Ambient Atmosphere

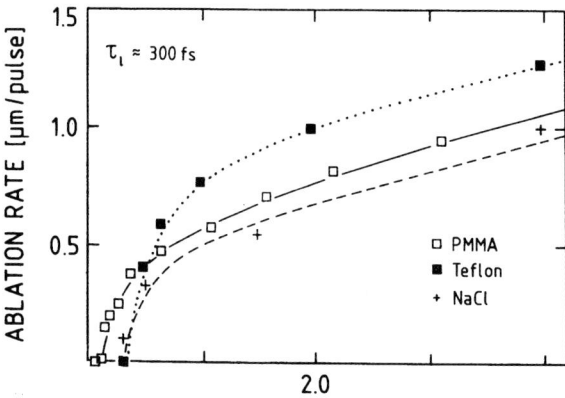

Fig. 12.5.3. Ablation rates for NaCl, PMMA, and PTFE (teflon) as a function of KrF-laser fluence ($\tau_l \approx 300\,\text{fs}$). For the high fluence range the ablation rates are similar. The lines are to guide the eye [*Küper* and *Stuke* 1989]

12.6 Influence of an Ambient Atmosphere

A *reactive* atmosphere can increase the rate of material removal. This is known as (dry) etching (Chaps. 14, 15). Additionally, such an atmosphere can change the physical and chemical properties of the ablated surface, which is the basis of many types of laser-induced surface modifications (Chap. 27). A reactive atmosphere can also change the chemical composition of ablated species. This is of particular importance in pulsed-laser deposition (Chap. 22).

A *non-reactive* atmosphere mainly influences the transport of species. This affects the ablation rate and the formation of debris.

Within any type of medium, pulsed-laser ablation can generate shock waves (Sect. 30.3).

12.6.1 Debris

Let us consider ablation within a non-reactive gaseous atmosphere. Due to collisions with gas-phase molecules, the transport of ablated species away from the irradiated surface area is hindered with respect to free expansion in vacuum. This favors the recondensation of product species on the substrate surface. As a consequence, a diminished ablation rate within the ablated region and the formation of debris outside of it is observed. Additionally, the confinement of the vapor plume increases the attenuation of the incident laser light and thereby diminishes the effective laser fluence. These effects become more pronounced with increasing gas pressure.

The condensation of product species (debris) on the ablated surface must be avoided with many applications. Here, it is very efficient to ablate the material in vacuum. This, however, is often impractical, in particular in production lines, or it is even impossible as for example with medical applications. Debris can often be diminished or even avoided by proper selection of the ambient atmosphere, e.g., a flow of low molecular weight species such as H_2, He, etc., or reactive species [*Küper* and *Brannon* 1992]; see also [*Kelly* et al. 1992; *Miotello* et al. 1992]. With certain conditions, debris can also be removed by subsequent laser treatment. For example, polyimide ablated in air by excimer-laser radiation can be cleaned by pulsed CO_2-laser irradiation [*Koren* and *Donelon* 1988].

After XeCl and KrF excimer-laser ablation of PET, PI, and PES (polyethersulfone) in air, a positive surface potential has been observed. This was ascribed to redeposited cationic fragments onto the ablated area; the process has been applied for selective electroless plating (Sect. 21.1.2).

13 Modelling of Pulsed-Laser Ablation

Pulsed-laser ablation has been analyzed on the basis of thermal, mechanical, photophysical, photochemical, and defect models. Almost all of these models try to describe ablation by a single dominating mechanism. For this reason, each of these models permits one to analyze experimental results only for a particular material and within a narrow range of parameters. A more general description requires simultaneous consideration of the different interaction mechanisms and the coupling between them. Let us discuss this in further detail by means of the block diagram shown in Fig. 13.0.1.

The process starts with single-photon or multiphoton material excitation. If the excitation energy is instantaneously transformed into heat, the increase in temperature changes the optical properties of the material and thereby the absorbed laser power. This coupling between the thermal field and the optical properties is indicated in the figure by a double-sided arrow. The temperature rise can result in (thermal) material ablation (vaporization) with or without surface melting. There is, however, another channel (dashed arrows) which may also result in ablation. The temperature rise induces stresses which can be so high that explosive-type ablation or, with thin films on thick substrates, material pop-off is observed. Stresses also change the optical properties of the material and thereby influence the laser-induced temperature rise. Another feedback could be related to thermally induced defects. Irrespective of whether thermally induced stresses or defects are important or not, we henceforth refer to this overall process as *thermal ablation*.

If the photon energy is high enough, laser-light excitation can result in direct bond breaking. As a consequence, single atoms, molecules, clusters or fragments desorb from the surface. Besides this direct channel, there is again an indirect channel (dashed arrows). Light-induced defects, for example photochemically dissociated bonds, can build up stresses which result in (mechanical) ablation. Both the direct and indirect path can take place, in principle, without any change in surface temperature. For this reason we term this process *photochemical ablation*.

Photophysical ablation shall describe a process in which both thermal and non-thermal mechanisms contribute to the overall ablation rate. An example would be a system in which the lifetime of electronically excited species or of broken bonds is so long that species desorb from the surface before the total excitation energy is dissipated into heat. The desorption process is enhanced

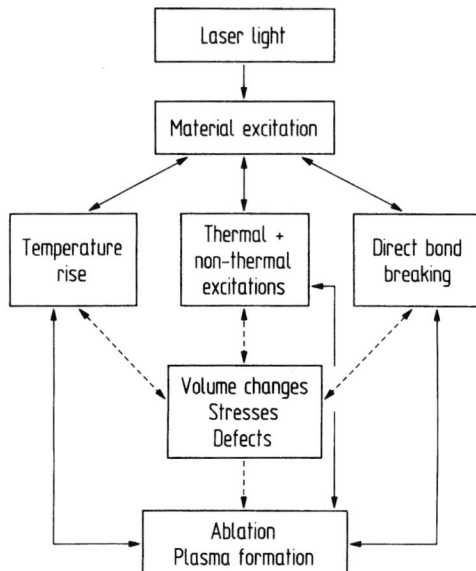

Fig. 13.0.1. Different interaction and feedback mechanisms involved in pulsed-laser ablation. Ablation can be based on thermal activation only (left path), on direct bond breaking (photo-chemical ablation; right path), or on a combination of both (photophysical ablation; intermediate path)

by the temperature rise. Thermally or non-thermally generated defects, stresses, and volume changes may again influence the overall process. Thermal ablation and photochemical ablation can be considered as limiting cases of photophysical mechanisms.

The different mechanisms and feedback channels included in Fig. 13.0.1 are by no means complete: Additional complexation arises from plasma formation, the ejection of electrons and ions which can build up surface electric fields, etc. Such electric fields, for example, may change activation energies for thermal desorption, for direct bond breaking, etc.

13.1 Model

One of the difficulties in modelling pulsed-laser ablation is related to the complexity of the optical excitation and energy dissipation mechanisms involved in the ablation process. Let us consider the electronic energy scheme shown in the schematic picture in Fig. 13.1.1. In organic polymers $S_0, S_1, S_2, \ldots$, and $T_1, T_2, \ldots$, denote singlet states and triplet states, respectively. In inorganic insulators or semiconductors, $S_0, S_1, S_2, \ldots$, indicate electronic energy bands. For example, S_0 would correspond to the highest valence band and S_1 to the lowest conduction band. $T_1, T_2, \ldots$, may be electronic defect states related to excitons, F-centers, surface states, etc.

13.1 Model

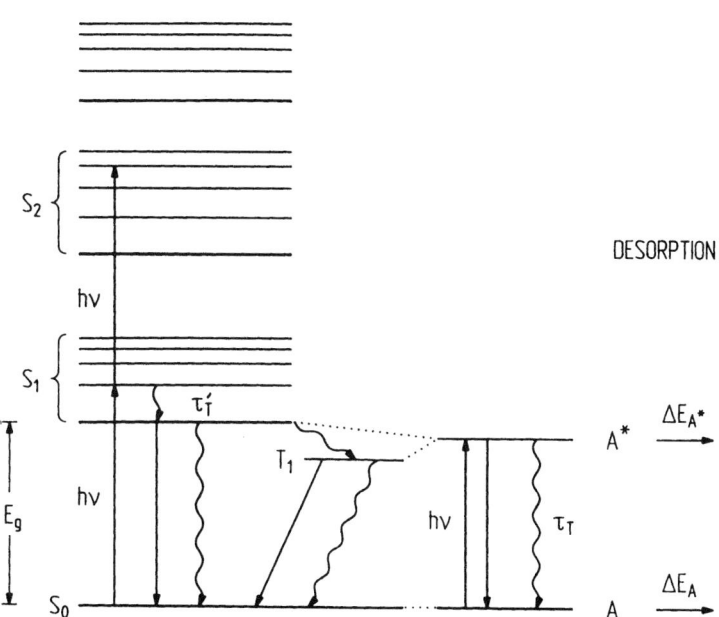

Fig. 13.1.1. Schematic picture which shows different electronic excitation and energy relaxation channels. Straight lines indicate the absorption or emission of photons while oscillating lines indicate non-radiative transitions. The photophysical model discussed in this chapter describes the electronic states of the material (left-hand side) by a two-level system, A, A* (right-hand side). ΔE_A and ΔE_A^* are activation energies for (thermal) desorption of ground state and excited state species, respectively

Infrared laser light excites electrons within the conduction band (intraband transitions), vibrations, etc. Here, the thermalization of the excitation energy is, in general, so fast that the laser can simply be considered as a heat source (Chap. 2).

The situation can be different with *ultraviolet* laser radiation which can induce single-photon or multiphoton interband transitions $S_0 \to S_1$, $S_0 \to S_2$,..., or excite defect states $T_1, T_2, ...$, etc. The excitation of defect states – or their generation by UV-laser radiation – is of particular importance when the photon energy is smaller than the band-gap energy, i.e., with $h\nu < E_g$. Non-radiative transitions between different electronic energy bands or defect levels shall be characterized by (thermal) relaxation times, τ_T, and those within energy bands by $\tau_T' \ll \tau_T$ (clearly, these times significantly differ for different energy bands). If τ_T becomes comparable to, or even exceeds, the characteristic time τ_R, for instance the time for activated desorption of 'excited' species, electronically excited states will play an important role in the ablation process (Sect. 2.1).

In order to investigate the role of electronic excitations, we consider, instead of the various excitation-energy relaxation channels, a two level system as shown on the right-hand side of Fig. 13.1.1. Species A refer to the electronic ground state and A* to the electronically excited state. A* may describe conduction band states, defect states, surface states, electronically excited atoms or molecules, dissociated bonds, etc. In any case, electronic excitation shall diminish the binding energy of species and thereby enhance their desorption from the surface. In the simplest case, the excitation and energy relaxation process can be described by single-photon transitions $A \rightarrow A^*$, stimulated emission, and thermal relaxation.

Within this simple model, the total ablation velocity, v, is determined by both 'ground-state species' A and 'excited species' A*, whose desorption rates depend on the respective activation energies, ΔE_A and ΔE_{A^*}, and on the local temperature rise which is controlled by the thermal relaxation time, τ_T. If A and A* desorb independently from the surface $z = 0$, v can be described by

$$v = k_A N_A^*(0) + k_{A^*} N_{A^*}^*(0), \qquad (13.1.1)$$

where $k_A = k_A^0 \exp[-\mathscr{E}_A/T(0)]$ and $k_{A^*} = k_{A^*}^0 \exp[-\mathscr{E}_{A^*}/T(0)]$ are rate constants for activated desorption. The coefficients k_A^0 and $k_{A^*}^0$ are related to the corresponding attempt (vibrational) frequencies of species. As before, we use the abbreviation $\mathscr{E}_A = \Delta E_A/k_B$ and $\mathscr{E}_{A^*} = \Delta E_{A^*}/k_B$. $T(0)$ is the surface temperature. $N_A^* = N_A/N$ and $N_{A^*}^* = N_{A^*}/N$ are the normalized number densities of species A and A*, respectively. $N = N_A + N_{A^*}$ is the total number density of optically active electronic states, defect states, surface states, chromophores in polymers, etc. For defect states, N may strongly depend on temperature, the number of laser pulses, stresses, etc. In such cases, in general, $N \ll N_t$ where N_t is the total number density of species within the solid. In organic crystals, polymers, etc., $N \approx N_t$ and almost independent of temperature. The coefficients k_A^0 and $k_{A^*}^0$ are then of the order of the sound velocity within the solid, v_A. For the activation energies we assume ΔE_{A^*} to be significantly smaller than ΔE_A. For small values of ΔE_{A^*}, the kinetic energy of ablation products should be taken into account.

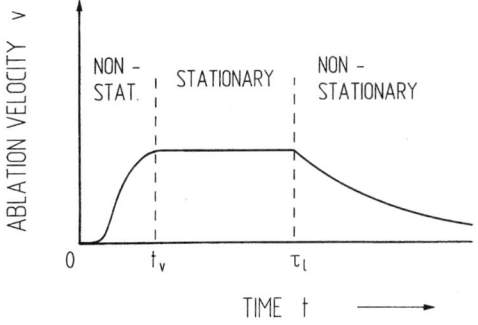

Fig. 13.1.2. Temporal dependence of the ablation velocity

13.2 Photothermal Ablation

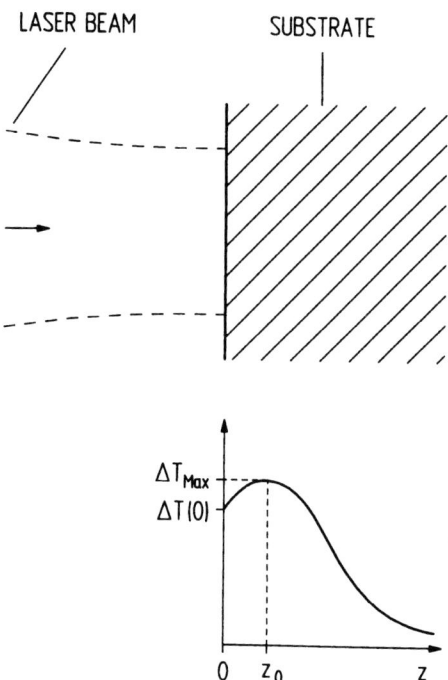

Fig. 13.2.1. Irradiation geometry and laser-induced temperature distribution for finite absorption

The behavior of the ablation velocity is schematically shown in Fig. 13.1.2. Stationary conditions shall be reached within a time t_v and they shall be sustained during the interval $\tau_l - t_v$. The energy stored near the surface during the laser pulse may be so high that ablation continues for times $t > \tau_l$. This effect is of particular importance with deep optical penetration depths and/or short laser pulses.

13.2 Photothermal Ablation

If the thermal relaxation time, τ_T, is very short compared to the time of activated desorption, laser ablation can simply be treated as a thermal process. This regime applies to pulsed-laser ablation by IR- and VIS-laser radiation and to many cases of UV-laser ablation.

Subsequently, we ignore any stress-related effects, the influence of the plasma plume, and any temperature dependences in parameters. The laser beam shall propagate in z-direction; the ablated surface shall be placed within the plane $z = 0$ (Fig. 13.2.1). The radius of the laser beam, w, shall be large compared to l_α and l_T. With this condition, the problem can be treated in one dimension.

We consider a single laser pulse of duration τ_l and a uniform laser-light intensity on the surface $z = 0$.

13.2.1 Stationary Conditions

The regime of stationary ablation $t_v < t \leqslant \tau_l$ (Fig. 13.1.2) can be described by the boundary-value problem

$$\frac{\partial^2 T}{\partial z^2} + \frac{v}{D}\frac{\partial T}{\partial z} + \frac{Q}{\kappa} = 0,$$

with the source term

$$Q = I_a \alpha \exp(-\alpha z),$$

and

$$\kappa \frac{\partial T}{\partial z}\bigg|_{z=0} = \rho v [\Delta H_t + c_p \Delta T(z=0)] \tag{13.2.1}$$

$$T(z \to \infty) = T(\infty)$$

$\Delta H_t = \Delta H_m + \Delta H_v$ is the sum of the latent heat of melting and evaporation, and c_p the specific heat per unit mass. $I_a = I_0(1 - R)$ is the laser-light intensity that actually reaches the surface and which is not reflected at $z = 0$. Thus, one has to correct the laser output intensity, I_l, for the attenuation by the expanding plasma (vapor) plume.

The difference between the present boundary-value problem and that treated in Sect. 11.1 is related to the *finite* penetration depth, l_α. The stationary velocity of the surface to be ablated is

$$v = k_A N_A^* = v_A \exp\left(-\frac{\mathscr{E}_A}{T(0)}\right), \tag{13.2.2}$$

where $v_A = k_A^0 N_A^*$. The solution of this problem is

$$\Delta T(z) = \frac{I_a}{\rho c_p (v - \alpha D)} \exp(-\alpha z)$$
$$- \frac{1}{2\rho c_p}\left(\rho \Delta H_t + \frac{I_a}{\rho c_p v}\frac{\alpha \kappa + \rho c_p v}{v - \alpha D}\right) \exp\left(-\frac{vz}{D}\right). \tag{13.2.3}$$

This temperature distribution is qualitatively shown in Fig. 13.2.1. The maximum temperature rise, ΔT_{max}, is obtained at a distance z_0 below the surface. At the surface,

$$\Delta T(0) = \frac{I_a - \rho \Delta H_t v}{2 \rho c_p v}. \tag{13.2.4}$$

Equation (13.2.4) is equal to (11.1.6). From (13.2.2) and (13.2.4) we can determine v and $\Delta T(0)$. Note that $\Delta T(0)$ does *not* depend on α. More detailed

13.2 Photothermal Ablation

considerations show that (13.2.4) is correct even if α is temperature dependent. The distance z_0 is given by

$$z_0 = \frac{D}{v - \alpha D} \ln\left[\frac{1}{2} + \frac{v}{2D\alpha} + \frac{\rho \Delta H_t v}{2\alpha I_a D}(v - \alpha D)\right]. \tag{13.2.5}$$

Substitution of (13.2.5) into (13.2.3) yields the maximum temperature rise, ΔT_{max}. For *surface* absorption we obtain $z_0(\alpha \to \infty) \to 0$. Thus, the maximum temperature rise occurs at the surface.

13.2.2 The Regime $t < t_v$

This regime can be described by the boundary-value problem

$$\frac{1}{D}\frac{\partial T}{\partial t} - \frac{\partial^2 T}{\partial z^2} = \frac{Q}{\kappa}$$

$$-\kappa \frac{\partial T}{\partial z}\bigg|_{z=0} = 0; \quad T(z \to \infty) = T(\infty); \quad T(t \leq 0) = T(\infty). \tag{13.2.6}$$

The solution of (13.2.6) is given by (7.5.8). For the surface temperature rise $\Delta T(0) = T_s - T(\infty) \equiv \Delta T(0, t_v)$ this yields

$$\Delta T(0) = \frac{I_a}{\kappa}\left\{\frac{2[Dt_v]^{1/2}}{\sqrt{\pi}} - \frac{1}{\alpha}[1 - \text{erfc}(\alpha^2 Dt_v)^{1/2} \exp(\alpha^2 Dt_v)]\right\}. \tag{13.2.7}$$

This equation permits one to calculate the latent time t_v by using for $\Delta T(0)$ the stationary value calculated from (13.2.2) and (13.2.4). For $\alpha \to \infty$ one finds $t_v \approx (c_p \Delta T/\Delta H_t)^2 D/v^2 \sim D/v^2$. This dependence can be derived from the condition that the velocity of the ablation front must become equal to the velocity of the heat front, i.e., $v_T = l_T/t \approx (Dt)^{1/2}/t$.

13.2.3 Average Ablation Velocity

The average ablation rate for a single pulse is

$$W_A \equiv \frac{\Delta h}{\tau_l} = \frac{1}{\tau_l}\int_0^\infty v(t)\,dt \approx \frac{\Delta h_1 + \Delta h_2 + \Delta h_3}{\tau_l}, \tag{13.2.8}$$

Where $v(t)$ denotes the *non*-stationary velocity (Fig. 13.1.2). $\Delta h_1 \equiv \Delta h_1(t < t_v)$ is the ablated layer thickness during the time interval required to reach stationary conditions. In most cases Δh_1 can be ignored, as in the preceding paragraph. $\Delta h_2 = v(\tau_l - t_v)$ is the ablated thickness within the regime of stationary ablation. Δh_3 refers to the time after the laser pulse, $t > \tau_l$. Here, ablation may continue for a certain time due to the energy stored within the irradiated surface layer.

Clearly, with picosecond or femtosecond pulses, the regime of stationary ablation may not be reached.

Subsequently, we consider the ablation rate for surface absorption and finite absorption. For $\alpha \to \infty$ significant ablation takes place only within the time interval $t_v \leq t \leq \tau_l$. This can be understood in the following way: According to (11.1.5) we can approximate the change in temperature at distance z by

$$\Delta T(z) \approx \Delta T(0)\left(1 - \frac{vz}{D}\right). \tag{13.2.9a}$$

After the end of the laser pulse, only the material layer whose temperature is high enough will ablate. The temperature corresponding to the 1/e decrease in ablation velocity (13.2.2) is

$$\Delta T(\Delta h_3) \approx \Delta T(0) - \frac{T^2(0)}{\mathscr{E}_A}. \tag{13.2.9b}$$

From (13.2.9) we obtain with $T(0) \approx \Delta T(0)$ for the layer ablated *after* the pulse

$$\Delta h_3 \approx \frac{\Delta T(0) D}{v \mathscr{E}_A}.$$

The thickness of the layer ablated *during* the pulse is $\Delta h_2 \approx v\tau_l$ and, therefore, $\Delta h_3 \ll \Delta h_2$ if

$$\frac{\Delta T(0)}{\mathscr{E}_A} \frac{D}{v^2 \tau_l} \ll 1.$$

This condition is in fact well fulfilled for the parameters commonly employed in ablation experiments with strongly absorbing materials.

The situation is different in the case of *finite* absorption. Here, the overheated layer of thickness, z_0, may have accumulated enough energy to be ablated after the pulse. This is the case if the dissipation of energy by heat conduction, characterized by $t_T \approx z_0^2/4D$, is slow compared to the ablation process, characterized by $t_A \approx z_0/v$. The condition $t_T > t_A$ yields $vz_0/4D > 1$. The total ablation rate per pulse is then

$$W_A \equiv \frac{\Delta h}{\tau_l} \approx \left[v\left(1 - \frac{t_v}{\tau_l}\right) + \frac{z_0}{\tau_l}\mathscr{H}\left(\frac{vz_0}{4D} - 1\right)\right]\mathscr{H}(\tau_l - t_v), \tag{13.2.10}$$

where $\mathscr{H}$ is the Heaviside function. The quantities t_v and z_0 can be calculated as outlined above. For *surface* absorption, (13.2.10) yields

$$W_A = v\left(1 - \frac{t_v}{\tau_l}\right)\mathscr{H}(\tau_l - t_v). \tag{13.2.11}$$

Figure 13.2.2 shows the normalized ablation rate, $W_A^* = W_A/v_A$, as a function of the normalized fluence $\phi^* = \phi/\rho\Delta H_t v_A \tau_l$ for three different cases characterized by the normalized absorption coefficient $\alpha^* = \alpha l_T$:

13.3 Photophysical Ablation

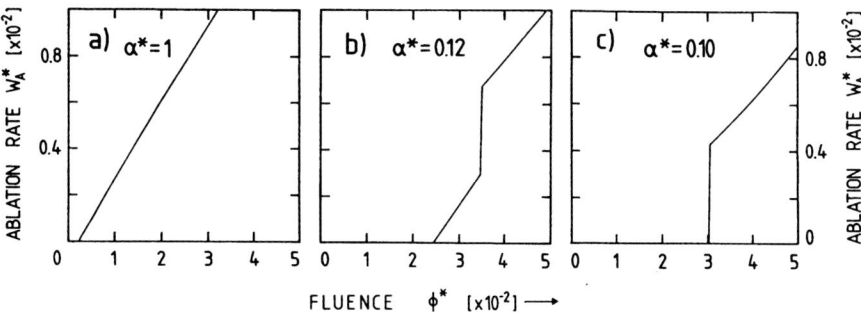

Fig. 13.2.2. Normalized ablation rate per pulse, W_A^*, versus normalized laser fluence, ϕ^*, calculated for three different values of α^*. The other parameters were $R = 0.5$, $c_p \mathscr{E}_A/\Delta H_t = 1$, and $v_A(\tau_l/D)^{1/2} = 300$ [Bäuerle et al. 1992]

- For large values of α^* the rate W_A^* increases continuously above a certain threshold fluence.
- For moderate values of α^* a step-like increase in W_A^* occurs at high fluences. This is due to the ablated layer thickness *after* the pulse, Δh_3. The experimental data shown in Fig. 12.4.2b may be an example of such behavior.
- For small values of α^* the 'tail' below the step is absent.

The results of this model are in agreement with many features observed experimentally (Chap. 12). Among these are:

- The overall dependence of the ablation rate on fluence. Equation (13.2.10) fits experimental data often much better than the logarithmic approximation (13.4.1) which is frequently employed.
- The occurrence of an ablation 'threshold'. In the present model, ϕ_{th} corresponds to a fluence where a fast rise in ablation rate is observed. In reality, the step-like increase in rate is smeared out.
- For fluences $\phi < \phi_{th}$, the ablation kinetics can be described by an Arrhenius-type law with $T \propto \phi$.
- The different ablation thresholds for metals and insulators, which are mainly due to the differences in heat conductivities.

If experimental data cannot, or not exclusively, be explained on the basis of this *simple* thermal model, it does not necessarily mean that non-thermal processes are important. It can equally well indicate the limits of validity in the standard description of laser-induced thermal processes (Chap. 2), and the approximations made in the present treatment.

13.3 Photophysical Ablation

Photophysical mechanisms become important if the second term in (13.1.1) cannot be ignored.

In order to elucidate the main ideas, we consider the simple two-level system shown on the right-hand side of Fig. 13.1.1. τ_T is taken to be independent of temperature. With the laser-light intensities under consideration, spontaneous emission can often be ignored. The decomposition of species shall occur simultaneously with their desorption from the material surface.

In a (moving) coordinate system that is fixed with the surface to be ablated, the density of species A* and A can be described by

$$\frac{\partial N_{A^*}}{\partial t} = v \frac{\partial N_{A^*}}{\partial z} + \frac{\sigma I}{h\nu}(N_A - N_{A^*}) - k_T N_{A^*}, \quad (13.3.1)$$

$$\frac{\partial N_A}{\partial t} = v \frac{\partial N_A}{\partial z} - \frac{\sigma I}{h\nu}(N_A - N_{A^*}) + k_T N_{A^*}, \quad (13.3.2)$$

where σ is the absorption cross section, and $k_T = \tau_T^{-1}$ the rate constant for thermalization of the excitation energy. Here, we ignore any diffusion processes which are slow compared to all other processes under consideration.

The propagation of the laser light within the substrate is given by

$$\frac{\partial I}{\partial z} = -\sigma(N_A - N_{A^*}) I. \quad (13.3.3)$$

The heating of the solid surface is described by the heat equation

$$\frac{\partial T}{\partial t} = v \frac{\partial T}{\partial z} + D \frac{\partial^2 T}{\partial z^2} + \frac{D}{\kappa} Q. \quad (13.3.4)$$

The heat source Q is determined by non-radiative transitions

$$Q = k_T h\nu N_{A^*}. \quad (13.3.5)$$

The heat flux at the surface $z = 0$ is dominated by the heat loss due to ablation of species A and A*

$$\kappa \frac{\partial T}{\partial z}\bigg|_{z=0} \approx \rho [\Delta H_A k_A N_A^*(0) + \Delta H_{A^*} k_{A^*} N_{A^*}^*(0)]. \quad (13.3.6)$$

Here, we have set $\rho = \rho_A = \rho_{A^*}$. This approximation does not hold for materials where stresses caused, e.g., by laser-induced bond breaking, significantly influence the ablation rate. For the transition enthalpies we set $\Delta H_A = \Delta E_A/m$ and $\Delta H_{A^*} = \Delta E_{A^*}/m$, where m is the (average) mass of ablated fragments A and A*. Besides (13.3.6) we employ the boundary conditions

$$N_A(z \to \infty) = N; \quad N_{A^*}(z \to \infty) = 0$$
$$T(z \to \infty) = T(\infty); \quad I(z=0,t) = I_a(t). \quad (13.3.7)$$

Due to the attenuation of the laser output intensity, I_l, within the plasma plume, we have $I_a(t) < I_l(t)(1-R)$. The initial conditions are

$$N_A(t=0) = N; \quad N_{A^*}(t=0) = 0; \quad T(t=0) = T(\infty). \quad (13.3.8)$$

13.3 Photophysical Ablation

Equations (13.1.1), and (13.3.1) to (13.3.8) characterize the boundary-value problem. Any temperature dependences in material parameters are ignored. For further details see *Luk'yanchuk* et al. (1993a).

13.3.1 Stationary Solutions

For organic polymers, a typical value of the ablation velocity is $v = 0.1$ μm/10 ns pulse; the thermal diffusivity is $D = 10^{-3}$ cm^2/s. The time to reach stationary conditions is then $t_v \approx D/v^2 \approx 1$ ns. For metals, on the other hand, t_v is of the order of 10^3 to 10^4 ns. Thus, with polymers and typical excimer-laser pulses (10 ns $\leqslant \tau_l \leqslant$ 40 ns) stationary conditions are really reached within times $t_v < \tau_l$.

Figure 13.3.1 shows the velocity v^* as a function of intensity I_a^* for different relaxation times τ_T^*. A purely thermal process is characterized by $\tau_T^* = 0$. With intensities $I_a^* = 1 - 10$ ($I_a = 10^7 - 10^8$ W/cm^2) and a thermal relaxation time $\tau_T^* = 0.01$ ($\tau_T \approx 10^{-10}$s), the interpretation of measured ablation rates would require surface temperatures $T(0) \approx (5-7) \times 10^3$ °C. With $\tau_T^* \geqslant 0.1$ the surface temperature near the ablation threshold becomes, however, only about 2000 °C. This temperature is in agreement with direct temperature measurements [*Brunco* et al. 1992] and with experimental data on the vibrational temperature of product species [*Srinivasan* and *Braren* 1989]. The value $\Delta E_A = 3$ eV employed in the calculations is realistic for PI (polyimide) where the dissociation energy of C—N bonds is about 3.15 eV. In many cases, for example in aromatic compounds, bond breaking energies may exceed this value considerably ($\Delta E \approx 4.8$ eV for C—H and C—O bonds, and 4.5 eV for C—N bonds [*Arjavalingam* et al. 1990]). With higher values of ΔE, the differences between thermal ablation and photophysical ablation become even

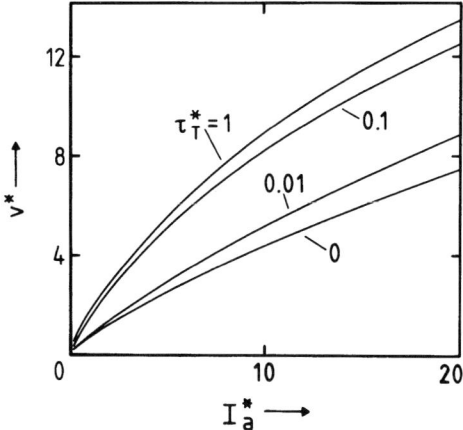

Fig. 13.3.1. Normalized stationary ablation velocity $v^* = v/v_n$ versus intensity $I_a^* = I_a/I_n$ for different relaxation times $\tau_T^* = \tau_T/t_n$ and $\Delta E_A = 3$ eV, $\Delta E_{A^*} = 0.3$ eV. The scaling factors are $v_n = l_\alpha/t_n$, $I_n = 10^7$ W/cm^2, and $t_n = h\nu/I_n \sigma = 10^{-8}$s [*Luk'yanchuk* et al. 1993a]

more pronounced. For $\tau_T^* = 0$ and an intensity $I_a^* = 10$, an increase in ΔE_A from 3 eV to 4.5 eV increases the surface temperature rise by a factor of about 1.4. Qualitatively, this is understandable: With increasing ΔE the ablation velocity decreases and the temperature increases because the energy loss due to evaporation is diminished.

The relative contribution of excited species to the overall ablation velocity can be characterized by the coldness

$$\text{cld} = \frac{k_{A*} N_{A*}^*(0)}{v}. \tag{13.3.9}$$

Purely thermal ablation is characterized by cld $= 0$. With increasing time τ_T, the concentration N_{A*} increases. The limiting value cld $= 1$ characterizes photochemical ablation.

Figure 13.3.2 shows the coldness as a function of intensity for different values of τ_T^*. For small values of τ_T^* the coldness increases with intensity. For $\tau_T^* > 0.05$, it saturates with very low values of I_a^*. With $\tau_T^* > 0.1$ ($\tau_T > 10^{-9}$s) and $I_a^* > 1$ ($I_a > 10^7$ W/cm^2) ablation is photophysical. Thus, even with nanosecond pulses ablation may be mainly photophysical, depending on τ_T^*. The relaxation times under consideration are realistic for polymers such as PET and PI, and photon energies 4.4 eV $\leq h\nu \leq$ 6.2 eV. With higher photon energies, photo-chemical decomposition may dominate.

Another interesting feature of the model is the behavior of the intensity distribution. From (13.3.3) it becomes evident that with increasing concentration, N_{A*} (increasing τ_T), the penetration depth of the laser light increases. The extension of this *bleaching zone* decreases with increasing ablation velocity v^*.

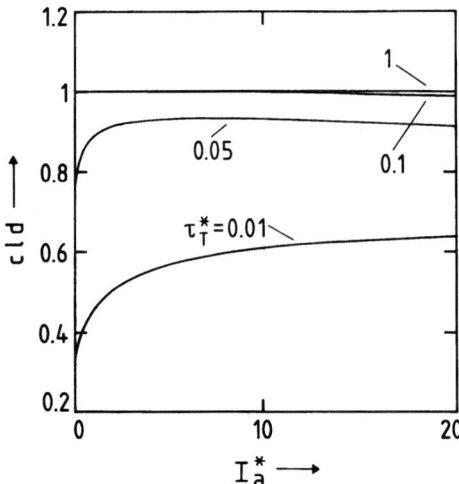

Fig. 13.3.2. Coldness cld [see (13.3.9)] as a function of intensity I_a^* for different values of τ_T^* [Luk'yanchuk et al. 1993 a]

13.3.2 Non-Stationary Ablation

With short high-intensity laser pulses, the dynamic ablation velocity, $v^*(t)$, may reach its maximum *after* the laser pulse [*Srinivasan* and *Braren* 1989]. Numerical calculations show that for very long relaxation times (for polymers this would mean $\tau_T \geqslant 10$ ns) damped oscillations in $T(0,t)$, $v(t)$, etc., may occur. The ablation kinetics is also very sensitive to the *shape* of the laser pulse. If the pulse is changed from rectangular to triangular (with the same duration and total energy) a significant increase in the ablated layer thickness is obtained.

Comparison of results

Figure 13.3.3 shows experimental data for PI together with numerical results. The full curve has been calculated by employing an activation energy $\Delta E_{A^*} \approx 1.7$ eV which is small compared to the photon energy $h\nu = 5.2$ eV (≈ 248 nm), and the bond breaking energy, 3.15 eV. If we assume that ablation is dominated by excited species, (13.2.8) can be written as

$$\Delta h \approx v_{A^*} \int_0^\infty dt \, \exp[F(t)], \tag{13.3.10}$$

where

$$F(t) = -\frac{\mathscr{E}_{A^*}}{T(0,t)} + \ln[N^*_{A^*}(0,t)].$$

The function $\exp[F(t)]$ shows a sharp maximum at $t = \bar{t}$ which corresponds to the maximum ablation velocity and almost coincides with the maximum in

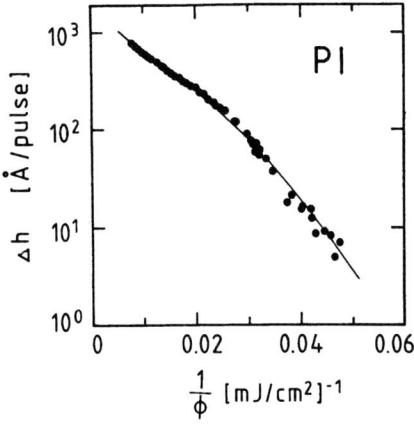

Fig. 13.3.3. Ablated layer thickness, Δh for PI derived from QCM measurements (data points), versus inverse laser fluence ϕ (248 nm KrF, $\tau_l \approx 15$ ns) [*Küper* et al. 1993]. The full curve is calculated by using $\alpha_p = 0.6\,\alpha$ [*Luk'yanchuk* et al. 1994]

$T(0)$. Integration yields

$$\Delta h \approx [2\pi]^{1/2} v_{A*} \frac{\exp[F(\bar{t})]}{(|d^2 F/dt^2|_{t=\bar{t}})^{1/2}} \approx A \exp\left(-\frac{\mathscr{E}_{A*}}{T_{\max}(0)}\right). \quad (13.3.11)$$

A is a slowly varying function of laser-light intensity which can be described by $A = a v_0 \tau_T$ where a is a constant. For low intensities, $T(0)_{\max}$ is proportional to the fluence ϕ. For higher fluences, the competition between bleaching and ablation leads to a sublinear dependence of $T(0)_{\max}$ on fluence. This yields a decrease in the slope in Fig. 13.3.3. Numerical integration of the original set of equations supports this interpretation. However, the resulting change in slope is not large enough to explain the experimental data. This discrepancy can be due to screening of the output intensity, $I_l(t)$, within the vapor plume, so that

$$I(0,t) = I_l(t) \exp\left[-\alpha_p \int_0^t v(t_1) dt_1\right]. \quad (13.3.12)$$

α_p is a coefficient which accounts for the absorption within the plume and which has been assumed to be a constant.

From (13.3.11 and 12) we find the interpolation formula for *photophysical ablation*

$$\Delta h = A \exp\left[-\frac{B}{\phi} \exp(\alpha_0 \Delta h)\right], \quad (13.3.13a)$$

or

$$\phi = B \exp(\alpha_0 \Delta h) \ln^{-1}\left(\frac{A}{\Delta h}\right). \quad (13.3.13b)$$

The exponential factor describes, heuristically, the attenuation of the laser light within the vapor plume. Because of (temporal) averaging, $\alpha_0 \neq \alpha_p$. B is described by $B = b\alpha^{-1} \mathscr{E}_{A*} \kappa/D$, where $b > 1$ is a constant. For low fluences (13.3.13) yields an Arrhenius-type dependence

$$\Delta h = A \exp\left(-\frac{B}{\phi}\right). \quad (13.3.14)$$

For high fluences we obtain within the range of realistic parameters with $A \gg \Delta h$ the logarithmic law

$$\Delta h = \frac{1}{\alpha_0} \ln\left(\frac{\phi}{\phi_{\text{th}}}\right), \quad (13.3.15)$$

where

$$\phi_{\text{th}} = \frac{B}{\ln(\alpha_0 A)}. \quad (13.3.16)$$

The interpolation formula (13.3.13) fits experimental results on polymer ablation with high accuracy (Fig. 12.4.4).

The photophysical model can explain many experimental features observed in UV-laser ablation of polymers:

- The Arrhenius-type behavior of the ablation rate for fluences $\phi < \phi_{th}$.
- Activation energies that are significantly smaller than bond-breaking energies.
- High ablation velocities at relatively low surface temperatures.
- The absence of thermal instabilities (Chap. 28).

The model also has a number of weak points:

- Because stimulated emission is not observed in ablation experiments, at least one additional energy level should be taken into account. This would describe single-photon excitation and relaxation processes as shown on the left-hand side of Fig. 13.1.1. Because $\tau'_T \ll \tau_T$, the energy $h\nu - E_g$ is dissipated into heat. Calculations based on a four-level system have been performed by *Luk'yanchuk* et al. (1996a). The ablation rates calculated from this model for wavelengths $\lambda = 351$ nm, 308 nm, and 248 nm coincide with the full curves in Fig. 12.4.4.
- Photophysical ablation mechanisms require relatively long relaxation times, τ_T.
- The activation energy ΔE_{A^*}, which strongly affects the results, is unknown.
- Thermo- and photomechanical contributions related to stresses, defects, etc., have been ignored. In particular, the model takes into account only desorption of species from the surface. In reality, volume effects often play an important or even dominating role in laser ablation. For example, when the maximum in the temperature distribution is located below the surface (Fig. 13.2.1) volatile products formed within the material volume may cause explosive-type ablation. This mechanism seems to be important with PMMA.

13.4 Photochemical Ablation

Experimental evidence for non-thermal processes in pulsed-laser ablation has been obtained with organic polymers and inorganic insulators and semiconductors.

Photochemical ablation rates are frequently described by a logarithmic law which follows, heuristically, from Beer's law

$$\Delta h = \frac{1}{\alpha} \ln\left(\frac{\phi}{\phi_{th}}\right), \tag{13.4.1}$$

where the absorption coefficient is often used as a fit parameter. In reality, α should be described by (12.3.1) and can depend on fluence itself. Equation (13.4.1) has the same form as (13.3.15). Clearly, the meaning of α and ϕ_{th} is different in both cases.

Within the model presented in Sect. 13.3, the purely photochemical channel can be described by $\Delta E_{A^*} \to 0$. For cw-laser irradiation this channel dominates if $I\sigma\tau_T/h\nu = \tau_T/\tau_0 > \exp(-\mathscr{E}_A/T)$. This holds also for pulsed irradiation if

$\tau_l \gg \tau_T \tau_0/(\tau_T + \tau_0)$. For short laser pulses the condition is $\tau_l(\tau_T + \tau_0)/\tau_0\tau_T > \exp(-\mathscr{E}_A/T)$.

Irrespective of the particular model and mechanisms considered, *purely* photochemical ablation is very rare, if possible at all.

13.4.1 Dissociation of Polymer Chains

Direct (non-thermal) bond breaking has been suggested to be the dominating mechanism in polymer ablation by means of ArF-laser radiation. Photo-dissociation of polymer chains into monomers leads to an expansion of the irradiated volume. The ejection of species from the surface was ascribed to built-up stresses related to this local volume increase [*Garrison* and *Srinivasan* 1985].

The fit of measured ablation rates by (13.4.1) is unsatisfactory even when ArF-laser radiation is employed. A much better approximation is obtained with the interpolation formula (13.3.13).

13.4.2 Defect-Related Processes, Incubation

Physical and chemical defects, including those that are generated by the laser radiation itself (incubation centers), can significantly influence microscopic interaction mechanisms which finally result in laser-induced emission of species from solid surfaces. Defects permit sub-band-gap electronic excitations in non-metals, enhance multiphoton band-gap excitations, alter binding energies of neighboring atoms and their coupling to the crystal lattice, trap electronic excitation energies, electrons, holes, etc. Energy trapping can have two consequences: It suppresses fast thermal relaxation and permits local energy transfer to a single or a few atoms only.

The mechanisms subsequently described are mainly discussed in connection with laser-induced-desorption (LID). Nevertheless, defect formation by these mechanisms may be responsible for material 'incubation' (Chap. 12), and for the 'tail' in the ablation curves, i.e., the observation that ablation starts for fluences $\phi \leqslant \phi_{th}$, even when $hv < E_g$.

With high fluences such mechanisms may initiate the ablation process.

The influence of lattice vacancies on photochemical ablation rates was discussed by *Okano* et al. (1993). In this model, ablation is based on the ejection of surface or near-surface atoms whose binding energy is diminished – with respect to those in the perfect crystal lattice – by nearby vacancies or vacancy clusters. With the emission of such weakly bound atoms, new vacancies are generated. In the initial stage of the process, this positive

13.4 Photochemical Ablation

feedback results in an exponential increase in the number of ejected atoms

$$N_A = N_A^0 \exp(\beta t), \tag{13.4.2}$$

where $\beta \propto I^n$. The exponent n describes multiple (electronic) excitation processes where, typically, $5 \leq n \leq 30$. For a given laser pulse duration, this mechanism predicts an apparent threshold fluence.

The calculations have been employed to qualitatively describe non-thermal emission of atoms from compound-semiconductor surfaces such as GaP [*Hattori* et al. 1992; *Nakai* et al. 1991] and GaAs under laser-light irradiation with photon energies that can cause electronic excitation of vacancies ($hv < E_g$).

Electron-hole pairs generated for photon energies $hv > E_g$ can become trapped at defect sites and produce Jahn–Teller distortions. Photo-excitation of such quasi-localized states may result in direct bond breaking, the desorption of species, and the generation of additional defects. The latter will increase optical absorption and thereby the efficiency for further bond breaking and surface heating. The mechanism is in agreement with the observation that the threshold fluence for perfect single-crystalline material is higher than for polycrystalline or ceramic material with many defects. A weak point of this model is the assumption of high order excitation processes required to explain the strong fluence dependence of measured ablation rates.

In wide band-gap insulators such as alkali halides, alkaline earth fluorides, SiO_2, etc., band-gap excitation requires, in general, multiphoton absorption. The probability of such transitions is strongly enhanced if there are 'intermediate' (defect) states within the band gap. In these materials, the generation of electron–hole pairs can result in the formation of *self*-trapped excitons (STE). An exciton in an alkali halide, for example, consists of a hole that is localized on a (negative) halogen ion – thereby forming a halogen atom – and an electron which is bound by the Coulomb potential of the surrounding (positive) alkali ions. Because of the strong electron-phonon coupling in these materials, such excitons can become self-trapped. The major contribution to the lattice relaxation energy (for one-center excitons) originates from the Jahn-Teller energy [*Itoh* et al. 1991]. STE states are located in the band gap and may be excited by laser light. Non-radiative decay of STE can result in the formation of F center – H center pairs (an F center is an electron in an anion vacancy; a H center denotes a molecular X_2^- ion, where X stands for a halogen atom) [*Haglund* et al. 1991; *Itoh* et al. 1991; *Jones* et al. 1989; *Matthias* and *Green* 1990]. X_2^- ions generated in this way are very mobile and diffuse over large distances. At the surface, they can dissociate and thereby lead to (preferential) emission of halogen atoms (the mobility of X_2^- ions depends on the crystal orientation and the emission of halogen atoms will therefore be anisotropic). What remains are F-centers and an alkali-rich surface layer (F-centers are not very mobile and neutralize metal ions only to some extent). At higher temperatures the metal atoms/ions are thermally desorbed from the surface. The defects generated in the halogen sublattice can be considered as 'incubation' centers. In any case, they increase the trapping of holes and the optical absorption coefficient, etc.

A different mechanism was suggested by *Wu* (1990). Here, laser-induced emission of neutrals and ions from Si and Ge surfaces was ascribed to energetic (electron) holes near defect sites. Such energetic holes which are generated via Auger recombination processes may induce bonding to antibonding electronic transitions and thereby contribute to the emission of species.

13.5 Thermo- and Photomechanical Ablation

Mechanical ablation is caused by built-up stresses generated by the laser light (Fig. 13.0.1). Depending on whether these stresses originate from thermal effects (thermal expansion, vaporization, thermally generated defects, etc.) or non-thermal effects (expansion due to direct bond breaking, non-thermal defect formation, etc.) we use the terms thermomechanical and photomechanical ablation, respectively. Such mechanisms are important in:

– Liquid-phase expulsion under the action of the recoil pressure of species evaporated from the surface (Sect. 11.1).
– Inorganic insulators and semiconductors where light-induced stresses change the absorptivity of the material.
– Polymer ablation where both thermal and non-thermal fragmentation of polymer chains leads to a volume increase.
– Cases where mechanical stresses change bond breaking energies.
– Strongly inhomogeneous systems that consist of different materials with different physical (thermal, optical, etc.) properties such as layered structures, certain types of ceramics, compound materials, etc.
– Pulsed-laser ablation of biological tissues [*Oraevsky* et al. 1991].

Thermal stresses, S, are particularly important at *low* laser-light intensities where $S \propto \Delta T \propto I$. At higher intensities the recoil pressure, p_{rec}, becomes of greater importance (Sect. 11.1.4). With even higher intensities, the (primary) shock related to the generation of a shock wave, $p_{sw} \propto (\phi/\tau_l)^{1/2}$ or, in dense media, in particular in liquids, the (secondary) shock related to bubble collapse, $p_{bubble} \propto E$, may dominate (Sect. 30.3).

Basic equations

Let us consider the influences of stresses in polymer ablation. While UV-laser ablation of PI can well be described within the frame of the photophysical model which ignores stresses (Sect. 13.3), the situation is different with PMMA. Here, stresses related to photochemical and thermal bond breaking seem to play an important role [*Garrison* and *Srinivasan* 1985].

13.5 Thermo- and Photomechanical Ablation

In a one-dimensional approximation, the rate of bond breaking can be described by

$$\frac{\partial N_b(z,t)}{\partial t} = \eta \frac{\sigma I(z,t)}{h\nu} N + k_0 N \exp\left(-\frac{\Delta E_b - \zeta S_{zz}(z,t)}{k_B T(z,t)}\right), \tag{13.5.1}$$

where N_b and N are the number density of broken bonds and of monomers, respectively. The first term describes direct (photochemical) bond breaking and the second term mechanical bond breaking. η is the quantum yield, and ζ a coefficient ($\zeta \approx 1.7 \times 10^{-21}$ cm^3 for PMMA). The stress component S_{zz} can be calculated from the equation of thermoelasticity [*Landau* and *Lifshitz: Theory of Elasticity*]

$$\frac{\partial^2 S_{zz}}{\partial t^2} = v_0^2 \frac{\partial^2 S_{zz}}{\partial z^2} - \frac{Y}{3(1-2\mu)} \frac{\partial^2}{\partial t^2}\left[\xi \frac{N_b}{N} + \beta_T(T-T_0)\right], \tag{13.5.2}$$

where v_0 is the sound velocity, Y Young's modulus, and μ the Poisson ratio. $\xi N_b/N = \Delta V/V$ describes the volume change caused by broken bonds. β_T is the coefficient of thermal (volume) expansion. Equations (13.5.1) and (13.5.2) must be solved together with the heat equation and Beer's law.

If we consider photochemical bond breaking only, the threshold fluence can be estimated from

$$N_b = \eta \frac{\phi \sigma}{h\nu} N > N_c = \frac{N}{Z}, \tag{13.5.3}$$

where Z is the number of monomers per persistent length (Kune segment; correlation length with respect to direction of polymer chain). For PMMA one finds $Z \approx 7$. This value is in reasonable agreement with the average length of monomers found in UV-laser ablation of PMMA [*Srinivasan* et al. 1986]. For efficient ablation, the length of fragments should be smaller than Z. From (13.5.3) we obtain

$$\phi_{th} \approx \frac{h\nu}{\eta \sigma} \frac{1}{Z}. \tag{13.5.4}$$

For PMMA and ArF-laser radiation with $\eta = 10^{-2}$, this value exceeds the experimentally derived ablation threshold [ϕ_{th}^{ex}(PMMA, 193 nm) ≈ 15 mJ/cm^2] by a factor of 10^3. If the decrease in activation energy, ΔE_b, due to mechanical stresses is taken into account, ϕ_{th} decreases by about two orders of magnitude [*Kitai* et al. 1990].

The remaining discrepancy to experimental findings can be overcome if we include in (13.5.1) the influence of *electronically* excited species, as in Sect. 13.3. This can be described by the additional term [*Luk'yanchuk* et al. 1996b]

$$k_0^* N^* \exp\left(-\frac{\Delta E_b^* - \zeta^* S_{zz}}{k_B T}\right), \tag{13.5.5}$$

where asterisks refer to excited species. As before, we assume $\Delta E_b^* < \Delta E_b$ and, additionally, $\zeta^* > \zeta$. Clearly, the number of broken bonds/defects will strongly influence the ablation rate. With the time scales involved in ns excimer-laser ablation, these defects appear at the surface mainly due to the moving ablation front.

It is evident that in most experiments *tangential* stresses cannot be ignored, in particular in laser micropatterning.

14 Etching of Metals and Insulators

Material removal by a gaseous, liquid, or solid etchant may be enhanced or only induced under the action of laser light. This is called laser-induced chemical etching. Symbolically, etching can often be described by the reversal of a corresponding deposition reaction, as already indicated in Fig. 1.2.1. Consider the deposition of Si according to $SiCl_4 \rightleftarrows Si(\downarrow) + 2Cl_2$. If the chemical equilibrium is shifted to the other side, the reaction describes the etching of Si in Cl_2 atmosphere. Another example is laser-enhanced electrochemical processing. Here, the course of the reaction can be turned around by simply changing the polarity of the substrate with respect to the counterelectrode.

With many materials, standard (dark) etching starts spontaneously when they become immersed in a gaseous or liquid etchant. With some systems, the reaction continues and the material becomes dissolved. With other systems there is an initial reaction which passivates the surface and thereby suppresses further reactions. Clearly, there are materials which are just inert against a particular medium.

Laser light can influence molecule–surface interactions by excitation of gas- or liquid-phase molecules, excitation of the solid surface or of adsorbate–adsorbent complexes. The etching mechanisms can be thermal or photochemical in nature. Laser-chemical etching may be classified into dry-etching which employs gaseous precursors, and wet-etching which is performed in liquids. Etching reactions using a solid etchant are quite rare.

In *dry-etching*, the precursor molecules most commonly used are halides, in particular Cl_2 and Br_2, and halogen compounds. The etching mechanisms are often based on the interaction between halogen radicals and charge carriers at or within the solid surface. These interaction mechanisms are, in many aspects, similar to those in surface oxidation (Chap. 26). The halogen radicals are formed spontaneously, for example by molecule–surface collisions, or only under the action of laser light. In a metal the free carriers are electrons, while in a semiconductor both electrons and holes are mobile. Dry-etching of metals and semiconductors is often classified into spontaneous etching, diffusive etching, and passivating reactions.

Wet-etching is mainly performed in aqueous solutions of acids like HCl, HNO_3, H_2SO_4, and H_3PO_4, or in lyes like NaOH and KOH, or in neutral salt solutions of NaCl, $NaNO_3$, K_2SO_4, etc. Also mixtures of different acids or of different lyes with or without other additives have been employed. Neutral salt solutions have the advantage of being less corrosive than acids or lyes.

The most important dry-etching techniques currently used in micromechanics and microelectronics are plasma-assisted etching (PE) and reactive ion etching (RIE). Basically, PE and RIE involve reactive radicals and charged particles which interact with the solid surface that is commonly kept at or near ambient temperature. Laser-induced etching can be considered as a *new* technique that not only permits direct maskless etching at high rates but also makes it possible to process a wide variety of materials that cannot be processed by standard techniques, or only very inefficiently. The general trend in semiconductor micro-fabrication is definitely towards dry-etching, because wet-etching, in general, introduces high levels of contamination. Nevertheless, wet-etching is still of great importance, mainly due to the high rates that can be achieved. A further advantage is its great versatility which is related to the large variety of reactants available without restrictions on volatility, etc. Laser-induced wet-etching is therefore a useful tool in *micro*machining such as cutting, drilling, shaping, etc.

An inspection of Appendix B.3 reveals that most of the work on laser-induced etching has concentrated on metals and semiconductors. Localized etching by direct writing, projection patterning, and laser-beam interference has been demonstrated with both gas-phase and liquid-phase precursors. Large-area etching with the laser-beam parallel to the substrate has been investigated mainly with gaseous etchants. Under otherwise identical experimental conditions, the etch rate strongly depends on the microstructure and morphology of the material, on crystal orientation, the type and concentration of admixtures, impurities, dopants, etc. Laser etching can be highly selective. This has been demonstrated for p-Si on SiO_2 and for W on Si substrates [*Loper* and *Tabat* 1984]. Here, the etch rates with respect to the substrate are, typically, $10^2 - 10^3$ times higher.

Etch rates achieved in laser-induced dry-etching can compete with those in conventional techniques. For example, with a 200 W excimer laser and Cl_2 atmosphere it is possible to remove Si from a 10 cm^2 SiO_2 wafer, or a Si wafer, with a rate of 20 Å/s. This rate is similar to typical removal rates obtained with RF-plasma etching (PE). For W in COF_2 atmosphere, ArF-laser radiation of comparable intensity would yield etch rates that are 3 to 4 times higher than those in PE. For many other metals, PE is even less efficient than for W. The high etch rates achieved with laser light, in particular at perpendicular incidence, are not only due to thermal or photochemical enhancement of the etching reaction itself, but also due to thermal or non-thermal desorption of non-volatile reaction products which are very often rate limiting in PE.

14.1 Photochemistry of Precursor Molecules

The precursor molecules mainly employed in laser-chemical etching are halides and halogen compounds. Overviews on the photophysics and photochemistry of these molecules can be found in [*Ben-Shaul* et al. 1981; *Calvert* and *Pitts* 1966].

14.1 Photochemistry of Precursor Molecules

14.1.1 Halides

Many materials are inert against Cl_2, Br_2, and I_2 molecules, but are heavily attacked by the (atomic) radicals. Halides show strong continua in the VIS and UV which result, at least in part, from optically active dissociative transitions $^1\Pi_u \leftarrow {}^1\Sigma_g$. The most detailed investigations on laser-induced photochemical etching have been performed with metals and semiconductors in chlorine atmosphere. For these reasons we will further outline the photochemistry of halides for the example of chlorine. The photodecomposition of Cl_2 can be described by

$$Cl_2 + h\nu(\lambda \leqslant 500\,\text{nm}) \rightarrow 2\,Cl. \tag{14.1.1}$$

The maximum in the dissociative continuum occurs at about 330 nm. At wavelengths $\lambda > 480$ nm, the continuum is very weak and has vibrational structure superimposed on it, resulting from transitions into the bound $^3\Pi(O_u^+)$ state. If absorption occurs at a wavelength short enough to break the bond ($\lambda < 498.9$ nm), this bound state predissociates with near unity yield by crossing over to a repulsive state [*Heaver* and *Clyne* 1982, and references therein]. For Br_2 and I_2 the corresponding wavelengths for dissociation are below 628.4 nm and 803.7 nm, respectively. Photoexcitation together with energy transfer via *collisions* permits predissociation at somewhat longer wavelengths. Furthermore, at elevated temperatures, excitation from higher vibrational levels ($v'' > 0$) becomes possible. These are the reasons why, for example, Cl_2 can even be photodissociated with 514.5 nm radiation, though with low efficiency (Table V).

Halogen radicals are very aggressive and strongly chemisorb on many material surfaces or even diffuse into these surfaces and break chemical bonds. Photochemical etching of silicon can be described by

$$Si + x\,Cl \rightarrow SiCl_x(\uparrow), \tag{14.1.2}$$

with $x \leqslant 4$. The product molecules desorb from the surface. In this reaction the etch rate is proportional to the radical concentration at the surface (Chap. 15). A quantitative analysis of such reactions thus requires the knowledge of the gas-phase chemistry that follows the photogeneration of radicals. For chlorine this can be described by

$$Cl_2 + h\nu \longrightarrow 2\,Cl$$

$$Cl + Cl + Cl_2 \xrightarrow{k_1} Cl_2 + Cl_2 \tag{14.1.3a}$$

$$Cl + Cl + Cl_2 \xrightarrow{k_2} Cl_2^* + Cl_2 \tag{14.1.3b}$$

$$Cl_2^* \xrightarrow{k_3} Cl_2 + h\nu' \tag{14.1.3c}$$

$$Cl_2^* + Cl_2 \xrightarrow{k_4} Cl_2 + Cl_2 \tag{14.1.3d}$$

$$Cl_2^* + Cl \xrightarrow{k_5} Cl_2 + Cl. \tag{14.1.3e}$$

The recombination of Cl atoms occurs via three-particle collisions (14.1.3a and 14.1.3b). The latter recombination channel has a quantum yield of a few percent only. Reaction (14.1.3c) results in chemiluminescence (afterglow) between about 450 nm and 1400 nm. This chemiluminescence permits one to measure the chlorine atom concentration in situ, e.g., during laser-induced chemical etching [*Kullmer* and *Bäuerle* 1988a]. The *total* chemiluminescence intensity observed during chlorine atom recombination is proportional to the density of excited Cl_2^* molecules

$$I \propto N_{Cl_2^*}. \tag{14.1.4}$$

With (14.1.3b–e) we obtain the rate equation

$$\frac{d}{dt} N_{Cl_2^*} = k_2 N_{Cl}^2 N_{Cl_2} - k_3 N_{Cl_2^*}$$

$$- k_4 N_{Cl_2^*} N_{Cl_2} - k_5 N_{Cl_2^*} N_{Cl}. \tag{14.1.5}$$

For stationary conditions one obtains

$$N_{Cl_2^*} = \frac{k_2 N_{Cl}^2 N_{Cl_2}}{k_3 + k_4 N_{Cl_2} + k_5 N_{Cl}}. \tag{14.1.6}$$

Thus, N_{Cl} can be determined from (14.1.4) and (14.1.6). Under certain circumstances, however, it is more appropriate to measure, instead of (14.1.4), the *spectrally resolved* intensity $I(\lambda)$. In the short wavelength region around 550 nm, this intensity can be described by [*Clyne* and *Stedman* 1968]

$$I(\lambda \approx 550 \text{ nm}) \propto N_{Cl_2^*}(v' \approx 12) \propto N_{Cl}^2 N_{Cl_2}^\gamma, \tag{14.1.7}$$

with $\gamma \approx 0.6$. Experimentally, the relative chlorine atom concentration, N_{Cl}, can then be determined from the measured intensity $I(550 \text{ nm})$ and the chlorine gas pressure (Fig. 30.1.1b).

14.1.2 Halogen Compounds

Laser-chemical etching by UV- and VIS-radiation or by IR-radiation is often performed by using halogen compounds such as HCl, XeF_2, NF_3, COF_2, CF_4, CF_2Cl_2, CF_3Cl, CF_2Br_2, CF_3Br, CF_3I, CCl_4, CF_3NO, SF_6, $CO(CF_3)_2$, etc.

Electronic excitations

The halogen compounds listed, can be photodissociated by UV- or VIS-laser radiation. The products are radicals such as F, Cl, CF_2, CF_3, etc. These radicals are highly reactive and therefore predestined for etching.

The pyrolytic and photolytic decomposition kinetics of many halides and halogen compounds has been reviewed by *Armstrong* and *Holmes* (1972).

IR Vibrational excitations

There are only a few examples for photochemical material processing with IR-laser radiation. The reason is that with the complex molecules and the high molecular densities employed in laser chemical processing, condition (2.3.9), and to an even greater extent (2.3.8), is difficult to fulfill.

Molecule-selective multiphoton vibrational excitation and dissociation (MPD) has been demonstrated for dry-etching with precursors such as SF_6, CF_3Br, CF_3I, and CDF_3. Here, SF_6 has been most extensively studied with respect to both its fundamental excitation mechanisms and its etching characteristics for pulsed CO_2 laser radiation. For *low* laser fluences (0.1 to 1 J/cm^2 or about 2 to 20 MW/cm^2), non-dissociative *coherent* excitation is observed

$$SF_6 + n_c h\nu(CO_2) \to SF_6^*, \qquad (14.1.8)$$

with $n_c \geqslant 3$. Asterisks now indicate vibrational excitation of the molecules. Because of the dense rotational structure in heavy polyatomic molecules such as SF_6 pumping into the quasi-continuum is possible *without* intermediate collisions (Sect. 2.3).

For fluences of 5 to 10 J/cm^2 a combination of coherent and sequential multiphoton dissociation of SF_6 is observed. This may be described, symbolically, by

$$SF_6 + nh\nu(CO_2) \to SF_5 + F, \qquad (14.1.9)$$

with $n \gtrsim 30$. SF_5 is unstable and further decomposes into SF_4 and another F atom.

The intensity of multiphoton absorption spectra depends on the laser fluence. Additional characteristic features are: A distinct resonance behavior and a broadening and shifting of the resonance to lower frequencies with increasing laser fluence. For SF_6, these characteristics have been studied by *Bagratashvili* et al. (1976). They compare favorably well with laser etching experiments performed in SF_6 atmosphere. A further point to consider is the dependence of the dissociation yield on gas pressure and composition [*Letokhov* 1983, and references therein]. For many monomolecular gases, the dissociation yield is independent of gas pressure, within a certain range. For SF_6 this range is 0.1 mbar $\leqslant p(SF_6) \leqslant 5$ mbar. This behavior is related to the fact that the v–v exchange between molecules of the same type can take place without a reduction of the average vibrational energy. Collisions between *different* types of molecules can result in a decrease or an increase in dissociation yield. For SF_6 [*Fuss* and *Cotter* 1977], CF_3I [*Bagratashvili* et al. 1978], etc., an admixture of monoatomic buffer gases decreases the dissociation yield. For other molecules such as CDF_3, C_2H_4, $C_2H_2F_2$, however, the dissociation yield shows a pronounced maximum when the buffer gas pressure is increased. In the case of CDF_3 with Ar, this maximum occurs at a pressure of $p(Ar) \approx 25$ mbar and exceeds the monomolecular yield by a factor of about 45. Vibrational energy transfer between vibrational modes will be more efficient with complex polyatomic molecules than with simple species (Sect. 2.3).

14.2 Concentration of Reactive Species

The time-dependent concentration of gas-phase radicals, A, generated in a (purely) photochemical reaction of the type

$$A_{\mu_A} B_{\mu_B} + h\nu \to \mu_A A + \mu_B B,$$

is given by the diffusion equation (3.3.2) which is approximated by

$$\frac{\partial N_A(\mathbf{x}, t)}{\partial t} \approx Q_{v,A}(\mathbf{x}, t) + D_A \nabla^2 N_A(\mathbf{x}, t) + f(N_{AB}, N_A, N_B). \tag{14.2.1}$$

The first term on the right-hand side describes the generation of species A within the gas (volume), and the second term the transport of these species by ordinary diffusion. The last term denotes the loss of radicals A by gas-phase recombination. The source term can be written as

$$Q_{v,A}(\mathbf{x}, t) = \mu \alpha_{AB}(\nu, N_{AB}(\mathbf{x}, t)) \frac{I(\mathbf{x}, t)}{h\nu}, \tag{14.2.2}$$

where $\mu \equiv \mu_A$. The boundary conditions are determined by the reaction (net) fluxes of species onto the various surfaces within the reaction chamber, including the substrate.

Henceforth, we assume cw-laser irradiation at normal incidence to the substrate and cylindrical symmetry (Fig. 14.2.1). In the simplest case, the reaction fluxes at the surfaces (normal components) can then be described by

$$-J_A = D_A \hat{\mathbf{n}}_r \nabla N_A = k_r N_A, \tag{14.2.3}$$

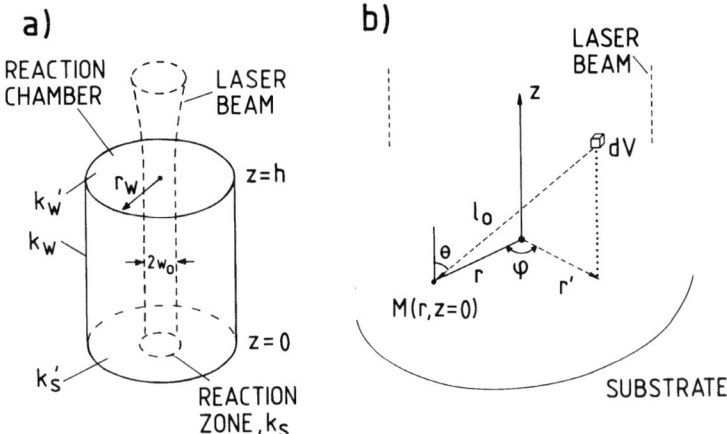

Fig. 14.2.1. (a) Reaction chamber and irradiation geometry employed in model calculations. The origin of coordinates is in the center of the basal plane. The radius of the reaction zone is $\sqrt{2}\, w_0$. (b) Magnification of laser-irradiated surface area. For notations see text

where $\hat{n}_r$ are unit vectors (directed into the reaction chamber) normal to the various surfaces (Fig. 14.2.1a); k_r are the corresponding reaction rate constants which are different for the different areas within the chamber: For the cylinder jacket defined by $r = r_w$ and $0 \leq z \leq h$, $k_r \equiv k_w$. The top surface (window) at $z = h$ with $r \leq r_w$ is characterized by $k_r \equiv k_w$. In the basal plane ($z = 0$) we define the reaction zone by $r \leq w_e = \sqrt{2}\,w_0$ and the corresponding rate constant by $k_r \equiv k_s$. Outside of this zone, the rate constant is $k_r \equiv k'_s$. Finally, we set up:

$$N_A(r, z, t) = 0 \quad \text{with} \quad t < 0. \tag{14.2.4}$$

The reaction rate in the center of the reaction zone is $W(0, 0, t) = -J_A(0, 0, t) = k_s N_A(0, 0, t)$.

Equation (14.2.1), together with (14.2.3 and 14.2.4), can only be solved numerically.

Some general aspects can be demonstrated by employing (crude) simplifications that permit an analytical treatment of the problem. Different approximations are discussed in the next subsections. Specific solutions of (14.2.1) for parallel laser-beam incidence are presented in Sect. 19.4.

14.2.1 Ballistic Approximation

An analytical expression for the reaction rate can be obtained if we assume a ballistic motion of A without recombination ($\lambda_m \gg l_0 \approx w_0$; Fig. 14.2.1b).

For low gas-phase absorption where $I(r, z) \approx I(r, h) = I(r)$ the rate of species A generated at a point (r', $z > 0$) is given by

$$\frac{\partial N_A}{\partial t} \approx \mu \sigma_{AB} N_{AB} \frac{I(r')}{h\nu}. \tag{14.2.5}$$

The number of species A generated per second in $dV = r'\,dr'\,d\varphi\,dz$ is then $[\partial N_A / \partial t]\,dV$. These species shall propagate isotropically out of dV. By integrating contributions from the entire gas volume above the substrate, the total flux at $M(r, z = 0)$ becomes

$$-J_A(r, 0) \approx \frac{\eta \mu \sigma_{AB} N_{AB} P}{\pi h\nu w_0^2} \int_0^{2\pi} d\varphi \int_0^h dz \int_0^{r_w} \frac{r' z}{4\pi l_0^3} \exp\left(-\frac{r'^2}{w_0^2}\right) dr'. \tag{14.2.6}$$

If *no* further reactions of A on the substrate take place as, for example, in some cases of photolytic LCVD, η has the meaning of a sticking coefficient. In the case of etching, η is the reaction probability that contains the sticking probability of A on the surface, the probability for further decomposition or reaction of A on the surface, the desorption of reaction products, etc. If both h, $r_w \gg w_0$ (Fig. 14.2.1a), equation (14.2.6) can be approximated by

$$-J_A(r,0) \approx \frac{\eta \mu \sigma_{AB} N_{AB}}{4\sqrt{\pi} h v w_0} P I_0\left(\frac{r^{*2}}{2}\right) \exp\left(-\frac{r^{*2}}{2}\right), \qquad (14.2.7)$$

where I_0 is the modified Bessel function. In the center of the laser focus this becomes

$$-J_A(0,0) \approx \frac{\eta \mu \sigma_{AB} N_{AB} P}{4\sqrt{\pi} h v w_0} \propto \frac{1}{w_0}. \qquad (14.2.8)$$

Thus, with the approximations made, the flux in the case of *gas-phase* photolysis at constant laser power, P, is $J_A \propto w_0^{-1}$. For *adsorbed-phase* photolysis we find $J_A \propto I \propto w_0^{-2}$. The reason is that species created at distances larger than w_0 are distributed over such a large area that they do not significantly contribute to the rate in the center of the laser focus. While these considerations are very simple and transparent, we should be aware of the many approximations made:

– The assumption of a constant coefficient η does not apply in many cases.
– With the gas pressures commonly employed in LCP, the assumption $\lambda_m \gg l_0$ does not hold.
– The assumption of a semi-infinite gas phase and the omission of product recombination is often inadequate. Their influence will be considered in the next two subsections.

14.2.2 Diffusion

In order to obtain further insight into the problem, we investigate the influence of diffusion on the number density of species A on the substrate surface. Here, we assume $k_s = k_s'$ and $h, r_w \gg w$ (Fig. 14.2.1a). In this limit, the interaction of species with the walls and the entrance window of the reaction chamber can still be ignored. With these approximations, an analytical solution of the stationary problem can be found. With $z > 0$ we have

$$D_A \nabla^2 N_A + Q_{v,A}(r) = 0, \qquad (14.2.9)$$

and with $z = 0$

$$D_A \frac{\partial N_A}{\partial z} = k_s N_A. \qquad (14.2.10)$$

with a Gaussian laser beam and low gas-phase absorption, the source term can be written as

$$Q_{v,A} = Q_0 \exp\left(-\frac{r^2}{w_0^2}\right), \qquad (14.2.11)$$

14.2 Concentration of Reactive Species

where

$$Q_0 = \frac{\mu \sigma_{AB} N_{AB} I_0}{h\nu}.$$

By using integral transforms, we find

$$N_A(r=0, z=0) = \frac{Q_0 w_0^2}{2D_A} \int_0^\infty \frac{\exp(-\zeta^2) d\zeta}{\zeta + \frac{k_s^*}{2}}, \tag{14.2.12}$$

where $k_s^* = k_s w_0 / D_A$. If $k_s^* \gg 1$ this yields

$$N_A(0,0) \approx \frac{\mu \sigma_{AB} N_{AB} P}{2\sqrt{\pi} h\nu k_s w_0}. \tag{14.2.13a}$$

This result differs from (14.2.8) by a factor 2 (in the ballistic approximation half of the species never reach the substrate). This approximation proves to be reasonable with $k_s^* \geq 5$ (see below). If $k_s^* \ll 1$ we obtain from (14.2.12)

$$N_A(0,0) \approx \frac{\mu \sigma_{AB} N_{AB} P}{2\pi h\nu D_A} \ln\left(\frac{2}{k_s^*}\right) \tag{14.2.13b}$$

14.2.3 Influence of Reaction Chamber

The concentration of photo-products impinging onto the substrate surface depends on the size of the reaction chamber and on the material from which it is fabricated. This has been demonstrated by solving (14.2.1) together with (14.2.2), (14.2.3) and (14.2.4) for the reactor geometry depicted in Fig. 14.2.1a. The calculations reveal that for constant laser power the exponent n, according to the ansatz

$$W(0,0) \propto \frac{1}{w_0^n}, \tag{14.2.14}$$

strongly depends on the rate constants $k_r \{k_s, k_s', k_w, k_w'\}$ and, under certain conditions, on the reactor size which is characterized by the dimensionless quantity $\Gamma = h/w_e$ where $w_e = \sqrt{2} w_0$ and $h = r_w$.

Let us first consider the situation where species A react at the *total* sample surface (inside and outside of the laser-irradiated area) and at the walls and windows of the reaction chamber with equal rate $k^* \equiv k_s^* = k_s^{*\prime} = k_w^* = k_w^{*\prime}$. Examples would be photolytic etching of metal and glass substrates by halogen radicals within metal and glass reactors, respectively.

Figure 14.2.2 shows the concentration $N_A^*(0,0)$ and the exponent n (right scale) as a function of k^*. Full and dotted curves belonging to values $\Gamma = 10^4$ and $\Gamma = 10^3$, respectively, almost coincide.

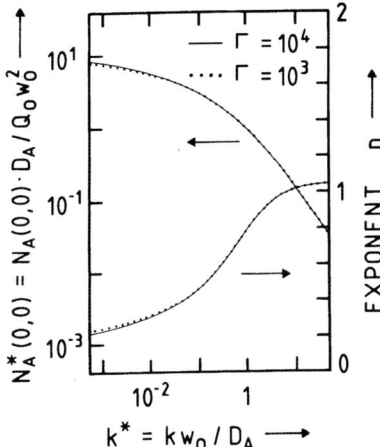

Fig. 14.2.2. Dependence of the stationary concentration of photogenerated species $N_A^*(0,0)$ (left-hand scale) and of the exponent n (right-hand scale) on the normalized rate constant $k^* \equiv k_s^* = k_s^{*'} = k_w^* = k_w^{*'}$ with $\Gamma = 10^4$ (full curves) and $\Gamma = 10^3$ (dotted curves) [*Piglmayer* and *Bäuerle* 1989]

If k becomes very large, i.e., $k \to k_m \equiv k_{max}(\eta \approx 1;$ Sect. 3.4.2) species A react with unit probability on all surfaces. This limit is described by (14.2.13a) (note that these calculations refer to a Gaussian beam while the present model uses a constant intensity). The numerical calculations yield $n \approx 1$ for $k^* \geqslant 5$ (the slightly larger value of n is a numerical artefact originating from the 'coarse' discretization).

If k decreases, i.e., $k < k_m$, the fraction $(1 - \eta)$ of molecules impinging onto the various surfaces is reflected into the volume of the reaction chamber, thereby increasing the concentration N_A. The dependence of $W(0,0)$ on w_0 can be described, approximately, by (14.2.14) where, however, the exponent n is not constant. With very small values of k^*, the density N_A^* becomes more uniform within the reaction chamber and is no longer affected by w_0. Therefore, n approaches zero for $k^* \to 0$.

We now assume that species A react with equal probability on the *total* substrate surface, while the walls of the reaction chamber are inert. This situation is described by $k^* \equiv k_s^* = k_s^{*'}$ and $k_w^* = k_w^{*'} \to 0$. It applies, for example, to photolytic dry-etching of metals in chlorine atmosphere, where the reaction chamber is fabricated out of glass. The results are similar to those shown in Fig. 14.2.2, as long as $k^* \geqslant 10^{-4}$. With $\Gamma = 10^4$, the deviations are negligible. With $\Gamma = 10^3$ and $k^* = 10^{-4}$, however, the deviations amount to a decrease of about 12% in n and an increase of about 25% in $N_A^*(0,0)$. The latter originates from the increase in back reflection of species A from the reactor walls into the gas volume.

Figure 14.2.3 shows the temporal dependence of W^* for different values of k^* where $k^* \equiv k_s^{*'} = k_w^* = k_w^{*'}$ and $k_s^* \to k_m^*$. The times necessary to reach steady-state conditions depend strongly on k^*. With very small values of k^*, these times can become longer or comparable to typical laser-beam dwell times involved in laser chemical processing.

14.2 Concentration of Reactive Species

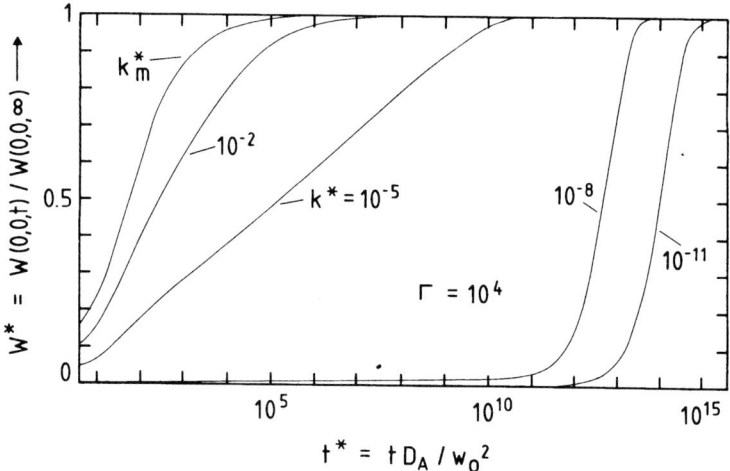

Fig. 14.2.3. Reaction rate W^* as a function of time t^* for various rate constants k^* and $\Gamma = 10^4$ [*Piglmayer* and *Bäuerle* 1989]

14.2.4 Gas-Phase Recombination

Let us consider the simplest case of gas-phase recombination where

$$f(N_A, N_B) = -k_A^{rec} N_A. \tag{14.2.15}$$

This ansatz permits an analytical solution of (14.2.1). The number density within the center of the laser beam is

$$N_A(0,0) = \frac{Q_0 w_0^2}{2 D_A} \int_0^\infty \frac{\zeta \exp(-\zeta^2) d\zeta}{\zeta^2 + k_A^{rec*} + \dfrac{k_s^*}{2}(\zeta^2 + k_A^{rec*})^{1/2}}, \tag{14.2.16}$$

where Q_0 and k_s^* are defined as in (14.2.12), and $k_A^{rec*} \equiv k_A^{rec} w_0^2/4 D_A$. With $k_A^{rec*} = 0$ equations (14.2.16) and (14.2.12) become identical. If the reaction is very fast and recombination very slow so that $k_s^* \gg 1$ and $k_A^{rec*} \ll 1$ we obtain

$$N_A(0,0) = \frac{\pi^{1/2} Q_0 w_0}{2 k_s}. \tag{14.2.17a}$$

This equation coincides with (14.2.13a). If, on the other hand, $k_A^{rec*} \gg 1$ we obtain

$$N_A(0,0) = \frac{Q_0 w_0}{w_0 k_A^{rec} + 4k_s(k_A^{rec*})^{1/2}}. \qquad (14.2.17b)$$

More general cases of gas-phase recombination introduce a nonlinearity in (14.2.1). Then, calculations can only be performed for specific systems. As an example, we study the photolysis of chlorine. The recombination of chlorine radicals (atoms) generated within the *gas phase* takes place via three-particle collisions (14.1.3a,b), which can be described by

$$f(N_{Cl_2}, N_{Cl}) = -2k_{Cl-Cl} N_{Cl}^2(r,z,t) N_{Cl_2}, \qquad (14.2.18)$$

where

$$k_{Cl-Cl} = k_1 + k_2 = 5.5 \times 10^{-32} \, cm^6/s. \qquad (14.2.19)$$

The substrate surface shall be characterized by $k_s \to k_m \approx \langle v_{Cl} \rangle/2 = 2.13 \times 10^4$ cm/s. We choose $k'_s = k_w = k'_w \equiv k = 0.1$ cm/s. This very small value must be considered as a *minimal* rate constant for reactions between chlorine radicals and material surfaces [*Clyne* and *Stedman* 1968].

Figure 14.2.4 shows the reaction rate, $W^*(0,0,t)$, as a function of time, t, for various chlorine pressures, $p(Cl_2)$. The time necessary to reach stationary conditions, t_∞, increases with decreasing chlorine pressure. The comparison of the 100 mbar curves shows that recombination diminishes t_∞. With 1 mbar, t_∞

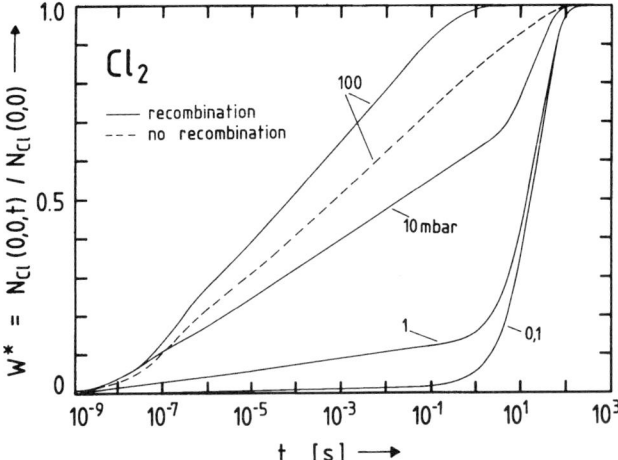

Fig. 14.2.4. Temporal dependence of the photolytic dissociation rate of chlorine. Full curves have been calculated by incorporating the proper recombination kinetics (14.2.18) into (14.2.1). The dashed curve shows the behavior without recombination. The parameters were α_{Cl_2}(514.5 nm, 1000 mbar) $I = 33$ W/cm^3, $w_0 = 5$ μm, $p_{Cl_2} D_{Cl} = 79$ cm^2 s^{-1} mbar, $k_s \to k_m = 2.13 \times 10^4$ cm/s, $k'_s = k_w = k'_w = 0.1$ cm/s, and $\Gamma = 10^4$ [*Piglmayer* and *Bäuerle* 1989]

is of the order of a hundred seconds. Such times can be longer or comparable to typical laser-beam dwell times employed in laser-induced dry-etching.

14.2.5 Gas-Phase Heating

The influence of gas-phase heating on the reaction flux of species can be calculated from the equations presented in Sect. 3.5. For pure etching we employ the boundary condition $x_{BC}(r, z \to \infty) = x_{BC}(\infty)$ and set $k_2 \neq 0$ and $k_1 = k_3 = 0$ in (3.5.1). The spatial distribution of the reaction rate is essentially the same as shown in Fig. 16.3.1, except that the height of the deposit is now replaced by the depth of the etched hole. At low temperatures, the depth of the hole decreases monotonically with increasing distance r^*. At higher temperatures, the maximum etch depth can occur at a certain distance $r > 0$.

14.3 Dry-Etching of Metals

Laser-enhanced dry-etching has been studied for many metals – both elements and compounds (Appendix B.3).

Subsequently, we will separately discuss spontaneous etching systems, diffusive etching systems, and passivating reaction systems.

14.3.1 Spontaneous Etching Systems

Spontaneous etching denotes a situation where the material dissolves within the ambient medium *without* any external influence. Among the examples are Al in Cl_2 atmosphere or Mo, Ta, Ti, W in XeF_2, etc. Small mass losses due to etching can conveniently be measured by means of a quartz crystal microbalance (QCM; Sect. 29.3). In most of these experiments, the metal film is directly evaporated onto the quartz.

Figure 14.3.1 shows measurements for the Al–Cl_2 system. Curve a shows the situation *without* laser light. After exposure of the Al film to Cl_2, the frequency of the QCM first decreases due to chlorine chemisorption and then increases due to etching. Etching continues until the chlorine gas is pumped off.

Light can significantly enhance the etch rate. This is shown in the lower part of Fig. 14.3.1 for (pulsed) N_2-laser radiation (note change in scales). The enhancement in etch rate depends on the laser parameters and the chlorine–gas pressure (curves b and c). With $\phi = 0.12$ J/cm^2 and $p(Cl_2) = 0.13$ mbar

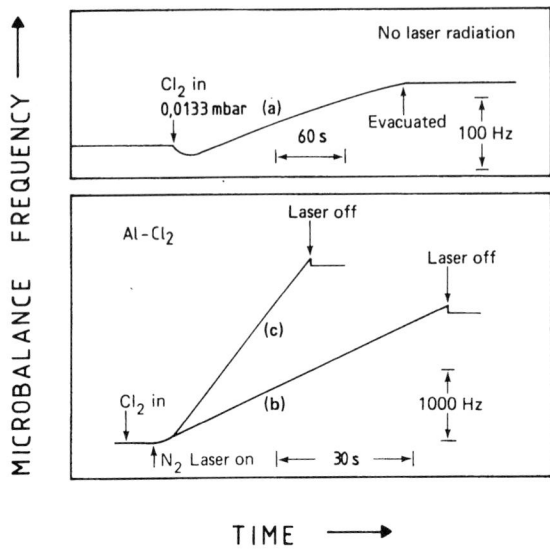

Fig. 14.3.1. Frequency responses of a quartz crystal microbalance (QCM) covered with a thin film of Al exposed to Cl_2 atmosphere. (a) Dark etching, $p(Cl_2) = 0.013$ mbar. (b) N_2 laser irradiation (0.12 J/cm², $\tau_l = 10$ ns, 30 pps), $p(Cl_2) = 0.013$ mbar. (c) As (b) but with $p(Cl_2) = 0.13$ mbar [*Sesselmann* and *Chuang* 1985]

the etch rate was $W_E \approx 11$ Å/pulse; with XeCl-laser radiation ($\phi = 0.3$ J/cm², $p(Cl_2) = 1.3$ mbar) it was $W_E \approx 30$ Å/pulse.

14.3.2 Diffusive Etching Systems

Diffusive etching systems are characterized by strong physisorption or chemisorption of the reactant *and* by the diffusion of corresponding radicals into the bulk. The main difference to spontaneous etching systems is related to the low vapor pressure of products. Among the model systems investigated are Ag and Cu in Cl_2 atmosphere. The degree of chlorination of the metal surface, $MeCl_x$, depends on the Cl_2 pressure, the time of exposure, and the distance from the surface. This has been evaluated from Auger depth profiles which were calibrated by means of X-ray photoemission spectroscopy (XPS) [*Sesselmann* and *Chuang* 1987]. For Cl_2 exposures lower than 10^7 L (1L = 1 Langmuir = 10^{-6} Torr s) the Cl concentration decreases rapidly with depth. In this regime, the chlorine uptake increases logarithmically with exposure time. This behavior can be explained on the basis of field-enhanced diffusion as described by Cabrera and Mott (Chap. 26). The strongly electronegative chlorine atoms adsorbed on the Ag surface become negatively charged by electron transfer from the metal. The resulting electric field across the surface layer causes diffusion of Cl^- ions into the metal and of Ag^+ ions towards the surface. Because of the smaller radius of Ag^+ (≈ 1.26 Å) compared to Cl^- (≈ 1.81 Å) Ag diffusion to the surface is dominating. In this surface-chlorination reaction, dissociative chemisorption of chlorine seems to be rate limiting. For exposures

14.3 Dry-Etching of Metals 243

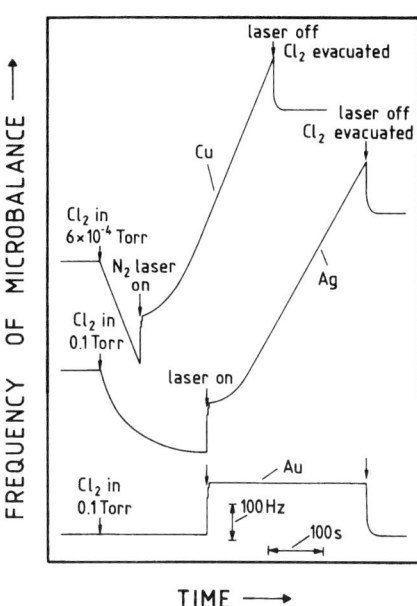

Fig. 14.3.2. Frequency response of QCM with evaporated Cu-, Ag- and Au-films after exposure to Cl_2- and N_2-laser irradiation ($\phi = 0.12$ J/cm^2, $\tau_l = 10$ ns, 30 Hz). The step-like frequency change observed when the laser is switched on and off are caused by temperature changes and *not* by desorption or adsorption of species [*Sesselmann* et al. 1986a]

$> 10^7$ L, the chlorinated layer becomes quite thick, up to 50 Å, and has a composition close to stoichiometric AgCl ($x > 0.8$). In this regime, bulk diffusion of ions becomes rate limiting and the chlorine uptake increases with $t^{1/2}$. Because both silver and copper chlorides have a very low volatility no spontaneous etching occurs.

Laser-light irradiation removes the chlorinated layer. Figure 14.3.2 shows the behavior of Cu, Ag, and Au in Cl_2 under N_2-laser radiation. No etching is observed with Au, and Cu etches faster than Ag ($W_E \approx 0.3$ Å/pulse). The desorption yields, as well as the mass and velocity distributions of species (Me, Cl, and Me$_y$Cl$_x$) for VIS and UV laser pulses, were determined by time-of-flight (TOF) mass spectrometry [*Sesselmann* et al. 1986a,b; *van Veen* et al. 1988]. The type and relative concentration of species desorbed from the surface depends on the laser parameters and differs significantly from that observed in standard thermal desorption experiments. The influence of gas-phase excitations of chlorine and of thermal and electronic excitations of the MeCl$_x$ layer has been studied by *Brannon* and *Brannon* (1989).

14.3.3 Passivating Reaction Systems

The class of passivating reaction systems investigated includes: Fe, Mo, Ni, W, and Ni$_x$Fe$_y$ alloys in Cl_2 atmosphere, and Ti in Br_2 and CCl_3Br. The situation is similar to surface passivation by native oxide formation (Chap. 26). The etchant chemisorbs on the metal surface and forms a stable and dense metal–halide layer which suppresses any further reaction. The layer is, typically, 10–30 Å thick and can be estimated from QCM measurements. The decrease in microbalance frequency observed directly after Cl_2 exposure of Ni, Fe, and Ni_xFe_y alloys can be seen in Fig. 14.3.3. In contrast to diffusive etching systems, no diffusion of the halogen into the bulk metal takes place.

Laser–light irradiation yields a local temperature rise which results in a step-like increase in QCM frequency. Subsequently, a continuous increase in QCM frequency due to laser-induced desorption of the metal-halide layer is observed. While the etch rate is almost negligible for pure Ni, it increases strongly with Fe alloying. With the parameters employed, the rates were $W_E(Ni) < 0.01$ Å/pulse, $W_E(Fe_{0.53}Ni_{0.47}) \approx 0.2$ Å/pulse, and $W_E(Fe) \approx 0.5$ Å/pulse. The differences in etch rates can be related to the volatilities of reaction products. $FeCl_3$ has a relatively high vapor pressure while Ni chlorides are essentially non-volatile below about 500 K.

Localized etching by means of focused cw Ar^+–laser radiation was investigated for Mo and W films on glass substrates immersed in Cl_2. Figure 14.3.4 shows etch rate measurements for various Cl_2 pressures. The full curves have been calculated on the basis of a thermally-activated process. The good agreement with experimental data suggests that etching at medium to high laser powers can be interpreted by (mainly) thermal desorption of reaction products. The influence of non-thermal effects is still under discussion.

14.4 Dry-Etching of Inorganic Insulators

Laser-induced dry-etching of inorganic insulators has been investigated for fused quartz, for various glasses of complex composition, for different ceramic, poly-, and single-crystalline oxides and nitrides, and some other materials. The

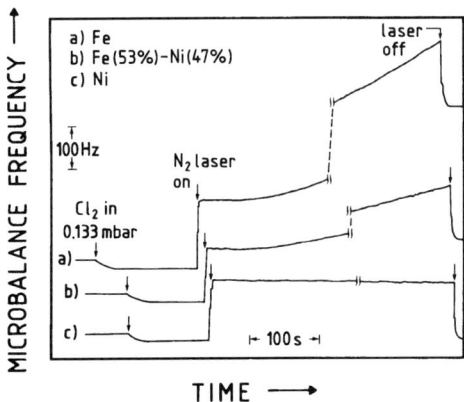

Fig. 14.3.3. Frequency response of quartz crystal microbalances covered with metal films exposed to a 0.13 mbar Cl_2 atmosphere and N_2-laser radiation (0.12 J/cm², $\tau_l = 10$ ns, 30 Hz) [*Chuang* et al. 1984]

14.4 Dry-Etching of Inorganic Insulators   245

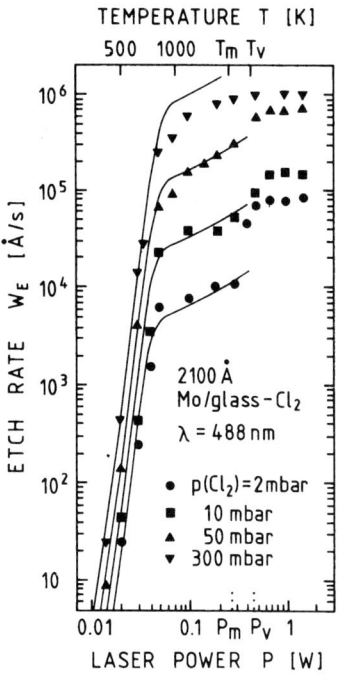

Fig. 14.3.4. Laser-induced etch rate for 2100 Å Mo films on glass as a function of 488 nm Ar$^+$-laser power ($w_0 \approx 5.7$ μm) and for various Cl_2 pressures. $T_m(P_m)$ and $T_v(P_v)$ indicate the temperatures (laser powers) for melting and evaporation of Mo, respectively. The full curves have been calculated [*Mogyorosi* et al. 1989a]

precursor molecules employed are mainly halides and halogen compounds. With the etching of oxides, H_2 is also frequently used as precursor. In this section we include some materials which are neither real insulators nor classified as semiconductors. References to the original literature can be found in Appendix B.3.

14.4.1 SiO$_2$ Glasses

Photochemical dry-etching of fused quartz (SiO_2) and of SiO_2-rich glasses has been investigated mainly with hydrogen and various halide radicals. The latter have been produced by both electronic and vibrational excitations of precursor molecules. Spontaneous etching can be ignored.

ArF– and KrF laser radiation has been used to photodissociate molecules like CF_2Cl_2, CF_2Br_2 and CF_3Br, CF_3I, CF_3NO, and $CO(CF_3)_2$ [*Brannon* 1984; *Loper* and *Tabat* 1984]. Etching was very efficient for CF_2Cl_2 and CF_2Br_2 where etch rates of, typically, 0.2 to 0.5 Å/pulse have been achieved. With the other compounds which mainly form CF_3 as photoproduct, etching was either inefficient or else did not take place at all. Apparently, CF_2 radicals interact with SiO_2 more strongly than CF_3 radicals – the reason for this observation remains unexplained. Experiments carried out under similar conditions, but with Br_2 and SF_6 did not etch SiO_2 significantly.

Hirose et al. (1985) have investigated etching of SiO_2 in a mixture of NF_3 and H_2 by ArF-laser radiation. The reaction has been qualitatively described by

$$SiO_2 + NF_3 + H_2 + h\nu(193 \text{ nm}) \rightarrow SiF_4(\uparrow) + NO_2 + N_2O + HF. \quad (14.4.1)$$

Etch rates of about 1 Å/s have been achieved. They were found to increase with H_2 partial pressure. In view of applications, it should be noted that ArF-laser light produces defects which act as electron traps.

Etching of SiO_2 in Cl_2 atmosphere by visible Ar^+-laser radiation has been investigated by *Chuang* (1982). Because SiO_2 is transparent within the visible spectral region, laser-induced heating of the substrate is of little importance. In fact, the etch rate has been found to be correlated to the concentration of chlorine radicals produced within the gas phase. For 457.9-nm radiation and a chlorine pressure of $p(Cl_2) = 133$ mbar, an etch rate of up to 3 Å/s was achieved. With the 514.5 nm laser line, under otherwise identical conditions, the etch rate was significantly lower. The lateral dimensions of etched features were much larger than the focal spot size. For example, for $2w_0 \approx 7$ μm the diameter of etched holes was 50 to 80 μm, depending on laser-beam illumination time. This can be explained by the diffusion of Cl radicals produced within the gas phase.

Large-area etching of SiO_2 activated by *multiphoton* vibrational dissociation (MPD) of CF_3Br and CDF_3 by means of pulsed CO_2-laser radiation has been reported by *Harradine* et al. (1981). Because of the high laser-light intensities, such experiments can be performed only with parallel laser-beam incidence. The etch rates achieved were around 0.3 Å/pulse.

Projection patterning of SiO_2-rich glasses by transient heating in H_2 atmosphere has been demonstrated with ArF-laser radiation. Gratings with a resolution of about 0.4 μm have been produced. Figure 14.4.1 shows the etch rate as a function of laser fluence for pyrex and thermally grown SiO_2 on Si.

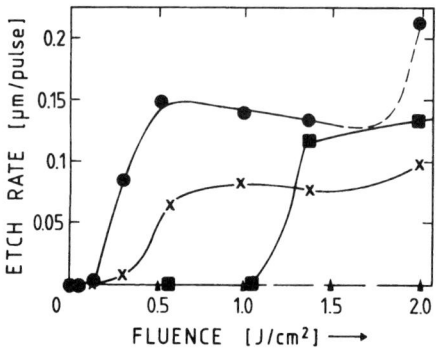

Fig. 14.4.1. Etch rate achieved for different glasses in H_2 atmosphere as a function of ArF-laser fluence. ● Pyrex, $p(H_2) = 266$ mbar. X Pyrex, $p(H_2) = 1333$ mbar. ■ SiO_2/Si, $p(H_2) = 266$ mbar [*Ehrlich* et al. 1985]

14.4.2 Oxides

Laser-induced dry-etching of various oxides and nitrides in ceramic, polycrystalline, and single-crystalline form has been studied for different ambient media. Among the materials investigated in most detail are oxidic perovskites and ferrites.

Oxidic perovskites

With short, high-intensity UV-laser pulses whose photon energies exceed the band-gap energy of the perovskite ($h\nu > E_g \approx 3$ eV), surface patterning is based on ablation and the results of such investigations are incorporated in Chap. 12.

Well-defined dry-etching with visible laser light ($h\nu < E_g$) of low to moderate intensity requires a reducing atmosphere. Here, laser-induced heating (initiated by an absorbing paint, defects, or by trapped radiation within ceramic samples) results in the formation of oxygen vacancies and quasi-free electrons (Sect. 27.1). With increasing laser power, metallization and, finally, etching occurs. The most detailed investigations have been performed for crystalline $BaTiO_3$ and $SrTiO_3$, and for ceramic $BaTiO_3$ and $PbTi_{1-x}Zr_xO_3$ (PZT) in H_2 atmosphere. SEM pictures of grooves produced in PZT using 647 nm Kr^+-laser radiation are shown in Fig. 14.4.2. Material removal in air is very irregular while, under otherwise identical experimental conditions, well-defined grooves can be produced in H_2. The threshold for material removal is considerably lower in H_2 than in air. The role of H_2 is interpreted by efficient local reduction of the material. This has two consequences:

- With the formation of oxygen vacancies, color centers, and quasi-free electrons, optical absorption within the laser-heated zone increases strongly and becomes spatially well defined.
- With increasing oxygen-vacancy concentration, a local collapse of the perovskite lattice occurs.

Temperature measurements using visual and photoelectric pyrometry revealed an Arrhenius-type behavior of the average etch depth. For laser-induced temperatures $T_s \leqslant 1600$ K, an apparent activation energy of $\Delta E = 41 \pm 8$ kcal/mol was derived. In this connection it seems interesting to note that the evaporation enthalpies for pure PbO [JANAF: Tables] and for PbO in $PbTi_{1-x}Zr_xO_3$ ($x = 0.65$) [*Northrop* 1968] are 56 kcal/mol and 39 kcal/mol, respectively. The evaporation enthalpies for pure TiO_2 and ZrO_2 are 137 and 170 kcal/mol, respectively. Thus, evaporation of PbO seems to be important in the etching process. This conclusion is supported by SEM and electron-beam-induced X-ray fluorescence studies. In a region of about 30–50 µm around groove edges, morphology changes due to a depletion of Pb, and sometimes even cracks are observed. Beyond this region, the morphology and chemical composition corresponds to that of bulk PZT. The thickness of the damaged

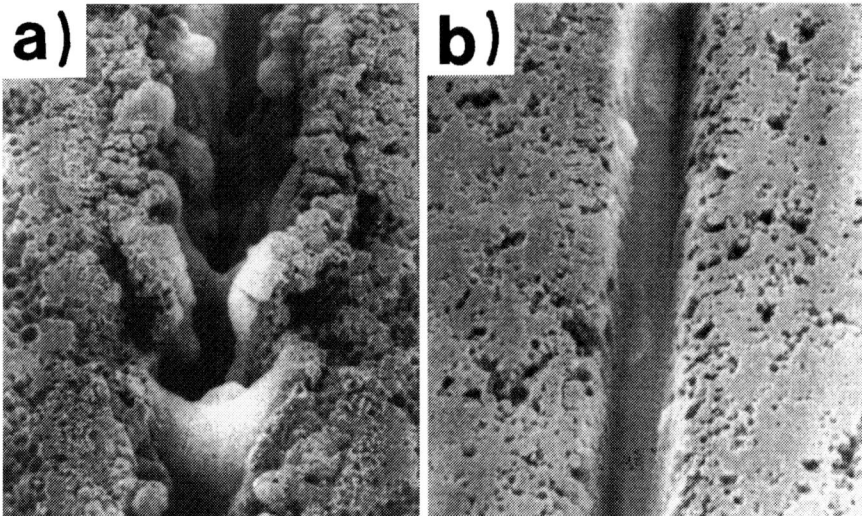

Fig. 14.4.2a,b. SEM pictures of grooves etched with 647 nm Kr^+-laser radiation in ceramic $PbTi_{1-x}Zr_xO_3$ ($P = 0.72$ W, $w_0 = 6.4$ μm, $v_s = 8.4$ μm/s). The ambient medium was (a) 90 mbar air; (b) 90 mbar H_2 [*Eyett* et al. 1986]

zone depends on the incident laser power and the scanning velocity. The etch rates achieved are, typically, between a few μm/s and some hundred μm/s. The results were quite similar for the other perovskites investigated. Note that patterning of PZT by ablation can be achieved without significant material damage (Chap. 12).

Ferrites

The most detailed investigations on (pyrolytic) laser-induced etching of single-crystalline (100) Mn-Zn ferrites (MnO:ZnO: $Fe_2O_3 = 31:17:52$) were performed by *Takai* et al. (1988a). The experiments employed Ar^+–laser radiation and gaseous CCl_4 or CCl_2F_2 as etchant. Crack-free etching with high aspect ratio and rates of up to 68 μm/s has been achieved (Fig. 14.4.3).

Similar experiments have been performed in aqueous solutions of KOH and H_3PO_4. With Fe:Al:Si (Sendust) etch rates up to 400 μm/s and aspect ratios of 40 have been achieved [*Takai* et al. 1994]. The technique is employed for the fabrication of magnetic head structures for computer, audio, and video recording applications. Single-sided MIG (metal-indium-gap) heads fabricated by LCP show higher performance than those fabricated by conventional machining.

14.5 Wet-Etching

Fig. 14.4.3. Magnetic head structure in MnO:ZnO:Fe$_2$O$_3$ etched by CCl$_4$ using a scanned Ar$^+$-laser beam ($P \approx 0.35$ W, $w_0 \approx 1.3$ μm, $v_s \approx 30$ μm/s, $p($CCl$_4) \approx 30$ mbar) [*Takai* et al. 1988a]

14.5 Wet-Etching

Laser-enhanced wet-etching has been studied for metals in the form of thin films, foils, and slabs. Only a few of such investigations have been performed for inorganic insulators. The microscopic mechanisms involved in liquid-phase reactions are described in Chap. 21.

Metals

Among the metals studied in most detail are Al, Cu, Fe, steel, and Ni. The etchants mainly employed were aqueous solutions of acids like HCl, HNO$_3$, H$_3$PO$_4$, etc., and neutral salt solutions of NaCl, KBr, NaNO$_3$, K$_2$SO$_4$, etc. Most of the experiments have been performed by means of Ar$^+$–laser radiation. In the absence of an external voltage, the etch rates achieved were, typically, between one and several μm/s. Such rates exceed the corresponding rates for dark etching by many orders of magnitude. Spatial resolution of better than 2 μm has been demonstrated.

Laser-enhanced electrochemical etching (LEE) is achieved by simply reversing the polarity of the cathode and anode in the experiments described for laser-enhanced plating (Sect. 21.2). LEE has been used to produce 50 μm diameter holes in stainless steel using aqueous NiCl$_2$ as an electrolyte and Ar$^+$–laser radiation [*von Gutfeld* 1984]. Etch rates of up to 10 μm/s have been obtained.

Wet-etching of Cu based on the photodecomposition of the etchant has been described by *Donohue* (1984). The etchant was an aqueous solution of Br_2 – sometimes with an admixture of KBr. XeCl- and XeF-laser radiation was used to photodissociate Br_2. The bromine radicals react with Cu to $CuBr_2$ and CuBr. While $CuBr_2$ is very soluble in water, CuBr is rather insoluble and may give rise to particulates that occur (mainly) at the edges of the illuminated area.

15 Etching of Semiconductors

This chapter deals with laser-induced etching of element and compound semiconductors. An inspection of Appendix B.3 shows that the most detailed investigations on dry-etching have been performed for Si, GaAs, and InP. The precursor molecules mainly employed include halides and halogen-compounds such as Cl_2, HCl, XeF_2, NF_3, CCl_4, CF_3Br, CF_3I, and SF_6. Laser-induced wet-etching has been studied mainly for compound semiconductors in aqueous solutions of H_2SO_4/H_2O_2, HNO_3, and KOH.

In *photothermal* dry-etching the rates typically achieved are between 0.01 μm/s and several μm/s, and about ten times higher in wet-etching.

Photochemical etching is based on the interaction between radicals and carriers within the semiconductor surface. Radicals can be formed spontaneously by molecule–surface interactions (Si-XeF_2 system), by selective electronic excitation (Si-Cl_2 system), or by vibrational excitation (Si-SF_6 system) of the etchant. The carriers can be incorporated into the semiconductor by doping or they can only be generated by photo-excitation. Thus, photochemical etch rates depend on the concentration of active species within the ambient medium, including the adsorbate, on the optical penetration depth of the laser light in the semiconductor surface, on carrier lifetimes, recombination processes, surface band bending, and on the concentration of impurities. The etch rates are typically between several Å/s and some 0.1 μm/s.

15.1 Dark Etching

Before discussing the mechanisms involved in light-enhanced etching of silicon in halogen atmosphere, it is useful to first consider some basic interactions between halogen radicals $X \equiv$ F, Cl, Br,... and silicon within the dark.

Chlorine molecules become dissociatively chemisorbed on clean Si surfaces. This process is mediated by charge (electron) transfer from silicon to physisorbed chlorine molecules (Fig. 15.1.1a). Clearly, the number of electrons and holes increases with temperature. Charge transfer results in the formation of a passivating $SiCl_x$ layer which suppresses any further reaction. This process is similar to the initial phase of surface oxidation.

The situation is different with fluorine which can etch Si *spontaneously*. The main differences to chlorine result from the higher electronegativity and the smaller radius of fluorine [atomic radius $r_a(F) \approx 0.64$ Å; (Pauling) ionic radius $r_i(F^-) \approx 1.36$ Å], compared to chlorine [$r_a(Cl) \approx 0.97$ Å; $r_i(Cl^-) \approx 1.81$ Å]. As a consequence, fluorine not only chemisorbs on Si but diffuses into the surface and forms a thick fluorosilyl layer (Fig. 15.1.1b). This becomes plausible from the inter-atomic distances for silicon (100), (110) and (111) surfaces. The Si radius in tetrahedral coordination is 1.17 Å. Thus, one finds, together with the lattice constants, that the radius of the 'hole' between Si atoms is about 1.54 Å for (100), 1.36 Å for (110) and 1.04 Å for (111) surfaces. It has in fact been confirmed by both model calculations [*Seel* and *Bagus* 1983] and experiments [*McFeely* et al. 1986; *Winters* and *Plumb* 1991] that fluorine can penetrate into all of the different Si surfaces. The fluorosilyl layer has a thickness of, typically, 10 to 30 Å and is dominated by SiF_3 groups. Spontaneous etching occurs from this fluorosilyl layer by desorption of SiF_x. The main volatile product at 300 K is SiF_4. To a minor extent, Si_2F_6 and Si_3F_8 are observed as well. Heating of Si to temperatures up to about 600 K does not change the distribution of reaction products, but only increases the overall reaction rate. The situation changes at higher temperatures. For example, thermal desorption studies of fluorinated Si have shown that at 800 K the primary etch product is SiF_2. The distribution of reaction products depends also on the orientation of the Si surface.

The main aspects of surface fluorination can be qualitatively understood along the lines of the Cabrera–Mott mechanism (Chap. 26). Consider the schematic picture shown in Fig. 15.1.1b. With stationary conditions, the thickness of the fluorosilyl layer remains constant. Thus, for compensation of desorbed SiF_x, transport of Si to the surface, or of F to the SiF_x-Si interface, is required. Because of the difference in radii, diffusion of fluorine through the SiF_x layer is expected to be dominating. The picture is similar to that of

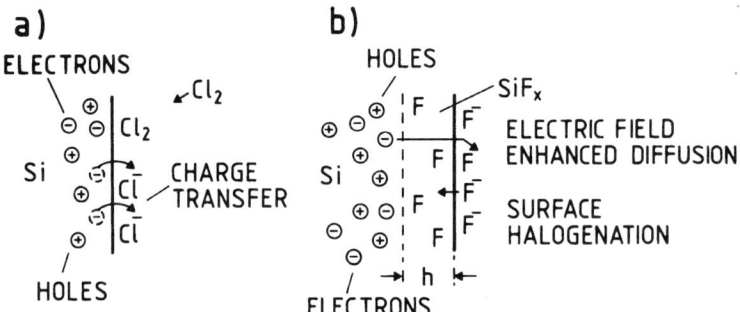

Fig. 15.1.1a,b. Schematic picture showing charge (electron) transfer and the generation of a surface electric field for Si in halogen atmosphere. (**a**) Chlorine molecules dissociatively chemisorb on clean Si surfaces. This process is mediated by electron transfer. (**b**) Fluorine chemisorbs on Si and diffuses into the surface. Diffusion is enhanced by the surface electric field

15.1 Dark Etching 253

Fig. 26.3.1. The main difference to surface oxidation is that many saturated halides are volatile while most oxides are not.

To achieve further insight, we consider the simple band model in Fig. 15.1.2. It schematically shows the highest valence band and the lowest conduction band of Si, the SiF$_x$ layer, and the vacuum level. At infinite distance, the affinity level of fluorine atoms is about 3.45 eV below the vacuum level. As fluorine approaches the surface, this value increases, mainly because of the image-charge attraction from the (bulk) Si and the SiF$_x$ layer. There are indications that the affinity level for fluorine is placed below the valence band of Si. When fluorine atoms become adsorbed, electrons will be transferred to them as long as the affinity level at the surface $z = 0$ is below the Fermi level. The electric field generated by F$^-$ ions on the surface of the SiF$_x$ layer and the (positive) space charge within the Si near the interface (Fig. 15.1.1b), causes the bands to bend upward. The degree of band bending is a measure of the field strength. The surface concentration of F$^-$ ions and the field strength are determined by the energy difference between the Fermi level, E_F, and the affinity level, E_a, at the Si–SiF$_x$ interface. This difference is denoted by $(E_F - E_a)_{z=h} = e\Phi$. The equilibrium (number) density of F$^-$ ions at the surface is, according to Gauss's law

$$N_{F^-} = \varepsilon_{SiF_x} \frac{\Phi}{eh}, \qquad (15.1.1)$$

where ε_{SiF_x} is the dielectric constant of the SiF$_x$ layer. h is the thickness of the SiF$_x$ layer which is assumed to be an insulator. The space charge accumulated within the bulk Si near the Si–SiF$_x$ interface must be equal (but of opposite sign) to the charge of F$^-$ ions at the SiF$_x$ surface (note that for a capacitor $C = \varepsilon F/h = Q/\Phi$ and $Q/F = eN_{F^-}$). This, together with (15.1.1), determines the potential difference Φ and the number density N_{F^-} for a given SiF$_x$ layer thickness. From qualitative arguments one finds that N_{F^-} increases with dopant concentration only with heavily doped (degenerate) n-type Si, but remains almost unaffected otherwise. With increasing layer thickness h, the

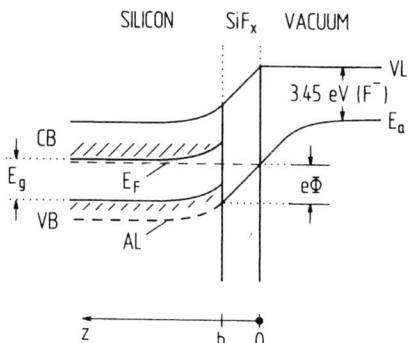

Fig. 15.1.2. Schematic band model which describes the influence of an electric field caused by chemisorbed F$^-$ ions on a silicon surface. CB stands for conduction band, VB for valence band, AL for affinity level, and VL for vacuum level. Adapted from [*Winters* and *Haarer* 1987]

density N_{F^-} decreases. Dark etching of doped (111) Si in XeF_2 atmosphere is consistent with this model; the measured etch rates are directly correlated with the surface concentration of fluorine ions, N_{F^-}. In a quantitative treatment of the problem, N_{F^-}, Φ, and h must be determined self-consistently from (15.1.1), the Poisson–Boltzmann equation, and a kinetic equation of the type (28.3.4) for $h = h(N_{F^-}, \Phi)$. The problem is, however, that the detailed surface chemistry and the corresponding kinetic coefficients are unknown.

15.2 Laser-Induced Etching of Si in Cl_2

The most detailed investigations on laser-induced dry-etching of element semiconductors have been performed for Si. Only a few papers have been published on Ge (see Appendix B.3).

15.2.1 Surface Patterning

High resolution etching of Si in chlorine atmosphere has been demonstrated by both direct writing and a special type of projection patterning. The spatial resolution achieved with both techniques is better than 1 µm. The chlorine gas pressures employed range from about 1 mbar to several hundred mbar.

Figure 15.2.1a shows a scanning electron micrograph of deep grooves obtained by Ar^+-laser direct writing. Here, line widths below 0.2 µm have been achieved by adjusting the laser-beam intensity in such a way that melting occurs only in the center of the laser-induced temperature profile. The increase

Fig. 15.2.1a,b. SEM pictures of patterns fabricated in c-Si by 488 nm Ar^+-laser direct writing in Cl_2 atmosphere. (**a**) Trenches ($P = 3$ W, $v_s = 1$ mm/s, 10 scans per line; $p(Cl_2) = 140$ mbar) [*Müllenborn* et al. 1995]. (**b**) Hollow with cone. The microscope objective was lowered in 1 µm increments after each plane is scanned at $v_s = 7.5$ mm/s [$2w_0 \approx 1$ µm, $p(Cl_2) = 133$ mbar] [*Ehrlich* 1993]

15.2 Laser-Induced Etching of Si in Cl_2

in resolution with respect to the diffraction-limited diameter of the laser beam is based on the much higher reactivity of liquid Si with respect to crystalline Si (Sect. 5.3.6; Fig. 15.2.2).

Figure 15.2.1b shows another example for three-dimensional patterning. Here, the laser beam was swept in a circular scan. The technique permits computer-aided design (CAD) or manufacturing (CAM) of silicon-based micro-electromechanical devices.

Projection patterning has mainly been investigated by means of excimer-laser radiation. In the simplest arrangement which just uses a projection mask and a Si substrate immersed in Cl_2 atmosphere, neither the contrast nor the resolution was satisfactory. This can be ascribed to the diffusion of both (gas-phase) Cl atoms and photo-carriers. These problems have been overcome by a technique developed by *Horiike* et al. (1987). Here, a permanent flow of Cl radicals generated in a microwave discharge tube and methyl-methacrylate (MMA) gas is used. The reaction between both produces a polymer film on the substrate surface. KrF-laser radiation removes the film from the illuminated region and thus permits etching by Cl radicals. Because of the perpendicular incidence of the laser light, the polymer film remains on the almost vertical side walls of the etched feature and thus avoids undercutting. This technique allows anisotropic etching and has been applied to n-type poly-Si. Here, satisfactory contrast has been achieved.

15.2.2 Photochemical and Thermal Etching

Silicon is quite inert to Cl_2 atmosphere at 300 K. Etching is observed only at higher temperatures, or in the presence of light.

Laser-induced etching of Si in Cl_2 atmosphere has been investigated by various groups. For *low* light intensities the molecule–surface interactions can be qualitatively explained on the basis of Figs. 15.1.1 and 15.1.2. The exact position of the chlorine affinity level is unknown – it is probably situated near the valence band edge. At higher laser-light intensities, thermal activation of the etching reaction is expected.

Figure 15.2.2 shows the etch rate achieved with pulsed-irradiation at three different wavelengths and constant Cl_2 pressure. The fluences $\phi(\lambda)$ were normalized to the fluences $\phi_m(\lambda)$ required for surface melting, $\phi^* = \phi/\phi_m$. The results obtained with 308 nm XeCl-laser radiation show three characteristic regimes:

– For low laser fluences which cause negligible surface heating (ϕ^*(308 nm) < 0.2), the etch rate increases linearly with ϕ^* (full line). Within this regime, etching is purely *photochemical* and based on both chlorine radicals produced within the gas phase and electron–hole pairs generated within the Si surface.

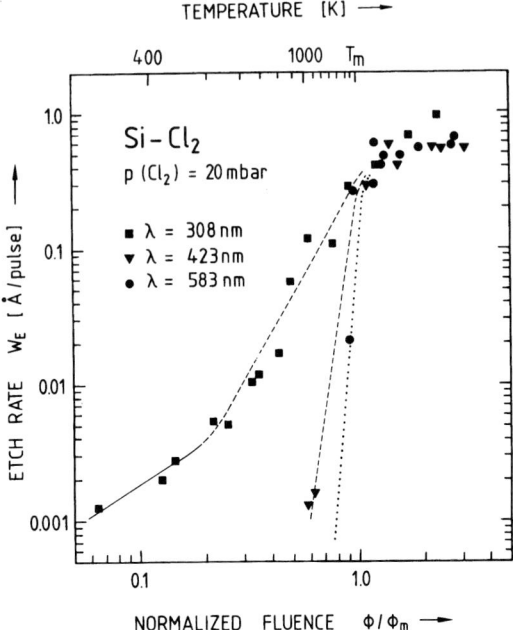

Fig. 15.2.2. Etch rate of (100)Si [slightly p-doped with $\approx 10^{14}$ B atoms/cm^3, $\rho = 100$ to $150\,\Omega$cm] as a function of (normalized) laser fluence, and for three different wavelengths (the 423 nm and 583 nm lines were obtained from a XeCl-laser-pumped dye laser) [*Kullmer* and *Bäuerle* 1987]

- At medium energy densities, i.e., between $0.2\,(\approx 150\,\text{mJ/cm}^2) < \phi^*$ (308 nm) $< 1\,(\approx 440\,\text{mJ/cm}^2)$, the etch rate increases nonlinearly with laser fluence. In this regime, thermal processes become important, but photo-generated Cl radicals are still required.
- At laser fluences that cause surface melting, i.e., with $\phi^* > 1$, the etch rate levels off. Such a behavior can be caused by the latent heat of melting, or the step-like increase in reflectivity at the melting point, or by mass transport limitations. With the Cl$_2$ pressures employed in this experiment the latter mechanism should be irrelevant. This could be proved by studying the dependence of the etch rate on the diameter of the laser focus. In any case, within this regime etching is mainly *thermally* activated.

Figure 15.2.2 shows that for $\phi^* > 1$ the etch rate depends only slightly on laser wavelength, while for $\phi^* < 1$ a strong decrease with increasing wavelength is observed. This wavelength dependence suggests that without surface melting, etching of Si requires the presence of Cl radicals which are generated only at wavelengths $\lambda \leqslant 500$ nm (Sect. 14.1). Photocarriers, on the other hand, are generated with all of the wavelengths investigated and their role cannot be revealed from these experiments because any changes in laser power or wavelength cause simultaneous changes in both the chlorine atom concentration within the gas phase *and* the concentration of photo-electrons within the Si surface.

15.2 Laser-Induced Etching of Si in Cl$_2$

To separate the influence of chlorine radicals and electron–hole pairs, one can use a *combined* irradiation scheme as shown in Fig. 30.1.1a. Such experiments have been performed for the Si-Cl$_2$ system by using 308 nm XeCl excimer-laser radiation at parallel incidence to the Si surface and 647.1 nm cw-Kr$^+$–laser radiation at perpendicular incidence. By this means, the concentration of Cl radicals within the gas phase, and of electron–hole pairs within the Si surface, can be controlled *independently* via the XeCl-laser fluence, ϕ, and the Kr$^+$-laser power, P, respectively. In brief, the main observation is: Significant etching of (p-doped) Si without or with negligible surface heating is observed only if both lasers are switched on. Thus, etching of p-type Si at laser powers that cause negligible surface heating requires both chlorine radicals *and* electrons.

15.2.3 Chlorine Radicals

The data points in Fig. 15.2.3a represent the etch rate measured as a function of XeCl-laser fluence. Both the chlorine gas pressure, and the Kr$^+$-laser power, were kept constant. The measured etch rate can be compared with the reaction flux of chlorine atoms onto the Si surface

$$J_{Cl}(r, z_s) = k N_{Cl}(r, z_s) \approx \frac{1}{4}\eta_{Si}(r) \langle v_{Cl} \rangle N_{Cl}(r, z_s). \tag{15.2.1}$$

The etch probability $\eta_{Si}(r)$ includes the sticking probability of Cl radicals, the conversion of Si to SiCl$_x$, and the desorption of SiCl$_x$ from the surface. Within

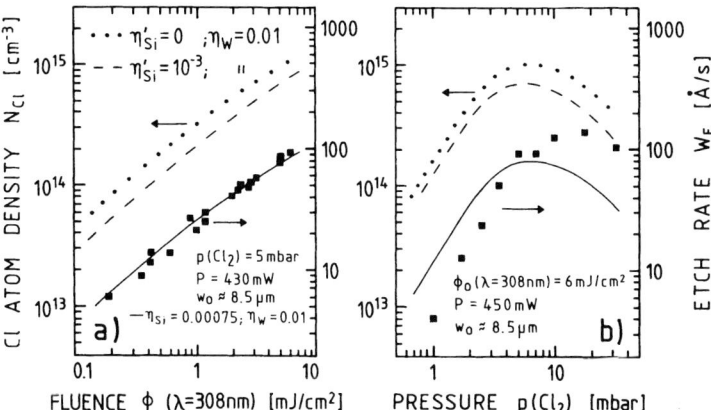

Fig. 15.2.3a,b. Data points show the measured etch rate W_E (right-hand scale) at constant Kr$^+$-laser power ($\lambda = 647$ nm; $P = 430$ mW; $w_0 \approx 8.5$ μm). The pulse repetition rate of the XeCl laser was 100 Hz. Dotted and dashed curves show the calculated chlorine atom density N_{Cl} on the Si surface within the center of the Kr$^+$-laser beam (left-hand scale). The full curves show the etch rates calculated with $\eta'_{Si} = 0, \eta_w = 0.01$, and $\eta_{Si}(r = 0) \equiv \eta_{Si} = 0.0075$. SiCl$_2$ was assumed to be the main reaction product. (a) Dependence on XeCl-laser fluence. (b) Dependence on chlorine pressure [*Kullmer* and *Bäuerle* 1988a]

the area of the (Gaussian) Kr^+–laser spot, it can be described by $\eta_{Si}(r) = \eta_{Si}(0)\exp(-r^2/l_c^2)$, where l_c is of the order of the free carrier diffusion length. $\langle v_{Cl} \rangle$ is the average Cl atom velocity.

The concentration of Cl atoms on the Si surface can be calculated as outlined in Sect. 14.2 (z_s is the distance of the Si surface from the center of the XeCl-laser beam; Fig. 30.1.1a). With $N_{Cl}(0, z_s)$ calculated for the reaction chamber employed in the experiments, and parameters $\eta'_{Si} = 0$ and $\eta_w = 0.01$ (η'_{Si} refers to non-irradiated Si and η_w to the walls of the reaction chamber; Sect. 14.2.3) the measured etch rate, $W_E(0)$, can be fitted with $\eta_{Si}(0) = 0.0075$ (this value is in good agreement with that reported by *Mogyorosi* et al. (1988); note that $W_E(0) = Zm_{Si}J_{Cl}(0)/\rho_{Si}$ where $Z = 1$, $1/2$, and $1/4$ with reaction products SiCl, $SiCl_2$, and $SiCl_4$, respectively). With pulsed UV- and VIS-laser radiation the main desorption products are SiCl and $SiCl_2$ [*Aliouchouche* et al. 1993; *Paulsen–Boaz* et al. 1992]. The result of the fit, which assumes $SiCl_2$ as reaction product, is shown in Fig. 15.2.3a by the full curve. The figure demonstrates that the observed dependence $W \propto \phi^{0.79}$ agrees very well with the calculated reaction flux of chlorine atoms. With the parameters employed, the reaction is *kinetically* controlled ($D_{Cl}/\eta_{Si}\langle v_{Cl}\rangle w_0 \gg 1$; Chap. 3). The data can also be compared with chemiluminescence measurements (Fig. 30.1.1b) from which we find $N_{Cl}(0,0) \propto \phi^{0.73}$. The consistency of results shows that the etch rate depends linearly on chlorine atom concentration

$$W_E \propto N_{Cl}. \tag{15.2.2}$$

The dependence of the etch rate on chlorine gas pressure (Fig. 15.2.3b) can be described by

$$W_E \propto p^\gamma(Cl_2), \tag{15.2.3}$$

where $\gamma \approx 2$ with pressures ≤ 3 mbar. The quadratic dependence of the etch rate on Cl_2 pressure can qualitatively be described by the full curve which was calculated by using $\eta'_{Si} = 0$ and $\eta_{Si} = 0.0075$, as in Fig. 15.2.3a. The discrepancies may originate from XeCl laser-induced gas-phase heating which is most pronounced in the regime of high Cl_2-gas pressures, additional mass transport by convection, other reaction products, in particular SiCl, etc. Such effects could explain the higher etch rates observed in the high pressure regime.

15.2.4 Electron–Hole Pairs

Figure 15.2.4 shows the etch rate versus Kr^+-laser power for two different sizes of the laser focus and two chlorine atom concentrations. The change in N_{Cl} was achieved by changing the distance between the excimer-laser beam and the silicon substrate (Fig. 30.1.1a). The full lines have equal slopes with $W_E \propto P^{0.7}$. The dashed curves are guides for the eye. The etch rate increases with Kr^+–laser power, and thus with the concentration of electron–hole pairs

15.2 Laser-Induced Etching of Si in Cl$_2$

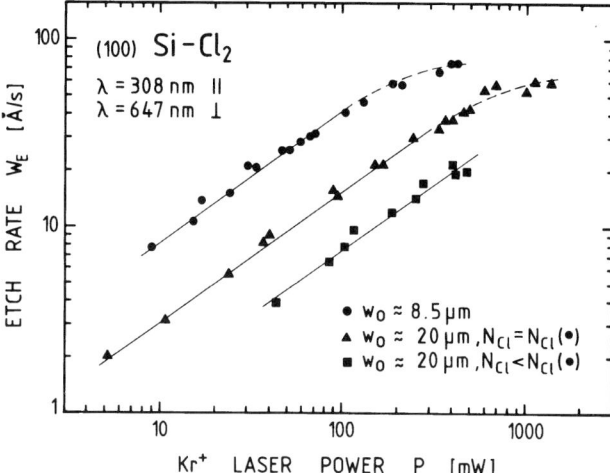

Fig. 15.2.4. Etch rate W_E versus Kr$^+$-laser power P for two different laser-spot sizes $w_0(1/e)$. The Cl atoms were generated by a XeCl-laser at parallel incidence to the Si substrate. The parameters employed were $p(Cl_2) = 5$ mbar and $\phi(308\,nm) = 4\,mJ/cm^2$. Full lines are fitted with equal slopes of 0.70 [*Kullmer* and *Bäuerle* 1988a]

within the Si surface. This observation can be described in (15.2.1) via the etch probability $\eta_{Si}(r)$. The simplest ansatz is

$$\eta_{Si}(r) \propto N_e(r, P) \propto P^\beta \quad (\lambda = 647\,nm). \tag{15.2.4}$$

The concentration of photo-electrons, N_e, calculated from (2.4.1) is shown in Fig. 15.2.5 for different laser wavelengths. The parameters employed refer to undoped c-Si with $D_e = 18.2\,cm^2/s$. Any surface recombination of electron–hole pairs or depletion of electrons at the surface due to surface reactions were ignored. With the wavelengths considered, N_e increases almost linearly up to $P \approx 300\,mW$ [$I(r = 0) \approx 10^5\,W/cm^2$]. Only for $N_e > 5 \times 10^{18}\,cm^{-3}$ does Auger recombination become effective and decrease the slope of curves.

From Fig. 15.2.5 and the ansatz (15.2.4) one would expect $\beta = 1$ and thus $\eta_{Si}(r) \propto P$ as long as $P < 300$ mW. This is in contrast to the measured power dependence of the etch rate, $W_E \propto \eta_{Si}(r) \propto P^{0.7}$. This discrepancy can be explained by surface effects which diminish the photoelectron concentration:

- Structural damages and, possibly, surface chlorination near the etched hole cause recombination or/and trapping centers for electrons.
- Chemisorbed Cl$^-$ ions produce a space charge layer at the silicon surface and thereby influence the carrier concentration within it, see Fig. 15.1.1a and [*Morrison* 1977].

The bending indicated by the dashed lines in the upper two curves of Fig. 15.2.4 is unexplained. From the model calculations on N_{Cl} we expect mass transport limitations to become effective only with etch rates that are one

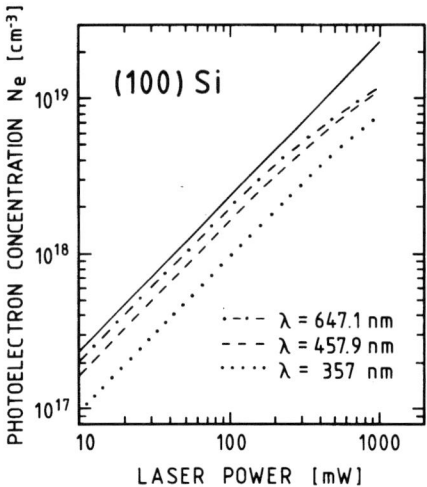

Fig. 15.2.5. Concentration of photo-electrons generated within a (100) Si surface as a function of laser power for different cw-laser wavelengths and $w_0 \approx 10.6\,\mu\text{m}$. The full line has been calculated for 514.5 nm radiation assuming $\alpha \to \infty$ and no Auger recombination [*Mogyorosi* et al. 1988]

order of magnitude higher than those observed in Fig. 15.2.4. In principle, such a bending could indicate the onset of Auger recombination.

The calculated diameters of the photoelectron distribution, $\text{FWHM}(N_e)$, and the measured diameters of holes, $2w$, defined by the hole depth at FWHM, are in good agreement and significantly larger than the spot size of the Kr^+-laser beam [see *Kullmer* and *Bäuerle* 1988a]. This loss in resolution produced by the diffusion of both photogenerated carriers within the Si surface and Cl atoms within the gas is a problem frequently encountered in photolytic processing (Sect. 5.3).

Calculations on the photochemical etching of Si in Cl_2 atmosphere have been performed by *Sytov* (1992, 1995).

15.2.5 Crystal Orientation and Doping

The etch rate of single-crystalline Si depends on crystal orientation. With low laser powers, corresponding to temperatures well below the melting point, the etch rate for (100) surfaces exceeds that for (111) surfaces by at least two orders of magnitude. The crystallographic orientation influences the etch rate via specific properties such as the degree of band bending, Fermi-level pinning, the geometrical arrangement of atoms, etc. The latter will strongly determine the diffusion of halogen ions into the Si surface (Sect. 15.1).

Because of the important role of electrons, photochemical etch rates in n-type Si exceed those in p-type Si. Clearly, the effect of doping can be observed only if the number of photoelectrons is small or comparable to the number of electrons originating from the dopant.

In heavily doped (degenerate) n-type Si, also denoted as n$^+$-type Si, the etch rate becomes almost independent of the irradiation geometry and spontaneous etching by chlorine radicals is observed [*Horiike* et al. 1987].

15.3 Si in Halogen Compounds

Light-enhanced etching of Si in XeF$_2$, NF$_3$, and SF$_6$ atmosphere has been investigated mainly by means of Ar$^+$- and CO$_2$-laser radiation.

15.3.1 Si in XeF$_2$

XeF$_2$ etches Si spontaneously (Sect. 15.1). Neither Ar$^+$– nor CO$_2$-laser radiation can excite gaseous XeF$_2$ in a single-photon process.

Band-gap excitation of Si with Ar$^+$– laser-light intensities $< 20\,\text{W/cm}^2$ changes the population of product species desorbing from the surface, but has almost no influence on the etch rate. For higher intensities, $W_E \propto I^n$ where $1 \leqslant n \leqslant 2$. The etch rate observed with low laser powers is somewhat higher with n-type Si than with p-type Si.

Qualitatively, the enhancement in W_E can be understood, at least in part, from the preceding outline. Band-gap excitation increases the potential difference Φ in (15.1.1). Additionally, laser-enhanced desorption of the etch products, SiF$_x$, decreases the thickness of the fluorosilyl layer below its equilibrium value. Both effects will increase the ratio Φ/h and thus W_E.

Figure 15.3.1 shows the (normalized) desorption fluxes $J(\text{SiF}_3)$ and $J(\text{SiF}_4)$ as a function of Ar$^+$–laser power. The figure suggests that only SiF$_3$ is a photoproduct. The selective nature of the etching reaction has been explained by F$^-$ ions that preferentially break those Si bonds to which the desorbing SiF$_3$ group is bound. Such a process can be described, symbolically, by [*Houle* 1989]

$$—\text{Si}^+—\text{SiF}_3 + \text{F}^- \to \text{Si}—\text{F} + \text{SiF}_3(\uparrow). \tag{15.3.1}$$

For pulsed-irradiation, the interaction mechanisms seem to be different from those involved in low power cw Ar$^+$–laser-enhanced etching. For example, irradiation with 532 nm frequency-doubled Nd:YAG-laser pulses results in the desorption of SiF$_x (x \leqslant 3)$ fragments and Si atoms [*Chuang* et al. 1984]. The yield for less F-coordinated species and Si atoms was found to increase with increasing light intensity. Similarly, with sub-band-gap irradiation using pulsed CO$_2$-lasers, the dominant species were also SiF$_x$ with $x \leqslant 3$. It was speculated that transient thermal electrons generated by single- or multiphoton processes create transient electric fields within the Si surface and thereby enhance the etch rate. However, with the fluences employed, the process may be entirely thermal

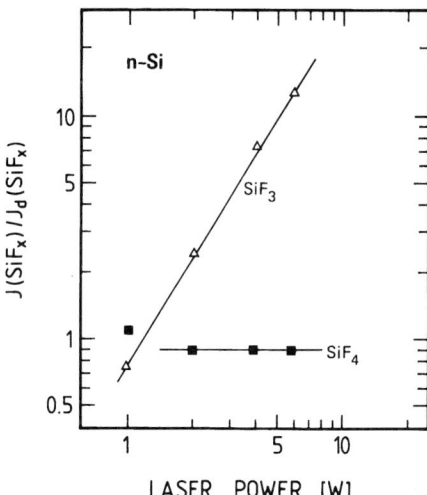

Fig. 15.3.1. 514.5 nm Ar$^+$–laser-induced desorption fluxes of SiF$_3$ and SiF$_4$ normalized to the respective dark fluxes [*Houle* 1989]

in nature, as discussed already with the Si-Cl$_2$ system. The estimated laser-induced temperatures are high enough to expect changes in the type of species desorbed from the surface (see also Sect. 15.1).

Ar$^+$–laser-enhanced etching of Si in a CF$_4$/O$_2$ plasma has been investigated by *Holber* et al. (1985).

15.3.2 Si in SF$_6$

A model system in which vibrational excitation has been demonstrated to enhance the molecule–surface reactivity is the etching of Si in SF$_6$ atmosphere (Sects. 2.3, 14.1).

Silicon at 300 K is almost inert against ground state SF$_6$. However, pulsed CO$_2$-laser excitation of SF$_6$ induces high etch rates, particularly at normal incidence. The major volatile reaction products observed are SiF$_4$ and SF$_4$. It has been suggested that gaseous or physisorbed SF$_6$ molecules are excited into higher vibrational states via coherent multiphoton excitation according to (14.1.8). In contrast to ground state SF$_6$, vibrationally excited SF$_6^*$ can dissociatively chemisorb on Si surfaces

$$SF_6^* \to 2F^- + SF_4(\uparrow) \ . \tag{15.3.2}$$

Part of the chemisorbed F$^-$ ions penetrate into the Si and form a fluorosilyl layer (Sect. 15.1). Via a number of subsequent surface processes, SiF$_4$ is formed and desorbs from the surface.

The importance of selective non-dissociative multiphoton vibrational excitation of SF$_6$ in the low intensity range has been derived from a number of

15.3 Si in Halogen Compounds

observations: The etch rate shows a pronounced wavelength dependence with a maximum that is much broader and lower in frequency than the single-photon absorption spectrum of the v_3 mode of SF_6 at 948 cm^{-1} (Fig. 15.3.2a). The dependence of the etch rate on laser fluence can be described by $W_E \propto \phi^{3.5}$ (Fig. 15.3.2b). This may indicate that the overall rate-limiting step is based on selective three- or four-photon excitations. The reaction yield increases monotonically with SF_6 gas pressure and saturates at about 2 mbar. However, in the intensity range considered, substantial substrate heating and, with the pressures employed, (non-selective) gas-phase heating is expected. In fact, gas-phase heating can also cause a broadening of the absorption spectrum and a shift to lower wavenumbers (longer wavelengths). Clearly, perpendicular ir-radiation will result in additional (indirect) gas-phase heating (Sects. 3.5 and 9.5).

With very high fluences and *parallel* incidence of the CO_2-laser beam, SF_6 molecules were decomposed into SF_5 and F atoms. This process was suggested to involve coherent and sequential multiphoton absorption resulting in the dissociation (MPD) of the molecule (Sect. 14.1.2). Unstable SF_5 decomposes into SF_4 and another F atom. The product species, SF_4 and F, diffuse to the Si surface and react to form SiF_4.

Similar experiments have been reported for the Si-NF_3 system [*Brannon* 1988]. Here, it was suggested that CO_2-laser radiation dissociates gaseous NF_3 via a collisionally enhanced multiphoton process generating NF_2 and F radicals which diffuse to the surface and cause etching. With the high pressures used in these experiments, up to $p(NF_3) > 300$ mbar, the formation of a fluorinated surface layer consisting of mainly SiF_3 and SiF_4 has been observed.

Due to the high laser-light intensities involved, material processing based on MPD can be applied only with parallel laser-beam incidence. Thus, the technique can be employed for large-area etching, unless a mask is used. This technique

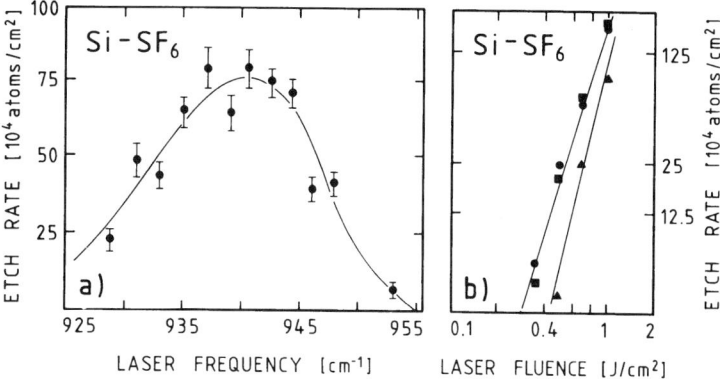

Fig. 15.3.2a,b. Etch rate for Si in SF_6 achieved with CO_2-laser radiation at perpendicular incidence. (a) Frequency dependence ($\phi = 0.9$ J/cm^2, $p(SF_6) = 3.3$ mbar). (b) Fluence dependence ($\lambda = 10.6$ μm ($\tilde{v} = 942.4$ cm^{-1}): ▲ $p(SF_6) = 1.1$ mbar; ● 2.7 mbar; ■ 6.7 mbar. Adapted from [*Chuang* 1981]

seems to be promising since many halogen-containing molecules, such as COF_2, CF_3X, CF_2X_2 (X = Cl, Br, I), N_2F_4, etc., can be readily decomposed by MPD to produce reactive radicals for surface reactions not only with Si but for many other materials as well.

15.4 Microscopic Mechanisms

In this section we will complete and summarize the microscopic mechanisms involved in the laser-induced etching of silicon in halogen atmosphere.

Photochemical etching

Photochemical etching of silicon in halogen atmosphere is consistent with a model in which the interaction of halogen radicals and carriers within the silicon surface plays a fundamental role. The different steps consist of the generation of halogen radicals, the formation of reaction products, and their desorption from the surface:

- Halogen radicals are formed spontaneously by molecule–surface collisions (Si-XeF_2 system), or by selective electronic (Si-Cl_2 system) or vibrational excitation (Si-SF_6 system) of the etchant.
- The radicals become adsorbed and, because of their strong electronegativity, capture an electron from the Si. Electron transfer is promoted by donator doping or by interband photo-excitation. With the formation of a *thin* SiX_x (X ≡ F, Cl, Br, ...) layer, tunneling of electrons through this layer will become important. This process is similar to surface oxidation (Chap. 26).
- X^- ions on the surface and positive holes within the silicon generate a surface electric field. This field causes a change in surface band bending and thus in charge transfer to the adsorbate. The strength of the surface electric field is proportional to the number of X^- ions adsorbed on the surface. However, even without charge transfer, laser irradiation of a semiconductor surface causes a surface electric field as high as 0.0001 to 0.1 V/Å. This field is related to the different mobilities of electrons and holes [*Gauthier* and *Guittard* 1976]. With localized irradiation, a separation of charges will also take place in radial direction (Dember effect; Fig. 15.6.4b). Clearly, any heating of the surface will generate additional electron–hole pairs.
- The surface electric field, and the noble-gas character of X^- ions, favor their diffusion into the Si. This mechanism is discussed in detail for surface oxidation (Chap. 26).
- With undoped, lightly-doped, and strongly p-doped materials, band-gap excitation will shift the Fermi level towards the conduction band and thereby increase the potential difference Φ in (15.1.1).

- Laser-light irradiation may change the thickness of the halogenated surface layer, h (Sect. 15.1).
- The etch rate depends on the density of X^- ions on the Si surface.
- The etch rate depends on the concentration of free carriers. This concentration is determined by the doping level, interband photo-excitation, electron–hole pair recombination, and electron trapping. In the Si-Cl$_2$ system, the sublinear increase in etch rate with Kr$^+$-laser power is related to electron traps or to a modified recombination kinetics caused by structural and chemical defects produced by the etching process itself.
- Defects influence the concentration of free carriers. Shorter lifetimes of free carriers in impure or heavily doped Si will diminish the etch rate but enhance the ultimate resolution via the decrease in carrier diffusion length.
- Adsorption of halogen radicals, charge transfer, and the penetration of species depend on the morphology, microstructure, and orientation of the Si surface.
- Reaction products must desorb from the surface. Desorption can take place spontaneously or it can be activated by the laser light. The composition of product species depends on the particular system and the parameters employed.

Combined photochemical and thermal etching

For laser-light intensities that generate high electron densities, a substantial amount of the light energy absorbed is directly converted into heat via Auger recombination (Sect. 2.4). In this intermediate regime (Fig. 15.2.2), surface etching is caused by both photochemical *and* thermal mechanisms. For the Si-Cl$_2$ system, chlorine radicals are still necessary to cause significant etching within this regime.

Thermal etching

With laser-light intensities that cause surface melting, or which are close to those, etching is mainly thermally activated. In this regime, photogenerated radicals play no, or only a minor role, at least, in the Si-Cl$_2$ system.

15.5 Dry-Etching of Compound Semiconductors

Laser-induced dry- and wet-etching of compound semiconductors has been investigated for the III–V compounds GaAs, InP, InSb, and the II–VI compounds Cds and CdSe. The microscopic etching mechanisms are similar to those described for silicon. Because of the lower thermal and chemical stability of compound semiconductors, photochemical etching is of particular importance. Surface patterning was demonstrated by direct writing, laser-beam interference, and projection. The precursor molecules most commonly used were Cl_2, Br_2, HBr, CCl_4, CH_3X, and CF_3X with $X \equiv Cl$, Br, I.

15.5.1 III–V Compounds

Thermal etching of GaAs, InP, and InSb in CCl_4 atmosphere using focused Ar^+-laser radiation has been studied by *Takai* et al. (1988b). Figure 15.5.1 shows an Arrhenius plot. The activation energies are quite similar and around 3.7 kcal/mol. The rate limiting step in the reaction is probably thermal desorption of the respective chlorides. The etch rates are 1 to 3 orders of magnitude higher than those observed in photochemical etching of these materials [*Brewer* et al. 1984]. The maximum resolution achieved with a 1.2 µm laser focus was about 0.6 µm. Scanning speeds of up to 60 µm/s have been employed. At medium to high laser powers, changes in the stoichiometry of the material surrounding the laser etched groove have been found.

ArF-laser-induced etching of (100) and (111) n-type GaAs and (100) p-type GaAs in CF_3Br and CH_3Br has been investigated by *Brewer* et al. (1984). Surface patterning has been demonstrated by direct masking and laser-light projection. Here, a resolution of about 0.2 µm was achieved.

ArF-laser radiation photodissociates both CF_3Br and CH_3Br. The radicals, Br, CF_3, and CH_3 react with GaAs and form various etch products that have been analyzed by laser-induced fluorescence (LIF). For low fluences,

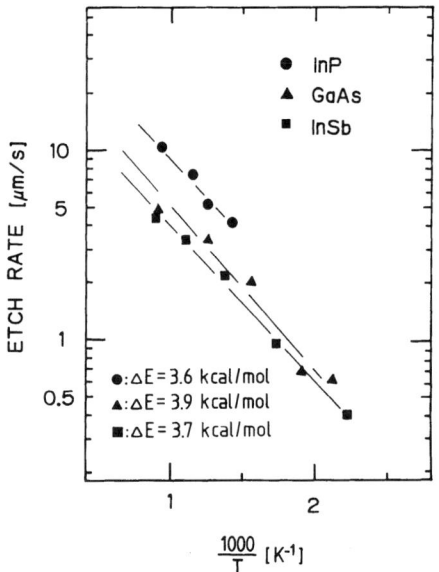

Fig. 15.5.1. Arrhenius plot of Ar^+-laser-induced etch rate achieved in CCl_4 with n-type (100) InP, (100) GaAs, and (111) InSb. Laser-beam scan speeds were 9 µm/s for InP and 3 µm/s for GaAs and InSb. The CCl_4 pressure was about 160 mbar [*Takai* et al. 1988b]

etching is mainly non-thermal. This interpretation is supported by experiments using XeF-laser radiation with otherwise identical experimental conditions. XeF-laser light does *not* photodissociate CF_3Br and CH_3Br but induces about the same surface temperature. No etching was observed in this case. Etch rates achieved with ArF-laser radiation at normal incidence were higher than those for parallel incidence. This difference was interpreted by photocarriers and laser-enhanced thermal desorption of nonvolatile products. Blocking of reactive surface sites by nonvolatile reaction products can be diminished by uniform substrate heating. As expected, an exponential increase in etch rate with substrate temperature was found. Etching is strongly anisotropic. For (111B), (100), and (111A) orientations, typical etch-rate ratios were 3:2:1. For fluences $\leqslant 35$ mJ/cm^2 the surface morphology is relatively smooth. The corresponding etch rates are about 0.01 µm/s. At higher fluences, ablation seems to be the primary mechanism. Then, the surface becomes rough and material damage is observed.

Laser-induced etching of n-type (100) InP has been investigated for excimer-laser radiation and Cl_2, HCl, and HBr as etchants. Figure 15.5.2 shows an etch pattern produced by ArF-laser-light projection in HBr atmosphere. The surface morphology is very smooth and well-defined edges down to a 10 µm scale have been observed. The etch rate, about 0.3 Å/pulse, seems to be determined by the desorption of $InBr_3$. Substrate heating to about 200°C increases the etch rate by almost a factor of ten, but it reduces the quality of the etch pattern, in particular near edges.

15.5.2 Laser Etching of Atomic Layers

Etching of atomic layers (EAL) also termed as *digital* etching is the inverse of atomic-layer epitaxy (ALE). Layer-by-layer etching requires chemisorption of a monolayer etchant on the substrate surface. Decomposition and product removal can be enhanced or only induced by laser light (Laser-EAL).

Figure 15.5.3 shows the etch rate of InP in Cl_2 atmosphere as a function of ArF-laser fluence. At $T_s \approx 140$ °C and a fluence of about 0.12 J/cm^2 the etch rate saturates at about 2.3 Å/pulse, which corresponds to about one monolayer of InP.

15.5.3 Dopants, Impurities, and Defects

Photochemical dry-etching of GaAs can be suppressed by impurities and ion-beam induced defects which serve as electron traps (Sect. 15.4) [*Ashby* et al. 1990, and references therein; *Houle* 1991]. The combination of ion-beam and laser-beam techniques permits selective high-resolution surface patterning.

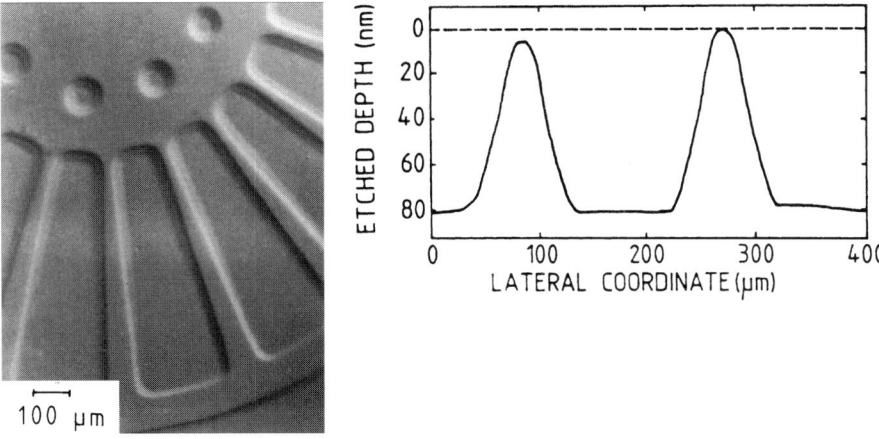

Fig. 15.5.2. Pattern etched into (100) InP at 300 K by means of ArF-laser-light projection using a stainless steel mask ($\phi = 120\,\mathrm{mJ/cm^2}$, 25 Hz). The precursor was HBr (30 sccm) with admixtures of H_2 (20 sccm) and Ar (150 sccm). The total pressure was 1.2 mbar and the processing time 2 min [*Matz* et al. 1990]

15.6 Wet-Etching

Laser-enhanced wet-etching of semiconductors has been performed in aqueous solutions of acids, lyes, and various mixtures. For high-resolution etching, laser-induced reactions within the bulk liquid must be avoided. Various mechanisms involved in liquid-phase processing are discussed in Chap. 21.

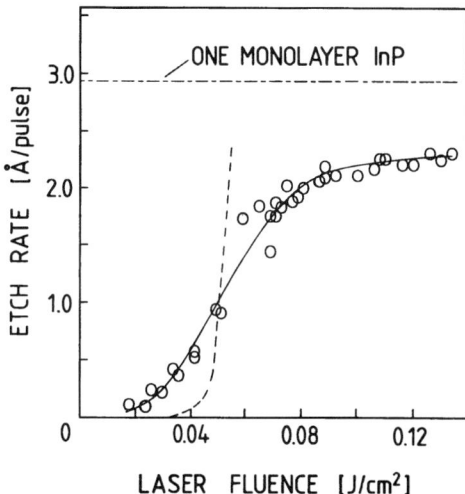

Fig. 15.5.3. InP etch rate as a function of ArF-laser fluence [10 Hz, $p(Cl_2) = 2.5$ mbar, $T_s \approx 140°C$]. The dashed curve represents the calculated etch rate near threshold if etching were limited by sublimation of $InCl_3$ [*Donnelly* and *Hayes* 1990]

15.6.1 Silicon

Laser-induced wet-etching of Si has been investigated in aqueous solutions of HF, NaOH, and KOH using cw Ar^+–, Nd:YAG–, and CO_2-laser radiation.

Figure 15.6.1 shows the volume etch rate as a function of Ar^+–laser power for holes etched in (111) Si wafers immersed in aqueous KOH. Corresponding experiments with ceramic Al_2O_3/TiC are included in the figure. Average (depth) etch rates up to 15 μm/s have been achieved. Nd:YAG-laser-induced etching of Si in KOH by rear-side irradiation (Fig. 9.5.1b) has been demonstrated by *Bunkin* et al. (1985).

At low to medium laser-light intensities non-thermal mechanisms seem to play an important role (Chap. 21). Etch rates in (100)-direction exceed those in (111)-direction by up to more than 2 orders of magnitude. For high intensities, melting and vaporization dominates. The high rates achieved in liquid-phase processing are related to the high density of reactive species within the liquid and the increase in mass transport by microstirring (Chap. 21).

15.6.2 Compound Semiconductors

Laser-enhanced wet-etching of compound semiconductors has been investigated mainly for GaAs, InP, and InSb.

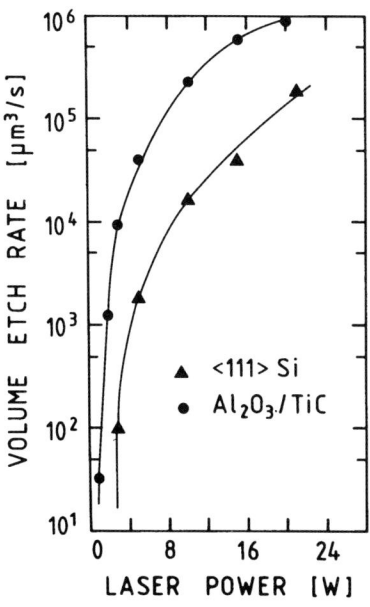

Fig. 15.6.1. Volume etch rate for (111) Si and ceramic Al_2O_3/TiC in KOH as a function of Ar^+–laser power ($\tau_l = 5$ s). Full curves are to guide the eye [*von Gutfeld* and *Hodgson* 1982]

Figure 15.6.2a shows a SEM picture of a via hole in GaAs. It demonstrates that liquid-phase etching permits one to fabricate deep high-quality holes with perfectly vertical walls. The high aspect ratio achieved can be attributed to waveguiding of the laser light (Sect. 15.6.4). Deep trenches have been produced by translating the substrate with respect to the laser beam. The gratings shown in b) have been etched by using laser-beam interference (Fig. 5.2.1). By varying the angle of incidence, different groove profiles with depth-to-spacing ratios between 0.2 and 0.8 were produced.

Figure 15.6.3a shows the etch rate for Si-doped (n-type) GaAs as a function of absorbed photon flux for two laser wavelengths. The HNO_3 solution employed was diluted in such a way that dark etching could be ignored. The etch rates are equal for 514.5 nm and 257 nm radiation. Initially, the rate increases almost linearly and saturates with fluxes above 10^{19} to 10^{20} photons/cm^{2}s $[I_a(257\,\text{nm}) \approx 10-10^2\,\text{W/cm}^2]$. The maximum temperature rise was calculated to be $< 1\,\text{K}$. The etch rate was independent of crystallographic orientation for UV illumination and only slightly dependent for VIS radiation. Figure 15.6.3b shows the etch rate achieved with 257 nm radiation in GaAs doped with Si, Cr (semi-insulating), and Zn (p-type). The rate increases with increasing n-type character of the material. Etching of p-type material was *not* possible with 514.5 nm laser light.

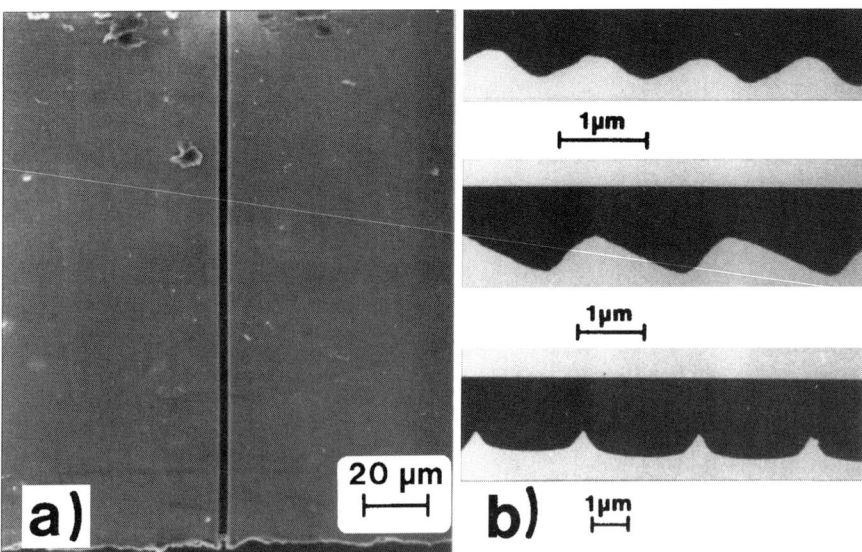

Fig. 15.6.2a,b. Ar^+-laser-induced photochemical etching of GaAs in aqueous $H_2SO_4 + H_2O_2$ (volume ratio H_2SO_4: H_2O_2: $H_2O = 1:1.3:25$). (**a**) Via hole, $\lambda(\text{SH}\,Ar^+) = 257$ nm [*Podlesnik* et al. 1984]. (**b**) Gratings produced by 514.5 nm Ar^+-laser-light interference. The different profiles were obtained by varying the angle of incidence, Θ [*Podlesnik* et al. 1983]

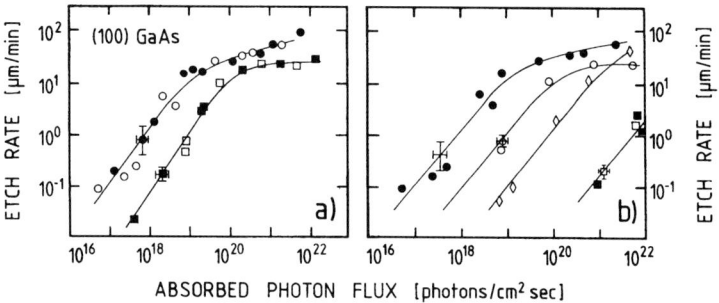

Fig. 15.6.3a,b. Ar$^+$-laser-induced etch rate in GaAs as a function of absorbed photon flux for different types of dopings. The etchant was aqueous HNO$_3$. (**a**) Si doped with $n = 3 \times 10^{18}$cm^{-3} (● $\lambda = 515$ nm, ○ $\lambda = 257$ nm) and with $n = 3 \times 10^{16}$cm^{-3} (■ 515 nm, □ 257 nm). (**b**) Doped with Si (● $n = 3 \times 10^{18}$cm^{-3}, ○ $n = 10^{16}$cm^{-3}), Cr(◇ SI, $\rho > 10^7 \Omega$cm), and Zn (□ $p = 10^{16}$cm^{-3}, ■ $p = 9.5 \times 10^{18}$cm^{-3}), and $\lambda = 257$ nm [*Ruberto* et al. 1991]

15.6.3 Interpretation of Results

With *low* laser-light intensities, the laser-induced temperature rise can be ignored. In this regime, the enhancement in etch rate can be interpreted by the generation of electron–hole pairs within the semiconductor surface and charge transfer at the liquid–solid interface.

Let us first consider the situation in the dark. The Fermi level in the semiconductor must match the redox level of the liquid. As a consequence, the semiconductor bands bend upwards or downwards, depending on whether the semiconductor is n-type or p-type (Fig. 15.6.4a) [*Gerischer* 1975; *Bockris* and *Reddy* 1977]. The band bending can be easily understood: The oxidizing agent will attract electrons. For n-type material, a negatively charged surface layer within the liquid is thereby formed. This causes an upward band-bending resulting in a barrier preventing any further flow of electrons towards the surface. In p-type material, on the other hand, electrons are the minority carriers. In this case, there is a "transfer of holes" (in reality, this means a transfer of electrons from the liquid side) and the (liquid) surface layer becomes positively charged. This causes downward band-bending, and thereby a barrier against any further migration of holes.

Low light intensities

Light with a photon energy that exceeds the band gap, $hv > E_g$, will generate electron–hole pairs. The different mobilities of electrons and holes cause spatial changes in their concentrations. For GaAs, e.g., the ratio of mobilities for electrons and holes is $\mu_e/\mu_h \approx 8$.

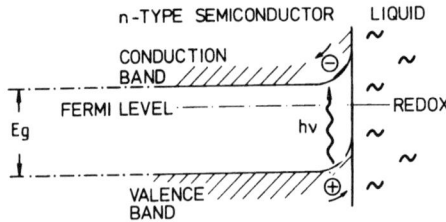

Fig. 15.6.4a. Band bending of n-type and p-type semiconductors at the interface with a liquid electrolyte. Conditions are similar at gas–semiconductor interfaces. The flow of electrons and holes under illumination with band-gap radiation is indicated

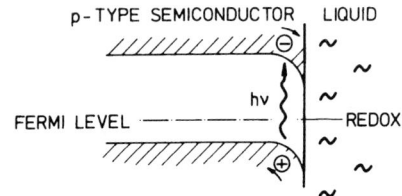

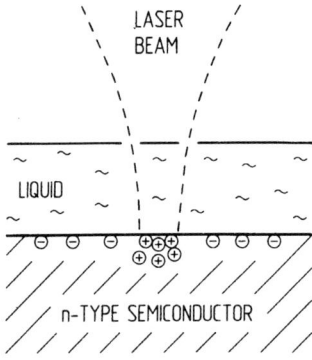

Fig. 15.6.4b. Generation of a local EMF by the Dember effect in an n-type semiconductor

With *uniform* (large-area) irradiation, the changes in carrier concentrations become effective only in the direction perpendicular to the semiconductor surface. In n-type material the holes will drift to the surface, the electrons further into the bulk. The holes can be considered as ionized or broken bonds. The disrupted lattice will strongly interact with the negatively charged surface species. This may result in oxide formation. If the solution contains an acid that dissolves this oxide, the semiconductor will dissolve into positive ions. For p-type material there is a depletion of holes near the surface. Thus, light-induced etching will occur at a slower rate or not at all. For particular liquid–solid interfaces, the process may even be reversed, resulting in material deposition. In this latter case, however, the change in physical properties at the interface will, in general, rapidly terminate the reaction (Chap. 21).

15.6 Wet-Etching

With *localized* irradiation, the different mobilities of carriers will generate a local electromotive force (EMF) in *radial* direction (Dember effect; Sect. 21.1). This is schematically shown in Fig. 15.6.4b for n-type material. Holes are enriched within the irradiated area.

n-type GaAs

Etching of n-type GaAs in a dilute acid can be understood as follows: The *anodic* reaction within the irradiated area can be described by

$$GaAs + 2H_2O + 6h \to Ga^{3+} + HAsO_2 + 3H^+$$
$$\to Ga^{3+} + As^{3+} + \cdots. \tag{15.6.1}$$

GaAs first reacts with water and forms an oxide which is subsequently dissolved by the acid. Ga^{3+} and As^{3+} ions go easily into solution. In electroless etching (no external EMF) the overall hole and electron currents must be equal. Thus, the consumption of holes requires the consumption of an equal number of electrons in the *cathodic* reaction outside the illuminated area. Here, electrons are transferred from the GaAs to the oxidizing agent in the solution.

Let us consider a simple estimation of the reaction rate: The band-gap energy of GaAs is $E_g(300\,K) \approx 1.43\,eV$ ($\lambda \approx 870\,nm$). In the absence of an external field and severe band-bending, which applies to semi-insulating (SI) material, in good approximation, the concentration of holes within the GaAs surface can be described by (2.4.1). If we consider steady-state conditions and assume at the liquid–solid interface the boundary condition

$$D_h \frac{\partial N_h}{\partial z}\bigg|_{z=0} = k N_h,$$

integration of (2.4.1) yields for the one-dimensional case

$$N_h(z) = \frac{\alpha I_a \tau_{rec}}{h\nu \left(\frac{\alpha^2 l_c^2}{4} - 1\right)} \left[\frac{kl_c + \alpha l_c D_h}{kl_c + 2D_h} \exp\left(-\frac{2z}{l_c}\right) - \exp(-\alpha z)\right], \tag{15.6.2}$$

where τ_{rec} is the time for carrier recombination, and $l_c \approx 2(D_h \tau_{rec})^{1/2}$ the diffusion length of holes. The reaction rate is thus

$$W = D_h \frac{\partial N_h}{\partial z}\bigg|_{z=0} = \frac{\alpha I_a l_c}{2h\nu \left(\frac{\alpha l_c}{2} + 1\right)\left(1 + \frac{2D_h}{kl_c}\right)} \approx \frac{I_a}{h\nu}. \tag{15.6.3}$$

The latter approximation refers to $\alpha l_c \gg 1$ and $D_h/kl_c \ll 1$. The linear increase in etch rate with $I/h\nu$ is consistent with the experimental observations at low to medium photon fluxes (Fig. 15.6.3).

The saturation in etch rate observed with higher laser-light intensities can be related to surface defects, impurities, ionic transport limitations within the liquid, etc.

p-type GaAs

With p-type GaAs the influence of VIS-laser light of low to medium intensity can be described by

$$GaAs + 3e \rightarrow Ga + As^{3-}. \tag{15.6.4}$$

As^{3-} goes into solution as AsH_3. On the other hand, metallic Ga produced on the surface is not easily dissolved and *passivates* the surface. Within the dark region slow anodic etching as described by (15.6.1) takes place. Thus, the sample is slowly dissolved, except within the illuminated region. The situation is different with UV light which induces etching also with p-type material, though at smaller rates (Fig. 15.6.3b). This can be explained by the shallow penetration depth of UV light which generates a much higher concentration of holes at the surface than VIS light $[l_\alpha(257 \text{ nm}) \approx 50 \text{ Å}, l_\alpha(514.5 \text{ nm}) \approx 1100 \text{ Å}]$. Thus, some holes may overcome the potential barrier imposed by the downward band-bending.

If a positive voltage is applied to p-GaAs, the concentration of holes at the interface is increased. Then, it is possible to oxidize and dissolve the Ga layer formed according to (15.6.4) via

$$Ga + 3h \rightarrow Ga^{3+}.$$

Thus, by switching the potential between a value at which photoreduction occurs, and a value at which oxidation occurs, it becomes possible to also etch p-type GaAs. This type of electrochemical etching has been demonstrated by means of He-Ne and Ar^+-laser radiation [*Ostermayer* and *Kohl* 1981]. Similar experiments have been performed for InP [*Bowers* et al. 1985]. Further details on electrochemical processes are discussed in Chap. 21.

Thermal activation

With *high* laser-light intensities, laser-induced heating will generate electron–hole pairs and result in effects similar to those discussed above. For n-type material the thermal EMF and the Dember EMF are oriented in the same direction, while in p-type material they are opposite.

The importance of thermal mechanisms follows from the observation that at high intensities the etch rate becomes independent of laser wavelength and material doping. In this regime, mass transport limitations may determine the etch rate.

15.6.4 Spatial Resolution, Waveguiding

For *low* laser-light intensities where the etch rate increases linearly with photon flux, the resolution is determined by carrier-diffusion and the intensity distribution of the absorbed laser light. Figure 15.6.5a shows the cross section of a shallow groove etched in n-type GaAs by means of 514.5 nm Ar^+-laser radiation. The carrier diffusion length depends on the number of physical and chemical defects. The dotted curves in Fig. 15.6.6 show measured etch-depth profiles for two doping levels. Here, the etching times were adjusted in such a way that the center etch depths were equal for both samples. With high doping, the etch profile reflects the Gaussian intensity distribution of the laser beam. With lower doping, the profile is bell-shaped and significantly wider. The narrower profile observed with the higher doping is a consequence of both the shorter diffusion length of holes and the smaller width of the depletion layer. The full curves represent the stationary surface density of photo-generated holes, $N_h(r, 0)$. This has been calculated from an equation of the type (2.4.1) by taking into account band-bending.

With *higher* laser-light intensities changes in etch profiles are observed. The groove shown in Fig. 15.6.5b exhibits a flattened bottom which is related to the saturation in etch rate observed in Figs. 15.6.3. It should be noted that with the laser-light intensities employed, the laser-induced temperature rise can still be ignored.

An estimation of the spatial resolution for grating formation on n-type semiconductors by light-enhanced electrochemical etching has been performed by *Ostermayer* et al. (1985).

Waveguiding

Optical waveguiding (light-guiding) is a well-known phenomenon which is described in standard textbooks. A hole, rod, or fiber of uniform cross section, for example in the form of a cylinder, can guide electromagnetic radiation in axial direction. This is mainly based on multiple reflections, as in metal tubes, or on total reflection.

In laser processing, this effect has been studied in connection with (thermal) laser machining of metals using CO_2-laser or Nd:YAG-laser radiation

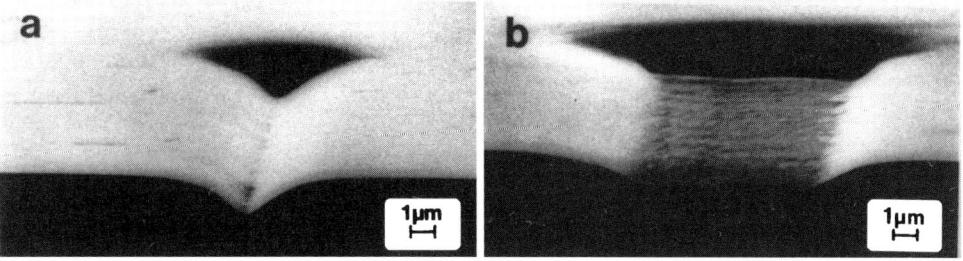

Fig. 15.6.5a,b. SEM pictures showing the relative widths of trenches in n-GaAs ($3 \times 10^{18}/cm^2$ Si) etched in HNO_3 by Ar^+-laser light ($\lambda = 515$ nm, $w_0 = 1.3$ μm). (**a**) 3 W/cm², $v_s = 0.43$ μm/s. (**b**) 150 W/cm², $v_s = 2$ μm/s [*Ruberto* et al. 1991]

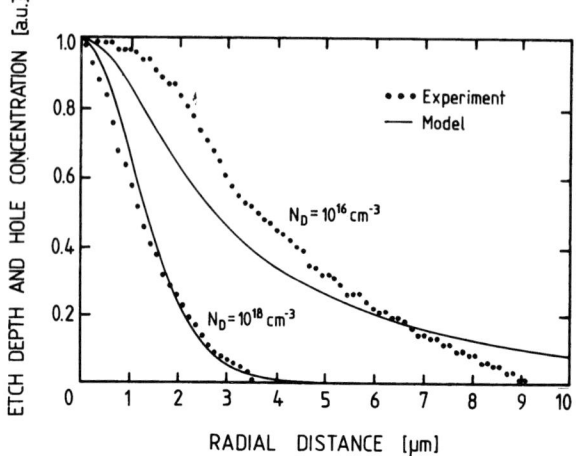

Fig. 15.6.6. Profiles of grooves in n-GaAs. Dotted curves: Experimental results achieved in HNO_3 with Ar^+-laser radiation. Except for the laser beam dwell time, the parameters employed were the same as in Fig. 15.6.5a. Full curves: Calculated surface distribution of photogenerated holes [*Ruberto* et al. 1991]

and with photochemical liquid-phase etching of semiconductors (Fig. 15.6.2). Consider Fig. 15.6.7. The etch velocity, v_E, is perpendicular to the surface being etched. Because of waveguiding, the wavefront near the bottom of the hole is almost plane. Therefore, the angle of incidence at the point $M \equiv M(z(t), r(t))$ can be approximated by

$$\Theta(r, z, t) = \arctan \left| \frac{dz}{dr} \right|. \tag{15.6.5}$$

The velocities of a surface element in axial and lateral directions are then given by

$$\frac{dz}{dt} = v_E \cos\Theta \quad \text{and} \quad \frac{dr}{dt} = v_E \sin\Theta. \tag{15.6.6}$$

For surface absorption and a purely photochemical process where the etch rate is proportional to the absorbed laser-light intensity, v_E can be described by

$$v_E = kI[1 - R(\Theta)] \cos\Theta, \tag{15.6.7}$$

where k is the rate constant and $R(\Theta)$ the reflectivity. The factor $\cos\Theta$ in (15.6.7) describes the change in incident laser power because of the surface tilt. Etch profiles can be calculated by integration of these equations.

For thermally activated reactions, including conventional laser processing of metals (drilling, cutting), the material removal rate depends exponentially on temperature. The temperature distribution, in turn, is a complicated function of the intensity distribution, the geometry of the hole, and the physical properties of the material.

15.6 Wet-Etching

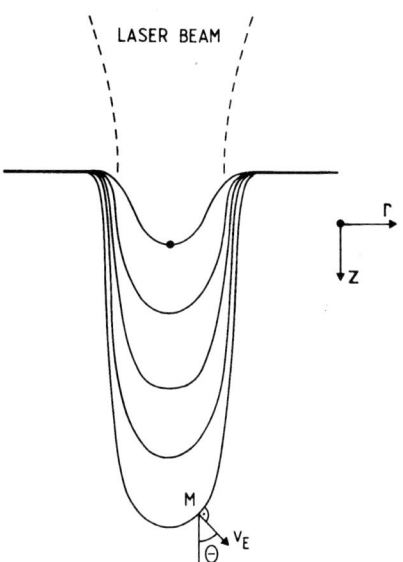

Fig. 15.6.7. Computer simulation of the temporal evolution of a tubular hole as shown in Fig. 15.6.2a. Adapted from [*Ruberto* et al. 1991]

A quantitative treatment must take into account changes in the intensity I due to multiple reflections, interference effects, etc. Thus, self-consistent calculations on the basis of Maxwell equations must be performed.

Part IV: Material Deposition

Laser-induced material deposition has been demonstrated from gases and condensed phases (Fig. 1.2.1).

Laser-induced chemical vapor deposition (LCVD) can be employed to fabricate *microstructures* of different types (Chaps. 16–18) and to grow large-area *thin films* (Chap. 19).

Adsorbates frequently play an important role in laser-CVD and in deposition techniques using a combination of laser and atomic (molecular) beams (Chap. 20). Material deposition from liquids has been demonstrated with ordinary liquids and with electrolytes with and without an external electromotive force (EMF) (Chap. 21). Thin films can also be fabricated from solid targets by pulsed-laser ablation or laser-induced evaporation and subsequent material condensation onto a substrate (PLD process). Laser-induced forward transfer (LIFT) is mainly employed for micropatterning (Chap. 22).

Nucleation, surface coating by laser surface cladding, and surface patterning by solid-phase transformation are discussed in separate chapters.

16 Laser-CVD of Microstructures

The decomposition of precursor molecules in laser-induced chemical vapor deposition (LCVD) can be activated thermally (*pyrolytic* LCVD) or non-thermally (*photolytic* LCVD) or by a combination of both (*photophysical* LCVD). The type of process activation can be verified from the morphology of the deposit and from measurements of the deposition rate as a function of laser power, wavelength, substrate material, etc; additional information is obtained from the analysis of data on the basis of theoretical models. The subsequent discussion concentrates on examples which were studied in most detail. Nevertheless, this discussion is very general in the sense that most of the trends, features, and results apply to all of the corresponding systems listed in Appendix B.4. Applications of laser-CVD in microfabrication are summarized in Sect. 18.5.

16.1 Precursor Molecules

The application of laser-CVD in micro-patterning requires proper selection of precursor molecules. An inspection of the bibliography reveals that the precursors most frequently employed are:

- Halogen compounds, hydrocarbons, and silanes.
- Alkyls, carbonyls, and various organometallic coordination complexes.

The first class of molecules possess electronic transitions in the near to deep UV where only a few or no adequate laser sources are available. Though the temperatures required for *thermal* decomposition are relatively high, these molecules are frequently used in pyrolytic LCVD because they permit the deposition of materials with high purity.

The second class of molecules possess electronic transitions in the VIS to near UV. Thus, they facilitate the matching of available laser wavelengths for (high-yield) *photolysis*. On the other hand, utilization of carbonyls and many organic precursors is often linked with the incorporation of large amounts of impurities into the deposit, in particular of carbon. Such impurities cause deterioration of the electrical properties of deposited materials. This problem is less pronounced if these molecules are decomposed either thermally – which

16.2 Pyrolytic LCVD of Spots

Investigations on the pyrolytic growth of spots allow one to test the adequacy of model calculations for pyrolytic laser-CVD. In such experiments, the substrate is immersed in a reactive gaseous ambient and perpendicularly irradiated by a focused laser beam. The setup typically employed is shown in Fig. 5.2.2. Both the laser beam and the substrate are at rest.

Henceforth, we assume that the laser light is exclusively absorbed on the substrate surface or on the already deposited material. Transparent substrates are frequently coated with a thin absorbing film or single absorption centers. By this means, latent times for nucleation can be almost avoided (Chap. 4), and a better reproducibility of data can be achieved.

16.2.1 Deposition from Halides

A model system that has been investigated in great detail is the deposition of W from WX_6 ($X \equiv F, Cl$) with or without H_2 or an inert carrier gas M. In analogy to (3.5.1) the overall reaction can be described by

$$WX_6 + 3H_2 + M \underset{k_2}{\overset{k_1, k_3}{\rightleftarrows}} W(\downarrow) + 6HX + M. \tag{16.2.1}$$

The rate constants k_1 and k_3 describe the decomposition of WX_6 at the surface and in the adjacent gas, respectively. k_2 describes etching of condensed W by HX.

Another practical example is the deposition of Si according to

$$SiCl_4 \underset{k_2}{\overset{k_1, k_3}{\rightleftarrows}} Si(\downarrow) + 2Cl_2. \tag{16.2.2}$$

Again, H_2 or an inert gas can be added. If the chemical equilibrium is shifted to the other side, this reaction describes the pyrolytic etching of Si in a Cl_2 atmosphere.

Morphology

The morphology of deposits depends mainly on the type of precursor molecules, the gas pressures, the laser-induced temperature distribution, and the illumination time.

16.2 Pyrolytic LCVD of Spots

Figure 16.2.1a–c shows SEM pictures of W spots deposited from $WF_6 + H_2$ by means of 514.5 nm Ar^+-laser radiation. The substrate employed was fused quartz covered with W absorption centers (radius $r_0 \approx 1.5\,\mu m$) produced by laser-induced forward transfer (LIFT).

The shape of spots obtained in a particular experiment is determined mainly by the ratio of partial pressures $\Gamma_p = p(H_2)/p(WF_6)$ and the laser-induced temperature distribution. The different types of spots shown in Fig. 16.2.1a–c are henceforth denoted by S_l, S_i, and S_d, respectively.

Spots of type S_l are spatially well *localized*. They consist of only a few single crystallites. The ratio of the spot diameter d and height h is small and, typically, within the range $2 \leq d/h \leq 3$. Such spots are obtained with small pressure ratios Γ_p, small focus diameters, and low laser powers. The (transparent) SiO_2 substrate permits inspection of the deposit by means of an optical transmission microscope (Fig. 16.2.1a'). This reveals that the spot is surrounded by a smooth, highly reflecting circular film which has a thickness of some ten Ångströms, and which consists of pure W. The spot is separated from the film by a transparent ring which shows the plain SiO_2 surface. These observations are interpreted by a combination of heterogeneous and homogeneous deposition reactions together with etching of condensed W by HF (Sect. 16.3.1).

Spots of type S_d (Fig. 16.2.1c) are fine grain, and very flat with large values of d/h. They are quite *diffuse* with no well-defined edge, and are also

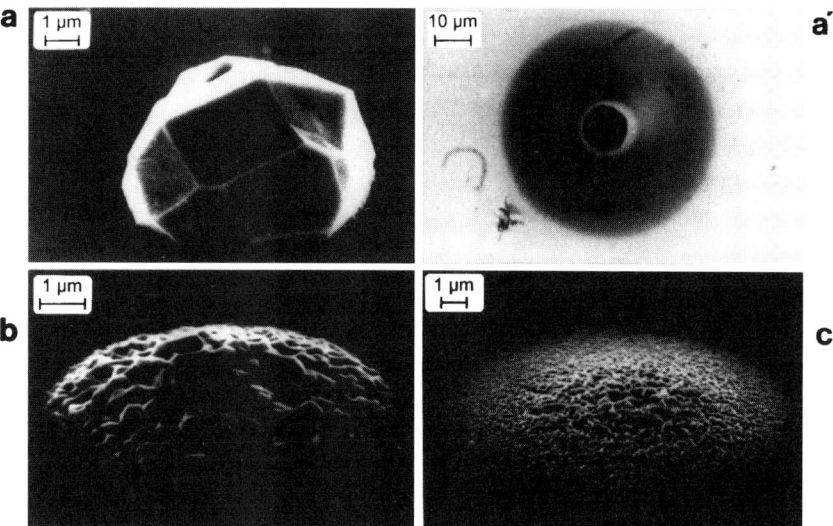

Fig. 16.2.1a–c. SEM pictures (**a–c**) and optical transmission microscope picture (**a'**) of W spots deposited from $WF_6 + H_2$ onto SiO_2 substrates by means of Ar^+-laser radiation [$\lambda \approx 515$ nm, $w_0(1/e) \approx 1.1\,\mu m$]. (**a, a'**) $P = 120$ mW, $\tau_l = 0.2$ s, $\Gamma_p = 2\,[10/5 \equiv 10\,\text{mbar}\,H_2 + 5\,\text{mbar}\,WF_6]$. (**b**) $P = 110$ mW, $\tau_l = 0.5$ s, $\Gamma_p = 5\,[25/5]$. (**c**) $P = 120$ mW, $\tau_l = 0.5$ s, $\Gamma_p = 50\,[250/5]$ [*Toth* et al. 1992]

surrounded by a thin film. Spots of this type are obtained with big ratios Γ_p or high laser powers.

Spots of type S_i (Fig. 16.2.1b) represent an *intermediate* situation. They are still well defined, have a coarse-grain morphology, and they are observed over a broad range of parameters. These spots are again surrounded by a thin film.

The diameter of W spots as a function of laser power is shown in Fig. 16.2.2 for different ratios Γ_p. Besides the change in diameter, a change in surface morphology is observed. This is indicated by full (type S_l or S_i) and open symbols (type S_d). In all cases, deposition starts only above a *threshold* power of about 40 ± 5 mW. Above this threshold, the diameter increases about linearly with laser power. With the laser powers investigated and with $1 \leq \Gamma_p \leq 3$ only spots of type S_l are observed. With higher pressure ratios, the shape changes with power. For example, with $\Gamma_p = 5$, spots of type S_i are observed up to powers of about 140 mW. Above about 160 mW, only spots of type S_d are observed. With increasing Γ_p, the transition in shape from type S_i to type S_d is shifted to lower laser powers. Spot shapes obtained with other precursor molecules and experimental parameters are shown in Figs. 16.2.5 and 16.5.2.

Dependence on illumination time

The time-dependent growth of W spots has been investigated with WF_6 and WCl_6 precursors. In such experiments, the laser beam illumination time is increased successively, while all other parameters are kept constant. The

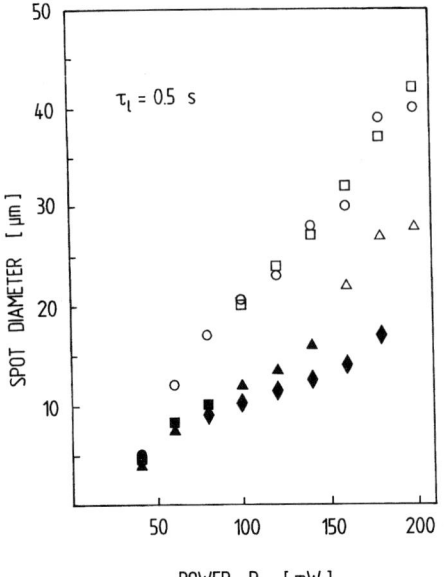

Fig. 16.2.2. Diameter of W spots as a function of laser power for fixed illumination time ($\tau_l = 0.5$ s). Full symbols refer to well-defined spots of type S_l or S_i (Figs. 16.2.1a, b). Open symbols refer to diffuse spots of type S_d (Fig. 16.2.1c). ◆: $\Gamma_p = 2(10/5)$, ▲△: $\Gamma_p = 5(25/5)$. ■ □: $\Gamma_p = 15(75/5)$, ● ○: $\Gamma_p = 25(125/5)$ [*Toth* et al. 1992]

16.2 Pyrolytic LCVD of Spots

growth is characterized by changes in both microstructure and morphology: The grain size, diameter, and maximum height of S_i-type spots increase with illumination time, while the ratio d/h decreases (Fig. 16.2.3a). For very long times τ_l and within certain parameter ranges, the growth of one or several large crystals is observed (Fig. 16.2.5a). Figure 16.2.3b shows similar investigations but for WF_6 and different absorbed laser powers, P_a. Within the parameter range investigated, these spots are of type S_i. Again, spot diameters first increase very rapidly and saturate for longer times, depending on the laser power.

Dependence on wavelength

For pyrolytic deposition, the growth rate is expected to be independent of the laser wavelength as long as the absorbed laser power, $P_a = P(1-R)$, is kept constant. This is confirmed by the experimental results in Fig. 16.2.3a.

Dependence on laser power

The diameter and height of W spots deposited from $WCl_6 + H_2$ is displayed in Fig. 16.2.4 as a function of laser power. Well above the threshold for

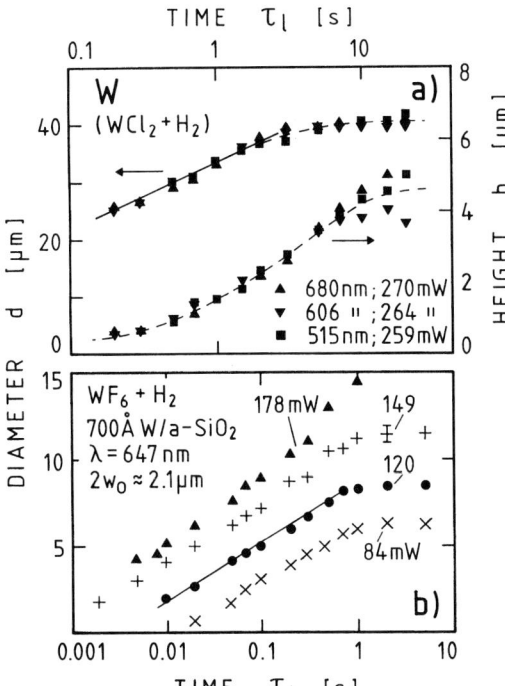

Fig. 16.2.3 a,b. Diameter/height of W spots as a function of laser-beam illumination time. The substrate material was SiO_2 covered with a 700 Å thick layer of sputtered W. (a) Deposited from 0.49 mbar WCl_6 + 50 mbar H_2 by using different laser wavelengths but constant absorbed laser power $P_a = P(1-R)$ with $2w_0 \approx 15\,\mu m$ [*Kullmer* et al. 1992]. (b) Deposited from 5 mbar WF_6 + 500 mbar H_2 at various Kr^+-laser powers ($\lambda \approx 647\,nm$; $2w_0 \approx 2.1\,\mu m$) [*Szörenyi* et al. 1988]

deposition, the spot diameter increases about linearly with power. A similar behavior has been found for the WF_6-H_2 system (Fig. 16.2.2) [*Szörenyi* et al. 1988] and for the deposition of Ni from $Ni(CO)_4$ [*Petzoldt* et al. 1984].

Dependence on gas pressure and composition

The dependence of the growth rate on partial pressures, pressure ratio Γ_p, and the admixture of noble gases has also been investigated.

The strong influence of Γ_p on the morphology of W spots deposited from $WF_6 + H_2$ has already been discussed. With the admixture of an inert gas such as Ar, the shape of the spots changes from S_l to S_i type and, with pressures $p(Ar) > 200$ mbar, to S_d type. With increasing Ar pressure, the ratio d/h increases while the diameter and the thickness of the thin circular film decreases (Fig. 16.2.1a').

Even when spots are spatially well-defined, considerable changes in shape may occur with pressure changes. Examples are illustrated in Figs. 16.2.5 for the WCl_6-H_2 system.

16.2.2 Deposition from Carbonyls

Pyrolytic deposition from metal carbonyls was studied in detail for Ni [*Petzoldt* et al. 1984], Cr, Mo, and W [*Singmaster* and *Houle* 1991]. The

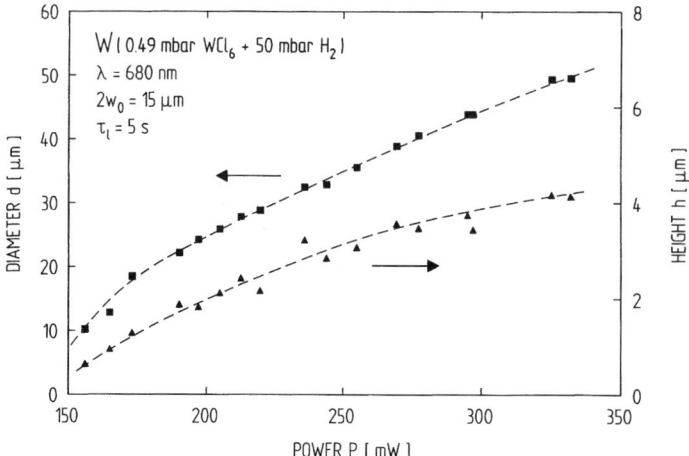

Fig. 16.2.4. Diameter d and height h of W spots as a function of (dye-) laser power ($\lambda = 680$ nm, $\tau_l = 5$ s). The substrate material was 700 Å W/a-SiO_2 [*Kullmer* et al. 1992]

16.3 Modelling of Pyrolytic LCVD

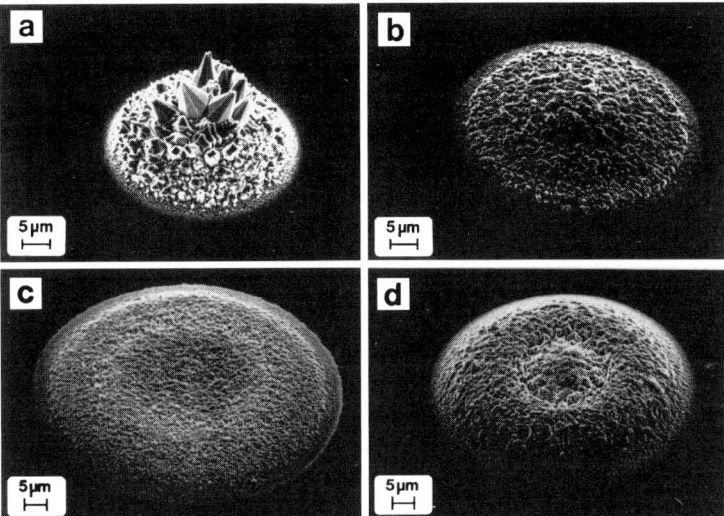

Fig. 16.2.5 a–d. SEM pictures of W deposited from WCl_6. (**a**) Pure WCl_6 (0.49 mbar) and 514.5 nm Ar^+-laser radiation ($P = 290$ mW, $\tau_l = 20$ s, $w_0 = 7.5$ μm). (**b**) Same conditions as in (**a**) but with an admixture of 50 mbar H_2. (**c**) 0.49 mbar WCl_6 + 400 mbar H_2 ; 680 nm dye-laser radiation ($P = 300$ mW, $\tau_l = 20$ s). Note the dip in the center. (**d**) Same as (**c**) but for $P = 240$ mW and $\tau_l = 40$ s [*Kullmer* et al. 1992]

concentration of carbon and oxygen impurities increases from the spot center towards the (colder) edge but is in total much lower than in photochemically deposited spots.

16.3 Modelling of Pyrolytic LCVD

The spot sizes observed under the conditions usually employed in laser-CVD can easily be measured and the high growth rates enable much data to be accumulated. Thus, various experimental dependences can be determined with high accuracy. Among those is the shape and size of spots, their axial and lateral growth rates as a function of laser power, gas pressures, etc. From these data, fundamental information on the gas-phase kinetics, the different activation mechanisms, and the influence of the substrate on the growth process can be derived, in principle. Such an analysis requires, however, a *self-consistent* treatment of the various gas-phase processes, the laser-induced temperature distribution, and the growth process itself. Besides the influence of trivial geometrical effects, axial and lateral growth rates may be determined by *different* apparent activation energies, sticking coefficients, etc. These may originate from the different physico-chemical surface properties of the deposit and the substrate, from temperature gradients on these surfaces, etc. Temperature gradients may influence growth rates via both surface diffusion of species and thermal diffusion within the gas phase. A self-consistent treatment of the general (coupled) problem is very complicated. For this reason, the analysis of data has been performed along two different lines:

- The gas-phase kinetics and transport is studied by assuming the geometry and temperature distribution of the deposit to be *fixed*. The basic equations and the various models employed in such calculations are presented in Chap. 3.
- The shape of the deposit is calculated by assuming a single, thermally activated purely heterogeneous reaction which is kinetically controlled. Here, gas-phase processes are ignored.

These different approaches will now be discussed in further detail.

16.3.1 Gas-Phase Processes

The gas-phase kinetics in pyrolytic LCVD has been investigated by describing the reaction zone either by a hemisphere (Fig. 3.4.1) or by a thin circular film placed on a semi-infinite substrate (Fig. 3.5.1). The hemispherical model is particularly suited for studying different types of transport processes (Sect. 3.4). The cylindrical model, on the other hand, permits one to include, in a simple way, the effect of volume (homogeneous) reactions and to describe qualitatively the influence of gas-phase processes on the shape of deposits. Certainly, this model applies to "flat" structures only. Nevertheless, it helps to interpret different spot shapes observed experimentally. Subsequently, we shall discuss some essential features of thermally activated reactions of type (3.5.1) which correspond to those described by (16.2.1), (16.2.2), etc. For simplification, we mainly assume the partial reaction orders $\gamma_i = \gamma_{v,i} = 1$. The flux of species onto the substrate surface $J(r) \equiv J(r, z = 0)$ can be calculated by employing the approximations made in Sect. 3.5. Let us consider different cases.

Case 1: Pure surface reaction $k_1 \neq 0$, $k_2 = k_3 = 0$

The *surface* (heterogeneous) reaction shall take place only in the forward direction. This case applies also to deposition from adsorbed species. Figure 16.3.1a depicts the normalized reaction rate $W^*(r^*) = J(r^*)/J(0)$ for different center temperatures T_c. At low temperatures, the concentration of reactants AB varies only slightly over the reaction zone and the shape of the deposit is determined mainly by the radial dependence of the rate constant $k_1(T(r, 0))$. The thickness of the deposit decreases monotonically with distance.

16.3 Modelling of Pyrolytic LCVD

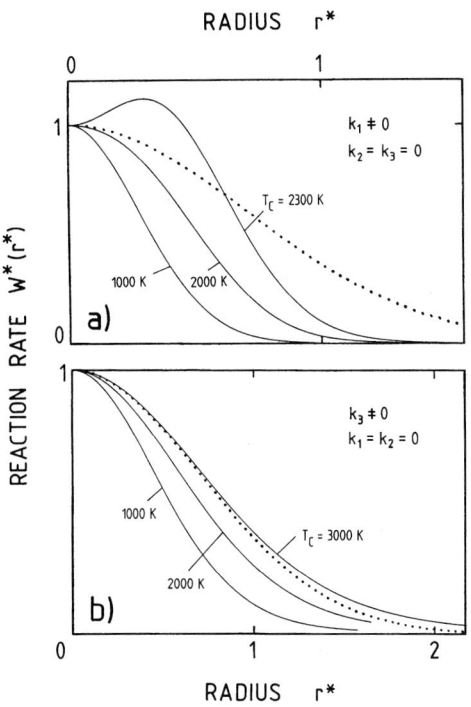

Fig. 16.3.1a,b. Normalized reaction rate, W^*, as a function of radius, r^*. Different curves refer to different laser-induced center temperatures T_c on the flat deposit/substrate (Fig. 3.5.1). Dotted curves represent the normalized laser-beam intensity. (a) Pure surface reaction with $\mathscr{E}_1 = 2 \times 10^4$ K. (b) Pure gas-phase reaction with $\mathscr{E}_3 = 8 \times 10^3$ K [Kirichenko and Bäuerle 1992]

At higher temperatures, the top of the deposit becomes flatter. When further increasing T_c, a dip near the center appears (note that the *absolute* value of $W^*(0)$ increases monotonically with T_c). This dip reflects gas-phase transport limitations of species AB towards the reaction zone. For all temperatures, the deposited spot is localized within the area $r \leqslant r_1$, see (3.5.8). The picture is in qualitative agreement with the spot shapes shown in Fig. 16.2.5b,c.

Case 2: Pure gas-phase decomposition $k_3 \neq 0$; $k_1 = k_2 = 0$

Figure 16.3.1b exhibits the rate for a pure *gas-phase* reaction. At low temperatures, $W^*(r^*)$ reflects the spatial behavior of the rate constant $k_3(T(r,0))$ given by (3.5.7b).

At high temperatures, W^* shows long tails. The shape of curves can be interpreted as follows. According to (3.5.8) the gas-phase reaction is localized within a volume of height h_v above the irradiated surface. At low temperatures, h_v is very small and atoms A generated within this volume directly diffuse to the surface and form the deposit. At higher temperatures, h_v increases and atoms A will diffuse over larger distances. As a consequence, deposition becomes less localized.

Thus, with low laser-induced temperatures, deposition is localized irrespective of whether decomposition of molecules takes place at the surface or within the gas. As a consequence, the growth of spots of type S_l or S_i is observed. With medium and high temperatures, the situation becomes more complicated. In any case, the "transition" from localized to diffuse deposits (Fig. 16.2.1) can, in principle, be understood by the increasing importance of homogeneous reactions. No new phenomena occur if $k_1 \neq 0$, $k_3 \neq 0$, $k_2 = 0$.

Case 3: Gas-phase decomposition, etching $k_2 \neq 0$, $k_3 \neq 0$, $k_1 = 0$

We now consider gas-phase pyrolysis of AB together with surface etching. Figure 16.3.2a shows three characteristic shapes of deposits corresponding to different temperatures T_c where $-J^*(r^*) \propto W^*(r^*)$. At low temperatures, the thickness of the deposit decreases monotonically with the distance r^*. Above a certain temperature, the thickness of the deposit approaches zero at $r^* \approx 1$. At

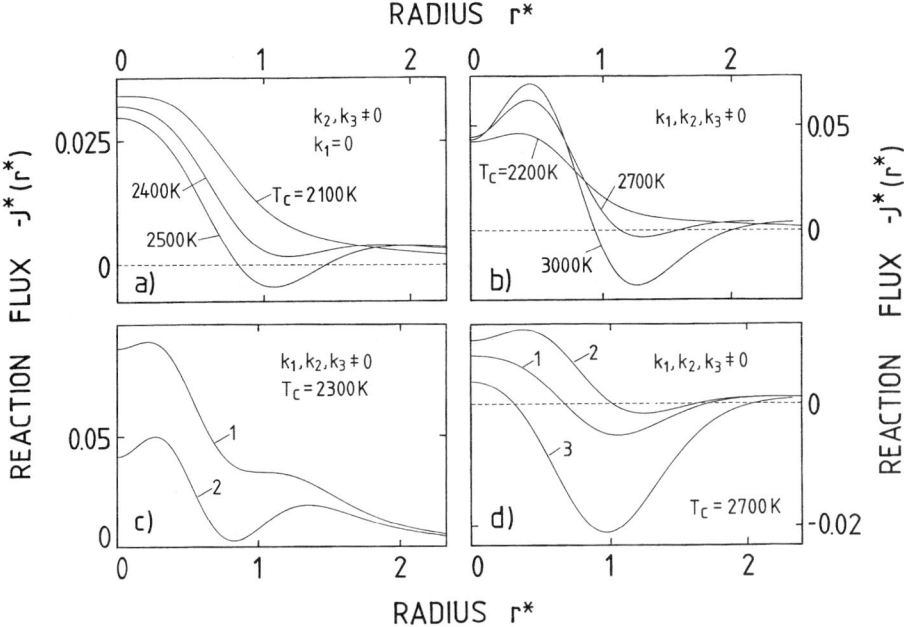

Fig. 16.3.2 a–d. Normalized reaction flux $J^*[\text{cm}^{-1}] = J\sqrt{\pi/2N(\infty)D_{BC}(\infty)}$ as a function of distance, r^*. (**a**) Pure gas-phase reaction and surface etching ($\mathscr{E}_2 = 8 \times 10^3$ K, $\mathscr{E}_3 = 10^4$ K, $D^*_{AB} = 0.1$). (**b**) Surface- and volume decomposition of precursors and surface etching ($\mathscr{E}_1 = 1.5 \times 10^4$ K, $\mathscr{E}_2 = 8 \times 10^3$ K, $\mathscr{E}_3 = 10^4$ K, $D^*_{AB} = 0.1$). (**c**) Same as (**b**) but for $\mathscr{E}_1 = 2 \times 10^4$ K, $\mathscr{E}_2 = 10^4$ K, $\mathscr{E}_3 = 7 \times 10^3$ K, $T_c = 2300$ K. Curve 1: $D^*_{AB} = 0.1$. Curve 2: $D^*_{AB} = 0.2$. (**d**) Same as (**b**) but for $\mathscr{E}_1 = 1.2 \times 10^4$ K, $\mathscr{E}_2 = 8 \times 10^3$ K, $\mathscr{E}_3 = 10^4$ K, $T_c = 2700$ K, $D^*_{AB} = 0.1$, $\gamma_{v,AB} = 1$. Curve 1: $\gamma_{AB} = \gamma_{BC} = 1$. Curve 2: $\gamma_{AB} = 0.5$, $\gamma_{BC} = 1$. Curve 3: $\gamma_{AB} = 1$, $\gamma_{BC} = 0.9$ [*Kirichenko and Bäuerle 1992*]

even higher temperatures, etching within a *ring* around the central deposit is observed – here we assume that the substrate consists of material A. At large distances, the reaction rates become small, for both deposition and etching. Here, a thin film is formed via diffusion of atoms A out of the reaction zone. The figure demonstrates an additional phenomenon. An increase in laser-beam intensity, and thus in temperature T_c, results in a *decrease* in the height at $r=0$. This reflects the competition between deposition and etching.

Case 4: Surface- and gas-phase decomposition, etching $k_1 \neq 0$, $k_2 \neq 0$, $k_3 \neq 0$

This is the most general situation which must, in fact, be considered with reactions of type (16.2.1). Figure 16.3.2 b–d depicts different shapes that can be obtained in this case. The profiles shown in (b) are quite similar to those in (a). Figure 16.3.2d demonstrates the strong influence of reaction orders on profiles.

Limits of validity

In spite of the good agreement between observed and calculated deposition/etch profiles, a direct correlation of results should be considered with care: First, the relative contributions of volume and surface reactions are unknown. Second, the model (Fig. 3.5.1) is adequate only for *flat* structures. Third, the assumption of a temperature profile which remains unaffected during deposition is a reasonable approximation only if $\kappa_D \approx \kappa_s$.

16.3.2 The Coupling Between $T(x)$ and $h(x)$

In Part II we investigated laser-induced temperature distributions for plane substrates that were infinitely extended in the lateral direction. In some kinds of laser microprocessing, however, these assumptions are invalid. Among the examples are the laser-induced deposition of *microstructures*, the transformation of spun-on metallopolymers, etc. Here, the width of patterns is of similar size as the diameter of the *focused* laser beam. This is schematically shown in Fig. 16.3.3. It is easy to demonstrate that in such cases the laser-induced temperature distribution, $T(x)$, depends on the geometry of the pattern, $h(x)$. We shall prove this for pyrolytic laser-CVD by assuming a *kinetically* controlled heterogeneous reaction ($k_2 = k_3 = 0$).

In the kinetically controlled regime, the growth rate depends exponentially on the temperature distribution induced by the absorbed laser light on the substrate or the already deposited material. This temperature distribution, in turn, depends on the geometry of the deposit and thereby changes during the growth process. This can be demonstrated by considering, for example, laser-induced temperature distributions for the (fixed) model structures depicted in Fig. 16.3.3a–d. Thus, even when gas-phase processes are ignored, any quantitative or semi-quantitative calculations on the shape of deposits require a *self-consistent* treatment of the laser-induced temperature distribution and the equation of growth.

In photolytic processing the situation is different. Here, the precursor molecules are decomposed within the ambient gas and the profile of patterns depends on the concentration of product species reaching the surface (Chaps. 14, 15 and 19).

Let us consider a deposit of arbitrary shape $h(x, y)$ within the region $0 < z < h(x, y)$ (Fig. 16.3.3d). The temperatures of the deposit and the substrate are denoted by T_D and T_s, respectively. The notations for the other quantities are analogous. The temperature distribution induced by the absorbed laser light can then be calculated from the boundary-value problem

$$c_D \rho_D \frac{\partial T_D}{\partial t} - \nabla[\kappa_D(T_D)\nabla T_D] = Q(x, y, z, t) \tag{16.3.1}$$

$$\kappa_D(T_D)\frac{\partial T_D}{\partial z}\bigg|_{z=0} = J_{loss}(z=0)$$

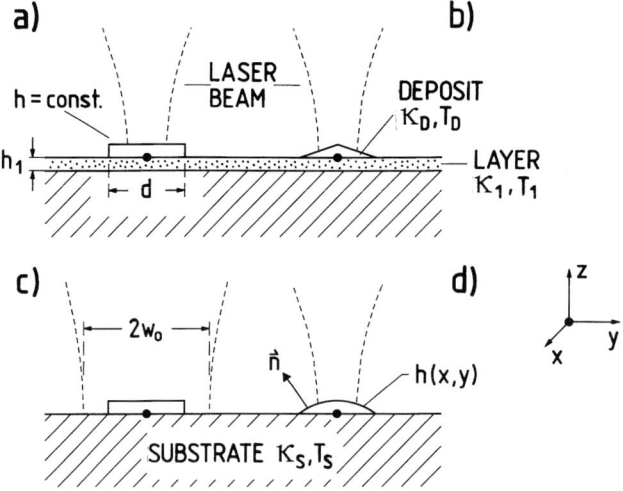

Fig. 16.3.3a–d. Deposits of various shapes with radius $r_D = d/2$ at the substrate surface and height h. In (d) the height is described by an arbitrary function $h(x, y)$. The intermediate layer has a thickness h_1. The radius r_D can be either larger or smaller than the radius of the laser focus, w_0. The origin of the coordinate system is indicated by the dot. If the substrate and the laser beam are fixed, these model structures describe the deposition of spots. If the laser beam is scanned, e.g., perpendicularly to the plane of the drawing, these structures describe laser direct writing of lines with different shapes

16.4 Temperature Distributions on Circular Deposits

$$-\kappa_D(T_D)\frac{\partial T_D}{\partial \hat{n}}\bigg|_{z=h} = J_{\text{loss}}(z=h).$$

Here, c_D is the specific heat and ρ_D the mass density of the deposit. The unit vector $\hat{n} = n/|n|$ is directed perpendicular to the surface $z = h(x, y)$ so that n has the components $(-\partial h/\partial x, -\partial h/\partial y, 1)$. Equation (16.3.1) together with the corresponding equations for the substrate and, if present, for the thin layer, can directly be solved numerically.

For the simulation of deposition profiles, (16.3.1) must be solved self-consistently together with the equation of growth. In a coordinate system that is fixed with the laser beam, the shape of the deposit is given by

$$\frac{\partial h}{\partial t} = W(T_D) \cdot |n| + v_s \frac{\partial h}{\partial x}, \tag{16.3.2}$$

The factor $|n| = [1 + (\nabla_2 h)^2]^{1/2}$ takes into account that growth takes place perpendicularly to the surface $h(x, y)$ (∇_2 is the gradient in the xy-plane). If $v_s \neq 0$, scanning of the laser beam is performed exclusively in the x-direction (Fig. 16.3.3). The growth rate can be described by

$$W(T_D) = k_0 \exp\left(-\frac{\mathscr{E}}{T_D}\right)\left[1 + \exp\left(\frac{T_{\text{th}} - T_D}{\delta T_{\text{th}}}\right)\right]^{-1}, \tag{16.3.3}$$

where $\mathscr{E} \equiv \Delta E/k_B$. According to our assumptions, the growth process shall be characterized by a *single* apparent activation energy ΔE. The additional factor in the Arrhenius-type law shall account for the "threshold" behavior observed with certain systems. Depending on the type of precursor molecules employed, the deposition rate either drops off very sharply for $T_D \approx T_{\text{th}}$ or it shows a smooth behavior. This "width" of the threshold is denoted by δT_{th}. In the *absence* of a threshold we set the last factor in (16.3.3) equal to unity. The boundary and initial conditions are usually characterized by $h = 0$. For the simulation of growth we can frequently employ the assumption $\partial T_D/\partial t = 0$. This is a good approximation if thermal equilibrium is reached within a time $\tau_T \ll [\partial (\ln h)/\partial t]^{-1}$. In many cases, we can also ignore heat losses to the gas phase and set $J_{\text{loss}}(z = h) = 0$.

Before we present self-consistent calculations, we consider temperature distributions induced on model structures of fixed geometry.

16.4 Temperature Distributions on Circular Deposits

In the present section we discuss *stationary* temperature distributions that have been calculated for combined structures consisting of a circular deposit, a plane semi-infinite substrate and, finally, a layer of thickness h_1 in between (Fig. 16.3.3). The geometry of the deposit is kept fixed. We consider cw-laser irradiation of Gaussian shape, if not otherwise noted. For circular spots, the maximum temperature occurs in the beam center. The source term in (16.3.1) can be written as

$$Q = \frac{P_a}{\pi w_0^2} \exp\left(-\frac{r^2}{w_0^2}\right) f(z),$$

where $P_a = P(1 - R)$ (Sect. 6.1; note the different definitions of the z-direction). With the geometries under consideration, the solution of the heat equation is most conveniently performed by employing finite element techniques.

The laser-induced temperature distribution is mainly characterized by the thermal conductivities κ_D, κ_1, and κ_s, by the radius of the deposited spot, $r_D = d/2$, and by the radius of the laser focus, w_0.

Subsequently, we discuss only a few important features of such temperature distributions.

Case 1: Cylinders with $r_D > w_0 = const$, $\alpha_D \to \infty$, $h_1 = 0$

In this case we approximate the deposit by a *circular cylinder* of (fixed) diameter d and height h (Fig. 16.3.3a). We set $h_1 = 0$ (this is identical with $h_1 \neq 0$ and $\kappa_1 = \kappa_s$). The normalized surface temperature rise is indicated in Fig. 16.4.1 as a function of the distance from the center. For $\kappa^* \equiv \kappa_D/\kappa_s = 1$ the temperature distribution approximates that of the plane substrate (dashed curve). Here,

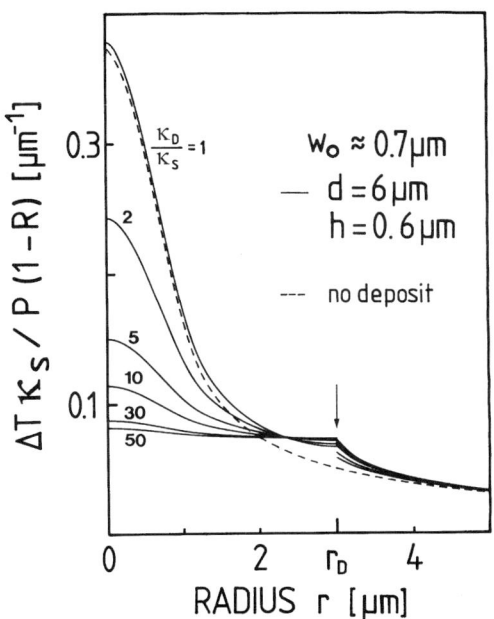

Fig. 16.4.1. Normalized laser-induced surface temperature rise calculated for a circular cylinder of radius $r_D = d/2$ and height h placed on a semi-infinite substrate. The arrow marks the edge of the cylinder. The dashed curve represents the temperature distribution calculated for a plane semi-infinite substrate [*Piglmayer* et al. 1984]

significant differences occur only near r_D. The center-temperature rise strongly decreases with increasing κ^*. In the limit $\kappa^* \gg 1$ the temperature is almost uniform over the surface of the cylinder. This latter case applies to metal deposits on thermally insulating substrates. The temperature distribution for radii $r \gg r_D$ is only slightly influenced by κ^*. A detailed analysis shows that the temperature at the edge of the deposit, $T(r_D)$, depends strongly on the diameter of the spot, but much less on its exact height and shape [Bäuerle 1986]. If $\kappa^* \gg 1$, the temperature rise $\Delta T(r_D, h)$ scales approximately inversely with d. This can easily be understood. With $w_0 < r_D$ the temperature gradient in the substrate is $\nabla T \approx \Delta T/r_D$. The energy balance yields $P_a \approx \pi r_D^2 \kappa_s \nabla T$. Thus, we obtain the simple relation $\Delta T(r_D, h) \approx P(1 - R_D)/\pi r_D \kappa_s$. An exact analytical solution yields for $\kappa^* \to \infty$

$$\Delta T(r_D, h) = \frac{P(1 - R_D)}{4 r_D \kappa_s}. \tag{16.4.1}$$

This temperature rise is not very sensitive to changes in κ_D as long as $\kappa^* w_0/r_D \gg 1$. For example, if κ^* is doubled, numerical calculations yield a decrease in $\Delta T(r_D, h)$ by a few percent only.

Case 2: Influence of a thin layer and the laser focus: Cylinders with
$r_D > w_0, \alpha_D \to \infty, h_1 \neq 0$

The *full* curves in Fig. 16.4.2 show temperature distributions for a cylinder of radius r_D and height $h = r_D/10, h_1 \neq 0$, and $w_0 = r_D/3$ (Fig. 16.3.3a). In Fig. 16.4.2 (a) and (c), the intermediate layer has no influence on the temperature

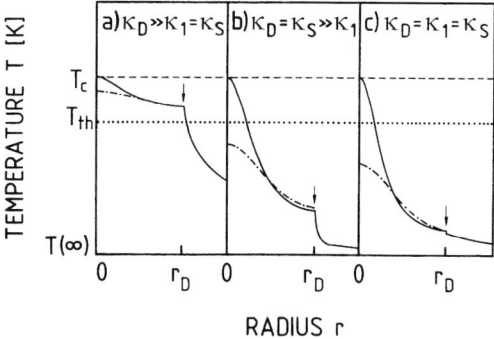

Fig. 16.4.2a–c. Temperature profiles calculated for the model structure depicted in Fig. 16.3.3a. Three different ratios of thermal conductivities and two different radii of the laser focus have been considered. The solid curves have been calculated for equal center temperature $T_c = 530$ K with $h_D = r_D/10$ and $w_0 = r_D/3$, and $T(\infty) = 300$ K. T_{th} schematically indicates a threshold temperature. (a) $\kappa_D = 0.7$ W/cm K, $\kappa_1 = \kappa_s = 1.3 \times 10^{-2}$ W/cmK. (b) $\kappa_D = \kappa_s = 0.7$ W/cm K, $\kappa_1 = 1.3 \times 10^{-2}$ W/cmK, $h_1 = 4000$ Å. (c) $\kappa_D = \kappa_1 = \kappa_s = 0.7$ W/cmK. Dash-dotted curves: $h = r_D/10$ and $w_0 = 2r_D/3$. All other parameters are the same as with the full curves [Bäuerle 1984]

distribution. In (b) the layer h_1 is thermally insulating. The parameters employed are appropriate to a situation where a metal spot is deposited onto a metal or a Si substrate that is covered with an oxide layer. The curves were calculated for equal center temperatures which require *different* absorbed laser powers.

Temperature profiles calculated for the same parameter sets, but with twice the diameter of the laser focus, $w_0 = 2r_D/3$, are presented by *dash-dotted* curves. The change in center temperature is more pronounced in (b) and (c) than in (a). As long as $r_D > w_0$ the temperature at the edge of spots $\Delta T(r_D, h)$ remains almost unaffected.

Temperature distributions calculated for the case shown in Fig. 16.3.3c were given in [*Piglmayer* and *Bäuerle* 1986].

16.5 Simulation of Pyrolytic Growth

We now consider the two-dimensional model (Fig. 16.3.3d) with $h = h(x, y)$. The problem is then described by (16.3.1 to 3) with $v_s = 0$. Henceforth we assume $\alpha_D \to \infty$ and a flat deposit with $\partial h/\partial x$, $\partial h/\partial y \ll 1$ so that $|n| \approx 1$.

Because changes in the laser-induced temperature distribution related to the geometry of the deposit are the more pronounced the more the ratio of thermal conductivities differs from unity, we consider $\kappa^* \equiv \kappa_D/\kappa_s \gg 1$. For flat structures we can write

$$\frac{\partial T_D}{\partial \hat{n}} = \frac{\partial T_D}{\partial z} - \nabla_2 h \cdot \nabla_2 T_D . \tag{16.5.1}$$

Furthermore, we assume $T_D \equiv T_D(x, y, z, t) \approx T_D(x, y, 0, t)$ which holds if $h/(r_D \kappa^*) \ll 1$ (this follows from $\kappa_D \partial T_D/\partial z = \kappa_s \partial T_s/\partial z$ and $\partial T_s/\partial z \approx [T_s(z=0) - T(\infty)]/r_D$). In this case we can expand T_D in a Taylor series and take into account only the first term

$$T_D(z) = T_D(z=0) + \mathcal{O}(z). \tag{16.5.2}$$

With these additional approximations the original boundary-value problem (16.3.1) can be written as [*Arnold* and *Bäuerle* 1993]

$$\theta_s = \theta_s^0 + \frac{\kappa^*(T(\infty))}{2\pi}\left[\nabla_2(h\nabla_2\theta_D)*\frac{1}{|r|}\right], \tag{16.5.3}$$

where

$$\theta_s^0 = \frac{1}{2\pi\kappa_s(T(\infty))}\left(I_a * \frac{1}{|r|}\right)$$

is the (linearized) temperature distribution without the deposit (Chap. 2). r is a two-dimensional radius vector within the xy-plane, and $*$ denotes the

16.5 Simulation of Pyrolytic Growth

convolution integral

$$f*g \equiv \int_{-\infty}^{+\infty}\int_{-\infty}^{+\infty} f(r')g(r-r')dx'dy'.$$

θ_s^0 remains unchanged during growth as long as the reflectivity R stays constant. If R changes (16.5.3) is still valid. In this case, however, the computational time will increase significantly. At $z = 0$ the temperatures θ_s and θ_D are related via the condition

$$T_D(\theta_D) = T_s(\theta_s) \quad \text{at } z = 0, \tag{16.5.4}$$

i.e., $\theta_s = \theta_s(\theta_D)$. This dependence can be fitted by the polynomial

$$\theta_s = \theta_D + \beta\theta_D^2 + \dots .$$

With this substitution, (16.5.3) becomes a nonlinear integro-differential equation for the determination of θ_D. The real temperature is obtained from the inverse Kirchhoff transform $T(x,y) = T_D(\theta_D)$. The advantages of solving (16.5.3) instead of (16.3.1) are:

- The problem is two-dimensional, but describes the temperature distribution in three dimensions.
- No boundary conditions need to be considered; they are included implicitly in θ_s and θ_D.
- Any temperature dependences in κ_D and κ_s can easily be taken into account.
- It is sufficient to calculate the convolution integrals within regions $h \neq 0$ and $I_a \neq 0$.

With (16.5.3) the shape of the deposits can be calculated from (16.3.2) with $v_s = 0$. The results of such calculations shall now be discussed for W and Ni spots deposited onto SiO_2 substrates. For convenience, we normalize h, x, y, and k_0 to the radius of the (Gaussian) laser beam, w_0. Temperatures and activation energies are normalized to $T(\infty)$. The normalized intensity is $I_0^* = I_0 w_0 / T(\infty)\kappa_s$.

Figure 16.5.1a shows the evolution of W spots together with surface temperature profiles. The kinetic data $k_0^* = 2.14$, $\mathscr{E}^* = 5.68$ were taken from experimental investigations on the deposition of W from $WCl_6 + H_2$ [Kullmer et al. 1992]. The other parameters are $T_{th}^* = 2.71$, $\delta T_{th}^* = 0.01$, $T(\infty) = 443$ K, $\kappa^*(T(\infty)) = 50.56$, $AI_0^* = 10.3$. The thermal conductivities were approximated by $\kappa_D(W) = c_1 + c_2/T - c_3/T^2$ with $c_1 = 42.65$ W/mK, $c_2 = 1.898 \times 10^4$ W/m and $c_3 = 1.498 \times 10^6$ WK/m and by κ_s (SiO_2) $= a_1 + a_2 T$ with $a_1 = 0.9094$ W/mK and $a_2 = 1.422 \times 10^{-3}$ W/m K^2 [Heraeus 1979]. The Kirchhoff transform permits one to find the approximations

$$\theta_s^* \approx \theta_D^* + 0.46\theta_D^{*2} \quad \text{and} \quad T_D^* \approx 1 + \theta_D^* + 0.12\,\theta_D^{*2}.$$

The saturation in width is reached faster than in height, mainly due to the small value of δT_{th}. This is in agreement with experimental observations (Fig. 16.2.3a). Within a short time, the temperature becomes almost uniform

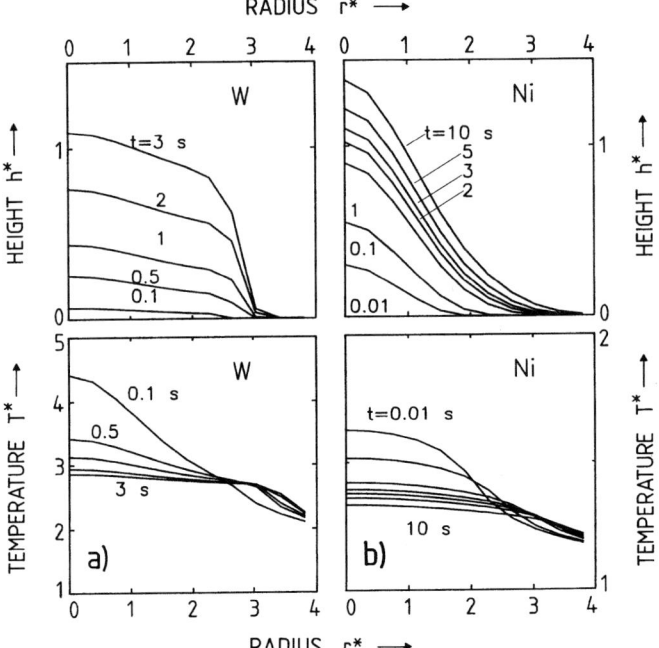

Fig. 16.5.1a,b. Normalized height of spots calculated for various stages of growth as a function of the (normalized) distance from the laser-beam center. The parameters employed are typical for laser-CVD. The lower part of the picture shows the evolution of the (normalized) surface temperature distribution. (a) Deposition of W from $WCl_6 + H_2$. (b) Deposition of Ni from $Ni(CO)_4$ [*Arnold* and *Bäuerle* 1993]

over the surface of the spot, as expected from the simple model developed in the previous section.

Figure 16.5.1b displays growth profiles for Ni spots. Here, no threshold was assumed and the thermal conductivities were kept constant [$\kappa^*(T(\infty) = 300\,\text{K}) = 30$]. The kinetic data, $k_0^* = 1.1 \times 10^{13}$ and $\mathscr{E}^* = 45$ (27 kcal/mol), correspond to pyrolytic decomposition of $Ni(CO)_4$. The other parameters employed were $T(\infty) = 300\,\text{K}$ and $AI_0^* = 1.33$. Due to the absence of a threshold, there is *no* abrupt saturation in width. Furthermore, the aspect ratio $\Gamma = h/d$ is bigger than for W. This is related to the higher activation energy which causes faster growth in the center than near the edge. The calculated shape is in qualitative agreement with experimental observations (Fig. 16.5.2).

16.6 Photolytic LCVD

Photolytic laser-induced chemical vapor deposition is based on *selective* excitations of precursor molecules. In many systems, decomposition of molecules

Fig. 16.5.2. SEM picture of a Ni spot deposited from Ni (CO)$_4$ onto a 1000 Å a-Si/glass substrate [*Piglmayer* and *Bäuerle* 1986]

takes place in the gas phase within the total volume of the laser beam (Fig. 1.2.2). The product species diffuse and condense, in part, on the substrate surface. Photolytic LCVD can therefore be employed for *thin-film* formation at relatively low substrate temperatures. Today, this is its main application (Chap. 19).

Single-step production of *microstructures* based on photolytic LCVD has also been demonstrated. With the systems investigated, the spatial confinement of deposits is closely related to physisorbed layers of parent molecules, and to nucleation processes. Here, adsorbed-layer photolysis significantly contributes or even dominates the deposition process. In systems where physisorption is weak, spatially well-defined direct patterning by photolytic LCVD is impossible.

Fabrication of microstructures based on selective *electronic* excitations has been studied mainly for metal-alkyls and metal-carbonyls. These compounds are readily available, have a relatively high vapor pressure at ambient or slightly elevated temperatures, and they can be decomposed by single-photon or sequential multiphoton excitation in the near to medium UV. This spectral range can easily be reached by frequency doubling of cw Ar^+- or Kr^+-laser lines, by excimer lasers, or by higher harmonics of Nd:YAG lasers. Other precursors frequently employed are organic coordination complexes, in particular various acetylacetonates, which can be designed for particular applications.

Photolysis by coherent multiphoton electronic excitations is only of limited value for controlled deposition of micrometer-sized patterns, mainly because of the high laser-light intensities required.

Up to now, there is *no* clear example for the deposition of microstructures based on selective multiphoton vibrational excitations using IR radiation. This is quite understandable from Sect. 2.3.2. In any case, this excitation mechanism would not permit one to achieve the resolution obtained with VIS or UV radiation.

It should be noted that in many of the experiments reported, the laser-light intensities employed are high enough to induce a significant temperature rise. Thus, many of the experimental data should be analyzed on the basis of a combined photochemical–photothermal (photophysical) model.

16.6.1 Metals

Photolytic LCVD of metals from *alkyls* has been most thoroughly studied for $Cd(CH_3)_2$ and $Al_2(CH_3)_6$. These (model) compounds can be dissociated by single-photon excitation in the near UV and their photochemistry is well known [*Price* 1972]. Dissociation of $Cd(CH_3)_2$ by single-photon excitation with 257 nm Ar^+-laser radiation can be described by

$$Cd(CH_3)_2 + hv\ (257\,nm) \rightarrow Cd(^1S_0) + 2CH_3\,. \tag{16.6.1}$$

The free methyl radicals subsequently react and form volatile hydrocarbons such as ethane. Photolysis can take place within the gas and on the surface. The latter mechanism requires adsorbed $Cd(CH_3)_2$ molecules. Adsorbed-layer photolysis determines the confinement of the deposition process. In other words, metal atoms generated within the gas condense, preferentially, within the irradiated area (see also Chaps. 4, 5, and 20).

Figure 16.6.1 shows the (thickness) deposition rate for Cd spots versus laser-light intensity. The linear increase reflects single-photon dissociation. At

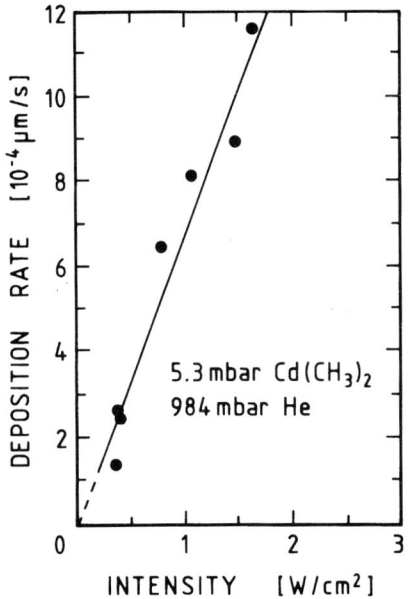

Fig. 16.6.1. Deposition rate for Cd spots versus Ar^+-laser light intensity ($\lambda = 257$ nm). Helium was used as buffer gas [*Ehrlich* et al. 1982]

very low laser fluences, the threshold for surface nucleation prevents deposition. At much higher fluences than those shown in the figure, the deposition rate saturates due to mass-transport limitations. This effect is more pronounced when a buffer gas is used. Mass transport could be increased by employing a gas flow. Deposition rates for Cd of up to 0.1 µm/s have been measured at UV intensities of about 10^4 W/cm^2. Direct writing of lines with widths as small as 0.7 µm has also been demonstrated.

Thermal and photochemical decomposition of $Al_2(CH_3)_6$ has been investigated by *Price* (1972), *Suzuki* et al. (1986), and others. Direct writing of Al lines with widths of 2 to 3 µm was demonstrated for different substrate materials. Deposition rates up to 0.1 µm/s and scanning speeds up to a few µm/s have been achieved.

Photolytic deposition was also studied with other precursors such as $Ga(CH_3)_3$, $In(CH_3)_3$, $Te_2(CH_3)_2$, $Zn(CH_3)_2$, $Zn(C_2H_5)_2$ ([*Eden* 1991] and references therein).

Deposition from carbonyls

Metal carbonyls such as $Ni(CO)_4$, $Fe(CO)_5$, $Cr(CO)_6$, $Mo(CO)_6$, and $W(CO)_6$ are used for metal deposition in the form of microstructures and extended thin films. For many metal carbonyls, molecular fragmentation begins to occur in the near UV at wavelengths $\lambda < 350$ nm. For cw- and pulsed-laser irradiation at low power densities, decomposition seems to be based on *sequential* elimination of CO ligands by single photon processes

$$Me(CO)_m + h\nu \rightarrow Me(CO)^*_{m-1} + CO, \qquad (16.6.2)$$

$$Me(CO)^*_{m-1} + h\nu \rightarrow Me(CO)^*_{m-2} + CO,$$

$$\vdots$$

$$Me(CO)^* + h\nu \rightarrow Me^* + CO,$$

where the asterisk indicates internal vibrational and possibly electronic excitations. Due to the high gas pressures employed in LCVD, or due to molecule-substrate interactions, stripping of the remaining ligands, for example after absorption of one or two photons, can also occur via collisions [*Ishikawa* et al. 1990; *Price* 1972; *Seder* et al. 1985]. Focused high power pulsed-laser excitation may favor *coherent* multiphoton rather than sequential single-photon photochemistry.

Laser-induced deposition of metals by photolysis of carbonyls has been studied by many researchers. The literature on the most important investigations is included in Appendix B.4. Recent investigations on the deposition of Cr, Mo, and W were performed by *Singmaster* et al. (1990). These experiments employed 257 nm Ar^+-laser radiation and Si substrates. With the laser powers employed, the estimated temperature rise on the substrate was < 50 K. The intrinsic compositions of deposits, as determined by scanning Auger microscopy, were found to be independent of the laser power. Auger analysis of deposits provides information on surface photodissociation and dissociative chemisorption of species. Chemical differences are observed only for the Cr system where the laser-light forces additional desorption of CO. Thus, metal-carbonyl fragments adsorbed on the spot surface lose CO groups spontaneously. The remaining CO forms metal-oxycarbide in the case of Mo and W, and an oxide mixed with amorphous or graphitic carbon in the case of Cr.

The deposits show columnar structure and, in the case of Mo and W, many cracks. For all three materials, deep ripples are observed (Sect. 28.2). Addition of N_2 buffer gas results in a delocalization of the deposit. This situation is quite similar to that described for pyrolytic deposition of W from $WF_6 + H_2$ in the presence of Ar (Sect. 16.2). The initial deposition rates achieved with laser-light intensities of 3×10^3 W/cm^2 are around 0.2 to 0.4 μm/s for Cr and Mo, and about one order of magnitude smaller for W. Without window purge, the rates decrease rapidly due to attenuation of the laser-light intensity by material deposited on the windows of the reaction chamber.

Organo-metallic coordination complexes

Various organic coordination complexes, and in particular different acetylacetonates, have been used to deposit Au, Cu, Ir, Pt, etc. *Baum* et al. (1987) demonstrated substrate patterning with Au by direct writing and laser-light projection. Typical deposition rates were a few Å/s. Feature sizes as small as about 2 μm were achieved.

16.6.2 Other Materials

There are only a very few investigations on photolytic LCVD of semiconductors and insulators as microstructures. Among them is the projection printing of 10 μm wide SiO_2 patterns by ArF-laser photolysis of $Si_2H_6 + N_2O/N_2$ [*Hiura* et al. 1991]. However, the real decomposition mechanism has not yet been investigated.

16.6.3 Process Limitations

Deposition rates that are attractive for applications can only be achieved when the laser wavelength matches a transition which results in efficient decomposition of the molecule. For many molecules which would be suitable for photolytic LCVD, such transitions are located in the medium to far UV where, at present, only a few powerful lasers are available. There are, however, more fundamental and thus more severe limitations: One is the tendency for homogeneous cluster formation at higher laser-light intensities or partial pressures of reactants. Clusters may condense everywhere on the substrate and on the walls of the reaction chamber, including entrance windows. Then, controlled deposition becomes impossible. For these reasons, deposition rates achieved in photolytic LCVD will always be lower than those in pyrolytic LCVD. Another limitation concerns the purity of deposits. Photoproducts, including incompletely decomposed precursors, are often incorporated into the deposit with high concentrations. These can only be diminished by substrate heating.

Controlled growth in UV-laser photodeposition of metals is observed for laser-beam intensities of, typically, $1-10^4$ W/cm^2, and for gas pressures between 0.1 and 100 mbar. Typical deposition rates are 0.001 to some 0.1 µm/s. These rates are by a factor of 10^2-10^4 smaller than those in pyrolytic LCVD. The lateral resolution achieved in photolytic deposition of stripes was of the order of 0.1 µm. It should be emphasized, however, that for the aforementioned reasons broad tails, thin films, or big clusters of material are frequently observed around the deposit. The main advantage of photolytic LCVD is the lower local processing temperature and the smaller influence of the physical properties of the substrate. On the other hand, without uniform substrate heating, the microstructure and purity of photodeposited materials, and thus their electrical properties, are unsatisfactory.

17 Growth of Fibers

As demonstrated in the preceding chapter, lateral growth of photothermally deposited spots saturates for long laser-beam illumination times τ_l. With certain systems and within certain parameter ranges, this saturation is accompanied by an increase in axial growth and the formation of a fiber along the axis of the laser beam. A typical fiber is shown in Fig. 17.0.1 with the example of Si. Two phases of growth can be observed. Near the onset, the deposition rate depends strongly on the physical properties of the substrate. Under quasi-stationary conditions, as characterized by a constant fiber diameter, the temperature in the tip becomes independent of substrate material and time. For this reason, the temperature can be measured *in situ* with high precision.

The maximum growth rate occurs at the center of the tip, and it is equal to the axial growth velocity v_a. Therefore, the deposition rate can be defined as the growth in length of the fiber per unit time

$$W_D(T) = v_a(T) \equiv v(r=0, T) = \frac{\Delta h(r=0, T)}{\Delta t}. \tag{17.0.1}$$

With partial pressures of precursor gases of up to 1 bar, the axial growth rates

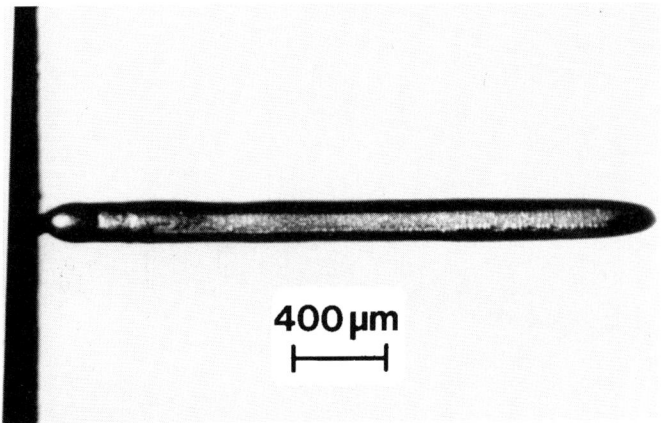

Fig. 17.0.1. Silicon fiber grown from SiH_4 by means of 488 nm Ar^+-laser radiation ($P = 400$ mW, $p \equiv p(SiH_4) = 133$ mbar) [Bäuerle 1983a]

are, typically, 10–100 µm/s. With higher pressures, growth rates up to 10^3 µm/s have been achieved.

Because of the high deposition rates and the possibility of in situ temperature measurements, the growth of fibers is a *unique* technique for both rapid determination of apparent chemical activation energies and investigations on gas-phase processes.

17.1 In Situ Temperature Measurements

An experimental setup employed for *in situ* temperature measurements during steady growth of fibers is schematically depicted in Fig. 17.1.1. The TEM_{00} beam of a cw Ar^+ laser operated at 488 nm was focused onto the tip of the growing fiber. The radius of the laser focus was, typically, $w_0 \approx 10$ µm. To achieve quasi-stationary conditions, the lens L_1 is moved with the velocity v_a as defined by (17.0.1). Thus, the tip of the fiber is always located within the focal plane.

For *monochromatic* pyrometry the thermal radiation emitted from the hot tip of the fiber is collected by the focusing lens L_1, transmitted through the beam-splitter (BS) and focused with a lens L_2 onto a pinhole (PH). The spatial resolution of the measurements is about 9 µm. The light transmitted by the pinhole and the interference filter (IF) is detected.

During deposition of pyrolytic carbon, temperature measurements between 2000 and 3000 K were performed at a center wavelength of 700 nm. With silicon fibers grown between 1100 and 1700 K, the center wavelength was 1000 nm. In both cases, a bandpass of 10 nm was employed. The radiation was detected by a high quality silicon photodiode whose output was measured by means of a lock-in amplifier. A tungsten band lamp was employed to calibrate the response of the detection system. From the measured spectral intensity of the thermal radiation, the local temperature was evaluated by using Planck's law (Sect. 29.4). The angular dependence of the emissivity was ignored. Because the radiation flux depends nonlinearly on temperature, the temperature evaluated in these measurements represents essentially the *maximum* temperature in the center of the fiber tip, i.e., $T \equiv T_D(r=0)$.

For measurements on carbon fibers, a constant emissivity $\varepsilon = 0.85$ was used. For silicon, the linear relation $\varepsilon(T) = 0.946 - (2.76 \times 10^{-4})\,T$ was employed within the range 1000 K $< T <$ 1688 K [*Lampert* et al. 1981; *Sato* 1967; *Allen* 1957].

The standard deviations in temperatures evaluated were, typically, about 10 K for C fibers and about 5 K for Si fibers.

For temperatures $T > 2300$ K, the axial growth rate of C fibers was measured by imaging the thermal radiation emitted from the hot tip of the fiber onto a position-sensing photodiode placed perpendicular to the fiber axis. The output of this diode was simultaneously used to control the distance between

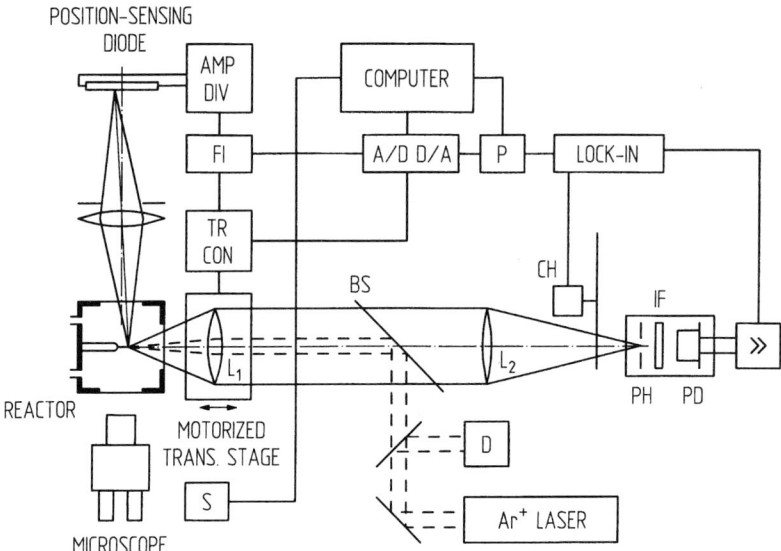

Fig. 17.1.1. Experimental setup employed for in situ temperature measurements during laser–CVD of fibers. AMP DIV: amplifier and analog divider, BS: beamsplitter, CH: light chopper, D: power meter, FI: electronic filter, P: peak detector, PD: Si photodiode, PH: pinhole, S: switch, TR CON: translation control [*Doppelbauer* and *Bäuerle* 1986]

the focusing lens and the fiber tip. In this way, quasi-stationary conditions are achieved. For temperatures below 2300 K, the growth rate could be measured more precisely by means of a microscope.

17.2 Microstructure and Physical Properties

The morphology and microstructure of fibers depend on the laser–induced temperature and on the gas pressure. They have been studied mainly by optical microscopy, scanning electron microscopy, X-ray diffraction, and Raman scattering. Fibers have been grown in amorphous (B, SiO_x, SiO_2, Si_3N_4), polycrystalline (Ni, C, Si, SiC), and single-crystalline form (B, Si, W).

Figure 17.2.1 exhibits SEM pictures of single-crystalline fibers of Si and W. The Si fiber was grown from SiH_4 at 1650 K with 530.9 nm Kr^+-laser radiation. The orientation of the fiber axis was found to be close to either $\langle 100 \rangle$ or $\langle 110 \rangle$ which are the fastest directions of growth in crystalline Si. With a silane pressure of $p \equiv p(SiH_4) = 133$ mbar, single-crystal growth was observed only above 1560 K. In this connection it is interesting to recall the microstructure of Si films grown on single-crystal Si substrates by standard CVD. Here, the regime of polycrystalline growth is separated from the regime

17.3 Kinetic Studies

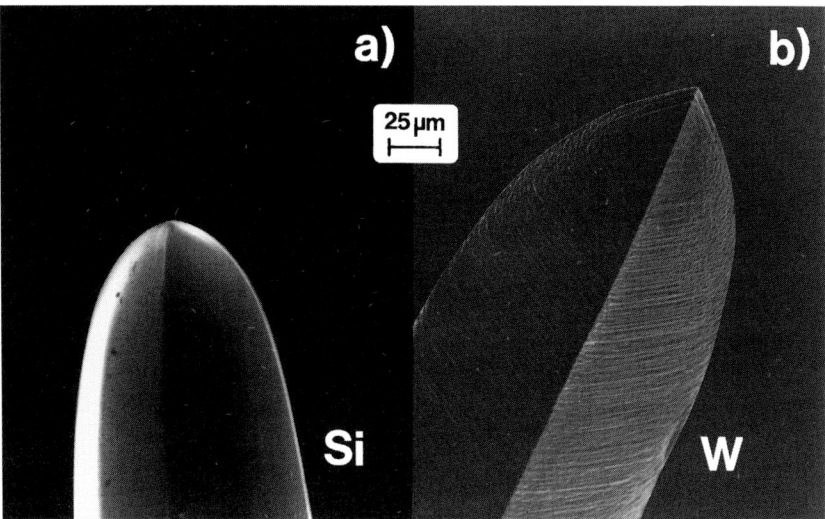

Fig. 17.2.1a,b. SEM pictures of the tip of laser-grown single-crystalline fibers. (**a**) Si grown from SiH_4 [*Bäuerle* et al. 1983]; (**b**) W grown from $WF_6 + H_2$

of single-crystalline growth by a border line (dashed line in Fig. 17.2.2). This line is determined by the flux of Si atoms giving rise to the observed growth rate, and the time for surface diffusion needed to arrange the arriving atoms on proper lattice sites. Linear extrapolation of this border line to higher temperatures yields an intersection point with the LCVD curve at about 1555 K. This value is in remarkable agreement with the temperature limit found for single-crystal growth of fibers.

The growth of fibers at gas pressures up to several bars has been investigated by *Wallenberger* et al. (1994). Here, growth rates up to several 10^2 μm/s have been achieved. The tensile strengths of fibers with about 15 μm diameter were about 7.6 GPa for B, 3 GPa for C, and about 2 GPa for SiC.

Potential applications of three-dimensional objects grown by laser-CVD are discussed in Sect. 18.5.

17.3 Kinetic Studies

Investigations on the steady growth of fibers yield accurate information on the chemical kinetics in pyrolytic LCVD. The systems investigated in most detail are listed in Table 17.3.1. Non-integral reaction orders indicate that the decomposition of precursor molecules includes different reaction channels, and that the activation energy depends on temperature (Sect. 3.1).

308   17 Growth of Fibers

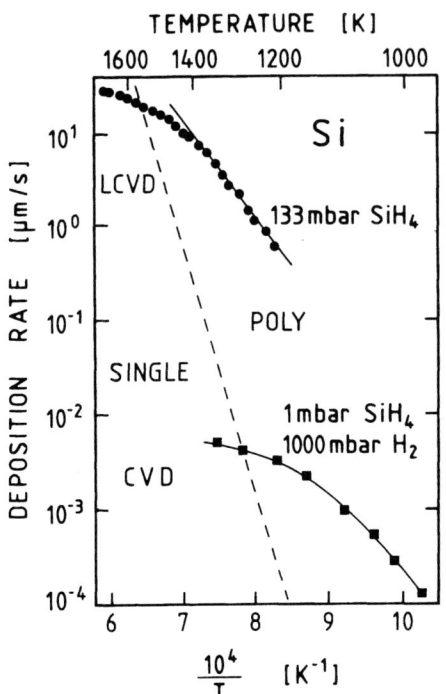

Fig. 17.2.2. Arrhenius plot for the growth of Si from SiH_4 by LCVD and standard CVD. The broken line separates regions of single- and polycrystalline growth; the intersection points with LCVD and CVD curves are at 1555 K and 1262 K, respectively [*Bäuerle* 1983b]

Table 17.3.1. Activation energies and reaction orders derived from steady growth of fibers

Material	Precursor	Temp. Range [K]	Pressure Range [bar]	Activation Energy [kcal/mol]	Reaction Orders	Reference
B	$BCl_3 + H_2$	1100–2300	(0.05–0.4) + (0.05–0.8)	26.5 ± 1	< 1	*Boman* and *Bäuerle* (1995)
C	CH_4	2850–3100	0.5–1	119 ± 2	1.25	*Leyendecker* et al. (1983b)
		2400–2750	0.5–1	43.5 ± 1.4	2.2	
	C_2H_2	1900–2450	0.05–1	47.3 ± 0.6	0.8	*Leyendecker* et al. (1983b)
	C_2H_4	2000–2250	0.3–1	58.3 ± 1.3	0.8	*Doppelbauer* (1987)
	C_2H_6	2200–2650	0.3–1	78.9 ± 4	2	*Doppelbauer* (1987)
Si	SiH_4	1150–1350	0.03–0.3	43.5 ± 1	0.6	*Bäuerle* et al. (1982)

Silicon

Figure 17.2.2 shows an Arrhenius plot for the growth of Si from SiH_4. The upper curve refers to data obtained for pyrolytic LCVD of Si fibers. The temperature has been measured in situ (Sect. 17.1). In the kinetically controlled regime, which reaches up to about 1400 K, the deposition rate increases exponentially with temperature and is characterized by an apparent chemical activation energy $\Delta E = 44 \pm 4$ kcal/mol (this value is not corrected for the temperature dependence of the preexponential factor [*Leyendecker* et al. 1983*b*]). The decrease in slope observed above 1400 K is probably due to mass transport limitation. The lower part of the figure shows the deposition rate obtained by standard CVD of Si from SiH_4 with H_2 as carrier gas [*v.d. Brekel* 1978]. A comparison of curves shows remarkable differences between localized and large-area CVD: The localization of the temperature distribution permits one to study the deposition process up to higher temperatures and much higher pressures of reactant molecules.

Carbon

The precursors employed for the growth of carbon fibers were C_2H_2, C_2H_4, C_2H_6 and CH_4. Figure 17.3.1a displays an Arrhenius plot of the axial growth rate for various pressures of C_2H_2. The diameter of fibers, $d = 2r_D$, increases almost linearly with laser power (Fig. 17.3.1b).

The activation energies derived from these investigations are also relevant to CVD and gas-phase epitaxy. The determination of ΔE via the standard techniques is very time-consuming and problematic because a number of parameters, such as the substrate temperature, gas velocity, and gas mixture must be kept constant over long time periods; for this reason, only a small number of data points can be obtained. However, because of the strong temperature gradients in LCVD, the activation energies are *not* necessarily equal to those derived from CVD experiments. Nevertheless, the deviations are probably small and below the inaccuracies in measurements.

17.4 Gas-Phase Transport

Investigations on the steady growth of fibers permit one to analyze contributions of different transport mechanisms to the growth rate. Because growth takes place mainly at the tip of fibers, the reaction zone can be described by a hemisphere (Sect. 3.4). In order to avoid confusion, we denote the temperature within the gas by T and the (uniform) temperature in the tip of the fiber by T_s.

17.4.1 The Coupling of Fluxes

In pyrolytic LCVD, strong temperature gradients near the deposit may strongly influence the gas-phase transport of species. Such temperature gradients together with temperature and concentration dependences of the transport

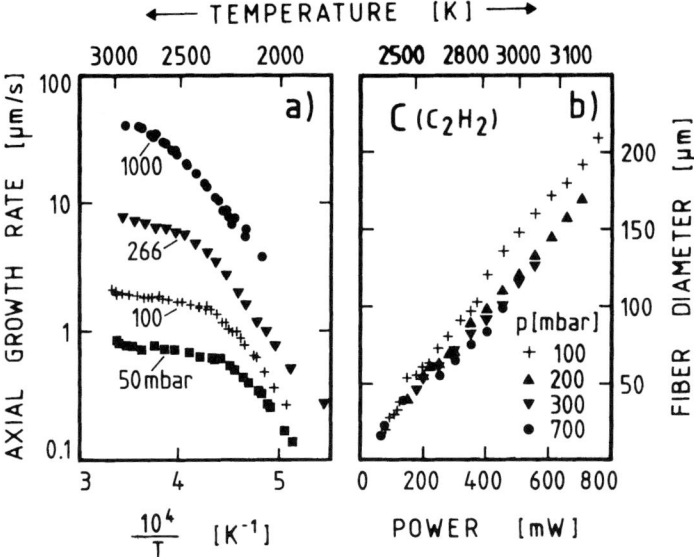

Fig. 17.3.1a,b. Ar$^+$-laser-induced growth of C fibers from (pure) C$_2$H$_2$ at various pressures. (**a**) Arrhenius plot ($\lambda = 488$ nm, $w_0 \approx 10\,\mu$m), adapted from [*Leyendecker* et al. 1983*b*, and *Doppelbauer* 1987]. (**b**) Diameter of fibers. Here, the temperature scale refers to the 300 mbar data only, adapted from [*Doppelbauer* 1987]

coefficients result in a coupling of fluxes. Let us consider a first-order non-equimolecular purely heterogeneous reaction of type (3.4.2) with $q = 1$, $\kappa =$ const, and $D_{AB} \propto T^n$ (Sect. 3.3). From (3.4.3 and 7) we obtain for the average velocity of the gas

$$v^*(r^*) = \frac{bx_{AB}(1)}{r^{*2}} \frac{T^*}{T_s^*} \exp\left(-\frac{\mathscr{E}^*}{T_s^*}\right). \tag{17.4.1}$$

For an equimolecular reaction ($b = 0$) chemical convection is absent and $v^* = 0$.

The temperature distribution within the gas is obtained from (3.4.5 and 8)

$$T^*(r^*) = 1 + \frac{T_s^* - 1}{1 - \exp(-\zeta)}\left[1 - \exp\left(-\frac{\zeta}{r^*}\right)\right], \tag{17.4.2}$$

where

$$\zeta = \frac{b\gamma k_0^* x_{AB}(1)}{T_s^*} \xi \exp\left(-\frac{\mathscr{E}^*}{T_s^*}\right), \tag{17.4.3}$$

where $\gamma \equiv c_p/c_v$. Henceforth, we set $\xi \equiv c_v D_{AB}(\infty) N(\infty)/\kappa = 1$. Figure 17.4.1 shows the gas-phase temperature calculated for $T_s^* = 10$, and for various values of b and the (normalized) thermal diffusion coefficient $\alpha_T^* = k_T/x_{AB}$. The

17.4 Gas-Phase Transport

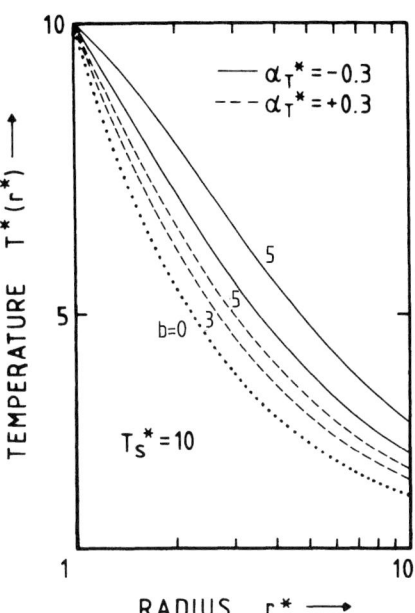

Fig. 17.4.1. Distribution of gas-phase temperature for $T_s^* = T_s/T(\infty) = 10$ with $T(\infty) = 300$ K and various values of b. For $b = 0$ the curvature is independent of α_T^* (dotted). The other parameters employed were $\mathscr{E}^* = 90$ (54 kcal/mol), $k_0^* = 5 \times 10^6$, $x_{AB}(\infty) = 0.1$, and $\gamma = 5/3$ [Luk'yanchuk et al. 1992]

parameter values employed for $\mathscr{E}^*$ and k_0^* describe, approximately, the deposition kinetics of carbon from various hydrocarbons of the type C_xH_y, and of silicon from SiH_4. Carrier gases investigated in this connection were mainly H_2, He, and Ar. For $b = 0$ (dotted), T is independent of α_T^*. The full and dashed curves, calculated for $b = 3$ and 5, refer to values of $\alpha_T^* = -0.3$ and $\alpha_T^* = +0.3$, respectively. In both cases, $\alpha_T^* > 0$ and $\alpha_T^* < 0$, chemical convection increases the heat flux to regions outside of the reaction zone. If $\alpha_T^* < 0$, the temperature distribution changes even its shape, in particular for big values of b. If $b < 0$, the curve is slightly below that for $b = 0$. The effect of additional gas-phase heating by chemical convection with $b > 0$ becomes more pronounced with decreasing values of α_T^*. This originates from the coupling between thermal diffusion and chemical convection. Negative values of α_T^* result in an enrichment of reactant molecules at the surface $r = r_D$ and thereby, in an enhancement of the reaction rate which, in turn, increases the flux of product molecules away from this surface. With decreasing temperature T_s^*, these phenomena become less important. With the same parameter values but temperatures $T_s^* \leqslant 5$, the temperature distribution becomes almost independent of b and α_T^* and close to the form $T^*(r^*) = 1 + (T_s^* - 1)/r^*$ which is obtained with $b = 0$.

Figure 17.4.2 exhibits the normalized ratio $x_{AB}^*(1) = x_{AB}(1)/x_{AB}(\infty)$ as a function of b for various values of α_T^*, $T_s^* = 10$ and $n = 3/2$. With a fixed value of b, $x_{AB}^*(1)$ increases with decreasing α_T^*. This is a consequence of thermal diffusion. The dependence of $x_{AB}^*(1)$ on b is more complex. Here, $x_{AB}^*(1)$ can

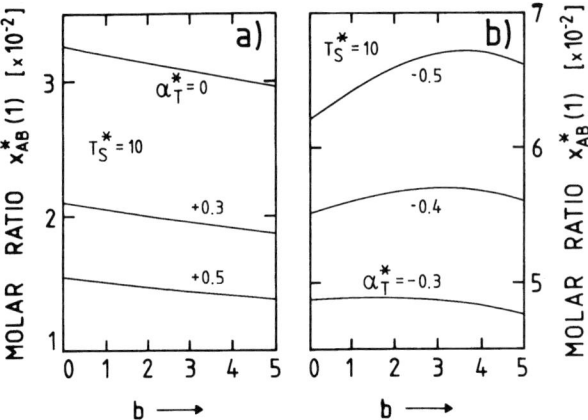

Fig. 17.4.2 a,b. Molar ratio at the surface of the reaction zone, $x^*_{AB}(1)$, as a function of b for $T^*_s = 10$ and various values of α^*_T [Luk'yanchuk et al. 1992]

decrease or increase with increasing b, depending on α^*_T. Thus, the chemical convection can decrease or increase the reaction rate. This behavior is by no means obvious. In any case, the effect of chemical convection is diminished by the coupling of fluxes. From a practical point of view the results demonstrate that for *high* substrate temperatures the influence of chemical convection is small for both $\alpha^*_T > 0$ and $\alpha^*_T < 0$. In other words, with molecules like C_2H_6, WF_6, $Mo(CO)_6$, etc., high deposition rates can be obtained in spite of strong chemical convection. The inverted behavior of $x^*_{AB}(1)$ observed for values $\alpha^*_T < 0$ is quite interesting. Here, chemical convection *increases* the reaction rate. This is a consequence of the spatial extension of the temperature profile and the temperature dependence of the molecular diffusion coefficient D_{AB}. This effect strongly increases with decreasing (more negative) values of α^*_T and increasing exponents n which, in certain cases, can exceed the value 3/2. For values of n close to 2 inversion exists even for $\alpha^*_T = 0$. In other words, for gas mixtures where the mass of carrier gas molecules, m_M, exceeds the mass of reactant molecules, m_{AB}, chemical convection can even increase the reaction rate. A practical example of such a situation is the deposition of C from mixtures of C_xH_y with Ar, Xe, etc. Thus, the coupling of ordinary non-isothermal diffusion, thermal diffusion, and chemical convection can result in non-trivial dependences of the temperature and concentration profiles. In particular, this coupling can result in inversion effects in the process kinetics.

17.4.2 Thermal Diffusion (Soret Effect)

The influence of thermal diffusion on the gas–phase distribution of constituents can most easily be studied for an equimolecular reaction ($b = 0$). The

17.4 Gas-Phase Transport

model employed is the same as in Sect. 3.4. The main features of the solution of the boundary-value problem are summarized in Figs. 17.4.3,4. Details of the calculations are presented in [Bäuerle et al. 1990a and Kirichenko et al. 1990].

Figure 17.4.3a shows the molar ratio $x_{AB}(r_D)$ as a function of the normalized temperature T_s^* with $x_{AB}(\infty) = 0.1$. The case $k_0^* \to 0$ characterizes the kinetically controlled regime where

$$x_{AB}(r_D) \approx \left(1 + \frac{1 - x_{AB}(\infty)}{x_{AB}(\infty)} T_s^{*\alpha_T}\right)^{-1}. \tag{17.4.4}$$

Comparison with the full curves shows that (17.4.4) is a good approximation for $T_s^* \leq 5$. Figure 17.4.3b depicts the normalized molar ratio $x_{AB}^*(r_D)$ as a function of T_s^* for various values of $x_{AB}(\infty)$.

The figures demonstrate to what extent the thermal diffusion flux $J_T \propto -\alpha_T \nabla T$ changes the surface concentration $x_{AB}(r_D)$. It is evident that with $\alpha_T < 0$ this concentration increases while with $\alpha_T > 0$ it decreases. In other words, addition of a carrier gas, M, will increase the surface concentration of species AB if $m_M > m_{AB}$ and it will decrease it if $m_M < m_{AB}$. The latter situation is also possible with $B \equiv M$. From Fig. 17.4.3b it becomes evident that the

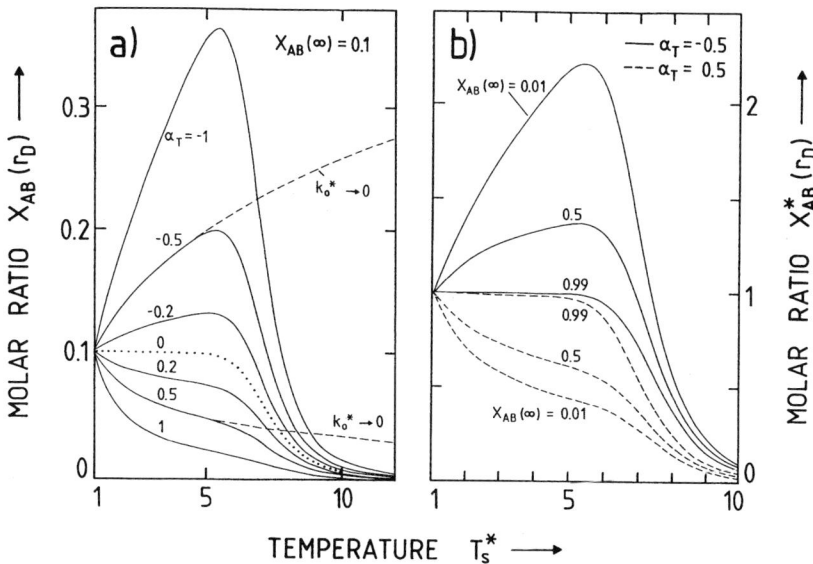

Fig. 17.4.3a,b. Influence of thermal diffusion on the concentration of species AB at the surface of the reaction zone as a function of surface temperature T_s^* for $b = 0$, $n = 2$, $\mathscr{E}^* = 90$, $k_0^* = 5 \times 10^6$. (**a**) $x_{AB}(\infty) = 0.1$. Full curves refer to various values of α_T. Dotted curve: $\alpha_T \to 0$. Dashed curves: $k_0^* \to 0$. (**b**) Different curves refer to different values $x_{AB}(\infty)$ with $\alpha_T = -0.5$ (full curves) and $\alpha_T = +0.5$ (dashed) [Bäuerle et al. 1990a]

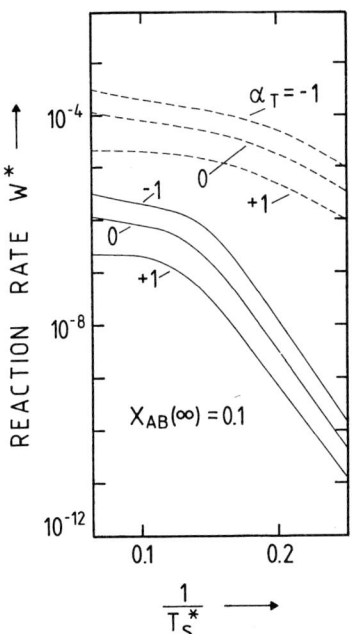

Fig. 17.4.4. Influence of thermal diffusion on the reaction rate W^* for $x_{AB}(\infty) = 0.1$, $b = 0$, and $\alpha_T = -1, 0, +1$. $T_s^* \equiv T_s/T(\infty)$ with $T(\infty) = 300$ K. Full curves: $\mathscr{E}^* = 90$ (54 kcal/mol), $k_0^* = 5 \times 10^6$. Dashed curves: $\mathscr{E}^* = 45$, $k_0^* = 5 \times 10^4$ [Bäuerle et al. 1990a]

effect of thermal diffusion becomes more pronounced with decreasing concentration $x_{AB}(\infty)$.

The influence of thermal diffusion on the reaction rate is shown in Fig. 17.4.4 for $x_{AB}(\infty) = 0.1$ and the values $\alpha_T = -1, 0$, and $+1$. Thermal diffusion influences *both* the kinetically controlled regime *and* the transport limited regime. With respect to the case $\alpha_T = 0$, the reaction rate is increased if $\alpha_T < 0$, and decreased in the opposite case.

The experimental results presented in Fig. 17.4.5 can tentatively be interpreted along these lines. The decomposition reaction

$$C_2H_2 + M \rightarrow 2C(\downarrow) + H_2(\uparrow) + M$$

is equimolecular. For the lighter carrier gases, $M \equiv He$ or H_2, we have $\alpha_T = \alpha_0$ $(m_{C_2H_2} - m_M)/(m_{C_2H_2} + m_M) > 0$. Thus, He and H_2 will accumulate near the hot tip of the fiber. As a consequence, the partial pressure of the reactant, and thereby the deposition rate, is lowered. For $M \equiv Ar$, the mass exceeds that of C_2H_2 and α_T becomes negative. Thus, Ar will be depleted near the hot surface. Therefore, the partial pressure of C_2H_2 near the surface of the reaction zone is increased and, consequently, the deposition rate. This behavior is in *qualitative* agreement with the theoretical results presented by the full curves in Fig. 17.4.4. However, these curves have been calculated for $r_D = $ const., while the radius of fibers increases with T_s (Fig. 17.3.1b). In any case, the experiments demonstrate that the selection of carrier gases has an important influence on the maximum deposition rates achieved in laser-CVD.

17.5 Simulation of Growth

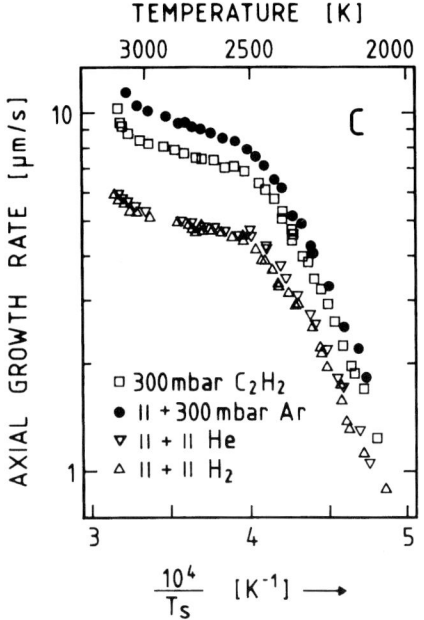

Fig. 17.4.5. Arrhenius plot for laser-induced deposition of carbon from pure C_2H_2 and from gas mixtures of C_2H_2 with H_2, He and Ar [*Doppelbauer* and *Bäuerle* 1986]

Gas-phase nucleation

The ultimate limit of controlled growth is set by gas–phase nucleation above the tip of the fiber (Chap. 4). This appears in Fig. 17.4.5 for temperatures above about 3000 K where the slope of curves starts to increase again. This limit decreases with increasing C_2H_2 pressure. Homogeneous gas-phase decomposition and nucleation can also result in spontaneous breakdowns due to autocatalyzation of the reaction. This has been observed during LCVD of Ni from $Ni(CO)_4$.

17.5 Simulation of Growth

In a simple, one–dimensional model quasi-stationary growth of fibers can be described by the energy balance and the Arrhenius law. Let us assume cylindrical symmetry. The laser-induced temperature can then be determined from the energy balance

$$\kappa_D \frac{\partial}{\partial z}\left[\pi r_D^2(z)\frac{\partial \theta_D}{\partial z}\right] - 2\pi r_D(z)\kappa_G \frac{\partial \theta_G}{\partial r}\bigg|_{r_D} = 0, \qquad (17.5.1)$$

and the boundary condition

$$\pi r_D^2 \kappa_D \frac{\partial \theta_D}{\partial z}\bigg|_{z=0} = P_a,$$

where z denotes the fiber axis. θ_D and θ_G are the linearized temperatures for the fiber and gas, respectively; κ_D and κ_G are the corresponding heat conductivities. We substitute $\partial \theta_G/\partial r|_{r_D} \approx \eta \theta_D/r_D$ where η is of the order of unity. In the simplest approximation we can calculate the temperature distribution from (17.5.1) and integrate the growth rate (normal to the surface) as given by the Arrhenius law. The radius of the fiber then becomes

$$r_D \approx \frac{P_a}{4\eta\kappa_G \mathscr{E}}, \qquad (17.5.2)$$

where $\mathscr{E}$ is the activation temperature. κ_D does not enter (17.5.2) because $\kappa_D \gg \kappa_G$. Both the linear increase in r_D with laser power and its only slight dependence on gas pressure are in agreement with the experimental results presented in Fig. 17.3.1b (in the elementary kinetic theory of gases, κ_G is independent of pressure). For a more detailed treatment see [*Arnold* et al. 1996].

18 Direct Writing

Laser direct writing by pyrolytic or photolytic decomposition of gas-phase precursors permits single-step surface patterning of planar and non-planar substrates. The process can most easily be studied by translating the substrate in one dimension perpendicular to the focused laser beam. By this means one obtains stripes (lines). The literature on such investigations is included in Appendix B.4.

18.1 Characteristics of Pyrolytic Direct Writing

The *morphology* and *geometry* of stripes strongly vary with laser power. This quite general phenomenon is illustrated in Fig. 18.1.1 with the example of Si deposited from SiH_4 onto Si wafers. At laser powers corresponding to center temperatures well below the melting point, a convex cross section is observed. When the laser power is increased, the cross section becomes mesa-type and, at even higher powers, a dip in the middle of the stripe occurs. This dip becomes more pronounced the higher the laser power. Such changes in surface morphology can be understood from the laser-induced temperature distribution and the transport of species. The latter includes both the transport of reactant and product molecules within the gas phase, and surface diffusion of species. Surface diffusion increases exponentially with temperature. Near melting, the surface tension strongly decreases, in general, and pulls off the soft or liquefied material from the valleys (Sect. 10.5).

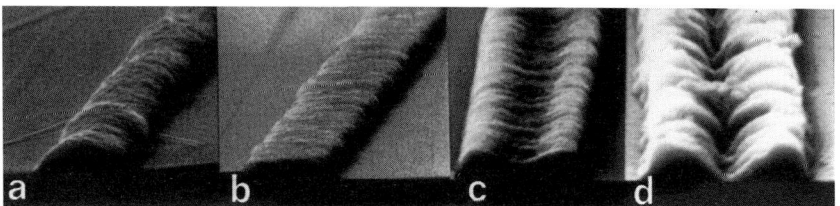

Fig. 18.1.1. Si stripes deposited onto Si wafers by Ar^+-laser pyrolysis of SiH_4 ($\lambda = 488$ nm, $v_s = 10$ μm/s, $p(SiH_4) \approx 40$ mbar). The laser-beam intensity increases from left to right

Additionally, high laser powers may induce gas-phase reactions (Sect. 3.5). The reaction products, such as polysilanes in the SiH_4 system, condense in the neighborhood of the stripe. Furthermore, coherent and non-coherent structure formation is frequently observed within certain ranges (Chap. 28).

The influence of melting and changes in the morphology of stripes complicate the understanding of the deposition process. The following analysis of direct writing will therefore be confined to laser powers where *no* dramatic changes in the shape of the cross section occur and where an unequivocal definition of a stripe width d and height h is possible.

Figure 18.1.2 exhibits SEM pictures of W lines deposited on a 1200 Å a-Si/SiO_2 substrate from $WF_6 + H_2$. The microstructure of the lines is polycrystalline. The grain size depends on the laser power and gas pressures. With this system, a thin tungsten film is initially formed by silicon reduction according to

$$2\ WF_6 + 3Si \rightarrow 2W + 3\ SiF_4\ (\uparrow). \tag{18.1.1}$$

This reaction is self-terminating at a film thickness of about 100–1000 Å [*Liu* 1986]. The slight deepening observed near the edges of lines is considered an indication for this process. Nevertheless, etching by HF may be important as well (Sect. 16.3.1). After the initial step of silicon reduction, deposition continues via hydrogen reduction of WF_6, as described by (16.2.1).

18.1.1 Dependence on Laser Parameters and Substrate Material

Figure 18.1.3 displays the height and width of lines as a function of laser power for two different substrate materials and various pressures of WF_6 and H_2.

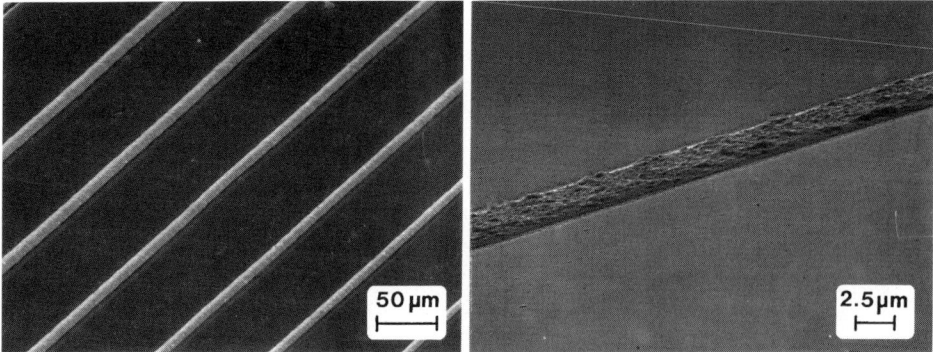

Fig. 18.1.2. SEM pictures of W stripes deposited from $WF_6 + H_2$ with Kr^+-laser light ($\lambda = 647$ nm, w_0 (1/e) ≈ 1.3 μm, $v_s = 100$ μm/s). Left picture: $p(WF_6) = 5$ mbar, $p(H_2) = 400$ mbar, $P = 146$, 132, 120, 107, 97, and 83 mW (left to right). Right picture: $p(WF_6) = 5$ mbar, $p(H_2) = 100$ mbar, $P = 135$ mW [*Zhang* et al. 1987]

18.1 Characteristics of Pyrolytic Direct Writing

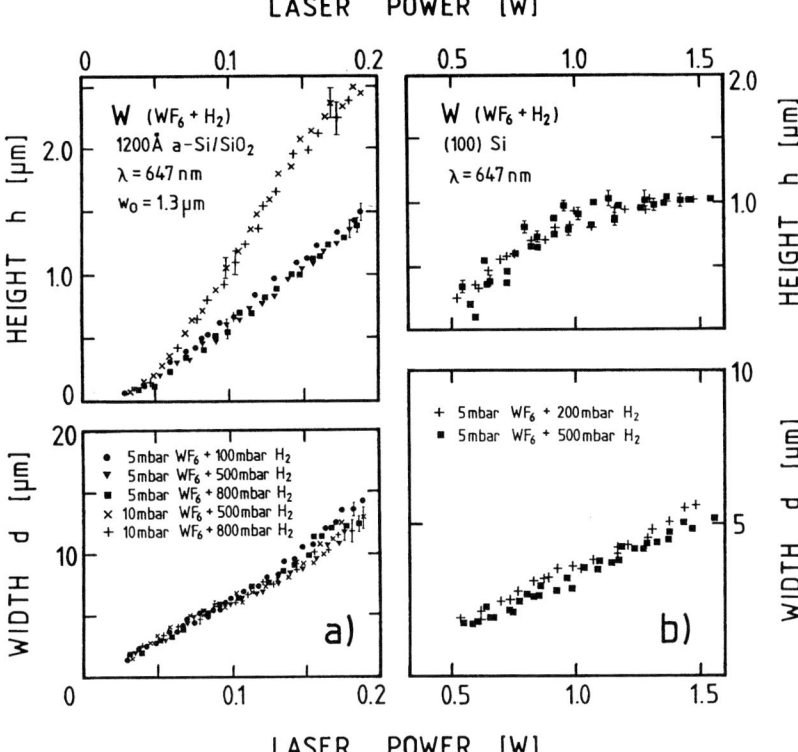

Fig. 18.1.3a,b. Height and width of W stripes as a function of Kr^+-laser power for two substrate materials ($\lambda = 647$ nm, $v_s = 100$ μm/s). (**a**) Fused quartz substrate covered with 1200 Å amorphous silicon. The laser focus was w_0 (1/e) ≈ 1.3 μm [*Zhang* et al. 1987]. (**b**) (100) Si-wafer substrate [*Zhang* et al. 1987, unpublished]

For the thermally insulating substrate (Fig. 18.1.3a) both d and h increase almost linearly with laser power. The width is independent of both the WF_6 and H_2 partial pressure. The height increases with WF_6 pressure but is independent of $p(H_2)$. The ratios d/h are approximately 10 and 5 for WF_6 pressures of 5 and 10 mbar, respectively. The mesa-type cross section (almost perfectly rectangular) observed with 5 mbar WF_6 changes to convex-type with 10 mbar WF_6. With $p(WF_6) > 5$ mbar, thin wings of W are frequently observed. Significant deposition takes place only above a certain threshold $\propto P/w_0$ which is related to the laser-induced temperature rise, see (7.1.4). From Fig. 18.1.3a we derive $P_{th} \approx 30$ mW; within the ranges investigated, this value is independent of the partial pressures of gases. Another feature is the increase in resolution which is observed for the lowest laser powers. Here, the width of lines is about half the diffraction-limited diameter of the laser focus. This effect originates mainly from the exponential dependence of the deposition rate on temperature (Sect. 5.3.6).

The situation is different with W lines deposited onto (100) Si wafers. Here, the laser beam was focused a few micrometers below the Si surface. In this way, the formation of periodic structures can be suppressed and the reproducibility of data improved. The laser powers required for direct writing are more than a factor of ten higher compared to SiO_2 substrates. This is mainly related to the high heat conductivity of Si [κ(c–Si; 300 K) $\approx$ 1.5 W/cmK; κ(a–Si/SiO_2; 300 K) $\approx \kappa$(SiO_2; 300 K) = 0.014 W/cmK]. The height of lines rapidly saturates with increasing power.

Similar experiments have been performed on the deposition of Ni from $Ni(CO)_4$ [Kräuter et al. 1983]. The laser wavelength has no or only little influence on the width and thickness of lines as long as the absorbed power remains constant. For thermally insulating substrates and medium to high powers, the width of metal lines is much larger than the laser focus ($d \gg 2w_0$) and independent of w_0. This is expected from Fig. 16.4.2a, where it has been shown that the temperature rise at the edge of the deposit remains almost unaffected if the laser focus is increased at constant power. On the other hand, deposition continues to lower laser powers as the diameter of the laser focus becomes smaller. Thus, the *smallest* widths of lines are obtained with the smallest focus. If the thermal conductivities of the deposit and the substrate are comparable, as in the case of the c–Si substrate, the width of lines remains of the order of the focus diameter.

The range of parameters and the maximum scanning velocities that can be employed in laser direct writing strongly depend on both the physical properties of the deposit *and* the substrate. While the possible range of variation in the width of stripes is very large for $\kappa_D \gg \kappa_s$, it is very small if $\kappa_D \approx \kappa_s$. The upper limit is essentially based on the maximum center temperature at which *controlled* deposition is possible, i.e, where no dramatic changes in the geometry of the deposit, no damaging of the substrate, and no triggering of a (uncontrolled) homogeneous gas-phase reaction above the surface of the deposit occur. Furthermore, small changes in w_0, or in the positioning of the substrate, will have a much stronger influence for systems where $\kappa_D \approx \kappa_s$ than for systems where $\kappa_D \gg \kappa_s$.

18.1.2 Electrical Properties

The electrical properties of W lines have been investigated for the same parameters and substrate as in Fig. 18.1.3a. Figure 18.1.4 shows the resistance of stripes per unit length and their resistivity, normalized to the bulk value of W, as a function of laser power and $p(WF_6) = 5$ mbar. No influence on H_2 pressure was observed. The strong decrease in resistance with increasing laser power is mainly due to the increase in cross section of stripes. The increase in resistivity with increasing laser power is ascribed to changes in the morphology and texture. Stripes deposited at low laser-induced temperatures are

18.2 Temperature Distributions in Direct Writing

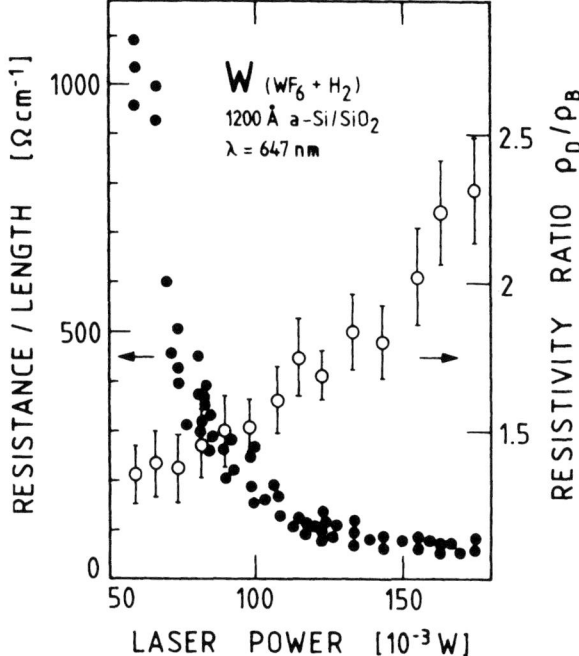

Fig. 18.1.4. Electrical resistance (left scale) and resistivity ratio (right scale) of W stripes as a function of laser power (647 nm Kr$^+$, $w_0 = 1.3\,\mu$m). The partial pressures of WF$_6$ and H$_2$ were $p(\mathrm{WF}_6) = 5$ mbar and 100 mbar $\leqslant p(\mathrm{H}_2) \leqslant 800$ mbar. For normalization we used ρ_B (W) $= 5.33 \times 10^{-6}\,\Omega$ cm [Zhang et al. 1987]

microcrystalline and have smooth surfaces. At higher temperatures, but otherwise identical experimental conditions, the surface becomes rougher and larger crystallites – up to several tenths of a micrometer – are formed. The larger grains may result in higher inter-grain resistances. Nevertheless, the resistivity values are only a factor of 1.5–2.5 larger than those of bulk W. Similar results have been achieved with Au and Cu (Appendix B.4).

18.2 Temperature Distributions in Direct Writing

The model employed in the calculations is depicted in Fig. 18.2.1. Scanning of the laser beam is performed in x-direction with velocity v_s. We consider quasi-stationary conditions with the coordinate system fixed to the laser beam. In this system, the geometry of the stripe remains unchanged. The precise temperature distribution induced within the deposited stripe can only be

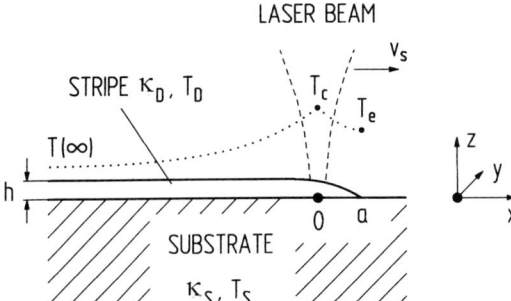

Fig. 18.2.1. Model for laser direct writing of stripes. The origin of the coordinate system (indicated by the dot) is fixed with the center of the laser beam. The forward edge of the stripe is at $x = a$. T_c and T_e are the temperatures at $x = 0$ and $x = a$, respectively. The temperature profile calculated analytically is schematically indicated by the dotted curve. The width of the stripe is $d = 2r_D$ and its height is h

calculated numerically. For many applications, however, a rough estimation of this distribution, or even of the center temperature only, is useful.

18.2.1 Center Temperature Rise

The simplest estimation of the temperature rise is based on the energy balance. Assume a stripe of rectangular cross section and uniform height within the region $-\infty < x \leq a$.

If $\kappa_D \leq \kappa_s$ the width of deposited stripes is comparable to the width of the laser spot, i.e., $w \approx r_D = d/2$. From simple energy considerations the center temperature rise is

$$\Delta T_c \equiv \Delta T(x=0) \approx \frac{P_a}{\pi w \kappa_s} \left(1 + \frac{v_s w}{D_s} + \frac{\kappa_D - \kappa_s}{\kappa_s} \frac{h d}{\pi w^2}\right)^{-1}, \qquad (18.2.1)$$

where $P_a = I_0 A \pi w^2$ is the absorbed laser power, and D_s the heat diffusivity of the substrate. The influence of scanning can be neglected as long as $v_s^* = v_s w / D_s \ll 1$. Because this estimation ignores temperature gradients in z-direction, it can be applied only if $h \ll d$.

If $\kappa^* \equiv \kappa_D / \kappa_s \gg 1$ the width of stripes can become large compared to the laser spot so that $w \ll r_D$. With the approximation $v_s^* = 0$ the energy balance for the tip of the stripe with cross section F can be written as

$$P_a \approx \pi r_D^2 \kappa_s \nabla T_s (x=0) + F \kappa_D \nabla T_D (x=0). \qquad (18.2.2)$$

The temperature gradient within the substrate surface can be approximated by $\nabla T_s(x=0) \approx \Delta T(x=0)/r_D \equiv \Delta T_c/r_D$. For an estimation of $\nabla T_D (x=0)$ we consider the energy balance for a stripe element between x and $x + dx$. If we ignore the source term, this yields $\nabla T_D (x=0) \approx \Delta T_c / l$ where $l = (F \kappa^*/\eta)^{1/2}$. l characterizes the drop in the laser-induced temperature rise in x-direction. If

18.2 Temperature Distributions in Direct Writing

we set $\eta = 2$ and substitute $\nabla T_s (x = 0)$ and $\nabla T_D (x = 0)$ in (18.2.2) we obtain

$$\Delta T_c \approx \frac{P_a}{\pi r_D \kappa_s} \left[1 + \frac{1}{\pi r_D} \left(2F \frac{\kappa_D}{\kappa_s} \right)^{1/2} \right]^{-1}. \tag{18.2.3}$$

With $F = 2r_D h = dh$ this yields

$$\Delta T_c \approx \frac{P_a}{\pi r_D \kappa_s} \left[1 + \frac{2}{\pi} \left(\frac{h}{r_D} \frac{\kappa_D}{\kappa_s} \right)^{1/2} \right]^{-1}.$$

From the comparison of terms $v_s \nabla T$ and $D_s \nabla^2 T$ in the heat equation (2.2.1) we find that the influence of scanning can be ignored as long as $v_s^* \equiv v_s r_D^2 / D_s l \ll 1$. It is evident that the previous equations represent only crude approximations. Nevertheless, they permit one to *qualitatively* understand some basic features observed in laser direct writing.

18.2.2 One-Dimensional Approach, $\kappa^* \gg 1$

With the one-dimensional model shown in Fig. 18.2.1 the laser-induced temperature distribution can also be calculated in a more sophisticated way from the energy balance

$$c_D \rho_D F \frac{\partial T_D}{\partial t} = \frac{\partial}{\partial x} \left(F \kappa_D \frac{\partial T_D}{\partial x} \right) - \int_{-r_D}^{r_D} \kappa_s \frac{\partial T_s}{\partial z} \bigg|_{z=0} dy + \mathscr{P}_a . \tag{18.2.4}$$

Here, the influence of scanning is again ignored. c_D is the specific heat, and ρ_D the mass density of the deposit. $\mathscr{P}_a \equiv \mathscr{P}_a(x)$ is the absorbed laser power per unit length of the stripe. The cross section F is assumed to be constant, because the heat loss along the stripe is dominated by the region where h and r_D are constant. This approximation is in good agreement with the exact solution of (18.2.4) with $F \equiv F(x)$. All material parameters have been assumed to be constants. If $\kappa^* \gg 1$ we can employ the approximation $T_s (z=0) \approx T_D$. We also approximate the integral in (18.2.4) by $\eta \kappa_s \Delta T_D$ where η is a dimensionless geometrical parameter which describes the heat flux from the deposit into the substrate by an effective (linear) heat exchange (Sect. 6.1.2). The value of η is near 2, because $\partial T_s / \partial z \approx \Delta T_D / r_D$ where $\Delta T_D = T_D - T(\infty)$. If $w < r_D$ we can replace $\mathscr{P}_a$ by $P_a \delta(x) = PA \delta(x)$ where A is the absorptivity.

With these simplifications and under stationary conditions (18.2.4) yields

$$l^2 \frac{\partial^2 T_D}{\partial x^2} - [T_D - T(\infty)] + \frac{P_a}{\eta \kappa_s} \delta(x) = 0, \tag{18.2.5}$$

where $l^2 = F\kappa^*/\eta$. The boundary conditions employed are

$$\left.\frac{\partial T_D}{\partial x}\right|_{x=a} = 0 \quad \text{and} \quad T_D(x=-\infty) = T(\infty).$$

The solution of this problem with $x < 0$ is

$$T_D(x) = T(\infty) + \Delta T_c \exp\left(\frac{x}{l}\right), \qquad (18.2.6)$$

where

$$\Delta T_c \equiv \Delta T(x=0) = \frac{P_a}{2\eta l \kappa_s}\left[1 + \exp\left(-\frac{2a}{l}\right)\right].$$

This relation coincides with (18.2.3) if we use expansions with respect to $a/l \ll 1$ and ignore coefficients of the order of unity.

With $0 < x < a$ we obtain

$$T_D(x) = T(\infty) + \Delta T_e \cosh\left(\frac{x-a}{l}\right), \qquad (18.2.7)$$

where

$$\Delta T_e \equiv \Delta T(x=a) = \frac{P_a}{\eta l \kappa_s}\exp\left(-\frac{a}{l}\right).$$

The unknown quantities ΔT_c, ΔT_e, a, F, l, h, and r_D must be calculated self-consistently together with the equation of growth.

18.2.3 Numerical Solutions

Numerical calculations of the laser-induced temperature distribution were performed for *rectangular* stripes of *uniform* thickness within the region $-\infty < x \leq a$ (Fig. 18.2.1). Heat losses to the ambient medium were ignored. The results obtained by employing the finite difference technique are presented in Fig. 18.2.2. If $\kappa_D = \kappa_s$, the temperature distribution remains essentially unaffected by the deposit and is almost symmetric. If, however, $\kappa_D \gg \kappa_s$, the general trend is the same as for the spots.

For realistic parameters, the temperature profiles are almost independent of scanning velocity. The temperature profiles shown in Fig. 18.2.2 are therefore very similar to those in Fig. 16.4.2. The main differences result from the heat transport along the stripe, which yields a reduction of the center temperature with increasing cross section (Fig. 18.2.3). This effect is important in particular if $\kappa_D \gg \kappa_s$.

18.3 Simulation of Direct Writing

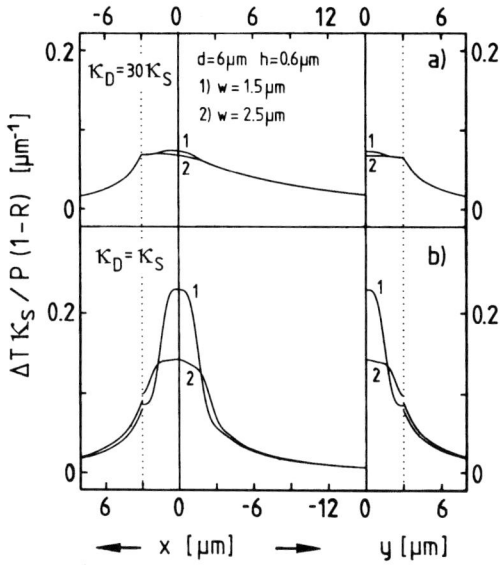

Fig. 18.2.2. Calculated temperature distributions for stripes with rectangular cross sections. The choice of coordinates is the same as in Fig. 18.2.1, adapted from [*Piglmayer* et al. 1984]

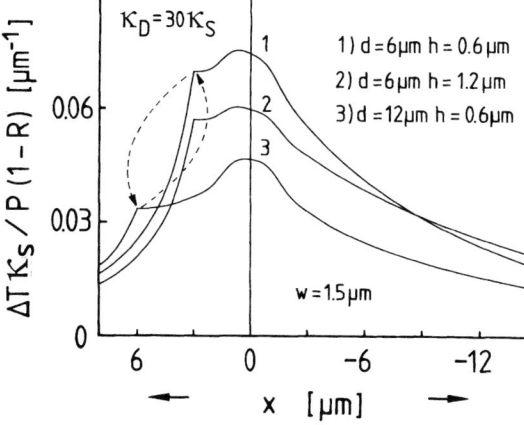

Fig. 18.2.3. Influence of the width and height of stripes on the temperature profile [*Piglmayer* et al. 1984]

For *arbitrary* cross sections with $h \ll \kappa^* r_D$ and $\kappa^* \gg 1$, the temperature distribution can directly be calculated from (16.5.3).

18.3 Simulation of Direct Writing

In this section we discuss self-consistent model calculations on pyrolytic direct writing for $\kappa^* \gg 1$. The theoretical results are compared with experimental data.

18.3.1 One-Dimensional Model

In a coordinate system that is fixed with the laser beam, the height of the stripe is given by (16.3.2) where $h = h(x, t)$. The stationary conditions are expressed by $\partial h/\partial t = 0$. The growth rate shall be described by the Arrhenius-type law (16.3.3). If we use for the position $x = 0$ the approximation $\partial h/\partial x \approx -h/(\gamma a)$, where h is the (constant) height behind the laser beam (Fig. 18.2.1), we obtain for lines with $|n| \approx 1$

$$W(T_c) - v_s \frac{h}{\gamma a} = 0, \qquad (18.3.1)$$

where γ is of the order of unity. To determine the unknown quantities we need three additional equations: The cross section of the stripe is

$$F \approx \zeta h r_D, \qquad (18.3.2)$$

where ζ is a dimensionless geometrical coefficient (for a rectangular stripe $\zeta \approx 2$, for a parabolic cross section $\zeta \approx 4/3$, etc.). The position of the laser beam with respect to the front edge of the stripe (Fig. 18.2.1) is characterized by

$$a \approx r_D/\xi. \qquad (18.3.3)$$

ξ is again dimensionless and of the order of unity. Ansatz (18.3.3) implies that the temperature distribution has almost axial symmetry near the position of the laser beam. This is confirmed by both experimental observations and numerical simulations of the growth process. Depending on the type of precursor molecule, we consider laser-induced deposition with or without a threshold temperature, see (16.3.3). *With* a threshold, T_{th} must be reached at the same distances from the center in x and y directions. The temperatures at $x = a$ and $y = r_D$ shall then be approximated by

$$T_e \approx T_{th}. \qquad (18.3.4a)$$

Without a threshold, we ignore deposition for temperatures $T < T_e$ where T_e is given by $W(T_e)/W(T_c) = \exp(-\beta)$ with $\beta \approx 1$. This yields

$$T_e = \frac{\mathscr{E} T_c}{\mathscr{E} + \beta T_c} \approx T_c\left(1 - \frac{\beta T_c}{\mathscr{E}}\right). \qquad (18.3.4b)$$

From (18.2.6 and 7) we obtain

$$\mu = \frac{a}{l} = \operatorname{arccosh}\left(\frac{\Delta T_c}{\Delta T_e}\right), \qquad (18.3.5)$$

where $\Delta T_e = T_e - T(\infty)$. From (18.2.7) we find

$$l = \frac{P_a}{\eta \kappa_s \Delta T_e} \exp(-\mu). \qquad (18.3.6)$$

Equations (18.2.6, 7), and (18.3.1 to 6) permit one to calculate all relevant dependences. If $v_s = $ const, T_c is independent of P_a while r_D, h, a, and l scale *linearly* with P_a. The dependence of these quantities on v_s is more complex. With (18.3.5 and 6) we can write (18.3.3) as

$$r_D = \xi a = \xi l \mu = \frac{\xi P_a}{\eta \kappa_s \Delta T_e} \mu \exp(-\mu). \tag{18.3.7}$$

From (18.3.2 and 7) we obtain with $l^2 = F\kappa^*/\eta$

$$h = \frac{\eta}{\zeta \xi \kappa^*} \frac{a}{\mu^2} = \frac{P_a}{\zeta \xi \kappa_D \Delta T_e} \mu^{-1} \exp(-\mu). \tag{18.3.8}$$

Equation (18.3.1) yields

$$v_s = \frac{\gamma \zeta \xi}{\eta} \kappa^* \mu^2 W(T_c). \tag{18.3.9}$$

Equations (18.3.7 to 9) are a parametric representation [with respect to T_c or μ in (18.3.5)] for the determination of h and $r_D = d/2$ as a function of v_s. Both, T_c and μ, increase with v_s. r_D depends only on κ_s while h depends on κ_D. The ratio h/r_D does *not* depend on P_a.

With very *low* laser powers, the width of stripes can become comparable to the laser focus so that $r_D \approx w_0$. Then, the point source approximation (18.2.5) becomes invalid. Heuristically, we can take into account the *finite* diameter of the laser focus by substituting $a = l\mu + \chi w_0$ where $\chi \approx 1$. Consideration of w_0 changes the power dependence of r_D and h within the range $r_D \approx w_0$ [*Arnold* et al. 1995].

18.3.2 Comparison with Experimental Data

With constant scanning velocity, the temperature T_c and thereby the rate $W(T_c)$ and $\mu(T_c)$ in (18.3.9) do not vary with laser power. Thus, according to (18.3.7 and 8) d and h increase *linearly* with laser power. This is in agreement with experimental data on pyrolytic direct writing for systems with $\kappa^* \gg 1$ (Fig. 18.1.3a). The constant value of T_c can be easily understood: With increasing power P_a, the height and width of stripes increase, and thereby the heat flow along the stripe.

Figure 18.3.1 shows the dependence of the normalized height, $h^* \equiv h\kappa_D T(\infty)/P_a$, the width $r_D^* \equiv r_D \kappa_s T(\infty)/P_a$ and the temperature $T_c^* \equiv T_c/T(\infty)$ as a function of scanning velocity. The curves were calculated from (18.3.7 to 9) together with (18.3.5). The height of lines decreases monotonically with increasing v_s because of the decrease in laser-beam dwell time $\tau_l \propto 1/v_s$. The increase in T_c with v_s can be understood from the overall decrease in the cross section F. The width of lines shows a more complex behavior:

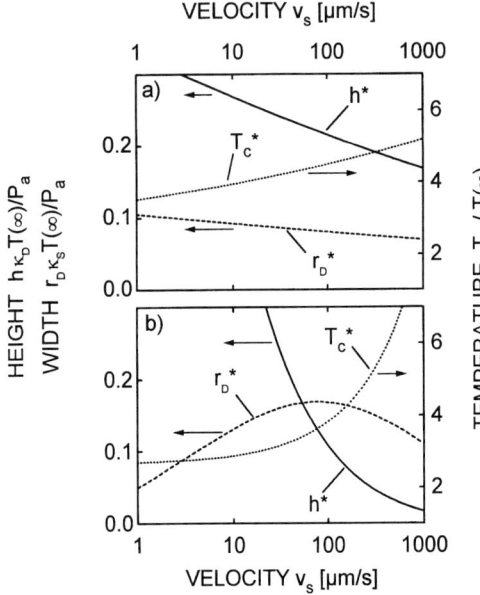

Fig. 18.3.1a,b. Normalized height, width, and center temperature for direct writing of stripes as a function of scanning velocity. The parameters employed correspond to SiO$_2$ substrates and deposits of: (a) Si from SiH$_4$ with $k_0 = 3 \times 10^8$ μm/s, $\mathscr{E}^* = 73$, $T(\infty) = 300$ K, $\kappa^* = 10$, $\eta = 1.6$, $\zeta = 4/3$, $\xi = 1.25$, $\gamma = 1.3$, $\beta = 1.8$; (b) W from WCl$_6$ + H$_2$ with $k_0 = 16$ μm/s, $\mathscr{E}^* = 5.7$, $T(\infty) = 443$, $T_{\text{th}}^* = 2.7$, $\kappa^* = 17$. η, ζ, ξ, γ are the same as in a), adapted from [*Arnold* et al. 1995]

For materials with *no* or low deposition threshold where T_e is given by (18.3.4b), r_D decreases monotonically with increasing v_s. The parameter set employed in Fig. 18.3.1a refers, approximately, to pyrolytic laser-CVD of Si from SiH$_4$. A similar dependence is found for the deposition of Ni from Ni(CO)$_4$ (Fig. 18.3.2).

In the *presence* of a threshold where T_e is given by (18.3.4a), the situation is different (Fig. 18.3.1b). The width first increases up to a maximum value r_D^{max} at

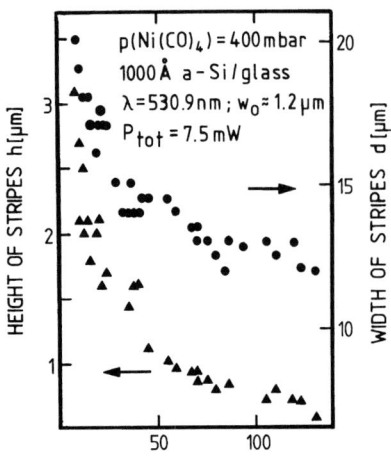

Fig. 18.3.2. Dependence of the height and width of Ni stripes on scanning velocity [*Kräuter* et al. 1983]

v_s^{max}, and then decreases for $v_s > v_s^{max}$. This behavior can be understood from the maximum in $r_D(v_s)$ at $\mu = 1$. For this point, we obtain

$$T_c^{max} = T(\infty) + \Delta T_{th} \cosh(1) \approx T(\infty) + 1.5 \Delta T_{th} \tag{18.3.10a}$$

$$r_D^{max} = \xi a = \xi l = \frac{\xi P_a}{\eta \kappa_s \Delta T_{th}} e^{-1} \tag{18.3.10b}$$

$$h^{max} = \frac{P_a}{\zeta \xi \kappa_D \Delta T_{th}} e^{-1} \tag{18.3.10c}$$

$$v_s^{max} = \frac{\gamma \zeta \xi}{\eta} \kappa^* W(T_c^{max}). \tag{18.3.10d}$$

The maximum in $r_D(v_s)$ is related to the threshold temperature. The increase in T_c is due to the decrease in heat flux along the stripe. Among the experimental examples that refer to this case is the deposition of W from $WCl_6 + H_2$.

Figure 18.3.3 shows the height and width of W stripes as a function of scanning velocity for two different WCl_6 pressures. The effective incident laser power was the same in both cases. The substrate was fused quartz (SiO_2)

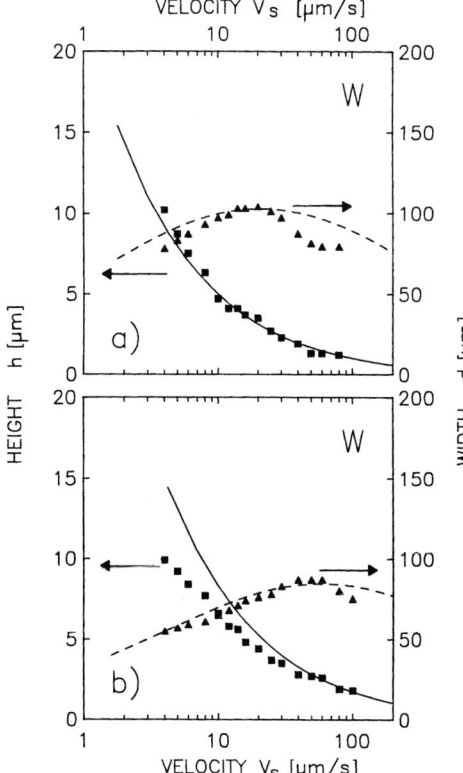

Fig. 18.3.3a,b. Height and width of Ar^+-laser deposited W stripes as a function of scanning velocity for two mixtures of $WCl_6 + H_2$ ($\lambda = 514.5$ nm, $P = 645$ mW, $w_0(1/e) = 7.5$ μm; $p(H_2) = 50$ mbar) [Kullmer et al. 1992]. The solid and dashed curves have been calculated as described in the text. The parameters were $A = 0.55$, $T(\infty) = 443$ K, $\mathscr{E}^* = 5.7$, $\kappa^* = 17$, $\kappa_s = 0.032$ Wcm^{-1}K^{-1}; $\eta = 1.6$, $\zeta = 4/3$, $\xi = 1.25$, $\gamma = 1.3$. (**a**) $p(WCl_6) = 0.49$ mbar, $k_0 = 7.15$ μm/s, $T_{th}^* = 2.4$. (**b**) $p(WCl_6) = 1.1$ mbar, $k_0 = 16.05$ μm/s, $T_{th}^* = 2.7$ [Arnold et al. 1993]

covered with a $h_1 \approx 700$-Å thick layer of sputtered W. Full and dashed curves were calculated from (18.3.7)–(18.3.9). The value $\xi = 1.25$ was derived from experimental data. The sputtered W layer has been ignored in the calculations; this is a good approximation if $\kappa^* h_1/r_D \ll 1$. The figure shows almost quantitative agreement between experimental data and theoretical curves. The maximum in d shifts to higher velocities with increasing WCl_6 pressure. This is expected from (18.3.10d) because of the preexponential factor k_0 in $W(T)$. For a partial reaction order of unity with respect to WCl_6, we expect $k_0 \propto p(WCl_6)$. From the figure we derive for the ratio of velocities $v_s(d^{max})$ a factor 2.5 which is in excellent agreement with the ratio of WCl_6 pressures employed, 2.25.

Figure 18.3.4 exhibits the width of W lines as a function of scanning velocity for different laser powers. The width increases more or less linearly with laser power; the position of the maximum width, however, remains unchanged. The dashed curves have been calculated by employing the same parameters as in Fig. 18.3.3a.

Equations (18.3.10b,c) suggest that d^{max} and h^{max} depend only on threshold temperature and increase linearly with laser power. This is confirmed by experiments which show, for example, that in the WCl_6–H_2 system d^{max} and h^{max} are almost independent of pressure [*Arnold* et al. 1993].

Within the approximations made, the agreement between theoretical predictions and experimental data is quite satisfactory. Besides the simplifying assumptions made, important parameters such as k_0, $\mathscr{E}$, T_{th}, and the reaction orders can only be estimated from experimental investigations.

The temperature dependences of κ_D, κ_s, and A can easily be taken into account. Even in this case a *linear* dependence of r_D, h, a, and l on absorbed laser power is obtained; the maximum center temperature, T_c^{max}, will still depend only on T_{th} and the materials' properties. With realistic parameters, the changes in quantities are below 30%.

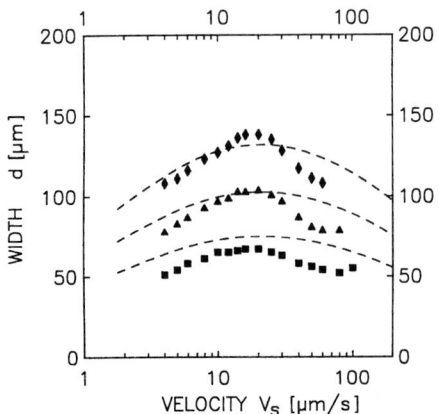

Fig. 18.3.4. Width of W stripes as a function of scanning velocity for three different laser powers. The precursors employed were 0.49 mbar WCl_6 + 50 mbar H_2. The parameters were the same as in Fig. 18.3.3a except for the laser power: ◆ $P = 825$ mW; ▲ $P = 645$ mW; ■ $P = 475$ mW [*Arnold* et al. 1993]

18.3.3 Two-Dimensional Model

Equations (16.3.1–3) permit a self-consistent calculation of line shapes and laser-induced temperature distributions. Here, we can ignore the time derivative in the heat equation and use the same approximations and normalizations as in Sect. 16.5.

Figure 18.3.5 shows contour lines (left-hand side) and isotherms (right-hand side) calculated for various stages of direct writing of W lines onto SiO_2. The laser beam is switched on at $t = 0$. Stationary conditions are achieved only after a relatively long time. The larger width of stripes observed with short times is related to the fact that energy losses due to heat conduction along the stripe are not yet effective. This behavior also becomes evident from the isotherms. It is in agreement with experimental observations.

Within the parameter range investigated, the calculated shape of W lines always remains uniform, i.e., it shows *no* oscillations in height or width. The calculations also permit one to derive the parameter $\xi = r_D/a$ for different scanning velocities; see (18.3.3). Here, we find $1.2 < \xi < 1.5$. The experimental value for the WCl_6–H_2 system is $\xi \approx 1.25$. The discrepancy between the one-dimensional and the two-dimensional model is, for realistic parameters, about 30% [*Arnold* and *Bäuerle* 1993].

Different morphologies of Cu lines including volcano shapes have been modelled by *Han* and *Jensen* (1994).

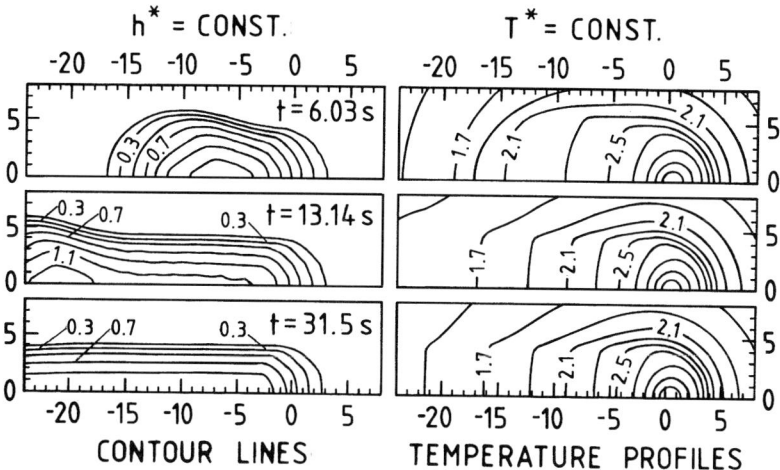

Fig. 18.3.5. Contour lines (left-hand side) and isotherms (right-hand side) calculated for laser direct writing of W onto SiO_2 ($WCl_6 + H_2$; $v_s^* = 2$, $AI_0^* = 20.6$; other parameters as in Fig. 16.5.1a). The coordinate system is fixed with the laser beam which is switched on at $t = 0$ [*Arnold* and *Bäuerle* 1993]

18.4 Photophysical LCVD

Up to now we have concentrated on pyrolytic direct writing. There are several reasons for this:

- *Pyrolytic* LCVD permits surface patterning with sub-micrometer dimensions. The deposition rates exceed those in photolytic LCVD by several orders of magnitude. Consequently, pyrolytic LCVD permits much higher scanning velocities in direct writing and also the production of three-dimensional structures. The microstructure and the electrical properties of pyrolytically deposited materials are superior to those deposited by photolysis. Pyrolytic reaction rates depend only slightly on the exact wavelength of the laser light. For this reason, a great variety of materials can be deposited with the same experimental setup. Disadvantageous are the high local laser-induced temperatures and their dependence on substrate material.
- The basic advantage of *photolytic* LCVD is the lower local processing temperature and the lower sensitivity to the physical properties of the substrate (Sect. 16.6). Therefore, direct writing of patterns onto heat-sensitive materials and more uniform writing onto combined materials with variations in thermal properties can be performed. Disadvantageous is the lower purity of the deposited material and, last but not least, the unsatisfactory localization of the deposit. For these reasons, purely photolytic LCVD *cannot* be applied in micropatterning, in general.

A possibility that makes the best of the advantages and disadvantages of pyrolytic and photolytic LCVD, is a twin-beam or a single-beam *combined* pyrolytic-photolytic reaction. This technique is also termed photophysical- or hybrid-LCVD. It should be emphasized, however, that in many systems classified as pyrolytic or photolytic, both mechanisms are important, at least during the phase of nucleation. Initial investigations of this type have been performed for the deposition of Ni from $Ni(CO)_4$ [*Kräuter* et al. 1983]. Here, it was demonstrated that for (visible) Ar^+- or Kr^+-laser radiation that is absorbed neither within gaseous $Ni(CO)_4$ nor by the substrate, the latent times for nucleation are significantly diminished when the UV plasma radiation of the laser tube, which is absorbed by $Ni(CO)_4$, is *not* blocked but focused onto the substrate together with the laser light. Short latent times have also been observed when using 356 nm Kr^+-laser light without the plasma radiation. This wavelength is slightly absorbed by $Ni(CO)_4$, but not by quartz substrates. When nucleation is started, deposition proceeds mainly thermally and depends only on the absorbed laser power and not on the presence of the plasma radiation. This can be understood from the strong absorption of the deposited Ni, which is approximately constant within this spectral range.

Photophysical LCVD was investigated in further detail for Mo, W, and Pt [*Gilgen* et al. 1987]. Here, the UV multiline Ar^+-laser output between 351 and

18.4 Photophysical LCVD

364 nm was used together with $Mo(CO)_6$, $W(CO)_6$ and Pt (hfacac) as parent molecules. With transparent substrates such as glass or sapphire, deposition is initiated by photolytic decomposition of the precursors. No delay or latent times were observed. After this initial step, absorption is determined by the deposited film and growth becomes dominated by pyrolysis. The deposition rates were, typically, 0.1–0.3 µm/s and the writing speeds some µm/s.

Figure 18.4.1 shows, for the example of W and two different substrates, the (normalized) resistivities of stripes versus laser power. The precursor was $W(CO)_6$ without any buffer gas. The most remarkable feature is the difference in resistivities of W stripes deposited on GaAs and glass substrates. This behavior originates from differences in the optical and thermal properties of these substrates. The thermal conductivity of GaAs exceeds that of glass by a factor of 10–30, depending on temperature. Therefore, the local temperature at a certain laser power will be considerably lower on GaAs than on glass. Consequently, the relative importance of pyrolytic and photolytic mechanisms will be quite different for these substrates. The lower laser-induced temperature on GaAs favors incorporation of photofragments such as C, O, CO, or CO_2 into the W films. The situation may even be more complicated. The higher film resistivities measured on GaAs substrates may also originate, in part, from the different metallurgical phases of W. Even in bulk W, the electrical conductivity

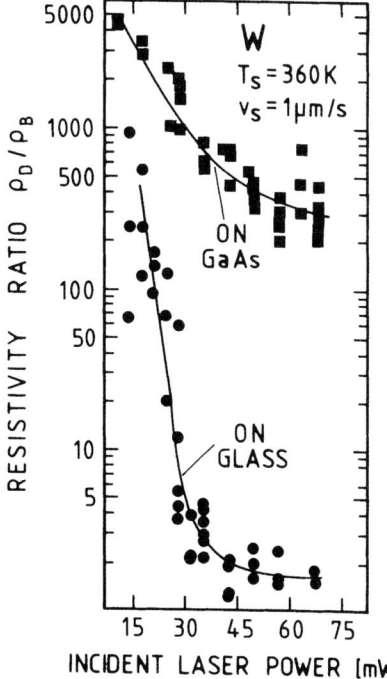

Fig. 18.4.1. Resistivities of laser-deposited W lines normalized to bulk values as a function of laser power (351–364 nm Ar^+-laser output). The pressure of the $W(CO)_6$ was 0.05 mbar. No buffer gas was used [*Gilgen* et al. 1987]

of the α-phase, which is formed between 300 and 650°C, exceeds that of the β-phase, which is formed at lower temperatures, by a factor of 100–300.

The examples for combined photolytic–pyrolytic LCVD mentioned in this section are by no means unique. In particular with transparent substrates, the initial phase of growth is quite frequently based on photolysis of gaseous or adsorbed precursors (Sects. 4.2, 20.2). Photophysical LCVD with single or dual laser beams thus permits one to pattern transparent substrates at technically relevant deposition rates.

18.5 Applications of LCVD in Microfabrication

From a technological point of view, the experimental results presented throughout Chaps. 16–18 demonstrate that laser-CVD enables one to fabricate smooth microstructures of good morphology and well-defined height-to-diameter ratio in a single-step maskless process. Subsequently, we summarize some of the real and potential applications of the technique in planar and non-planar materials processing (Sect. 1.2).

18.5.1 Planar Substrates

In microelectronic fabrication, laser-CVD has been found to be an invaluable technique for increasing product yields and testing early engineering designs. Among the present-day applications are ohmic contacts, interconnects for circuit and mask repair, device restructuring and customization, design and fault correction, etc. [see, e.g., *Baum* et al. 1991; *Nassuphis* et al. 1994]. Another interesting application is the fabrication of microlenses of SiO_2 on flat quartz substrates [*Kubo* and *Hanabusa* 1990].

Contacts

Spot-like deposits can be used for the fabrication of contacts. As demonstrated in Sect. 16.2, the smallest diameters of well-defined W spots produced from $WF_6 + H_2$ on 700 Å W/a–SiO_2 substrates were around 0.6 µm. The fabrication of a similar feature by standard CVD or plasma CVD (PCVD) in combination with photolithographic techniques would require many production steps. The sticking of W spots on amorphous SiO_2 substrates (with or without the 700 Å W layer) has been studied by the Scotch-tape test. The different types of spots have all passed this test.

Photodeposited Cd has been employed to form high Schottky-barrier-height contacts on InP and $In_{0.53}Ga_{0.47}As$ substrates [*Licata* et al. 1991].

18.5 Applications of LCVD in Microfabrication

Circuit repair

The requirements for the repair of electrical open-circuits in microelectronic applications are quite strict: The repair must be of good electrical integrity, be reliable to electrical and environmental stressing, withstand subsequent chemical and physical treatments, and match the dimensions of the existing circuit. All these requirements are fulfilled by LCVD of gold using $Au(CH_3)_2$ (tfacac) and Ar^+-laser radiation. Figure 18.5.1 shows a SEM picture of a copper circuit on a polyimide substrate after repair by LCVD of gold. The repaired defect was about 250 µm long, 15 µm wide, and 5–10 µm high. The electrical resistivity was, typically, between 3 and 10 times higher than for bulk gold (2.4 µΩcm).

Interconnects

Similarly to circuit repair, LCVD can be used to interconnect discrete regions on modules or integrated circuits (IC). The process has been used to rewire defective regions of a circuit, to customize components for specific designs or applications, to form interconnects on IC gate arrays, to interconnect interlevel metal planes by filling vertical vias, etc. The materials mainly employed were Ni, Pt, and W.

Mask repair

Pyrolytic and photolytic LCVD has been employed to repair clear defects on lithographic masks, e.g., chrome-on-glass masks, where the metal is missing. Here, the only demands are to match the resolution and to completely block any light transmission.

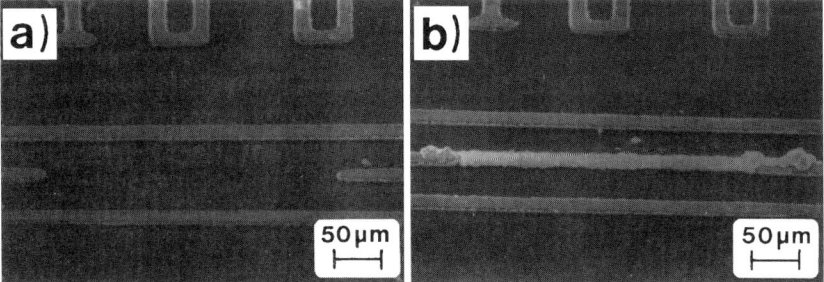

Fig. 18.5.1a,b. SEM picture demonstrating the repair of a defective copper circuit on a polyimide substrate. (**a**) Defect prior to repair ($d \approx 15$ µm). (**b**) Similar feature after repair by LCVD of gold [*Baum* et al. 1991]

18.5.2 Non-planar Substrates, 3-D Objects

Laser-CVD offers unique possibilities for the coating, patterning, and fabrication of non-planar three-dimensional (3-D) objects.

For example, it is possible to produce uniform protective coatings on mechanical workpieces of complex geometry, or to *selectively* increase the thickness of protective coatings (produced by standard techniques) on the edges or sides of tools or devices where mechanical or chemical requirements are particularly severe.

One can also conceive of many applications in chemical technology. For example, for catalytic purposes, production of metal coatings on cheap substrates, such as glass wool, could become of importance. Here, high porosity sponge-like deposition would be possible by appropriate control of processing parameters.

Laser-CVD makes it possible to directly write virtually any pattern onto three-dimensional devices. An example is shown in Fig. 18.5.2a for the case of a W helix deposited by means of a focused Ar^+-laser beam on a (rotating) Si core. Here, patterning by mechanical masking or lithographic techniques together with large-area processing would be extremely difficult and in most cases even impossible. Laser-CVD also enables one to fabricate free-standing 3-D patterns by dissolving the substrate after the direct-write process (Fig. 18.5.2b).

Millimeter-high wall-like patterns with lateral dimensions of only a few micrometers can be fabricated for various different materials by repetitive

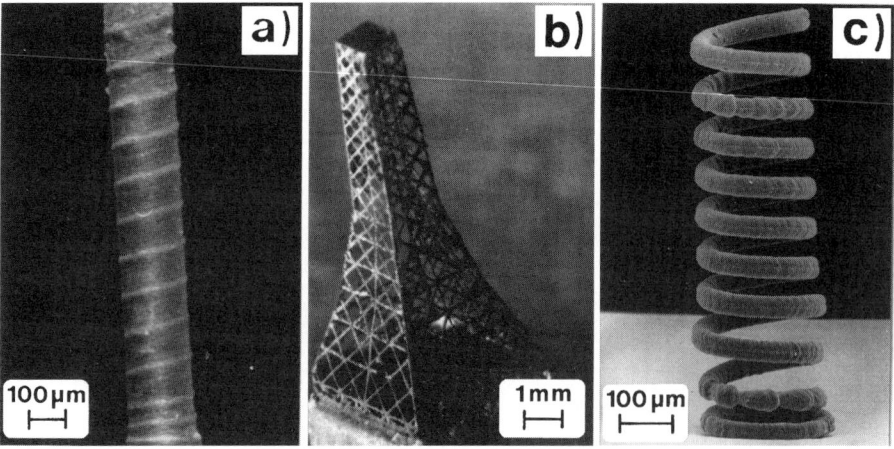

Fig. 18.5.2a–c. Examples of non-planar processing and fabrication of 3-D objects by LCVD. (**a**) Microsolenoid with silicon core and tungsten helix [after *Westberg* et al. 1993]. (**b**) Al-grid structure fabricated by laser direct writing onto poly-carbonate which was dissolved after LCVD [*Lehmann* and *Stuke* 1991]. (**c**) Free-standing boron microspring [*Johansson* et al. 1992]

18.5 Applications of LCVD in Microfabrication

direct writing. LCVD can also be used to fill up deep grooves or holes with almost any type of material.

Fibers of almost any length can be grown, even as single crystals, without any crucible and in an otherwise completely cold atmosphere. This technique not only permits one to grow materials with higher purity, but also to grow new materials, quite possibly under extreme physical conditions, for example at very high temperatures, pressures, or within electric or magnetic fields. Here, applications can only be speculated on.

Laser-CVD allows direct three-dimensional growth and packaging by means of pattern projection, interference techniques, etc. An example of the *direct* growth of a free-standing boron microspring by means of a focused laser beam is shown in Fig. 18.5.2c. Here, the spring was moved in such a way that the laser focus was always positioned at its end.

19 Thin-Film Formation by Laser-CVD

Light-assisted CVD (chemical vapor deposition), and in particular laser-CVD, open up new possibilities in thin-film fabrication. Laser light permits one to *selectively* generate high concentrations of atomic or molecular intermediate species that are present either not at all or only in small equilibrium concentrations in standard CVD using the same precursor molecules. Thus, laser-CVD enables one to study *new* reaction pathways and altered kinetics in thin-film growth. Lasers are often preferred over high-intensity lamps, at least in fundamental investigations, because of their high experimental versatility related to their intensity, monochromaticity, tunability, and directionability. In particular, at parallel incidence to the substrate surface, lasers permit pure gas-phase excitation. With perpendicular laser-beam irradiation, gas *and* surface excitations, including adsorbed layers, are important. The different irradiation geometries employed in large-area laser-CVD have been described together with Fig. 5.2.3.

From a practical point of view, light-assisted film growth is mainly studied with the intention of fabricating high-quality films at *lower* substrate temperatures. Laser-CVD based on gas-phase heating (parallel incidence only; Fig. 1.2.2b) or laser-induced photolysis of gas- or adsorbed-phase precursors (Fig. 1.2.2a,b) permits one to deposit thin films without significant heating. Thus, elevated substrate temperatures are employed only, if at all, for improving film morphologies. This is one of the main advantages over conventional CVD. Furthermore, in contrast to plasma-CVD, there are no problems with VUV radiation or particle bombardment (Sect. 1.2). Finally, laser-CVD permits monolayer control of film thicknesses. This is a prior condition for the well-defined fabrication of heterostructures consisting of multiple thin-film layers with different material properties.

As before, we concentrate on model systems for which the most complete data are available. For additional information and a list of systems investigated so far, the reader is referred to the Appendix B.5. This Appendix includes results on thin film formation using lamps for either direct or indirect (via photosensitization) decomposition of parent molecules [*Eden* 1991]. Systems for which adsorbed-layer decomposition has been proved to be decisive, are discussed in Chap. 20.

19.1 Direct Heating

Subsequently we consider systems where gas-phase molecules *directly* absorb the laser light. The excitation energy shall be locally randomized via collisions within a time that is short compared to any non-thermal reaction step. The laser beam shall propagate *parallel* to the substrate surface and thermally activate a (purely) homogeneous reaction (Figs. 19.1.1 and 1.2.2b). This situation applies to many types of large-area laser-CVD but also to some types of laser-induced etching.

The temperature distribution induced within the ambient medium can be calculated from the heat equation (2.2.1). In the case of uniform substrate heating, the temperature T_s enters the problem via the boundary condition $T(h_s) = T_s$. General solutions of (2.2.1) have been presented in Sect. 6.2. For the present problem, we can directly employ specific solutions given in Chap. 7. Here, the window of the reaction chamber (at $z = 0$) can be taken into account by employing either $\eta \approx \kappa_w/w$ ("semi-infinite" windows) or $\eta \to \infty$ ("cold" windows). Solutions with $\eta = 0$ ignore the influence of the window; nevertheless, such solutions can be employed for $z > z_0 \approx \kappa/\eta \approx w\kappa/\kappa_w$ and weak absorption with $\alpha^* \ll 1$ and $z_0 < l_\alpha$. κ and κ_w are the heat conductivities of the gas and the window, respectively.

19.1.1 Stationary Solutions

For many estimations it is appropriate to find stationary solutions of the heat equation. For a laser beam with cylindrical symmetry, this can be written as

$$\frac{1}{r}\frac{\partial}{\partial r}\left(r\frac{\partial T}{\partial r}\right) + \frac{\partial^2 T}{\partial z^2} + \frac{\alpha I_0}{\kappa} g(r) \exp(-\alpha z) = 0, \tag{19.1.1}$$

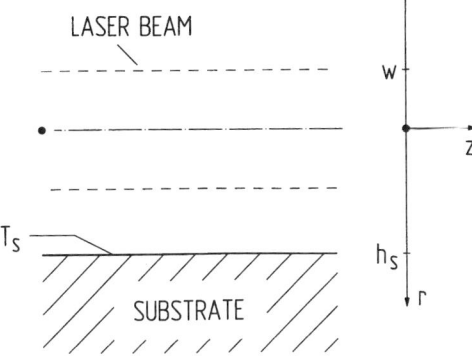

Fig. 19.1.1. Direct heating of an ambient medium by laser-light irradiation at parallel incidence to the substrate. The origin of the coordinate system, indicated by a dot, is in the center of the laser beam on the left-hand side. The laser beam has a radius w and propagates in z-direction. The distance from the beam center to the substrate surface is h_s

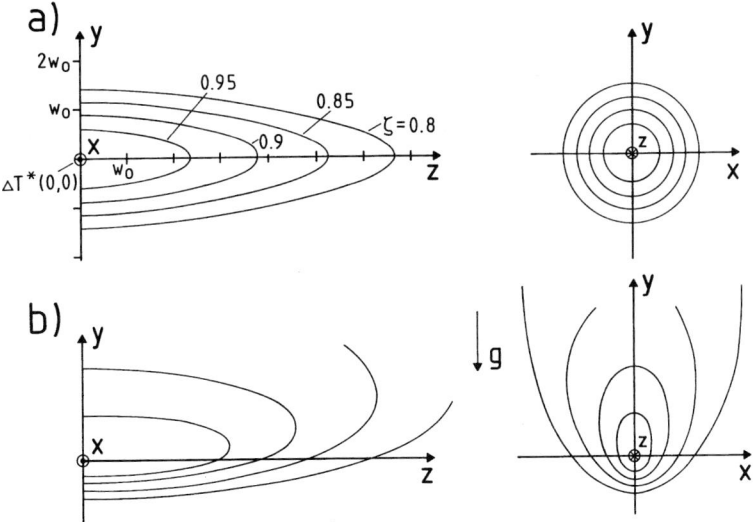

Fig. 19.1.2 a,b. Isotherms $\Delta T^*(r, z) = \zeta \Delta T^*(0, 0)$ plotted in planes $x = 0$ and $z = 0$. (a) Gaussian beam, no convection. (b) Qualitative influence of free convection; g is the vector of gravity

where α and κ are constants, $r \geq 0$, $z > 0$ and $T = T(r, z)$. In the *absence* of the substrate, the boundary conditions are

$$T(z \to \infty) = T(r \to \infty) = T(\infty); \quad \left.\frac{\partial T}{\partial r}\right|_{r=0} = 0, \qquad (19.1.2a)$$

and

$$\left.\frac{\partial T}{\partial z}\right|_{z=0} = 0. \qquad (19.1.2b)$$

The boundary condition for $z = 0$ ignores the influence of the chamber window. In many cases it is more realistic to use instead

$$T(z = 0) = T(\infty). \qquad (19.1.2c)$$

In any case, by employing Hankel transforms, the solution of the problem yields the temperature rise

$$\Delta T(r, z) = \frac{\alpha I_0}{\kappa} \int_0^\infty dq \, J_0(qr) \, g(q) \, \frac{q \exp(-\alpha z) - \xi \exp(-qz)}{q^2 - \alpha^2}, \qquad (19.1.3)$$

where $\xi \equiv \alpha$ with (19.1.2b) and $\xi \equiv q$ with (19.1.2c). J_0 is the Bessel function of order zero and $g(q)$ is given by

$$g(q) = \int_0^\infty r J_0(rq) \, g(r) \, dr. \qquad (19.1.4)$$

19.1 Direct Heating

Subsequently, we consider only solutions for $\zeta \equiv \alpha$ and for various intensity distributions.

Case 1: Gaussian laser beam

For a Gaussian laser beam $g(r) = \exp(-r^2/w_0^2)$ we obtain

$$g(q) = \frac{1}{2}w_0^2 \exp\left(-\frac{q^2 w_0^2}{4}\right). \tag{19.1.5}$$

Substituting (19.1.5) into (19.1.3) yields the temperature rise $\Delta T(r, z)$ whose maximum is given by

$$\Delta T(0, 0) = \frac{\alpha I_0 w_0^2}{2\kappa} \int_0^\infty \frac{dq}{\alpha + q} \exp\left(-\frac{w_0^2 q^2}{4}\right). \tag{19.1.6}$$

We introduce the same normalization as in (7.2.8). The dependence of $\Delta T^*(0, 0)$ on α^* is identical to that shown in Fig. 7.2.2. Isotherms $\Delta T^*(r, z) = \text{const} = \zeta \Delta T^*(0,0)$ within planes $x = 0$ and $z = 0$ are plotted in Fig. 19.1.2a.

Case 2: Rectangular intensity distribution

A circular beam with constant intensity over its cross section is described by $g(r) = \mathcal{H}(w - r)$ where $\mathcal{H}$ is the Heaviside function. With (19.1.4) we obtain

$$g(q) = \int_0^w r J_0(qr) \, dr = \frac{w}{q} J_1(qw). \tag{19.1.7}$$

If we still ignore the substrate, the temperature distribution $T(r, z)$ can be calculated by substituting (19.1.7) into (19.1.3). The maximum temperature rise is

$$\Delta T(0,0) = \frac{\alpha I_0 w}{\kappa} \int_0^\infty \frac{J_1(qw)}{q(q + \alpha)} \, dq. \tag{19.1.8}$$

The dependences $\Delta T(0,0) = f(\alpha)$ and $\Delta T(r,z) = \text{const.}$ are qualitatively similar to those plotted in Figs. 7.2.2 and 19.1.2, respectively.

Case 3: Influence of the substrate

If we assume *cylindrical* symmetry where the substrate has the form of a tube, the condition $T(r \to \infty) = T(\infty)$ in (19.1.2a) must be replaced by $T(r = h_s) = T_s$ where T_s is the temperature, and h_s the radius of the tube (Fig. 19.1.1). This geometry can be used for inside coatings of tubes by LCVD. For a weakly absorbing medium we can set $\partial T/\partial z \approx 0$ and $\exp(-\alpha z) \approx 1$. The solution of

(19.1.1) is then

$$T(r) = T_s + \frac{\alpha I_0}{\kappa} \int_r^{h_s} \frac{dr'}{r'} \int_0^{r'} g(r'')r'' dr''. \qquad (19.1.9)$$

For a *Gaussian* beam we obtain

$$T(r) = T_s + \frac{\alpha I_0 w_0^2}{4\kappa}\left[\operatorname{Ei}\left(-\frac{r^2}{w_0^2}\right) - \operatorname{Ei}\left(-\frac{h_s^2}{w_0^2}\right) + 2\ln\left(\frac{h_s}{r}\right)\right], \qquad (19.1.10)$$

where Ei denotes the exponential integral (Appendix A.3.3). This equation can be applied only if $h_s < (w_0 l_\alpha)^{1/2}$ or $h_s < \kappa/\eta$. If both h_s and the length of the reactor become very large, the temperature distribution is given by (19.1.3).

For a circular beam with constant intensity distribution, (19.1.9) can be written for $r < w$ as

$$T(r<w) = T_s + \frac{\alpha I_0 w^2}{2\kappa}\left[\ln\left(\frac{h_s}{w}\right) + \frac{1}{2}\left(1 - \frac{r^2}{w^2}\right)\right], \qquad (19.1.11a)$$

and for $r > w$

$$T(r>w) = T_s + \frac{\alpha I_0 w^2}{2\kappa}\ln\left(\frac{h_s}{r}\right). \qquad (19.1.11b)$$

The range of validity is similar to (19.1.10).

In pyrolytic laser-chemical processing with *plane* substrates, the assumption of cylindrical symmetry is certainly a crude approximation. This is true, however, even in the absence of a substrate or reaction chamber, because of convection (Fig. 19.1.2b). In gases, convection can be diminished by decreasing the pressure [note that $\mathcal{R}a \propto p^2 \Delta T/T$; see (9.5.12)] or by changes in the geometry of the reaction chamber. In some cases, the homogeneity of surface processes can be improved by increasing the laser power so that convective flows become turbulent.

At medium to high gas pressures and $\Delta T_c \geq T_s$ convection becomes very important. The characteristic length in (9.5.9), is of the order $l \leq h_s$. Typical flow velocities, v_c, are 0.2–5 cm/s. In the case of turbulence, v_c can exceed these values by several orders of magnitude.

19.1.2 Non-stationary Solutions

For a Gaussian beam, the center temperature rise derived from (2.2.1), (19.1.2), and the initial condition $T(t=0) = T(\infty)$ is given by (7.5.1). For very short pulses where $t \ll w^2/D$ one can employ the energy balance and obtain for weak absorption, in analogy to (7.5.6)

$$T(r,t) = T(r,0) + \frac{\alpha I(r)}{\rho c_p} t. \qquad (19.1.12)$$

Influence of substrate

For cylindrical symmetry we can use the solution (19.1.12) in the initial phase and (19.1.9) for $t > h_s^2/D$.

For a *plane* substrate with $T_s \equiv T(y=0) = T(t=0)$ (Fig. 19.2.1), weak absorption, and a Gaussian beam with $w_0 < h_s$, the center temperature rise is

$$\Delta T(0,0,t) = \frac{\alpha I_0 w_0^2}{4\kappa} \left[\ln(1+4t^*) + \mathrm{Ei}\left(-\frac{4h_s^{*2}}{1+4t^*}\right) - \mathrm{Ei}(-4h_s^{*2}) \right], \qquad (19.1.13)$$

where $t^* = tD/w_0^2$ and $h_s^* = h_s/w_0$. This solution can be applied only if $l_\alpha \gg h_s$. For $t^* \to \infty$ we obtain

$$\Delta T(0,0) = \frac{\alpha I_0 w_0^2}{4\kappa} [C - \mathrm{Ei}(-4h_s^{*2}) + \ln(4h_s^{*2})], \qquad (19.1.14)$$

where $C = 0.577$ is Euler's constant.

19.2 Pyrolytic Processing Rates

We now calculate the reaction flux of product species, and thereby the processing rate, for homogeneous pyrolysis of precursor molecules. The laser beam shall have either cylindrical symmetry with the radius $w \equiv w_y$ or rectangular symmetry with the widths $2w_x$ and $2w_y$ (Fig. 19.2.1). The latter situation applies to many cases of excimer-laser processing. For a first-order reaction of the type

$$AB \to A(\downarrow) + B(\uparrow) \qquad (19.2.1)$$

the number density $N_A \equiv N_A(x,t)$, can be calculated from (14.2.1). Henceforth, we assume $N_{AB} \gg N_A, N_B$. The generation of species A within the *volume* heated by the laser beam can be described by

$$Q_{v,A} = k(T)N_{AB} = k_0 N_{AB}\exp\left(-\frac{\mathcal{E}}{T}\right) \equiv W_v(T), \qquad (19.2.2)$$

where $T \equiv T(x,t) = T(\infty) + \Delta T(x,t)$ is the temperature within the medium *above* the substrate (Sect. 19.1). With stationary conditions and a number of additional assumptions, analytic solutions of (14.2.1) can be found.

19.2.1 Diffusion

We consider AB and B as ideal gases and set $D_A \propto T$, so that $N(T)D_A(T) = N(\infty) D_A(\infty)$ with $N \approx N_{AB}$. Gas-phase recombination shall be ignored. Molecules impinging onto the surface shall stick on it with unit probability, i.e., $N_A(x, 0, z) = 0$. All quantities shall be independent of z. With these assumptions, (14.2.1) can be written as

$$N_{AB}(\infty) D_A(\infty) \nabla^2 x_A + W_v(T) = 0. \tag{19.2.3}$$

The flux of species A on the surface $y = 0$ is

$$J_A(x) = - N_{AB}(\infty) D_A(\infty) \nabla x_A \bigg|_{y=0}. \tag{19.2.4}$$

By employing the Green's function technique, we obtain

$$- J_A(x) = \frac{1}{\pi} \int_{-\infty}^{\infty} \int_0^{\infty} W_v(T) \frac{y'}{(x-x')^2 + y'^2} dy' dx' = W_s(x), \tag{19.2.5}$$

where $W_s(x)$ is the surface reaction rate. Note that $T = T(x', y')$. Let us consider various cases.

Case 1: w or w_x, $w_y \ll h_s$

The laser beam is circular or rectangular and concentrated near $x = 0$, $y = h_s$. Substitution of $x' = 0$ and $y' = h_s$ in (19.2.5) yields

$$W_s(x) = - J_A(x) \approx \frac{1}{\pi} \frac{h_s}{x^2 + h_s^2} \Phi. \tag{19.2.6}$$

The function Φ describes the total number of species A produced within the

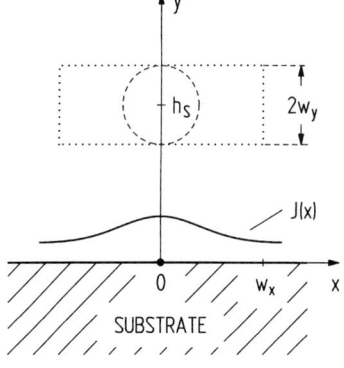

Fig. 19.2.1. Laser-chemical processing by means of a parallel laser beam propagating in the z-direction (normal to drawing plane) at height h_s above the substrate surface ($y = 0$). The beam profile is either circular (dashed) or rectangular (dotted). The flux profile of species is shown by the solid curve

19.2 Pyrolytic Processing Rates

volume heated by the laser beam (per unit time and length in z-direction)

$$\Phi = \int\int W_\mathrm{v}(T(x',y'))\,\mathrm{d}x'\,\mathrm{d}y' \approx F W_\mathrm{v}(T_\mathrm{c}). \tag{19.2.7}$$

F is the cross section of the reaction zone. We then obtain

$$W_\mathrm{s}(x) = \frac{1}{\pi}\frac{h_\mathrm{s} F}{x^2 + h_\mathrm{s}^2} W_\mathrm{v}(T_\mathrm{c}). \tag{19.2.8}$$

If we introduce the width of the temperature distribution w_T (Sect. 6.5) we can make the approximation $F \approx \pi w_\mathrm{T}^2 T_\mathrm{c}/\mathscr{E}$. The normalized *film thickness* is

$$h^*(x^*) = \frac{h(x)}{h(0)} = \frac{W_\mathrm{s}(x)}{W_\mathrm{s}(0)} = \frac{1}{1 + x^{*2}}, \tag{19.2.9}$$

where $x^* = x/h_\mathrm{s}$. This profile can be compared with experimental data obtained for a-Si:H films deposited from SiH_4 by means of CO_2-laser radiation. The data points in Fig. 19.2.2 represent average values obtained for distances $h_\mathrm{s} = 6,7,8$ and 9 mm and a reaction zone radius $w \approx w_\mathrm{T}(T_\mathrm{c}/\mathscr{E})^{1/2} \leqslant 0.5\,\mathrm{mm} \ll h_\mathrm{s}$. The full curve represents the (average) thickness profile calculated for a point source from (19.2.9).

Case 2: $w_\mathrm{y} \ll h_\mathrm{s}$

With the approximation $W_\mathrm{v}(x',y') = 2w_\mathrm{y} W_\mathrm{v}(T(x'))\,\delta(y' - h_\mathrm{s})$ integration over y' in (19.2.5) yields

$$-J_\mathrm{A}(x) = \frac{2w_\mathrm{y}}{\pi} \int_{-\infty}^{+\infty} W_\mathrm{v}(T(x')) \frac{h_\mathrm{s}}{(x-x')^2 + h_\mathrm{s}^2}\,\mathrm{d}x'.$$

If we assume $W_\mathrm{v}(T(x')) = W_\mathrm{v}(T_\mathrm{c})$ with $|x'| \leqslant w_\mathrm{x}$ and $W_\mathrm{v}(T(x')) = 0$ with $|x'| > w_\mathrm{x}$ we obtain

$$W_\mathrm{s}(x) = \frac{\Phi}{2\pi w_\mathrm{x}}\left[\arctan\left(\frac{w_\mathrm{x}-x}{h_\mathrm{s}}\right) + \arctan\left(\frac{w_\mathrm{x}+x}{h_\mathrm{s}}\right)\right], \tag{19.2.10}$$

with $\Phi = F W_\mathrm{v}(T_\mathrm{c}) \approx 4 w_\mathrm{x} w_\mathrm{y} W_\mathrm{v}(T_\mathrm{c})$ where w_x and w_y characterize the width of the reaction zone in x and y direction. With $w_\mathrm{x} \to 0$ we obtain (19.2.6). The normalized film thickness can be calculated in analogy to (19.2.9).

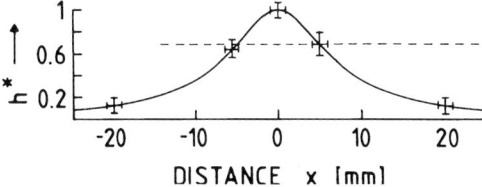

Fig. 19.2.2. Data represent the (average) thickness profile of an a-Si:H film deposited from SiH_4 by means of a CO_2-laser beam at parallel incidence to a vertical substrate [$w(1/e) \approx 5\,\mathrm{mm}$; $T_\mathrm{s} \approx 350\,°C$] [*Golusda* et al. 1992]. The solid curve has been calculated

Case 3: $w_y \ll h_s$, $w_x \to \infty$

If we introduce the total number of species produced within the heated volume per unit length in the z- and x-directions, $\tilde{\Phi} = \Phi/2w_x$ [species/cm²s], we obtain from (19.2.10)

$$-J_A(x) \approx \tilde{\Phi}. \tag{19.2.11}$$

In this approximation all species condense on the substrate.

19.2.2 Recombination

We now estimate the influence of recombination with

$$f(N_A, N_B) = -\frac{N_A(\infty)}{\tau_A^{rec}} x_A = -k_A^{rec} N_A(\infty) x_A = -C x_A, \tag{19.2.12}$$

where $C = $ const. In the absence of gas-phase heating, this ansatz coincides with (14.2.15). Together with (19.2.3) and $x_A(y=0) = 0$ we obtain for a point source instead of (19.2.6)

$$-J_A(x) = \frac{2\Phi}{\pi} \frac{h_s}{l l_A^{rec}} K_1\left(\frac{2l}{l_A^{rec}}\right), \tag{19.2.13}$$

where $l_A^{rec} = 2(D_A \tau_A^{rec})^{1/2}$ is the recombination length, $l = (h_s^2 + x^2)^{1/2}$, and K_1 the modified Bessel function. The (normalized) thickness profile is

$$h^*(x^*) = \frac{1}{l^*} \frac{K_1(2\beta l^*)}{K_1(2\beta)}, \tag{19.2.14}$$

where $l^* = l/h_s = (1 + x^{*2})^{1/2}$ and $\beta = h_s/l_A^{rec}$. Comparison of (19.2.9 and 14) shows that recombination leads to a better localization of the deposit (Fig. 19.2.3). The analysis of the shape of the deposit permits one to investigate experimentally the recombination length l_A^{rec} from β. For a point source, the

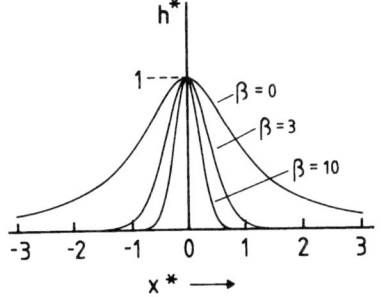

Fig. 19.2.3. Influence of recombination ($\beta > 0$) on the (normalized) thickness profile of the deposit

total flux to the surface is

$$J_{tot} = \frac{2\Phi}{\pi} \int_0^\infty \frac{2\beta}{l^*} K_1(2\beta l^*) dx^*. \qquad (19.2.15)$$

If $\beta \gg 1$ this yields

$$J_{tot} = \Phi \exp(-2\beta). \qquad (19.2.16)$$

With $\beta \ll 1$ the flux is $J_{tot} = \Phi$. For $w_x \to \infty$ we obtain (19.2.15), except that Φ must be replaced by $\tilde{\Phi} = \Phi/2w_x$.

19.3 Photolytic Processing Rates

The density of product species $N_A(x,t)$ generated in a photochemical reaction

$$AB + h\nu \to A(\downarrow) + B(\uparrow) \qquad (19.3.1)$$

can be calculated from (14.2.1 and 2) together with the appropriate boundary conditions.

Subsequently, we discuss analytical solutions for the simple reaction (19.3.1) and the irradiation geometry in Fig. 19.2.1. With the assumptions made in Sect. 19.2.1 and *continuous* irradiation, the reaction rate on the surface can be described by (19.2.5). If w or w_x, $w_y \ll h_s$ we can use the approximate solution (19.2.6). The only difference is that W_v in (19.2.7) is now the *photolytic decomposition rate*

$$W_v(I) = \sigma_{AB} N_{AB} \frac{I}{h\nu},$$

where σ_{AB} is the excitation/dissociation cross section. With $\Phi \approx F W_v(I)$ we obtain

$$W_s(x) = -J_A(x) = \frac{1}{\pi} \frac{h_s F}{x^2 + h_s^2} W_v(I), \qquad (19.3.2)$$

where F is now the cross section of the laser beam. The other cases discussed in Sect. 19.2 can be employed in analogy. For *pulsed* irradiation, the average rate is

$$\langle W_s(x) \rangle \approx \frac{1}{\pi \tau_i} \frac{h_s F}{x^2 + h_s^2} \sigma_{AB} N_{AB} \frac{\phi}{h\nu}, \qquad (19.3.3)$$

where $1/\tau_i$ is the pulse repetition rate, and ϕ the laser fluence per pulse. The normalized thickness profiles are equal to (19.2.9). The effect of recombination can be taken into account as in Sect. 19.2.2.

19.4 Metals

Extended thin films of metals have been fabricated with areas up to about $10\,\text{cm}^2$. The precursors mainly employed were metal halides, metal alkyls, and metal carbonyls. The preferred light sources are excimer lasers and harmonics of Nd:YAG lasers. In many systems, decomposition of precursors is mainly photolytic at low laser-light intensities and mainly pyrolytic at higher intensities.

19.4.1 Deposition from Metal Halides

Detailed investigations on large-area film deposition have been performed in particular for W using mixtures of WF_6 and H_2 [*Deutsch* 1984; *van Maaren* et al. 1991]. Here, ArF-laser radiation at *parallel* incidence has been employed. The substrates, mainly silicon wafers with and without a thermally grown SiO_2 layer, were preheated to a certain temperature, T_s. The surface morphology of films, and in particular the conformal coverage of the substrate, is excellent (Fig. 19.4.1).

Figure 19.4.2 shows an Arrhenius plot of the deposition rate. T_s is the (uniform) substrate temperature. No experimental information on the laser-induced gas-phase temperature exists. For comparison, corresponding results obtained by standard CVD are included in the figure. Here, it should be noted that with the present systems and the parameters under consideration, standard CVD is *selective*, i.e., W deposition is observed only on Si but not on SiO_2. Laser-CVD is *non-selective* and permits deposition at temperatures $T_s < 600\,\text{K}$. With many applications, however, selective substrate coating is desirable. Another important feature seen in the figure is the strong decrease in

Fig. 19.4.1. SEM pictures showing a side and a top view of a W film over a SiO_2 step on a Si substrate. Deposition was performed with ArF-laser radiation from $WF_6 + H_2$ with $T_s = 285\,°\text{C}$ [*Deutsch* 1984]

19.4 Metals

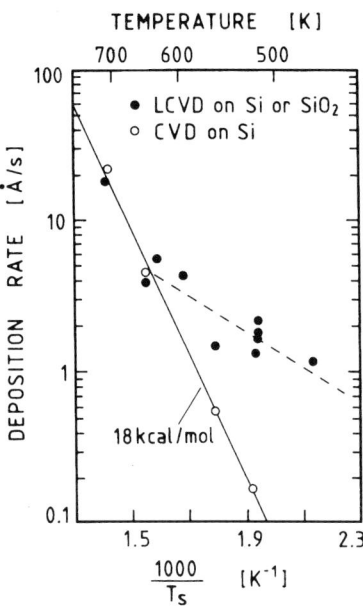

Fig. 19.4.2. Arrhenius plot of the (thickness) deposition rate of W films. The precursor was WF_6 diluted in H_2. ●: Deposition by ArF-laser radiation on Si and SiO_2/Si substrates. ○: Deposition by standard CVD on Si substrates, adapted from [*Deutsch* 1984]

the apparent chemical activation energy which is characterized by the slope of the dashed line. This may demonstrate that the laser light changes the chemical reaction pathways involved in the deposition process. There are, however, uncertainties in this interpretation because of laser-induced gas-phase heating.

The overall decomposition of WF_6 in the presence of H_2 is described by (16.2.1). With Si substrates, film growth may be initiated by thermally activated reduction of WF_6 according to (18.1.1).

Photochemical activation by ArF-laser radiation ($\lambda = 193$ nm; 6.4 eV) produces radicals, probably via the reaction

$$WF_6 + h\nu (ArF) \rightarrow WF_x + (6-x)F, \qquad (19.4.1)$$

with $x = 4$ or 5. This process is followed by reactions with H_2 and, in the case of Si substrates, with Si. The latter reaction channel seems to be of minor importance (Figs. 19.4.1 and 19.4.2). Fluorine atoms can react with H_2 in a strongly exothermic reaction

$$H_2 + F \rightarrow HF + H. \qquad (19.4.2)$$

Atomic H formed in the gas can react with WF_6 and with WF_x radicals via a complex series of reactions that result in W deposition and in gaseous HF. Such a reaction pathway could qualitatively explain the strong change in activation energy with respect to standard CVD, where the rate-limiting step in the kinetically controlled region is probably determined by the formation of atomic H via dissociative adsorption of H_2 on the W surface [*Hitchman* et al.

1989]. It should be emphasized that these mechanisms are still under discussion.

Laser-deposited W films are of high purity with a concentration of $F < 1\%$. Good adherence and low electrical resistivity of films was achieved only at elevated substrate temperatures. At $T_s = 400\,°C$, the electrical (sheet) resistivity of LCVD films exceeds that of pure bulk W by less than a factor of two. While this T_s seems to be rather high, standard CVD of W is frequently performed at even higher temperatures. In particular with SiO_2 substrates, reasonable deposition rates are only obtained with $T_s \geqslant 700\,°C$.

Figure 19.4.3 compares electrical resistivities of films deposited by laser-CVD and standard CVD on Si substrates at various temperatures, T_s. The film resistivity decreases strongly with increasing T_s. This behavior is mainly ascribed to the grain size of films which increases with T_s. With temperatures as low as $250\,°C$ the resistivity of laser-deposited films is only $20\,\mu\Omega\mathrm{cm}$. At this temperature, films produced by standard CVD are strongly non-uniform. Laser-deposited films show, in general, significantly smaller thickness variations than standard CVD films. This was revealed from Rutherford back-scattering (RBS) [van Maaren et al. 1989, 1991].

Laser-CVD of circular Ti films with diameters of a few mm has been studied both experimentally [Chou et al. 1989] and theoretically [Kar et al. 1991]. Here, CO_2-laser radiation at perpendicular incidence has been used to thermally decompose $TiBr_4$. The substrate was stainless steel and the deposition rate about $0.05\,\mu\mathrm{m/s}$.

19.4.2 Deposition from Alkyls and Carbonyls

Intense investigations have been performed on the deposition of Al from $Al_2(CH_3)_6$ and of Cr, Mo, and W from the hexacarbonyls on Pyrex, SiO_2 and

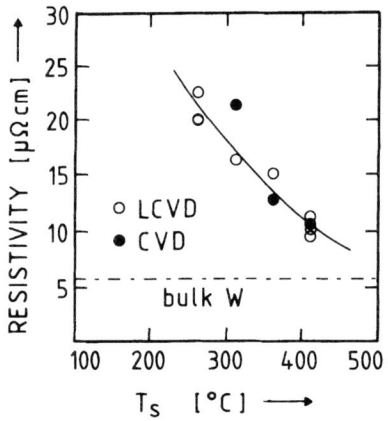

Fig. 19.4.3. Resistivity of W films deposited from a mixture of WF_6 and H_2 on Si substrates by laser-CVD (open symbols) and standard CVD (full symbols) at various substrate temperatures T_s, adapted from [van Maaren et al. 1989]

19.4 Metals

Si substrates. Most of these experiments were performed with ArF and KrF lasers.

With *perpendicular* laser-beam incidence, bright silvery metallic films of reasonable quality were obtained. Typical deposition rates are between a few Å/s and some ten Å/s. The adhesion of films depends strongly on the substrate material, its cleaning prior to deposition, and on the laser power and wavelength. With SiO_2 substrates good film adhesion has been achieved ($> 5 \times 10^8$ dynes/cm^2). The tensile stresses were between 10^9 and 7×10^9 dynes/cm^2. Films of Cr and Mo deposited at room temperature had a tendency to peel when exposed to air. This could be avoided by heating the substrate to about 150 °C either during deposition or prior to removal from the cell.

With *parallel* laser-beam irradiation and low substrate temperatures, grey or black particulate films with high impurity contents were obtained.

Films deposited photolytically without significant substrate heating have resistivities that exceed the corresponding bulk values by a factor of 10 to 10^4. This large variation in ρ originates from differences in grain sizes and, more importantly, in impurity concentrations which strongly depend on surface temperature. The concentrations of C_xH_y or C_xO_y fragments incorporated, decrease with increasing T_s. This explains the superior electrical properties of films deposited with perpendicular incidence $[\rho(Cr, Mo, W) \approx 20 \rho_B(Cr, Mo, W)]$. With Al the resistivity ratio was significantly lower.

There are several possible ways to improve the morphology and physical properties of photodeposited films. Among these are:

- Uniform substrate heating to temperatures where no significant dark reactions take place.
- Combined twin-beam or single-beam photophysical LCVD (Sect. 18.4) using irradiation geometries as shown in Figs. 5.2.3b,c.
- Post-deposition treatment. For example, in Al films deposited from $Al_2(CH_3)_6$ the contamination with CH_3 can be significantly diminished by subsequent ArF-laser irradiation in vacuum [*Higashi* and *Rothberg* 1985]. By using pulsed optoacoustic IR spectroscopy it has been demonstrated that high-yield non-thermal photodesorption of CH_3 groups incorporated in the films can be achieved with ArF laser, but not with KrF-laser radiation. Such post-treated films had electrical and optical properties similar to those deposited at elevated substrate temperatures. If these observations hold more generally, this technique would be a unique tool to improve the quality of both photolytically and pyrolytically deposited films. It would also be relevant in high-resolution pyrolytic patterning where low (local) laser-induced temperatures are employed. The improvement of other laser-deposited materials, such as Al_2O_3, by UV-laser irradiation, can eventually be explained, at least in part, along similar lines. Inhomogeneous laser-beam illumination and structure formation (Chap. 28) can result in inhomogeneous impurity incorporation [*Houle* et al. 1986].

19.5 Semiconductors

Light-enhanced/induced growth of thin films of semiconductors has been studied mainly with excimer lasers, frequency-doubled Nd:YAG lasers, cw- or pulsed-CO_2 lasers, and with lamps. In the latter case, the precursor molecules were either directly excited or indirectly excited via Hg photosensitization.

19.5.1 Photodecomposition of Silanes

Amorphous, polycrystalline, and epitaxial films of Si and Ge have been deposited mainly from SiH_4, Si_2H_6 and GeH_4. The photodecomposition channels can most easily be understood from the UV absorption spectra and energy-level diagrams shown in Fig. 19.5.1 for SiH_4 and Si_2H_6. Arrows indicate

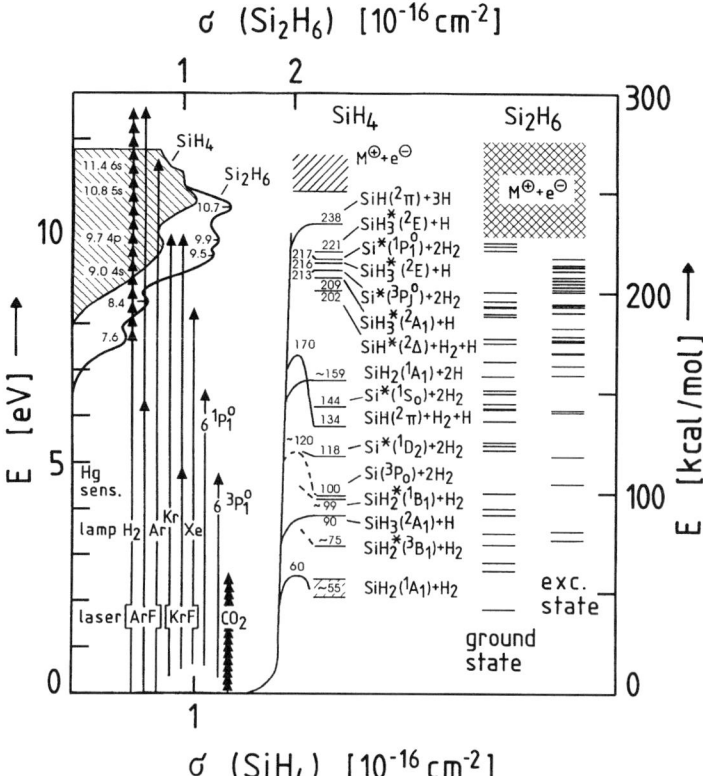

Fig. 19.5.1. Energy level diagrams of SiH_4 and Si_2H_6 including absorption spectra, optical excitation processes, and primary decomposition pathways, adapted from [*Stafast* 1988]

19.5 Semiconductors

various optical transitions induced by lasers, lamps, or photosensitization. Single-photon decomposition of SiH_4 and Si_2H_6 can be achieved only with energies $h\nu \geqslant 8\,eV$ ($\lambda \leqslant 155\,nm$) and $\geqslant 6.2\,eV$ ($\lambda \leqslant 200\,nm$), respectively. Thus, decomposition of SiH_4 by 193 nm (6.4 eV) ArF- or 248 nm (5 eV) KrF-excimer laser radiation requires coherent absorption of at least two photons. This process is quite efficient even with relatively low intensities ($< 10^6\,W/cm^2$). Si_2H_6 can be decomposed by single-photon absorption of either ArF- or 157 nm F_2-laser radiation. This is the reason why Si_2H_6 is the preferred precursor for selective photochemical LCVD. With high laser-light intensities, two-photon absorption and optical activation of photofragments seems to become important also with Si_2H_6.

The primary *photochemical* decomposition processes of SiH_4 and Si_2H_6 have been investigated by means of a 147 nm Xe lamp [*Perkins* et al. 1979]. The main reaction steps with SiH_4 can be described by

$$SiH_4 + h\nu \to SiH_2 + 2H \tag{19.5.1a}$$

and

$$SiH_4 + h\nu \to SiH_3 + H, \tag{19.5.1b}$$

where the relative quantum yield of (19.5.1a) is by about a factor of 5 larger than that of (19.5.1b). Corresponding experiments with Si_2H_6 have revealed three primary reaction channels

$$Si_2H_6 + h\nu \to SiH_2 + SiH_3 + H \tag{19.5.2a}$$

$$Si_2H_6 + h\nu \to Si_2H_4 + 2H \tag{19.5.2b}$$

$$Si_2H_6 + h\nu \to Si_2H_5 + H, \tag{19.5.2c}$$

with relative quantum yields 0.61:0.18:0.21.

Mercury-photosensitized decomposition of SiH_4 and Si_2H_6 has also been extensively studied (Sect. 2.3.1). The most probable channel for fragmentation due to energy transfer from $Hg(^3P_1)$ to SiH_4 can be described by

$$Hg(^3P_1) + SiH_4 \to Hg(^1S_0) + SiH_3 + H. \tag{19.5.3}$$

Thermal activation by either standard heating or by IR-laser radiation permits decomposition along the most *energetically* favorable pathway

$$SiH_4 \to SiH_2 + H_2. \tag{19.5.4}$$

Other decomposition pathways are characterized by higher activation energies and primary fragments such as SiH_3, SiH_2, and SiH radicals, or diradicals and Si atoms.

With Si_2H_6, the lowest activation energy is required for the fragmentation reaction

$$Si_2H_6 \to SiH_4 + SiH_2, \tag{19.5.5}$$

while the thermodynamically most stable products are formed according to

$$Si_2H_6 \rightarrow Si_2H_4 + H_2. \tag{19.5.6}$$

For further details on the photochemistry of SiH_4 and Si_2H_6, see e.g., [*Stafast* 1988] and references therein.

The amount of hydrogen incorporated in deposited films depends on the type and density of precursor molecules, the particular decomposition channel involved, and on the substrate temperature.

19.5.2 Crystalline Ge and Si

The growth of Si and Ge films has been investigated for various substrate materials by employing excimer-laser radiation and mixtures of SiH_4/N_2 and GeH_4/He [*Eden* 1991]. With *parallel* laser-beam incidence, the growth rates were, typically, between 0.1 and 1Å/s, depending on the (uniform) substrate temperature and gas pressure. These rates are 10–100 times lower than those achieved at normal laser-beam incidence. The average grain sizes of deposited films depend on the specific parameter set employed.

Epitaxial growth of Si on (100) Si wafers at (moderate) temperatures ($600\,°C \leqslant T_s \leqslant 650\,°C$) was demonstrated with ArF- and XeF-laser radiation at perpendicular incidence using Si_2H_6 diluted in H_2 [*Yamada* et al. 1989]. The good crystallinity and electrical properties of films were interpreted by laser-enhanced surface migration of Si atoms.

Epitaxial growth of Ge onto (100) GaAs has been demonstrated with ArF-laser radiation at parallel incidence and substrate temperatures $T_s \geqslant 285\,°C$ [*Kiely* et al. 1989a]. In the absence of laser light, films are amorphous ($T_s \approx 305\,°C$) or polycrystalline ($T_s \approx 415\,°C$). This clearly demonstrates the change in deposition kinetics due to the excitation of GeH_4. The initial step is ascribed to photochemically generated GeH_2 and GeH_3 radicals that are transformed into Ge_2H_6 via collisions. These species diffuse to the substrate surface where they subsequently pyrolyse.

19.5.3 Amorphous Hydrogenated Silicon (a-Si:H)

The enormous interest in a-Si:H arises from various applications of this material. Among these are low-cost solar cells, electrophotographic plates, thin-film transistors, optical sensors, etc. [*Fritzsche* 1984; *LeComber* et al. 1981; *Takahashi* and *Konagai* 1986]. Hydrogen incorporation into a-Si is necessary because it saturates Si dangling bonds and relieves strains, resulting in a reduced defect level and the ability to modulate the Fermi level by substitutional doping. Currently, the preferred technique of producing a-Si:H films is plasma-CVD (PCVD) using SiH_4 as a precursor. Structurally superior amorphous films are produced by standard CVD. However, to obtain reasonable deposition rates, CVD requires substrate temperatures of at least 600 °C. With such high temperatures, the films contain an insufficient amount of hydrogen (< 1 at.%) to achieve good electronic properties.

Photoenhanced deposition of thin films of a-Si:H has been demonstrated with lasers, lamps [*Kim* et al. 1989; *Mizukawa* et al. 1989; *Nishida* et al. 1986], and by Hg photosensitization [*Kamimura* and *Hirose* 1986]. The precursors most commonly employed are SiH_4, Si_2H_6, and Si_3H_8 diluted in appropriate buffer gases.

Excimer-laser-induced deposition

Thin films of a-Si:H can be deposited efficiently from Si_2H_6 with ArF-laser light (Fig. 19.5.1). Deposition from SiH_4 was observed only if significant substrate heating takes place [*Murahara* and *Toyoda* 1984].

Figure 19.5.2 compares film growth rates obtained by conventional CVD using Si_2H_6 diluted in H_2 and He with those obtained under otherwise identical conditions in the presence of an ArF-laser beam at *parallel* incidence. For conventional CVD the apparent activation energy is $\Delta E \approx 35 \pm 2$ kcal/mol. With laser-CVD and temperatures $T_s < 400°C$, this energy is only $\Delta E = 2.1 \pm 0.5$ kcal/mol. This weak dependence on T_s suggests that deposition is dominated by gas-phase photolysis. The small increase in slope observed

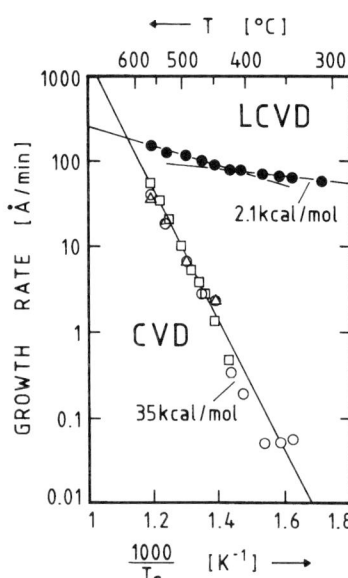

Fig. 19.5.2. Arrhenius plot for the deposition of a-Si:H films on thermally oxidized (100) Si ($p = 5$ Torr; 40 sccm of 10% Si_2H_6/90% H_2 and 390 sccm of He window flush). T_s is the (uniform) substrate temperature. Open symbols refer to standard CVD and full symbols to laser-CVD (ArF laser at parallel incidence; 20 Hz, 750 mW/cm²) [*Eres* et al. 1988]

near 450 °C reflects the increasing importance of thermal activation. At a laser-pulse repetition rate of 20 Hz, growth rates between 1 and 3Å/s have been obtained. The high deposition rates achieved at relatively low temperatures enable one to optimize the physical properties of films with each substrate material.

The film-growth rate increases linearly with laser power (Fig. 19.5.3). Thus, with the laser powers employed, disilane photolysis is dominated by a single-photon process. No film growth was observed with KrF-laser radiation.

The influence of buffer gases (H_2, He, and Ar) was studied in detail by *Dietrich* et al. (1989).

CO_2-laser-induced deposition

For CO_2-laser light at *perpendicular* incidence, deposition of Si onto fused quartz is dominated by pyrolysis of precursors at the gas-solid interface (SiO_2 strongly absorbs CO_2-laser light). Nevertheless, *gas-phase* heating can be important as well. This has been demonstrated with SiH_4 where the wavelength dependence of the deposition rate correlates with the absorption spectrum of gaseous SiH_4. Maximum deposition rates over areas of 1 cm^2 were about 20 Å/s with SiH_4 and 200Å/s with Si_2H_6 [*Hanabusa* et al. 1984].

Fabrication of high-quality a-Si:H films using *parallel* CO_2-laser-beam incidence requires uniform substrate heating to, typically, 200–400 °C. Deposition rates are much lower than for perpendicular incidence, even when significantly higher laser powers are used. With SiH_4, typical growth rates are 2–5 Å/s. However, the uniformity of films is superior to those produced at perpendicular incidence. Films with excellent adherence were deposited over areas up to 80 cm^2 [*Curcio* et al. 1986; *Pauleau* et al. 1984].

Decomposition of SiH_4 seems to occur in two steps [*Newman* et al. 1979]. In the first step, which is considered as the rate-limiting process, SiH_2 radicals

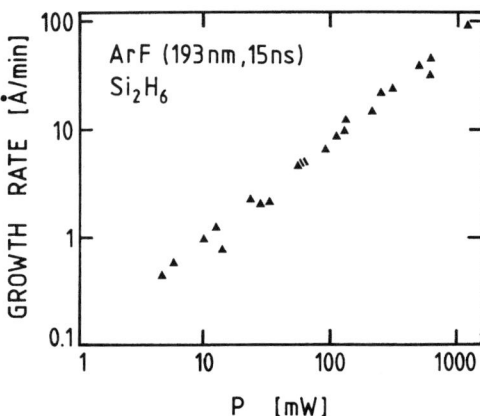

Fig. 19.5.3. Growth rate of a-Si:H films as a function of ArF-laser power [*Eres* et al. 1988]

19.5 Semiconductors

and H_2 are produced by gas-phase heating according to (19.5.4). Here, SiH_4 molecules vibrationally excited in a *single-photon* absorption process (indicated by *) transfer their energy among each other via collisions (Sect. 2.3) and decompose according to

$$SiH_4 + h\nu \to SiH_4^* \to \cdots \to SiH_2 + H_2. \tag{19.5.7}$$

This process is most efficient if the CO_2-laser frequency is tuned to a strong vibrational transition of the SiH_4 molecule. The second step depends on SiH_4 pressure and laser fluence. At high pressures or/and high fluences, SiH_2 radicals react homogeneously with other SiH_2 or SiH_4 molecules to produce particulates within the gas [*Sladek* 1971]. At lower pressures and fluences, diffusion of SiH_2 radicals to the surface dominates and a thin film of a-Si:H will grow, following the reaction

$$SiH_2 \to SiH_y(\downarrow) + (1 - y/2)H_2(\uparrow), \tag{19.5.8}$$

where y denotes the hydrogen content within the film.

Figure 19.5.4 shows an Arrhenius plot of the film growth rate obtained with different samples, laser powers, substrate temperatures, gas compositions, and gas pressures. T_G is the maximum gas-phase temperature calculated from (19.1.1). Data obtained under different experimental conditions can be described by a single activation energy. This supports the interpretation that deposition is controlled by *single-photon* vibrational absorption of SiH_4 and fast energy relaxation via collisions within the volume of the laser beam. This

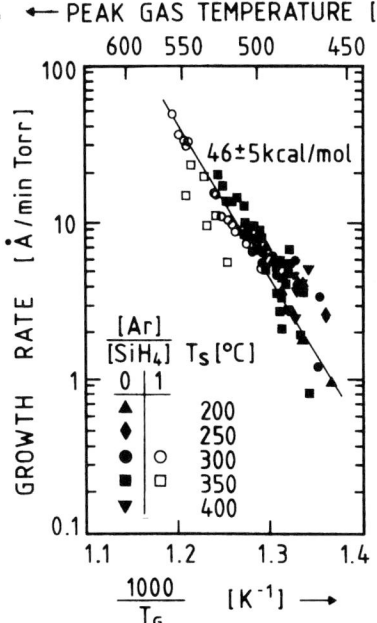

Fig. 19.5.4. Arrhenius plot of the growth rate (Å/min per torr of SiH_4) of a-Si:H films deposited by means of CO_2-laser radiation at parallel incidence. T_G is the calculated maximum laser-induced gas temperature, adapted from [*Meunier* et al. 1987a]

is in agreement with simple estimations: For typical conditions [an absorbed laser power of 1 W/cm, $p(SiH_4) = 10$ mbar, and $T_G = 10^3$ K] the excitation rate is $W_{exc} \approx 10^3/s$, and the average time between collisions about 10^{-8} s. The vibrational–translational relaxation rate is $1/\tau_{v-T} \approx 10^4/s$. Therefore, (2.3.10) is readily fulfilled. This example shows that even *non-selective* LCVD opens up new possibilities: With parallel laser-beam incidence high temperatures, and thereby high reaction rates, can be induced in the immediate neighborhood of heat-sensitive substrates.

The fabrication of doped a-Si:H films using SiH_4 diluted in Ar together with either B_2H_6 or PH_3 has also been investigated [*Curcio* et al. 1986; *Branz* et al. 1986]. The substrates (SiO_2, Si) were uniformly heated to, typically, 200–400 °C. The growth rates were 2–5 Å/s.

The composition, morphology, optical and electrical properties of CO_2-laser-deposited a-Si:H films have been analyzed in detail [*Curcio* et al. 1986; *Branz* et al. 1986; *Meunier* et al. 1987b; *Metzger* et al. 1988]. The main techniques employed were: X-ray diffraction, UV, VIS and IR spectroscopy, hydrogen effusion, and electron spin resonance (ESR). The magnitude of the unpaired spin concentration (this is a measure of the concentration of dangling bonds, which should be as low as possible for high-quality films) in LCVD films was about 400 times lower than in conventional CVD films and similar to that of PCVD and HOMOCVD films (Table 19.5.1). The main parameter that determines the physical properties of LCVD films is the substrate temperature which controls the total concentration of hydrogen. The films have a predominantly $Si-H_2$ structure and an optical band gap similar to HOMOCVD films. Undoped LCVD and HOMOCVD films are almost intrinsic, while PCVD films are usually somewhat n-type, probably due to defects generated by ion bombardment. A high ratio of the photo-conductivity, σ_p, and the dark conductivity, σ_d, is an indication for a low overall defect density.

Table 19.5.1. Physical properties of a-Si:H films deposited by UV-laser CVD [*Dietrich* et al. 1989; *Zarnani* et al. 1986], CO_2-laser CVD [*Golusda* et al. 1993; *Metzger* et al. 1988], HOMOCVD [*Scott* 1984], and PCVD [*Hirose* 1984]

Material constants	LCVD		HOMOCVD	PCVD
$T_s[°C]$	280	300	250	230
	193 nm ArF	10.6 μm CO_2		
Optical band gap [eV]	1.82	1.6–2.2	1.8	1.7–1.8 (H ≈ 18%)
Refractive index	3.20	–		3.43
Dark conductivity				
$\sigma_d[\Omega\,cm]^{-1}$	$10^{-11}-10^{-8}$	10^{-12}	$\approx 10^{-9}$	$10^{-13}-10^{-9}$
Photoconductivity (AM1 100 mW cm^{-2})				
$\sigma_p[\Omega\,cm]^{-1}$	$10^{-7}-10^{-5}$	$10^{-7}-10^{-4}$	$\approx 10^{-4}$	$10^{-7}-10^{-3}$
σ_p/σ_d	$3 \times 10^4 - 5 \times 10^5$	$10^3 - 10^5$	10^5	3.3×10^5
Density of gap states [cm^{-3}]	3×10^{16}	$10^{16}-10^{18}$	3×10^{15}	$< 10^{16}$

The high hydrogen content of cw CO_2-laser-deposited films (at $T_s = 300\,°C$ about 20–25%) is one of the main drawbacks to better film qualities. This problem can be overcome by SF_6 sensitization [*Golusda* et al. 1993]. Contrary to SiH_4, vibrational excitation of the quasi-continuum of SF_6 is possible without intermediate collisions (Sects. 2.3 and 14.1). This is the reason why SF_6 strongly absorbs CO_2-laser light and permits one to deposit from SiH_4 films with a more optimal hydrogen content ($\approx 10\%$). The photoconductivity of such films is about 10^{-4} $[\Omega cm]^{-1}$. Similar experiments have been performed to fabricate a-Ge:H films [*Barth* et al. 1994].

19.5.4 Compound Semiconductors

The most detailed investigations on photo-enhanced CVD of compound semiconductors have been performed for GaAs, GaP, InP, CdTe, HgTe, and some oxides. In the presence of light, films can be grown with lower substrate temperatures/higher deposition rates. Frequently, the surface morphology of films is significantly improved with respect to those grown by standard metal-organic CVD (MOCVD). Clearly, with lower substrate temperatures, interdiffusion of film and substrate material is diminished. The precursors most frequently employed are alkyls (Sects. 16.1,6).

III–V Compounds

Light-enhanced MOCVD of GaAs, GaAlAs, GaP, and InP has been demonstrated with excimer lasers, frequency-doubled Nd:YAG lasers and mercury lamps (Appendix B.5).

GaAs films were deposited from $Ga(CH_3)_3$ and AsH_3 diluted in H_2. Figure 19.5.5 shows the growth rate achieved with and without laser-light versus substrate temperature. In the presence of light, growth can be extended to

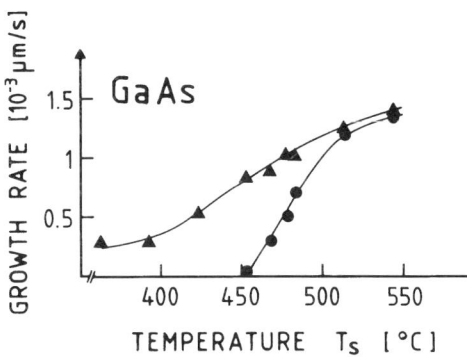

Fig. 19.5.5. Growth rate of an epitaxial (100) GaAs film as a function of GaAs-substrate temperature, T_s. ●: Conventional MOCVD; ▲: MOCVD enhanced by 532 nm pulsed Nd:YAG-laser radiation at perpendicular incidence [*Beneking* 1984]

much lower temperatures. The morphology of films depends on laser fluence. Smooth monocrystalline layers were obtained with a fluence $\phi_\perp \approx 120 \text{ mJ/cm}^2$. At a substrate temperature of 450 °C (this does not account for the laser-induced temperature rise which cannot be ignored with these fluences) these layers were p-type ($3 \times 10^{17} \text{ cm}^{-3}$), probably due to incorporated carbon.

The precursor molecules employed for the deposition of InP were $(CH_3)_3$ InP$(CH_3)_3$ [this adduct decomposes into In$(CH_3)_3$ and P$(CH_3)_3$ above approximately 80 °C] and P$(CH_3)_3$ diluted in He and H_2. Stoichiometric films with thicknesses of up to more than 1 µm were deposited in the presence of ArF-laser radiation with rates of up to 0.4 Å/pulse. Their microstructure was amorphous, polycrystalline or epitaxial, depending on the incident laser fluence and the substrate temperature, which was typically around 320 °C. At this temperature, the dark growth rate is negligible.

The laser-power density on the substrate/film surface is a very critical parameter that affects crystallization, film and substrate damage, and impurity incorporation. ArF-laser irradiation of the surface is very efficient in suppressing contaminations of C and C_xH_y (see also Sect. 19.4.2). For example, for properly deposited InP films, the C contamination was below detectability by AES. *In situ* gas-phase fluorescence measurements give further insight into photodecomposition mechanisms and also allow one to monitor metal-organic precursor concentrations. The comparison of results achieved at different wavelengths suggests that laser-enhanced growth of III-V compounds is controlled by pyrolytic and photolytic gas-phase decomposition of precursors, charge-transfer at the gas-solid interface (Chap. 15), enhanced surface diffusion of species, and surface heating–depending on the intensity and photon energy of the light. For low substrate temperatures and UV-laser light, adsorbed-layer photolysis plays an important or even dominant role (Sect. 20.2).

II–VI Compounds

Light-enhanced growth of ZnTe, CdTe and HgTe films has been studied with lasers and lamps.

Films of CdTe have been grown on (100) GaAs substrates in the presence of 257 nm Ar$^+$-laser radiation. The precursors were Cd$(CH_3)_2$ and Te$(C_2H_5)_2$ or Te$(CH_3)_2$. For epitaxial growth, photo-enhancement factors (ratio of growth rates for illuminated and non-illuminated regions) of about 8 have been found for laser-light intensities of 0.16 W/cm^2, a (uniform) substrate temperature of 300 °C, and Te$(C_2H_5)_2$ as precursor [*Irvine* et al. 1989]. At *low* light intensities, different surface processes control the growth kinetics. Among those are: Adsorption of Te atoms, photolytic decomposition of adsorbed radicals or undecomposed alkyls, and desorption of C_2H_5 or CH_3 groups. At *higher* intensities, photolytic gas-phase decomposition becomes important as well.

Photo-assisted epitaxy of CdTe on GaAs substrates has also been investigated for KrF-laser radiation at *parallel* incidence [*Zinck* et al. 1988]. With a mixture of $Cd(CH_3)_2 + Te(CH_3)_2$ and substrate temperatures as low as 165 °C growth rates of about 2 µm/h have been achieved. The experimental results have been analyzed on the basis of model calculations based on gas-phase photolysis [*B. Liu* et al. 1992]. As expected from (19.3.3) the growth rate increases linearly with laser-light intensity and gas pressure.

The growth of HgTe films was studied with both Hg-lamp photosensitization [*Irvine* et al. 1984] and ArF- and KrF-laser radiation at parallel incidence [*Fujita* et al. 1989]. In the latter case, the growth rate was, typically, 1 µm/h.

Oxides

Light-enhanced growth of ZnO, In_2O_3, and SnO_2 has been studied with lasers and lamps.

ZnO films are of relevance for acoustoelectric and acoustooptic applications that are based on the large piezoelectric and optical coupling constants of this material. Additionally, its high band gap energy ($E_g \approx 3.3$ eV) has attracted interest in using ZnO as a coating for solar cells because it is less expensive than indium-tin-oxide (ITO). SnO_2 combines optical and electronic properties that are interesting in optoelectronics, solar cell manufacture, heating element fabrication, and gas monitoring.

ZnO is an n-type semiconductor. It has been deposited from $Zn(CH_3)_2$ and NO_2 or N_2O by means of ArF- or KrF-laser radiation. With deposition rates of up to 18 µm/h, films were of good optical quality, low pinhole density, good adherence and good thickness uniformity (better than 5%) over areas of 10 cm^2. At higher deposition rates the adherence and quality of films deteriorated. The stoichiometry revealed by ESCA was 49% Zn and 51% O. The concentrations of impurities, mainly C and N, were below 1%. By varying the ratio of partial pressures, the stoichiometry of films can easily be changed. Resistivities may then range 10^3-10^{-1} Ωcm for 0.5 µm thick films [*Solanki* and *Collins* 1983].

Thin films of In_2O_3 have been deposited with ArF-laser radiation at glancing (5°) incidence [*Donnelly* et al. 1984]. SiO_2, GaAs, and InP served as substrate materials. The parent molecules $(CH_3)_3$ $InP(CH_3)_3 + P(CH_3)_3 + O_2$ or H_2O were diluted in H_2 and He. High-quality, nearly stoichiometric [In:O = 2:(3.2 ± 0.3)] films have been obtained at elevated temperatures, $T_s \approx 330$ °C. No C or P impurities were detected by sputtering AES. Perpendicular irradiation under otherwise identical conditions resulted in shiny films containing In and O in a ratio of 1:1 with no C or P.

19.5.5 Carbon

Classification of carbon is somewhat arbitrary because its electrical and thermal properties depend on whether its microstructure is amorphous, graphitic, or diamond-like.

Carbon films have been deposited by photodissociation of C_2H_3Cl and CCl_4 using ArF-laser radiation at perpendicular incidence [*Tachibana* et al. 1988]. The ten times higher deposition rate achieved with C_2H_3Cl (about 2 Å/s) is mainly due to its higher absorption cross section. Because of the higher deposition yields, C_2H_3Cl is also to be preferred to CH_4, C_2H_2, and C_2H_4.

Films deposited from C_2H_3Cl at $T_s < 300\,°C$ did consist of amorphous carbon and graphite; their hardness was about 7 Mohs. With $T_s > 400\,°C$ the graphite structure dominates.

Laser-CVD of diamond-like carbon (DLC) was demonstrated by *Kitahama* et al. (1986). ArF-laser radiation and C_2H_2 diluted in H_2 have been employed. The substrate temperature was varied between 40 and 800 °C. DLC films fabricated by pulsed-laser deposition are discussed in Sect. 22.3.3.

19.6 Insulators

Photo-assisted growth of (mainly) insulating films permits one to combine good step coverage, low mechanical stresses and defect concentrations with high deposition rates at low to moderate substrate temperatures.

Direct laser-enhanced surface oxidation/nitridation in pure oxygen/nitrogen is described in Chap. 26.

19.6.1 Oxides

The oxidant most frequently employed in photo-assisted CVD of oxide films is N_2O. This molecule can be dissociated within the range $138 \leq \lambda \leq 210$ nm with almost unity quantum yield. Thus, ArF lasers are the preferred sources in this application. Dissociation can be described by

$$N_2O + h\nu\,(193\,\text{nm}) \to N_2 + O(^1D), \tag{19.6.1}$$

where excited *atomic* oxygen is the primary reactive product. Atomic oxygen reacts with the precursor of the other component, or with fragments of it, and forms the oxide.

Al_2O_3

Al_2O_3 is an attractive substitute for SiO_2, mainly in microelectronics; it also has many other applications which benefit from a low-temperature, high-rate deposition process.

19.6 Insulators

Thin films of Al_2O_3 were deposited by means of ArF and KrF lasers at *parallel* incidence using a mixture of $Al_2(CH_3)_6$ and N_2O. The substrates employed were glass, and wafers of Si and III-V compounds. By moving the substrate relative to the laser beam, films of uniform thickness ($\pm 5\%$) across 3 inch wafers were deposited at rates of 30 Å/s. With higher deposition rates, the films were of bad adherence [*Solanki* et al. 1983].

Table 19.6.1 facilitates comparison of high-quality films fabricated by LCVD and RF sputtering. The mechanical, optical and electrical properties of LCVD films can be further improved by using *combined parallel/perpendicular* laser-beam incidence [*Deutsch* 1984].

A better understanding of the photodeposition process must await further investigations. Certainly, excited atomic oxygen generated according to (19.6.1) plays a dominant role. It reacts with fragments of the Al precursor and thereby forms Al_2O_3.

SiO_2

The fabrication of thin films of (amorphous) SiO_2 by laser-CVD has been studied in great detail. Films of SiO_2 are an integral part of every semiconductor device, providing electrical insulation and passivation as well as masking layers for both the diffusion of dopants and pattern transfer. Additionally, SiO_2 layers are often used for materials passivation against corrosion.

Table 19.6.1. Physical and chemical properties of Al_2O_3 films deposited by ArF-laser-CVD [*Emery* et al. 1984] and RF sputtering [*Novicki* 1977]. T_s is the substrate temperature, p the total gas pressure, W_D the deposition rate, n the index of refraction, ρ the resistivity, and ε the dielectric constant

	LCVD	RF sputtered
T_s [°C]	150 – 400	150 – 400
p [mbar]	1.3	1.3×10^{-2}
W_D [Å/s]	30	5
	$\lambda = 193$ nm	
	$\tau_l \approx 10$ ns	
	100 Hz	
Adhesion [10^8 dyne/cm^2]		
(1000 Å on Si)	> 6.5	Strongly adherent
Pinholes/cm^2	< 1	31
	(1000 Å)	(2500 Å)
Stress [dyne/cm^2]	$< 6 \times 10^9$ (tensile)	2.8×10^9 (compressive)
n	1.63	1.66
ρ [Ωcm]	10^{11}	10^{12}
ε (1 MHz)	9.74	9.96
Etch rate in 10% HF [Å/min]	100	–
Stoichiometry	Al_2O_3	$Al_{2.1}O_3$
Impurities [%]	C < 1	Ar $\approx$ 5

Laser-assisted growth of SiO_2 was mainly investigated by means of ArF-laser radiation. Here, mixtures of 1–5% SiH_4, 89–85% N_2O and 10% N_2 with a total pressure of $\leqslant 10$ mbar have been employed [*Emery* et al.]. Mainly Si wafers were used as substrates.

For *parallel* incidence, deposition rates of 10 to 50 Å/s were obtained (the upper value seems to be too high for good quality films). The deposition rates were proportional to gas pressure and laser intensity, but independent of T_s within the range $300\,°C \leqslant T_s \leqslant 600\,°C$. The quality of films, however, did depend sensitively on T_s. For $T_s \geqslant 250\,°C$, optically clear, stoichiometric, scratch-resistant, and adherent films of good uniformity over areas of 10 cm^2 were obtained. For $T_s < 200\,°C$, the films were milky.

The main physical and chemical properties of films produced by laser-, plasma-, and electron-beam-induced CVD are compared in Table 19.6.2. The pinhole densities, dielectric breakdown voltages, and step-coverage abilities of LCVD films were comparable to thermally grown native oxides.

The dependence of growth rates and film properties on partial pressures of gases ($SiH_4 + N_2O$ diluted in Ar) and on the distance between the (parallel) ArF-laser beam and the substrate was investigated by [*Szörenyi* et al. 1994, 1990].

Deposition of SiO_2 using *combined* parallel and perpendicular ArF-laser beam incidence has also been studied [*Emery* et al. 1984]. The parallel beam creates a thin sheet of photofragments in a localized region just above the substrate with minimal interaction with the surface. The perpendicular beam creates transient heating of the SiO_2/Si interface. This dual beam configuration allows for in situ annealing by surface heating, and photo-stimulation of surface reactions. As a result, an increase in the deposition rate (10–20%), a reduction in chemical etch rate, and an increase in refractive index were observed. Simultaneously, a decrease of SiH and SiOH bonds, and a lowering of the nitrogen content was found. Good-quality films with deposition rates of about 15 Å/s were obtained at substrate temperatures of only 250 °C.

Photo-assisted growth of SiO_2 layers by means of different types of lamps have been performed, e.g., by *Boyd* (1995), *Inushima* et al. (1988), *Marks* and *Robertson* (1988), *Scoles* et al. (1988).

The detailed photodeposition process is not well understood: The deposition kinetics is controlled by a competition between quenching and recombination of atomic oxygen, oxidation of silicon hydrides, creation of reactive nitric oxide species, and substrate reactions. From the low nitrogen concentration detected within films, it was concluded that only $O(^1D)$ and O_2^* are important in the growth kinetics (see also Sect. 22.2).

The growth of SiO_2 films based on *gas-phase heating* via vibrational excitation of SiH_4 by CO_2-laser radiation at parallel incidence was studied by *Fernandez* et al. (1994). The results have been analyzed on the basis of calculated gas-phase temperatures.

19.6 Insulators

Table 19.6.2. Physical and chemical properties of SiO_2 films deposited by various techniques [*Emery* et al. 1984; *Schuegraf* 1983]

	LCVD	PCVD	EBCVD
T_s [°C]	100–400	380	250–400
p [mbar]	0.4–10	1.4	0.35
$N_2O/SiH_4/N_2$	80/1/40	33/1/10	75/1/75
W [Å/s]	3–17	5	9
CVD parameters	$\lambda = 193$ nm	450 kHz	4.7 kV
	$\tau_l \approx 10$ ns		
	100 Hz		12 mA
	Hg photosens.		
ρ [g/cm^3]	2.1		
Adhesion [10^8 dyne/cm^2]			
(1000 Å on Si)	> 7	> 7	> 7
Pinholes/cm^2			
(1000 Å on Si)	1–100	25–100	100–700
(2000 Å on Si)	≈ 1	≈ 1	10–100
n (632.8 nm)	1.45–1.49	1.46	1.46
Stress on Si			
[10^9 dyne/cm^2]			
(compressive)	1.5	3.6	9.4
Breakdown voltage			
[MV/cm]			
(1000 Å on Si)	6–8	–	2–6
(2000 Å on Si)	–	10	–
ρ [Ω cm] at 5 MV/cm	10^{13}–10^{14}	10^{16}	10^{14}–10^{16}
Flat band voltage [V]	2–10	< 0.2	0.5–3
ε (1 MHz)	3.9–4.6	4.6	3.5
	(thermal oxide = 3.9)		
Stoichiometry	SiO_2	$SiO_{1.94}N_{0.06}$	SiO_2
N [%]	< 1	3	< 1
Hydrogen bonding			
(2270 cm^{-1} SiH [%])	2.3	2	< 1
(3380 cm^{-1} H$_2$O [%])	≈ 0.01	< 0.001	< 0.001
(3650 cm^{-1} OH [%])	0.6	0.002	< 0.01
Etch rate [Å/s] in			
7:1 buffered HF			
(as deposited)	30	20	30–60
(after 60 min			
1220 K N$_2$ anneal)	≈ 18	–	–

19.6.2 Nitrides

Large-area LCVD of nitrides has been investigated mainly for Si_3N_4 and TiN. Here, NH_2 radicals generated by photodissociation via

$$NH_3 + h\nu \,(193 \text{ nm}) \to NH_2 + H \qquad (19.6.2)$$

play a similar role as atomic oxygen in oxide formation. With ArF-laser light ground-state NH_2 is formed with nearly unit efficiency. The absorption cross section of NH_3 [$\sigma(193$ nm$) \approx 10^{-17}$ cm^2] exceeds that of N_2O by a factor 10^2. For this reason, the partial pressure of NH_3 generally employed is much lower than that of N_2O used for the deposition of oxides.

Si_3N_4

Silicon nitride films are used for various applications, for example as interlayers or capacitor dielectric in microelectronics; as gate insulators for active matrix liquid-crystal displays; as gate insulators in power field-effect transistors; and for final encapsulation of components, IC chips, etc.

Si_3N_4 has been deposited mainly by means of ArF lasers from mixtures of SiH_4 or Si_2H_6 with NH_3/N_2 or He [*Sugii* et al. 1988; *Emery* et al. 1984]. The substrates, mainly Si, thermally oxidized Si wafers, and ZnSe, were uniformly heated (50 °C $\leqslant T_s \leqslant$ 600 °C). With combined parallel/perpendicular ($\phi_\parallel \approx 10\phi_\perp$) incidence, deposition rates of up to 12 Å/s were achieved.

The physical and chemical properties of LCVD, PCVD and EBCVD films are similar (Table 19.6.3). Good-quality LCVD films possess good adherence, low pinhole density, low compressive stress, and excellent step coverage. The etch rate of films fabricated by combined $\parallel$ and $\perp$ irradiation is diminished by more than a factor of five. The dielectric properties of LCVD films are not yet satisfactory.

Low-temperature photo-deposition of silicon nitride by means of Hg lamps [*Inushima* et al. 1988; *Petitjean* et al. 1992] and Hg-photosensitization [*Schuegraf* 1983] was also demonstrated.

TiN

Films of TiN with diameters of a few mm have been deposited from an atmosphere consisting of $TiCl_4$, H_2, and N_2 by using CO_2-laser radiation at perpendicular incidence [*Conde* et al. 1992]. The deposition profiles have been analyzed on the basis of model calculations for pyrolytic LCVD.

19.7 Heterostructures

Laser-CVD with *parallel-beam* incidence has several advantages in low-temperature fabrication of amorphous multilayer structures. Among those are:

– High-resolution control of the film thickness. At moderate laser fluences less than one monolayer can be deposited per pulse.
– High deposition rates.
– Well-defined layer boundaries with minimal impurity or dopant diffusion.

19.7 Heterostructures

Table 19.6.3. Physical and chemical properties of Si_3N_4 films deposited by various techniques [*Emery* et al. 1984; *Petitjean* et al. 1992; *Schuegraf* 1983]

	LCVD	PCVD	EBCVD
T_s [°C]	50–400	380	50–400
p_{tot} [mbar]	0.4–3	3	0.5
$NH_3/SiH_4/N_2$	1/1/40	7/1/0	60/1/44
W_D [Å/s]	1–10	5	3
CVD parameters	$\lambda = 193$ nm	450 kHz	4.2 kV
	$\tau_l \approx 10$ ns		
	100 Hz		25 mA
	185 nm Hg lamp photosens.		
ρ [g/cm^3]	1.8–2.4	2.8	
Adhesion [10^8 dyne/cm^2]			
(1000 Å on Si)	> 6	> 6	> 5.5
Pinholes/cm^2			
(1000 Å on Si)	< 10–100	2–3	5–100
(2000 Å on Si)	< 1	< 1	–
n (632.8 nm)	1.80–2.4	2	1.85
Stress [dyne/cm^2]	4×10^9 compr.	4.7×10^9 compr.	
Breakdown voltage [MV/cm]	3–8	4	6
ρ [Ωcm] at 4 MV/cm	$10^{14} - 6 \times 10^{15}$	$10^{15} - 10^{16}$	$10^{12} - 10^{14}$
ε (1 MHz)	5.5–7.1	7	7.1
Stoichiometry	$Si_3N_{4.3}$	Si_3N_4	Si_3N_4
Hydrogen bonding			
SiH [%]	12	12–16	< 0.1
NH [%]	11	2–7	8–10
Etch rate [Å/s]			
in 5:1 buffered HF	20–100	1.7	3–20

These advantages have been demonstrated for heterostructures of a-Si/a-Ge and a-Si/a-Si_3N_4 [*Lowndes* et al. 1988]. The precursors (Si_2H_6, GeH_4, and mixtures of $Si_2H_6 + NH_3$) were photochemically decomposed by ArF-laser radiation. Figure 19.7.1 shows a laser–deposited heterostructure of a-Si/a-Ge. The respective layer thicknesses were 10.7 ± 0.4 nm and 5.4 ± 0.2 nm.

Similar experiments have been performed to fabricate heterostructures of a-Si:H/a-$Al_{1-x}O_x$ from Si_2H_6 diluted in H_2, and $Al_2(CH_3)_6 + O_2$ [*Uwasawa* et al. 1991]. The growth rates of a-Si:H and a-$Al_{1-x}O_x$ layers were around 0.3 Å/s and 0.8 Å/s, respectively ($\phi_\parallel$(ArF) ≈ 35 mJ/cm^2, $\tau_l \approx 17$ ns, 40 Hz), and almost independent of substrate temperature (200 °C $\leqslant T_s \leqslant$ 350 °C). The film properties were analyzed by SIMS, TEM, XPS, and optical spectroscopy. Good thickness uniformity and sharp interfaces have been obtained.

The low substrate temperatures that can be employed in photo-assisted CVD permit one to fabricate well-defined heterostructures of heat sensitive materials such as II-VI compounds. This has been demonstrated by means of Hg/Xe-arc-lamps for HgTe/CdTe [*Ahlgren* et al. 1988].

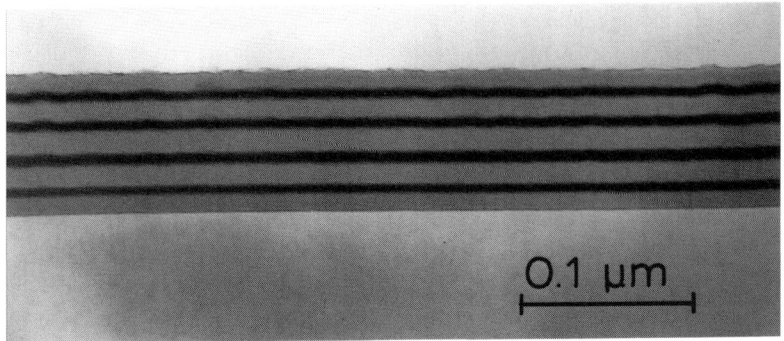

Fig. 19.7.1. TEM cross-section of a a-Si/a-Ge (dark) heterostructure deposited on (100) Si by ArF-laser photolysis at $T_s = 250\,°C$ [*Lowndes* et al. 1988]

The fabrication of heterostructures by pulsed-laser deposition (PLD) is described in Chap. 22.

19.8 Comparison of Laser-CVD and Standard Techniques

While standard CVD requires *uniform* substrate heating, typically, 400–1500 °C, LCVD can be performed at considerably lower temperatures and in some cases even at room temperature. Thus, LCVD permits deposition on temperature sensitive materials such as polymer foils, ceramics and compound semiconductors. These materials would melt, crack or decompose at the temperatures required for many conventional CVD systems. Even without such problems, lower temperatures can be desirable because they avoid or diminish material warpage, heat-induced mechanical stresses, diffusion or mixing between materials, and side reactions of the material surface with the ambient medium. An example where elimination or reduction of diffusion is a prior condition is the coating and patterning of prefabricated Si surfaces. Here, lower processing temperatures can increase the yields achieved in device fabrication, both in microelectronics and micromechanics. There are many materials that can be deposited with high quality only at relatively low substrate temperatures. Among these are compound semiconductors and hydrogenated amorphous Si. With other materials, the purity, morphology and crystallinity may deteriorate with decreasing film/substrate temperature and one has to strike a balance between the film quality and the heat sensitivity of the substrate. Another advantage of LCVD is the excellent step coverage. The process enables one to uniformly coat irregularly shaped substrates.

In plasma-CVD (PCVD), a plasma is used to generate reactive species. The processing temperatures for thin film deposition can thus be significantly lower than those used in standard CVD. However, plasma techniques have a number of other inherent properties which may be disadvantageous or even

19.8 Comparison of Laser-CVD and Standard Techniques

make the technique inadequate for a particular application. Among these are heavy ion bombardment, vacuum ultraviolet (VUV) irradiation of the substrate, loading effects, and contamination of the deposit by impurities originating directly from the reactant, or from carrier gases, or from sputtering of the reaction chamber. Further problems arise from the difficulty to control processing parameters. The RF power and frequency, the discharge geometry, electrode configuration, gas flow, total pressure, substrate temperature, etc., are all interrelated; thus, it is almost impossible to control and characterize effects due to *single* parameter variations. Above all, a stable discharge can be maintained only for a very narrow range of operating parameters and this limits the versatility of the technique even further. However, plasma-CVD is cheap, relatively simple, and permits large-area processing with high throughputs.

In laser-CVD, radiation damage, impurity sputtering, loading effects, etc., are absent or cause only minor problems. Laser-CVD also allows one to vary the laser power and wavelength, the spatial location, gas flow, total pressure, substrate temperatures, etc., *independently*, i.e., without affecting any one of the other parameters. Lasers permit selective processing. For example, it is conceivable that alternating layers of elements or compounds can be deposited from admixtures of gases by simply changing the laser wavelength and thereby the decomposition yield of particular species.

The deposition rates achieved in laser-CVD are comparable to, or even higher than, those achieved in PCVD, at least for areas up to about 10 cm^2. The maximum deposition rates can be determined by the available laser power or the mismatch of the laser wavelength and the maxima in the absorption cross section of reactant molecules, by transport limitations or gas-phase nucleation. The physical and chemical properties of laser-deposited films are already satisfactory in many respects.

Photochemical processing with lamps is often disadvantageous because large volume irradiation produces photofragments in regions which do not contribute to film growth. Consequently, reactants are not only lost, but they may also lead to unwanted reactions on reactor walls, etc. The reaction rates achieved with lamps are, in general, very low. For example, in the most efficient Hg-photosensitized reaction for the deposition of SiO_2, a deposition rate of 3 Å/s has so far been achieved.

Among the most serious disadvantages of laser-CVD is the low throughput that can be achieved in production lines (Sect. 1.2).

20 Adsorbed Layers, Laser–MBE

Adsorbed precursor molecules, reaction products, or impurities play an important role in various types of laser surface processing. They may control reaction rates, the spatial resolution in surface patterning, nucleation times in material deposition or synthesis, concentration profiles in surface doping, etc. Adsorbates are of importance also in new techniques that use a combination of laser *and* molecular/atomic beams.

Static and dynamic interactions between molecules/atoms and solid surfaces have been studied extensively. Most of the experimental investigations were performed under ultrahigh-vacuum (UHV) conditions at pressures below 10^{-10} mbar. Here, molecule–surface interactions are studied for low surface coverages or for particle beams and physically well-defined solid surfaces.

The number of molecules being adsorbed depends on their binding energy to the surface, their interactions between each other (e.g., dipole–dipole or direct Coulomb interactions), on the microstructure, roughness and temperature of the surface, and on the density of species within the ambient medium. The strength of these interactions also determines the extent of changes in the electronic and vibrational properties of adsorbed molecules with respect to free molecules. Adsorption may result in shifts and broadenings of electronic and vibrational energy levels and in the relaxation of selection rules for the interaction with light. As a consequence, the cross section for photoexcitation at a particular wavelength may differ significantly for adsorbed molecules and free gas-phase molecules. Contrary to investigations on basic molecule–surface interactions, LCP is, in general, performed only in high-vacuum (HV) reaction chambers that can be pumped out to, typically, 10^{-6} mbar. Therefore, substrate surfaces are covered by rest-gas molecules. Additionally, the substrates are, in most cases, only chemically cleaned according to standard procedures; such surfaces may be contaminated with water molecules, organic molecules, etc.

Laser light changes the adsorption behavior via substrate heating, and via selective electronic or vibrational excitations of gas-phase molecules, the solid surface, or the adsorbate–adsorbent complex. Due to interactions with light, adsorbed molecules may desorb from the surface, migrate across the surface, change the nature of bonding to the surface (e.g., from physical to chemical), diffuse into the bulk, or decompose at or react with the solid surface (Fig. 1.2.1a, b).

20.1 Fundamental Aspects

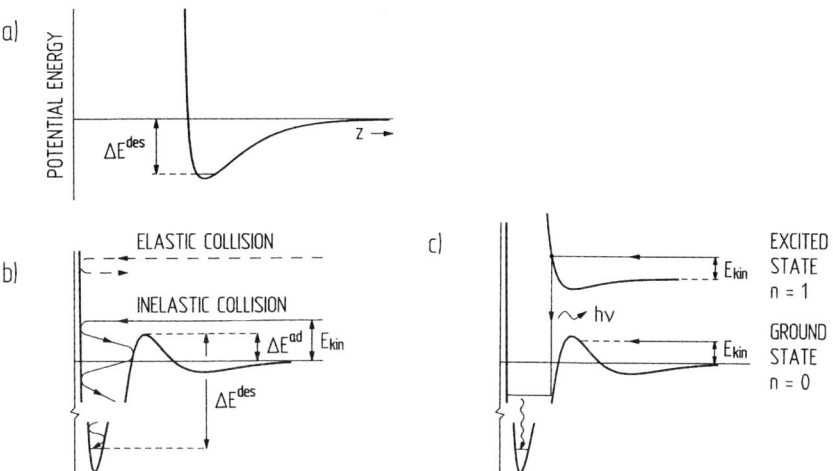

Fig. 20.1.1a–c. Molecule–surface interaction potentials for simple demonstration of adsorption processes. z is the distance from the surface. (**a**) Physical adsorption. (**b**) Activated chemical adsorption; the break in scale shall indicate that ΔE^{des} is much larger than in case (**a**). (**c**) Chemical adsorption via electronically excited molecule. Energy dissipation by emission of a photon is only one of various possible mechanisms

The various systems investigated are listed in Appendix B.6. This also contains materials that have been deposited by laser-MBE or laser-ALE.

20.1 Fundamental Aspects

Molecules on solid surfaces may be adsorbed either physically or chemically. *Physical adsorption* (physisorption) is, in general, a non-activated process with binding energies of, typically, 0.05–0.5 eV (1–10 kcal/mol). The adsorption process is reversible, i.e., the adsorbate can be removed without any chemical changes by increasing the temperature or by lowering the (number) density of species within the ambient medium.

Chemisorption is, in general, an activated process. The binding energies are similar to those of intramolecular bonds, typically, 0.5 to 5 eV (10–100 kcal/mol). Thus, chemisorbed species cannot be desorbed from the surface without chemical changes.

If we assume a surface coverage of at most one monolayer, only *one* type of species, AB, and first order kinetics, the density of adsorbed species N_{AB}^{ad} (molecules per unit area), can be described by

$$\frac{\partial N_{AB}^{ad}}{\partial t} = k^{ad}\left(1 - \frac{N_{AB}^{ad}}{N_{as}}\right)N_{AB} - k^{des} N_{AB}^{ad} - W + \nabla_2\left(D_{AB}^{ad} \nabla_2 N_{AB}^{ad}\right) - J_s, \quad (20.1.1)$$

with

$$J_s = -D_{AB} \frac{\partial N^s_{AB}}{\partial z}\bigg|_{z=0}$$

where N^s_{AB} [species/cm^3] is the number density of species within the substrate ($z > 0$). In the most general case, the various quantities depend on surface coordinates and time, for instance, $N^{ad}_{AB} \equiv N^{ad}_{AB}(x_s, t)$ and, if relevant, on laser parameters. The first three terms on the right-hand side describe changes in N^{ad}_{AB} due to adsorption, desorption, and surface reactions. Note that N_{AB} (first term) is the density of molecules within the gas. The fourth term describes (ordinary) diffusion of species along the surface, and J_s the flux of species into the bulk substrate. Thermal diffusion has been ignored. N^{ad}_{AB} must be calculated self-consistently together with the 3D-diffusion equation for N^s_{AB}, and the boundary conditions $N^s_{AB}(z=0) = N^{ad}_{AB}/h_l$ and $N^s_{AB}(z \to \infty) = 0$. h_l is the thickness of the adlayer. Single terms will now be discussed in further detail.

According to the Langmuir theory, adsorption occurs at active adsorption sites whose density N_{as} is, in general, smaller than the total atom density of the substrate surface, N_s. The number of species adsorbed is

$$N^{ad}_{AB} \leq N_{as} \leq N_s . \tag{20.1.2}$$

The limiting case $N^{ad}_{AB} = N_{as}$ characterizes a monolayer coverage. The rate constant k^{ad} can often be described by the Arrhenius law

$$k^{ad}(T) = k^{ad}_0 \exp\left(-\frac{\Delta E^{ad}}{k_B T_G}\right), \tag{20.1.3}$$

where ΔE^{ad} is the activation energy for adsorption. T_G is the temperature of gas-phase molecules, which shall be equal to the substrate temperature, $T_G = T_s$. For physical adsorption we can set $\Delta E^{ad} \approx 0$. A typical potential energy curve for this case is shown in Fig. 20.1.1a. It is often described by a Lennard–Jones potential. *Chemical* adsorption is described by the potential energy curve (b). It can be constructed by assuming, for example, a Lennard–Jones potential for physisorbed species and, for example, a Morse potential for chemisorbed species. In the case of *molecular* adsorption, the intersection of potential curves occurs at energies $E \leq 0$. *Dissociative* adsorption is activated with $\Delta E^{ad} > 0$ (it results from an energy separation between the two potentials for $z \to \infty$ which corresponds to the dissociation energy of the gas-phase species). ΔE^{ad} can be as high as several tenths of an eV [for GaAs we find $\Delta E^{ad}(N_2O) \approx 0.15$ eV and $\Delta E^{ad}(O_2) \approx 0.47$ eV]. If we describe adsorption by "active collisions", the kinetic energy of species impinging onto the surface with velocity component v_z, must exceed the activation energy, i.e., $E_{kin} = m v_z^2 / 2 > \Delta E^{ad}$. For a Maxwell distribution we can write

$$k^{ad}_0 \approx \frac{1}{4} \langle v_{AB} \rangle , \tag{20.1.4}$$

20.1 Fundamental Aspects

where $\langle v_{AB} \rangle = (8k_B T_G / \pi m_{AB})^{1/2}$ is the arithmetic mean velocity of gas molecules. However, even if $E_{kin} > \Delta E^{ad}$, species may be reflected from the surface. This process is indicated in (b) by the broken trajectory. For adsorption to take place, collisions between incoming species and the surface must be inelastic (full trajectory). Among the mechanisms for inelastic molecule–surface interactions are the transfer of kinetic energy (excitation of phonons), the transformation of translational energy into vibrational/rotational energy of the molecule (excitation of internal degrees of freedom), etc. If the time of energy conversion is long compared to molecule–surface interactions, the adsorption rate is small.

The second term in (20.1.1) describes desorption of molecules from the surface. The rate constant for *thermally* activated desorption can be described by

$$k^{des}(T) = k_0^{des} \exp\left(-\frac{\Delta E^{des}}{k_B T_s}\right). \tag{20.1.5}$$

ΔE^{des} is the activation energy for desorption. k_0^{des} is sometimes termed the attempt frequency with which adsorbed species try to escape from the surface. This frequency is of the order of molecular vibrational frequencies, typically some 10^{11}–10^{14}/s. Because of anharmonicity, k_0^{des} depends on temperature and can be described, approximately, by $k_0^{des} \propto T_s$. In the presence of an atmosphere, k_0^{des} may become dependent on gas pressure, mainly due to "sputtering" of adspecies via impinging molecules. From (20.1.3 and 5) it follows that the enthalpy of adsorption

$$\Delta H^{ad} = \Delta E^{ad} - \Delta E^{des} \tag{20.1.6}$$

is negative if $\Delta E^{des} > \Delta E^{ad}$ (exothermal process) and positive if $\Delta E^{des} < \Delta E^{ad}$ (endothermal process). The latter case requires an interaction potential with a metastable state. Parameters that characterize adsorption/desorption processes can be found for various systems, e.g., in [*Kreuzer* and *Gortel* 1986].

The third term in (20.1.1) describes the change in the density of adsorbed species due to surface reactions. If the reaction of adsorbed molecules AB is of first order

$$W = k N^{ad}_{AB}. \tag{20.1.7}$$

The fourth term in (20.1.1) describes ordinary diffusion of adsorbed molecules along the surface with

$$D^{ad}_{AB}(T_s) = D^{ad}_0 \exp\left(-\frac{\Delta E^{ad}_d}{k_B T_s}\right) \tag{20.1.8}$$

and $T_s \equiv T_s(x_s, t)$ [for Xe on (110) W, $D^{ad}_{AB}(T_s = 400 \text{ K}) \approx 10^{-7}$ cm^2/s and $\Delta E^{ad}_d \approx 0.05$ eV $\approx \Delta E^{des}/4$]. The surface diffusion length can be estimated from

$$l^{ad}_{AB} \approx 2\left(D^{ad}_{AB}\tau\right)^{1/2}, \tag{20.1.9}$$

where τ is the average lifetime of species on the surface which, in general, is mainly determined by the desorption process. Because $\Delta E_d^{ad} < \Delta E^{des}$, l_{AB}^{ad} decreases with increasing temperature, T_s.

The last term in (20.1.1) describes ordinary diffusion of adsorbed species *into* the bulk substrate with

$$D_{AB}(T_s) = D_0 \exp\left(-\frac{\Delta E_d}{k_B T_s}\right) \tag{20.1.10}$$

and $T_s \equiv T_s(z,t)$. ΔE_d is the activation energy for (bulk) diffusion (for B in solid Si, for example, $\Delta E_d = 3.69$ eV). This term is important in surface doping from adsorbed layers, and in surface oxidation, mainly of semiconductors.

The surface density of adsorbed molecules is often written as

$$N_{AB}^{ad}(x_s, t) = \Theta_{AB}(x_s, t) N_{as}, \tag{20.1.11}$$

where Θ_{AB} is the coverage. The number density within *one* monolayer can be estimated, if we assume that the mass density of the adsorbed film is equal to that of a liquid, i.e.,

$$N_{AB}^{ad}(\Theta_{AB} = 1) = N_{as} \approx \left(\frac{\rho}{m_{AB}}\right)^{2/3} \approx 10^{14} - 10^{15} [\text{species/cm}^2], \tag{20.1.12}$$

where ρ is the mass density of liquid AB, and m_{AB} the mass of molecules.

The preceding discussion simplifies the real situation considerably. In reality

- The assumption of first-order kinetics does not hold in many cases.
- Activation energies often change with coverage. The simplest ansatz is $E^{ad}(\Theta) = E^{ad}(\Theta = 0) + \zeta\Theta$.
- Adsorption processes strongly depend on the distribution function of gas-phase species. This holds also for desorption, mainly due to "sputtering". For this reason, molecular/atomic beams are frequently employed for adlayer formation or surface cleaning.
- Surface defects may play an important or even dominating role in adsorption/desorption processes.

Influence of laser light

Laser light may thermally or non-thermally excite the solid surface, the incoming molecules, and the adsorbate–adsorbent system. The influence of laser-induced heating on the adsorption process is quite clear from the preceding equations. Additionally, laser light can increase or decrease the number of active adsorption sites, for example by the generation of vacancies, electron–hole pairs, etc.

Selective electronic excitations of molecules change the molecule–surface interaction potential. For the example depicted in Fig. 20.1.1c the kinetic energy of incoming species in the electronic ground state is smaller than the activation energy for chemical adsorption, $E_{kin} < \Delta E^{ad}(n=0)$. Thus, ground-state molecules will mainly be physisorbed (trapped by the shallow minimum in the potential curve). For electronically excited molecules ($n = 1$) there is no such barrier and they can directly chemisorb via transitions $n = 1$ to $n = 0$. Among the various mechanisms for energy conversion is the emission of photons. This process is extremely fast and one of the simplest to explain the increase in sticking coefficients frequently observed with excited species.

Changes in the adsorption behavior originating from *vibrational* excitations of molecules can be based on various mechanisms [*Benedek* 1987]. For example, vibrational excitation can change or only generate a dipole moment of a molecule which, in turn, can directly interact with the solid surface via electrostatic or dynamic forces. A strong increase in sticking coefficient has been observed with vibrationally excited SF_6^* on semiconductor surfaces (Sect. 15.3) and with BCl_3^* on metals.

Laser light can directly excite adsorbed species (intramolecular bonds or adsorptive bonds), cause charge-transfer reactions within adsorbate–adsorbent complexes, enhance diffusion of adspecies along the surface and into the bulk of the substrate, etc.

In total, laser light increases or decreases adsorption/desorption rates, causes an inhomogeneous distribution of adspecies on the surface and, eventually, thermal or non-thermal decomposition of species.

The situation is, however, even more complex. For a particular species, the excitation probability is determined not only by the type of substrate material, but also by its surface properties, e.g., its microscopic structure (amorphous, crystalline), its roughness, degree of contamination, oxidation, etc. For example, the optical excitation cross section may strongly increase when a molecular resonance overlaps with optically active surface resonances such as surface plasmons [*Nitzan* and *Brus* 1981]. The influence of the surface morphology on light–molecule–surface interactions has been clearly demonstrated, e.g., in surface enhanced Raman scattering (SERS).

20.2 Deposition from Adsorbed Layers

Physisorbed or chemisorbed molecules strongly influence laser-induced reaction rates on substrate surfaces, in particular in cases where the light is neither absorbed by the substrate nor by the gaseous or liquid ambient medium. Among the effects that result in rate changes are:

– The high concentration of molecules within adsorbed layers.
– Spectral shifts which change absorption cross sections.
– Electric field enhancements at the solid surface.

Photodecomposition of adlayers has been investigated in experiments performed in vacuum (Fig. 1.2.1a) and in atmospheres consisting of the gaseous form of the adspecies (Fig. 1.2.1b).

Some systems for which the role of adsorbed layers has been investigated, in particular with respect to material deposition, are listed in Appendix B.6. It should be emphasized, however, that adsorbed layers may play an important or even decisive role in nucleation (Chap. 4) and in many of the other systems employed in laser-CVD (Chaps. 16–19). The extent to which adlayers contribute to laser-induced material deposition can be discovered most unambiguously by investigating deposition rates as a function of surface temperature and reactant-gas pressure; the dependence on laser-beam spot size does *not* permit unique conclusions (Chap. 14).

From a practical point of view, adsorbed-layer processing permits high-resolution patterning because it avoids pattern smearing by diffusion of gas-phase species. Photodissociation of adsorbed layers can be used to *prenucleate* condensation areas that are filled up later by large-area standard CVD, pyrolytic LCVD, etc. Such combined techniques permit maskless formation of microstructures and higher overall throughputs in production lines.

20.2.1 Vacuum

Laser processing with adsorbates in vacuum is possible only if the molecules are strongly physisorbed or chemisorbed on the substrate surface. In such investigations the substrate is first cleaned, for example by a single or a few UV laser pulses, and then exposed to the gas AB. During exposure, an adsorbate is formed. After exposure, the gas AB is pumped off. Even when multilayers have been formed, only the *first* layer which is most strongly bound to the surface remains, in general. Thus, besides a few exceptions, the surface coverage in a vacuum is $\Theta_{AB} \leq 1$.

Langmuir equation

Let us first consider the exposure cycle and a system where the equilibrium coverage (in the presence of the gas AB) is $\Theta_{AB} \leq 1$. If we consider only adsorption and desorption processes in (20.1.1), the change in coverage is

$$\frac{d\Theta_{AB}}{dt} = k^{ad}\left(1 - \Theta_{AB}\right)\frac{N_{AB}}{N_{as}} - k\Theta_{AB} = s\frac{J_{AB}}{N_{as}}\left(1 - \Theta_{AB}\right) - k\Theta_{AB}, \quad (20.2.1)$$

with $k \equiv k^{des}$. J_{AB} is the flux of molecules AB onto the surface. The sticking coefficient, s, is assumed to be a constant. In reality, s depends on Θ. The fraction $(1 - s)$ of impinging molecules is directly reflected from the surface. In

20.2 Deposition from Adsorbed Layers

the simplest kinetic theory, the flux of impinging molecules is

$$J_{AB} = \frac{1}{4}\langle v_{AB}\rangle N_{AB} = \frac{p_{AB}}{(2\pi m_{AB}k_B T_G)^{1/2}}, \quad (20.2.2)$$

where $p_{AB} = N_{AB}k_B T_G$. The steady-state coverage obtained from (20.2.1) yields the *Langmuir isotherm*

$$\Theta_L \equiv \Theta_{AB}(t \to \infty) = \frac{1}{1 + \Gamma_{AB}^{ad}} = \frac{bp_{AB}}{1 + bp_{AB}}, \quad (20.2.3)$$

with $\Gamma_{AB}^{ad} = \tau_{AB}^{ad}/\tau_{AB}^{des}$. Here, $\tau_{AB}^{ad} = N_{as}/k^{ad}N_{AB} = N_{as}/sJ_{AB}$ and $\tau_{AB}^{des} = 1/k^{des}$ are the time constants for adsorption and desorption, respectively. The adsorption coefficient is given by

$$b = \frac{s\tau_{AB}^{des}}{N_{as}\left(2\pi m_{AB}k_B T_s\right)^{1/2}} = \frac{\exp\left(-\frac{\Delta H^{ad}}{k_B T_s}\right)}{k_0^{des}N_{as}\left(2\pi m_{AB}k_B T_s\right)^{1/2}}, \quad (20.2.4)$$

where $T_s = T_G$. $1/b$ is sometimes denoted as adsorption pressure which, at room temperature, has typical values of the order of 10^{-6} bar. We obtain $\Theta_L = bp_{AB}$ if $bp_{AB} \ll 1$ and $\Theta_L = 1$ if $bp_{AB} \gg 1$. With (20.2.1) and the initial condition $\Theta_{AB}(t=0) = 0$ the coverage increases as

$$\Theta_{AB}(t) = \Theta_L\left[1 - \exp\left(-\frac{t}{\tau_{AB}}\right)\right] \approx 1 - \exp\left(-\frac{t}{\tau_{AB}^{ad}}\right). \quad (20.2.5)$$

$\tau_{AB} = \tau_{AB}^{ad}\Theta_L$ is the characteristic time for surface coverage which decreases with increasing temperature. The latter approximation in (20.2.5) assumes $\tau_{AB}^{ad} \ll \tau_{AB}^{des}$ and thus $\Theta_L \approx 1$.

Desorption, laser irradiation

In the next step of the (idealized) experiment, the gas pressure is reduced to zero, i.e., $p_{AB} \propto N_{AB} = 0$, and the laser is switched on. For simplicity, we assume that this takes place at time $t = 0$. The coverage will now decrease due to both desorption *and* laser-induced decomposition of species. If we assume first order kinetics, the change in Θ is still given by (20.2.1) with $J_{AB} = 0$ and $k = k^{des} + k^{dec}$ where k^{dec} is still the rate constant for laser-induced adlayer decomposition. The coverage then decreases as

$$\Theta_{AB}(t) = \Theta_{AB}(0)\exp(-kt), \quad (20.2.6)$$

where the initial value is $\Theta_{AB}(0) = \Theta_L$. If, during exposure, a multilayer coverage is formed and reduced to a monolayer due to pumping, we can still employ this equation, but with $\Theta_{AB}(0) = 1$.

Let us now consider the reaction rate for adlayer decomposition. For a focused laser beam with intensity $I(x_s, t)$ the coverage, and thereby the rate, becomes dependent on surface coordinate

$$W(x_s, t) \approx k^{dec}(T_s) N_{AB}^{ad}(x_s, t) = k^{dec} N_{as} \Theta_{AB}(x_s, t) \ . \tag{20.2.7}$$

For *pyrolytic* activation k^{dec} is given by (3.1.2). For single-photon *photolytic* activation, k^{dec} is given by (3.2.4a) where σ_{AB}^{ad} is the dissociation cross section of adspecies. In many systems, both thermal and non-thermal contributions will determine the processing rate.

Qualitatively, the kinetics for adlayer decomposition under the action of *focused* laser light can be described as follows: Initially, the laser light simply decomposes the adlayer. If molecules AB can diffuse into the sink produced by photodecomposition, a localized surface reaction can be sustained. This reaction is initially *kinetically* controlled. At later times, molecules must diffuse over larger distances and the reaction becomes *transport limited* and, finally, tends to zero [*Freeman* and *Doll* 1983; *Zeiger* et al. 1989].

Experimental examples

In the experiments described in this paragraph, the number of molecules adsorbed on the surface is, in general, of the order of one monolayer, i.e., 10^{14}–10^{15} molecules/cm^2. Therefore, the amount of deposited, etched, or doped material is very small, and even further decreases with increasing temperature.

Laser-induced adlayer photolysis has been investigated mainly for $Ni(CO)_4$, $Cd(CH_3)_2$, and $Al_2(CH_3)_6$ adsorbed on glass, SiO_2, and Al_2O_3 substrates. For example, for the deposition of nickel, the glass or quartz substrate was first exposed to a $Ni(CO)_4$ atmosphere of some millibars for several minutes. Then, after several cycles of pumping and purging with He or H_2, the substrate was irradiated with 476 nm or 356 nm Kr^+-laser light. As expected from the optical absorption spectra of $Ni(CO)_4$ the latent time for photodecomposition was much shorter for 356 nm than for 476 nm radiation. Experiments on adsorbed $Cd(CH_3)_2$ and $Al_2(CH_3)_6$ were mainly performed with 193 nm ArF-, 257 nm frequency-doubled Ar^+-, and 356 nm Kr^+-laser radiation.

The metal deposits are less than a monolayer thick. They can be used as nucleation centers for subsequent film growth by standard techniques.

20.2.2 Gaseous Ambient

The role of adsorbed layers is quite different in gas-phase LCP which is usually performed at reactant (partial) pressures ranging from 10^{-4} mbar up to more than 1 bar. In this case, the adsorbed layer is in a *dynamic equilibrium*

20.2 Deposition from Adsorbed Layers

with the surrounding atmosphere and its thickness is determined by the pressure p_{AB}. Laser radiation may interact with the adsorbate–adsorbent system and with gas-phase species and thereby increase or decrease the sticking coefficient of species and thus the thickness of the adlayer (note that s may also change due to laser-induced material deposition, etching, etc; Sect. 4.2.1).

Low coverages

At *low* pressures p_{AB} where $\Theta_{AB} \leqslant 1$ the change in coverage for a first order reaction can be calculated from (20.2.1) with $k = k^{des} + k^{dec}$. The steady-state reaction rate is

$$W = k^{dec} N_{as} \frac{\Theta'_L}{1 + \Gamma \Theta'_L}. \tag{20.2.8}$$

Due to the modified sticking coefficient, $s = s(I)$, Θ'_L can differ from Θ_L in (20.2.3). $\Gamma = k^{dec} N_{as}/k^{ad} N_{AB} = k^{dec} N_{as}/s J_{AB}$. If $\Gamma \Theta'_L \ll 1$, the rate is

$$W = k^{dec} N_{as} \Theta'_L. \tag{20.2.9a}$$

In this regime W changes *nonlinearly* with pressure p_{AB} as in (20.2.3). If $\Gamma \Theta'_L \gg 1$

$$W = k^{dec} N_{as} \frac{1}{\Gamma} \propto p_{AB} \tag{20.2.9b}$$

and the rate becomes *linearly* dependent on p_{AB}.

Multilayer coverages

At *high* pressures p_{AB}, multilayer molecular films may be formed. The upper layers are weakly bound by van der Waals forces to the more strongly bound first monolayers. The total number of adsorbed molecules depends on the strength of bonding, the temperature, and the gas pressure. Figure 20.2.1 shows isotherms for adsorbed $Cd(CH_3)_2$ as a function of pressure. The low-pressure knee occurs at a coverage of about one monolayer. The curves measured at 34 and 43 °C almost saturate at this coverage. Only the 25 °C curve shows two further steps related to the second and third monolayer. When $p_{AB} \equiv p[Cd(CH_3)_2]$ approaches the vapor pressure, p_v, the coverage can become very high, 10 monolayers or more. This can be seen from the strong decrease in QCM frequency.

A good fit to the isotherms of many physisorbed systems is provided by the BET theory (*Brunauer, Emmett,* and *Teller* 1938)

$$\Theta_{BET} \equiv \Theta(p_{AB}, T) = \frac{C p_{AB}}{(p_v - p_{AB})[1 + (C-1)p_{AB}/p_v]} \approx \frac{C p_{AB}}{p_v} \tag{20.2.10}$$

where Θ_{BET} describes the amount of adsorbed species with respect to one monolayer, and may therefore become *larger* than unity. The latter

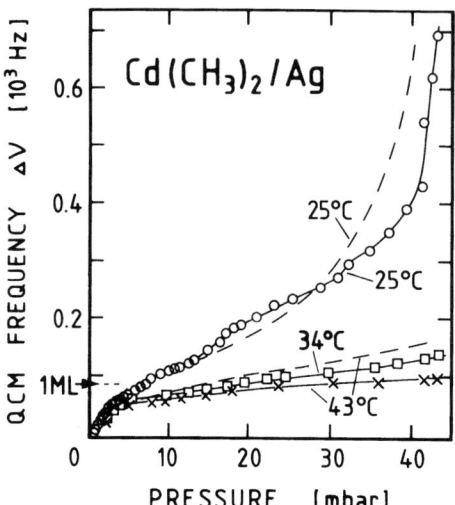

Fig. 20.2.1. Isotherms of $Cd(CH_3)_2$ adsorbed on an Ag film evaporated on a quartz-crystal microbalance (QCM) as a function of $Cd(CH_3)_2$ gas pressure. The dashed curves have been calculated from the BET theory ($E_{ms} \approx 0.44$ eV, $E_{mm} \approx 0.37$ eV) [Ehrlich et al. 1982]

approximation refers to systems with $(C-1)p_{AB}/p_v \ll 1$. The coefficient C characterizes the particular adsorbate–adsorbent system, and its temperature dependence is approximately given by $\exp[(E_{ms} - E_{mm})/k_B T]$; E_{ms} and E_{mm} are molecule–surface and molecule–molecule interaction energies, respectively. The total (equilibrium) number density of adsorbed species is

$$N_{AB}^{ad}(p_{AB}, T) = \Theta_{BET} N_{as}, \qquad (20.2.11)$$

The surface coverage Θ_{BET} varies *nonlinearly* with pressure. This can be seen from the dashed curves included in Fig. 20.2.1. The theoretical curve reproduces the first knee in the isotherms which is due to the strongly bound first monolayer. The additional steps observed in the 25° C curve at integral multiples of a monolayer are not reproduced; this is expected because the BET theory assumes a common binding energy for all the upper layers. The BET theory becomes inaccurate with pressures $p_{AB}/p_v > 0.5$. Θ_{BET} depends sensitively on temperature, primarily via the Arrhenius factor in p_v. The change in Θ_{AB} with even slight changes in temperature is a consequence of the (weak) van der Waals binding of the upper layers. This can be seen from the isotherms measured at higher temperatures (Fig. 20.2.1). Only the first strongly bound layer remains. In fact, this layer could not be removed even after 1 hour of pumping at room temperature. Similar results have been obtained with $Al_2(CH_3)_6$ and with Au and oxidized Si-layer substrates.

If we ignore the role of reaction products and surface diffusion of species, the coverage in the *presence* of *light* can be described by

$$\frac{d\Theta_{AB}}{dt} = \frac{1}{\tau'_{AB}}(\Theta'_{BET} - \Theta_{AB}) - k^{dec}\Theta_{AB}. \qquad (20.2.12)$$

Note that all quantities refer to *multilayer* molecular films. τ'_{AB} is the (pressure-dependent) relaxation time in the BET theory. Θ'_{BET} takes into account the change in coverage caused by the change in sticking coefficient due to the laser light [in (20.2.10) s is included in the coefficient C]. The steady-state coverage becomes

$$\Theta_{AB}(\infty) = \frac{\Theta'_{BET}}{1 + \Gamma'} \tag{20.2.13}$$

where $\Gamma' = \tau'_{AB} k^{dec}$. The reaction rate is then

$$W = k^{dec} N_{as} \Theta_{AB}(\infty). \tag{20.2.14}$$

Experimental examples, separation of mechanisms

Adsorbates that are in a dynamic equilibrium with the corresponding gas-phase molecules contain a high density of relevant species. Their influence on reaction rates in laser-chemical processing may thereby become important or even dominating. Decomposed adlayer molecules are continuously replenished by gas-phase molecules and this permits rapid growth of films that are thick compared to a monolayer.

The variation of the substrate temperature is the most simple and transparent way to investigate the relative importance of adsorbed-phase, gas-phase, and solid-surface excitations. Figure 20.2.2 displays the thickness of Zn lines deposited from $Zn(C_2H_5)_2$ by means of 257 nm Ar^+-laser radition. The deposition rate *decreases* almost exponentially with temperature – between 20 °C and 60 °C by about a factor of 100. This reflects the strong decrease in coverage with temperature – mainly via the increase in p_v. This is a clear indication of adsorbed-layer-controlled deposition. For higher substrate temperatures, gas-phase deposition becomes dominating. The situation is, however, more complicated. A change in surface temperature does not only alter the surface coverage but also the surface mobility of species, recombination efficiencies of photoproducts, etc. The interrelation between such mechanisms may explain why the photodeposition rate of Al observed with $Al_2(CH_3)_6$ on quartz substrates first increases and then decreases with increasing substrate temperatures [*Tsao* and *Ehrlich* 1984].

The relative importance of adsorbed-phase and gas-phase photodecomposition can also be studied via the dependence of the reaction rate on pressure p_{AB}. In *gas-phase* processing the reaction rate is $W \propto p_{AB}^\gamma$ where, for first order kinetics, $\gamma \approx 1$. For *adsorbed-layer* photolysis we have $W \propto \Theta_{AB}$. Furthermore, the reaction zone in *photolytic* gas-phase processing is, in general, much wider than the laser-beam spot size. In adsorbed-phase photolysis the profile of the deposit is similar to the intensity distribution of the laser beam. Further insight into excitation mechanisms can often be obtained also from the different wavelength dependences of gas-, adsorbed-, and solid-phase excitations (Sect. 2.1).

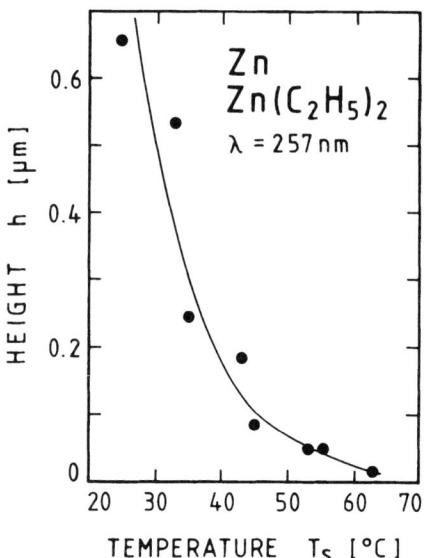

Fig. 20.2.2. Height of Zn lines photodeposited from $Zn(C_2H_5)_2$ on glass by means of 257 nm Ar^+-laser radiation ($v_s = 1.4\,\mu m/s$). The strong dependence on substrate temperature, T_s, is a clear indication of adsorbed-layer photolysis [*Krchnavek* et al. 1987]

Fig. 20.2.3. SEM picture of an Al pattern produced by photolysis of adsorbed $Al(C_4H_9)_3$ by KrF-laser light projection and subsequent standard CVD [*Higashi* 1989]

A clear demonstration of the application of adsorbed-layer photolysis is given in Fig. 20.2.3: Excimer-laser-light projection has been used to photolyze adlayers of $Al(C_4H_9)_3$ on substrate materials such as Al_2O_3, SiO_2, GaAs, and Si. The reactive Al sites produced in this way act as nucleation centers for (spatially) selective growth by standard CVD. The figure demonstrates that 4 μm holes are accurately reproduced at pressures where the mean free path of (gas-phase) species is $\lambda_m \approx 100\,\mu m$. Features of this type would be difficult to reproduce if gas-phase photoreactions would dominate. In adsorbed-layer processing, the resolution of features is controlled mainly by the laser wavelength and the imaging optics. With ArF-laser radiation and optimized imaging, it should be possible to improve the resolution achieved in Fig. 20.2.3 by more than a factor of 20. The electrical resistivity of features exceeds that of bulk Al by only a factor of two, although the deposits were very rough.

Adsorbed-layer photolysis often plays an important role in the initiation of the deposition process (Chap. 4). A few nuclei or monolayers of photodeposited material may already absorb enough light to cause significant laser-induced heating; film growth then continues thermally.

The influence of local electric field enhancements on adsorbed-layer photodecomposition yields has been studied for $Cd(CH_3)_2$ adlayers in dynamic equilibrium with gaseous $Cd(CH_3)_2$ and Ar [*Chen* and *Osgood* 1983]. The substrate was a C film with predeposited Cd spheres. Cd possesses a plasmon resonance which can be excited with 257 nm laser light. The observed growth kinetics is modified by the plasmon-excitation and in agreement with model calculations.

20.3 Combined Laser and Molecular/Atomic Beams

There is considerable interest in the growth of epitaxial layers, mainly of semiconductors, by molecular beam epitaxy (MBE) and atomic layer epitaxy (ALE). Lasers have been incorporated in these techniques to enhance surface diffusion and photodecomposition of adsorbed species, to replace standard effusion cells, to generate species of variable kinetic energies for surface doping, etc.

20.3.1 Laser-MBE

The term laser-assisted MBE (in short, laser-MBE) shall henceforth be used to describe *all* cases of MBE where lasers have been incorporated in the deposition process.

Standard MBE with substrate illumination

Laser-assisted growth of p-type CdTe:Sb films has been reported by *Bicknell* et al. (1986). Here, a standard MBE apparatus was used together with an Ar^+-laser for substrate illumination (Fig. 20.3.1). Films grown under similar conditions in the absence of light were insulating. The same technique has been used to grow n-type layers of $Cd_{1-x}Mn_xTe$.

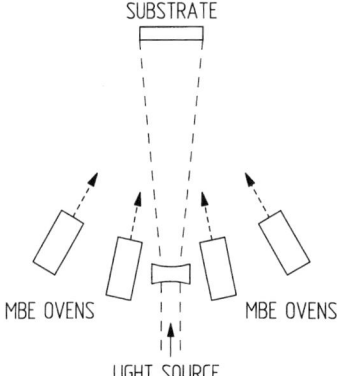

Fig. 20.3.1. Apparatus for film growth by light-assisted molecular beam epitaxy (MBE). Single material components are provided from MBE ovens (effusion cells)

Laser-assisted gas-source MBE

In gas-source MBE (also denoted as chemical-beam epitaxy, CBE) a molecular beam is used instead of a standard effusion cell.

Laser-assisted MOMBE (metalorganic-MBE) of GaAs using $Ga(C_2H_5)_3$ and elemental As_2 and As_y [formed by standard pyrolysis from $As(CH_3)_3$ or $As(C_2H_5)_3$] has been demonstrated by *Donnelly* et al. (1988). With ArF-laser radiation and substrate temperatures $T_s < 450\,°C$ very smooth films can be grown selectively (Fig. 20.3.2). The measured deposition rates cannot be explained on the basis of gas-phase photodecomposition, mainly because of the low (gas-phase) density of precursor molecules ($\approx 10^{-6}$ mbar) and the low laser pulse repetition rate (20 Hz). The observations can be understood, however, by adlayer decomposition of $Ga(C_2H_5)_3$, if desorption of species between laser pulses is ignored. Decomposition of the $Ga(C_2H_5)_3$ adlayer can be based on direct (non-

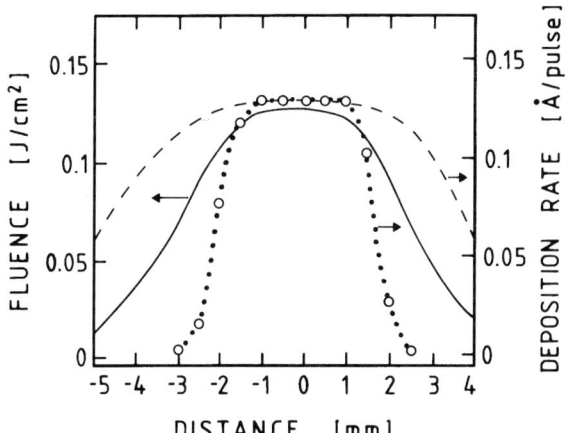

Fig. 20.3.2. ArF-laser-induced deposition rate of GaAs as a function of distance from the beam center (data and dotted curve; $T_s = 400\,°C$). The full curve represents the laser-beam intensity ($\phi_{max} \approx 0.12\,J/cm^2$, $\tau_l = 15\,ns$, 20 Hz). The dashed curve is the deposition rate expected for adsorbed-layer photolysis [*Donnelly* et al. 1988]

thermal) photodissociation of molecules, on interactions with electron–hole pairs generated within the GaAs surface, and on transient heating. The comparison of the measured film profile (dotted curve) with the laser-light intensity (full curve) suggests a strongly nonlinear decomposition mechanism. Both adsorbed-layer photolysis (dashed curve) and charge transfer mechanisms cannot explain the rapid drop-off in deposition rate. Therefore, laser-enhanced growth has been ascribed to transient heating and thermal decomposition of *adsorbed* $Ga(C_2H_5)_3$. The analysis of data on the basis of calculated temperature distributions yields reasonable agreement with the measured deposition profile ($\Delta E[Ga(C_2H_5)_3 \approx 18\text{--}25$ kcal/mol). This interpretation is supported by XPS studies which show that adsorbed $Ga(C_2H_5)_3$ is not efficiently photolyzed at 193 nm [*McCaulley* et al. 1989]. By means of excimer-laser-light projection, selective-area growth of GaAs has been demonstrated [*Donnelly* and *McCaulley* 1989].

MBE using laser-induced material ablation

In another type of laser-assisted MBE, one or several standard MBE ovens (effusion cells) are replaced by solid or liquid sources (targets) from which the material is ablated under the action of laser light (Chaps. 11–13). The technique has been employed for the fabrication of thin films and for band-gap engineering (Sect. 22.3).

20.3.2 Laser-ALE

Atomic-layer epitaxy (ALE) and atomic-layer etching require self-limiting surface processes. Here, the binding energy of the first monolayer adsorbed on the substrate should considerably exceed that of subsequent layers, if existent. This can be controlled by proper selection of the substrate temperature or by laser-light irradiation. Light can enhance the decomposition of these strongly adsorbed species or it can desorb reaction products.

Laser-assisted ALE (laser-ALE) has been demonstrated for semiconductors, and in particular for GaAs [*Isshiki* et al. 1993]. A typical setup is shown in Fig. 20.3.3. The precursors, AsH_3 and $Ga(C_2H_5)_3$ or $Ga(CH_3)_3$ are introduced into the reaction chamber in alternating "pulses". The latter were synchronized with the laser pulses, as shown in b). Ideal ALE was observed with 355 nm or 515 nm pulsed Ar^+-laser radiation but not with 1.064 μm Nd:YAG-laser radiation [*Meguro* et al. 1988a]. This wavelength dependence was ascribed to the photolysis of adsorbed $Ga(C_2H_5)_3$. Photodecomposition seems to be self-limiting. If $Ga(C_2H_5)_3$ is adsorbed on the As layer of the GaAs surface, it is decomposed by Ar^+-laser radiation. If, however, $Ga(C_2H_5)_3$ is adsorbed on a Ga layer, no significant decomposition takes place.

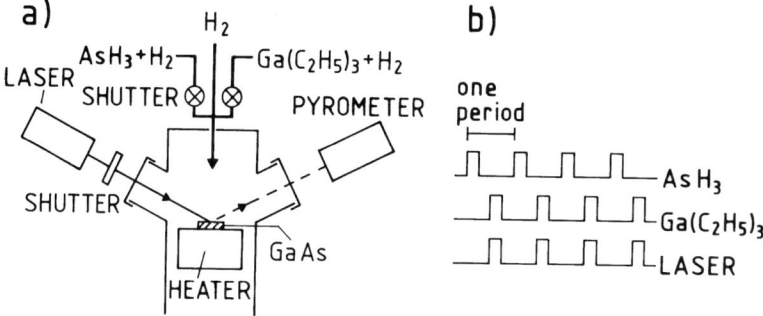

Fig. 20.3.3. (a) Setup employed in laser-ALE. (b) Typical time sequences of gas flow and laser irradiation, adapted from [*Aoyagi* et al. 1987]

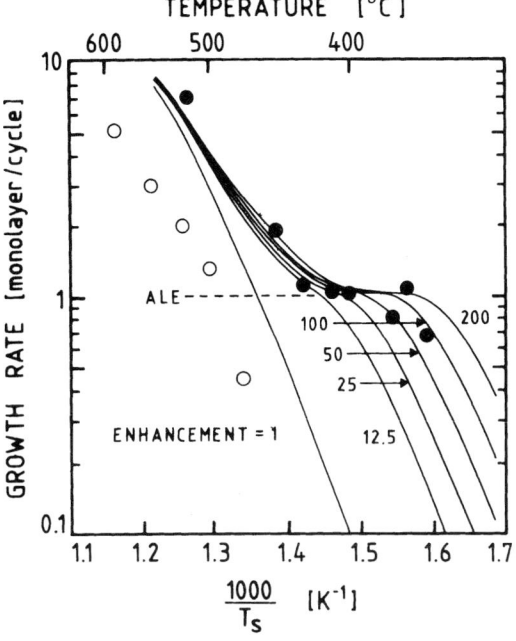

Fig. 20.3.4. Arrhenius plot for the growth of GaAs. ●Ar$^+$-laser irradiation (laser-ALE). ○ Standard growth (no light). Full curves have been calculated for various ratios of decomposition rates of $Ga(C_2H_5)_3$ on As and Ga layers [*Aoyagi* et al. 1990]

Figure 20.3.4 shows the growth rate in monolayers per cycle as a function of substrate temperature, T_s. The full curves have been calculated by assuming different ratios of decomposition rates of $Ga(C_2H_5)_3$ on As and Ga layers. Without laser light, the growth rate increases continuously (open symbols, enhancement factor = 1). In the presence of laser light, the growth rate shows a plateau at one monolayer per cycle and only increases at higher temperatures. The fabricated films are of high quality. The same technique has been employed for the fabrication of AlAs and $Ga_{1-x}Al_xAs$ films.

21 Liquid-Phase Deposition, Electroplating

Laser-enhanced/induced liquid-phase processing is mainly applied for material deposition and etching. Both electrolytic solutions and, to a smaller extent, ordinary liquids (non-electrolytes) are employed. The precursor molecules are dissolved in the liquid (solvent) and possibly dissociate. The dipoles or ions formed in this way interact with each other, with the solvent, and with the substrate. Thus, electrochemical effects will often play an important or even decisive role. They may arise from the illumination of the liquid–solid interface, or from an external electromotive force (EMF).

The high number densities of species involved in liquid-phase processing cause significant changes in transport properties with respect to gases. These rely on both changes in transport coefficients and additional transport mechanisms. For example, the shorter mean free path of molecules makes diffusion in liquids much slower than in gases (Sect. 3.3.2). Therefore, convection, turbulence, and bubbling related to strong temperature gradients in pyrolytic liquid-phase processing can exceed the transport of species by diffusion by several orders of magnitude.

Laser-induced processes in electrolytes are quite different for metals, semiconductors, and insulators. Some of these differences have already been discussed in connection with laser-induced wet etching (Chaps. 14, 15).

21.1 Liquid-Phase Processing Without External EMF

In the absence of an external EMF, the fundamental mechanisms in laser-induced liquid-phase processing are based on thermal, electrothermal, or electrochemical effects.

21.1.1 Thermal Decomposition

In the simplest case, electrical interactions can be ignored and the precursor molecules are just thermally decomposed at the liquid–solid interface. This situation frequently applies to laser processing using ordinary liquids. Because of strong changes in optical properties and boiling, processing can be performed in a relatively small temperature interval only. The difficulty with quantitative calculations arises from the paucity of information on kinetic constants and hydrodynamic effects. From the reaction rates achieved in pyrolytic liquid-phase processing, (normalized) activation energies $\mathscr{E}^* \equiv \Delta E/k_B T(\infty)$ in the range, 10–30 can be estimated. Let us consider a reaction of the type

$$AB_\mu + M \rightarrow A(\downarrow) + \mu B(\uparrow) + M,$$

where M now denotes the liquid solvent. Here, the total particle number density N can be considered to be independent of temperature so that $q = 0$ in (3.3.11).

To demonstrate some essential features, we consider transport by (ordinary) diffusion only. In this case we can directly calculate the (normalized) density, N_{AB}^*, and the reaction rate, W^*, from the equations given in Chap. 3. Here, we assume a spherical reaction zone (Fig. 3.4.1) and $\mathscr{E}^* = 10$. The dashed curves in Fig. 21.1.1a have been calculated for temperature-independent transport coefficients, i.e., $n = 0$ and $m = 0$ (Sect. 3.3.2). The results show that transport limitations start with relatively low temperatures. An Arrhenius plot of the normalized reaction rate is shown in (b) for various values of $k_0^* \equiv k_0 r_D/D_{AB}(\infty)$. The dashed curves apply to $n = 0$ and $m = 0$. As in the case of gases, N_{AB}^* and $W^* \equiv W/k_0 N(\infty) x_{AB}(\infty)$ are strongly affected by the temperature dependence of D_{AB} but not by the heat conductivity κ (compare full and dotted curves). The temperature distribution does not depend on q and n.

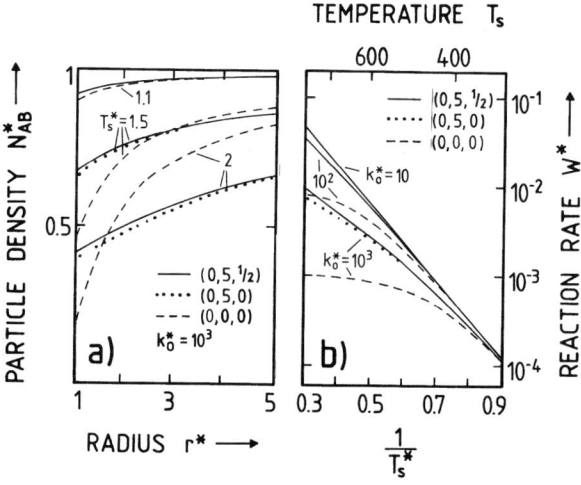

Fig. 21.1.1a,b. Normalized concentration of species AB and reaction rate in *liquid-phase processing*. Different curves belong to various parameter sets $(q = 0, n, m)$ with $T_s^* \equiv T_s/T(\infty)$ and $\mathscr{E}^* = 10$. q, n, m describe the respective temperature dependences in particle number density, molecular diffusion coefficient, and heat conductivity [Bäuerle et al. 1990a]

Convection and bubble formation

Convective flows are of particular importance in liquid-phase processing. Even at moderate laser-light intensities, Rayleigh numbers $\mathcal{R}a \approx 10^3$ to 10^4 are obtained. In many systems, convection determines the transport of species and the temperature distribution near the reaction zone. In the simplest approximation, the convective flux can be estimated from

$$J_c \approx v_c N_{AB}. \qquad (21.1.1)$$

For a spherical reaction zone with radius r_D, the diffusion flux can be estimated from (3.4.17)

$$J_d \approx \frac{D_{AB} N_{AB}}{r_D} = v_d N_{AB}, \qquad (21.1.2)$$

where v_d is the diffusion "velocity". The ratio of fluxes is

$$\frac{J_c}{J_d} = \frac{v_c r_D}{D_{AB}} = \frac{v_c}{v_d}. \qquad (21.1.3)$$

With $v_c = 1$ cm/s, $r_D = 10$ μm, and $D_{AB} = 10^{-5}$ cm^2/s, we obtain $J_c/J_d = v_c/v_d \approx 10^2$. The influence of convection on the processing rate can be estimated from the (modified) Smoluchowski equation (3.4.14), with $k^* = k/(v_c + D_{AB}/r_D) \approx k/v_c$. The latter approximation refers to strong convection. In this case, reaction rates within the transport-limited regime may be enhanced by several orders of magnitude. Additionally, convective flows increase the temperature further away from the laser-heated area and thereby favor homogeneous reactions and cluster formation within the liquid. The clusters may absorb the laser radiation and, in turn, further heat the liquid. Such a positive feedback may cause instabilities and structure formation (Chap. 28). Cluster formation may result in porous deposits with weak adhesion to the substrate. Convective flows can be diminished or even avoided by using short laser-beam dwell times, see (9.5.15).

With higher laser-light intensities, bubble formation above the irradiated surface may occur [*Yavas* et al. 1994]. In this regime controlled surface processing becomes very difficult, or even impossible.

Pyrolytic laser-induced metal deposition has also been demonstrated with organic solutions containing compounds with zero (metal) valency. Among those are triphenyl-phosphine complexes of Au and complexes of Cr, Fe, Mo, and W ([*Brook* et al. 1991] and references therein). Materials such as C, Si, etc., can be deposited from liquid hydrocarbons, silanes, etc. Clearly, this process permits deposition onto both conducting and insulating substrates.

21.1.2 Electroless Plating

In electroless plating the charge balance is maintained via a reducing agent, Re, incorporated into the solution. Thus, plating can be performed without simultaneous etching. The process can be described by the redox equations

$$Re^{z_0+} \to Re^{(z_0+z)+} + ze^-$$

$$Me^{z+} + ze^- \to Me \, , \qquad (21.1.4)$$

where Me^{z+} stands for z-fold charged metal ions within the solution. The metal ions are reduced by capturing electrons from the reducing agent Re which itself becomes oxidized. Laser light can thermally or non-thermally enhance or initiate this process. When the reaction has been started, the deposited metal may act as a catalyst and the reaction becomes self-sustaining. Laser-induced/enhanced electroless plating has been employed for fast localized deposition of metals. In general, virtually no plating takes place outside the laser-illuminated region. The technique can be applied also to electrically insulating substrates. Among the experimental examples is the electroless plating of premetallized glass substrates with Ni from solutions containing sodium hypophosphate (I) as reducing agent. With Ar^+-laser radiation local deposition rates up to 0.1 µm/s have been achieved. The background plating rates were about 5 Å/s [*von Gutfeld* 1984].

Other types of electroless plating have been demonstrated for laser pretreated surfaces. Among the examples are Au films on n-doped GaAs [*Sugioka* and *Toyoda* 1992], Ni and Cu films on modified [*Niino* and *Yabe* 1993a] or ablated [*Niino* and *Yabe* 1993b] polymer surfaces, etc.

21.1.3 Metal–Liquid Interfaces

Reactions at interfaces between metals and electrolytic solutions may be enhanced or initiated by laser light via changes in the electrochemical (Nernst) potential or via the external photoeffect. For this latter mechanism there is, up to now, no experimental evidence.

The (equilibrium) electrochemical potential of a metal with respect to the electrolyte can be described by

$$\Phi_e(T, N_i) = \Phi_0(T) + \frac{R_G T}{zF} \ln\left(\frac{a_i}{a_{Me}}\right), \qquad (21.1.5)$$

where $F = eL$ is the Faraday constant, and L the Avogadro number. a_i and a_{Me} are the activities of ions i within the solvent and the metal, respectively. For low ion concentrations $x_i = N_i/N \ll 1$ we have $a_i \propto x_i$. For pure metals, the activity is $a_{Me} = 1$, per definition. The potentials Φ_e and Φ_0 cannot be measured directly. The potential difference between a metal in an electrolyte with $a_i = 1$ and a standard hydrogen electrode, measured at $T = 25\,°C$, is called the standard electrode potential [*Bockris* and *Reddy* 1977].

21.1 Liquid-Phase Processing Without External EMF

Laser-induced changes in Nernst potential

Laser-light irradiation of a metal generates a local electric cell via changes in the potential (21.1.5). If this change in Φ_e is based on a local laser-induced temperature difference, $\Delta T = T_c - T(r)$, a *thermobattery* with an EMF

$$U_{EMF}(\Delta T) \leqslant [\Phi_e(T_c) - \Phi_e(T(r))]_{N_i = \text{const}} \tag{21.1.6}$$

is generated. $T_c \equiv T(r=0)$ is the laser-induced center temperature, and $T(r)$ the temperature at distance r. The EMF originates mainly from the temperature dependence of $\Phi_0(T)$. For most metals, $\partial\Phi_0/\partial T \approx \text{const} > 0$ and has typical values of some 10^{-3} V/K. The situation is schematically shown in Fig. 21.1.2. The potential at the center of the laser beam will be higher (anode) than outside (cathode). For a temperature rise of $\Delta T = 100$ K an EMF of some 0.1 V is generated. While these voltages are very small, the electric field strengths are very high due to the small dimensions of the battery. Because the metal itself acts like an external load resistance in a standard Galvanic element, the current within the metal flows from the center to peripheral regions. In order to maintain the overall charge neutrality, the (positive ion) current within the solution must flow towards the center. Thus, we observe deposition (plating) in the heated center and simultaneous etching outside. This situation is exactly opposite to that in standard electroplating using an *external* battery. If $\partial\Phi_0/\partial T < 0$ etching occurs in the center and deposition outside. Because of the finite current, the real EMF, (21.1.6), is below the value estimated from the right side of the formula. The current $j(\xi)$ and the overpotentials $\xi(T_c)$ and $\xi(T(r))$ can be estimated from equations of the type (21.2.1) and Ohm's law for the microgalvanic element in Fig. 21.1.2. The plating rate is proportional to the normal component of the current at the electrolyte-metal interface. Because etching (plating) takes place on a large area outside the laser focus, the change

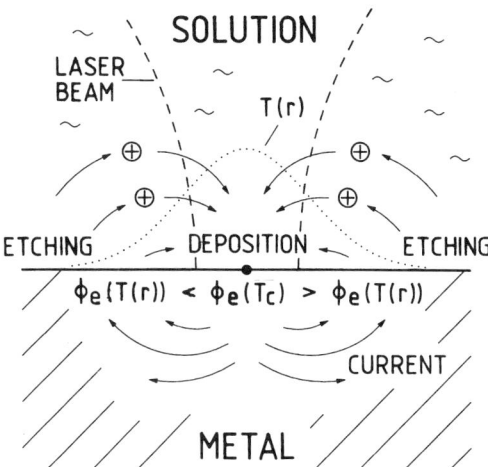

Fig. 21.1.2. Laser-induced thermobattery on a metal surface

in film thickness can be ignored, in general. This type of liquid-phase processing which is related to a laser-generated thermobattery is also termed *laser-enhanced exchange plating* (etching).

According to (21.1.5) an EMF can also be generated via a local laser-induced change in concentration N_i

$$U_{EMF}(\Delta N_i) = [\Phi_e(N_i(0)) - \Phi_e(N_i(r))]_{T=const} . \qquad (21.1.7)$$

This is referred to as a *concentration battery*. The change in N_i can be related to the reaction itself (this is important within the transport-limited regime only), to thermal diffusion of species [*Bunkin* et al. 1986], etc. In general, the thermobattery effect exceeds the concentration battery effect, though the two effects cannot be separated.

Up to now we have considered a situation where both the metal surface and the metal ions in solution are of the same nobility as, for example, with a Cu substrate immersed in aqueous $CuSO_4$. Laser-induced exchange plating enables one, however, to plate a less noble metal, Me_2, onto a more noble metal, Me_1. For a uniform temperature T_0 we have $\Phi_{e2}(T_0) < \Phi_{e1}(T_0)$ and no plating takes place. With a focused laser beam, however, we can generate a local thermobattery so that $\Phi_{e2}(T_c) > \Phi_{e1}(T(r))$. Thus, plating becomes possible above a certain threshold intensity (temperature).

Laser-enhanced exchange plating has been studied for premetallized substrates (glasses, etc.) and for bulk metals such as Cu, Ni, W, etc. These substrates have been plated with Cu from aqueous $CuSO_4$, and with Au from various solutions (Appendix B.7). At constant laser-light intensity, the deposition rate for Cu spots was found to decrease with increasing thickness of the metal film covering the glass substrate. This is in agreement with the corresponding decrease in temperature rise.

21.1.4 Semiconductor–Liquid Interfaces

Laser-light with photon energies $h\nu > E_g$ generates electron–hole pairs (Fig. 2.1.1). Because of the different mobilities of electrons and holes (for most semiconductors $\mu_e > \mu_h$), spatial changes in carrier concentrations take place (Dember effect). These, in turn, alter the charge transfer rates at the semiconductor–electrolyte interface. This (non-thermal) mechanism is of particular importance in wet-etching of semiconductors at low laser-light intensities (Sect. 15.6). Nevertheless, this mechanism is also important within the *initial* phase of plating. With localized irradiation, the Dember effect generates a local EMF. If we consider an n-type semiconductor, a depletion of electrons within the irradiated area and an enrichment outside will take place (Fig. 15.6.4b). Thus, the laser-irradiated surface acts like an *external* battery. Positive ions within the solution move from illuminated to dark regions, exactly opposite to the situation shown in Fig. 21.1.2. As a consequence, etching within the center and ring-shaped (metal) plating outside is observed. For an estimation of the current, electrical conductivities are required. For a 1-molar electrolyte solution the conductivity is, typically, about 10^{-1} $(\Omega\text{cm})^{-1}$; for an n-type semiconductor, it is between 10^4 and 10^{-2} $(\Omega\text{cm})^{-1}$, depending on the degree of doping.

Electron–hole pairs are also generated if the absorbed laser light just heats the surface. A local temperature rise results in a depletion of the major carriers (Sect. 15.6.3). The thermal EMF for semiconductors is about 10^3 times higher than for metals.

21.1.5 Further Experimental Examples

In most cases of laser-induced/enhanced liquid-phase processing, different effects will simultaneously contribute to the overall reaction rate and the dominating mechanism may even change during the reaction.

Pt, Au, and Ni have been deposited from aqueous solutions of H_2PtCl_6, $HAuCl_4$ (also used as methanolic solution) and $NiSO_4$ by means of pulsed dye-laser radiation (580 nm $\leq \lambda \leq$ 720 nm; within this range, the solutions are transparent) [*Karlicek* et al. 1982]. The substrates employed were mainly doped and undoped InP. In the initial phase, a thermally activated chemical reaction between the InP and the metal salt leads to the formation of PtP_2, NiP or similar compounds within the interfacial layer. Deposition seems to proceed via thermal decomposition of precursors at the metal–liquid interface. Smooth platinum films up to a thickness of 0.5 μm were deposited. Within the platinum deposits, no solution contaminants were found. Deposits of Pt and Au on undoped InP exhibited ohmic behavior. Pt has been deposited sucessfully also on n-type GaAs, but not on Si.

21.2 Electrochemical Plating

In laser-enhanced electrochemical plating, an *external* battery is applied in such a way that, in general, the substrate is negatively biased with respect to a counterelectrode. The applied voltages are, typically, 1–2 V. Detailed experiments have shown that the enhancement of the reaction rate is based on local laser-induced heating [*von Gutfeld* 1984]. By reversing the polarity of the electrodes, the same process can be employed to *etch* material surfaces (Chaps. 14, 15). With the systems investigated, photochemical effects within the liquid have been excluded. We shall therefore concentrate on the effect of local heating on charge- and mass-transfer rates within an electrochemical system. Here, we have to consider the temperature dependence of the current density. In the kinetically controlled regime, i.e., at low overpotentials, the current density j [A/cm^2] is given by the Butler–Volmer equation [*Bockris* and *Reddy* 1977]

$$j = j_0 \left[\exp\left(\frac{\beta' \xi z F}{R_G T}\right) - \exp\left(-\frac{\beta \xi z F}{R_G T}\right) \right], \qquad (21.2.1)$$

where j is directed from the electrode to the electrolyte. β and β' are the so-called symmetry factors (transfer coefficients). In general, $\beta' \approx 1 - \beta$ and $\beta \approx 1/2$. $\xi = \Phi - \Phi_e$ is the overpotential. If $\xi < 0$ we obtain plating with $j < 0$, and etching if $\xi > 0$ ($j > 0$). Equation (21.2.1) can be derived in analogy to the Frenkel–Wilson law (10.1.9) by taking into account the distortion of the potential barrier by the external electric field. The exchange current density j_0 (also termed charge transfer rate) is given by

$$j_0 = z N_c F k_c \exp\left(-\frac{\beta z F \Phi_e}{R_G T}\right). \qquad (21.2.2)$$

j_0 describes the equilibrium flux of charges through the interface in one direction. Clearly, in equilibrium the *total* flux is zero. N_c[mol/cm^2] is the ion concentration near the electrode; $N_c L$ is the number of ions per square centimeter. For concentrated solutions we must introduce, instead of N_c, the molar activity/cm^2 which takes into account ion–ion interactions. The rate constant is given by

$$k_c = \frac{k_B T}{h} \exp\left(-\frac{\Delta E}{R_G T}\right). \qquad (21.2.3)$$

ΔE is the activation energy for ions within the solution to become incorporated into the metal in the absence of an external field. The temperature dependence of j_0 is dominated by the Arrhenius term, since $\Delta E > \beta z F |\Phi_e|$. Even when the temperature dependence of Φ_e is taken into account, the rate always increases with temperature.

At higher overpotentials, transport of ions becomes rate limiting. In addition to diffusion and forced convection, transport due to gradients in the potential Φ must be taken into account. According to *W. Nernst* and *M. Planck*, the flux of ions i within the solution can be written in the form

$$J_i = -D_i \nabla N_i + N_i v_c - \frac{z_i F}{R_G T} D_i N_i \nabla \Phi . \qquad (21.2.4)$$

Forced convection (second term) has been estimated for liquid-phase processing by solving the 3D-Navier-Stokes equations. In the general case one has to solve the total magnetohydrodynamic problem. The influence of *focused* laser-light irradiation in electrochemical plating (etching) can be summarized as follows:

– Localized heating causes a positive shift in the rest potential and thereby permits localized plating on large-area electrodes. Because there is no back-

ground plating, the technique can be employed for single-step fabrication of microstructures.
– Localized heating results in 3D-diffusion of ions and, at higher laser-light intensities, in convection. In any case, the rate achieved within the transport-limited range is increased; with tight focusing, the current density increases as $j \propto w_0^{-1}$. This has been verified experimentally [*Puippe* et al. 1981].

Laser-enhanced electrochemical plating (etching) has been studied most extensively with Ar^+- and Kr^+-lasers. The power densities employed were between 10^2 and 10^6 W/cm^2. Both continuous and pulsed plating (etching) were demonstrated by modulating the external voltage source, the laser output power, or both synchronously. Plating has been studied in detail for Au, Cu and Ni. The substrates were glass and c-Al$_2$O$_3$, both covered with 0.1µm thick films of Au, Cu, Ni, Mo, or W. The resolution achieved in these experiments was a few micrometers.

The plating mechanism was investigated by illuminating the metallized glass surface with Ar^+-laser radiation either from the front through the solution, or from the back through the optically transparent glass. In the latter geometry, no photons reach the electrolyte. The deposition rate was found to be equal in both cases. This is expected for a thermally activated reaction. Hence, photochemical processes cannot play an important role. Further support for the thermal character of the process was obtained from the comparison of plating rates achieved with premetallized c-Al$_2$O$_3$ ($\kappa \approx 0.2$ W/cmK) and glass ($\kappa \approx 0.01$ W/cmK) substrates. Under otherwise identical experimental conditions, the rates on c-Al$_2$O$_3$ substrates were found to be much lower than on glass, as expected from the lower laser-induced center temperature.

Detailed investigations on electrochemical Au plating have revealed that dense, small-grained, crack-free, and uniform deposits of good adhesion are formed at elevated temperatures and high concentrations of gold within the electrolyte. Here, the operating potential should be *below* the mass-transport limit. Near this limit, Au of good morphology was deposited over areas of 500 µm in diameter with rates of up to 1 µm/s. Direct writing of Cu lines on premetallized glass substrates was possible with widths of ≥ 2µm.

Jet-plating

Laser-enhanced jet-plating permits one to achieve significantly higher deposition rates. Here, the mass transport to the substrate is increased by a jet (Fig. 21.2.1; the flow velocities are, typically, 10^3 cm/s). The laser beam is focused to the center of the orifice of the jet and is maintained within the liquid column by total internal reflection until impingement on the cathode occurs. The potentiostat is set to deliver constant current, i.e., to plate galvanostatically.

Jet plating permits high-quality, rapid, localized plating. The electrochemical and hydrodynamical parameters determining the mechanical and metallurgi-

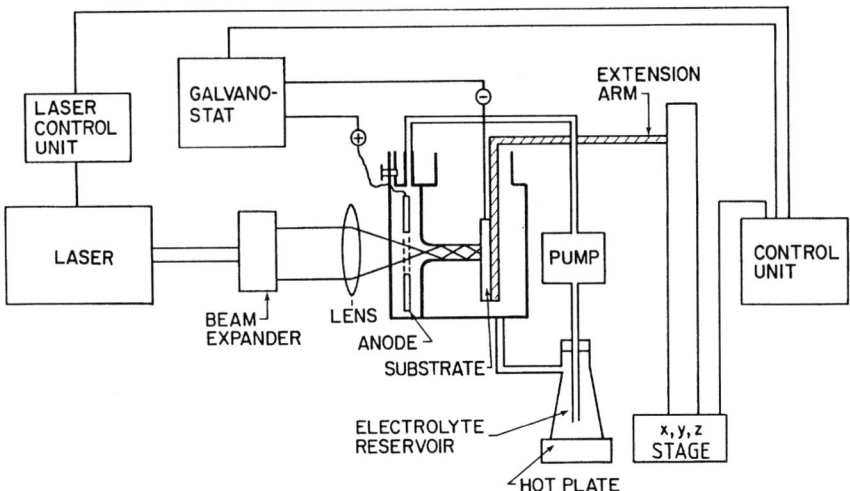

Fig. 21.2.1. Experimental setup for laser-induced electrochemical jet plating. The laser-beam is focused to the center of the jet orifice. The substrate can be moved via the extension arm [*von Gutfeld* 1984]

cal properties of deposits have been investigated, in particular for Au. Here, plating rates of up to 12 µm/s have been achieved. The surface smoothness of Au films increases with laser-light intensity. Simultaneously, their nodularity decreases and voids disappear. The Knoop hardness of films was between 20 and 90 kg/mm^2, which is characteristic for soft gold.

Laser-enhanced plating can be applied for circuit and mask repair [*Jacobs and Nillesen* 1990], the fabrication of interconnects, in customization and ohmic contact formation, etc. (Sect. 18.5).

22 Thin-Film Formation by Pulsed-Laser Deposition and Laser-Induced Evaporation

Lasers can be used to fabricate thin extended films by condensing on a substrate surface the material that is ablated from a target under the action of laser light. Depending on the specific laser and material parameters, ablation takes place under quasi-equilibrium conditions, as in laser-induced thermal vaporization (Chap. 11), or far from equilibrium, as in pulsed-laser ablation (Chap. 12). In the latter case, thin-film formation is termed pulsed-laser deposition (PLD). Instead of PLD, terms such as laser-sputter deposition (LSD), pulsed-laser evaporation (PLE), laser-induced flash evaporation (LIFE), and others, are also used in the literature. Pulsed-laser deposition is of particular interest because it enables one to fabricate multicomponent stoichiometric films from a *single* target.

From the aspect of film formation, the detailed ablation mechanism (Chap. 13) is of minor relevance. It is only important that ablation takes place on a time scale that is short enough to suppress the dissipation of the excitation energy beyond the volume ablated during the pulse. Only with this condition, can damage of the remaining target and its segregation into different components be largely avoided. In this regime of interactions, the relative concentrations of species within the plasma plume remain almost unchanged for successive laser pulses and they are almost equal to those within the target material. This is the main reason why pulsed-laser deposition has been found to be useful; in particular, for the deposition of thin films with complex stoichiometry. Among the materials studied in most detail are compound semiconductors, insulators, and high-temperature superconductors (Appendix B.8).

The short response times are also the reason why it is useful, for certain materials, to integrate pulsed-laser ablation into molecular beam epitaxy (MBE; Sect. 20.3).

Pulsed-laser deposition is a very reliable technique. It offers great experimental versatility, it is fairly simple, and fast – as long as small-area films of up to several square-centimeters are to be fabricated. For these reasons, PLD is particularly suitable in materials research and development. The strong non-equilibrium conditions in PLD allow, however, some *unique* applications:

– The synthesis of metastable materials that cannot be produced by standard techniques.

- The fabrication of films from species that are generated only during pulsed-laser ablation. With certain systems, the physical properties (microstructure, morphology, adhesion, optical, electrical, etc.) of such films are superior to those fabricated by standard evaporation, electron-beam evaporation, etc.

Besides restrictions with respect to film areas, the major disadvantage of PLD is related to particulates on the substrate and film surface. Clearly, other thin-film techniques have their peculiarities as well. For example, RF sputtering enables one to produce large-area films with good thickness uniformity and small surface roughness (typically < 100 Å with 1000 Å thick films). Here, the control over the correct stoichiometry is, however, much more problematic. Furthermore, sputtering requires large targets and longer preparation cycles, and it affords less experimental versatility.

In the present chapter we put special emphasis on thin film formation by PLD. The diagnostics and expansion of laser-induced vapor and plasma plumes is incorporated into Chap. 30.

22.1 Experimental Requirements

A typical setup employed for film deposition is schematically shown in Fig. 22.1.1. It essentially consists of a laser, a reaction chamber, a target, and a substrate. The material ablated from the target is condensed on the substrate and forms a thin film. Ablation can take place in either a vacuum or a low pressure inert or reactive atmosphere. The latter technique is termed *reactive laser ablation* (reactive laser sputtering).

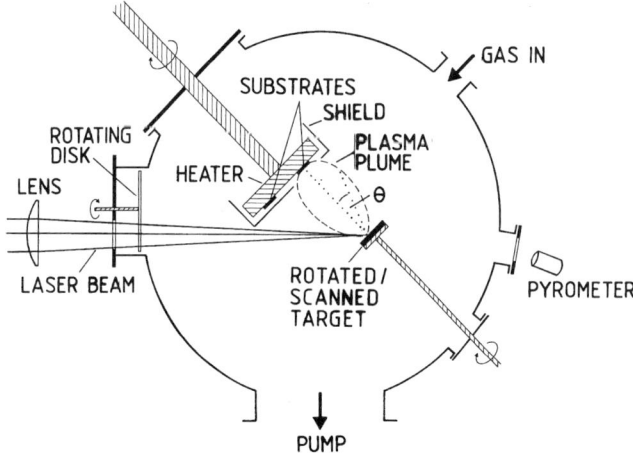

Fig. 22.1.1. Experimental setup employed in thin-film formation by PLD

22.1 Experimental Requirements

It is evident that the setup depicted in Fig. 22.1.1 is considerably simplified and shows only the main components. The proper choice of the laser depends on the physical properties of the target which, in any case, should strongly absorb the laser light. The attenuation of the incident light by ablated material condensed at the entrance window of the chamber may become a problem, in particular when deposition is performed in a vacuum. This can be diminished by a rotating disc, as shown in the figure, or by a light pipe, etc. The fluences typically employed in PLD, $\phi \approx 0.1\text{--}10 \, \text{J/cm}^2$, generate a plasma plume in front of the target. Uniform ablation is achieved by rotation and scanning of the target with respect to the laser beam. The distance between the target and the substrate should match, approximately, the length of the plasma plume (Chap. 30) and it is, typically, 3–8 cm. The uniformity in film thickness can be improved by moving the substrate relative to the plasma plume, for example by excentric rotation of the substrate holder. With some materials, the throughput in film preparation can be enhanced by mounting several substrates. The substrate temperature determines, to a large extent, the morphology and microstructure of films. Proper temperature control requires shielding of the sample holder, thermometers mounted on the substrate surface and the sample holder and, if possible, pyrometric measurements. Special optical arrangements permit laser-induced substrate cleaning prior to film deposition and in situ laser annealing of the deposited film. For stoichiometric deposition and the suppression of particulates, different types of shutters, masks, and apertures are introduced in the chamber. Further experimental aspects are outlined throughout this chapter.

22.1.1 Congruent and Incongruent Ablation

The proper choice of laser parameters is of great importance since they determine the type and the relative concentrations of species leaving the target surface, their degree of ionization, and their spatial and temporal distribution.

Congruent ablation

High laser-power densities and short pulses (dwell times) cause short interaction cycles resulting in (almost) congruent ablation of small material volumes. Here, the thickness of the ablated layer per pulse must fulfil the condition $\Delta h \approx \max (l_\text{T}, l_\alpha)$, see (12.0.1). With this condition, the heat loading of the target material is small and no material segregation takes place. This is the reason why the ablated material has essentially the same composition as the original target, even after many laser pulses. The angular distribution of ablated products is strongly forward-directed and can be described, in good approximation, by a $\cos^n \Theta$ law, where Θ is the angle between the surface normal and the direction of propagation of species within the plasma plume. The forward orientation of the plume becomes more pronounced with increasing laser fluence. With the laser parameters typically employed in this regime, the exponent has values $n \geqslant 10$. Congruent ablation is a prerequisite for the synthesis of thin films of multicomponent materials from *single* targets. High-intensity laser pulses also permit one to partially or completely dissociate materials that ordinarily evaporate in molecular form only.

Incongruent ablation

In the case of equilibrium or quasi-equilibrium laser heating, the target surface is melted and vaporized in a similar way as in conventional thermal evaporation (Chap. 11). Here, the relative concentrations of species leaving the surface, in general, differ significantly from those of the original target. Components with high vapor pressure leave the target before those with low vapor pressure. The angular distribution of species can be described by a $\cos \Theta$ law, as expected for thermal (equilibrium) evaporation. This parameter range is not appropriate for stoichiometric deposition of compounds from single targets.

With materials that consist of a *single* component, the terms congruent and incongruent ablation become meaningless. However, the type of ablated species (atoms, molecules, clusters), their density, degree of ionization, and their velocity are all dependent on the specific laser parameters employed. These properties of species strongly influence the microstructure, morphology, and quality of films.

Simple estimates

With strongly absorbing inorganic materials, the requirements for congruent laser ablation are reasonably well fulfilled with nanosecond laser pulses and fluences of, typically, 1–10 J/cm². This becomes plausible from a simple estimate of the characteristic times involved in the process. Let us assume purely thermal ablation and one–dimensional heat flow. The time t_v to reach the effective vaporization temperature on the target surface can then be estimated from (7.5.8). For surface absorption we obtain

$$t_v \approx \frac{\pi}{4D}\left(\frac{\kappa \theta_v^{\text{eff}} \tau_l}{\phi(1-R)}\right)^2. \tag{22.1.1}$$

The effective temperature rise $\theta_v^{\text{eff}} \equiv \Delta T^{\text{eff}} = T_v^{\text{eff}} - T(\infty) = T_v - T(\infty) + \Delta H_v/c_p$ takes into account, in a crude way, the enthalpy of vaporization. With $\phi \approx 5 \,\text{J/cm}^2$ and $\tau_l \approx 20\,\text{ns}$, we find for a Cu target $t_v \approx 10^{-8}\,\text{s}$, and for a ceramic target of a high temperature superconductor about $10^{-11}\,\text{s}$. Clearly, this estimate ignores laser–plasma interactions. At least for non-metals, these

22.1 Experimental Requirements 401

heating times are too short for significant material segregation via convection (Sect. 10.4). With certain materials and laser parameters, stoichiometric ablation can even be achieved with rapidly scanned cw-lasers or fast rotating targets. In this case τ_l corresponds to the dwell time of the laser beam.

If (12.0.1) is fulfilled, the heat load of the target is minimized. Thin-film formation, however, requires multiple-pulse irradiation and target heating can become significant. This can be estimated from the average absorbed intensity. In any case, one has to compromise between the pulse repetition rate which determines the overall growth rate of the film, and the maximum tolerated temperature of the target. Consideration of these conditions is very important beause material segregation may also take place *between* laser pulses.

22.1.2 Targets

The targets mainly employed in PLD are solids in single crystalline, polycrystalline, ceramic, powdery, or amorphous form. In some cases, liquid targets are also used.

With single-crystalline and coarse-grain polycrystalline targets uniform deposition and low concentrations of particulates within the film can frequently be achieved only during the first few laser pulses. Subsequent fracturing of the target material due to thermal shocks can result in the ejection of large fragments which deteriorate the uniformity and quality of the deposited film.

Small-grained dense ceramic targets permit more uniform conditions during longer sputtering times. An additional advantage of such targets is the enhanced extinction due to radiation trapping. This is of particular importance with materials that are otherwise transparent at the laser wavelength under consideration. The diminished thermal conductivity originating from thermal barriers between grains is also an advantage. Both properties considerably enhance the surface temperature with respect to that induced in single- or polycrystalline material.

The deposition of uniform films also requires good surface morphology and smooth ablation of the target. Independent of the angle of incidence, the vapor plume has its preferred direction perpendicular to the target surface. With very high fluences the plume bends towards the incident laser beam. With an uneven surface the direction of plume expansion will continuously change. This is another reason why targets should be rotated and/or scanned during ablation.

Liquid or surface-molten targets remain smooth at all times and permit one to efficiently reduce or almost avoid particulates on the film surface. This technique, however, can only be applied with single component materials or with materials that melt congruently.

For the fabrication of heterostructures, multiple targets consisting of the individual elements, compounds, or various types of mixtures are used.

Multiple targets are also used for the synthesis of compounds for which the requirements for stoichiometric ablation cannot be fulfilled, or if single stoichiometric targets are not available, or if the sticking coefficients of single constituents on the film surface are very different. Here, ablation can be achieved by sequential or simultaneous exposure of the individual targets. In any case, ablation from multiple targets is more difficult to handle because the rate of ablation will vary with each source. The composition of the film can then be controlled via the dwell time or the power of the laser beam on each source. Nevertheless, such arrangements are used in laser molecular beam epitaxy (LMBE). With this technique, epitaxial layers of binary and ternary compounds with predesigned compositional profiles and varying band-gap energy have been produced. One of the important advantages of LMBE is the instantaneousness of the ablation process. This enables one to properly control the stoichiometry in the deposited film.

22.2 Volume and Surface Processes, Film Growth

Figure 22.2.1 shows, schematically, different volume and surface processes in reactive PLD. The laser-induced plasma plume consists of UV radiation, electrons, ionized or neutral atoms, molecules, clusters, and fragments (Sect. 30.2). The different species may react with each other or with a background atmosphere and impinge onto the film/substrate surface. Adsorbed species diffuse on the surface and contribute to film growth or they desorb. If the grown film does *not* require any post-deposition treatment, the process is termed *in situ* fabrication.

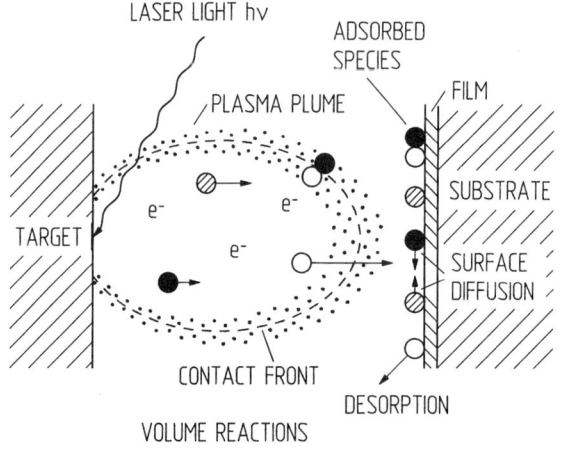

Fig. 22.2.1. Volume and surface processes relevant in thin-film formation by PLD. Reactions (collisions) between ablated species and the ambient atmosphere take place, in particular, near the contact front

22.2.1 Plasma and Gas-Phase Reactions

On their way from the target to the substrate, reactive and non-reactive collisions of ablated species take place within the plasma plume, at the contact front between the plume and the ambient gas, and near the substrate surface. Such volume reactions are strongly influenced by the UV plasma radiation, by free electrons within the plasma and, near the target, by the laser radiation itself. In the absence of an ambient atmosphere, the number of collisions between species is largest close to the target. In any case, collisions and chemical reactions of species within the volume between the target and the substrate determine, to a large extent, the type, energy, and flux of species at the substrate surface.

Reactive ambient atmospheres frequently employed are oxidizing gases such as O_2, O_3, N_2O, or NO_2. They allow one to fabricate stoichiometric oxide films by ablation of single-component or multicomponent targets. Here, the exact oxygen content within the film can be tuned via the background gas pressure.

Volume excitation/dissociation of species can be thermally or non-thermally activated. Comparison of bond dissociation energies shows that the formation of atomic oxygen from O_2 requires about 5.11 eV while with O_3 only 1.05 eV, and with N_2O about 1.67 eV are needed.

Thermal activation is most pronounced near the target and substrate surface. Non-thermal processes take place, in particular, within the plasma plume. Laser-induced volume excitations play a significant role only near the target because of the oblique beam incidence. The photochemistry of the oxidizing gases mentioned above can be described by

$$XO + h\nu \rightarrow X + O(^1D, {}^3P) \,, \tag{22.2.1}$$

where $X \equiv O, O_2, NO, N_2$. Thus, oxygen atoms are mainly in either the ground state (^3P) or the first excited state (^1D), depending on the precursor molecule, the laser wavelength, and the spectral intensity distribution of the plasma radiation. The absorption cross sections for different wavelengths can be found in Table V. At very high laser fluences, multiphoton dissociation of species near the target surface may become important.

Electronic excitation and dissociation of species can also be mediated via electron impact. For example

$$O_2 + e \rightarrow O_2^* + e \rightarrow 2O(^1D, {}^3P) + e \,, \tag{22.2.2}$$

or

$$N_2O + e \rightarrow [N_2O]^* + e \rightarrow N_2 + O(^3P) + e \,, \tag{22.2.3}$$

and

$$O(^3P) + e \rightarrow O(^1D) + e \,. \tag{22.2.4}$$

Similar processes can take place with the other precursors mentioned above. With the laser fluences typically employed in PLD, the average energy of electrons is between 2 and 4 eV. The rate constants for these processes are then $k(O_2) \leqslant 5 \times 10^{-11}$ cm^3/electron s and $k(O_3) \approx k(N_2O) \approx 10^{-8} - 10^{-9}$ cm^3/electrons [*Kline* et al. 1991; *Cleland* and *Hess* 1989; *Eliasson* and *Kogelschatz* 1986].

Energetic neutrals, ions, and clusters cause similar excitation/dissociation processes. Unfortunately, very little is known about the yield related to such collisons. In any case, with the background gases under consideration, the species ablated from the target will react with oxygen and form oxide molecules.

Dissociation of species i by UV radiation and electron impact is particularly strong near the target surface where the flux $J_i(0)$ is generated. For *low* densities of background species, N_j, and strong forward direction (one-dimensional propagation) the flux at the substrate at distance l can be written as

$$J_i(l) = J_i(0) \exp(-\sigma_i N_j l), \qquad (22.2.5)$$

where σ_i is the cross section for both non-reactive and reactive collisions; any electromagnetic interactions are ignored. Within this approximation, the film growth rate should exponentially decrease with increasing background pressure. If, however, species i are only generated from gas-phase molecules j, for example by photodissociation or reactive collisions with ablated species, the flux J_i will *increase* with increasing pressure p_j, at least close to the target surface. The validity of (22.2.5) has been proved for pulsed-laser ablation of Y-Ba-Cu-O targets in O_2 and Ar-atmosphere [10^{-5} mbar $\leqslant p_j(O_2, \text{Ar}) \leqslant 0.4$ mbar] by means of ion-probe measurements [*Geohegan* 1992]. The scattering cross sections derived from these experiments were around 2.5×10^{-16} cm^2. For *high* number densities, a hydrodynamic description of the problem is required.

22.2.2 Substrate Temperature, Laser-Pulse Repetition Rate

Species impinging onto the substrate need a certain time for surface diffusion and incorporation at proper lattice sites. Surface diffusion is a thermally activated process which increases with increasing temperature. This is the reason why high-quality crystalline films can be deposited at reasonable growth rates only at elevated temperatures. The time necessary for surface diffusion is also the reason why laser pulse repetition rates must be adapted to the particular material under investigation, the substrate temperature, and the flux of species onto the surface. Due to the dissipated kinetic energy of impinging species, their internal excitation energies, and their heat of condensation and desorption, the temperature on the substrate surface changes during and between laser pulses. Such (rapid) changes in temperature which significantly influence surface processes cannot, at present, be detected with the setups employed in PLD. The total (additional) energy input by the impinging species depends on the laser parameters, and in particular on the laser fluence and pulse repetition rate, and on the distance between the target and the substrate. Thus, there is a complex interrelation between the laser parameters, the properties of the target, the volume expansion and reaction of ablated species, and the different processes on the substrate surface. Therefore, non-monotonic dependences in film growth with laser fluence, etc., may be observed.

Another problem related to high laser pulse repetition rates is the overall increase in target temperature which favors material segregation. If, on the other hand, the laser-pulse repetition rate is too low, the relative importance of thermal desorption of volatile components from the film increases. Additionally, in the case of non-reactive deposition, incorporation of gas-phase impurities takes place at a higher relative rate.

22.2.3 Energetic Species

The energy of ablated species is between 10 and several 10^3 eV, depending on the laser parameters employed. The bombardment of the growing film by high-energy species can improve or deteriorate its overall morphology, stoichiometry, and microstructure. This is well known from ion-beam techniques. High-energy species may break atomic bonds, generate subsurface vacancies and displacements of atoms, induce recoil implantation and self-sputtering, cause thermal spikes, etc. However, energetic species also enhance surface diffusion of adsorbed atoms and thereby permit film growth at lower (average) substrate temperatures. Thus, the film quality achieved with a fixed parameter set, depends on the properties of the film itself. The energy of species leaving the target can be tuned only within a limited region via the laser fluence, simply because of the requirements for stoichiometric ablation. The energy of species impinging onto the substrate can, however, be controlled via the distance between the substrate and the target, and via the pressure of the background atmosphere. The optimal position of the substrate for high-quality film growth is often found to be close to the tip of the visible plasma plume (Chap. 30).

22.2.4 Particulates

Characteristic for pulsed-laser deposited films is the appearance of particulates and of various other features on the film/substrate surface (Fig. 22.2.2). These can be classified according to different main types:

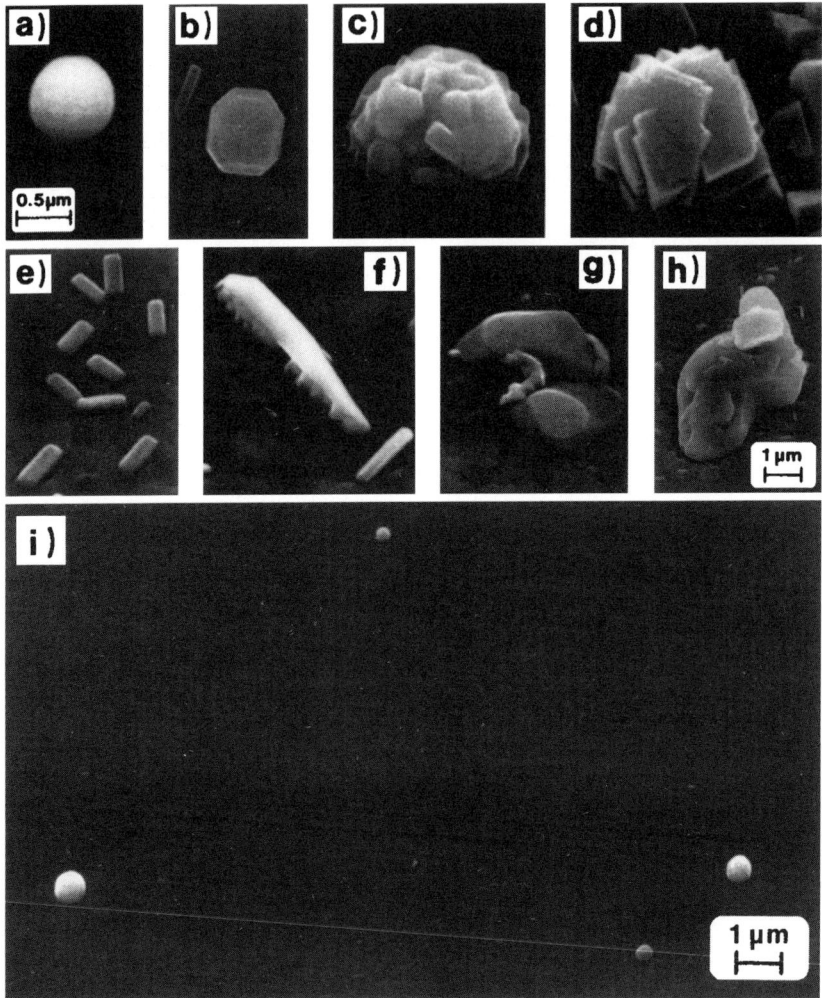

Fig. 22.2.2a–i. Most prominent types of particulates and other features observed on KrF-laser deposited c-axis oriented YBCO films on MgO. The magnification in (a to g) is the same. (**a**) Droplet with smooth surface. (**b**) Crystallized droplet. (**c, d**) Droplets with overgrowths. (**e**) Outgrowth due to a-axis misorientation. (**f, g**) Cu-enriched needles and platelets. (**h**) Big fragment directly ejected from target. (**i**) Film deposited under optimized experimental conditions; the remaining particulates are (mainly) droplets, adapted from [*Proyer* et al. 1996]

- Solidified melt drops with diameters between 0.1 and 3 µm. These *droplets* may be related to vapor condensate and/or hydrodynamic instabilities at the target surface. During film growth, droplets may change their shape due to crystallization, material coverage, etc. (Fig. 22.2.2 a–d)
- Irregularly shaped solid grains with diameters between 1 µm and more than 10 µm. These grains are directly ejected from the target, e.g., due to thermal

22.2 Volume and Surface Processes, Film Growth

stresses, without any vaporization or total melting taking place (Fig. 22.2.2h).
- Solidified splash drops with diameters of up to more than 10 µm, originating from superheating and liquid-phase expulsion.
- With organic polymers and biological materials high molecular-weight fragments and clusters are observed (Sect. 30.2.5).

Besides these main types, film surfaces frequently show additional features and particulates such as outgrowths, needles, platelets, etc. (Fig. 22.2.2e–g), which are specific to the particular material. Some of these features, e.g., the outgrowth, are probably formed at the substrate surface only, and they are also observed on films deposited by other techniques. Subsequently, the term "particulates" shall also include all types of these other features.

Particulates reduce the film quality in many respects. They locally destroy the microstructure, diminish the surface smoothness, reduce the minimum width of features in microfabrication, degrade the electrical properties, etc. For these reasons, great efforts have been made to eliminate the particulates or, at least, reduce their size and density. In fact, with proper experimental handling, the density of the *coarse* particulates can drastically be diminished. This can be achieved by:

- Selecting the proper target material (Sect. 22.1.2).
- Employing smooth target surfaces and uniform target ablation. Only by this means can the ablation conditions be kept stable.
- Outgassing of the target prior to ablation. In particular with pressed-powder or ceramic targets rapid expansion of trapped gas bubbles beneath the target surface may result in explosive-type ejection of large particulates.
- Combined rotation and scanning of the target with respect to the laser beam.
- Optimization of the laser parameters (Chap. 12). Material damages such as microcracks, exfoliation, etc., result in loosely attached fragments which can be ejected from the target during multiple-pulse irradiation. In the case of laser-induced evaporation where a molten surface layer is important, optimization of the laser parameters is necessary even with single component material, because of droplet formation due to hydrodynamic instabilities, or splashing of the melt due to overheating. With finite absorption, subsurface superheating (Sect. 13.2) may result in explosive-type ablation of large fragments. With amorphous carbon and silicon films, the density of particulates can be significantly lowered when using fs instead of ns laser pulses. Copper films, on the other hand, show less particulates when grown with ns pulses [*Müller* et al. 1993].
- Optimization of the distance between the target and the substrate surface, l. If we assume that particulates that originate from the target are homogeneously distributed, their number density decreases with increasing distance according to $N_i \propto l^{-\zeta}$ where $\zeta \approx 2$. Thus, a strongly forward-directed plume would be favorable.

Among the other techniques which prevent particulates from reaching the substrate are:

- Variation of the substrate orientation with respect to the (main) direction of plasma-plume expansion, and off-axis deposition [*Trajanovic* et al. 1995; *Holzapfel* et al. 1992].
- Multiple vane wheels [*Barr* 1969], choppers [*Pechen* et al. 1995], electromagnetic shutters [*Lubben* et al. 1985], etc., which are placed between the target and the substrate and which deflect or filter the (slower) particulates but transmit the fast moving atoms and molecules.
- Shadow masks [*Iwabuchi* et al. 1994].
- Ablation from the edge of disc-shaped targets rotating at high speed.
- Fragmentation of particulates by means of an additional laser beam propagating in parallel to the substrate surface.
- Dual-laser ablation from a single target [*Witanachchi* et al. 1995], or from two targets [*Strikovsky* et al. 1993].
- Deflection of particulates by means of a supersonic gas jet [*Murakami* 1992].

Most of these techniques increase the complexity of the experimental setup and reduce the deposition rate. While a particular technique, e.g., off-axis deposition, permits one to suppress some types of particulates, it may favor the formation of others, e.g., outgrowths in HTS films, or it may cause the film stoichiometry to deteriorate, lead to thickness inhomogeneities, etc. For these reasons, the conditions can be optimized only for each particular system and for the specific requirements.

Droplets

While the coarse particulates can be largely avoided by selecting the proper experimental conditions and, if necessary, by employing the additional techniques described, the finer particulates, the droplets, are more difficult to suppress. They seem to be inherently linked with PLD. Figure 22.2.2i shows a typical example of a thin film on which only the droplets can be seen. The formation of droplets may have different origins.

If ablation is performed in a vacuum, vapor cooling due to the expansion of the plasma plume may result in gas-phase condensation. This effect is particularly important for big adiabatic exponents, γ. In the presence of a background gas, condensation becomes more pronounced with increasing pressure. Nucleation and droplet formation can also occur near or at the substrate due to interactions between impinging vapor species and species desorbing from the substrate surface. In any case, with the parameters typically employed, the size of such droplets should be in the submicrometer region.

Droplets can also be formed at the target due to explosive-type material removal related to strong superheating or due to Kelvin–Helmholtz or

22.2 Volume and Surface Processes, Film Growth

Rayleigh–Taylor instabilities (Sect. 28.5). Different mechanisms related to surface instabilities are schematically shown in Fig. 22.2.3. The typical radii of droplets have been estimated by *Brailovsky* et al. (1995). The characteristic time for droplet formation is of the order of some ns (cases a to c). Let us first consider Fig. 22.2.3a. The recoil pressure and the expansion of the vapor plume generate a (relative) *lateral* velocity, v_{vl}^l, at the vapor–liquid interface. This may cause Kelvin–Helmholtz instabilities and the excitation of surface capillary waves. If the laser fluence is high enough, droplets with characteristic radii of 10–100 nm may be formed, even with single-pulse irradiation.

The surface corrugation due to capillary waves solidifies after the first laser pulse. With *multiple-pulse* irradiation and low to medium laser fluences, only a thin layer of the corrugated surface becomes molten with each pulse (Fig. 22.2.3b). The motion of the liquid out of the (hotter) valleys increases the amplitude of the corrugation and causes centrifugal forces near the hills. The latter can result in Rayleigh–Taylor instabilities and the formation of droplets. Their radii are again, typically, 10–100 nm.

If the amplitude of the corrugation becomes very large (Fig. 22.2.3c) the velocity of the vapor along the walls, v_{vl}^a, may cause shear flow (Kelvin–Helmholtz) instabilities and droplets with radii between 10 nm and 1 μm. This range of radii is determined by the velocity v_{vl}^a which depends on the spacing Λ. If some of the Λ^a become comparable to h and if the corresponding amplitudes are large enough, this mechanism may cause necking and the formation of solid particulates with radii of 0.1 to several μm (Fig. 22.2.3d).

There is, in fact, experimental evidence that droplets originate from hydrodynamic instabilities at the target surface [*Bennett* et al. 1995; *Proyer* et al. 1996].

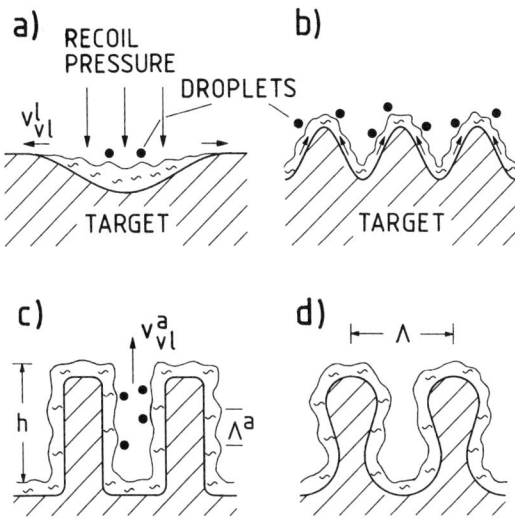

Fig. 22.2.3a–d. Droplet formation due to hydrodynamic instabilities generated during laser-induced melting and ablation. For simplicity, the laser beam and vapor plume are not drawn (Fig. 11.1.1). (**a**) The (relative) velocity v_{vl}^l excites surface capillary waves via a Kelvin–Helmholtz instability. (**b**) With multiple-pulse irradiation, the motion of the liquid from the valleys of the capillary waves increases the corrugation. Centrifugal forces near the hills may cause Rayleigh–Taylor instabilities. (**c**) Shear flow (Kelvin–Helmholtz) instability. (**d**) Necking and formation of solid particulates

22.2.5 Chemical Composition of Films

With certain compounds, PLD can result in films which are sub-stoichiometric with respect to a particular constituent, even when ablation is, in total, congruent. There can be various reasons for this:

- Differences in the transport of species between the target and the substrate.
- The sticking coefficients of different species impinging onto the substrate surface are not equal.
- The desorption enthalpies of adspecies differ from each other.
- Differences in binding energies, outdiffusion, and desorption of species that were already built-in on proper lattice sites.

Compensation can be achieved in various ways, depending on the particular material:

- A uniform background pressure yields an additional reactive flux onto the substrate surface. Furthermore, ablated species may react with gas-phase molecules *before* they condense onto the substrate.
- Via a nozzle which directs a gas stream onto the substrate.
- By means of another source, for example an ion gun.
- By using a target that contains an excess of the particular constituent.

22.3 Overview of Materials and Film Properties

The large variety of different materials deposited as thin films by laser ablation is listed in Appendix B.8. The discovery of high-temperature superconductors (HTS) in 1986 has had a strong impact on pulsed-laser deposition. Here, the possibility to synthesize high-quality HTS films from *single* targets has made PLD a widely used technique. The great versatility of PLD and the short turn-around times for film fabrication enable one to efficiently study a great variety of different compounds and film dopings. This success has stimulated activities to deposit other materials, e.g., dielectric and ferroelectric oxide films, compound semiconductors, etc. PLD permits a precise and fast control of the material composition. This capability is of particular importance in band-gap engineering and the fabrication of heterostructures.

The lasers mainly employed in film fabrication are excimer lasers, Nd:YAG or Nd:glass lasers, and pulsed CO_2 lasers. Some of the experiments have been performed with scanned cw-laser beams.

Films deposited by laser ablation are amorphous, polycrystalline, or single crystalline. Higher substrate temperatures favor, in general, crystalline growth. With multiphase materials, the particular crystalline phase formed during deposition is determined by both the substrate temperature and the gas pressure. The film growth rates achieved range from a few Å/s to some ten μm/s. The most important techniques to characterize film qualities are summarized in Sect. 30.4.

22.3.1 Metal Targets

Non-reactive PLD from metal targets results in the formation of metallic films. *Reactive* deposition in oxygen atmosphere yields semiconducting or insulating metal oxides. In a similar way, metal nitrides, halides, and other metal compounds can be synthesized (Appendix B.8).

22.3.2 Semiconductors

Pure Si and hydrogenated Si films have been fabricated by ablation of Si targets in vacuum and H_2 atmosphere, respectively (see also Sect. 19.5). With frequency-doubled Q-switched Nd:YAG-lasers, deposition rates of 10 to 100 µm/s were obtained [*Hanabusa* et al. 1983].

Ge films have been grown on (100) Si and (100) NaCl at temperatures as low as 250–300 °C by using molten Ge targets and pulsed CO_2-laser radiation [*Cheung* and *Sankur* 1992]. Films grown on Si substrates were single crystalline, had an index of refraction $n \approx 4.08$ and an intrinsic stress $S \approx 7.5 \times 10^3$ N/cm². In contrast, films grown by standard evaporation at the same substrate temperature were polycrystalline with $n \approx 3.85$ and $S \approx 3 \times 10^4$ N/cm². The higher quality of PLD films was ascribed to hyperenergetic species generated during pulsed-laser ablation.

Compound semiconductors

Epitaxial layers of compound semiconductors have been fabricated by employing either multiple targets for the individual elements, or single stoichiometric targets. In the latter case, additional sources must frequently be employed to obtain stoichiometric films. For example, with GaAs even congruent ablation results in As-deficient films. This originates from the lower sticking coefficient of As with respect to Ga. Similar observations have been made with $Hg_{1-x}Cd_xTe$. Here, the loss of Hg was compensated by an additional Hg source or a Hg background pressure.

The dependence of film properties on laser fluence and dwell time has been investigated for different materials.

PbF_2 films deposited with high-intensity pulsed CO_2 lasers show a superior crystallinity with respect to those deposited with cw CO_2-lasers [*Sankur* 1986].

The surface smoothness and the electrical properties of CdTe films grown from CdTe targets with high laser fluences are superior to those grown at lower laser fluences. This was ascribed to an increasing fraction of *atomic* Te among the ablated species, Cd, Te, and Te_2. With the highest laser fluences, only atomic Te has been detected. Surface recombination between Cd and Te atoms is considered to be the rate limiting step during film growth. The presence of Te atoms therefore enhances film growth rates, and it also suppresses the formation of Te clusters. Resistive heating of CdTe results in the evaporation of atomic Cd and molecular Te_2 only. This shows how laser ablation enables one to change the type of species within the vapor plume and their influence on film growth.

Heterostructures, band-gap engineering

Heterostructures are of increasing interest for both fundamental investigations and materials device engineering. They permit one to study carrier transport across layered planes, quantum-size effects, etc.

Among the systems fabricated by pulsed-laser deposition are InSb/CdTe, InSb/PbTe, InSb/CdTe, Bi/CdTe, etc. [*Gaponov* et al. 1980]. The single layers were polycrystalline or epitaxial with thicknesses down to less than 50 Å.

The capability of pulsed-laser ablation in combination with MBE (Sect. 20.3) has been demonstrated for $Hg_{1-x}Cd_xTe$ [*Cheung* and *Sankur* 1992]. Here, Knudsen effusion cells are used to produce Hg and Te beams with constant flux intensities. Pulsed-laser ablation is used to produce a *modulated* flux of Cd and Te, simply by changing the pulse repetition rate. In this way, the composition of the material, and thereby its band-gap, can be controlled. The fabrication of tailored band-gap profiles is denoted as band-gap engineering. The resolution in composition profile achieved by this combined technique is by at least one order of magnitude higher than in MBE.

Superlattices of $Hg_{1-x}Cd_xTe/Hg_{1-y}Cd_yTe$ grown on (100) CdTe substrates show excellent interfacial abruptness, uniformity, and electrical characteristics. By continuously varying the laser pulse repetition rate during deposition of single layers, multiple quantum wells with triangular, trapezoidal, and sawtooth-shaped composition profiles were grown.

Heterostructures are important components in a number of applications, in particular in junction devices. Among the novel devices using $Hg_{1-x}Cd_xTe/Hg_{1-x}Cd_yTe$ are staircase solid-state photomultipliers, optical modulators [*Chang* et al. 1989], and second-harmonic generators for long wavelength radiation.

22.3.3 Diamond-Like Carbon

Diamond-like carbon (DLC) is of increasing interest for technical applications. Its hardness is attractive for coatings of specific mechanical tools; its high thermal conductivity together with its insulating properties are interesting in semiconductor device technology, etc.

DLC consists of a mixture of sp^2 (graphite) and sp^3 (diamond) bonds. The relative concentration of these bonds depends on the energy of the impinging species during film growth.

DLC films were produced by excimer laser [*Murray* and *Peeler* 1993; *Rengan* and *Narayan* 1992; *Sato* et al.1988] and Nd:YAG laser [*Scheibe* et al. 1992] ablation of graphite. With uniformly heated substrates, mainly Si and SiO$_2$ ($T_s \approx 25$–$500\,°$C), growth rates between 3 and 10 Å/s were achieved. The films were amorphous, had a hardness of up to 33 GPa, a refractive index $n(633\,\text{nm}) = 2.2 \pm 0.2$, a resistivity around 10^8 Ωcm, and an optical band-gap of 1.1 ± 0.2 eV. Such films could not be dissolved in solutions of HF + HNO$_3$.

DLC films of high optical quality with almost no particulates have been grown by 500 fs KrF-laser ablation of graphite targets [*Müller* et al. 1993]. These properties have been ascribed to both the high degree of atomization/ionization of species and to their high kinetic energies which were, typically, in the 10^3 eV range.

22.3.4 Insulators

Among the dielectric materials deposited as thin films are SiO$_2$, ZrO$_2$, Al$_2$O$_3$, Si$_3$N$_4$, and many fluorides such as MgF$_2$, CaF$_2$, SrF$_2$, etc. The deposition rates achieved vary, typically, between 1 Å/s and 50 Å/s. The depletion of oxygen observed for many oxides can be compensated by reactive ablation in oxygen atmosphere at pressures between 10^{-4} and 1 mbar [*Slaoui* et al. 1992].

ZrO$_2$ films deposited by using an optimized laser fluence and pulse duration have a refractive index $n\,(632.8\,\text{nm}) \approx 2.15$ which is typical for bulk material. With films produced by cw-laser or electron-beam evaporation one finds $n \approx 1.89$ [*Sankur* 1984; *Gaponov* et al. 1977]. When increasing the laser fluence from 28 to 128 J/cm^2 the average size of crystallites increases from 200 to 2200 Å at fixed substrate temperature. This effect was ascribed to the higher energy of species impinging onto the film/substrate surface.

Ferroelectric materials

Among the ferroelectric films fabricated by PLD are oxides of BaTiO$_3$, KaTa$_{1-x}$Nb$_x$O$_3$ (KTN), PbTi$_{1-x}$Zr$_x$O$_3$ (PZT), SrTiO$_3$ (incipient ferroelectric; see, e.g. *Migoni* et al. 1976), and perovskite-like materials such as Bi$_4$Ti$_3$O$_{12}$, LiNbO$_3$, etc.

Epitaxial films of BaTiO$_3$ have been grown on (100) LiF substrates by XeCl-laser ablation of sintered BaTiO$_3$ pellets [*Davis* and *Gower* 1989]. In a similar way, polycrystalline BaTiO$_3$ was deposited on (100) Si substrates. The dielectric constant of these films was $\varepsilon \approx 200$ and their dielectric strength 1MV/cm [*Yeh* et al. 1993].

Crystalline stoichiometric films of $PbTi_{1-x}Zr_xO_3$ ($x \approx 0.48$) were deposited on bare and Pt coated Si substrates by means of KrF-laser radiation ($\varepsilon \approx 850$, remanent polarization ≈ 22 μC/cm^2, coercive field ≈ 40 kV/cm, $\rho \approx 10^{13}$ Ωcm; *Roy* et al. 1992).

Oriented films of $Bi_4Ti_3O_{12}$ (BTO) have been grown on $SrTiO_3$, $LaAlO_3$, MgO, etc., mainly by means of KrF-laser radiation. The fabrication of heterostructures consisting of BTO/YBCO/YSZ/Si has also been demonstrated [*Ramesh* et al. 1991].

Potential applications of ferroelectric films are found in dynamic random access memories (DRAMs).

Polymer targets

Most of the experiments on thin films generated by pulsed-laser ablation of organic polymer targets were performed for mechanistic studies on the degradation and decomposition kinetics [*Hansen* and *Robitaille* 1988]. There is, however, an important exception, PTFE (Teflon). High-quality PTFE films are desirable for many applications. Because of the insolubility of PTFE, the conventional coating techniques can be applied only in very special cases. Here, PLD seems to be a promising technique. Films produced by 266 nm Nd:YAG-laser radiation are composed of amorphous and highly crystalline material [*Blanchet* et al. 1993].

22.4 High-Temperature Superconductors

Thin high-temperature superconductor (HTS) films have been fabricated by pulsed-laser deposition in both reactive and non-reactive atmosphere (Appendix B.8). The material investigated in most detail is $YBa_2Cu_3O_{7-\delta}$ which is also denoted YBCO or Y-123. Among the other materials investigated are different RE-123 compounds, and different phases of $Bi_2Sr_2Ca_{n-1}Cu_nO_{2(n+2)\pm\delta}$ [$n=1$ semiconducting, $n=2$ and $n=3$ superconducting with $T_{c0} = 85$ K and 110 K, respectively], $Tl_2Ba_2Ca_{n-1}Cu_nO_{2(n+2)\pm\delta}$ [$T_{c0} = 80$ K ($n=1$), 110 K ($n=2$), 125 K ($n=3$)], and $HgBa_2Ca_{n-1}Cu_nO_{2(n+1)\pm\delta}$ [$T_{c0} = 95$ K ($n=1$), 120 K ($n=2$), 130 K ($n=3$)].

Non-reactive deposition was mainly performed in vacuum. It requires post-annealing for establishing the superconducting phase. Reactive deposition in O_2, O_3, N_2O, or NO_2 permits *in situ* fabrication of high-quality superconducting films.

In most of the experiments, UV excimer-laser radiation and high density ceramic targets are used. The preference for excimer lasers is related to their short wavelengths, high pulse energies, and short pulse lengths, typically, 10–40 ns (Table I). These properties favor strong light absorption and congruent target ablation. In this parameter regime, the depth of laser energy deposition can be estimated from the heat diffusion length, since $l_T \gg l_\alpha$. For YBCO and KrF-laser radiation, e.g., $l_T \approx 0.6\,\mu\text{m}$ [$D(T_v) \approx 0.05$ cm^2/s, $\tau_l = 20$ ns] while $l_\alpha \approx 0.04$ μm.

Stoichiometric laser ablation of (ceramic) targets requires energy densities of, typically, 1–5 J/cm^2 (Fig. 12.2.1). With low energy short pulses and high energy long pulses, compositional changes have been observed in the targets, and thus in deposited films [*Auciello* et al. 1988; *Heitz* et al. 1990]. Stoichiometric target ablation can still result in films that are sub-stoichiometric with respect to a particular constituent. Surface desorption and outdiffusion from the bulk are probably responsible for the loss of Bi, Tl, and Pb in the respective HTS systems.

Among the substrate materials investigated for thin film deposition were SrTiO$_3$, LaAlO$_3$, MgO, YSZ, ZrO$_2$, Al$_2$O$_3$, LiNbO$_3$, NdGaO$_3$, GaAs, Si, thermally oxidized Si, and different metal sheets such as Ag foils and stainless steel. With substrates that do not match the lattice spacing of the film or which favor strong interdiffusion of species, a buffer layer becomes inevitable. The buffer layer can be fabricated by standard techniques or it is laser-deposited prior to the HTS film by employing an additional target. The (uniform) substrate temperatures used during film deposition are between 20 °C and 850 °C. In most publications, the temperatures quoted were measured at the substrate holder.

22.4.1 Non-reactive Deposition

For YBa$_2$Cu$_3$O$_{7-\delta}$ targets the threshold fluence for stoichiometric ablation with 248 nm KrF radiation is around $\phi_{th} \approx 1$ J/cm^2 (Fig. 12.2.2). The laser fluences commonly employed in film deposition are, however, well above this value. The substrate temperature, T_s, is, typically, 300–400 °C. At such (low) temperatures, deposited films are amorphous and non-superconducting. In order to form the desired superconducting phase with the correct oxygen concentration ($0 \leqslant \delta \leqslant 0.2$), films must be post-annealed. Typically, annealing is performed in O$_2$ atmosphere for 30–60 minutes at temperatures between 800 and 950 °C. In such films the transition to the superconducting state is around $T_{c0} \approx 91$ K (zero-resistance temperature in the absence of a magnetic field; the criterion typically employed is 1–10 μV/cm with bridges of several μm in width). Mainly due to grain boundaries, the critical current densities are by about a factor of 10 smaller than those found with in situ fabricated films.

The synthesis of stoichiometric single-phase films of Bi-Sr-Ca-Cu-O is more complicated because of the multiphase behavior of this material. With Tl-Ba-Ca-Cu-O and Hg-Ba-Ca-Cu-O an additional problem is the incorporation of the correct amount of Tl and Hg. This has been achieved by sputtering sequential layers, e.g., of Ba-Ca-Cu-O and HgO, and/or post-annealing of films in the presence of precursor pellets [*Krusin–Elbaum* et al. 1995].

22.4.2 Reactive Deposition

Post-annealing of laser-deposited films improves their electrical properties, but it is time consuming and, more importantly, it damages the surface morphology and the film-substrate interface, mainly due to material interdiffusion. The high annealing temperatures limit the choice of substrate materials and are incompatible with many potential applications, e.g., in semiconductor device technology, the fabrication of multilayer structures, etc. Therefore, great efforts have been made to produce high-quality films *without* external post-annealing and at substrate temperatures which are as low as possible. Considerable advances in this direction have been achieved by reactive PLD.

An ambient atmosphere attenuates the propagation of the plasma plume (Sect. 30.3). For this reason, the laser fluences are somewhat higher than those employed in vacuum deposition, typically between 3 and 10 J/cm^2. The background pressure of the oxidant is, typically, between 0.01 and 1 mbar. Subsequent to deposition, films are cooled to room temperature in a well-defined way. This is essential for incorporating the correct oxygen content, because oxygen out-/indiffusion is significant at temperatures $T_s > 300\,°C$. Deposition rates for high-quality films are between 1 Å/s and 150 Å/s (about 0.1–6 Å/pulse).

Y-Ba-Cu-O

In situ deposited YBa$_2$Cu$_3$O$_7$ films are black with a smooth mirror-like appearance, good surface hardness, good chemical stability to air, and metallic resistance behavior within the normal conductive state [ρ (100 K) $\approx$ 60 $\mu\Omega$cm on SrTiO$_3$]. Films with thicknesses between 300 Å and 1 μm were produced over areas of up to 5 cm in diameter. With the largest areas, film thickness variations of up to a factor of 2 have been observed. These thickness variations can be diminished by moving the substrate holder. The microstructure of films is crystalline [for a detailed discussion see, e.g., *Eibl* and *Roas* 1990].

The best quality films with respect to the transition temperature, T_{c0}, and the critical current density, j_c (T), have been obtained with (100) SrTiO$_3$ substrates (the mismatch with the b-axis of orthorhombic YBa$_2$Cu$_3$O$_7$ is < 1%). At temperatures $T_s \approx 750\,°C$ epitaxial c-axis oriented (c $\approx$ 11.68 Å $\perp$ to substrate surface) films have been grown. Their T_{c0} values are between 90 and 92 K while j_c (77 K) is within 10^6 and 10^7 A/cm^2. The microwave surface resistance of these films is one to two orders of magnitude lower than for copper [measured at 77 K and frequencies between 1 and 100 GHz; *Inam* et al. 1990].

With other crystallographic orientations of $SrTiO_3$ and other substrate materials, lower values of both T_{c0} and j_c (T) have been obtained. For example, with (100) MgO substrates the best values achieved are $T_{c0} \approx 89$ K and j_c (77 K) $\approx 4 \times 10^6$ A/cm^2.

Figure 22.4.1a shows the temperature dependence of the resistivity of a film deposited by means of KrF-laser radiation onto (100) $SrTiO_3$. The width of the transition was defined by the 10% to 90% resistance interval. A high-resolution transmission-electron microscope (TEM) picture of a similar film taken at the film–substrate interface is shown in b). No grain boundaries can be observed. The dependence of j_c (20 K) on the angle between the (a, b)-plane and magnetic fields of different strengths has been investigated by *Samadi* et al. (1994).

Deposition of Y-Ba-Cu-O in N_2O atmosphere

For substrate temperatures *below* about 650 °C, the superconducting properties of $YBa_2Cu_3O_7$ films deposited in N_2O are superior to those achieved in O_2 (Fig. 22.4.2). With samples prepared in O_2 at temperatures $T_s \leqslant 570$ °C no transition to the superconducting state was observed.

The superior superconducting properties of films deposited in N_2O are tentatively ascribed to atomic oxygen generated by *thermal* dissociation of N_2O at the film surface [*Schwab* et al. 1992]. This becomes plausible from the lower dissociation energy of N_2O with respect to O_2 (Sect. 22.2.1). Nevertheless, higher concentrations of atomic oxygen generated within the plasma plume, and a higher reactivity near the contact surface (Fig. 22.2.1) can also be responsible for this behavior.

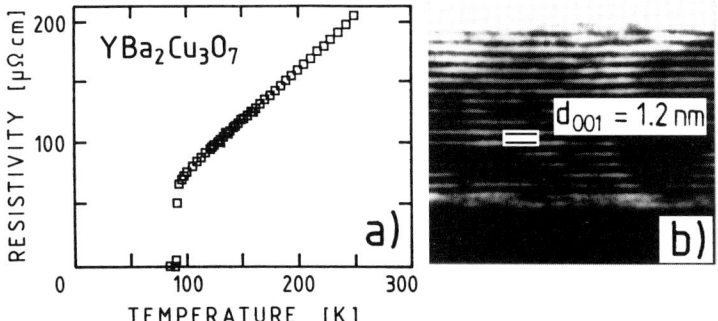

Fig. 22.4.1. (a) Temperature dependence of the resistivity of an $YBa_2Cu_3O_7$ film deposited by KrF-laser radiation ($\phi \approx 3.5$ J/cm^2, $\tau_l \approx 17$ ns, 10 Hz) onto (100) $SrTiO_3$ ($T_s \approx 725$ °C, $p(O_2) \approx 0.4$ mbar). The transition temperature is $T_{c0} \approx 92$ K and the transition width $\Delta T_c \approx 0.6$ K [*Schwab* 1988, unpublished]. (b) TEM picture taken at the $YBa_2Cu_3O_7$/(100) $SrTiO_3$ interface [ϕ (308 nm XeCl) ≈ 4.5 J/cm^2, $\tau_l = 60$ ns, 5 Hz; $T_s \approx 750 \pm 30$ K, $p(O_2) \approx 0.3$ mbar] [*Roas* et al. 1988]

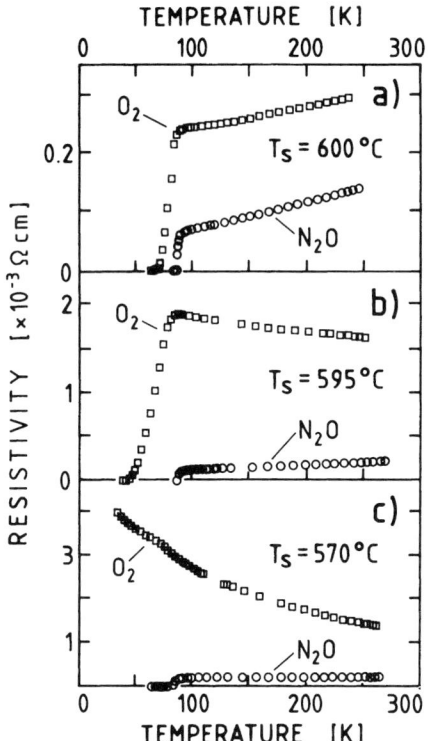

Fig. 22.4.2a–c. Temperature dependence of the resistivity of $YBa_2Cu_3O_7$ films deposited by means of KrF-laser radiation onto (100) MgO substrates ($\phi \approx 4$ J/cm^2, $\tau_l \approx 34$ ns, 10 Hz). The gas pressures were equal in all cases, $p(N_2O) = p(O_2) = 0.4$ mbar [Schwab and Bäuerle 1991]

22.4.3 Buffer Layers, Technological Aspects

For the integration of thin films in device technology, the substrate material that yields the best quality films is not always the most suitable one. An example is $SrTiO_3$ which is an almost ideal substrate for $YBa_2Cu_3O_7$ films. On the other hand, single crystalline $SrTiO_3$ is very expensive and its RF properties are inadequate for applications presently under consideration. With MgO, cubic YSZ, and $LaAlO_3$ substrates the transition temperatures and critical current densities achieved are somewhat lower with respect to $SrTiO_3$. However, these substrates, and, in particular, MgO and YSZ, are much cheaper and have significantly lower RF losses.

For an integration of HTS films in microelectronics, the important substrate materials are Si, SiO_2, Si_3N_4, etc. Si is also one of the preferred materials with other applications, for example, sensors and microwave devices (because of its low RF losses). All of these materials, however, do not fulfil the requirements necessary for high-quality film growth. Besides the mismatches in lattice parameters and thermal expansion coefficients, these materials, and in particular Si, react strongly with YBCO at elevated temperatures. These problems can be overcome, in part, by employing buffer layers. Ideal buffer layers should favor epitaxial growth and block interface diffusion. Here, YSZ seems to be one of the best choices. With YSZ (200–1500 Å thick) fabricated by PLD on (100) Si, c-axis oriented YBCO films with a transition temperature $T_{c0} \approx 90$ K and a critical current density around 10^6 A/cm^2 have been obtained [*Habermeier* 1992]. YSZ has also been used for growing YBCO films on silicon on sapphire (SOS) and metallic substrates, e.g., stainless steel (Appendix B.8).

22.4.4 Heterostructures

Heterostructures allow one to study coupling mechanisms across layered planes of varying compositions, thicknesses, etc.

In-situ epitaxial growth of multiple layers of $YBa_2Cu_3O_7/Y_{1-x}Pr_xBa_2Cu_3O_7$ has been demonstrated over the entire range $0 \leqslant x \leqslant 1$ [*Venkatesan* et al. 1990; *Lowndes* et al. 1992]. C-axis oriented layers with periodicities between some 10 Å and several 100 Å were grown on MgO, SrTiO$_3$ and LaAlO$_3$ substrates. AES depth profiles and cross-sectional TEM indicate abrupt Pr/Y interfaces within one unit cell and virtually no disruptions of layers at interfaces. For device applications the demands on the crystallinity and interface properties are quite stringent. This is related to the short coherence length in HTS (typically a few Å along the c-axis and 20 Å within a,b-planes). Among the various systems under consideration, $Y_{1-x}Pr_xBa_2Cu_3O_7$ is particularly suitable. First, the orthorhombic structure is maintained with only small changes in lattice constants ($< 1.5\%$ for all axes) over the entire range of compositions. Second, this system is superconducting for $x < 0.6$ and semiconducting otherwise.

22.5 Metastable Compounds, Mixed Systems

Pulsed-laser deposition permits one to synthesize compounds that can be prepared either not in single-phase or not at all by solid-state reactions or by standard evaporation techniques. This unique possibility is related to the lower substrate temperatures that can be employed in PLD, to the type and energy of species involved in this process, and to the short turn-around times. Synthesis can be achieved by ablation of either multiple targets or a single target consisting of non-reacted or multiphase *mixtures* of the individual components. We mention two model systems:

Y-Ba-Sr-Cu-O

An example which demonstrates that pulsed-laser deposition enables the extension of the solid–solution range for substitutions in HTS compounds is

$YBa_{2-x}Sr_xCu_3O_7$. Standard ceramic techniques permit one to stabilize the (pure) 123-phase only for Sr contents $x < 1.2$. With pulsed-laser deposition this range was extended up to $x = 1.8$ [*Schwab* et al. 1993]. The chemical composition of films was equal to that of the targets, as proved by electron microprobe (EDX) analysis.

Lu-Ba-Cu-O, Lu-Ba-Sr-Cu-O

Single-phase $LuBa_2Cu_3O_7$ films were successfully prepared on (100) MgO and (100) $SrTiO_3$ substrates ($T_s \approx 690\,°C$) by using O_2 atmosphere and stoichiometric multiphase ceramic targets [*Schwab* et al. 1992]. $LuBaSrCu_3O_7$ could only be synthesized at $T_s \approx 600\,°C$ in N_2O atmosphere.

With (100) $SrTiO_3$ substrates $LuBa_2Cu_3O_7$ and $LuBaSrCu_3O_7$ films were c-axis oriented with T_{c0} values of 90 K and 54 K, respectively. With $LuBa_2Cu_3O_7$, the transition width was $\Delta T_c < 1K$, and the critical current density j_c (83 K) $\approx 10^6$ Å/cm^2.

22.6 Laser-Induced Forward Transfer

The laser-induced forward transfer (LIFT) technique employs laser radiation to transfer a film (target) initially precoated on an optically transparent support onto a substrate (Fig. 22.6.1). With certain systems, good results can also be achieved by laser irradiation of the target film via the (transparent) substrate. Patterning is achieved by direct writing, projection, or laser-beam interference (Fig. 5.2.1).

Among the prerequisites for good morphology, spatial resolution, and adherence of patterns are:

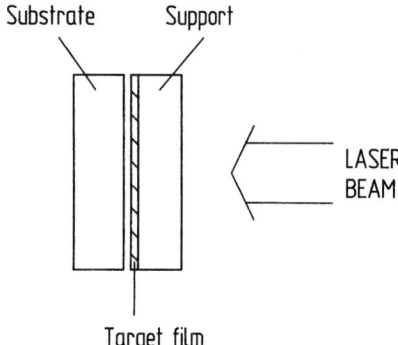

Fig. 22.6.1. Schematic picture demonstrating the LIFT technique. The target film which is precoated on a transparent support is brought into close proximity to the substrate. Irradiation can take place also from the left side, if the substrate is transparent at the laser wavelength

22.6 Laser-Induced Forward Transfer

- The laser fluence employed should just exceed the threshold fluence for removal of the target-film from the support.
- Target films should not be too thick, typically less than a few 1000 Å.
- The target film should be in close contact to the substrate.
- The absorption of the target film should be high and that of the support as low as possible.

The mechanisms of film removal depend on the laser-beam intensity. Near threshold, elastic forces related to thermal expansion of the heated film are important. With increasing intensity, the formation of a vapor or a plasma at the target-support interface dominates the transfer process. Laser-induced modification of the material can proceed after its transfer onto the substrate.

The LIFT technique is simple and can be employed with a wide variety of target films. The film thickness on the substrate can be controlled by repetitive transfer of thin films. In a similar way, multilayer structures can be produced. Problems may occur with the uniformity, morphology, and adhesion of films, and with material implantation into the substrate surface. At low laser fluences transfer may be incomplete, while at high fluences droplet formation or substrate doping/alloying is observed. Additionally, plasma formation at the film-support interface may result in surface ablation of the support. This introduces impurities into the transferred material and, eventually, into the substrate.

Pattern formation by LIFT has been demonstrated mainly for metals [*Kantor* et al. 1994; *Toth* et al. 1993; *Mogyorosi* et al. 1989b; *Baseman* et al. 1988; *Adrian* et al. 1987], but also for some semiconductors and high-temperature superconductors [*Fogarassy* 1990]. Film supports frequently employed are standard glass, SiO_2, Al_2O_3, $SrTiO_3$, etc. The substrate materials investigated include various metals, semiconductors, ceramics, glasses, and polymers.

The smallest pattern sizes presently achieved with metal films are several μm wide. These widths exceed the laser foci employed in laser direct writing, mainly due to lateral heat conduction within the target film. The fluence necessary for material transfer increases with increasing film thickness. Practical metal film thicknesses are between several 100 Å and 3000 Å. The corresponding fluences employed with ns pulses are typically $0.1-10$ J/cm^2.

Material transfer from targets consisting of stacked layers of elements has also been demonstrated. By tuning the laser fluence, multilayer deposits with low interlayer mixing, or metal alloy patterns can be produced.

Part V: Surface Transformations, Synthesis and Structure Formation

Laser-induced surface modifications can be classified into physical, chemical, and physicochemical transformations. *Physical* surface transformations can be performed in an inert atmosphere and they take place without any changes in the overall chemical composition of the material. Here, no real chemical reaction between the different constituents, if present, is activated. *Chemical* transformations are characterized by an overall change in the chemical composition of the material or the activation of a real chemical reaction. This shall include material synthesis, decomposition, and some types of surface modifications. The latter can take place either in a chemically reactive ambient medium or they are performed by adding a new material to the surface, or by depleting a certain constituent of the material from the surface. With *physicochemical* transformations both physical and chemical processes are important.

Large-area surface modifications are performed with excimer lasers, Nd:YAG or Nd:glass lasers, and with high power CO_2 lasers. For localized processing, low-power cw lasers such as Ar^+ or Kr^+ lasers are employed.

23 Structural Transformations

Laser-induced structural transformations such as transformation hardening, annealing, recrystallization, glazing, shock hardening, etc., are based on the high processing temperatures that can be reached during *short* heating and cooling cycles under high-power pulsed laser or rapidly scanned cw-laser irradiation. Short processing cycles permit material transformations within thin films and surfaces without significant influence on the substrate or the underlying bulk material. In the case of surface absorption, which is a good approximation with many applications, the thickness of the heated zone is approximately described by the heat diffusion length, l_T. The time for heating the material to a certain temperature and depth, and the time for cooling can both be calculated from the equations given in Chaps. 6–9. The thickness of the modified layer, $\Delta h \approx l_T$, decreases with decreasing pulse length. With ultrashort laser pulses Δh becomes so small that cooling rates up to more than 10^{12} K/s can be achieved. If $\tau_l < D/v_0^2 \approx 10^{-12}$ to some 10^{-14} s (v_0 is the sound velocity), the finite velocity of the heat front must be taken into account (Sect. 2.2). In any case, with such cooling rates it is possible to freeze *non-equilibrium* phases, suppress nucleation, etc. There is, however, a limitation. The transformation temperature must be sustained during a time which is longer or at least comparable to the time required for the phase transformation to take place. Furthermore, with many systems, successful laser processing is related to strong temperature gradients which induce internal stresses, redistributions of defects, different types of transport phenomena, etc.

23.1 Transformation Hardening

Transformation hardening is performed by heating the material under investigation above a certain transformation temperature and subsequent rapid cooling. The transformation temperature is, in general, *below* the melting temperature. This type of heat treatment permits one to freeze non-equilibrium phases with modified physical properties.

A well-known example is the transformation hardening of steel. With carbon steel the austenitizing temperature is above 723 °C (eutectic), depending on the carbon content. At this temperature, the (soft) pearlite phase with non-dissolved carbon transforms into austenite which is a solid solution of carbon in γ-Fe. Rapid cooling at rates $> 10^3$ K/s results in the formation of a (hard) metastable martensite. Frozen-in thermal stresses, micro-cracks, segregation phenomena, etc., are often of importance in hardening processes as well.

With many applications it is desirable to harden only the material surface and otherwise retain the bulk properties. In such cases, lasers are particularly useful because they permit rapid heating and cooling of the material surface without affecting the bulk; for the same reason, material distortions are significantly diminished.

Experiments were mainly performed with CO_2 lasers at powers exceeding 10^3 W. Because CO_2-laser radiation is strongly reflected by metals (Fig. 7.2.4; for steel and perpendicular incidence $A(\Theta = 0) \leqslant 0.2$) surfaces are often coated with absorbing layers consisting of graphite, sulfides such as MoS_2 or Fe_2S_3, metal oxides, or black inks. Absorption can also be increased by employing the Brewster effect. With metals for which $A(\Theta = 0) \ll 1$, irradiation with π-polarized light under the Brewster angle

$$\Theta_B \approx \frac{\pi}{2} - \frac{A(\Theta = 0)}{2\sqrt{2}}$$

yields an absorptivity $A(\Theta_B) \approx 2/(1 + \sqrt{2}) \approx 0.83$ which is *independent* of the specific material.

In a typical experiment, a laser beam with an intensity of about 10^4 W/cm^2 is scanned over the surface with a dwell time $\tau_l = 2w/v_s$ of 0.01 to 1 s. τ_l determines the depth of the heat affected zone (HAZ). Laser-hardened surface layers are, typically, some 0.1 to a few mm thick, depending on the material properties and the laser parameters employed. Because of the shorter cooling cycles with respect to standard processes (induction heating, flames, arcs, etc.) the martensite formed in laser hardening is finer and the residual stress more compressive. Thus, laser-treated surfaces have improved fatigue and wear properties with reduced friction coefficients. For steel, the (Vickers) hardness is, typically, increased by a factor of 3–5.

For materials in which solid-phase diffusion is *not* fast enough to obtain complete solution of the relevant species, laser transformation hardening can be achieved by melting and resolidification (Chap. 10). In such cases, surface roughening and subsequent machining are, in general, inevitable.

The criteria for laser hardening and the depth of the hardened zone have been investigated both experimentally and theoretically for various types of steel [*Davis* et al. 1986; *Ashby* et al. 1985; *Gregson* 1983; *Wissenbach* et al. 1985]. The depth of the hardened zone can be estimated from the laser-induced temperature distribution, the phase diagram of the material, and the related kinetic constants.

23.2 Laser Annealing, Recrystallization

Laser annealing shall denote the epitaxial regrowth of thin defective or amorphous layers which are formed during a particular processing step on otherwise crystalline bulk material. The term laser-induced recrystallization often refers to the transformation of small grain polycrystalline or amorphous films or slabs into large-grain crystalline material. This includes the crystallization of films deposited on substrates by standard evaporation, electron-beam evaporation or other techniques. In the literature, the terms laser annealing and laser recrystallization are often used as synonyms.

The advantages of laser annealing and recrystallization over standard (oven) annealing are related to the short processing cycles that can be achieved with lasers. The good surface crystallinity of laser-annealed/recrystallized materials is often related to laser-induced melting and regrowth of the material. Here, solidification velocities can reach several meters per second (Chap. 10). Solid-phase transformations where the laser-induced surface temperature stays *below* the melting temperature have been demonstrated as well. In any case, preheating of the substrate is advantageous in most of these applications because it allows processing at lower laser-light intensities and better process control. Most of the investigations on laser annealing have concentrated on ion implanted silicon surfaces. Laser-induced recrystallization of films has been demonstrated for a large number of materials. Laser- and electron-beam remelting of ion-implanted metal surfaces has been demonstrated for Al, Cu, Ni, etc. [*Hirvonen* et al. 1980; *Buene* et al. 1981; *Picraux* and *Follstaedt* 1983]. The motivation for such experiments is the improvement of surfaces against corrosion and wear.

23.2.1 Ion-Implanted Semiconductors

One of the transformations studied in most detail is the recrystallization of amorphous and polycrystalline semiconductor surfaces, and in particular of Si [*Cullis* 1985; *Celler* 1984; *Brown* 1983]. Thin amorphous films are generated on single-crystal Si wafers during surface doping by ion implantation which is widely used for the fabrication of thin p- or n-type layers in semiconductor device technology. The amorphized surface layers, which are typically some 0.1 μm thick, must be recrystallized in order to restore the physical properties of crystalline Si and to electrically activate the dopant atoms by incorporating them into proper (substitutional) lattice positions.

Pulsed-laser annealing

Pulsed-laser annealing (PLA) offers new possibilities in comparison to oven annealing. With laser-beam dwell times of, typically, some ps to some 100 ns, diffusion broadening or redistribution of dopant profiles, or the precipitation of dopants, can widely be suppressed. This permits one to generate very high concentrations of *active* dopants. Figure 23.2.1 demonstrates some of the main features for the example of ion-implanted As and B in (100) Si. Similar results have been obtained with In, Ga, P, Sb, etc. The solid solubility limit in Si can be exceeded by a factor of up to 100. More than 95% of the implanted dopant atoms can be (electrically) activated. Pulsed-laser annealing involves, in general, surface melting. Almost defect-free crystallization of the melt on pure (100) Si has been found for solid–liquid interface velocities, v_{sl}, of up to about 15 m/s (Fig. 23.2.2). The maximum interface velocities for *defect-free* crystal growth on pure Si lie in the order $(001) > (011) > (112) > (111)$.

Pulsed lasers are the superior sources for the fabrication of very steep and ultra-shallow pn junctions, good ohmic contacts or regions with very low sheet resistivity. The lasers most commonly employed are excimer lasers and Q-switched Nd:YAG lasers. In any case, the photon energy should exceed the band-gap energy, i.e., $hv > E_g$. Otherwise, process control becomes very difficult due to the strong feedbacks in the absorption process (Sect. 7.6). All of the experimental results achieved with ns and ps laser pulses can be explained on the basis of purely thermal processes [*Malvezzi* et al. 1985].

Cw-laser annealing

Cw-laser annealing does *not* involve surface melting, in general. Here, the laser-beam dwell times are, typically, some ms. Today, surface annealing of

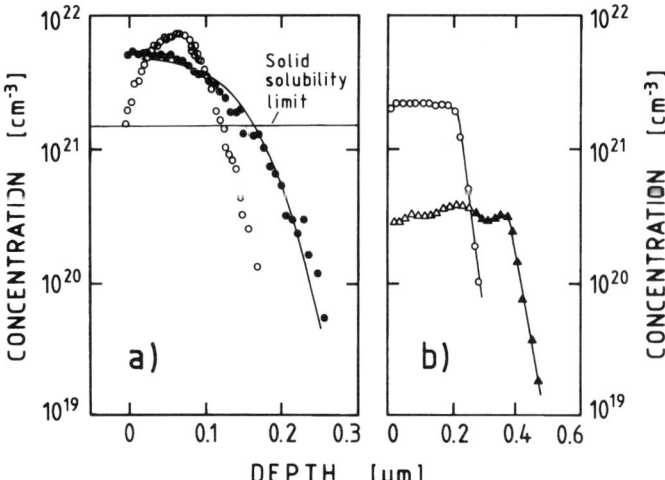

Fig. 23.2.1a,b. Dopant profiles generated in (001) Si. (a) ○ after implantation with 100 keV As$^+$ ions (6.4×10^{16}/cm^2). ● after laser annealing, adapted from [*White* et al. 1980]. (b) Implantation of B$^+$ (△, ▲) and As$^+$ (○) ions followed by multiple-pulse laser annealing [*Hill* et al. 1982]

23.2 Laser Annealing, Recrystallization

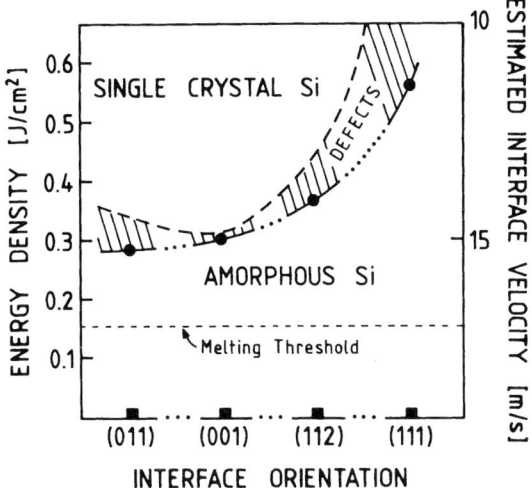

Fig. 23.2.2. Maximum (estimated) interface velocities for defect-free crystalline growth (dashed curve) and amorphization on Si for 347 nm frequency-doubled ruby-laser irradiation ($\tau_l \approx 2.5$ ns) and for different crystallographic orientations [*Cullis* 1985]

ion-implanted Si is mainly performed with intense lamps. This widely used technique is termed rapid-thermal annealing (RTA).

Laser annealing of compound semiconductors such as GaAs, InP, etc., has *not* achieved real importance. Here, higher processing temperatures result in defect formation and material decomposition.

23.2.2 Thin Films

Laser-induced recrystallization of thin films of silicon on insulators (SOI) has been studied by *Fan* and *Zeiger* (1975), *Nguyen* et al. (1984), and others. Among the systems investigated in most detail are silicon on sapphire (SOS) and silicon on quartz (SOQ). Pulsed-laser crystallization of amorphous Si (a-Si) films on quartz substrates for the fabrication of high mobility poly-Si thin-film transistors (TFT) has been investigated by *Fogarassy* et al. (1993). With ArF excimer-laser pulses which cause full melting of the as-deposited layer, field-effect mobilities up to 140 cm^2/Vs have been achieved.

Among the other experimental examples are the transformation of (insulating) diamond-like carbon (DLC) films into graphite (this is associated with a change in resistivity from 10^6 Ω cm to 0.1 Ω cm [*Prawer* et al. 1986]), the transformation of C$_{60}$ [*Phillips* et al. 1993a], the surface annealing of silicon nitride films on Si substrates [*Donnelly* and *Mucha* 1989], the fabrication of micron-sized magnetic domains in amorphous (magnetic) films of TbFe [*Suits* et al. 1986], etc.

New phenomena may occur during surface patterning by laser direct writing of crystalline lines into amorphous films. With the transformation of amorphous Si and Ge the heat of crystallization can become comparable to the laser power absorbed. As a consequence, explosive crystallization which gives rise to the formation of periodic structures is observed (Chap. 28).

While the incorporation of laser annealing and recrystallization in real device fabrication processes is still questionable, these investigations have stimulated rapid thermal annealing with lamps and also revealed many fascinating fundamental aspects. Among the latter are rapid solidification processes with liquid–solid interface velocities of several meters per second. Such investigations provide insights into a new regime of crystallization. Time-resolved measurements with picosecond and femtosecond laser pulses open up new regimes of photon–electron–phonon interactions [*Lompré* et al. 1983]. Studies on dopant segregation, supercooling, the formation of metastable phases, instabilities, and structure formation have elucidated many new phenomena at liquid–solid and solid–solid interfaces.

23.3 Glazing

Just as amorphous layers can be crystallized under the action of laser light, crystalline surfaces can often be transformed into the amorphous state. This is achieved if the cooling rate exceeds a critical velocity, v_{ls}^{cr}, which strongly depends on the type of material. v_{ls}^{cr} can vary over many orders of magnitude.

As already mentioned, amorphization of Si takes place above interface velocities of 11–15 m/s, depending on crystal orientation (Fig. 23.2.2). With high dopant concentrations, amorphization thresholds can significantly change [*Campisano* et al. 1984].

Laser treatment of *metals* often results in the formation of fine grain polycrystalline material and not in real metallic glasses. Glass formation requires cooling rates in excess of 10^6–10^{12} K/s. The critical interface velocities are typically 10^2–10^3 m/s.

With *insulators*, liquid phases can be frozen-in with cooling rates between 10^{-4} and 10^2 K/s.

The thickness of amorphized layers is typically in the range 0.1–10^3 µm. Laser-induced glass formation is applied for improved corrosion resistance and surface hardening of metals. Problems may arise with crack formation, recrystallization, etc. Investigations of this type have been extensively reviewed [*von Allmen* and *Blatter* 1995]. The kinetics of non-equilibrium phase transformations has been reviewed by *Chvoj* (1993).

23.4 Shock Hardening

In contrast to the structural transformations discussed in the preceding sections, laser-induced shock hardening is performed at typical laser-light intensities of $I > 10^8\,\text{W/cm}^2$ (Sect. 11.2.3). Shock hardening is based on the densification observed with some materials under the action of shock waves [*Fairand* et al. 1972, 1974]. An example is the transformation of some fourfold coordinated non-metals into more densely packed metallic phases. Shock hardening can also be related to the generation of very high densities of nonequilibrium defects (vacancies, dislocations, etc.).

24 Doping

As with laser-induced structural transformations, laser-induced surface doping takes advantage of the high heating and cooling rates that can be achieved with lasers. The short temperature cycles enable one to produce very shallow, heavily doped layers within solid surfaces. Pulsed lasers are mainly used for both large-area doping and local doping by projection. Cw lasers allow local doping by direct writing. In any case, the absorbed light intensity must be high enough to substantially heat or even melt the sample surface in order to allow dopant incorporation by high-temperature diffusion or liquid-phase transport. The dopant source may be an adsorbate, a gas, a liquid, or an evaporated film.

Doping from *gas-phase* precursors requires thermal or photochemical decomposition of parent molecules. With surface melting, gas-phase transport of species is usually rate-limiting, in particular if convective fluxes within the molten layer become effective. Efficient surface doping based on solid-phase diffusion of species was observed for some types of polycrystalline materials, and for III-V semiconductors with dopants such as Cd, S, Se, and Zn.

The *thickness* of the doped layer can be controlled via the laser-beam dwell time. With fixed dwell time, the dopant density within the solid can be tuned via the concentration of precursors within the gas or liquid, via the thickness of the evaporated film, etc.

This chapter deals with the incorporation of *electrically active* dopants into semiconductor surfaces and, in particular, with Si. This is based on the special importance of Si and also on its excellent stability which allows one to study wide parameter ranges.

The deposition of doped films is discussed in Chaps. 19, 20, and 22. Surface doping by ion implantation in combination with laser annealing is described in Chap. 23. Sheet doping in connection with thin film coating has been studied for metals, particularly for stainless steel. This is included in Chap. 25.

24.1 Solid-Phase Diffusion

Surface doping by solid-phase diffusion is schematically shown in Fig. 24.1.1. The starting material shall consist of a semi-infinite substrate S that is covered

24.1 Solid-Phase Diffusion

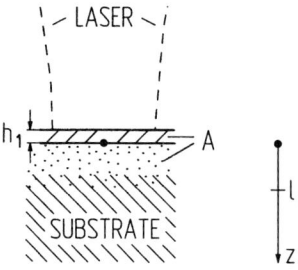

Fig. 24.1.1. Surface doping by laser-induced solid-phase diffusion. The semi-infinite substrate is covered with a dopant layer A

with a layer of the dopant, A. Due to laser heating, species A will diffuse into S and vice versa. In the one-dimensional case, the density profile of A in S can be described by $N_A \equiv N_A(z,t)$. Because of the temperature gradients in z-direction, N_A must be calculated by solving the diffusion equation

$$\frac{\partial N_A(z,t)}{\partial t} = \frac{\partial}{\partial z}\left(D_s(T(z,t))\frac{\partial N_A(z,t)}{\partial z}\right) \tag{24.1.1}$$

simultaneously with the heat equation. $D_s(T(z,t))$ is the diffusion coefficient for species A in S (we denote the coefficient for solid-phase diffusion by D_s and for liquid-phase diffusion by D_l). The temperature dependence of D_s can be approximated by

$$D_s(T(z,t)) = D_0 \exp\left(-\frac{\mathscr{E}_d}{T(z,t)}\right). \tag{24.1.2}$$

The initial condition shall be characterized by

$$N_A(z, t=0) = 0 \quad \text{with } z > 0. \tag{24.1.3}$$

The concentration of species A at the interface $z=0$ shall be constant, and it shall vanish far away from the surface, i.e.,

$$N_A(0,t) = N_A^0 \quad \text{and} \quad N_A(\infty, t) = 0. \tag{24.1.4}$$

The heat equation can be written in analogy to (2.2.1).

Let us consider surface doping where we can set $h_1 \approx 0$. Because the heat diffusion length, l_T, or the optical penetration depth, l_α, is much larger than the effective diffusion length of dopants, i.e., $l_s \ll \max\{l_\alpha, l_T\}$, the temperature within the depth l_s can be assumed to be uniform and the diffusion coefficient becomes independent of coordinate z. The solution of the boundary-value problem is then

$$N_A(z,t) = N_A^0 \operatorname{erfc}\left(\frac{z}{l_s}\right). \tag{24.1.5}$$

The spatial variation of N_A is schematically shown in Fig. 24.1.2 by the solid curve. The effective diffusion length can be approximated by $l_s \approx 2(\langle D_s \rangle t)^{1/2}$

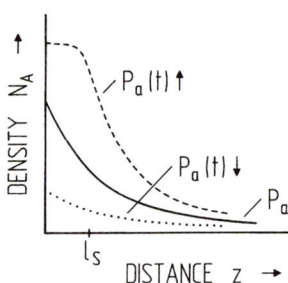

Fig. 24.1.2. Concentration profiles observed in laser-enhanced solid-phase diffusion. z is the distance from the substrate surface (Fig. 24.1.1). The absorbed laser power P_a can be constant during the laser-beam dwell time τ_l (solid curve), it can increase (dashed), or decrease (dotted curve)

where $\langle \rangle$ denotes the time-average, i.e.,

$$\langle D_s \rangle = \frac{1}{t} \int_0^t dt' \, D_s(T(0, t')) . \tag{24.1.6}$$

For pulsed-laser irradiation with uniform intensity and constant absorptivity the diffusion length l_s at time t can be estimated from [*Libenson* and *Nikitin* 1973]

$$l_s \approx 2 \left[\frac{[\Delta T(t)]^2}{\mathscr{E}_d \dot{T}(t)} D_s(T(t)) \right]^{1/2} . \tag{24.1.7}$$

For *surface absorption* this yields

$$l_s \approx 2 \left[\frac{2\Delta T(0)}{\mathscr{E}_d} D_s(T(0)) t \right]^{1/2} , \tag{24.1.8a}$$

where $T(0) \approx \Delta T(0) = I_0 A \, l_T / \sqrt{\pi} \kappa$ and $l_T \approx 2(Dt)^{1/2}$. For *finite absorption* with $l_\alpha \gg l_s$ we obtain

$$l_s \approx 2 \left[\frac{\alpha I}{\mathscr{E}_d \rho c_p} D_s(T(t)) t^2 \right]^{1/2} . \tag{24.1.8b}$$

For Si and dopant atoms such as B, Bi, Ga, In, P, and Sb, typical values of D_s at temperatures well below melting are within 10^{-12} cm²/s $< D_s(T \ll T_m) < 10^{-10}$ cm²/s. For temperatures near T_m typical values are 10^{-6} cm²/s $\leq D_s(T \leq T_m) \leq 10^{-5}$ cm²/s [*Kimerling* and *Benton* 1980]. With $\Delta T(0) = 10^3$ K, $D_s = 10^{-6}$ cm²/s, $\mathscr{E}_d = 10^4$ K and $\tau_l = 10^{-3}$ s we obtain $l_s \approx 0.3$ μm.

The preceding treatment ignores a number of important effects:

- Interdiffusion of A and S and temperature dependences in R, α, etc., change the optical properties of the solid. Thus, depending on the particular system, the absorbed laser power will increase (positive feedback) or decrease (negative feedback) during the laser-beam dwell time. Changes in absorbed laser power result in changes in the temperature distribution and thereby in the diffusion profile of species. This is schematically shown in Fig. 24.1.2 by dashed and dotted curves for positive and negative feedback, respectively.

- Thermal diffusion related to strong temperature gradients, in particular in pulsed-laser doping.
- Stress gradients related to thermal expansion increase, in general, solid-phase diffusion [*Shewmon* 1963]. They can be estimated by solving the problem of thermoelasticity. Stress gradients can also decrease the activation energy for diffusion, change the optical properties of the material, etc.
- Defects such as vacancies, dislocations, microcracks, etc., may significantly enhance solid-phase diffusion of dopants. Defects can be generated by laser-induced heating, stresses, shock waves, etc.

  Thermally induced defects are localized near the solid surface. Near melting, the concentration of vacancies can be very high, in metals and semiconductors up to 10^{19} or 10^{20} cm^{-3}. Due to thermal diffusion and stress effects, the real concentration of vacancies can be much higher than their equilibrium concentration. The diffusion coefficient can then be approximated by

$$D_s^{ne} = D_s \frac{N^{ne}}{N},$$

where N^{ne} and N denote non-equilibrium and equilibrium concentrations of defects, respectively. N^{ne}/N can be of the order of 10^5 to 10^6. When the density of vacancies exceeds a critical value, condensation and droplet formation may occur. This is well known from radiation damage caused by particle bombardment. It is often termed *cold melting*.
- Laser-induced shock waves can produce a high density of defects within a *large* volume of the solid.
- Electronic [*Wautelet* et al. 1988] or vibrational [*Karlov* et al. 1992] excitations can induce charge-transfer effects, generate electron-hole pairs, vacancies, etc., and may thereby enhance diffusion of host or impurity atoms/ions.
- Compound formation in laser alloying and synthesis can result in the formation of various kinds of defects which also enhance diffusion.

Because of the numerous contributions to solid-phase diffusion, an estimate of the density N_A, as presented at the beginning of this section, will hardly pretend to be quantitative.

24.2 Liquid-Phase Transport

Surface melting considerably increases processing rates in surface doping and alloying. This is due to enhanced transport by both diffusion and convection (Chap. 10). For surface alloying from thin evaporated layers, the melt depth h_l must exceed h_1.

The diffusion length of species within the melt can be estimated from $l_l \approx 2(D_l \Delta t_m)^{1/2}$ where Δt_m is the time during which the surface stays molten.

The laser fluences commonly employed in these processing applications are between $0.1\,\text{J/cm}^2$ and $10\,\text{J/cm}^2$. In the simplest approximation, h_l can be estimated from (10.1.3), and Δt_m from (10.2.1).

Liquid-phase diffusion coefficients exceed solid-phase diffusion coefficients near melting by one to three orders of magnitude. In molten metals, typical values of D_l are between $10^{-5}\,\text{cm}^2/\text{s}$ and $10^{-4}\,\text{cm}^2/\text{s}$. In Si, liquid-phase diffusion coefficients for dopant atoms such as $\text{B}[D_l(\text{B}) \approx 2.5 \times 10^{-4}\,\text{cm}^2/\text{s}]$, $\text{Bi}, \text{Ga}, \text{In}, \text{P}[D_l(\text{P}) \approx D_l(\text{B})]$ and Sb are almost equal and typically in the range $10^{-4} \leqslant D_l \leqslant 10^{-3}\,\text{cm}^2/\text{s}$.

For Si and excimer-laser radiation, the estimated melt depth achieved with the parameters typically employed is between $0.1\,\mu\text{m}$ and $10^3\,\mu\text{m}$. Thus, l_l is significantly smaller than h_l. These simple estimates are supported by numerical calculations [*Sameshima* and *Usui* 1987].

Experimental investigations have demonstrated, however, that in surface doping and alloying species A and S are often well mixed within the *total* melt depth. This fast mixing can only be explained by convective fluxes or surface instabilities within the liquid layer. Remnants of convective fluxes and instabilities frequently appear as cellular structures on the resolidified surface [*Fogarassy* et al. 1985]. The flow velocities, v_c, are between $1\,\text{cm/s}$ and some m/s (Sect. 10.4). For a homogeneous distribution of A in S, convective mixing must take place within a depth $l_c \approx v_c \Delta t_m \approx h_l$.

24.3 Sheet Doping

Large-area thin-layer doping (sheet doping) has mainly been performed with (pulsed) UV-laser radiation which is strongly absorbed in semiconductors and also photodissociates most of the relevant precursor molecules.

24.3.1 Silicon

Surface doping of Si has been demonstrated with adsorbed layers, gases, liquids, and solid films.

Adsorbed layers, no ambient medium

Adsorbed layers provide a finite source of dopant and permit extremely shallow doping profiles to be generated. Because the precursor gas is pumped off prior to laser-light irradiation, efficient adlayer doping requires strong adsorption of parent molecules (Sect. 20.2.1). Most of the experiments were performed with excimer lasers using BCl_3, B_2H_6, BF_3, and PCl_3 as precursors [*Deutsch* 1984].

24.3 Sheet Doping

Gaseous Ambient

Gas-phase doping of Si from AsH_3, BCl_3, BF_3, B_2H_6, PCl_3, and PH_3 has been demonstrated mainly with ArF-laser radiation. BF_3, which does not absorb 193 nm radiation, is thermally decomposed at the gas-solid interface. The other molecules are directly photodissociated by ArF-laser radiation [*Clark* and *Anderson* 1978; *Slaoui* et al. 1990].

Figure 24.3.1a exhibits SIMS profiles of the B concentration obtained with (100) n-type Si wafers. With the laser parameters employed, the Si surface is melted [$\phi_m(197 \text{ nm}) \approx 0.25 \text{ J/cm}^2$]. For a *single pulse*, the melt depth estimated from (10.1.3c) with $\phi_a \to (\phi - \phi_m)(1 - R)$ and $R = 0.7$ is $h_l^{max} \approx 0.54 \mu m$. From (10.2.6b) we find the solidification time $\tau_s \approx 60$ ns. The time during which the surface is molten is $\Delta t_m \approx \tau_l + \tau_s \approx 80$ ns. The estimated diffusion length within the liquid, $l_l \approx 2(D_l \Delta t_m)^{1/2} \approx 0.09 \mu m$, is in good agreement with the doping depth derived from the single-pulse data. Because the observed doping depth agrees well with l_l but not with h_l^{max}, convection seems to be unimportant in these experiments. Thus, we can calculate the doping profile from (24.1.5). The result is shown in Fig. 24.3.1b. It is in excellent agreement with the experimental data. For *multiple-pulse* (mp) irradiation we estimate $\Delta t_m(\text{mp}) \approx N \Delta t_m$. Thus, dopant atoms further diffuse into the molten layer and $l_l(\text{mp}) \approx l_l \sqrt{N}$. The agreement between the experimental data obtained for 5 pulses and the corresponding theoretical curve is still reasonable. When further increasing the

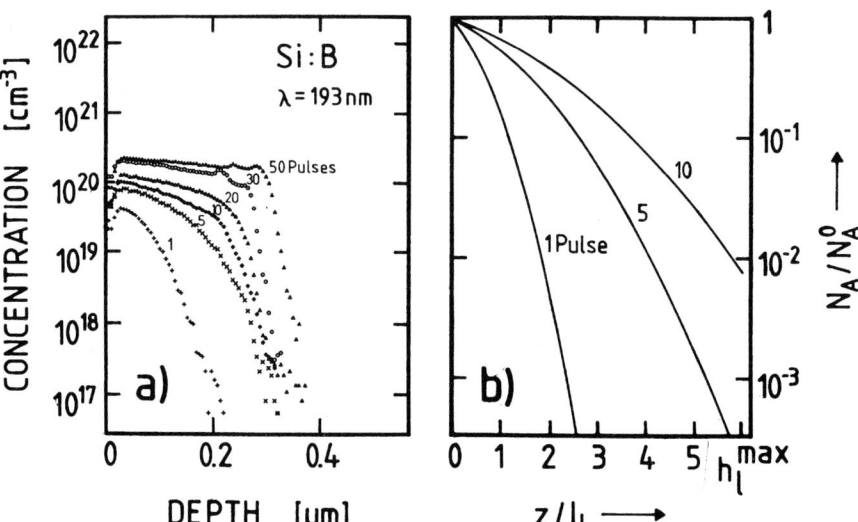

Fig. 24.3.1a,b. ArF-laser-induced doping of (100) Si with B. (a) SIMS profiles of the B concentration ($\phi \approx 1 \text{ J/cm}^2$; $\tau_l \approx 21$ ns; 2 Hz; $p(BCl_3) \approx 6.7$ mbar) [*Slaoui* et al. 1990]. (b) Concentration profile of B calculated from (24.1.5) with l_s replaced by l_l

number of pulses, the concentration of dopants becomes almost uniform within the molten layer and drops off sharply beyond it because $D_s \ll D_l$. This transition is observed when $l_t\sqrt{N} \approx h_l^{max}$, i.e., with $N \approx 36$ pulses. This is in qualitative agreement with the experimental observation. Due to the cut-off in diffusion with $z \approx h_l^{max}$, significant differences between the data and calculated curves occur already with $N > 5$ pulses.

Figure 24.3.2 shows the sheet resistance of (100) Si versus the number of ArF-laser pulses for BF_3 and B_2H_6. The initial drop-off in sheet resistance is much slower for BF_3 compared to B_2H_6, in spite of the much higher pressure employed. This demonstrates that photolysis of B_2H_6 substantially enhances the doping efficiency. With B_2H_6 the surface carrier concentration saturates at about 10^{21} B atoms/cm^3. This value exceeds the equilibrium solubility of boron in silicon at 1300 °C by about a factor of two. From electrical measurements and SIMS profiles, it was estimated that about 70% of the B atoms incorporated are electrically active. With very high doping densities, the mobility of carriers is diminished due to scattering.

Adsorbed molecules play an important role in systems such as Si:BCl_3. This follows from the pressure dependence of the sheet resistance which was found to be in qualitative agreement with adsorption isotherms [*Deutsch* 1984].

Liquid ambient

Liquid-phase doping of Si with P and Sb has been demonstrated by *Stuck* et al. (1981). The doping solutions were either tributyl-phosphate ($C_{12}H_{27}O_4P$)

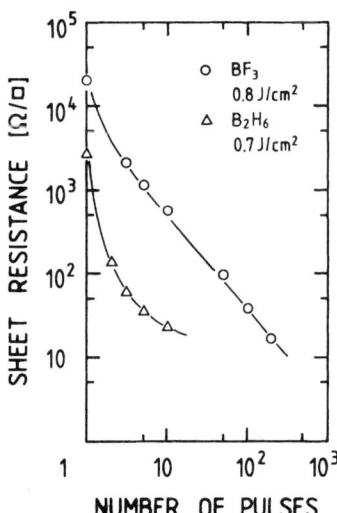

Fig. 24.3.2. Sheet resistance versus number of laser pulses. The partial pressures of BF_3 and B_2H_6 precursors were about 67 mbar and 6.7 mbar, respectively [*Matsumoto* et al. 1990]

24.3 Sheet Doping

with 2.5×10^{21} P atoms/cm^3, or SbCl$_3$ with 1.5×10^{21} Sb atoms/cm^3 in ethanol. Irradiation was performed with Q-switched ruby laser light ($\phi \approx 1-2$ J/cm^2, $\tau_l \approx 20$ ns) which is either only very slightly or else not at all absorbed by these liquids. The sheet resistances of P and Sb doped layers were $120\,\Omega/\square$ and $60-80\,\Omega/\square$, respectively.

Solid films

Doping of Si with Al, B, Bi, Ga, In, P, and Sb was also performed by laser-induced heating of spun-on or evaporated dopant films [*Inui* et al. 1991; *Fogarassy* et al. 1985; *Deutsch* 1984; *Narayan* et al. 1978]. Some of the experimental results are in reasonable agreement with model calculations and similar to those obtained by laser annealing of ion-implanted Si surfaces.

Applications

Laser-induced sheet doping of Si can be used to produce very shallow junctions [*Bollmann* et al. 1993; *Kramer* et al. 1993; *Matsumoto* et al. 1990]. The I-V characteristics of pn-junction diodes fabricated by ArF-laser doping from B$_2$H$_6$ is depicted in Fig. 24.3.3.

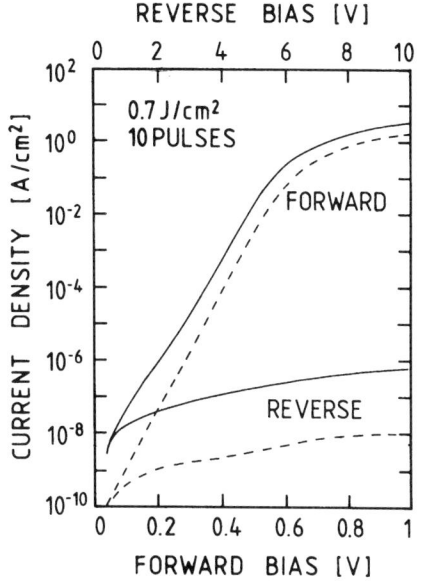

Fig. 24.3.3. I-V characteristics of pn-junction diodes fabricated by ArF-laser doping of Si in B$_2$H$_6$. The surface carrier concentration is $\approx 10^{21}$ cm^{-3} and the junction depth $\approx 0.1\,\mu$m. The solid and dashed curves refer to non-annealed and annealed (800 °C, 30 min) samples, respectively [*Matsumoto* et al. 1990]

24.3.2 Compound Semiconductors

Laser-induced sheet doping of compound semiconductors is more problematic. With the laser fluences necessary for efficient in-diffusion of dopants, the surface may decompose via selective evaporation of the more volatile component. Additionally, laser-induced heating can introduce slip planes and other defects which degrade the electrical properties of these materials. Doping was mainly studied for GaAs and InP by employing gas-phase precursors.

Doping of GaAs was performed with frequency-doubled Nd:YAG lasers (with and without blocking the fundamental line) and excimer lasers.

The formation of n-type layers was demonstrated with dopants such as Si [*Sugioka* and *Toyoda* 1994; *Bentini* et al. 1988], Se [*Kräutle* et al. 1985], S [*Zhang* et al. 1994; *Deutsch* 1984], and Ge [*Garcia* et al. 1988]. The Se doping source was gaseous H_2Se diluted in $H_2 + AsH_3$. During the 3 ns Nd:YAG-laser pulse employed in these experiments, the surface is heated to temperatures near melting. H_2Se molecules adsorbed on the GaAs wafer thermally decompose and Se diffuses into the surface. The loss of As during heating is compensated by the supply of As from adsorbed AsH_3. With 532 nm radiation, surface layers with a thickness less than 0.02 μm and with more than 10^{20} Se atoms/cm^3 have been produced. Thicker dopant profiles have been obtained by simultaneous illumination with the 1064 nm fundamental line. Similar experiments have been performed with $Zn(C_2H_5)_2$.

Acceptor doping of GaAs was demonstrated for C [*Sugioka* et al. 1989] and Zn [*Beneking* 1984]. KrF-laser-induced doping from gaseous CH_4 did yield a sheet resistance as low as 165 Ω/□ and a surface carrier density of 8×10^{14} cm^{-2}. The doped layer was about 0.05 μm thick and had a surface concentration of about 10^{21} atoms/cm^3. The maximum efficiency of active dopants was 69%. In this way, non-alloyed ohmic contacts can be produced. No significant deterioration of the GaAs surface was observed. Similar results have been obtained for Si doping from SiH_4.

Formation of p-type layers in InP was demonstrated by ArF-laser photolysis of $Cd(CH_3)_2$ [*Deutsch* et al. 1984].

24.4 Local Doping

Local doping of semiconductors by cw Ar$^+$-laser direct writing has been demonstrated for lateral dimensions down to submicrometer levels. Boron doping patterns in (111) and (100) Si have been produced by local pyrolysis or photolysis of BCl_3 at gas pressures of about 10^2 mbar [*Ehrlich* and *Tsao* 1982]. For 515 nm Ar$^+$-laser radiation *below* the melting intensity, the doped region was about 0.6 μm wide. This width is considerably smaller than the laser focus employed, $2w_0 \approx 1.8$ μm. The increase in spatial resolution is consistent with

24.4 Local Doping

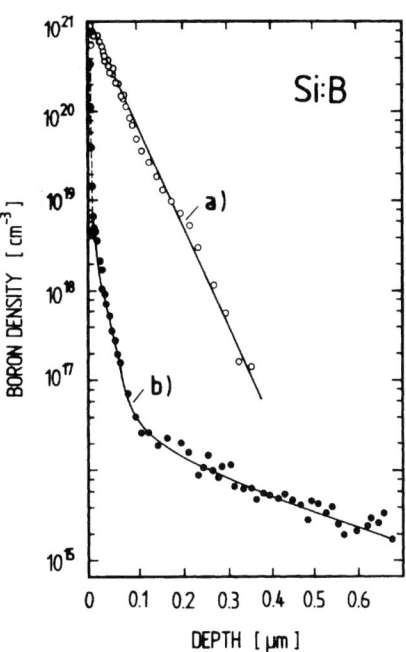

Fig. 24.4.1. Boron depth profiles determined by SIMS analysis of a raster-scanned 2000 Ωcm Si (111) sample. Scan speeds were (a) 19 μm/s, (b) 440 μm/s [*Ehrlich* and *Tsao* 1982]

the exponential dependence of the diffusion coefficient on temperature (Sect. 5.3.6; the activation energy for isothermal diffusion of B in Si is $\Delta E_d \approx 3.69$ eV). A depth profile of the boron density determined by SIMS analysis is shown in Fig. 24.4.1.

Direct cw-laser doping of InP with Cd and Zn was also demonstrated [*Ehrlich* et al. 1980]. In these experiments 257 nm frequency-doubled Ar$^+$-laser radiation was used collinearly with the 515 nm fundamental output. By controlling the individual beam intensities it was possible to vary independently the flux of Cd atoms produced by photodissociation of Cd(CH$_3$)$_2$ at or near the gas-solid interface, and the surface temperature. In fact, electrical measurements have revealed a linear increase in dopant concentration (up to $> 10^{19}$ Cd atoms/cm^3) with the 257 nm light flux. The 515 nm radiation controls the surface temperature and thereby the spatial width of the doped region. The dominant mechanism for dopant incorporation seems to be *solid-phase* diffusion due to thermal, stress, and concentration gradients.

Local doping of GaAs with Si has been demonstrated by KrF-laser light projection [*Sugioka* and *Toyoda* 1990] and with Zn by Ar$^+$-laser direct writing [*Licata* et al. 1990]. Linewidths between about 0.3 and 3 μm have been achieved.

25 Cladding, Alloying, and Synthesis

The emphasis of this chapter is on laser-assisted cladding, alloying, and synthesis of stoichiometric compounds. The main objective of laser-assisted cladding is the coating of substrates with thick films of metals or alloys. Laser-assisted alloying is applied for both surface modification and compound formation.

In the experiments discussed throughout this chapter, the material to be added to the substrate, or the starting material, is mainly in *solid* form. It can be a powder, a single layer, a stack of several layers, a solid solution, etc. In most of these applications, the laser powers employed cause surface melting. With cw-laser irradiation and long laser-beam dwell times, materials alloying and synthesis may take place via solid-phase diffusion. Laser-induced synthesis of organic polymers from monomers is of mainly photochemical nature.

25.1 Laser-Assisted Cladding

Laser-assisted cladding denotes the coating of a substrate with a thick metal or alloy film (clad). The coating material is either predeposited on the substrate surface in the form of a powder or a solid layer, or it is only fed in during the process. An arrangement for the latter technique is schematically shown in Fig. 25.1.1a. The technique is based on laser-induced melting of both the coating material and a *shallow* layer of the substrate surface. Fusion bonding guarantees good adhesion of the clad to the substrate. In contrast to surface alloying, intermixing of materials is minimized in order to avoid degradation of the physical and chemical properties of the clad and the substrate. The lasers mainly employed are CO_2 and Nd:YAG lasers with powers up to 4×10^3 W and beam radii between 0.5 and 3 mm. Substrate velocities are between 0.5 and 5 cm/s. Figure 25.1.1b shows the cladding rate W_c [cm^2/s] for stainless steel on a mild steel substrate as a function of laser power. Typical thicknesses of the clad are between 0.1 and 2 mm. Thicker layers can be built up by repetitive scans. Among the cladding materials most frequently used are alloys of B, C, Co, Cr, Ni, Fe, Si, and W.

If the coating material is predeposited in the form of a solid layer, the temperature rise induced by a Gaussian laser beam and $v_s w_0/D \ll 1$ can be

25.2 Alloying

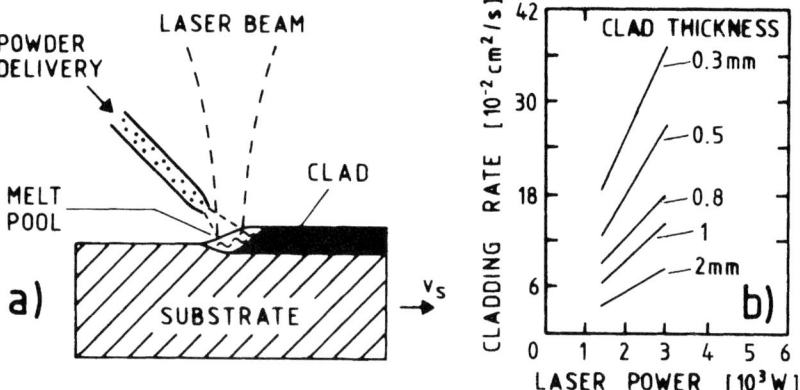

Fig. 25.1.1. (a) Laser cladding using powder delivery. (b) Cladding rate, W_c, for stainless steel (316 L) on a mild steel substrate as a function of CO_2-laser power ($w = 2.5$ mm) and for various thicknesses of the clad. The coating material was fed in as powder (powder flow ≈ 0.2 g/s, particle velocity ≈ 1.4 m/s) [*Weerasinghe* and *Steen* 1984]

estimated from (9.4.2) and (9.4.3), or from Fig. 9.4.2. With $v_s w/D \geqslant 1$ and arbitrary beam shapes, the Green's function given by *Burgener* and *Reedy* (1982) can be used.

One of the main advantages of laser cladding is the low thermal damage caused to the substrate material. The technique can be employed for applications involving wear, erosion, abrasion, corrosion, impact, etc. It competes with other techniques such as plasma spraying, plasma-CVD (PCVD), and electrochemical plating. For further details see [*Steen* 1991; *Li* and *Mazumder* 1985; *Juch* and *Löschau* 1994; *Ollier* et al. 1995].

25.2 Alloying

Laser alloying can be applied for modifying the physical and chemical properties of solid surfaces and for fabricating new metastable materials.

25.2.1 Laser–Surface Alloying

The objective in laser-surface alloying (LSA) is the complete mixing of the added material within the *molten* substrate surface. This is an important difference to laser cladding. Within the alloyed layer, the concentrations of added material and substrate material are of the same order of magnitude–in contrast to surface doping. Many of the investigations have been performed on the alloying of steel with C, Cr, Mn, Ni, V, and W. Of increasing interest is the hardening of Al by alloying with Ni or Si. Mainly CO_2 and Nd:YAG lasers have been employed in these experiments.

LSA enables one to modify surface properties of materials within depths of, typically, 1–2000 µm without affecting the bulk. In order to achieve uniform mixing of the added and the substrate material within the molten layer, the time Δt_m must be sufficiently long (Chap. 10). Scanning velocities employed are up to some 10 cm/s. With the high cooling rates that can be achieved, material segregation can be widely suppressed. Surface alloying with expensive precursor elements is considerably cheaper than bulk alloying, for details see, e.g., [*von Allmen* and *Blatter* 1995].

High concentration doping and coating of stainless steel (SUS 304) with Si by KrF-laser irradiation in SiH_4 atmosphere has been studied by *Jyumonji* et al. (1995). The hardness and chemical stability of the surface with respect to etching in aqueous HCl and H_2SO_4 was significantly improved. A new field is surface alloying and coating of ceramics, mainly with metals.

25.2.2 Formation of Metastable Materials

The formation of metastable materials is mainly based on the high heating and cooling rates achieved with lasers. For scanned cw-laser-heated surfaces the cooling rates are comparable to those achieved in standard splat cooling or melt spinning techniques (about 10^6 K/s). With Q-switched and mode-locked pulsed lasers, rates of $10^{10} - 10^{12}$ K/s can be obtained. Such cooling rates make it possible to produce glassy alloys which have never before been available. Detailed investigations have been performed for binary alloys consisting of two transition metals, and for combinations of metals with group-IV elements. Among the binary transition metal systems, glasses of Au-Ti, Cu-Ti, Co-Ti, Cr-Ti, V-Ti and Ag-Cu have been produced [*von Allmen* and *Blatter* 1995]. Glass formation seems to fail if single-phase crystallization is possible from the melt, or if the glass is unstable at ambient temperatures. The most extensive investigations on binary alloys consisting of metals and group-IV elements have been performed for silicides.

25.2.3 Silicides

Lasers enable one to synthesize silicides with various compositions, simply by varying the laser fluence. This is quite different to conventional synthesis within a furnace where sequential phase formation is observed. Most of the experiments start out with a metal film deposited on the Si substrate. The main analytical techniques employed for product characterization are summarized in Sect. 30.4.2.

Silicide formation by means of scanned *cw lasers* has been demonstrated for Co, Nb, Pd, and Pt. Uniform layers of MeSi, Me$_2$Si, or MeSi$_2$ (Me = metal) consisting of essentially a single phase, have been produced [*Shibata* et al. 1980]. The process seems to be dominated by solid-phase diffusion.

Silicide formation using *pulsed-laser* irradiation has been investigated for Au, Co, Cr, Mo, Ni, Pd, Pt, Ti, and W. Excimer, ruby, Nd:YAG, and Nd:glass lasers were mainly employed in these experiments. Here, silicide formation is mainly based on surface melting and liquid-phase mixing of the metal and Si. Due to rapid solidification, amorphous films, or films with different Me$_x$Si$_y$ precipitates surrounded by Si, are formed. Suitable variation of the laser fluence, pulse length, and thickness of the metal film, makes it possible to produce uniform layers with a predetermined single phase or average composition $x:y$ [*Bohac* et al. 1993; *D'Anna* et al. 1988; *Baeri* et al. 1985]. An example for which both multiphase and single-phase silicide formation has been demonstrated is the Ni-Si system. With relatively thick metal layers, typically 0.1 µm, evaporated on a Si wafer, the quantity of material liquefied by the laser fluences employed was so large that nucleation and growth of different phases took place during solidification. Simultaneous formation of Ni$_2$Si, NiSi, NiSi$_2$, and NiSi$_3$ was observed [*Bentini* et al. 1982]. On the other hand, with thin evaporated metal layers ($h_1 < 0.02$ µm) and with certain processing conditions, single-phase NiSi$_2$ can be grown even epitaxially onto (100) and (111) Si wafers [*Grimaldi* et al. 1983]. The physical properties of these films are very similar to those produced by standard furnace annealing.

The results obtained by different groups can be summarized as follows: Depending on the laser fluence and dwell time, the film thickness, and the optical and thermal properties of the system, it is possible to
- synthesize different silicides.
- completely react the metal film with Si.
- react only part of the film at the Me/Si interface.
- produce *single-phase* silicides, e.g., Me$_2$Si if Me is a noble metal (Pt, Pd, etc.) and MeSi$_2$ if Me is a refractory metal (Cr, Mo, Ti, W). Here, the reaction starts when the interface temperature, T_i, and the eutectic temperature, T_e, of the Me/Si system are equal, i.e., $T_i = T_e$. Single-phase silicide formation is observed if $T_e < T_i < T_m(\text{Me})$.
- simultaneously form silicides with complex composition. This situation is observed with *thick* Me layers and high fluences that cause deep melting.
- widely suppress surface instabilities (Chap. 28).

25.3 Synthesis

Lasers have been used to synthesize stoichiometric compounds, mainly in the form of thin films and fibers. In the experiments subsequently described, the starting material did consist of a single layer, a stack of several layers, or a rod.

25.3.1 Thin Films

Laser-induced synthesis of thin crystalline films has been investigated in particular for compound semiconductors [*Laude* et al. 1986]. The systems studied in detail include binary compounds such as III–V (AlSb, AlAs), II–VI (CdTe, CdSe, ZnSe), IV–VI (GeSe$_2$, GeSe) and IV–IV (Si$_{1-x}$Ge$_x$) semiconductors, and even ternary compounds, for example CuInSe$_2$. Synthesis has been achieved by irradiating solid solutions or alternating multiple layers consisting of appropriate proportions of the elements. In order to avoid oxidation or other contaminations during processing in air, films were often encapsulated between SiO$_x$ layers.

Cw-laser synthesis

Cw-laser synthesis has been demonstrated for AlSb, GeSe$_2$, and CuInSe$_2$. The size of crystallites formed is, typically, in the µm range. Both free-standing films and films on glass or NaCl substrates exhibit good stoichiometry and optical properties comparable to those of single crystals. For the Ge-Se system different compositions and scanning velocities ranging from 1 cm/s to 20 m/s, have been investigated. Well-defined periodic structures were observed within certain parameter ranges (Sect. 28.3.4).

Pulsed-laser synthesis

Synthesis by pulsed-laser radiation requires melting, or at least partial melting, of the elemental constituents. The dependence of the threshold fluence on pulse duration and substrate temperature was investigated for AlSb.

An interesting point is the possibility to select particular phases by changing the laser-beam illumination time. For example, with dye-laser pulses of $\tau_l = 10^{-6}$ s, irradiation of Cd-Te and Cd-Se free-standing films did yield a mixture of hexagonal and cubic CdTe and hexagonal CdSe, respectively. Chopped Ar$^+$-laser irradiation with pulse durations of $\tau_l = 3.6 \times 10^{-2}$ s, on the other hand, revealed films of *cubic* structure only. Thus either type of phase can be produced by proper control of the laser beam dwell time (note that II–VI single crystals are cubic if grown at low temperatures and hexagonal for high temperatures). The physical properties of CdTe films are very similar to those of single crystals grown by standard techniques. In particular, the laser-grown material is very pure with electrical impurity concentrations ranging from 10^{15} to 10^{17} defects/cm^3 only. The deviation from the correct stoichiometry is 1–2%.

Microcrystalline films of semiconducting Ge$_{1-x}$Sn$_x$ ($x \approx 0.22$) have been synthesized from amorphous RF-sputtered films by means of ArF- and KrF-laser radiation [*Oguz* et al. 1983].

25.3.2 Fibers

Lasers permit one to synthesize polycrystalline and single-crystalline fibers, some of which cannot be fabricated by conventional techniques. This has been demonstrated for various materials by the float-zone technique (Fig. 25.3.1). Here, in contrast to the technique described in Chap. 17, the source material is solid, frequently a ceramic in the form of a rod. Fiber growth is achieved by moving the source rod and the fiber through the laser-heated zone. Here, CO_2 lasers are most commonly employed. If the fiber diameter is smaller than the diameter of the source rod, the technique is also termed laser-heated pedestal growth (LHPG) [*Feigelson* 1988]. Similarly to the growth of fibers by laser-CVD, the technique does not require any crucibles and therefore avoids contaminations and frozen-in stresses. The composition of fibers can easily be controlled via the starting material. With the fiber diameters under consideration and focused laser-beam irradiation, strong temperature gradients, and hence rapid pulling rates, typically 10–100 µm/s, can be achieved. The heat source in pedestal growth must have dimensions comparable to the diameter of the fiber and source rod. This requirement is hard to realize with resistance heating. Induction heating would require either a conducting sample or a susceptor. Electron-beam (EB) techniques can be applied only in vacuum. Thus, with many systems, lasers are the ideal sources. Laser light can be tightly focused, can readily melt any known material, and it can be applied in conjunction with inert or reactive atmospheres. With appropriate optics, uniform ring-shaped illumination of the molten zone is possible.

Laser-heated pedestal growth has been employed to grow single-crystal fibers of halides, oxides, borides, carbides, metals, semiconductors, and high-temperature superconductors. Most of the materials so far investigated are listed in Table 25.3.1.

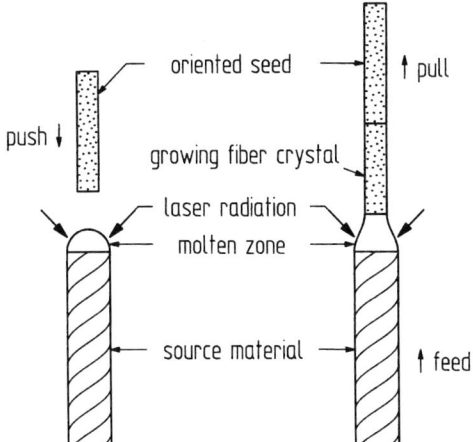

Fig. 25.3.1. Growth of fibers by pedestal growth, adapted from [*Feigelson* 1988]

Table 25.3.1. Fibers grown by laser-heated pedestal growth, adapted from [*Feigelson* 1988]. T_m is the melting temperature

Material	$T_m[°C]$	Orientation	Diameter [μm]
Fluorides			
BaF_2	1280	[110]	200–600
CaF_2	1360	[111]	600
Eutectics			
$Li_2O\text{-}GeO_2$	1106	–	–
LiF-NaF	676	–	40–500
NaF-NaCl	640	–	–
Oxides			
$Bi_2Sr_2CaCu_2O_8$			30–400
Nd:YAG	1940	[111], [100]	6–1000
YAG	1940	[111], [100]	100–1000
Al_2O_3	2045	a, c	55–600
$Ti:Al_2O_3$	2045	c	200–800
$Cr:Al_2O_3$	2045	c	3–170
$LiNbO_3$	1260	a, c	20–800
$Nd:LiNbO_3$	1260	c	800
$LiTaO_3$	1650	[110]	600
Li_2GeO_3	1170	a, c	100–600
$Gd_2(MoO_4)_3$	1157	[110]	200–600
$CaSc_2O_4$	2200	a, b, c	100–600
$Nd:CaSc_2O_4$	2200	c	600
$SrSc_2O_4$	2200	–	600
YIG	1555	[110]	100–600
$Eu:Y_2O_3$	2410	c	500–800
$Ti:MgAl_2O_4$	2105	–	1000
$Ti:YAlO_3$	1875	–	1000
BaB_2O_4	1095	–	500
Nd_2SiO_5	1980	–	750
$SrTiO_3$	1860	–	600
Nb_2O_5	1495	–	700–1700
$BaTiO_3$	1618	c*	300–800
$SrBaTi_2O_6$	1700	a, c	200–600
$ScTaO_4$	2300	a	200–600
$ScNbO_4$	2100	–	200–600
$SrBaNb_2O_6$	1500	a, c	600–1700
$Cr:Y_2O_3$	2400	c [110]	600
$Cr:Sc_2O_3$	2400	c [110]	600
$Cr:Lu_2O_3$	2400	c [100]	600
Semiconductors and Metals			
LaB_6	2715	–	200
Nb	2468	–	200
Si	1420	[111]	200
Ge	960	–	200
B_9C	2400	–	200
Co	1495	–	100–600
Fe-Co	1500	–	100–600
Fe	1539	–	100–600

* Hexagonal and cubic phases.

25.3.3 Polymerization, Waveguides

Ultraviolet laser light can be used for polymerization by photoexcitation of monomers or sensitizer molecules. An example is the photopolymerization of MMA (methylmethacrylate) to PMMA (poly-MMA).

Laser-induced photopolymerization of spun-on polymer films (Fig. 27.3.1) was used to fabricate optical waveguides on thermally oxidized Si wafers [*Krchnavek* et al. 1989]. The width and thickness of such waveguides is of the order of a few μm. Their optical attenuation, measured at $\lambda = 632.8$ nm was, typically, below 1 dB/cm.

Optical waveguides have also been produced in $LiNbO_3$ by KrF-laser-induced decomposition of $TiCl_4$ and Ti indiffusion [*Lavoie* et al. 1991*a*].

26 Oxidation, Nitridation

Surface oxidation of metals and semiconductors in an oxidizing agent is a well-known phenomenon. Clean surfaces of many materials like Al, Nb, Si, etc., *spontaneously* react in air, even at room temperature, to form thin native oxide layers. With the materials under consideration the native oxide layer is very dense and terminates further oxidation. Native oxide layers are, typically, 10–100 Å thick.

For many applications such as local hardening, chemical passivation, electrical insulation, etc., it is desirable to increase the thickness of the oxide layer, or to stimulate oxidation on material surfaces that do *not* spontaneously oxidize in an oxygen containing environment. The latter includes the reoxidation of oxygen-deficient oxide layers and the transformation of oxides, for example, of M_xO_y into $M_{x\pm\varepsilon}O_{y\pm\delta}$.

Among the many techniques investigated for solid-surface oxidation, the most commonly used ones are thermal oxidation by uniform substrate heating in an oxygen-rich atmosphere, and plasma oxidation. Both techniques have their characteristic advantages and disadvantages (Chap. 1).

Light-enhanced, and in particular laser-enhanced materials oxidation is based on thermal or non-thermal molecule-surface excitations. Single-photon dissociation of O_2 starts at around 5.1 eV ($\lambda \leqslant 240$ nm).

Nitridation of solid surfaces is generating increasing interest as an alternative to oxidation. The fundamental aspects and the techniques employed are similar to those for surface oxidation. Photodissociation of gaseous N_2 starts at around 9.8 eV.

Among the advantages of laser-enhanced materials oxidation and nitridation are the lower substrate temperatures, the short processing times, and the possibility to form *new* types of oxides. The main disadvantages are the low total throughput, the roughness of the oxide surfaces, and the tendency of crack formation.

Laser-enhanced surface oxidation was mainly performed in air and oxygen atmosphere. Some experimental results achieved with liquids have also been reported. An overview on the various systems so far investigated is given in Appendix B.9. Laser-induced chemical vapor deposition of oxides and nitrides is included in Chaps. 16–19. Thermochemical instabilities observed during laser-enhanced oxide formation are discussed in Chap. 28.

26.1 Basic Mechanisms

Native oxide formation involves a number of consecutive steps:

- Transport of oxygen from the ambient medium to the solid surface.
- Adsorption of molecular oxygen.
- Electron transfer to adsorbed O_2.
- Electric-field-enhanced diffusion of species through the oxide layer.

Adsorption of (strongly electronegative) O_2 on a metal or semiconductor surface favors electron transfer. The situation is similar to that shown in Fig. 15.1.1a. Dissociative chemisorption is enhanced for O_2^- because it requires only 3.8 eV compared to 5.1 eV for O_2. Chemisorbed oxygen reacts with surface ions/atoms and forms an ultrathin oxide layer. Further oxide growth can proceed via electron tunneling and diffusion of species through the growing layer. This process is self-terminating, because it becomes less likely with increasing layer thickness. Clearly, the surface morphology, the microstructure of the substrate material, the concentration of physical and chemical defects, etc., will strongly influence this *initial* phase of oxide growth. While the fine details certainly depend on the particular system under consideration, the main aspects of this scenario can be applied to many materials.

Oxide growth beyond a single or a few monolayers is controlled by the transport of electrons, ions, atoms, or molecules through the oxide layer (Fig. 26.1.1). With *thin films* and charged species, this transport is enhanced/suppressed by the electric field which builds up within the oxide due to electron transfer from the metal to oxygen. This process takes place until a quasi-equilibrium state is reached. The related potential difference across the oxide layer is, typically, of the order of $\Phi \approx 1$ V and almost independent of layer thickness. With *thick films*, the transport of species becomes dominated by ordinary diffusion. For most metal oxides such as those of Cu, Fe, Pb, Zn, etc., diffusion of metal cations dominates. The metal ions reaching the surface react with adsorbed oxygen. Thus, oxide growth takes place at or near the interface between the gas and the oxide surface. For the oxides of Hf, Nb, Ta, W, and Zr diffusion of oxygen seems to dominate and growth takes place mainly at the metal-oxide interface [*Roberts* and *MacKee* 1978]. With some systems, the species which predominantly diffuses even changes with temperature.

The growth in oxide layer thickness can often be described by the Cabrera-Mott theory which was originally developed for metal oxidation only. Here, three different regimes of oxidation are distinguished:

Very thin films

In the initial phase of growth the oxide layer is very thin and the electric field set up by electron tunneling, $E \approx \Phi/h$, is of the order of 10^7 V/cm or more. In this regime the transport of ions is dominated by this strong surface electric field and can be described by a Butler-Volmer type equation (Sect. 21.2), which leads to the kinetic law

$$\frac{\partial h}{\partial t} = k'_0 \exp\left(-\frac{\mathscr{E}}{T}\right) \exp\left(\frac{h_M}{h}\right), \qquad (26.1.1)$$

where $k'_0 \approx 10^4$ cm/s. This equation holds for $h \approx (20-100\,\text{Å}) \ll h_M = qa\,\Phi/k_B T \approx 100-1000\,\text{Å}$. q is the ionic charge, and a the distance between ion jumps which is of the order of the oxide lattice constant. Henceforth we assume $q = e$.

Thin films

After the initial phase of oxidation, i.e., for film thicknesses $h_M < h \leqslant h_D$ the (very mobile) electrons can still pass easily through the oxide layer. However, the drift of ions can now be assumed to be proportional to the electric field $E = \Phi/h$, as with Ohm's law. The oxidation behavior can then be described by the transport equations and the Poisson equation. The flux of electrons is given by

$$J_e = -D_e \frac{dN_e}{dz} - N_e \mu_e E. \qquad (26.1.2)$$

The ion flux is

$$J_i = -D_i \frac{dN_i}{dz} + N_i \mu_i E. \qquad (26.1.3)$$

N_e and N_i are the number densities of electrons and ions within the oxide layer. μ_e and μ_i are the respective mobilities where $\mu_e \gg \mu_i$. The Poisson equation can be written as

$$\frac{d^2 \Phi}{dz^2} = -\frac{4\pi e}{\varepsilon} [N_i(z) - N_e(z)], \qquad (26.1.4)$$

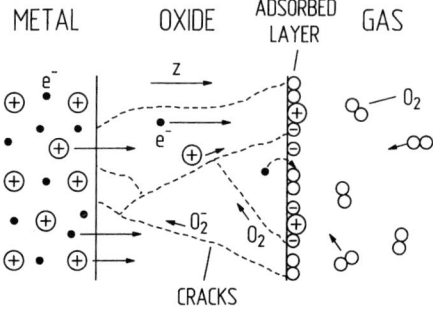

Fig. 26.1.1. Model for surface oxidation of metals. With most metals, the metal ions diffuse to the surface to react with oxygen. Diffusion is enhanced if cracks within the oxide layer are formed

26.1 Basic Mechanisms

where ε is the dielectric constant. From these equations and the appropriate boundary conditions one can derive the stationary ion flux

$$J_i = J_i(h) \propto N_i E \propto \frac{N_i \Phi}{h}. \tag{26.1.5}$$

With $\partial h/\partial t \propto J_i$ we find the kinetic law

$$\frac{\partial h}{\partial t} = \frac{k_{0n}}{h^n} \exp\left(-\frac{\mathscr{E}_{ox}}{T}\right) = \frac{k_n}{h^n}, \tag{26.1.6}$$

where k_{0n} is a constant and $\mathscr{E}_{ox} \equiv \Delta E_{ox}/k_B$. Here ΔE_{ox} includes the activation energy for diffusion of species through the oxide layer and some energies which characterize interface processes. The exponent n has values $0, 1, 2$, etc., depending on the material and the regime of surface oxidation. Let us consider three different systems.

– For *interstitial* diffusion which is typical for oxidation of metals such as Al, Zn, etc., $N_i = $ const. From (26.1.5) we find $\dot{h} \propto h^{-1}$ which corresponds to $n = 1$ in (26.1.6). This yields the parabolic law

$$h^2 = 2 k_1 t. \tag{26.1.7}$$

– For *substitutional* diffusion which is typical for oxidation of Cu, Fe, etc., the density $N_i \propto \Phi/h$ (note that the oxide layer can be considered as a capacitor with fixed potential difference Φ; see also (15.1.1)). This yields $\dot{h} \propto h^{-2}$ which corresponds to $n = 2$ in (26.1.6). Integration results in the cubic law

$$h^3 = 3 k_2 t. \tag{26.1.8}$$

– For a *non-compact* layer, or the case where k_1 increases with h, one often observes an "apparent" linear behavior

$$h = k_0 t. \tag{26.1.9}$$

This situation is typical for the oxidation of Mg.

Thick films

The surface electric field drops off within the distance $h_D \approx (\varepsilon k_B T/8\pi N_e e^2)^{1/2}$ which is the thickness of the double layer related to the space charge built up near the oxide surface or near the oxide–metal interface, depending on the particular system under consideration (h_D is also known as screening or Debye length). Thus, for thick films with $h \approx 10^4$ Å $\gg h_D$ the electric field can be ignored and ordinary diffusion of species determines the rate. Then, $J_i \propto dN_i/dz \approx N_i/h$ so that $\dot{h} \propto h^{-1}$. This again yields a parabolic law (Wagner relationship)

$$h^2 = 2k_1' t, \tag{26.1.10}$$

where $k'_1 = k'_{01} \exp(-\mathscr{E}_d/T) \propto D$. Note that the physical origin of this law is quite different from (26.1.7).

Parameters derived from oxidation experiments performed in furnaces can be found for various materials in the Oxide Handbook [*Samsonov* 1973]. For example, the growth kinetics of Cu_2O can be described by this parabolic law with $k'_{01} = 4.38 \times 10^{-5}$ cm^2/s and $\mathscr{E}_d = 9645$ K.

In the preceding treatment a single chemically uniform oxide layer has been assumed. With many materials, however, the concentration of oxygen within the layer is *not* constant but varies with depth. Such chemical inhomogeneities can be described by multilayer structures as schematically shown in Fig. 26.1.2 for Cu and Fe. In such situations, the oxidation kinetics becomes much more complex. For example, for two-layer systems such as $CuO/Cu_2O/Cu$ the kinetics of thick films can be described by the Wagner-Valency theory [*Karlov* et al. 1992].

Influence of laser light

Laser-enhanced *pyrolytic* (photothermal) oxidation can be understood along similar lines as ordinary thermal oxidation. The influence of the oxide-layer thickness, h, on the laser-induced temperature rise, ΔT, is related to changes in the thermal and optical properties of the irradiated material (Chap. 9). Changes in the absorptivity due to laser-beam interferences yield oscillations in the laser-induced temperature distribution and thus in the oxidation rate. Instead of considering single well-defined layers, one can solve the Maxwell equations for (smoothly) varying concentrations of oxygen in z-direction. In any case, because of the temporal changes in temperature, and therefore in coefficients k_n in (26.1.6), laser-enhanced oxidation can be modelled by means of the kinetic laws in their differential form only.

The laser-induced temperature rise enhances the diffusion flux and the reaction rate of species within the irradiated area. This enhancement is based on various different mechanisms: The temperature dependence of ordinary

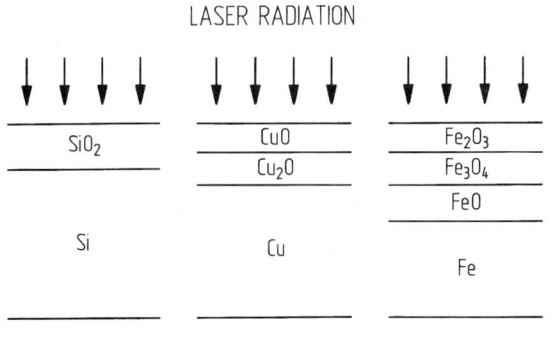

Fig. 26.1.2. Laser-induced oxide formation on different material surfaces. The change in oxygen concentration with depth can be described by multiple layers. The composition and thickness of these layers depends on the laser parameters

diffusion, the thermal generation of defects such as vacancies, etc., and the thermal excitation of electrons. The latter increases the rate of electron transfer to oxygen and thereby the electric field which enhances the diffusion flux. Furthermore, strong temperature gradients induced by the laser light will enhance the transport of species via thermal diffusion and via the formation of stresses, strains, cracks, and other defects (Fig. 26.1.1). Finally, with the dwell times involved, in particular in pulsed-laser oxidation, thermodynamically unstable phases may remain, while other phases cannot nucleate or form larger crystallites within such short times. It is evident that the formation of different oxides follows different kinetic laws and that optical excitations can modify activation energies, etc. Thus, it is not very astonishing that in spite of the thermal character of the interaction mechanisms, the growth rate, composition, thickness, morphology, and microstructure of laser-fabricated oxide layers differ significantly from those observed with films fabricated under equilibrium conditions in an oven. For example, CO_2-laser-induced surface oxidation of Cu slabs in air results in the formation of CuO and Cu_2O, as shown in Fig. 26.1.2. On the other hand, the oxidation of thin Cu films on glass or sapphire substrates under the action of visible Ar^+-laser radiation yields mainly Cu_2O.

Photolytic (photochemical) mechanisms have been proved to be important, in particular in cases where the photon energy matches the energy for selective excitation of a particular transition:

- Far-UV radiation results in gas-phase formation of atomic oxygen and ozone which can efficiently react with many material surfaces. The mechanisms are similar when, instead of molecular oxygen, other oxygen containing species are employed.
- If $h\nu$ exceeds the band-gap energy, E_g, electron–hole pairs are generated within the oxide layer or within the (non-metallic) substrate. Photoelectrons and holes can modify oxygen adsorption on the surface, its reaction with surface atoms, or its migration into the surface, etc. UV-laser light may directly excite adsorbed oxygen and thus promote its dissociation.
- UV- and far-UV-laser light can generate high concentrations of (mainly) oxygen vacancies within the oxide layer. These vacancies increase the mobility of species, lower the temperature for surface melting, etc.
- For both metals and non-metals, photoelectrons may directly be ejected into the oxide layer and modify the electric field.
- In cases where $h\nu < E_g$, excitation of band-gap states may become important. These states are related to impurities, lattice imperfections, surface states, localized states at the substrate–oxide interface, etc. Even if the density of such localized states is low, they may efficiently contribute to the overall oxidation process because of their long lifetimes.

At high intensities, laser light initiates a breakdown at or near the solid surface. This is the regime of *pulsed-laser plasma chemistry* (PLPC; in the literature this is also termed laser-pulsed plasma chemistry and abbreviated by

LPPC). The laser-generated plasma contains excited reactants which can combine with substrate atoms to form oxide overlayers.

26.2 Metals

Investigations of laser-enhanced oxidation of metals have been performed with essentially two types of samples: metal plates and thin metal films. With plate-like samples, the laser-oxidized surface layer is very thin in comparison to the sample thickness. This case will be termed laser-induced surface oxidation. Here, the oxidized layer contains, in general, various types of oxides that can be described by Me_xO_y, where Me stands for metal (Fig. 26.1.2). With thin metal films, having thicknesses of, typically, $h \leqslant 0.1$ μm, oxidation can succeed throughout the whole film. In this case, oxidation often results in the synthesis of a single stoichiometric oxide. Very thin metal-oxide layers have been produced by PLPC.

26.2.1 Photothermal Oxidation

Photothermal oxidation of metal plates and thin metal films has been studied mainly with CO_2 lasers, Nd:YAG lasers, Nd:glass lasers, and Ar^+ lasers. Many of the systems investigated are listed in Appendix B.9. With many metals, laser-enhanced surface oxidation can be described along the lines of the Cabrera–Mott theory.

Among the systems so far investigated in detail, is the oxidation of thin Cu films on sapphire substrates by means of cw Ar^+-laser radiation. The laser-grown oxide consists mainly of Cu_2O. Figure 26.2.1a depicts the time-resolved reflectivity measured *in situ* with a HeNe-laser probe beam for different Ar^+-laser powers. The reflected light intensity shows damped oscillations due to interference. The decrease in time between oscillations observed with increasing incident laser power is due to the increase in film growth rate. However, even with constant laser power, the absorbed power increases with film thickness (Fig. 9.2.2). For a quantitative description of the results, the oscillations in the absorbed Ar^+-laser power are of relevance. Figure 26.2.1b displays the oxide-layer thickness as a function of irradiation time for constant incident laser power. The stepwise growth is related to the oscillations in absorbed laser power.

For an analysis of results, the exothermal energy release during oxidation can be ignored. The additional source term can be approximated by $I_{ox} \approx \Delta H_{ox} \, v$ with $\Delta H_{ox} \approx 7 \times 10^3 \, \text{J/cm}^3$. For an average oxidation velocity $v \approx 10^3 \, \text{Å}/10 \, \text{ms}$ (corresponding to a laser power of about 4.4 W) we have $I_{ox} \approx 10^{-3} I_0$.

26.2 Metals

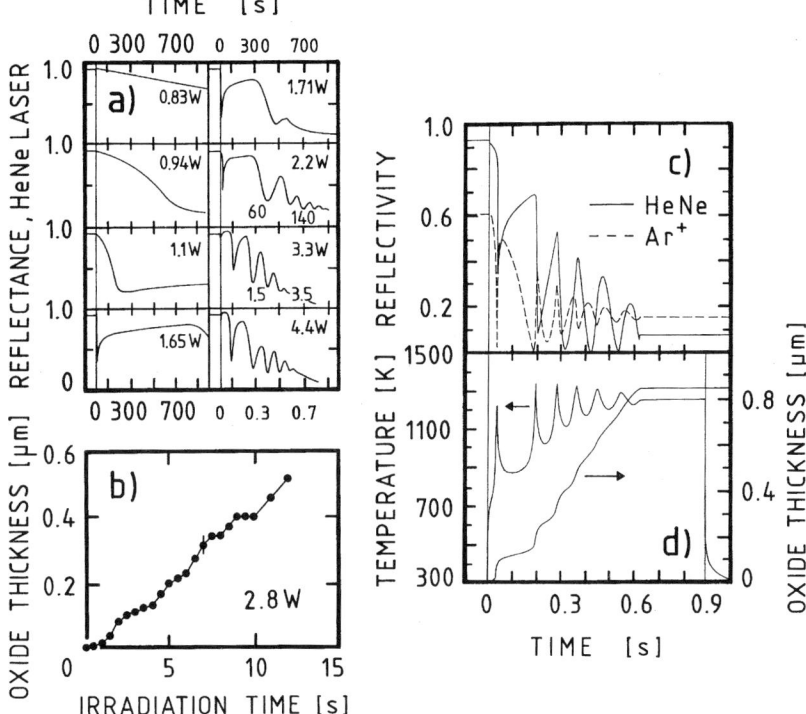

Fig. 26.2.1a–d. Ar$^+$-laser-induced oxidation of 0.05 μm thick Cu films on sapphire ($\lambda = 514.5$ nm, $w_0 \approx 80$ μm, $p(O_2) \approx 200$ mbar). (**a**) Time-resolved reflectance measured by HeNe-laser probe beam. Vertical lines indicate beginning of Ar$^+$-laser irradiation. Note changes in time scales with 2.2, 3.3, and 4.4 W curves. (**b**) Oxide-layer thickness as a function of irradiation time. (**c,d**) *Calculated* temporal dependence of the reflectivity of HeNe- and Ar$^+$-laser beams, of the film thickness, and of the temperature for $P = 4.4$ W. Oscillations end when the film is oxidized over the whole thickness, adapted from [*Baufay* et al. 1987]

The experimental data presented in Fig. 26.2.1a,b have been simulated on the basis of different models [*Baufay* et al. 1987]. Figure 26.2.1c,d exhibits the results of calculations where the thermally well-conducting copper film is placed on a semi-infinite (sapphire) substrate. The main features observed in the experiments are reproduced by the calculations. If the thermal properties of the sapphire substrate are adapted, almost quantitative agreement with the experimental data is achieved.

The good agreement between experimental and theoretical results achieved for the preceding example does *not* mean that laser-enhanced metal oxidation can, in general, be described on the basis of such a simple model with similar success. There are many different reasons for this:

- For *surface* oxidation, the measurement and definition of the oxide-layer thickness is by no means trivial, in particular when the surface oxide is chemically inhomogeneous.
- There are no reliable measurements of the laser-induced temperature distribution within the oxide layer. Thus, the temperature that enters the kinetic equations can only be estimated from model calculations. Such calculations require, however, accurate knowledge of the optical and thermal material parameters and their spatial and temporal changes during the oxidation process.

A more sophisticated model has also to account for:
- The transport of oxygen or oxygen containing species from the gas-phase to the solid surface and the adsorption kinetics of these species on the surface.
- Non-thermal oxidation mechanisms.
- The transport of metal ions and oxygen through the oxide layer with highly non-uniform properties related to strong temperature gradients.
- Diffusion of species along grain boundaries and cracks within the oxide (Fig. 26.1.1).
- Nucleation and growth of crystallites.
- Effects due to macroscopic non-equilibrium in the system when temperature variations are faster than the chemical relaxation of the medium, the relaxation of the charge distribution, etc.
- Laser-induced thermal or non-thermal decomposition, transformation or ablation of oxides that have already been formed.

Some of these effects and their relative influence on the laser-induced temperature distribution and the transport of species have been considered [*Karlov* et al. 1992; *Wautelet* 1990]. The oxidation kinetics depends sensitively on the specific material and laser parameters. This is the reason for the significant differences in apparent kinetic constants reported by different groups for similar materials.

The influence of an *external* electric field on oxide formation was investigated for cw CO_2-laser enhanced oxidation of Cu and V in air [*Nanai* et al. 1993].

26.2.2 Photochemical Contributions

For certain systems, surface oxidation cannot, or only poorly be described by (26.1.6). Certainly, some of the discrepancies may simply be related to the difficulties in the measurements and analysis of data, as mentioned above. There are, however, a number of systems where *non-thermal* effects seem to be of importance. An example is the oxidation of Ni in O_2 atmosphere which was studied for laser wavelengths 308 nm $\leqslant \lambda \leqslant$ 1064 nm [*Mesarwi* and *Ignatiev* 1989]. Here, the oxidation rate was found to increase with decreasing wavelength. This was interpreted by photoexcitation of molecularly adsorbed oxygen at the NiO surface.

26.2.3 Oxidation by Pulsed-Laser Plasma Chemistry

Surface oxidation based on pulsed-laser plasma chemistry (PLPC) was studied in detail for Nb films (≈ 130 μm thick) in O_2 atmosphere and pulsed CO_2-laser radiation [*Marks* et al. 1983]. Single-pulse laser-activated oxidation was found to produce thicker films than multiple-pulse irradiation. For a single pulse with a fluence of $\phi = 0.75$ J/cm^2, the thickness of the native oxide consisting of $Nb_2O_{5-\delta}$ was increased by 18 Å, while 3 pulses, each having a comparable fluence, yielded a net increase of only 11 Å. This is interpreted in terms of competing mechanisms: Oxidation by PLPC, and oxide ablation due to absorption of CO_2-laser radiation within the oxide layer. As revealed by XPS, the valence defect, δ, decreases monotonically with increasing layer thickness. A similar behavior has been found for films produced by standard plasma oxidation. However, for a given layer thickness, δ is 3 to 5 times smaller for PLPC oxides ($0.02 \leqslant \delta \leqslant 0.04$) than for plasma oxides ($0.1 \leqslant \delta \leqslant 0.2$). In other words, PLPC yields more complete oxidation. Furthermore, in comparison with laser-enhanced photothermal oxidation or conventional oxidation, PLPC diminishes the formation of suboxides. Niobium oxide layers (18–40 Å with 0.24–0.79 J/cm^2) produced by PLPC may be applied for the fabrication of tunnel barriers in tunneling devices. It is difficult to produce such well-defined dielectric layers with comparable thickness control and quality by standard techniques.

26.2.4 Nitridation

Nitridation of metal surfaces has been studied in detail for Ti and Zr. Pulsed and cw CO_2 and Nd:YAG lasers, and excimer lasers were employed. The ambient media were gaseous and liquid N_2, NH_3, and air. Nitridation is extremely sensitive to traces of oxygen. Nitrified Ti and Zr surfaces produced by means of CO_2-laser radiation in N_2 atmosphere were *non*-stoichiometric, containing varying mixtures of MeN, MeN$_x$O$_y$, and MeO$_2$ (Me $\equiv$ Ti or Zr). The composition, thickness (typically 10–100 μm), microhardness, and adherence of layers depend strongly on laser-pulse repetition rate, pulse duration, and gas pressure. The microhardness of films in the center of the spot was typically increased by a factor of 2-3 with respect to bulk Ti and Zr. The best nitrified layers on Zr substrates reached approximately the hardness of ZrN. Very pure, hard, and adherent layers of TiN have been produced by multiple pulse XeCl-laser irradiation of Ti in N_2 and NH_3 atmosphere [*D'Anna* et al. 1991].

Thick TiN overlayers have been fabricated by CO_2-laser-induced surface melting in combination with an N_2/Ar-gas jet. With laser-beam dwell times $\tau_l < 1$s, layer thicknesses of 50–500 μm have been produced (oxide layers have been fabricated in a similar way; *Gasser* et al. 1989).

Excimer-laser-enhanced nitridation of Ti and Fe immersed in *liquid* N_2 or NH_3 has been studied by *Ogale* et al. (1987). RBS revealed large quantities of nitrogen and oxygen incorporated into the metal surface over depths of some 10^3 Å. Stable nitrides and oxides such as TiN and TiO have been formed. Stresses built up within the surface layer result in crack formation.

26.3 Elemental Semiconductors

Surface oxidation and nitridation of elemental semiconductors is of basic importance in semiconductor-device technology. The investigations have concentrated on silicon and, to a smaller extent, on germanium.

Surface oxidation of crystalline and amorphous silicon has been investigated with cw and pulsed lasers and various atmospheres. Air and O_2 were most commonly employed, sometimes with admixtures of other species that catalyze the overall oxidation process. Oxide formation by annealing of oxygen-ion (O^+) implanted Si surfaces has also been demonstrated. In many experiments, the Si substrate was preheated to several hundred degrees. For photon energies $h\nu < E_g$ the enhancement in oxidation rate is primarily due to laser-induced surface heating (Sect. 7.6). The band-gap energy [E_g(Si, 300 K) ≈ 1.12 eV $\hat{=} 1107$ nm] decreases with increasing temperature. Excitation by visible and ultraviolet laser radiation results in the emission of electrons from Si into the SiO_2 layer. These electrons enhance the dissociation of molecular oxygen and thus the oxidation rate (Fig. 26.3.1).

26.3.1 Photothermal Oxidation of Si

During thermal oxidation of Si in O_2 atmosphere the oxide-layer thickness increases linearly for short oxidation times (thin layers) and follows a parabolic

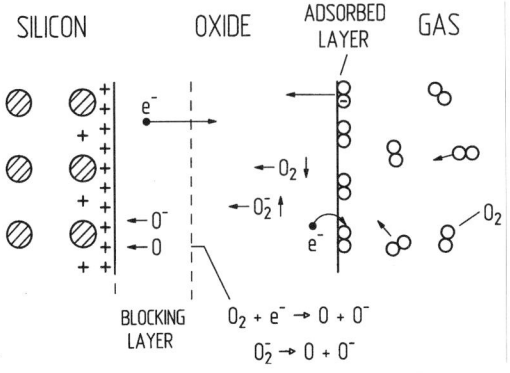

Fig. 26.3.1. Model for laser-induced surface oxidation of Si. Arrows ↑ and ↓ denote increasing and decreasing concentrations, respectively

26.3 Elemental Semiconductors

law for longer times [*Deal* and *Grove* 1965]. These growth regimes can be described by (26.1.6) with $n = 0$ and $n = 1$, respectively.

Within the *linear* regime and for temperatures above 600 °C, activation energies between 1.7 and 2 eV have been measured. These activation energies are close to the Si–Si bond breaking energy which is around 1.8 eV. Thus, the oxidation rate within this regime can be related to the formation of Si radicals.

The *parabolic* regime can be described along similar lines as metal oxidation. With Si – and this holds also for Ge – mainly oxygen diffuses through the oxide layer. Thus, oxide growth takes place at the interface between the semiconductor and the oxide. With Si, the activation energy measured within this regime is around 1.23 eV. This compares favorably well with activated diffusion of O_2 through SiO_2 for which $\Delta E_d \approx 1.17$ eV ($\mathscr{E}_d \approx 13600$ K).

The microscopic mechanisms considered are schematically depicted in Fig. 26.3.1. Native oxide formation proceeds as described in Sect. 26.1. Electrons are transferred from Si to adsorbed O_2. Thus, physisorbed O_2 becomes chemisorbed and forms an ultrathin oxide layer. Up to a thickness of about 10 Å, rapid oxide growth due to electron tunneling and oxygen diffusion through the thin oxide layer is observed.

A further increase in oxide layer thickness requires both the diffusion of adsorbed oxygen into the oxide and the ejection of "hot electrons" from the Si. Activated diffusion of oxygen is possible because the free volume within the SiO_2 lattice is about 45 Å^3 and thus of similar size as the O_2 molecule. On its way towards the Si/SiO_2 interface, oxygen will pick up electrons which are trapped in the oxide; thus, the concentration of O_2 decreases while that of O_2^- increases. Diffusion of O_2^- ions is enhanced by the electric field related to the positively charged holes ("broken bonds") at the Si surface. At distances of 10–20 Å from the interface, diffusion of O_2^- and O_2 becomes blocked. This is related to the reduction in lattice parameters caused by the mismatch between the (smaller) lattice constant of Si and that of SiO_2. However, *hot* electrons emitted into the SiO_2 conduction band can easily penetrate the blocking layer. These hot electrons promote dissociation of oxygen into O and O^-. Atomic oxygen has a volume of only about 5 Å^3 and can easily diffuse through the blocking layer to the Si/SiO_2 interface, react with silicon and increase the oxide-layer thickness. Clearly, the oxidation rate can be limited by diffusion of oxygen towards the blocking layer or by dissociation of oxygen molecules near the blocking layer. The latter is related to the flux of electrons penetrating the blocking layer.

The enhancement of the oxidation rate by CO_2-laser radiation is due to the heating of both the oxide and silicon. SiO_2 strongly absorbs CO_2-laser radiation via vibrational excitations (Si-O stretching mode at around 1000-1100 cm^{-1}; see also Fig. 7.2.4). Heating of the oxide layer, in turn, favors oxygen diffusion. CO_2-laser heating of Si occurs via free carrier excitations (Sect. 7.6). This increases the density of hot electrons and thereby favors oxygen dissociation. Clearly, preheating of the substrate has the same effect.

With CO_2-laser radiation, films up to a thickness of about 0.2 µm have been grown at an average rate of about 3×10^{-5} µm/s. Films with thicknesses > 0.04 µm exhibited an average dielectric breakdown of about 6.5×10^6 V/cm, for further details see [*Boyd* 1987].

A sharp increase in oxide formation is observed with laser fluences that cause surface melting. In this regime, however, strong surface degradation and structural damage is observed.

26.3.2 Photochemically Enhanced Oxidation of Si

The enhancement of the oxidation rate by VIS or UV laser light can be qualitatively understood with the same model (Fig. 26.3.1). Besides thermal stimulation, direct band-gap excitation increases the density and average energy of electrons emitted into the oxide. Thus, electron capture and dissociation of oxygen molecules becomes more likely and takes place further away from the Si/SiO_2 interface.

The enhancement in oxide growth rate in the presence of Ar^+-laser radiation (about 40% with $I = 10^2$ W/cm²) was found to be *linearly* proportional to the photon flux [*Young* 1988]

$$J = \frac{I_a}{h\nu} = \frac{I_a}{hc}\lambda, \qquad (26.3.1)$$

and only slightly dependent on crystal orientation [*Massoud* and *Plummer* 1987]. According to (26.3.1) the enhancement increases with laser wavelength – as long as I_a stays constant and the oxidation mechanism remains unchanged. When the photon energy exceeds 3.15 eV, a step-like increase in oxidation rate is observed. This is related to direct electronic transitions from the Si conduction band into the SiO_2 conduction band. For example, with 308 nm (4.03 eV) XeCl laser radiation, an enhancement with respect to visible laser light by a factor of 10 was obtained. For photon energies above 4.25 eV, electrons can directly be excited from the Si valence band into the SiO_2 conduction band. This gives an additional 20% increase in oxidation rate. When increasing the photon energy above 5.1 eV, photodissociation of molecular oxygen begins. The effect of atomic oxygen can directly be seen in Fig. 26.3.2a which shows the oxide-layer thickness as a function of fluence for 193 nm (6.42 eV) ArF-and 248 nm (5eV) KrF-laser light. The maximum fluence employed is just below the melting threshold (≈ 0.35 J/cm²). The high oxidation rate observed with ArF-laser radiation is mainly ascribed to direct photodissociation of O_2 (Table V). With respect to visible laser light, the enhancement observed with ArF-laser radiation is about a factor of 300. Figure 26.3.2b shows the dependence of the oxide-layer thickness on the number of laser pulses.

26.3 Elemental Semiconductors

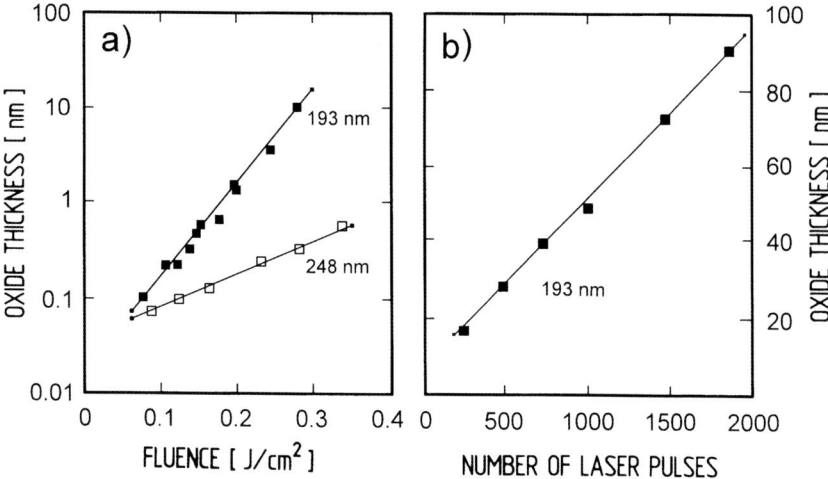

Fig. 26.3.2a,b. Thickness of oxide layer formed on (100) Si ($T_s = 300$ K) in oxygen atmosphere. (a) 20 ArF- or KrF-laser pulses, $p(O_2) \approx 13$ mbar. (b) $\phi(\text{ArF}) \approx 0.16$ J/cm^2, $p(O_2) \approx 366$ mbar, adapted from [*Orlowski* and *Mantell* 1988]

Surface oxidation of Si has also been demonstrated with twin-beam irradiation where, for example, Ar$^+$-laser light produces excess free carriers in the conduction band so that CO$_2$-laser radiation can be absorbed more efficiently. This technique permits localized surface oxidation.

Surface oxidation of Si and Si$_{0.8}$Ge$_{0.2}$/Si enhanced by Hg-lamp ($\lambda = 254$ nm, 185 nm) irradiation has been studied by *Boyd* et al. (1993). At 550 °C enhancements in growth rates of up to 50 times have been observed.

Film properties

Characterization of oxide films was performed by IR spectroscopy, XPS, and capacitance-voltage (CV) measurements. The latter experiments revealed fixed oxide charge densities of $(3\text{-}8) \times 10^{11}$/cm^2 eV and surface state densities of the same magnitude. A considerable improvement in the electrical quality of as-grown films can be achieved by means of a short (about 20 min) anneal at 900 °C in 1000 mbar O$_2$. The fixed charge density was then about 6×10^{10}/cm^2 eV, and the breakdown voltage $> 5 \times 10^5$ V/cm.

26.3.3 Nitridation of Silicon

Surface nitridation of Si has been investigated mainly for N$_2$ and NH$_3$ atmospheres and excimer-laser radiation. Efficient nitridation was observed only in

NH$_3$. With ArF-laser radiation, nitridation is characterized by an initial phase of rapid growth (during about 2000 pulses with $\tau_l = 12$ ns and $\phi \approx 15$ mJ/cm^2) followed by inhibited growth similar to thermal nitridation [*Sugii* et al. 1988]. The maximum film thickness achieved was ≈ 25 Å. On the basis of AES studies, laser-grown films are very similar to those thermally grown at $T_s \approx 1000$ °C. Nitridation seems to be related to photogenerated NH$_2$ radicals which easily react with Si. Nitridation may proceed similarly to oxidation.

26.4 Compound Semiconductors

The oxidation of compound semiconductors and its photon enhancement has extensively been studied. The most detailed experiments have been performed for the III-V compounds GaAs, InAs, and InP. With these materials, surface oxidation proceeds in three successive steps [*Mönch* 1986]:

- Oxygen adsorption at cleavage-induced defects.
- Activated adsorption and dissociation of oxygen followed by breaking of adjacent III-V surface bonds.
- Field-assisted diffusion of oxygen and film growth.

Laser-enhanced oxidation of compound semiconductors, in particular of GaAs and InP, has been demonstrated with VIS- and UV-laser light. The experiments were mainly performed in O$_2$ and N$_2$O atmosphere (Appendix B.9).

The (dark) sticking coefficient of O$_2$ on GaAs depends on coverage and has values in the range $10^{-5} \leqslant s \leqslant 10^{-9}$. Light with photon energies $h\nu \geqslant E_g$ (GaAs, 300 K) ≈ 1.43 eV $\cong 867$ nm increases s by about a factor of 10^3. This is related to the increase in electron-transfer which favors chemisorption of O$_2$. With VIS and near UV light, the enhancement in oxidation rate saturates at around 1 ML (monolayer) [*Lu* et al. 1990; *Bertness* et al. 1987].

At a photon energy of about 4.1 eV (≈ 302 nm) a sharp increase in oxidation rate due to photoemission of electrons into the GaAs oxide was observed. Another increase in rate observed with 193 nm ArF-laser radiation is related to photodissociated O$_2$.

The stoichiometry of oxides depends on the laser parameters. After laser irradiation, the oxide layer mainly consists of Ga$_2$O$_3$ and variable amounts of As$_2$O$_3$ and As$_2$O$_5$. These amorphous oxide mixtures may crystallize and form stable Ga$_2$O$_3$ and GaAsO$_4$ [*Schwartz* 1975].

X-ray-induced low-temperature oxidation of (110) GaAs in N$_2$O has been studied by *Seo* et al. (1990).

Laser-enhanced oxidation of compound semiconductors can be described along similar lines as silicon oxidation. Important differences are related to the differences in band-gap energy and the type of surface oxides. For elucidating finer details, more experimental data are required.

Summary

Semiconductor oxidation takes place mainly at the semiconductor-oxide interface while metal oxidation frequently proceeds at the oxide surface. In semiconductors, the oxidation rate depends *strongly* on wavelength, mainly due to electron-hole pair generation for photon energies $h\nu > E_g$. Laser radiation may generate defects within the oxide layer and at the oxide-substrate interface. Defects such as oxygen vacancies, cracks, etc., significantly enhance the transport of species.

Laser-induced surface oxidation and nitridation of metals and semiconductors results in the formation of films with thicknesses of, typically, 10 Å to 1 µm. Such films can be produced either by single-step direct writing, or over extended areas. The technique is *complementary* to laser-CVD which permits the growth of films with typical thicknesses of 1–100 µm.

26.5 Oxide Transformation, Reoxidation

Laser irradiation of an oxide M_xO_y in oxygen or an oxygen containing atmosphere can result in a new oxide of composition $M_{x\pm\varepsilon}O_{y\pm\delta}$. Experiments of this type have been performed, in particular, for metal oxides.

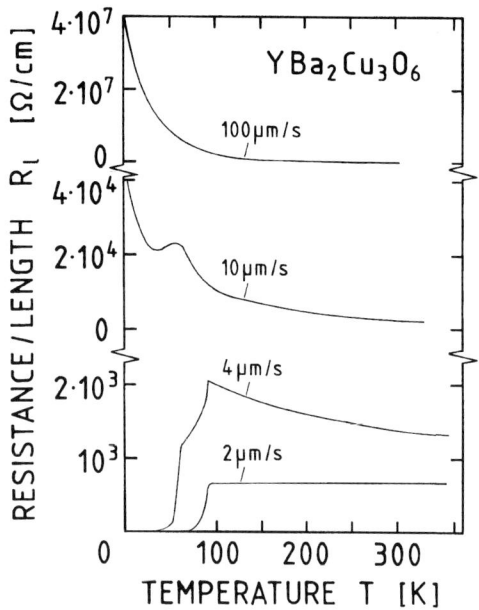

Fig. 26.5.1. Temperature dependence of the resistance R_l [Ω/cm] of Ar^+-laser fabricated lines in ceramic $YBa_2Cu_3O_6$ for various scanning velocities ($P = 500$ mW, $\lambda = 488$ nm, $2w_0 = 35$ µm, $p(O_2) = 1800$ mbar) [*Liberts* et al. 1988*a*]

Laser-induced reoxidation permits direct writing of superconducting $YBa_2Cu_3O_7$ lines into semiconducting $YBa_2Cu_3O_6$ [*Sobolewski* et al. 1994; *Shen* et al. 1991; *Liberts* et al. 1988a]. Experiments have been performed with both ceramic pellets and thin films. The oxygen uptake within the material is induced by cw Ar^+- or Kr^+- laser heating in O_2 atmosphere. The electrical properties of lines depend on the laser-beam intensity, the scanning velocity, and the oxygen pressure. The temperature dependence of the line resistance per unit length is shown in Fig. 26.5.1 for various scanning velocities. For $v_s = 2\,\mu m/s$ the onset to superconductivity occurs at around 88 K. The large transition width of about 10 K is probably related to chemical inhomogeneities within such lines. Similar experiments have been performed for oxidic perovskites [*Bäuerle* 1985; *Otto* et al. 1984].

27 Depletion and Exchange of Species

This chapter deals with laser-induced surface modifications based on the depletion or exchange of one or several components of a material; it includes cases where the solid surface is decomposed so that only a single component or a mixture of components of the original material remains. Among the many examples found in the literature (Appendix B.10), we shall discuss in further detail the reduction and metallization of oxides, surface modifications of polymers, and chemical transformations of spun-on films.

27.1 Reduction and Metallization of Oxides

Just as laser-induced heating of certain materials in oxygen-rich atmosphere permits one to incorporate oxygen into the lattice or to form a surface oxide layer, specific materials can give up their oxygen to a reducing environment under the appropriate laser-heating conditions. This has been demonstrated for oxidic perovskites and perovskite-related oxides such as $BaTiO_3$, $PbTiO_3$, $PbTi_{1-x}Zr_xO_3$ (PZT), $Pb_{1-3y/2}La_yTi_{1-x}Zr_xO_3$ (PLZT), $SrTiO_3$, and $LiNbO_3$, for high-temperature superconductors, and for different metal and semiconductor oxides.

27.1.1 Oxidic Perovskites and Related Materials

Oxidic perovskites are insulators with a band gap of, typically, 3 eV. Most of them are ferroelectric and thereby piezoelectric. It is well known that the physical properties of these materials can be dramatically changed by reducing the bulk material at elevated temperatures, e.g., in H_2 atmosphere at 500–1500 K. This treatment results in the formation of oxygen vacancies and free or quasi-free electrons. The concentration of oxygen vacancies and free electrons increases with increasing temperature and with decreasing oxygen partial pressure. The oxygen vacancies act as shallow donor levels and the insulating material becomes an n-type semiconductor. The originally transparent material changes to blue or black, depending on the concentration of vacancies. Because of the fundamental role of the oxygen ion in connection with the dynamical properties of perovskites, oxygen vacancies strongly influence the structural phase transitions (ferroelectric and non-ferroelectric) observed in these materials [*Migoni* et al. 1976; *Wagner* et al. 1980; *Bäuerle* et al. 1980].

Laser-light irradiation of oxidic perovskites in a reducing atmosphere can result in *local* reduction of the material surface. While for sub-band-gap radiation ($hv < E_g$) the reduction mechanism is mainly thermal, UV and far-UV radiation ($hv > E_g$) directly generates quasi-free electrons and oxygen vacancies. The reduction process is reversible, i.e., on heating the material in O_2 atmosphere or air, the reduced (blue to black) regions vanish and only small changes in surface morphology remain. With increasing laser-light intensity, the degree of reduction increases and the electrical properties of laser-treated regions change from semiconducting to metallic. Beyond a certain threshold intensity, etching or cutting of the material is observed.

Laser-induced reduction and metallization of oxidic perovskites allows single-step conductive patterning of the otherwise insulating material surface. Metallization has been studied in detail for hot-pressed optically transparent ferroelectric PLZT ceramics [*Kapenieks* et al. 1986a, b]. The electrical resistivity of the bulk material was $> 10^{14}$ Ωcm. Figure 27.1.1a exhibits the resistance per unit length of lines, R, produced by UV-laser direct writing as a function of laser power in H_2 atmosphere. Above about 180 mW, microcracks

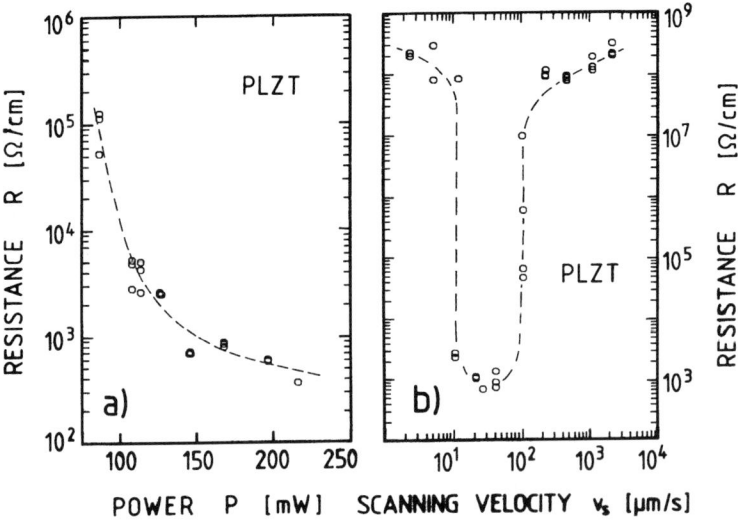

Fig. 27.1.1. (a) Resistance per unit length of lines produced by laser-direct writing on ceramic PLZT as a function of laser power (337–356 nm cw Kr$^+$laser, $w_0 \approx 0.9$ μm, $v_s = 25$ μm/s; $p(H_2) = 500$ mbar). (b) Same as (a) but for constant laser power and different scanning velocities ($P = 190$ mW, $p(H_2) = 500$ mbar) [*Kapenieks* et al. 1986a]

are occasionally observed in the region adjacent to the metal line. These cracks have no influence on the electrical conductivity of lines. However, with $P > 250$ mW, cracks penetrate deeply into the bulk material. The resistance of lines as a function of scanning velocity shows a pronounced minimum between about 10 and 100 µm/s (Fig. 27.1.1b). Very slow scanning velocities favor evaporation of Pb and the formation of cracks. For high velocities, the laser-beam dwell time is too short for oxygen out-diffusion. The location of the minimum and the overall change in R depend on the laser parameters and the H_2 pressure. For the dependence of R on H_2 pressure a similar behavior has been found.

The experiments show that there are ranges of optimal parameters where metallic lines with $R < 10^3$ Ω/cm ($\rho \approx 10^{-4}$ Ωcm) can be fabricated with good reproducibility.

The conductivity of metallized regions within PLZT surfaces produced by *UV-laser* radiation in H_2 atmosphere is essentially determined by the reduction of the material to metallic Pb, Ti, and Zr, the evaporation of Pb and, for certain parameters, the cracking of lines. The evaporation of Pb results in the occurrence of a shallow groove in the middle of the metal line. Local depletion of Pb is consistent with X-ray microanalysis and similar investigations performed on $PbTi_{1-x}Zr_xO_3$.

Electrodes of areas up to 0.5×0.5 cm^2 have been fabricated on PLZT surfaces in a similar way [*Kapenieks* et al. 1986b]. They were characterized by temperature-dependent dielectric measurements on samples with different thicknesses (0.1–2 mm) at frequencies between 10 kHz and 10 MHz. Below 400 K, laser-fabricated contacts led to higher dielectric constants than conventional evaporated Au electrodes. The difference is most pronounced for small sample thicknesses. In combination with the observed increase in adherence, laser-processed electrodes could be superior to conventional electrodes in microprocessing.

Other materials

Ar_2 excimer-laser radiation ($hv \approx 9.8$ eV $\cong 126.5$ nm) has been used to generate spherically shaped crystalline Si precipitates with submicrometer diameter in SiO_2 surfaces. Direct writing of p-Si lines has also been demonstrated [*Kurosawa* et al. 1993].

27.1.2 Superconductors

The physical properties of high-temperature superconductors (HTS) depend sensitively on the oxygen content.

The depletion of oxygen under laser-light irradiation has been investigated for ceramic platelets and thin films of $YBa_2Cu_3O_{7-\delta}$ [*Shen* et al. 1991; *Liberts* et al. 1988b; *Rothschild* et al. 1988]. Here, the oxygen content within the material surface is diminished by local heating under cw Ar^+- or Kr^+-laser irradiation in vacuum, H_2 or N_2 atmosphere. This technique permits one to locally diminish the transition temperature from the normal to the superconducting state, or to direct write semiconducting ($\delta > 0.5$) or metallic patterns into the otherwise superconducting material ($0 \leq \delta < 0.5$). Figure 27.1.2 shows the temperature-dependent resistance of a strip line before laser treatment. It shows a clear transition from the normal to the superconducting state. After laser-induced reduction in N_2, the strip line is semiconducting. The process is reversible, i.e., laser-induced reoxidation recovers the superconducting state. The technique can be employed for surface patterning, trimming, tuning of critical currents, the fabrication of weak links for SQUIDs (superconducting quantum interference devices), etc.

27.1.3 Qualitative Description

The laser-induced depletion of species can be described by the diffusion equation, similarly to solid-phase doping. The main difference is that (24.1.3,4) must be replaced by

$$N_A(z, t=0) = N_A^0 ; \quad N_A(0, t) = 0 ; \quad N_A(\infty, t) = N_A^0 , \qquad (27.1.1)$$

where N_A^0 is the initial concentration of those species that are depleted. When species A reach the surface it is assumed that they immediately desorb. With (27.1.1) the solution of (24.1.1) becomes

$$N_A(z,t) = N_A^0 \left[1 - \mathrm{erfc}\left(\frac{z}{l_s}\right) \right] = N_A^0 \, \mathrm{erf}\left(\frac{z}{l_s}\right), \qquad (27.1.2)$$

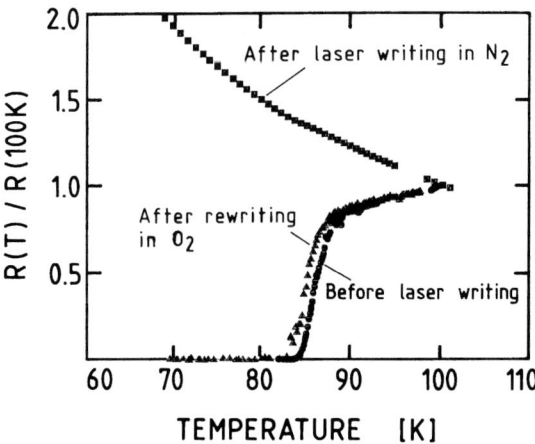

Fig. 27.1.2. Normalized resistance of a strip line in $YBa_2Cu_3O_{7-\delta}$ as a function of temperature. After Ar^+-laser-induced reduction in N_2 atmosphere the material is semiconducting ($I = 6 \times 10^5$ W/cm^2, $v_s = 5$ µm/s, 48 scans, $p(N_2) = 1$ bar). Superconductivity can be recovered with 2 scans in O_2 atmosphere. For the three data sets $R(100\,K)$ was 62 Ω, 1850 Ω, and 85 Ω [Shen et al. 1991]

27.1 Reduction and Metallization of Oxides

where the effective diffusion length is given by $l_s \approx 2(\langle D_s \rangle t)^{1/2}$. Thus, the spatial variation (27.1.2) can directly be derived from Fig. 24.1.2. Equations (24.1.7) and (24.1.8) can be employed without any changes.

Metallization

Let us now consider a somewhat different situation where a metal oxide in the form of a slab with area F and thickness h_s is reduced according to the reaction

$$\text{Me}_x\text{O}_y + y\text{H}_2 \rightarrow x\,\text{Me} + y\,\text{H}_2\text{O} + \Delta H \, . \tag{27.1.3}$$

In the simplest approximation, this process can be described by the energy balance

$$\rho c_p F h_s \dot{T} = PA(h_1) - P_{\text{loss}}(T) - \Delta H \dot{h}_1 F \, , \tag{27.1.4}$$

and the equation for the growing metal film

$$\dot{h}_1 = v_0 \exp\left(-\frac{\mathscr{E}}{T}\right), \tag{27.1.5}$$

where $h_1 \ll h_s$. In general, the absorptivity of an oxide covered with a thin metal film is lower than that of the oxide, i.e., $A_{\text{Me}} < A_{\text{ox}}$ (Fig. 7.2.4), and decreases further with increasing film thickness h_1. This results in a *negative* feedback.

Let us first ignore the reaction enthalpy, ΔH. Figure 27.1.3 shows, qualitatively, the temporal behavior of the surface temperature (full curves) for two different laser powers P_1 and P_2 with $P_1 < P_2$. Initially, the *absorbed* laser power is determined by the absorptivity of the oxide and a rapid rise in surface temperature is observed. With the formation of a metal film, the absorptivity decreases. When $h_1 > l_\alpha$ (metal), the absorbed laser power is determined by the absorptivity of the

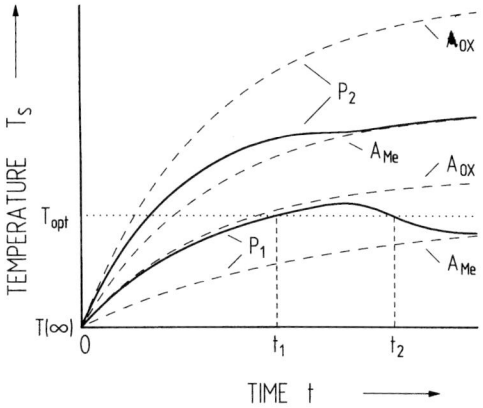

Fig. 27.1.3. Qualitative behavior of surface temperature T_s (solid curves) during laser-induced metallization of a metal oxide. The laser is switched on at $t = 0$ (power $P_1 < P_2$). The dashed curves show the behavior of T_s for a pure oxide with absorptivity A_{ox} and an oxide covered with a metal film with $A_{\text{Me}} < A_{\text{ox}}$

metal and T_s increases more slowly, or it may even decrease. This may explain, in part, the behavior of the surface resistance in Fig. 27.1.1a. Figure 27.1.3 reveals another interesting feature. For low laser powers, P_1, the surface temperature shows a maximum and exceeds the optimal temperature, T_{opt}, for controlled surface reduction only within a time interval $t_1 \leqslant t \leqslant t_2$. The final layer thickness is then almost independent of the total illumination time. This self-stabilization permits efficient and reproducible surface processing (without crack formation, etc.) by selecting the optimal laser power.

If ΔH cannot be ignored, the additional source term in (27.1.4) may cause strong transient changes in temperature and uncontrolled processing. An example would be the reduction of Cu_2O according to (27.1.3) by CO_2-laser light where $A_{ox} \equiv A(Cu_2O, 10.6\,\mu m) \approx 0.82$ and $A_{Me} \equiv A(Cu, 10.6\,\mu m) \approx 0.02$, and $\Delta H \approx -73.5$ kJ/mol (Fig. 3.1.1).

27.2 Surface Modification of Polymers

Laser fluences below or around the ablation threshold can modify the surface morphology, crystallinity, chemical composition, reactivity, and resistivity of polymer surfaces.

27.2.1 Laser-Enhanced Adhesion

Successful material coating and bonding is closely related to the adhesion forces between the materials under consideration. With many materials, adhesion must be increased by modifying the surfaces prior to the coating or bonding process. Besides standard chemical and physical (e.g., RF plasma) techniques, UV-laser light has been proved to be a powerful tool which opens up new possibilities. Laser-enhanced adhesion may be based on different types of surface roughening, the scission of surface chemical bonds and their saturation with other atoms/radicals, etc. Systematic investigations of this type have been performed for polymers.

Figure 27.2.1 shows the surface of a PET foil after single-pulse KrF-laser irradiation. While the initial roughness of this (ultra-flat) foil was $\leqslant \pm 2.5$ nm the roughness of the irradiated area due to dendrite formation was between ± 5 and ± 30 nm. Dendrites grow during many hours after laser-light irradiation (Fig. 4.2.2).

The adhesion of metal films evaporated onto foils preirradiated with UV light is significantly increased with respect to non-irradiated foils. This is shown in Fig. 27.2.2 for a 200 nm thick CoNi film deposited by means of electron-beam evaporation. The adhesion force was measured by a standard peel test (EAA test) [*De Puydt* et al. 1988] and a modified test [*Hagemeyer*

27.2 Surface Modification of Polymers

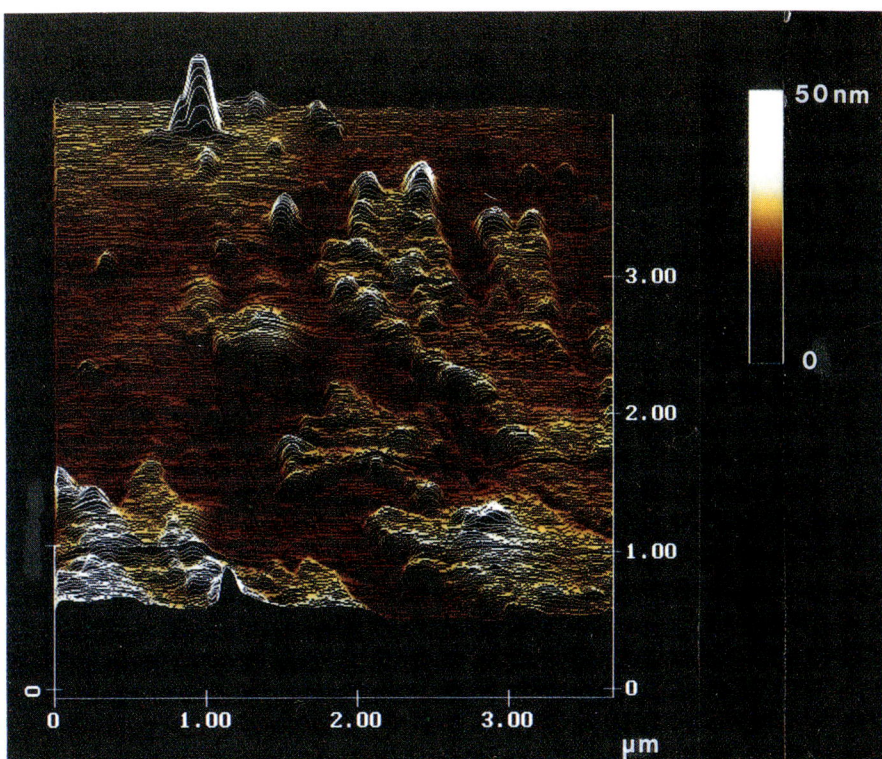

Fig. 27.2.1. AFM (atomic force microscope) picture showing the surface roughening of an ultraflat PET foil after single-pulse KrF-laser irradiation ($\phi \approx 41$ mJ/cm^2)

et al. 1994]. The peel strength rapidly increases with laser fluence and reaches the limit of the standard test (≈ 4 N/cm) at about 15 mJ/cm^2. This fluence is considerably *below* the threshold fluence for ablation [ϕ_{th} (PET) ≈ 40 mJ/cm^2]. Similar results have been achieved with other Co alloys, .eg., CoCr, CoNi-oxide, etc. These types of metal coatings on PET are applied for video tapes, floppy disks, etc. Here, the adhesion of the metal film on the polymer foil determines, to a large extent, the lifetime of the product.

Epoxy bonding of Teflon (PTFE) to stainless steel was demonstrated by *Murahara* (1994). If Teflon is irradiated with ArF-laser light prior to bonding, the shearing tensile strength is increased, typically from 20 to 1280 N/cm^2.

27.2.2 Changes in Crystallinity

Laser-light irradiation with fluences $\phi \leq \phi_{th}$ can significantly enhance the relative amount of amorphous material within polymer surfaces. Recrystallization

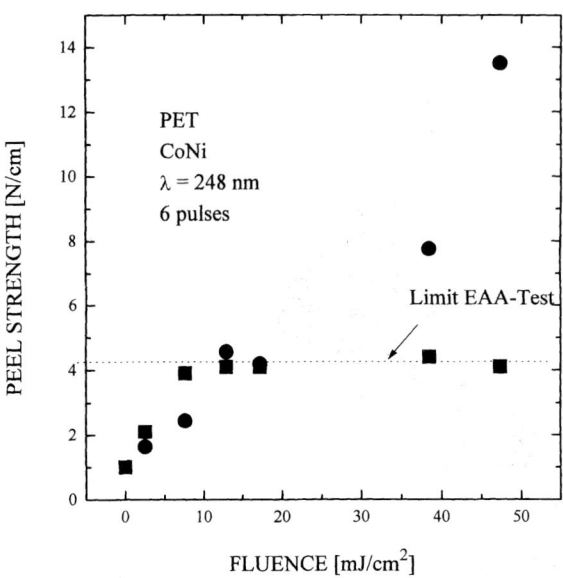

Fig. 27.2.2. Adhesion force of 200 nm thick $Co_{80}Ni_{20}$ films on PET foils (thickness ≈ 50 μm) irradiated with 248 nm KrF-laser light prior to film deposition. Two different types of peel tests have been employed (■ EAA test, ● modified test), adapted from [Hagemeyer et al. 1994]

may result in the formation of dendrites (Fig. 4.2.2). In oxygen or nitrogen atmosphere, dendrite formation is suppressed. The maximum growth velocity of dendrites is around the glass temperature $T_g(\text{PET}) \approx 67\,°\text{C}$.

27.2.3 Chemical Degradation

UV-laser radiation with fluences $\phi \approx \phi_{th}$ may cause drastic changes in the composition of polymer surfaces.

ArF- or KrF-laser irradiation of PI significantly decreases the surface concentration of oxygen and nitrogen with respect to carbon. This process results in a dramatic increase in surface conductivity (Fig. 27.2.3). The sheet conductance has been defined by $\sigma_s = 1/\mathscr{R}$ where $\mathscr{R}[\Omega/\square]$ is the sheet resistance (Sect. 30.4.3). After about 500 laser pulses, the conductance is increased by about 15 orders of magnitude. This effect is ascribed to the formation of a carbon layer on the polymer surface. If instead of air the ambient medium is pure O_2, the conductance is lowered, while in pure Ar or N_2 it is enhanced. The maximum in the conductance appears for fluences near ϕ_{th} (Fig. 27.2.3b). By using laser-beam interference (Fig. 5.2.1c) line patterns with widths of 0.5 μm and periods of 0.9 μm have been fabricated [Phillips et al. 1993b; Feurer et al. 1993].

The temperature dependence of σ_s is shown in Fig. 27.2.4 for different laser fluences. Open symbols refer to samples which were exposed to aqueous HCl

27.2 Surface Modification of Polymers

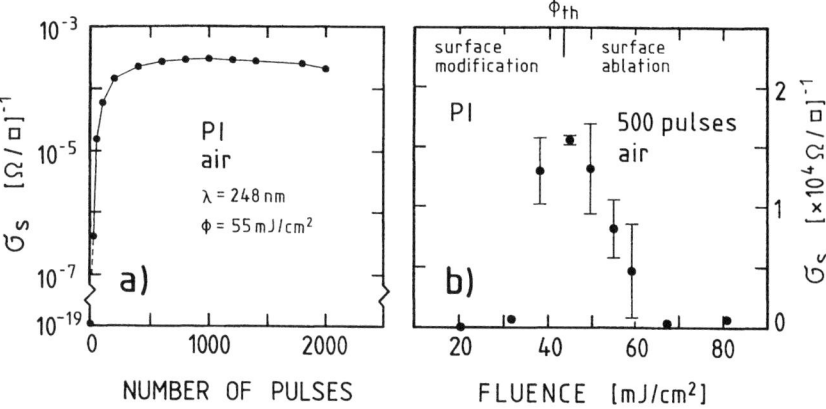

Fig. 27.2.3a,b. Dependence of the sheet conductance of PI (Upilex R) foils on KrF-laser treatment in air (τ_l(FWHM) ≈ 28 ns, 5 Hz, $w_x \times w_y \approx 2 \times 20$ mm^2). The threshold fluence for ablation is ϕ_{th}(PI, 248 nm) ≈ (44 ± 3) mJ/cm^2 [*Arenholz* et al. 1993]

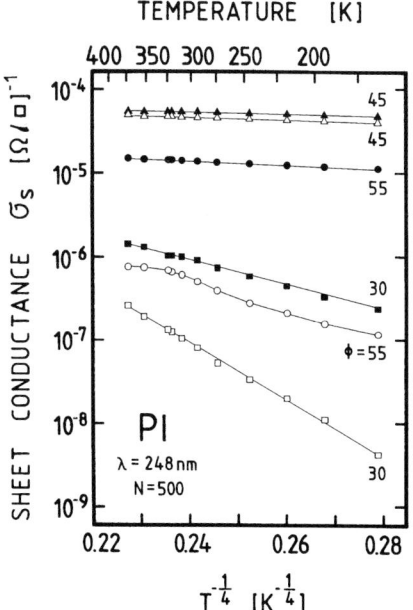

Fig. 27.2.4. Temperature dependence of the sheet conductance for PI irradiated with KrF-laser light at different fluences (full symbols: after irradiation; open symbols: after irradiation and treatment in aqueous HCl) [*Bäuerle* et al. 1996]

subsequent to irradiation. This treatment destroys carbon–carbon double bonds formed at the surface. With fluences both below and above the ablation threshold, the conductivity can be described by variable range hopping [*Mott* and *Davis* 1979]

$$\sigma = \sigma_0 \exp\left[-\left(\frac{T_0}{T}\right)^{1/4}\right]. \tag{27.2.1}$$

For $\phi \approx \phi_{th}$ the temperature dependence of the conductivity shows "metallic-like" behavior.

Laser-induced surface modifications of polymers in combination with electroless plating permit selective area metallization (Chap. 21). Among the materials investigated are PET, PI, and PES (polyethersulfone).

27.2.4 Photochemical Exchange of Species

ArF-laser irradiation of Teflon in $B(CH_3)_3$ atmosphere permits one to exchange F atoms for methyl groups [*Murahara* 1994]. By this means, the oleophobic surface changes into oleophilic. In similar investigations with $B(OH)_3$, fluorine atoms have been replaced by OH molecules. The modified surface shows hydrophilic properties. Artificial blood vessels of Teflon were locally transformed from hydrophobic to hydrophilic by ArF-laser irradiation in an atmosphere of NH_3 and B_2H_6.

27.3 Chemical Transformation of Solid Films

The fabrication of microstructures by laser-induced transformation and decomposition of thin solid films is shown schematically in Fig. 27.3.1. In the first step, the substrate is coated with a thin film which contains the precursor molecules. Frequently, such films are fabricated by spin-on, spray-on, or paint-on techniques. Here, the precursor is dissolved in an adequate solvent, coated onto the substrate, and baked for excess solvent removal. For laser-induced transformation the remaining solid film should have a thickness of typically 0.1–10 µm. In the second step, the film is transformed by laser light, for example in a direct-write process. This step goes along, in general, with a strong shrinkage in height. In the third step, non-illuminated film material is dissolved. This method is compatible with photolithographic techniques, it is simple, and it does not require any vacuum equipment. Patterning by direct writing has been performed with scanning speeds up to 10 cm/s. Presently, the main drawbacks for applications in microfabrication are related to the porosity of the resulting material, the incorporation of impurities originating from non-volatile decomposition products, and the tendency for (non-coherent) structure formation.

Thin films can be fabricated by using high-power unfocused or defocused laser beams, or a line focus. The thickness of deposits can subsequently be increased by employing standard techniques, e.g., CVD, electrochemical plating, etc.

Among the precursor films investigated in most detail are organometallic compounds, metal acetates, organosilicates, and composites.

27.3 Chemical Transformation of Solid Films

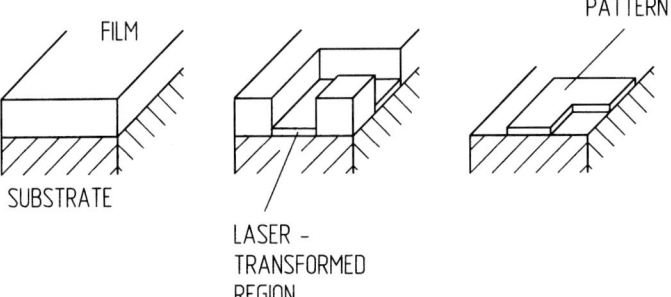

Fig. 27.3.1. Main steps of pattern formation by laser-induced solid-phase transformation. Note the difference in height between the precursor film and the resulting pattern

Metal patterns

For the fabrication of metal patterns, one of the most interesting classes of precursor compounds are metallopolymers that are stable at room temperature. In the first detailed experiments in this area Au and Pd patterns were produced on SiO_2 and Si substrates by using Ar^+-laser radiation for (overall exothermal) material transformation [*Gross* et al. 1986, 1987]. Subsequent to laser direct writing, non-transformed parts of the precursor film were dissolved in CH_2Cl_2.

Figure 27.3.2a displays the average width of Au lines as a function of Ar^+-laser power for two scanning velocities. The width of lines increases approximately linearly, as in LCVD. Lines produced with powers < 10 mW were incompletely reacted and did not survive rinsing in a mixture of aqueous HNO_3, H_2CrO_4, and H_2SO_4. Rinsing removes regions of lines having a high carbon and sulfur content. With increasing thickness of the metallopolymer film, the average linewidth was found to increase for all laser powers and scan speeds. Complete transformation was achieved in films with an initial thickness of 1.7 μm. The (normalized) electrical resistivity of Au lines is shown in Fig. 27.3.2b as a function of laser power. For $P > 25$ mW, the resistivity is about five times that of evaporated Au films. The large error bars in the low power range are related to periodic non-coherent structures (Sect. 28.3.4).

The same technique has been used for a variety of other materials (Appendix B.10).

The fabrication of metal patterns by excimer-laser light projection or by means of a contact-mask, has also been demonstrated [*Esrom* and *Wahl* 1989]. For example, Pd patterns were fabricated from palladium (II) acetate films (≈ 0.1 μm thick) which were spin-coated onto SiO_2 or ceramic Al_2O_3 substrates, illuminated by excimer-laser light, and developed in CCl_4. Subsequently, the thin Pd patterns (≈ 60 Å thick) were electroless plated with Cu ($h_1 \approx 0.1$ to 6 μm; plating rate about 15 Å/s). In this way, *regular* patterns with well-defined sharp edges have been obtained. With ceramic Al_2O_3 substrates and special surface pretreatment, the film adhesion was typically

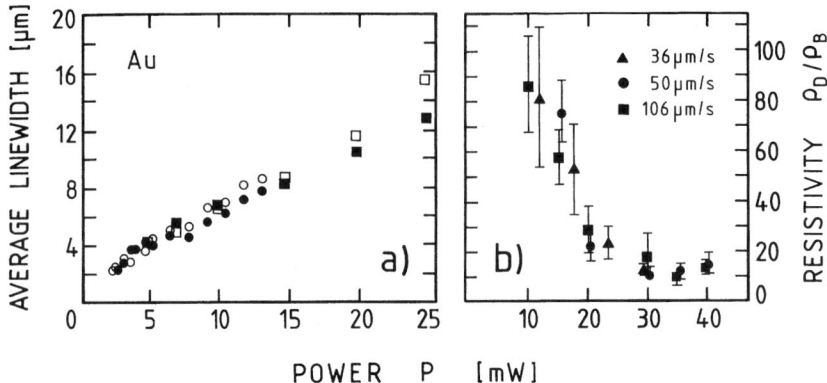

Fig. 27.3.2a, b. Properties of Au lines fabricated by 515 nm Ar$^+$-laser-induced decomposition of a metallopolymer film (thickness after pre-baking $\approx 1.7\,\mu$m). (a) Average width: ●, ■: $v_s = 36\,\mu$m/s, ○, □: $v_s = 206\,\mu$m/s (different symbols designate data from different samples) [*Fisanick* et al. 1985]. (b) Electrical resistivity of lines normalized to the resistivity of bulk Au ($\rho_B \approx 2.44\,\Omega$ cm) [*Fennell* et al. 1985]

$\geqslant 30$ N/mm^2. Contact metallization of laser-drilled via-holes using the same process was also demonstrated. The mechanism of Pd-acetate film decomposition is mainly thermal.

SiO$_2$ patterns

Another example of solid-phase transformation is the generation of SiO$_2$ patterns by laser pyrolysis of organosilicate films produced by spin-coating from Si(OR)$_x$(OH)$_{4-x}$ [*Krchnavek* et al. 1984]. Smooth, continuous lines were fabricated on Si-wafers by employing 515 nm Ar$^+$-laser radiation and scanning speeds up to 100 µm/s. Linewidths as small as 1 µm were obtained. The film thicknesses were, typically, 0.5 µm but could be varied by varying the spin-on speed or the viscosity of the organosilicate solution. Material not exposed to the laser beam was rinsed off with methanol. The laser-induced reaction probably consists of an initial elimination of the organic liquid and a subsequent release of OH groups. The quality of the SiO$_2$ films was at least as good as that obtained in thermally cured spun-on glasses. The breakdown strength was about 1.6×10^6 V/cm.

Photoresists for UV-laser lithography

There are numerous investigations on the exposure of photoresist layers by (mainly) excimer-laser radiation for applications in high-resolution imaging. For example, with single-layer resists based on methacrylate-terpolymers and ArF-laser radiation a resolution of 0.25 µm has been achieved [*Wallraff* et al. 1993].

28 Instabilities and Structure Formation

Instabilities are observed in many different types of laser processing. They are related to positive feedbacks between different characteristics (degrees of freedom) in laser–matter interactions. The appearance of an instability means that a small perturbation of the system initially grows up exponentially with time. Its further development can lead to different self-organization phenomena like spontaneous formation of periodic or stochastic dissipative structures, spiral waves, and many other structures with broken or unbroken symmetry ("unbroken" refers to transformations from an unstable structure to a stable or metastable structure without changes in initial symmetry).

From a *scientific* point of view, the analysis of instabilities yields fundamental information on laser–matter interactions. From a *technical* point of view, many applications of laser materials processing require spontaneous structure formation to be suppressed or even avoided altogether.

28.1 Coherent and Non-coherent Structures

Structures that develop on solid or liquid surfaces under the action of laser light can be classified into coherent structures and non-coherent structures.

Coherent structures are *directly* related to the coherence, the wavelength, and the polarization of the laser light. For non-coherent structures such a direct relation to these laser parameters is absent[1].

The feedback that causes coherent or non-coherent structure formation can originate from different mechanisms such as local thermal expansion, changes in optical or thermal properties, surface tension effects, surface acoustic waves (SAW), capillary waves, melting, vaporization, transformation energies, chemical reactions, etc.

Coherent structures have a common origin: The oscillating radiation field on the material surface which is generated by the *interference* between the

---

[1] In the traditional theory of self-organization (synergetics) the term "coherent" is sometimes used in a different way. It denotes cooperative structure formation with an "internal" period that does *not* depend on the initial and boundary conditions, any external disturbances, etc.

incident laser beam and scattered/excited surface waves. The spatial periods of such structures are therefore proportional to the laser wavelength.

Non-coherent structures are *not* directly related to any spatial periodicity of the energy input caused by interference phenomena. Here, the feedback results in either spontaneous symmetry breaking or in a non-trivial spatio-temporal ordering of the (whole) system [*Gaponov-Grekhov* et al. 1989; *Haken* 1978; *Nicolis* and *Prigogine* 1977].

An example where coherent *and* non-coherent structure formation can be seen *simultaneously* is shown in Fig. 28.1.1a. The slow oscillation (long spatial period) is not related to the laser wavelength and polarization. It depends only on the laser-beam intensity, scanning velocity, and the gas pressure. Superimposed on these slow oscillations are fine periodic structures. The distance between these ripples is proportional to the laser wavelength. Their orientation is perpendicular to the electric vector of the incident light. Non-coherent structure formation can be suppressed by changing the experimental conditions. This is demonstrated in (Fig. 28.1.1b). The ripples are still present. Their orientation can be changed by changing the polarization of the incident light (Fig. 28.1.1c). Ripple formation can also be suppressed, to a large extent, by an appropriate selection of experimental parameters.

Equations

A theoretical description of coherent and non-coherent structure formation is quite complex. It requires one to solve the electrodynamic, heat, and kinetic equations simultaneously in three dimensions by taking into account the different *feedback* mechanisms involved. The (partial) differential equations have an infinite number of degrees of freedom (here, this term is used in a way

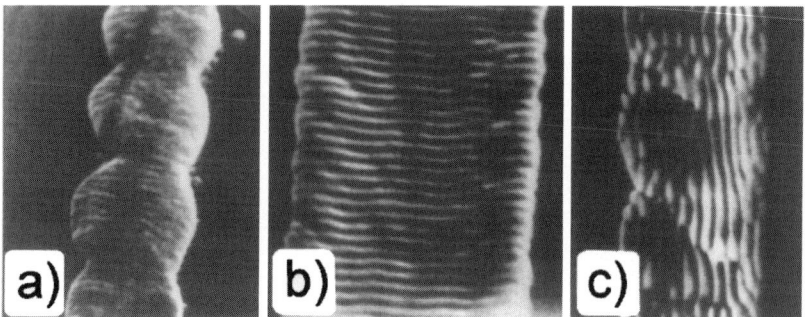

Fig. 28.1.1a–c. Coherent and non-coherent structures observed during 488 nm Ar⁺-laser direct writing of Si lines (from SiH$_4$) on Si wafers. The slow oscillation in (**a**) is superimposed by ripples whose periods are of the order of the laser wavelength. The orientation of ripples depends on the polarization of the electric vector of the radiation. This was parallel to the scanning direction in (**b**) and perpendicular to it in (**c**) [*Bäuerle* 1984]

28.1 Coherent and Non-coherent Structures

different to classical mechanics). Therefore, simplified models which provide further insight into the problem are very useful. Following synergetic methods, one can restrict the problem to the "most important" degrees of freedom which are termed *order parameters*, O_i. Their choice is sometimes a question of physical intuition.

From a mathematical point of view, the evolution of order parameters can be described by differential equations of the type

$$\left(\frac{\partial}{\partial t} + \boldsymbol{v}_s \nabla\right) O_i = \sum_j \nabla(\mathscr{D}_{ij} \nabla O_j) + \mathscr{F}_i, \quad (28.1.1)$$

with $i = 1, 2, \ldots, n$, where n is the number of order parameters. $\mathscr{D}_{ij} \equiv \mathscr{D}_{ij}(\boldsymbol{O})$ are "diffusion" coefficients and $\boldsymbol{O} \equiv \{O_1, O_2, \ldots, O_n\}$. $\boldsymbol{v}_s$ is the velocity of the moving medium as, for example, the moving ablation front, the scanning velocity of the laser beam, etc. For non-autonomous systems the source term depends explicitly on time so that $\mathscr{F}_i = \mathscr{F}_i(\boldsymbol{O}; \boldsymbol{x}, t)$. For autonomous non-homogeneous systems $\mathscr{F}_i$ depends on the (spatial) coordinates $\boldsymbol{x}$ only, i.e., $\mathscr{F}_i = \mathscr{F}_i(\boldsymbol{O}; \boldsymbol{x})$. If $\mathscr{F}_i = \mathscr{F}_i(\boldsymbol{O})$ the system is termed homogeneous and autonomous.

In some cases (28.1.1) can be reduced to a set of *ordinary* differential equations. Some examples will be discussed in Sect. 28.3.

Stability

There are different approaches to investigate the stability of solutions [Haken 1978]. Let us consider *small* perturbations, $O_i^p(\boldsymbol{x}, t)$, of the stationary solution $O_i = O_i^s(\boldsymbol{x})$ so that

$$O_i(\boldsymbol{x}, t) = O_i^s(\boldsymbol{x}) + O_i^p(\boldsymbol{x}, t).$$

The relevant equations can then be linearized with respect to O_i^p by the ansatz

$$O_i^p(\boldsymbol{x}, t) = O_i(\boldsymbol{x}) \exp(\Gamma t). \quad (28.1.2)$$

This ansatz enables one to determine the eigenvectors $\boldsymbol{O}^{(j)}(\boldsymbol{x}) = \{O_1^{(j)}, \ldots, O_n^{(j)}\}$ and the dispersion relations for the increments $\Gamma^{(j)} = \gamma^{(j)} - i\Omega^{(j)}$ where $j = 1, 2, \ldots, n$.

For distributed systems, only a part of the spatial coordinates $\boldsymbol{x}$ explicitly enter the linearized equations. These coordinates shall be denoted by $\boldsymbol{x}_1$. $\boldsymbol{x}_2$ stands for all other coordinates, i.e., the linearized problem is *homogeneous* with respect to $\boldsymbol{x}_2$. Then, $O_i^{(j)}(\boldsymbol{x})$ can often be written in the form

$$O_i^{(j)}(\boldsymbol{x}) = O_i^{(j)}(\boldsymbol{x}_1) \exp(i\boldsymbol{q} \cdot \boldsymbol{x}_2).$$

All $\Gamma^{(j)}$, $\gamma^{(j)}$ and $\Omega^{(j)}$ are functions of $\boldsymbol{q}$. If for some value of $\boldsymbol{q}$ one of the $\gamma^{(j)}(\boldsymbol{q}) > 0$, the particular stationary solution $O_i^s(\boldsymbol{x})$ is unstable. If for all values of $\boldsymbol{q}$ all $\gamma^{(j)}(\boldsymbol{q}) < 0$, the solution is stable.

Coherent structures

For coherent structures, the source term depends on order parameters in such a way that it becomes a periodic function of some coordinates, for example,

$$\mathscr{F}_i \propto \mathscr{R}e\{\exp(i\boldsymbol{q}\cdot\boldsymbol{x})\} = \cos \boldsymbol{q}\cdot\boldsymbol{x}, \qquad (28.1.3)$$

where $\boldsymbol{q}$ is the wavevector of the surface corrugation. The dispersion relation has the form

$$\Gamma = \Gamma(\boldsymbol{q}, \boldsymbol{k}_i, \varepsilon), \qquad (28.1.4)$$

where $\boldsymbol{k}_i$ is the wavevector of the incident laser light, and ε the permittivity of the material. The orientation and dominant period of the structure that develops most rapidly is given by the maximum value of the increment, γ_{max}.

Non-coherent structures

For non-coherent structures, $\mathscr{F}$ is *not* periodic in $\boldsymbol{x}$ and sometimes even independent of $\mathcal{O}_i$. Nevertheless, the increment Γ depends on some internal period, $\boldsymbol{q}_{nc}$. The $\boldsymbol{q}_{nc}$ related to the maximum increment $\gamma_{max}(\boldsymbol{q}_{nc})$, determines the period of the structure in the initial phase.

With increasing amplitude of the (coherent or non-coherent) structure, nonlinearities become important. In this regime, the generation and coupling of different harmonics, the interference of surface electromagnetic waves oriented in different directions, etc., make the problem more complex.

28.2 Ripple Formation

Frequently observed coherent structures in laser–surface processing are the ripples. These are spatially periodic structures (Figs. 28.1.1, 28.2.1).

Ripple formation was first observed by *Birnbaum* (1965) after ruby-laser irradiation of various semiconductor surfaces. Further investigations have shown that ripple formation is a quite general phenomenon which is practically always observed on solid or liquid surfaces within certain ranges of laser parameters. Ripples originate from the interference between the incident laser light and the scattered/excited wave along the interface.

Scattering of the incident light can be caused by microscopic roughnesses of the surface, by defects, spatial variations in the dielectric constant, etc. The interference between the incident and the scattered radiation field leads to an inhomogeneous energy input which, in turn, can cause surface instabilities. Such structures are termed non-radiative "waves" because they are not true surface modes. The incident laser light can, however, excite true surface electromagnetic waves (SEW) if there is a polariton-active mode at the laser

28.2 Ripple Formation

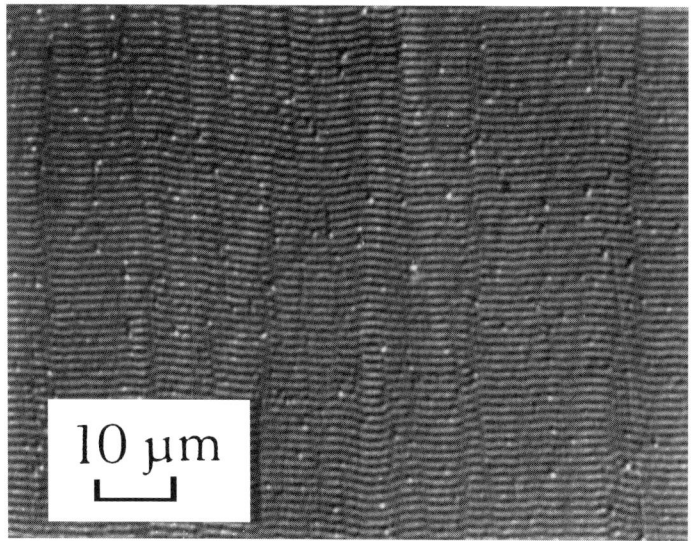

Fig. 28.2.1. Ripples generated on Ge with linearly polarized 1.06 μm Nd:YAG-laser radiation at perpendicular incidence ($q \parallel E_i$) [*van Driel* et al. 1985]

wavelength under consideration. In contrast to non-radiative structures, SEW can propagate energy along the surface. In any case, the period of the emerging structure (distance between ripples) depends on the wavelength ($\Lambda \propto \lambda$) and the angle of incidence, Θ_i, of the laser beam. The orientation of ripples is determined by the polarization of the incident light and, in some cases, by Θ_i. With metals and semiconductors the ripples are mainly oriented perpendicularly to the electric vector of the incident light. With dielectrics, both perpendicular and parallel orientations are frequently observed.

28.2.1 Interference Pattern

Let us first consider the interference pattern of the radiation field on the surface. The solid shall be characterized by a uniform dielectric constant ε. For the linear problem it is sufficient to study the diffraction of the incident laser light, $E_i = E_0 \exp[i(k_i \cdot r - \omega t)]$, by the Fourier component of the surface roughness which is characterized by the wave vector q and frequency Ω_q. The vectors k_i and q are oriented, in general, in arbitrary directions. Henceforth, we assume k_i and q to lie within the xz-plane. Let us consider the situation shown in Fig. 28.2.2. Here, the "roughness" of the *rigid* surface ($\Omega_q = 0$) is described by

$$z(x) = -\zeta \cos(qx), \tag{28.2.1}$$

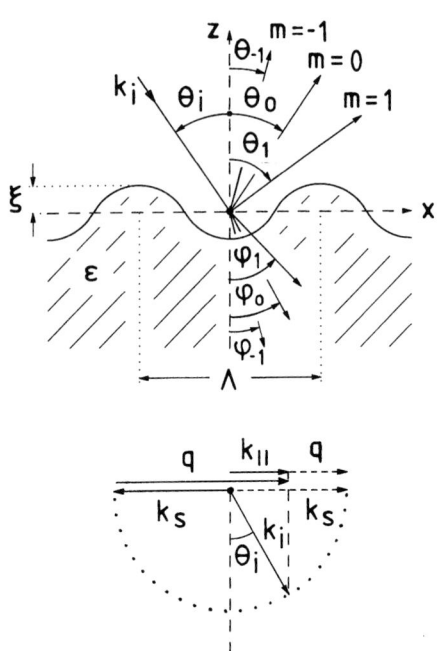

Fig. 28.2.2. Reflection and refraction at a corrugated surface. The lower part illustrates a wave vector diagram for the diffracted surface wave. k_i is the wave vector of the incident laser light and $k_\parallel$ its projection onto the surface. q is the grating vector which characterizes the surface corrugation. The scattered waves can have wave vectors $k_s = k_\parallel - q$ (Stokes wave; full arrows) or $k_s = k_\parallel + q$ (anti-Stokes wave; k_s, q dashed)

where $q = 2\pi/\Lambda$. With σ-polarization $E_i \parallel y$ (TE wave) and with π-polarization $H_i \parallel y$ (TM wave) (instead of σ and π one sometimes uses the letters s and p, respectively). Due to diffraction, we obtain a set of reflected and refracted waves. The reflected waves ($z > 0$) are given by the grating equation

$$k_i \sin \Theta_m = k_i \sin \Theta_i + mq. \tag{28.2.2}$$

Θ_m is the angle of the reflected beam of order m with $m = 0, \pm 1, \pm 2, \ldots$ If the medium with $z > 0$ is vacuum, the wave vector of the laser light is given by $k_i = \omega/c = 2\pi/\lambda$.

Correspondingly, the refracted waves ($z < 0$) are described by

$$k \sin \varphi_m = k_i \sin \Theta_i + mq, \tag{28.2.3}$$

where $k = n\omega/c$ and $n = \mathcal{R}e\{\sqrt{\varepsilon}\}$ is the refractive index. We now consider the situation where the first-order scattered (reflected or refracted) waves are along the x-axis. The wave vector of the scattered wave is either $k_s = k_\parallel - q$ (Stokes wave) or $k_s = k_\parallel + q$ (anti-Stokes wave); $k_\parallel$ is the wave vector component of the incident laser light on the surface $z = 0$. For the *reflected* wave we obtain

$$q = k_i(1 \pm \sin \Theta_i). \tag{28.2.4}$$

The plus sign refers to the Stokes wave ($m = -1$, $\sin \Theta_{-1} = -1$) and the minus sign to the anti-Stokes wave ($m = 1$, $\sin \Theta_1 = 1$). With $q = 2\pi/\Lambda$ the

28.2 Ripple Formation

spatial period of the interference pattern becomes

$$\Lambda = \frac{\lambda}{1 \pm \sin \Theta_i}. \tag{28.2.5}$$

For the *refracted* wave we obtain

$$q = k_i(n \pm \sin \Theta_i). \tag{28.2.6}$$

The period of the interference is then

$$\Lambda = \frac{\lambda}{n \pm \sin \Theta_i}. \tag{28.2.7}$$

The surface waves k_s related to (28.2.4, 6) are often denoted as Rayleigh waves. Ripples with periods equal to (28.2.5, 7) have been found experimentally.

28.2.2 Distribution of Energy

We now consider the distribution of the absorbed energy. For strong absorption, the (heat) source term related to the interference field is determined by the normal component of the Poynting vector on the surface, S_n. This can be calculated from the Maxwell equations. In first order with respect to the corrugation ξ/λ we obtain [*Guosheng* et al. 1982]

$$S_n = I_a[1 + \mu \cos(qx - \psi)]. \tag{28.2.8}$$

Thus, in addition to the average laser-light intensity absorbed on a flat surface, there is a spatially sinusoidal intensity distribution with period Λ. In general, there is a phase shift ψ between S_n and the surface corrugation $z(x)$. The coefficient μ determines the energy transformation into the periodic structure. It is given by

$$\mu = \frac{\xi}{\lambda} F(\boldsymbol{q}, \boldsymbol{k}_i, \varepsilon).$$

F describes the (dimensionless) interference flux which depends on the polarization of light, on the orientations of $\boldsymbol{q}$ and $\boldsymbol{k}_i$, and on the permittivity $\varepsilon = \varepsilon' + i\varepsilon''$ (*Akhmanov* et al. 1985; *Guosheng* et al. 1982 and references therein).

If $\mu < 0$ and $\psi = 0$ the amount of energy absorbed on the hills of ripples is larger than that in the valleys; for $\mu > 0$ the situation is opposite.

The inhomogeneous part in the absorbed energy distribution (28.2.8) can generate a modulation of thermal and non-thermal surface excitations. The modulation in surface temperature is

$$\Delta T(x) \propto I_a \mu \cos(qx - \psi). \tag{28.2.9}$$

If the incident light is π-polarized and $\varepsilon' < 0$ the incident wave excites a surface electromagnetic wave which propagates along the interface. If $|\varepsilon'| \gg \varepsilon''$, $|\varepsilon'| \gg 1$, and $\Delta k_s \ll k_i/|\varepsilon'|$ the flux $F(\boldsymbol{q}, \boldsymbol{k}_i, \varepsilon)$ becomes

$$F \propto \frac{k_i}{|\varepsilon'|^{1/2}|(\Delta k_s - i\Pi)|}, \tag{28.2.10}$$

where

$$\Delta k_s = k_s - k_i \left(\frac{|\varepsilon'|}{|\varepsilon'|-1}\right)^{1/2}, \tag{28.2.11}$$

and

$$\Pi = \frac{k_i \varepsilon''}{2|\varepsilon'|^2} \ll k_i. \tag{28.2.12}$$

k_s is the wave vector of the scattered Stokes and anti-Stokes surface waves. From (28.2.10, 11) we directly find that for low damping the function F, and thereby μ, shows a *resonance* for $\Delta k_s \approx 0$, i.e.,

$$k_s \approx \frac{\omega}{c} \left(\frac{|\varepsilon'|}{|\varepsilon'|-1}\right)^{1/2} \approx k_i. \tag{28.2.13}$$

Note that SEWs exist only for π-polarization and decay exponentially in both media (which have dielectric permittivities of opposite sign). They can be described by $I_s = I_s(z) = I_0 \exp[-2(k_s^2 - k_i^2)^{1/2} z]$ with $z > 0$ and $I_s = I_0 \exp[2(k_s^2 + k_i^2 |\varepsilon'|)^{1/2} z]$ with $z < 0$. Among the types of surface polaritons, phonon polaritons are observed in insulators and semiconductors, and plasmon polaritons in metals.

28.2.3 Feedback

We now discuss the mechanisms for the change in surface corrugation. The phase shift ψ is given in many papers [*Guosheng* et al. 1982]. It depends on Θ_i and q. Let us consider, qualitatively, some typical cases as depicted in Fig. 28.2.3 for $\mu < 0$. The full curves represent the surface profile $z(x)$ and the dashed curves the temperature modulation $\Delta T(x)$. With $\psi = 0$ the profiles z and ΔT are in phase and with $\psi = \pi$ in opposite phase. The response depends on the particular feedback mechanism:

- If thermal expansion dominates, the feedback is positive with $\psi = 0$. The maximum temperature occurs on the hills and increases the amplitude $\xi \propto \Delta T$; the increase in ξ yields an increase in $\Delta T \propto \mu \propto \xi$, etc.
- If material evaporation dominates, the feedback becomes positive with $\psi = \pi$.
- For surface capillary waves, which arise from a thermal modulation of the surface tension coefficient, the feedback is positive for $\psi = \pi$ if $d\sigma/dT < 0$ (Sect. 10.4).

28.2 Ripple Formation

- Temperature modulations cause a modulation in the dielectric permittivity which, in principle, can cause ripple formation on its own [*Guosheng* et al. 1982].

There are a large number of other feedback mechanisms which may be related to different types of phase transitions, the generation of defects, or electron-hole pairs in semiconductors, to plasma formation, changes in surface chemistry, etc. However, in none of the many experiments performed was the exact feedback mechanism responsible for the observed surface grating unequivocally elucidated. In any case, the important parameters are the surface corrugation ξ and the temperature T which can be considered as order parameters within the framework of a more general theory (Sect. 28.1).

Ripple movement

For phase shifts $0 < \psi < \pi$ the right side of hills (Fig. 28.2.3b) and for $\pi < \psi < 2\pi$ the left side of hills in d) are heated to different temperatures. This may result in ripple movement along the x-axis, i.e.,

$$z = -\xi_0 \mathcal{R}e\{\exp(\Gamma t + iqx)\},$$

with $\Gamma = \gamma - i\Omega_q$ where γ is the increment of instability. With a scanned laser beam, positive feedback may occur at a particular velocity, v_s. Here, an intense amplification of surface acoustic waves may be observed [*Dykhne* and *Rysev* 1986].

The preceding examples show the importance of the phase ψ and the type of feedback mechanism. Ripple formation will be most pronounced with optimal values $\mu = \mu_{opt}$ and $\psi = \psi_{opt}$ which correspond to $\gamma_{max} > 0$. Here, μ and ψ depend differently on q and Θ_i and the optimal conditions cannot be obtained in a simple form.

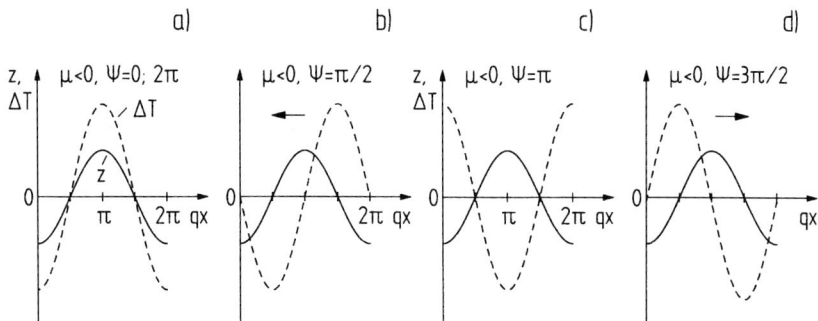

Fig. 28.2.3 a–d. Modulation of temperature (dashed curves) and surface profile (full curves) for $\mu < 0$ and different values of ψ

28.2.4 Comparison of Experimental and Theoretical Results

The orientation and period of ripples for *arbitrary* wave vectors k_i and q can be found from the dispersion equation (28.1.4). They are summarized in Table 28.2.1 for ripples caused by evaporation. The situation is similar for other feedback mechanisms. Terms "normal" and "anomalous" refer to ripples with $q \| E_i$ and $q \perp E_i$, respectively. The table illustrates that there are many coherent structures with different orientations and periods. Under certain conditions, some of these periods coincide with the Rayleigh waves related to (28.2.4,6). For example, (28.2.5) coincides with the period given for $\varepsilon' < -1$ and π-polarization, if we consider a metal or semiconductor with $|\varepsilon| \gg 1$ where $n^* \approx 1$. The period (28.2.7) becomes equal to that obtained for $0 < \varepsilon' < 1$ with σ-polarization and weak absorption where $n \approx \sqrt{\varepsilon'}$. Let us now consider the orientation and period of ripples found *experimentally*.

Metals

For metals and linearly polarized light with angles of incidence $\Theta_i < 35°$ the ripples are perpendicular to the projection of the electric vector E_i onto the surface $z = 0$. In other words, the orientation of the grating vector is $q \| E_i$. The periods observed were $\Lambda \approx \lambda/\cos\Theta_i$ with σ-polarization and $\Lambda \approx \lambda/(1 \pm \sin\Theta_i)$ with π-polarization [*Isenor* 1977; *Jain* et al. 1981]. This is in agreement with the theoretical results for $\varepsilon' < -1$.

Semiconductors

For semiconductors, the situation is similar to metals and the ripples are mainly oriented with $q \| E_i$ (Figs. 28.1.1, 28.2.1). With the experimental

Table 28.2.1. Orientation and period, Λ, of ripples as a function of Θ_i and $\varepsilon = \varepsilon' + i\varepsilon''$ with $|\varepsilon'| \gg \varepsilon''$. $\Lambda(\sigma)$ and $\Lambda(\pi)$ refer to σ- and π-polarization, respectively. The mechanism considered is based on laser-induced evaporation, adapted from [*Akhmanov* et al. 1985]

$\mathscr{R}e\,\varepsilon$ Type	$\varepsilon' < -1$ Normal	$-1 < \varepsilon' < 0$ Anomalous	$0 < \varepsilon' < 1$ Anomalous	$\varepsilon' > 1$ Normal
Orientation $\Lambda(\sigma)$	$q \| E_{iy}$ $\dfrac{\lambda}{[n^{*2} - \sin^2\Theta_i]^{1/2}}$	$q \perp E_{iy}$ $0 < q < k_i$	$q \perp E_{iy}$ $\dfrac{\lambda}{|\sqrt{\varepsilon'} \pm \sin\Theta_i|}$	$q \| E_{iy}$ $\dfrac{\lambda}{[n^{*2} - \sin^2\Theta_i]^{1/2}}$
Orientation $\Lambda(\pi)$	$q \| E_{ix}$ $\dfrac{\lambda}{n^* \pm \sin\Theta_i}$	$q \perp E_{ix}$ $0 < q < k_i$	$q \perp E_{ix}$ $\dfrac{\lambda}{[\varepsilon' - \sin^2\Theta_i]^{1/2}}$	$q \| E_{ix}$ $\dfrac{\lambda}{n^* \pm \sin\Theta_i}$

The following values for n^* are used
$n^{*2} = 1 + (n+\kappa)^2/(n^2 + \kappa^2)^2$ for $|\varepsilon| \gg 1$ with $\sqrt{\varepsilon} = n + i\kappa$
$n^{*2} = |\varepsilon'|/[|\varepsilon'| - 1] \approx 1$ for $\varepsilon' < -1$ (SEW)

28.2 Ripple Formation

conditions typically employed, surface melting is frequently observed. The fact that ripples are formed on semiconductor surfaces at laser-beam intensities that are typically one order of magnitude lower than those employed with metals can be interpreted by the lower melting threshold.

On the basis of these observations it has been suggested that the (irreversible) formation of ripples on metal and semiconductor surfaces is mainly due to surface capillary waves (Chap. 10).

Dielectrics

Ripple formation has also been studied for dielectric materials. For example, ripples generated on fused and crystalline quartz by 10.6 μm ($\tilde{v} = 943.3$ cm^{-1}) CO_2-laser radiation show the "normal" orientation $q \parallel E_i$. With σ- and π-polarization a single period and two periods, respectively, were observed ([*Keilmann* 1983] and references therein). If, however, 9.33 μm ($\tilde{v} = 1072$ cm^{-1}) radiation is used, the situation changes. For all angles $\Theta_i > 0$ gratings with "anomalous" orientation $q \perp E_i$ occur. In this case two spacings were found with σ-polarization and a single spacing with π-polarization. An inspection of the dielectric function of quartz reveals that $|\varepsilon'(10.6~\mu m)| > 1$ while $|\varepsilon'(9.33~\mu m)| < 1$. Thus, the normal and anomalous behavior observed in these experiments is consistent with the results listed in Table 28.2.1. It should be noted that anomalous behavior ($|\varepsilon'| < 1$, $|\varepsilon'| \gg \varepsilon''$) is also observed with surface capillary waves and surface acoustic waves.

On polymers, ripple formation under the action of UV excimer-laser radiation has been observed for fluences well below and well above the threshold for ablation, ϕ_{th} [*Bäuerle* et al. 1995; *Bolle* and *Lazare* 1993; *Dyer* and *Farley* 1990]. Figure 28.2.4 exhibits experimental data for PET and 308 nm laser radiation with $\phi \approx 26$ mJ/cm$^2 < \phi_{th} \approx 170$ mJ/cm^2. For σ-polarization the data can be fitted by $\Lambda = \lambda/(\sqrt{\varepsilon'} - |\sin \Theta_i|)$ and for π-polarization by $\Lambda = \lambda/(\varepsilon' - \sin^2 \Theta_i)^{1/2}$ with $\sqrt{\varepsilon'} = 1.17 \pm 0.01$ (solid curves). For 248 nm KrF-laser radiation $\sqrt{\varepsilon'} \approx 1.22$ is obtained (for untreated PET $n \approx 1.6$).

It is interesting to note that ripple formation is extented over areas that greatly exceed the spatial coherence area of the laser. This can be related to repeated scattering during multiple-pulse irradiation which can act to propagate the pattern. Here, local spatial coherence must exist, at least, over distances of the order of Λ.

Ambient medium with $\varepsilon_M \neq 1$

Up to now we have assumed that the laser light is incident from vacuum. If the permittivity of the ambient medium is $\varepsilon_M \neq 1$ the previous equations must be modified. In particular, the denominator in (28.2.10) becomes (in the general

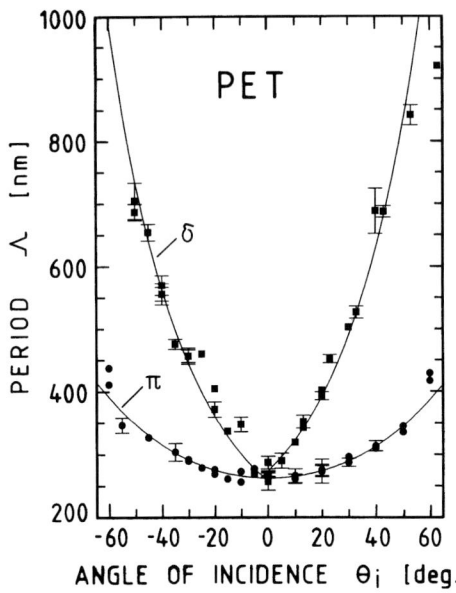

Fig. 28.2.4. Period of ripples observed on PET versus angle of laser-beam incidence, Θ_i (308 nm XeCl laser, $\phi < \phi_{th}$). Different branches refer to π- and σ-polarization. The grating vector of ripples, is $\mathbf{q} \perp \mathbf{E}_i$. Solid curves have been calculated [Bäuerle et al. 1995]

case) proportional to

$$D(k_s) \propto |\varepsilon'(k_s^2 - \varepsilon_M' k_i^2)^{1/2} + \varepsilon_M'(k_s^2 - \varepsilon' k_i^2)^{1/2}|\ .$$

For non-radiative modes with $|\varepsilon'| \gg |\varepsilon_M'|$ and $|\varepsilon'| \gg 1$ resonance occurs at $k_s \approx \varepsilon_M'^{1/2} k_i$. Thus, for materials like Si ($\varepsilon' \approx 12$) or Ge ($\varepsilon' \approx 16$) in air, the ripple period is determined by the vacuum wavelength, i.e., $\Lambda = \lambda/\sqrt{\varepsilon_M'}(1 \pm \sin\Theta_i)$, see (28.2.5). On the other hand, with dielectrics, like NaCl, MgF$_2$, etc., where $\varepsilon' \gtrsim \varepsilon_M'$ components with $k_s = \sqrt{\varepsilon'}\, k_i$ can have a large amplitude and the period becomes $\Lambda = \lambda/(\sqrt{\varepsilon'} \pm \sqrt{\varepsilon_M'}\sin\Theta_i)$, see (28.2.7).

If the laser light excites surface polaritons, remnant fields still exist if $\varepsilon_M > 0$. However, their amplitude can be ignored in comparison to the polariton field structures for which we find

$$k_s = \left(\frac{\varepsilon_M' \varepsilon'}{\varepsilon_M' + \varepsilon'}\right)^{1/2} k_i\ , \qquad (28.2.14)$$

where ε_M'', $\varepsilon'' < \varepsilon_M'$, ε'. The preceding discussion has shown that most of the features observed with coherent structures can be quantitatively described by theoretical models.

28.3 Spatio-temporal Oscillations

The formation of non-coherent structures due to spatio-temporal ordering is observed with laser-induced oxidation, explosive crystallization, exothermal

28.3 Spatio-temporal Oscillations

reactions, direct writing, etc. They are related to changes in absorptivity, the release of latent heat, spatial inhomogeneities induced by localized laser-beam irradiation, etc.

Order parameters most commonly used in such problems are:

- The temperature near the position of the laser beam.
- The temperature at the crystallization or reaction front.
- The velocity of a moving front.
- The width or height of the deposit near the laser beam.
- The concentration of species in the reaction zone.
- The width of the temperature or concentration distribution.

For the problems under consideration, the evolution of these order parameters can be described by ordinary differential equations.

From a physical point of view, the volume around the laser beam where transformations and reactions take place, can often be described by a "chemical reactor" [*Karlov* et al. 1992]. If such a reactor is closed with respect to mass and energy transfer, undamped oscillations are forbidden. This follows directly from the second law of thermodynamics. In laser processing, however, this reactor is an *open* system. The input of energy is provided by the laser light and, if relevant, by the latent heat of chemicals supplied to the reaction zone from outside. The energy output includes heat transport into the surrounding medium, enthalpy changes in endothermal processes, etc. The input and output of matter (precursor molecules, reaction products, transformed materials, etc.) can take place via diffusion, convection, laser-beam scanning, etc.

The appearance of oscillations requires, in general, *at least* two degrees of freedom.

To demonstrate the situation in further detail, let us assume a system where these two degrees of freedom are the temperature T, and a variable h which describes the thickness of a transformed layer, or the height of a deposit, or the concentration of a particular species, etc. In any case, T shall be the fast variable and h the slow variable. The fast variable shall reach its equilibrium value for each value of h very fast. Qualitatively, such a system can be described by

$$c_p \rho \dot{T} = P_{input}(T, h; P_a, v_s, \ldots) - P_{loss}(T, h; v_s, \ldots) \equiv f(T, h; P_a, v_s, \ldots) \quad (28.3.1)$$

and

$$\dot{h} = g(h, T; P_a, v_s, \ldots). \quad (28.3.2)$$

The behavior of the system can most conveniently be understood from the shape of zero isoclines.

28.3.1 Zero Isoclines

The zero isocline of a particular variable is given by the condition that the time derivative of this variable is zero. The zero isocline for the temperature, $f(T, h; P_a, v_s, \ldots) = 0$, is shown schematically in Fig. 28.3.1 by the solid curve. The branches which attract adjacent trajectories are drawn by thick lines and the branch which repels them by the thin line. The dashed curves represent zero isoclines for the slow variable for different values of parameters. The intersection points of the zero isoclines for T and h characterize stationary states of the system.

Oscillatory behavior of systems where a clear separation into fast and slow variables is possible, requires the zero isocline of the fast variable to be non-monotonic, as drawn in the figure. Such a behavior is obtained if, within a certain temperature range, the supply of energy [first term in (28.3.1)] increases more rapidly than the heat losses. Oscillations will take place only if the zero isocline of the slow variable varies as shown by curve 2. The stationary state (intersection point 0) is unstable. The limit cycle indicated by the dotted trajectory *abcd* represents (stable) oscillatory behavior.

The case of curve 1 corresponds to the *latent regime*. If the intersection point is close to the minimum of the solid curve, a small deviation from equilibrium may result in a large *single* (non-oscillatory) response.

If the zero isocline of h behaves as shown by curve 3, the system is bistable.

If a clear separation into "fast" and "slow" variables is *not* possible, oscillations can occur for almost all shapes of zero isoclines.

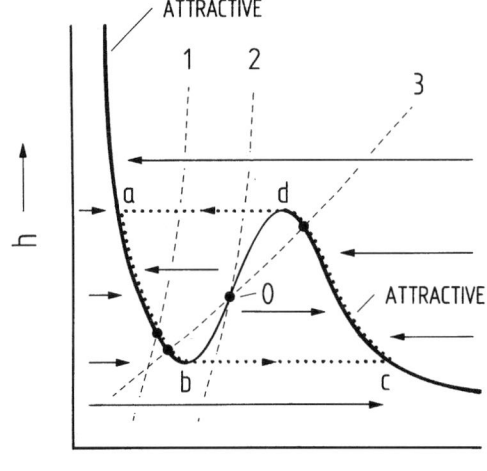

Fig. 28.3.1. Typical shape of zero isocline for the fast variable T (solid curve; thick lines indicate attractive branches, the thin line the repulsive branch). The dashed curves show three different cases of zero isoclines for the slow variable h. Curve 1 is characteristic for the latent regime, curve 2 for the oscillatory regime, and curve 3 for the regime of bistability. The dotted trajectory shall demonstrate oscillations (limit cycle). Arrows indicate the evolution of the sytem upon disturbances

28.3.2 Instabilities in Laser-Induced Oxidation

As an example of a thermochemical instability we consider laser-induced surface oxidation of a metallic slab of thickness h_s. If $h_s \ll l_T$, the temperature in z-direction can be considered as uniform. The heat conduction problem can then be described by

$$h_s \rho c_p \frac{\partial T}{\partial t} = h_s \kappa \nabla^2 T + I_a + \rho \frac{k_0}{h} \Delta H_{ox} \exp\left(-\frac{\Delta E_{ox}}{k_B T}\right)$$

$$- \rho \Delta H_v v_0 \exp\left(-\frac{\Delta E_v}{k_B T}\right) - \eta [T - T(\infty)], \qquad (28.3.3)$$

where h is the thickness of the oxide layer, $I_a = I_a(x,y)$, and $T = T(x,y)$. The third term on the right-hand side describes the heat release due to oxidation and the fourth term the heat loss due to evaporation (sublimation) of the oxide. ΔH_{ox} and ΔH_v are the corresponding enthalpies. The last term stands for heat losses by ordinary convection. The change in oxide-layer thickness is given by

$$\frac{dh}{dt} = \frac{k_0}{h} \exp\left(-\frac{\Delta E_{ox}}{k_B T}\right) - v_0 \exp\left(-\frac{\Delta E_v}{k_B T}\right). \qquad (28.3.4)$$

Even with *uniform* laser-beam irradiation [first term on the right-hand side of (28.3.3) equal to zero] the growth in oxide-layer thickness can be stationary or it can oscillate, depending on the parameter values and the laser-beam intensity. In this case, the behavior of the system is described by two ordinary differential equations, the modified equation (28.3.3) and equation (28.3.4). Thus, the system possesses two degrees of freedom and the situation is similar to that described in Fig. 28.3.1. Oscillations of this type have been observed within certain parameter ranges during (large-area) CO_2-laser-induced oxidation of metals, e.g., Mo and W. Figure 28.3.2 shows such oscillations in the laser-induced temperature for Mo. The temporal dependence becomes more evident from the derivative dT/dt. These oscillations are *not* related to interference effects. The main oxidation product, (liquid) MoO_3, strongly absorbs CO_2-laser radiation so that $l_\alpha \leqslant \lambda \ll h$.

Besides these homogeneous stationary and oscillating structures, *inhomogeneous* stationary[1], oscillating, and spatio-temporal structures [autowaves such as rotating spiral waves, circular waves (pacemakers, leading centers), etc.] can be observed. For *non-uniform* irradiation, different spatio-temporal behavior may coexist within different areas. For instance, with a Gaussian beam the period of structures varies with radius r.

---
[1] The situation is somewhat similar to the formation of Benard cells.

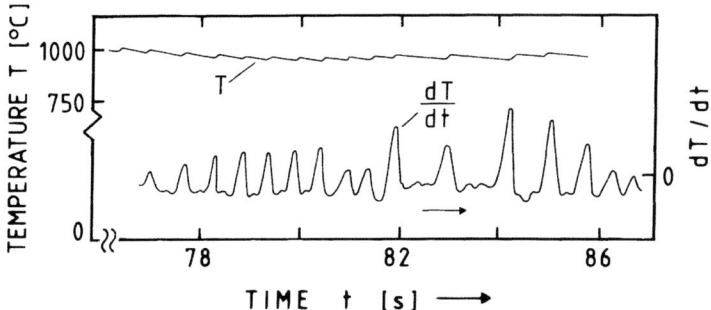

Fig. 28.3.2. Temporal dependence of temperature, T, and derivative dT/dt during CO_2-laser-induced oxidation/evaporation of Mo/Mo_xO_y. The disturbances of the periodicity reflect experimental difficulties, mainly related to the liquid oxide layer, adapted from [Karlov et al. 1992]

28.3.3 Explosive Crystallization

Explosive crystallization is observed during laser-induced crystallization of amorphous Si and Ge films on thermally insulating substrates [Chapman et al. 1980]. In these systems, the latent heat release related to the (structural) transformation from amorphous to crystalline exceeds the absorbed laser power and crystallization becomes self-promoting. If the laser beam is scanned with respect to the substrate, large-grained crescent-like periodic structures are observed within certain ranges of scanning velocities, v_s. This is shown in Fig. 28.3.3. for plasma-deposited Si on Si_3N_4 coated glass substrates. The period of the structure increases with scan speed; it also depends on the type of material, the layer thickness, the laser power, and the ambient temperature.

Explosive crystallization can be understood along similar lines as discussed in the previous two paragraphs.

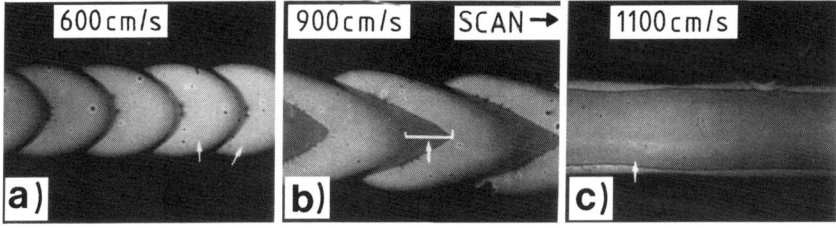

Fig. 28.3.3. Optical transmission photographs of laser-crystallized Si lines for different scanning velocities, v_s. With the parameters employed, explosive crystallization takes place without melting. The pictures are similar when melting takes place [Nguyen et al. 1984]

28.3 Spatio-temporal Oscillations

Let us first consider explosive crystallization in the absence of the laser beam. In this case, our "chemical reactor" is the area near the crystallization front. It can be characterized by two dynamic variables, the velocity of the interface crystalline-amorphous, v_{ca}, and the corresponding temperature, T_i. Here, T_i is not necessarily the fast variable; in some cases, it may even be v_{ca}.

The zero isocline for v_{ca} is given by the Frenkel–Wilson law (Sect. 10.1.1) which describes the equilibrium front propagation

$$v_{ca} = v_0 \left[1 - \exp\left(\frac{\Delta \mathcal{H}_c^a \Delta T_i}{T_m \, T_i}\right)\right] \exp\left(-\frac{\mathscr{E}_{ac}}{T_i}\right), \tag{28.3.5}$$

where $\mathscr{E}_{ac} = \Delta E_{ac}/k_B$ is the activation temperature for the transformation; $\Delta \mathcal{H}_c^a = \Delta H_c^a(T_m)/k_B < 0$ is the normalized heat of crystallization per atom at the melting temperature T_m (for the definitions of ΔE_{ac}^a and ΔH_c^a see Fig. 10.1.5).

Another relation between T_i and v_{ca} follows from the heat balance. Consider a film of thickness h on a semi-infinite substrate. In the simplest model we can assume the propagation of the crystallization front to be stationary. The temperature T_i is then determined only by the heat of crystallization and the heat losses. The latter are mainly due to heat conduction into the crystallized film and the substrate. With these approximations the zero isocline for T_i has the form [*Shklovskii* and *Kuz'menko* 1989]

$$T_i = T(\infty) + \frac{\Delta H_c}{c_p} \Psi(v_{ca}^*), \tag{28.3.6}$$

with $v_{ca}^* = v_{ca} h/D$ where D is the thermal diffusivity of the film. Ψ is a monotonically increasing function which depends on the particular experimental conditions. The rate of relaxation of T_i to the equilibrium value is determined by the heat conductivity.

The zero isocline (28.3.5) is strongly non-monotonic and the system can become bistable. This corresponds to a situation where the intersection of the two isoclines is of the type displayed in Fig. 28.3.1 (case 3). Thus, within a certain range of parameters D, h, and $T(\infty)$, there are *two* stable velocities of the propagation front.

In *laser* direct writing, the absorbed power ignites the crystallization process whenever it has stopped, and a new cycle starts near the edge of the crystallized material. If $v_s < v_{ca}$ periodic structures are observed. If, however, $v_s > v_{ca}$ the line of crystallized material becomes uniform (Fig. 28.3.3c).

28.3.4 Exothermal Reactions

Non-coherent structure formation can also be related to latent heats in laser-induced chemical reactions. Examples are periodic structures observed during laser direct writing of metal lines, mainly of Au and Pd from metallopolymer films (Sect. 27.3) or of Fe lines from $Fe(CO)_5$ [*Jackman* et al. 1986], or during rapid-scan synthesis of Ge_xSe_{1-x} films from sandwich layers of Ge and Se [*Laude* et al. 1986], etc. In all of these examples, the spatial periods of structures are much larger than the laser wavelength.

If the overall exothermal energy exceeds the absorbed laser-light energy, the periodic structures observed in chemically reactive systems can be described in analogy to explosive crystallization [*Kurtze* et al. 1984]. If, however, the latent heat release is of comparable size only, a description in analogy to that outlined in the next paragraph is more adequate. Here, a term proportional to $\exp(-\Delta\mathcal{H}/T)$, where $\Delta\mathcal{H} = \Delta H/k_B$ is the exothermal energy release, must be added in (28.3.7). The important point is that the overall energy input must be strong enough to produce a non-monotonic shape of the zero isocline for the temperature (solid curve in Fig. 28.3.1).

28.3.5 Instabilities in Direct Writing

Stable oscillations with (spatial) periods much longer than the ripples, have been observed in different LCVD systems during pyrolytic direct writing (Figs. 28.1.1a, 28.3.4). These oscillations are neither related to the wavelength and polarization of the laser light nor to latent heat effects. Their period has been found to increase with laser power, scanning velocity, size of focus, and the pressure of the reactant gas.

Figure 28.3.5 shows the dependence of Λ on laser power and scanning velocity for W. It should be noted that in this system periodic structures were observed only in the presence of small amounts of O_2 and only within a certain range of laser powers and scanning velocities (shaded area in Fig. 28.3.6). The concentrations of tungsten oxychlorides related to these traces of O_2 cannot, however, contribute significantly to the total W deposition rate. Additionally, on the basis of available thermodynamic data, the heat of chemical reaction can be ignored in comparison to the absorbed laser power. An important point seems to be the oscillating behavior observed in the surface absorptivity, A [*Kargl* et al. 1993a]. A change in absorbed laser power changes the surface temperature and thus the growth rate which, in turn, changes the surface morphology, and thereby A.

Interpretation of oscillations

A phenomenological description of the oscillations has to consider the interdependence between the geometry of stripes, the laser-induced temperature distribution which depends on the absorptivity, and the kinetics of the growth process. Let us consider the most simple model. The width of the deposited line shall be $d = 2r_D$ and the reaction shall take place only in a region of radius

28.3 Spatio-temporal Oscillations

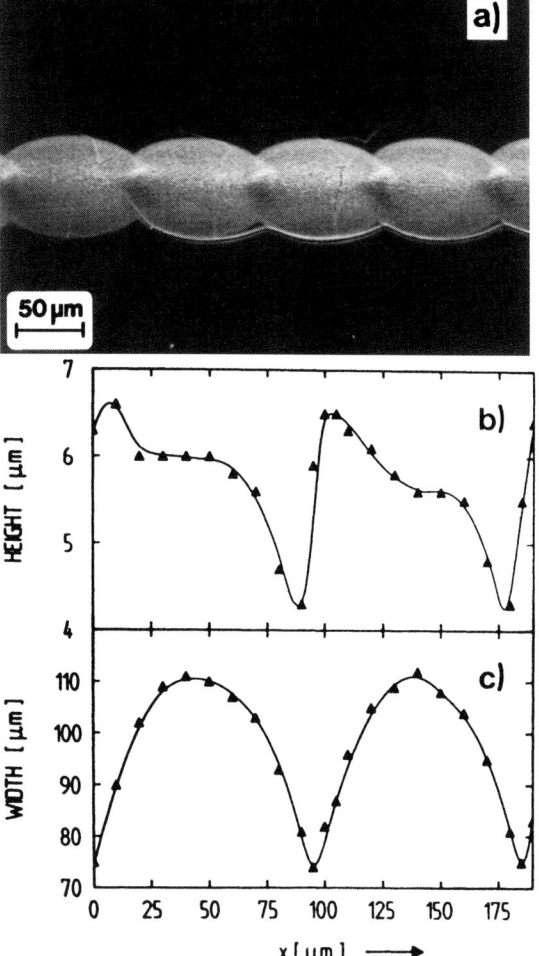

Fig. 28.3.4. (a) Periodic structure observed for a W line deposited from 1.1 mbar WCl_6 + 50 mbar H_2 + 15 μbar O_2 by cw-Ar^+ laser direct writing ($\lambda = 514.5$ nm, $P = 650$ mW, $2w_0 = 15$ μm, $v_s = 15$ μm/s; the SEM picture was taken under an angle of 45°). (b) and (c) show the height and width along the scanning direction [*Kargl* et al. 1993a]

r_D around the center of the laser beam. Our "chemical reactor" is then determined by the area πr_D^2, the height of the deposit, h, at the position of the laser beam, and the average temperature at this position, T. For certain systems, for example for W deposited from $WCl_6 + H_2$, the changes in r_D are much faster than in h (Fig. 16.5.1a). We choose as order parameters the (fast) variable T and the slow variable h. The energy balance yields

$$c_p \rho \pi r_D^2 h \dot{T} = P A(T) - P_{\text{loss}} . \tag{28.3.7}$$

$PA(T) = P(1 - R)$ is the energy input. If $\kappa_D \gg \kappa_s$, the heat losses can be

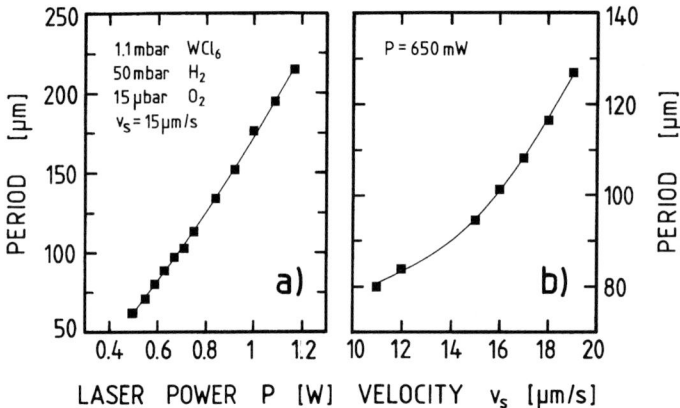

Fig. 28.3.5a,b. Period of oscillations observed in direct writing of W lines. (**a**) Dependence on laser power for a fixed scanning velocity, v_s. (**b**) Dependence on velocity v_s at a fixed laser power [*Kargl et al. 1993a*]

approximated by

$$P_{loss} \approx \left(\pi r_D^2 \kappa_s + h r_D \kappa_D\right)\frac{\Delta T}{r_D} = \kappa_s \Delta T\left(\pi r_D + h\frac{\kappa_D}{\kappa_s}\right), \tag{28.3.8}$$

where $\Delta T = T - T(\infty)$. The first term describes heat losses to the substrate, and the second term the losses due to heat transport along the stripe. h is given by the equation of growth (16.3.2)

$$\dot{h} = v_0 \exp\left(-\frac{\mathscr{E}}{T}\right) - v_s \frac{h}{r_D}. \tag{28.3.9}$$

Here, the "removal" of material from the reactor due to scanning has been approximated by $v_s \partial h/\partial x \approx -v_s h/r_D$. If the increase in absorptivity with temperature is sufficiently strong, the zero isoclines are similar to those shown in Fig. 28.3.1 by the full curve and the dashed curve 2. Thus, oscillations are expected.

It is evident that this model oversimplifies the real situation. However, it outlines the essential features and explains the main relationships. For example, this model predicts that oscillations will exist only within a certain range of scanning velocities and laser powers.

Self-consistent calculations based on the 1D model discussed in Chap. 18 have been performed by taking into account the temperature-dependent absorptivity. They describe the main features observed in the experiments, as for example the range of oscillations in Fig. 28.3.6 (solid curves).

The growth of even isolated islands observed within certain parameter ranges during direct writing of Ni "lines" (Fig. 28.3.7a) can be tentatively

28.3 Spatio-temporal Oscillations

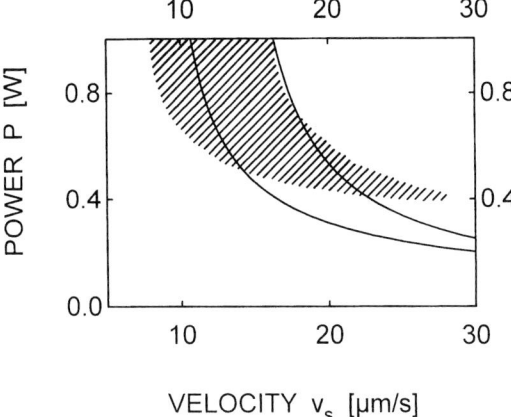

Fig. 28.3.6. The shaded area indicates the range of laser powers and scanning velocities where oscillations in Ar$^+$-laser direct writing of W have been experimentally observed [$p(WCl_6) = 1.1$ mbar, $p(H_2) = 50$ mbar, $p(O_2) = 15$ µbar]. The solid curves have been calculated [*Arnold* et al. 1995]

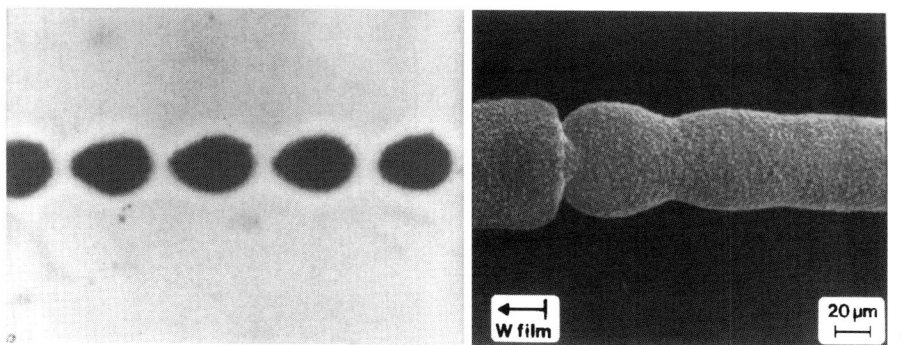

Fig. 28.3.7. (a) Periodic structures observed in pyrolytic direct writing of Ni deposited from Ni(CO$_4$) onto 1000 Å a-Si/glass substrates [*Kräuter* and *Bäuerle* 1982, unpublished]. (b) Strongly damped oscillations observed during Ar$^+$-laser direct writing (from left to right) of W (WCl$_6$ + H$_2$). The substrate was quartz which, in part, was covered with a 700-Å W film [*Kargl* et al. 1993a]

understood in the following way: The laser light absorbed within the a-Si layer induces a temperature distribution with a center temperature barely exceeding the threshold for deposition. This results in Ni deposition and a concomitant decrease in temperature. Once deposition has ceased, it cannot start again until the overlap of the laser focus with the spot-like well-reflecting heat sink has decreased sufficiently for the threshold temperature to be attained again.

Figure 28.3.7b shows *damped* oscillations which start at the edge between the covered and uncovered substrate.

28.3.6 Discontinuous Deposition and Bistabilities

The first clear observation of discontinuous growth was made during pyrolytic direct writing of Si lines. Figure 28.3.8 demonstrates the essential features. In the lower part of the figure the laser power was continuously increased when scanning from left to right. The scanning velocity and the gas pressure were kept constant. For low laser powers a uniform line with a height of, typically, a few μm is observed. When a certain laser power is reached, the line becomes non-uniform and consists of single tiny rods which are in close contact to each other. When further increasing the power, a discontinuous change occurs. The deposit consists now of single, almost equidistant rods tilted into the scanning direction. The height of these rods increases continuously with laser power. The upper part of the figure shows the result of a similar experiment where the laser power was decreased with scanning from right to left. Here, a transition from rod-type to line-type growth is observed.

Figure 28.3.9a exhibits the height of deposits as a function of laser power. Full triangles and full circles refer to data obtained with increasing and decreasing laser powers, respectively. It becomes evident that there is a well-pronounced hysteresis (bistability). With increasing power the transition occurs at about 59 mW, and with decreasing power at about 55 mW. The *critical power* P^{cr} increases with v_s. A similar behavior has been observed when the scanning velocity is varied at otherwise constant parameters (Fig. 28.3.9b).

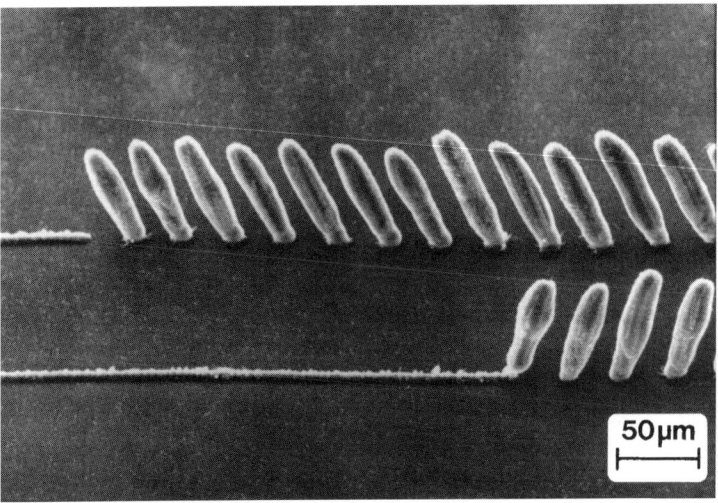

Fig. 28.3.8. SEM picture of Si deposited by Ar$^+$-laser direct writing ($\lambda = 514.5$ nm, $2w_0 = 3$ μm, $p(SiH_4) = 500$ mbar). Lower trace: Constant scanning ($v_s = 15$ μm/s) from left to right with continuously increasing laser power from 53 mW to 63 mW. Upper trace: Scanning from right to left with decreasing laser power [*Kargl* et al. 1993b].

28.3 Spatio-temporal Oscillations

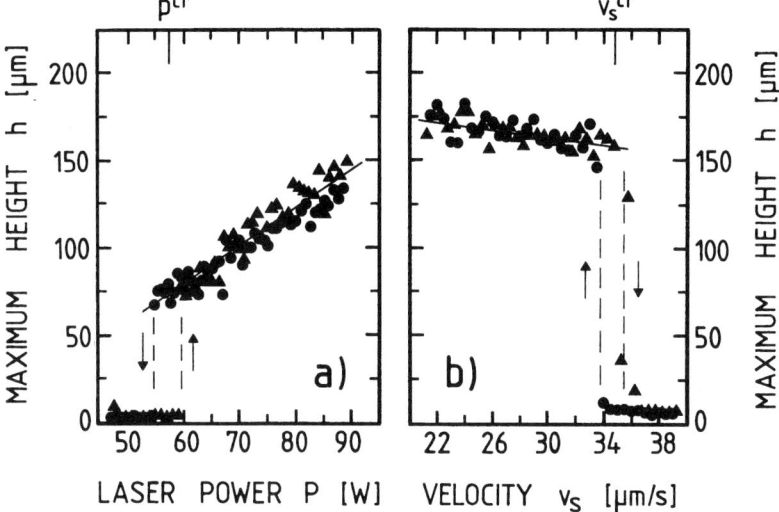

Fig. 28.3.9a,b. Maximum height of deposits as a function of: (a) Laser power at constant scanning velocity $v_s = 15$ µm/s. (b) Scanning velocity at constant laser power (136 mW; the other parameters are the same as in Fig. 28.3.8). ▲ and ● refer to increasing and decreasing laser power/scanning velocity [*Kargl* et al. 1993b]

The experimental results can be interpreted, qualitatively, along the lines of the one-dimensional model in Chap. 18. For "thick" deposits, however, the temperature gradient in z-direction (Fig. 18.2.1) cannot be ignored any further. In the simplest approximation we can estimate this gradient by employing a Taylor expansion

$$T_D(z=h) \approx T_D(z=0) + h \frac{\partial T_D}{\partial z}\bigg|_{z=0} . \tag{28.3.10}$$

From the continuity of the heat flux at $z=0$ and the approximation $\partial T_s/\partial z|_{z=0} \approx \theta_s/r_D$ with $\theta_s = T_s(x=0, z=0) - T(\infty)$ we obtain for the center temperature rise at the surface of the deposit

$$\theta_c(z=h) \approx \left(1 + \frac{h}{\kappa^* r_D}\right) \theta_c(z=0) . \tag{28.3.11}$$

If we set $h/r_D \approx 1$, $\kappa^* = \kappa_D/\kappa_s = 10$, and $\theta_c \approx 2 \times 10^3$ K, the difference between temperatures at $z=h$ and $z=0$ is 200 K. Because of the exponential increase in growth rate with temperature, a dramatic change can occur. Let us consider the surface of the stripe at $x=a$, and $x=0$ (Fig. 18.2.1). If v_s decreases, both $T_c(z=0)$ and T_e decrease because of the increasing cross section of the stripe. On the other hand, with increasing h, the difference $T_D(x=0, z=h) - T_D(x=0, z=0)$ increases. Because T_e is *not* affected by the temperature gradient in

z-direction, the ratio of growth rates at $x = 0$ and $x = a$ will increase with decreasing v_s. Therefore, there exists a minimum velocity v_s below which continuous deposition of a stripe with finite slope near its edge becomes impossible. As a consequence, the stripe will turn over like the front of a nonlinear wave and a transition to rod-type growth is observed. If we incorporate this physical picture into the model, we obtain the following results: Line-type growth is possible only above a critical scanning velocity

$$v_s^{cr} \approx W(T_c^{cr}(z=h)). \tag{28.3.12}$$

The transition occurs when the maximum height of stripes is of the order of the width, i.e., if $h^{cr}/r_D^{cr} \approx 1$. The critical temperature is $T_c^{cr}(z=0) \approx \mathscr{E}/\kappa^*$. For further details see [*Arnold* et al. 1995].

28.4 Instabilities in Laser Ablation

For some types of non-coherent instabilities, the picture of the "chemical reactor" described in Sect. 28.3 is inadequate. A typical example is the instability of the ablation front which is related to temperature gradients at the material surface.

Consider a planar surface from which material is ablated under the action of laser light (Fig. 28.4.1). Here, small perturbations of the surface may give rise to instabilities which result in spontaneous symmetry breakings and structure formation [*Anisimov* and *Khokhlov* 1995]. The laser-induced temperature

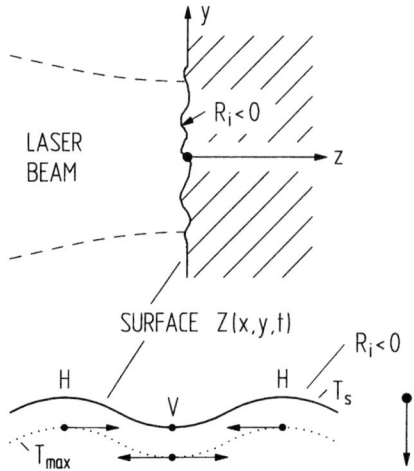

Fig. 28.4.1. Schematic picture of a surface $Z(x, y, t)$ irradiated by a laser beam. The lower part demonstrates how a surface deformation in z-direction generates a heat flux in x, y-directions. H and V stand for "hills" and "valleys", respectively. R_i denote the principal radii

28.4 Instabilities in Laser Ablation

distribution with $z \geq 0$ is given by the heat equation

$$\rho c_p \frac{\partial T}{\partial t} - \kappa \nabla^2 T = Q, \tag{28.4.1}$$

with

$$Q = \alpha I_0 (1 - R) \exp\{-\alpha[z - Z(x, y, t)]\}, \tag{28.4.2}$$

where $Z = Z(x, y, t)$ describes the (moving) surface. We assume $I_0(1 - R) =$ const. The boundary conditions can be written as

$$\kappa \frac{\partial T}{\partial z}\bigg|_{z=Z} = \rho \Delta H_v \dot{Z} \quad \text{and} \quad T(z \to \infty) = T(\infty), \tag{28.4.3}$$

where ΔH_v [J/g] is the enthalpy of vaporization and $\dot{Z} \equiv dZ/dt$. The curvature of the surface shall be described, at a given point, by the principal radii R_1 and R_2. The sign of R_i is defined in the picture. This yields

$$\dot{Z} = v_0 \exp\left[-\frac{\Delta E_v(\infty)}{k_B T_s} - \frac{\sigma}{N_0 k_B T_s}\left(\frac{1}{R_1} + \frac{1}{R_2}\right)\right], \tag{28.4.4}$$

where $T_s = T(z = Z)$ and σ the surface tension coefficient. $\Delta E_v(\infty) \approx \Delta H_v^a$ is the enthalpy of vaporization per atom/molecule from a *plane* surface, and $N_0 \equiv V_n^{-1}$ the number density of atoms/molecules within the surface of the material. The second term in the exponent describes the change in activation energy due to the surface corrugation, see also (4.1.13). Equations (28.4.1 to 4) have a stationary solution in the form of a plane ablation front (Sect. 13.2.1). Here, z in (13.2.3) must be substituted by $z^* = z - vt$. The maximum temperature, T_{max}, occurs at a distance $z^* = z_0$ *below* the surface. Thus, the temperature gradient at the surface is positive, $\partial T/\partial z|_{z=Z} > 0$. Due to the inhomogeneity in the temperature distribution, a small deformation of the surface will generate heat fluxes *perpendicular* to the z-direction. For *long* wave perturbations with $q < q_{min}$ (Fig. 28.4.2) the overall heat flux will decrease the temperature within the valleys, V (Fig. 28.4.1). Thus, material evaporation will be slower in the valleys than on the hills, i.e., the feedback is negative and $\gamma < 0$.

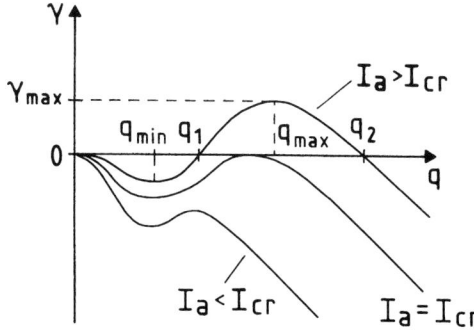

Fig. 28.4.2. Dependence of the increment γ on the wave vector of the surface perturbation, q. With $\gamma < 0$ surface perturbations are damped out. The instability which develops most rapidly is characterized by γ_{max} and q_{max}. The decrease in γ for $q > q_{max}$ is due to surface tension effects

For *short* wave perturbations with $q > q_{min}$ the situation is opposite. This is schematically shown in the lower part of Fig. 28.4.1. The resultant heat flux to the valleys, V, is positive and the surface corrugation is increased. With $q > q_1 \approx 2\pi/z_0$, $\gamma > 0$ and the surface becomes unstable. However, this positive feedback competes with a negative feedback related to the change in vaporization energy with curvature [see exponent in (28.4.4)]. This negative feedback increases with increasing wave vector of the perturbation.

Let us now continue the calculations and assume $\boldsymbol{q} \parallel y$. The perturbation in the temperature distribution can then be described by

$$T = T_0(z^*) + T_1(z^*) \mathscr{R}e \{\exp(iqy + \gamma t)\}, \tag{28.4.5}$$

where $T_0(z^*)$ corresponds to the stationary solution (13.2.3). The shape of the surface shall be described by

$$Z = vt + \xi,$$

where v is the stationary velocity of the ablation front and

$$\xi = \xi_0 \mathscr{R}e \{\exp(iqy + \gamma t)\}. \tag{28.4.6}$$

With the approximation $R_1^{-1} \approx -\partial^2 Z/\partial y^2$ which holds for $\xi/\Lambda \ll 1$, the boundary-value problem can be solved. Figure 28.4.2 displays, qualitatively, the dependence of γ versus q. If the absorbed laser-light intensity is smaller than some critical intensity, i.e., $I_a < I_{cr}$, surface perturbations will be damped out because $\gamma < 0$. If, however, $I_a > I_{cr}$, there exists a region $q_1 < q < q_2$ in which $\gamma = \gamma(q) > 0$. Perturbations with such wave numbers grow exponentially with time and the surface becomes unstable. For wavevectors $q > q_2$ the surface is stabilized again due to surface tension effects.

The threshold for the appearance of an instability is determined by the dimensionless parameter

$$\Xi \equiv \frac{\alpha \sigma}{\Delta E_v(\infty) N_0} \approx \alpha N_0^{-1/3}. \tag{28.4.7}$$

The dependence $I_{cr}^* = I_{cr}/I_k = I_{cr}(1-R)/v_0 \rho \Delta H_v$ is shown in Fig. 28.4.3 as a function of Ξ. Vaporization is stable only within the shaded region. For a material with an absorption coefficient $\alpha = 10^3$ cm^{-1} the characteristic value I_{cr} is 10^6–10^8 W/cm^2.

The period of the structure which develops most rapidly is given by

$$q_{max} \approx \left(\frac{\alpha I_0 (1-R) N_0 \Delta E_v(\infty)}{2\sigma \kappa} \frac{}{T_s} \right)^{1/3}. \tag{28.4.8}$$

Note that q_{max} is *not* directly related to the laser wavelength, except via the absorption coefficient α. Typical values of q_{max} are between a few μm and some ten μm. The time for the development of this instability is

$$\gamma_{max}^{-1} \approx \frac{D}{v^2} \left(\frac{T_s}{\Delta E_v(\infty)} \right)^2. \tag{28.4.9}$$

28.4 Instabilities in Laser Ablation

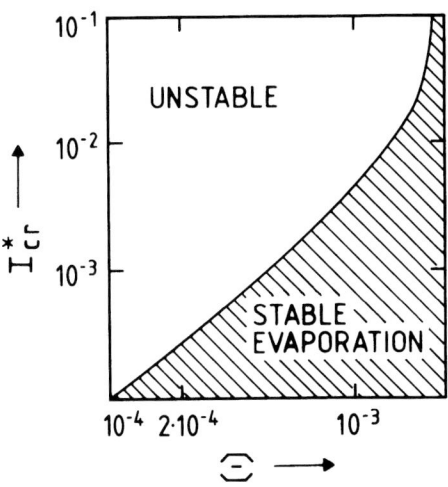

Fig. 28.4.3. Stability boundary. Within the shaded region evaporation is stable [*Anisimov* et al. 1980]

For a metal, e.g., with $D = 1$ cm^2/s, $v = 0.1$ μm/pulse, $\tau_l \approx 10^{-8}$ s and $[T_s/\Delta E_v(\infty)] \approx 0.1$, we obtain $\gamma^{-1} \approx 10^{-8}$ s. Thus, in metals this type of instability can develop during ns pulses but not during ps and fs pulses. From the preceding discussion it becomes evident that, with the mechanism under consideration, a positive feedback arises only if $\partial T/\partial z|_{z^*=0} > 0$ where $z^* = 0$ denotes the position of the (plane) ablation front.

Similar considerations for UV-laser (photophysical) ablation of dielectrics, and in particular of organic polymers, show that the ablation velocity is a function of temperature, the concentration of excited species, N_{A^*} (Chap. 13), and some other parameters, Ψ_i, if relevant. The necessary condition for the development of a surface instability is then

$$\frac{\partial v}{\partial z} = \frac{\partial v}{\partial T}\frac{\partial T}{\partial z} + \frac{\partial v}{\partial N_{A^*}}\frac{\partial N_{A^*}}{\partial z} + \sum_i \frac{\partial v}{\partial \Psi_i}\frac{\partial \Psi_i}{\partial z} > 0, \qquad (28.4.10)$$

where all derivatives with respect to z should be taken at $z^* = 0$. The second term originates from activated desorption of excited species. The third term describes the influence of mass density changes, stresses, shielding by impurities, debris, etc. Let us first ignore Ψ_i. Because $dN_{A^*}/dz|_{z^*=0} < 0$ the second term *stabilizes* the ablation front [*Luk'yanchuk* et al. 1993b]. This stabilization becomes more pronounced with increasing thermal relaxation time, τ_T. For many polymers, the surface is stable if $\tau_T > 10^{-11}$ s. The typical length of structures calculated for the unstable regime is of the order of 10^3 Å.

These considerations allow one to understand why it is difficult to obtain smooth surfaces during ns UV-laser ablation of metals or during IR-laser ablation of polymers. Instabilities will *not* develop during laser pulses with $\tau_l \ll \gamma_{max}^{-1}$. This is one of the reasons why polymer surfaces ablated by ps or fs laser pulses are smoother than those ablated by ns laser pulses [*Küper* and *Stuke* 1989].

A different type of surface instability occurs if the temperature $T_{max}(z_0)$ becomes so high that liquid or gaseous bubbles are formed *below* the surface. In such cases, explosive-type ablation will be observed. For metals, this mechanism seems to be unimportant because the time for bubble formation is very long compared to γ_{max}^{-1} in (28.4.9). The situation may, however, be different with non-metals.

Conical structures

For many types of materials such as ceramics, organic polymers, etc., conical structures develop on the ablated surface [*Krajnovich* and *Vazquez* 1993; *Kullmer* and *Bäuerle* 1988b; *Dyer* et al. 1986]. The formation of such cones is *not* necessarily related to surface melting. This can be seen from Fig. 28.4.4 for the example of PI which can only sublimate. The cone axes are oriented along the direction of the incident laser beam. The number density of cones increases with the number of laser pulses.

The formation of conical structures has often been explained by shielding effects related to local enrichments of photofragments or material impurities, to debris condensing on the surface between laser pulses, etc. At least within the initial phase, structure formation may also be ascribed to spontaneous symmetry breaking which develops within the plain ablation front and which can be described in analogy to (28.4.10) with $\Psi_i \neq 0$.

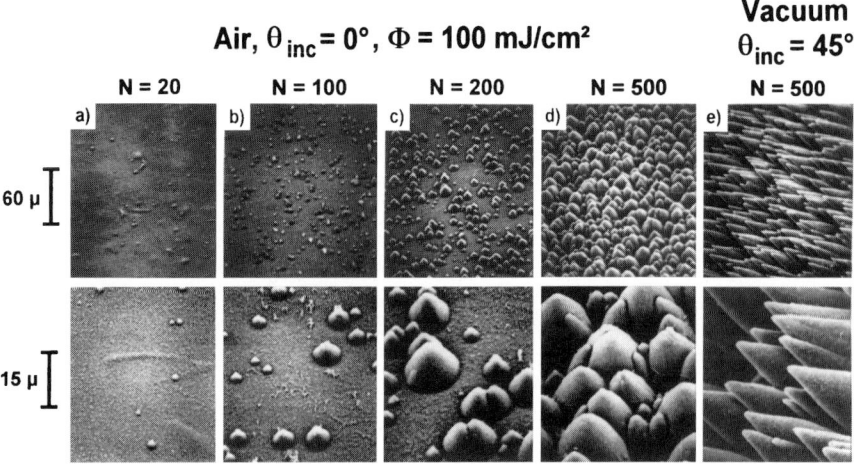

Fig. 28.4.4 a–d. SEM photographs of polyimide ablated in air at $\phi = 100 \, mJ/cm^2$ for different numbers of pulses N. The laser beam was normally incident on the sample. (**e**) Same as (**d**) but irradiated in vacuum under $\Theta = 45°$ [*Krajnovich* and *Vazquez* 1993]

Let us consider PI where laser irradiation results in an enrichment of carbon within the surface (Sect. 27.2). If we denote the density of carbon atoms by $\Psi \equiv N_c$ and if $\partial v/\partial N_c < 0$, the third term in (28.4.10) is positive because $\partial N_c/\partial z|_{z^* = 0} < 0$. Such a mechanism would *destabilize* the surface. Instabilities of this type can be described, in principle, by localized perturbations of the form (28.1.2) with $\Omega = 0$ and $\mathcal{O}_i(\mathbf{r}) \equiv J_0(qr)$ where J_0 is the Bessel function so that

$$\mathcal{O}^p(\mathbf{r}, t) \propto J_0(qr)\exp[\gamma(q)t]. \tag{28.4.11}$$

28.5 Hydrodynamic Instabilities

Hydrodynamic instabilities can be classified according to Kelvin–Helmholtz and Rayleigh–Taylor type instabilities [*Chandrasekhar* 1961].

Kelvin–Helmholtz instabilities are excited, under certain conditions, at interfaces of stratified heterogeneous or even homogeneous liquids when different layers of the liquid are in relative (horizontal) motion, characterized by a velocity v_l. In laser processing the velocity v_l can be related to:
- Surface tension effects (Sect. 10.4).
- The expansion of the plasma plume (Sect. 30.3.1).
- The recoil pressure in laser ablation (Sect. 11.1.4).
- The gas jet in liquid-phase expulsion (Sect. 10.7), etc.

Rayleigh–Taylor instabilities arise at interfaces between liquids of different densities that are superimposed over one another and which are in an external field (gravity, centrifugal forces, etc.). In laser processing, such instabilities can cause:
- Fast mixing of gases at the contact front between the plasma plume and the ambient medium in reactive laser ablation (Sect. 30.3.1).
- The formation of droplets in laser-surface melting and vaporization (Sect. 22.2.4).

Different types of hydrodynamic instabilities have been discussed in various chapters. For example, the laminar convection due to non-uniform surface melting under the action of focused laser-beam irradiation (Fig. 10.4.1) loses its stability at some critical intensity. As a consequence, axial symmetry is broken and the formation of rotating spiral waves is observed [*Arnold* and *Kirichenko* 1992].

28.6 Stress-Related Instabilities

Spontaneous breakings in axial symmetry imposed by laser beams can result in instabilities, as shown in Fig. 28.6.1. With low laser-light intensities, etching of regular holes in Mo films immersed in Cl_2 atmosphere is observed

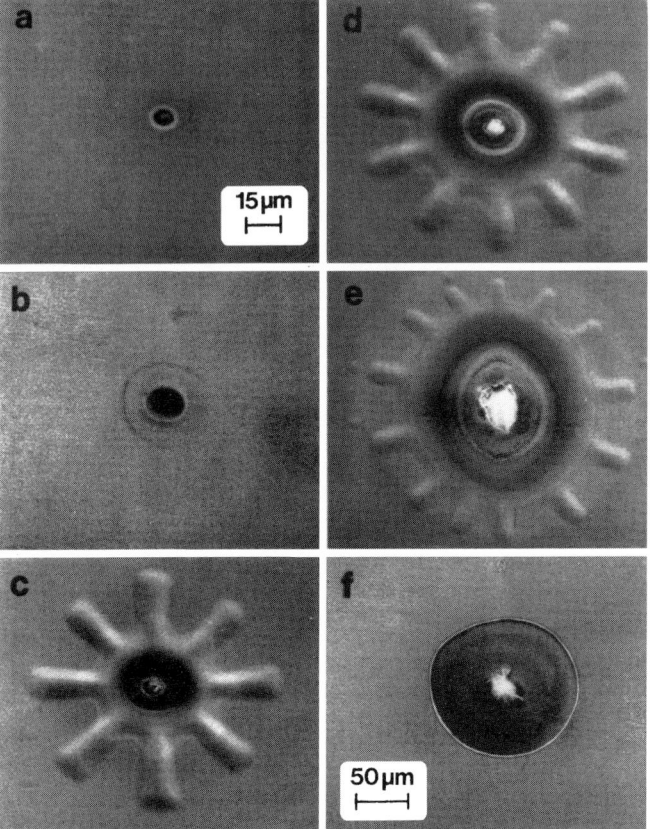

Fig. 28.6.1. SEM pictures of holes and stars formed in 2500 Å Mo/glass films during Ar$^+$-laser-induced etching in Cl$_2$ atmosphere ($\lambda = 488$ nm, $w_0(1/e) \approx 5.7\,\mu$m, $p(\text{Cl}_2) = 50$ mbar). The laser powers employed were: **a)** $P = 10$ mW, **b)** 20 mW, **c)** 50 mW, **d)** 100 mW, **e)** 500 mW, **f)** 150 mW [*Mogyorosi* et al. 1989a]

(Sect. 14.3). With medium intensities a "star-like" structure develops. The number of "rays" increases with intensity. The formation of stars is probably related to built-up stresses caused by laser-induced heating. Beyond a certain laser-light intensity, the stresses become so high that a large area of the film pops off. This area has about the same diameter as the ring surrounding the holes. This behavior was only observed with films which were not strongly adherent on the substrate surface. From a mathematical point of view, the appearance of these stars can be described by a perturbation

$$\mathcal{O}_i^p \propto \exp\left[im\varphi + \gamma(m)t\right] \, .$$

The number of rays is given by the particular m for which the increment ($\gamma > 0$) becomes maximum.

28.6 Stress-Related Instabilities

Different types of non-coherent structures have also been observed with polymers [*Arenholz* et al. 1993; *Yabe* and *Niino* 1992; *Bahners* et al. 1989; *Novis* et al. 1988]. For example, polymer surfaces ablated with UV-laser light frequently show wall- or nap-type structures, as shown for PI in Fig. 28.6.2. These structures are related to internal stresses within the polymer foil. Uniaxial stretching prior to laser-light irradiation results in wall-type structures, while biaxial stretching gives rise to nap-type structures. The surface of annealed (unstretched) foils is smooth. Naps seem to be formed by "superposition" of walls aligned perpendicularly to the respective stresses [*Arenholz* et al. 1991]. The average distance between walls or naps increases with both the number of laser pulses (Fig. 28.6.3) and the laser fluence.

A tentative explanation for the formation of these structures is based on a stress-release model. Here, craze formation is considered to be due to an externally applied or frozen-in stress, S, parallel to the surface (Fig. 28.6.4). Formation of a craze significantly changes the stress field. In the simplest approximation, stress release takes place within a half cylinder (shaded) of volume $V_{el} \approx \pi a^2 / 2$ [per unit length] where a is the depth of the craze. The elastic energy release is $E_{el} \approx V_{el} \, S^2/2Y$ where Y is Young's modulus. For narrow cracks where $r_c \ll a$ the surface energy can be approximated by

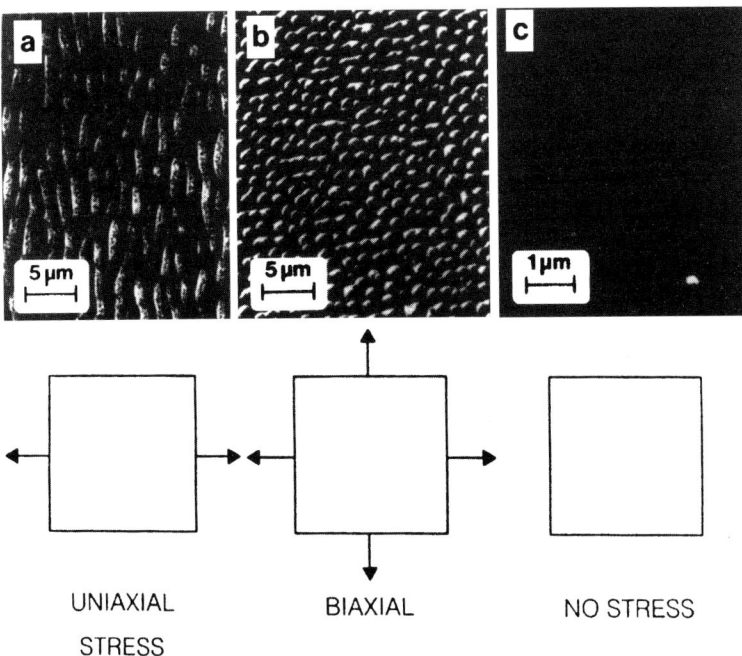

Fig. 28.6.2a–c. SEM micrographs of different types of structures observed on PI foils after ablation with KrF-laser radiation. (**a**) Foil was uniaxially stretched before ablation ($\phi \approx 136$ mJ/cm^2, 30 pulses). (**b**) Foil was biaxially stretched before ablation ($\phi \approx 115$ mJ/cm^2, 50 pulses). (**c**) Annealed foil ($\phi \approx 120$ mJ/cm^2, 50 pulses) [*Arenholz* et al. 1992]

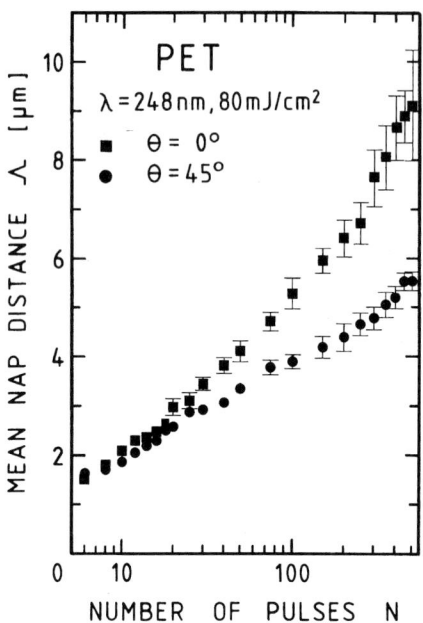

Fig. 28.6.3. Mean nap distance versus number of laser pulses observed on untreated PET foils [*Arenholz* et al. 1994]

$E_s = 2a\sigma$. The overall energy decreases after craze formation if

$$a \geqslant a_c = \frac{2Y\sigma}{\pi(1-\mu^2)S^2}.$$

The factor $2/\pi(1-\mu^2)$, instead of $8/\pi$, is obtained in a self-consistent treatment of the problem where μ is the Poisson ratio ($0 < \mu < 0.5$) [*Landau* and *Lifshitz* VII *Theory of Elasticity*; *Griffith* 1920; *Williams* 1980]. For polymers Y depends strongly on temperature and has values of about 10^5 N/cm^2 for 20 °C and 10^2 N/cm^2 for 150 °C. Typical values of σ are between 0.02 J/cm^2 (PMMA) and 0.15 J/cm^2 (PS). Frozen-in stresses in polymers are of the order of 10^2–10^4 N/cm^2. Thus, we find as "average" value $a_c \approx 1$ μm. With $a > a_c$, the craze will propagate, because E_{el} increases faster than E_s. It is well known that polymer foils possess internal stresses due to the fabrication process. Fast heating of the foil by UV-laser radiation further increases this stress. If S exceeds a critical value, craze formation takes place and relaxes the stress field. The depth of the craze, which determines the volume of stress relaxation, is related to the thickness of the laser-heated surface layer. Thus, a quasiperiodic surface structure with a mean period $\Lambda \approx 2a$ develops. The model also explains, qualitatively, the increase in Λ with the number of laser pulses. Irradiation of a quasiperiodic structure leads to preferential ablation of top-layers where the stress fields are relaxed. At the bottom both ablation and further extension of the craze increase the depth of the structure. Simultaneously, the relaxed volume increases and the stabilization of the structure

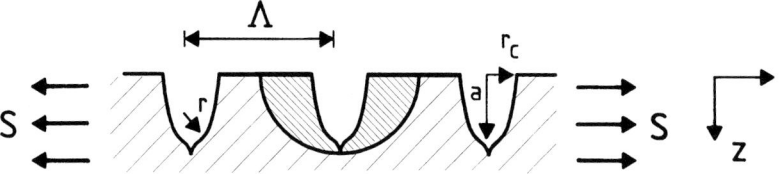

Fig. 28.6.4. Schematic picture of quasiperiodic structures originating from craze formation in materials with frozen-in or externally applied stress fields

by the stress field becomes lost. Naps or walls drop over in random directions so that, for example, a new nap is formed out of two. Thus, the average period increases.

28.7 Technological Aspects

For applications it is often desirable to suppress structure formation because it influences the smoothness of the surface, the ultimate resolution in micropatterning and laser lithography, etc.

Ripple formation can be suppressed by diminishing the correlation time of the polarization, for example by using circularly polarized light, and by avoiding good spatial coherence. Another possibility is to use fluences which result in plasma formation—as long as this is compatible with the particular processing application. With the generation of a plasma, interference effects on the surface are significantly diminished.

On the other hand, ripple formation may be applied for the fabrication of gratings [*Kim* et al. 1995], for well-defined orientation of molecules, e.g., in liquid-crystal displays, for surface roughening and the improvement of surface adhesion, e.g., to evaporated layers, etc.

The combination of different techniques opens up a great variety of possibilities. An example is the periodic submicrometer-dot structure shown in Fig. 28.7.1. The period in *y*-direction, generated by laser-beam interference (Fig. 5.2.1c), is given by (5.2.1)

$$\Lambda_h = \frac{\lambda}{2\sin\Theta_i} \,. \tag{28.7.1}$$

The grating in *x*-direction is formed by ripples with period

$$\Lambda_r \approx \frac{\lambda}{(n^{*2} - \sin^2\Theta_i)^{1/2}} \,, \tag{28.7.2}$$

where n^* is given in Table 28.2.1. With $\lambda = 355$ nm, $\Theta_i = 19°$, and $\varepsilon = 6.42 + i\,12.52$ we obtain $\Lambda_h \approx 545$ nm and $\Lambda_r \approx 333$ nm which is in reasonable

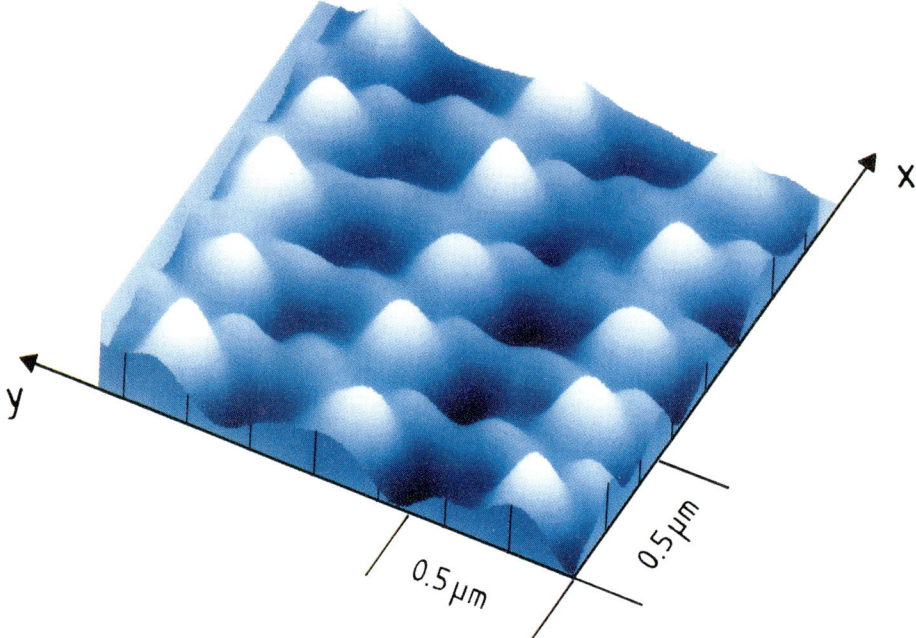

Fig. 28.7.1. AFM picture of dot structures fabricated on n-type (100) InP by 355 nm Nd: YAG-laser etching in CH_3Br/He [*Kumagai* et al. 1992]

agreement with the experimental observations. The conical dots have a diameter of 200 nm at FWHM. The depth of valleys between dots is 60 nm and 30 nm in y- and x-direction, respectively. Because of their unique physical properties, such small dots have potential applications in electronic and optical device fabrication.

Non-coherent structure formation can be suppressed for certain ranges of process parameters which must be determined experimentally for every system of interest. Here, even the very simple theoretical models discussed in the preceding sections can help to find these ranges. With some systems, and in particular with stress related instabilities, structure formation can be diminished or even avoided altogether by proper pretreatment of the sample prior to laser processing.

Part VI: Measurement Techniques, Diagnostics

The characterization of experimental tools, and *in situ* measurements of processing rates, of laser-induced temperatures, and of product species is of importance for both an understanding of fundamental laser–matter interactions and for process optimization. The next two chapters give an overview on the most important techniques and their application in various types of laser processing.

29 Measurement Techniques

29.1 Characterization of Laser-Beam Profiles

The intensity distribution of laser beams with diameters in the range 0.1–5 cm, can conveniently be measured in real time by employing a linear or a two-dimensional array of photodiodes, a vidicon (a semiconductor or pyroelectric detector array), or a CCD (charge coupled device) camera.

For *focused* laser beams with spot sizes of some tenths of a micrometer to a few micrometers a simple and cheap method is the scanning knife-edge technique [*Suzaki* and *Tachibana* 1975]. Here, the transmitted (reflected) laser power P_T is measured during scanning a sharp edge, e.g., a razor blade, across the laser spot. For a Gaussian laser beam P_T is given by

$$P_T(x) = \frac{P}{\pi^{1/2} w_0} \int_x^\infty \exp\left(-\frac{x'^2}{w_0^2}\right) dx' = \frac{P}{2} \mathrm{erfc}\left(\frac{x}{w_0}\right), \tag{29.1.1}$$

where P is the total power. The fit of the experimental curve $P_T(x)/P$ yields the beam radius, w_0.

29.2 Homogenization of Laser Beams

The production of a spatially uniform energy density in a plane perpendicular to the optical axis of the laser beam is termed beam homogenization. Here, single "rays" of the incident beam which propagate along different paths are superimposed in the exit plane.

Homogenization of highly coherent lasers which have a Gaussian or Gaussian-like beam profile is mainly performed by diffractive methods.

Highly non-coherent lasers such as excimer lasers, have a top-hat profile. Here, beam homogenization is, in general, based on reflective and refractive methods. The beam profile of excimer lasers is often optimized inside the resonator, for example by X-ray preionization of the laser-active gases or by special geometries of the electrodes.

Other requirements on homogenizers include: Low energy losses, transmission of high energy densities, compact design for applications with small working distances, and long lifetimes.

Many techniques employed for laser-beam homogenization are also suitable for other light sources such as lamps.

Diffractive methods

Laser beams with Gaussian or Gaussian-like profiles can be homogenized by diffractive methods employing metallic grids, optical gratings, phase plates, or holographic techniques [*Jain* et al. 1984; *Possin* et al. 1983; *Veldkamp* 1982; *Han* et al. 1983]. The application of these methods requires a long term stability of the position, the (resonator) mode, and, with pulsed lasers, of subsequent pulses of the incident beam.

Reflective methods

Homogenization by multiple reflections is achieved by focusing the laser beam into a light guide of rectangular or circular cross section. This is schematically shown in Fig. 29.2.1a. The light guide can consist of highly reflecting mirrors, a quartz rod, or a fiber inside which the light is totally reflected. High energy densities at the entrance plane can cause material damages. For a given length of the light guide, the number of reflections is determined by the beam divergence in the entrance plane. Single rays which undergo a different number of reflections inside the light guide, are superimposed at the exit plane. The homogeneity can be improved by inserting a diffuser plate in front of the light guide. By means of an imaging optics, the homogenized beam can be concentrated onto a mask for substrate patterning (Fig. 5.2.1).

Other arrangements for beam homogenization use a *curved* quartz rod [*Cullis* et al. 1979] or multifaceted mirrors which divide the laser beam into rectangular segments that are superimposed at the image plane by means of an imaging mirror [*Dickey* and *O'Neil* 1988].

Refractive methods

Particularly suitable for homogenization of large-area high power laser beams, for example excimer-laser beams, is the fly's eye homogenizer which is frequently used in lithography (Fig. 29.2.1b). It consists mainly of crossed anamorphic-lens arrays, in most cases cylinder lenses, and an imaging lens [*Kahlert* et al. 1993; *Ozaki* and *Takamato* 1989]. The telescope is used to match the diameter of the (excimer) laser beam to the size of the homogenizer entrance aperture. The size of the homogenized beam in the exit plane is

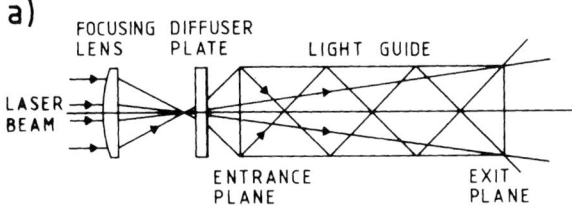

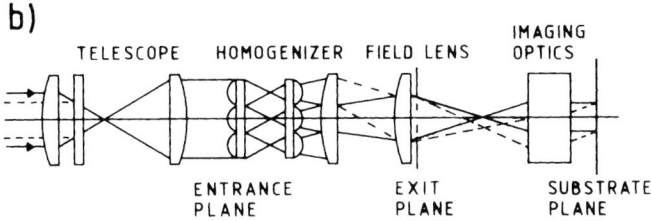

Fig. 29.2.1a,b. Laser-beam homogenization by means of (**a**) multiple reflections within a light guide, and (**b**) a fly's eye homogenizer

determined by the width of the anamorphic lenses, by the focal lengths of the second lens array, and the image formation lens. For projection patterning, a mask is positioned at the exit plane. The field lens in front of the mask reduces vignetting. The imaging optics transfers the image produced by the mask onto the substrate. The fly's eye homogenizer permits one to employ relatively large cross sections of optical components (quartz or glass) and thereby high energy throughputs.

29.3 Deposition, Etch, and Ablation Rates

This section deals with the most commonly used techniques for *in situ* measurements of deposition, etch, and ablation rates.

Speckles

When irradiating a substrate with a cw-laser beam, a characteristic speckle pattern is observed, in general. With a fixed laser beam and substrate, the beginning of deposition or etching can be detected by the onset of uniform speckle movement. The observation of the speckles is a qualitative but very sensitive method and it is mainly used in gas-phase processing.

Interference fringes

The growth of thin films can be monitored from the thin-film interference pattern of a probe beam, for example a HeNe-laser beam. The distance between subsequent maxima in the oscillations of the reflected beam intensity (perpendicular incidence) corresponds to a change in layer thickness $\Delta h = \lambda/2n$ where n is the refractive index of the layer. The detailed equations that apply to different cases are given in Sect. 9.2. Interference fringes are also used for in situ measurements of etch depths.

Transmission measurements

Deposition and etch rates can sometimes be estimated from the transmitted (reflected) intensity of either the laser beam employed in processing, or of a probe beam, through (from) the deposited or etched film. This technique is primarily used in photolytic processing with low processing rates. An estimation of film thicknesses, however, requires a knowledge of the optical constants of the film and the substrate, if present. For thin metal films, whose thickness is comparable to the optical penetration depth ($l_\alpha = \alpha^{-1}$) the optical behavior is dominated by surface effects. The optical constants of thin films may differ significantly from the bulk values and, additionally, they may change with film thickness. As a consequence, estimates of film thicknesses from transmitted light intensities often involve tremendous inaccuracies.

If we ignore these effects, the film thickness can be estimated from the transmittivity D which is given for different cases in Sect. 9.2.

Microbalances

A very sensitive method for measuring mass deposition, etch, and ablation rates over larger areas uses a microbalance. This is a piezoelectric slab-shaped crystal or ceramic with metal electrodes. The materials most commonly used for microbalances are crystalline quartz and ceramic $PZT(PbTi_{1-x}Zr_xO_3)$. For etch- and ablation-rate measurements the material to be investigated must first be evaporated as a thin film or bonded as a thin platelet onto the microbalance surface. Detailed investigations on quartz crystal microbalances (QCM) including different mass load ranges and measurement techniques have been described by *Benes* (1984). When the electrodes of the microbalance are attached to an oscillating circuit, resonance occurs at a (fundamental) frequency v. For *small* mass changes originating from laser-induced deposition ($\Delta m > 0$) or material removal ($\Delta m < 0$) the frequency change can be approximated by

$$\Delta v \approx -\frac{v^2 \Delta m}{C\rho F}, \qquad (29.3.1)$$

where C is denoted as frequency constant. ρ is the mass density of the piezoelectric slab and F its *active* area on which material is deposited or from which it is removed. For AT-cut quartz crystals $C \approx 1.668 \times 10^5$ cm/s and $\rho \approx 2.648$ g/cm^3. Equation (29.3.1) is valid for mass changes $\Delta m/m \leqslant 2\%$. Since the resonant frequency can be monitored continuously, material deposition or removal rates can directly be measured. Microbalances are commercially available and can routinely measure mass load changes of 10^{-9} g/cm^2 and, if special care is taken, changes down to 10^{-12} g/cm^2.

Chuang (1981) has used QCMs to investigate laser-induced etch rates. For evaporated Si films sensitivities of about 2×10^{14} atoms/Hz ($\leqslant 0.5$ ML/Hz) have been achieved. A typical frequency response of a QCM covered with a thin evaporated Si film to pulsed CO_2-laser radiation is illustrated in Fig. 29.3.1. The microbalance was positioned with the active area perpendicular to the unfocused laser beam. The momentary increase in microbalance frequency during each laser pulse is due to the temporary temperature rise caused by the absorbed laser light. In the absence of any reactive gas, the microbalance returns to its original frequency (curve a). Thus, under the experimental conditions used, the laser radiation by itself did not cause any significant removal of Si. In a reactive gaseous atmosphere, e.g., in SF_6, the microbalance frequency increases (curve b). The detected frequency change of $\Delta v = 52$ Hz/20 pulses, corresponds to a removal rate of 4.4×10^{14} Si atoms/pulse. In a similar way, excimer-laser-induced ablation rates of organic polymers [*Küper* et al. 1993; *Lazare* and *Granier* 1989] and 257 nm Ar^+-laser-

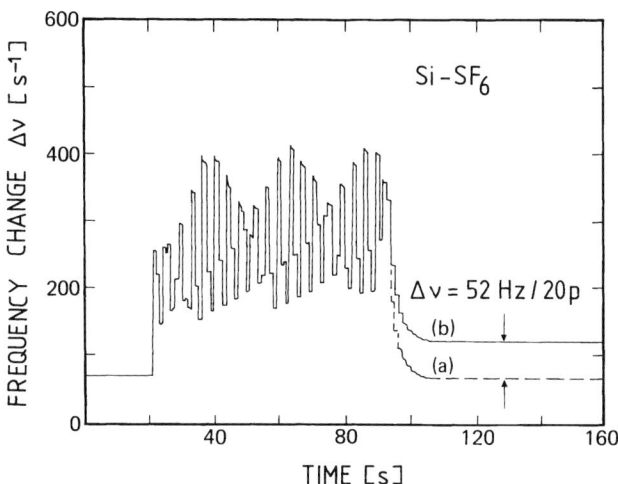

Fig. 29.3.1. Frequency response of a quartz crystal microbalance (QCM) covered with a Si film, to pulsed CO_2-laser radiation ($\lambda = 942.4$ cm^{-1}; $\phi = 1$ J/cm^2). (**a**) Vacuum, (**b**) $p(SF_6) = 2.7$ mbar [*Chuang* 1981]

induced deposition rates of Cr, Mo, and W from the corresponding hexacarbonyls [*Jackson* and *Tyndall* 1988] have been measured.

It should be emphasized that (*mass*) deposition, etch, and ablation rates and the corresponding threshold intensities measured by QCMs can differ significantly from those determined by means of optical techniques, mechanical profilometers, etc. In material deposition, these differences can be related to the incorporation of impurities, porosities, etc. On the other hand, mass losses related to the depletion of single species can take place without significant material etching or ablation.

Thin film acoustic and pyroelectric monitors

For detection of elastic (acoustic) waves emitted from the laser-processed region, one frequently uses, instead of discs of crystalline quartz or ceramic PZT [*Melcher* 1984], piezoelectric materials in the form of films or foils. Ideal for this application are foils of (ferroelectric) PVDF (polyvinylidene fluoride). Here, the sample to be investigated is either bonded or, in the case of thin films, directly evaporated onto the PVDF foil.

Piezoelectric materials that possess a unique polar axis (dipole moment) in the absence of a stress, are also pyroelectric. Pyroelectric materials develop a polarization charge that is proportional to the (uniform) temperature rise (note that a piezoelectric material that is *not* pyroelectric can generate an electric charge if it is *non*-uniformly heated; this is simply due to piezoelectric stresses created by thermal expansion). Because all ferroelectric materials are piezoelectric *and* pyroelectric, PVDF foils can also be used as pyroelectric detectors.

Thin film acoustic and pyroelectric monitors have been employed during pulsed-laser ablation of various materials. The technique has been described by many authors [*Dyer* et al. 1992a].

Photoelectric methods, laser-beam deflection

Deposition rates achieved during steady growth of fibers can be measured by imaging the hot tip of the fiber onto a position-sensing diode (Fig. 17.1.1).

The *deflection* of a probe-laser beam propagating in parallel to the substrate surface can be used to detect ablation thresholds [*Hunger* et al. 1992; *Sell* et al. 1989]. The experimental setup is similar to that shown in Fig. 30.1.1a. The beam at perpendicular incidence to the substrate, for example an excimer-laser beam, causes ablation. The probe beam, in general a HeNe-laser beam, becomes deflected due to transient changes in refractive index above the laser heated/ablated surface.

29.4 Temperature Measurements

The knowledge of the laser-induced temperature distribution, or at least of the maximum temperature rise, is of great importance for fundamental investigations, and for process control and optimization.

Temperature measurements can be classified according to optical techniques (photoelectric pyrometry, Raman spectroscopy, photothermal deflection, etc.) and other (non-optical) techniques (pyroelectric calorimetry, time-of-flight techniques, etc.).

29.4.1 Photoelectric Pyrometry

In photoelectric pyrometry the local temperature is derived from the optical radiation emitted from the laser-heated surface. The spectral radiance emitted from a solid depends on temperature and is given by

$$L(v, T) = \varepsilon(v, T) L_{b0}(v, T) , \tag{29.4.1}$$

where $\varepsilon(v, T)$ is the spectral emissivity and

$$L_{b0}(v, T) = \frac{2hv^3}{c^2} \frac{1}{\exp\left(\frac{hv}{k_B T}\right) - 1} = \frac{2hv^3}{c^2} \langle n \rangle$$

the spectral radiance of the blackbody radiation in vacuum (radiant power of the non-polarized radiation emitted per unit projected area of the surface per unit solid angle within the interval between v and $v + dv$). If ε is known, measurements of $L(v, T)$ permit direct determination of the temperature. The situation is more complicated if the material is irradiated with a focused laser beam which results in spatially localized heating. In this case, the total emitted radiation is a mixture of radiation emitted from different elements at different temperatures. This has led to the concept of apparent emissivities, ε_a [*Zeldovich* and *Raizer* 1966]. For the (simpler) case of large-area laser-beam irradiation (one-dimensional problem), the apparent emissivity is about equal to the emissivity of a uniformly heated solid only if the variation in temperature is small over the distance $l_{\alpha e} = \alpha_e^{-1}$ (α_e is the absorption coefficient at the wavelength employed in the pyrometric measurements).

The main advantages of photoelectric pyrometry are the high temperature sensitivity, the relatively low sensitivity to surface properties – as compared to other techniques – and the suitability for in situ measurements. The spatial resolution is limited by diffraction and, at low temperatures, by the sensitivity of the detection system. The temperature can be derived from the emitted radiation in various ways:

- From its intensity at a certain wavelength $\lambda_1 \equiv \lambda_e$ (monochromatic pyrometry).
- From its intensity integrated over a broader spectral region.
- From its spectral dependence, including temperature determination from the ratio of the radiances at two wavelengths.

For *monochromatic* pyrometry, we obtain from the law of error propagation in Wien's approximation

$$\delta T = \frac{\lambda_1 T^2}{C_2}\left[\left(\frac{\delta I}{I}\right)^2 + \left(\frac{\delta \varepsilon}{\varepsilon}\right)^2\right]^{1/2}, \tag{29.4.2}$$

where δT is the uncertainty in temperature. $C_2 = hc/k_B = 1.4388 \times 10^7$ nm K is the second radiation constant, and I the measured intensity. A similar expression applies to broad-band pyrometry.

If the temperature is evaluated from the ratio of radiances at *two wavelengths*, we obtain

$$\delta T = \frac{\lambda_{\text{eff}} T^2}{C_2}\frac{\delta R_i}{R_i}, \tag{29.4.3}$$

with $\lambda_{\text{eff}} = (1/\lambda_2 - 1/\lambda_1)^{-1}$ and $R_i = (I_1/\varepsilon_1)/(I_2/\varepsilon_2)$ where I_1 and I_2 are the intensities, and ε_1 and ε_2 the emissivities at wavelengths λ_1 and λ_2, respectively. In large-area pyrometry of uniformly heated materials, the error in temperature measurements is mainly determined by uncertainties in the emissivity, so that two-wavelength pyrometry gives the best results [*Battuello* and *Ricolfi* 1980]. The situation is different in laser *microprocessing* where temperature distributions are strongly localized. Here, the light fluxes to be measured are rather small and the uncertainties in the measured intensities can exceed those in the emissivities. If inaccuracies in the intensity measurements dominate, we find from (29.4.2 and 3) that, irrespective of the difference between the two wavelengths, the evaluation of the temperature from the radiance at the smaller of the two wavelengths yields a smaller error than the evaluation from the ratio of radiances. Additionally, chromatic errors in the imaging system become more serious for small spot sizes. Monochromatic pyrometry is therefore the superior technique in most cases of laser microchemical processing. Additional advantages are the simplicity of the experimental setup and calibration (via a radiance standard such as a tungsten band lamp). However, in many cases sufficient accuracy can only be achieved if the influence of the laser-induced temperature distribution on the apparent emissivity is known.

The sensitivity of the detector required for photoelectric pyrometry can be estimated from the spectral radiance of the blackbody in the temperature range under consideration. Figure 29.4.1 exhibits the wavelength dependence of the spectral radiance [W/nm μm² sr] of the blackbody radiation for various temperatures; the range of applicability for different detectors has also been included. The intensities that can be measured in a real experiment are diminished by a factor of $10-10^2$ with respect to the ideal values, mainly due to the

29.4 Temperature Measurements

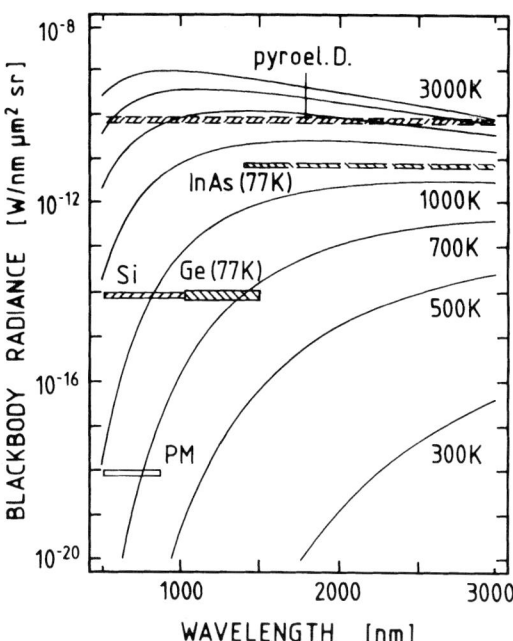

Fig. 29.4.1. Blackbody radiances that can be measured with various detectors in photoelectric pyrometry. The solid curves are isotherms for various temperatures. The bars indicate the noise equivalent power (NEP) in $Ws^{1/2}$ and the range of spectral sensitivity. PM: photomultiplier

lower emissivity, $\varepsilon < 1$, the detected solid angle which is often < 1 sr, and losses in the optical components. At temperatures above 1000 K, high quality silicon photodiodes (noise equivalent power NEP $\approx 10^{-14}$ W/Hz$^{1/2}$) can often be employed. Lower temperatures, down to about 700 K, can be measured by using cooled Ge diodes and a spectral bandwidth of several nanometers.

An experimental example

Typical spectra of the emitted radiation measured during steady growth of silicon fibers are shown in Fig. 29.4.2a for various laser powers. The experimental arrangement employed was similar to that in Fig. 17.1.1. The depth of the temperature distribution along the axis of the glowing fiber was estimated to be larger than about 100 µm, even when the reactor was filled with H_2. Since the energy of the bandgap of Si decreases with increasing temperature, the condition $w_T^a \gg l_{\alpha e}$ (Sect. 6.5) is satisfied over the spectral range investigated. The fit of the spectra by Planck's law (Wien's approximation; Fig. 29.4.2b) permits one to derive the corresponding temperatures. The results are quite consistent and demonstrate that the influence of excess carriers at temperatures above 900 K can be ignored. The agreement with measured

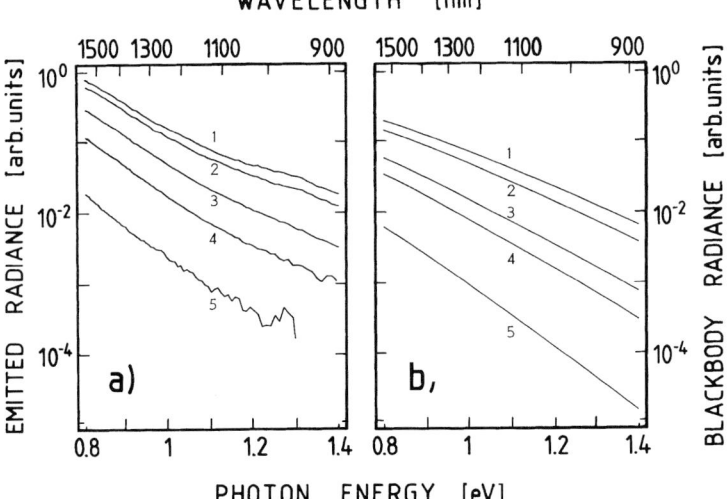

Fig. 29.4.2(a). Spectral dependence of the radiation emitted from laser heated tips of Si fibers grown by LCVD from SiH$_4$ at various laser powers P. Curve 1: $P = 150$ mW (1360 K); 2: 140 mW (1300 K); 3: 120 mW (1150 K); 4: 100 mW (1080 K); 5: 80 mW (900 K). The temperature values given in parentheses were derived from fits to Planck's law (Wien's approximation). (**b**) Blackbody radiances for the temperature values given in (**a**) [*Doppelbauer* and *Bäuerle* 1987]

spectra can even be improved if we take into account the (extrapolated) emissivity of Si, $\varepsilon(v, T)$ [*Jellison* and *Modine* 1983].

29.4.2 Other Optical Techniques

Raman spectroscopy allows one to determine temperatures from the relative intensities of Stokes and anti-Stokes components which are given by

$$\frac{I_s}{I_{as}} = f(\alpha_i, v_i) \frac{\langle n \rangle + 1}{\langle n \rangle} \, , \qquad (29.4.4)$$

where $\langle n \rangle \equiv \langle n(v_M) \rangle$ is the Bose–Einstein factor for the particular (vibrational, rotational, etc.) mode under consideration, and $f(\alpha_i, v_i)$, a factor which corrects the measured Raman intensities for the actual absorption coefficients, the frequency dependence of Raman efficiencies, the spectral sensitivity of the experimental setup, etc. In a crude approximation, $f(\alpha_i, v_i) \approx 1$. Alternatively, the temperature can be derived also by analyzing the shifts and shapes of Stokes Raman lines only. The accuracy achieved in such measurements with Si at 1200 K is about ±100 K [*Compaan* 1985; *Balkanski* et al. 1983]. With strongly inhomogeneous temperature distributions, similar problems as in photoelectric pyrometry arise. Local laser-induced temperature distributions were derived from Raman spectra during cw-laser heating of Si [*Pazionis* et al. 1989] and laser-CVD of Si from SiH_4 [*Magnotta* and *Herman* 1986]. A clear disadvantage of the Raman technique is the high cost of the experimental setup.

Laser-induced temperatures have also been derived from time-resolved *reflectivity* and *transmittivity* measurements [*Lompré* et al. 1983] and from changes in optical *absorbance* [*Lee* et al. 1992].

Photothermal deflection

Localized heating causes thermal expansion of the sample surface. The surface deformation can be measured via the deflection of a probe beam. For modulated or pulsed-laser irradiation, the deflection angle is, within a certain range [*von Gutfeld* et al. 1986; *Karner* et al. 1985; *Melcher* 1984], given by

$$\varphi \propto \frac{\beta_T P_a \tau_l}{\kappa}, \qquad (29.4.5)$$

where β_T is the thermal expansion coefficient [*Vicanek* et al. 1994]. Because of the difference in the thermal and elastic (acoustic) response, the deformation is *not* a direct indicator of the temperature. Furthermore, the technique can be employed only in very special types of laser processing.

29.4.3 Other Techniques

Thin foils (10–30 µm) of PVDF (Sect. 29.3) have been employed to measure average surface temperatures [*Emmerich* et al. 1992] and phase changes [*Coufal* 1984] in thin evaporated films.

Time of flight

Time-of-flight (TOF) techniques permit one to estimate the surface temperature from the velocity of desorbed/ablated species. This technique has been used during pulsed-laser annealing [*Stritzker* et al. 1981] and laser-induced deposition, ablation, and etching [*Baller* 1990]. Because the evaluated temperature depends on the model employed in the analysis of the TOF spectra, the temperature values are not very accurate (Sect. 30.1.2).

Thermistors

Thin film thermistors of NiSi placed between thin polymer foils and quartz substrates were used to directly measure laser-induced temperature profiles with ns resolution [*Brunco* et al. 1992].

30 Analysis of Species, Plasmas, and Surfaces

30.1 Precursor and Product Species

Optical and mass spectroscopy have been applied to analyze product species in laser annealing, ablation, etching, and laser-CVD.

30.1.1 Optical Spectroscopy

The analysis of fundamental interaction mechanisms in laser processing (LP) requires one to separate excitations within the ambient medium from those at the substrate surface. Among the different optical techniques are LIF (laser-induced fluorescence), optical emission spectroscopy, Raman spectroscopy, UV-, VIS-, and IR-absorption spectroscopy, optical deflection, etc.

Figure 30.1.1a illustrates an optical setup which uses combined laser-beam irradiation. The beam at perpendicular incidence excites the substrate. The beam propagating parallel to the substrate surface is used for thermal or non-thermal excitation of the ambient medium, or as a probe beam, as for example in laser ablation. This irradiation geometry has been applied, for example, in investigations on photolytic *etching* of Si in Cl_2 atmosphere (Sect. 15.2). Here, XeCl-laser radiation at parallel incidence has been used to photodissociate Cl_2 molecules, while Kr^+-laser radiation at perpendicular incidence has been employed to *only* generate electron-hole pairs within the Si surface. The number density of Cl atoms can be derived from the measured chemiluminescence intensity and (14.1.7). This is shown in Fig. 30.1.1b together with the density calculated from (14.2.1).

A similar experimental setup is used for the analysis of laser-induced plasmas. Here, the beam at normal incidence is used for ablation. The plasma plume is then viewed perpendicularly to its expansion direction and investigated by means of a spectrometer, or a polychromator in combination with a CCD camera, a streak camera, etc. The detected species are excited during the ablation process itself, or via inelastic collisions or chemical reactions within the plume, or via a second (parallel) laser beam. The spatial distribution of species can be investigated by imaging different volumes of the plume; here, the

30.1 Precursor and Product Species

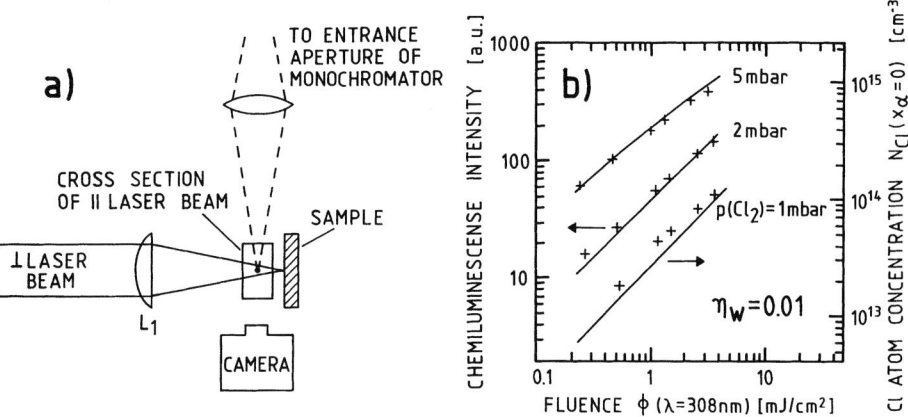

Fig. 30.1.1. (a) Schematic picture of an experimental setup used for *in situ* measurements during LP. The optical emission of species is analyzed by means of a spectrometer. Spatially or temporally resolved measurements can be performed by imaging different volumes onto the entrance slit of the monochromator, or by using a polychromator together with a CCD camera. (b) Chemiluminescence intensity measured *during* etching of Si in Cl_2 atmosphere as a function of XeCl-laser fluence [parallel incidence to sample surface in (a)]. The solid curves (right-hand scale) have been calculated. For details see Sect. 15.2.3 and [*Kullmer* and *Bäuerle* 1988a]

resolution achieved along the target normal is typically 100 μm. By time-resolved detection of single emission lines, expansion velocities of species can be deduced as well. The type of product species observed in such experiments are classified in Sect. 30.2.

LIF has been employed to study species desorbed from Si and Al surfaces during laser-induced etching [*Herman* et al. 1991], and during LCVD of W from WF_6 [*Heszler* et al. 1993].

30.1.2 Mass Spectrometry

An experimental setup that permits *in situ* identification of species taking part in laser-driven surface reactions is shown in Fig. 30.1.2. Here, the surface reaction is modulated via the laser-light intensity. The modulation causes transients in the concentrations of gas-phase precursor and product species. These species are sampled through an orifice in the center of the reaction zone and identified by phase-sensitive detection using a mass spectrometer. Thus, product species can be differentiated from those formed by ionization and fragmentation of precursor molecules. This and similar setups have been used for gas-phase analysis during LCVD of Au from dimethyl-gold-hfacac [*Kodas* and *Comita* 1989], W from WF_6 [*van Maaren* et al. 1990], a-Si:H from SiH_4 and Si_2H_6 [*Golusda* et al. 1991], etc.

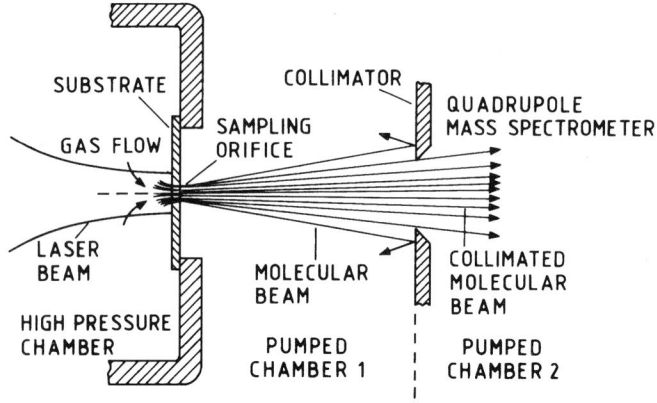

Fig. 30.1.2. Experimental setup employed for the diagnostics of species [*Kodas* and *Comita* 1989]

Mass spectrometers coupled with time-of-flight (TOF) techniques permit one to determine the chemical nature of product species and their desorption dynamics and energy distribution (Fig. 30.1.3). The experimental apparatus shown in the figure contains two stages of differential pumping which separate the quadrupole mass spectrometer (QMS) from the sample chamber. The QMS is mounted with its axis perpendicular to the molecular flight path. The sample is aligned with its surface normal in the direction of observation. Illumination of the substrate occurs at an angle of 45°. TOF data are collected by setting the QMS to a desired mass and recording the number of counts during a certain integration time. Such setups have been used for the *in situ*

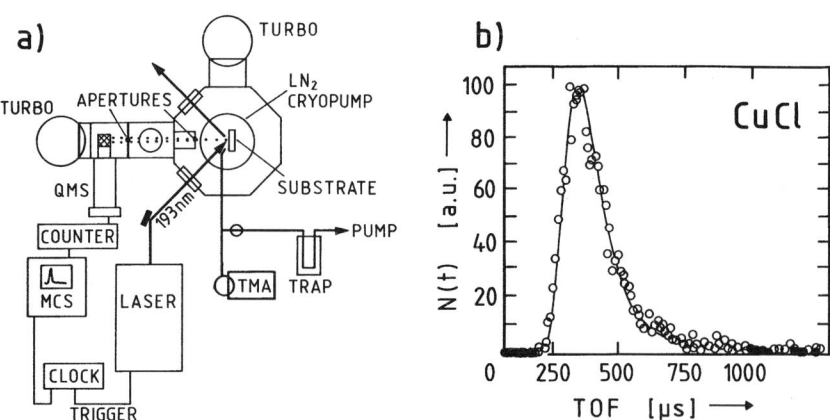

Fig. 30.1.3. (a) Schematic picture of an apparatus to measure photoproducts, here of $Al(CH_3)_3$ (TMA), by TOF [*Orlowski* and *Mantell* 1989]. (b) TOF distribution for XeCl-laser-induced etching of (100) Cu in Cl_2 atmosphere ($\lambda = 308$ nm, 90 mJ/pulse, 1Hz). The mass spectrometer setting was $m/e = 98$ [*Kools* et al. 1992]

diagnostics of species generated during laser-induced decomposition of Al(CH$_3$)$_3$ [*Orlowski* and *Mantell* 1989], Ga(CH$_3$)$_3$ and Ga(C$_2$H$_5$)$_3$ [*Donnelly* 1991], during etching of Cu and Si in Cl$_2$ [*Kools* et al. 1992; Baller 1990; *van Veen* et al. 1988], and during ablation of organic and inorganic materials (Sect. 30.2). Care must be taken that the observed fragments actually originate from the sample surface, since larger mass fragments can also create smaller fragments within the QMS ionizer. This can be accomplished by investigating correlations in the transit times of species.

An example of a typical TOF distribution is shown in Fig. 30.1.3b. The full curve is a fit to experimental data by an elliptical Maxwell–Boltzmann distribution

$$N(t) = \frac{\zeta}{t^3} \exp\left\{ -\frac{m}{2k_B T_{xy}} \left[\left(\frac{x}{t}\right)^2 + \left(\frac{y}{t}\right)^2 \right] - \frac{m}{2k_B T_z} \left(\frac{z}{t} - u\right)^2 \right\}, \qquad (30.1.1)$$

where t is the flight time, ζ a scaling factor, m the fragment mass, and u the stream velocity. The values of the fit parameters employed were $T_{xy} = 1400$ K, $T_z = 2400$ K, and $u = 1200$ m/s. From T_{xy} and T_z the (thermodynamic) temperature at the surface has been calculated by assuming *isothermal* expansion [*Kools* et al. 1992].

It should be noted that TOF spectra can be fitted equally well by *adiabatic* solutions of the gas-dynamic equations [*Anisimov* et al. 1996]. The temperatures derived in these two ways may differ significantly.

30.2 Species in Vapor and Plasma Plumes

The analysis of species in vapor/plasma plumes during pulsed-laser ablation yields important information on both the ablation process itself and the properties of films fabricated by PLD.

The types of species observed by different techniques can be classified into atomic and molecular neutrals, ions, electrons, photons, large molecular fragments, and clusters. Here, we do *not* consider macroscopic particulates (Sect. 22.2.4). Atomic and molecular neutrals, and large molecular fragments are the dominating species observed with laser fluences below and around the ablation threshold, ϕ_{th}. With increasing fluence, the number of electronically excited species, ions, and lower molecular weight species increases. Changes in the type of ablated species with laser parameters can originate from changes in the ablation mechanism, from secondary photolysis (Sect. 12.4.1) or from reactions with the ambient medium, if present. *Primary* ablated products can only be analyzed for femtosecond laser ablation in a vacuum.

30.2.1 Species at Subthreshold Fluences

With both inorganic and organic materials, the ejection of species has been detected well below ϕ_{th}, for example by means of acoustic measurements. Figure 30.2.1 demonstrates that for YBCO irradiated by KrF-laser light, material removal takes place for fluences $\phi < \phi_{th} \approx 0.75\,\mathrm{J/cm^2}$. This is consistent with the results presented in Sect. 12.2. Similar experiments have been performed with organic polymers such as PMMA, PET, and PI. Here, the ejection of molecular species under UV-laser irradiation with fluences $\phi < \phi_{th}$ has also been observed. For PI, these products consist mainly of CO and CN [*Srinivasan* 1994].

KrF-laser irradiation of Si and Ge [*Bialkowski* et al. 1991] and of MgO [*Dickinson* et al. 1994] with fluences below the ablation threshold results in the emission of charged species.

30.2.2 Atomic and Molecular Neutrals

Even for *quasi-equilibrium* evaporation using cw lasers or "long" laser pulses, the species within the vapor may significantly differ from those in standard evaporation. This is based, in most cases, on the higher temperatures involved in laser ablation and on the interactions between the laser light and the plume.

For example, the vapor-phase species observed in standard evaporation of compound semiconductors are mainly atoms for group-II and -III elements, and mainly molecules and clusters for group-V and -VI elements. With appropriate parameters, lasers enable one to produce also the latter in atomic form only. This can significantly modify the growth kinetics and properties of films (Sect. 22.3.2).

With *short* laser pulses, the differences in the evaporation characteristics become more pronounced (Sect. 22.1.1). The most prominent features are the

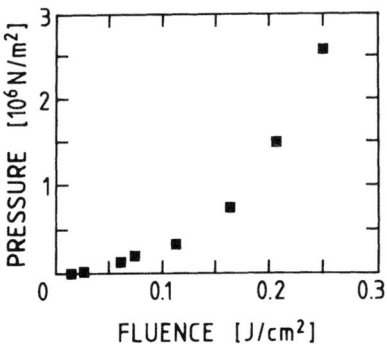

Fig. 30.2.1. Peak surface pressure measured by means of a PVDF foil during KrF-laser irradiation of $YBa_2Cu_3O_7$, adapted from [*Dyer* et al. 1992b]

congruent (stoichiometric) ablation of multi-component targets and the forward direction of the vapor plume. With moderate laser fluences which, however, well exceed the threshold for stoichiometric ablation, the ablated species consist of ground state and electronically excited neutrals and a *small* fraction of ions, depending on fluence. This is the regime employed in many cases of PLD. Here, optical spectroscopy permits one to directly correlate emission spectra with excited neutrals and ionized species.

Atomic and molecular neutrals and ions have been detected during excimer-laser-induced ablation of Au [*Bennett* et al. 1995], Al_2O_3 [*Walkup* et al. 1986], graphite [*Krajnovich* 1995; *Chen* et al. 1991; *Koren* and *Yeh* 1984], CaF_2 and SrF_2 [*Mitzner* et al. 1993; *Kreitschitz* et al. 1994], Cu and CuO [*Saenger* 1989], for various high-temperature superconductors [*Otis* and *Goodwin* 1993; *Geohegan* 1992; *Chrisey* et al. 1991; *Dyer* et al. 1992b; *Girault* et al.; *Wu* et al. 1989], organic polymers [*Hansen* 1989; *Dyer* et al. 1988; *van der Wel* and *Lub* 1993; *Ventzek* et al. 1991; *Feldmann* et al. 1987; *Davis* et al. 1985], etc.

Vapor/plasma plumes generated by excimer-laser ablation of Y-Ba-Cu-O in *vacuum* contain neutral and ionized atomic Y, Ba, and Cu, molecular O_2 ions, and diatomic molecules of BaO, CuO, and YO. The fractional ionization of ablated species for fluences $\phi \leqslant 4$ J/cm^2 is $\leqslant 4\%$. At 1 J/cm^2, typical velocities of species derived from optical emission spectra were around 10^6 cm/s and somewhat smaller than those measured with an ion-probe. The plume temperature has been estimated to be 5×10^3 to 10^4 K. With an O_2 background pressure, the optical emission intensity increases. This is mainly related to inelastic collisions between elemental species and electronically excited oxygen atoms, to collisions between electrons and ions, and to the recombination of species (Sect. 22.2.1).

With many materials and a broad range of laser parameters, the velocity of neutral atoms and molecules can be fitted by a Maxwell–Boltzmann distribution. The kinetic energy of species in excimer-laser ablation is typically in the range of a few eV to several tens of eV. Hyperthermal neutrals corresponding to the "hot" tail of the distribution function can significantly influence the film properties (Sect. 22.2.3).

30.2.3 Ions

With increasing light intensities, the fraction of ions within the plasma plume increases. Above a certain threshold, avalanche ionization is observed and the fraction of ions becomes almost 100%. This threshold for optical breakdown depends on the target material, the ambient medium, and the laser parameters (Sect. 11.2). If photochemical ablation mechanisms are important, this threshold is one to two orders of magnitude *lower* than with purely thermal ablation. In any case, with laser-beam intensities exceeding some 10^9 W/cm^2, electrons and ions are the dominant species. With inorganic materials, the ions are mainly positively charged and consist of atoms, molecules, and clusters. The fragmentation and degree of ionization of species increases with laser fluence.

The kinetic energy distribution of ions has been investigated for different systems, mainly by means of TOF techniques. The multiple peak structure observed with multicomponent targets is related to different ion velocities. The faster ions are more concentrated around the centerline of the plume.

30.2.4 Electrons, X-rays

The photoemission of electrons and their interaction with laser-ablated species has been investigated for some metals, semiconductors and insulators [*Cronberg* et al. 1991; *Honig* 1963]; see also Sect. 22.2.1. At laser-light intensities $> 10^{10}$ W/cm^2, soft X-ray emission was observed during ablation of various metal targets, and of carbon [*Murakami* 1992; *Dyer* et al. 1976].

30.2.5 Fragments and Clusters in Polymer Ablation

High molecular-weight fragments and clusters are observed during laser ablation of compound materials, organic polymers, and biological tissues. At least with the latter, ablation often seems to be related to volume explosions where gaseous products have built up a sufficiently high pressure to break-up the irradiated volume. This results in the ejection of large fragments.

Experiments on the analysis of product species in polymer ablation have been reviewed by *Srinivasan* (1994). Besides using optical and mass spectroscopy, ablated species are often analyzed from the material condensed on a substrate. Among the materials studied in most detail are PI, PET, and PMMA.

With PI and *ns excimer-laser* radiation the main ablation products are CO, HCN formed from CN, hydrocarbons such as CH, C_2H_2, C_3H_2, and carbon in the form of C, C_2, C_3, and large clusters [*Illmer* et al. 1990; *Campbell* et al. 1990]. The large carbon clusters seem to be formed in the vapor plume only. In oxygen atmosphere an increasing amount of CO_2 at the expense of C_2H_2 and carbon particulates was observed with increasing pressure [*Singleton* et al. 1989]. The composition of ablation products changes fundamentally with *long UV-laser pulses* ($\tau_l > 10$ µs) and *IR-laser pulses* of ns duration or longer [*Srinivasan* 1993]. In both of these cases the ablation products are essentially equal to those observed in standard pyrolysis of PI, namely C_6H_5CN, C_6H_5OH, $C_6H_5NH_2$, $C_6H_4(CN)_2$, etc. [*Wright* 1981].

The main volatile products observed with *CO_2-laser* ablation of PET, analyzed by gas chromatography and mass spectroscopy, were CO, CO_2, CH_4, C_2H_2, C_2H_4, C_6H_6, and CH_3CHO. These products are similar to those observed in standard thermal decomposition of PET. The relative yield of these products depends strongly on laser fluence. For $\phi \approx \phi_{th}$, the ablated material consists mainly of involatile species of relatively high molecular weight [*Dyer* et al. 1989]. With ns XeCl-laser pulses almost the *same* products are observed [*Dyer* et al. 1991]. Thus, with PET rapid thermal relaxation of the excitation energy and pyrolysis of the polymer seems to be the dominating mechanism in *both* regimes. This situation is quite different from that observed with PI where the ablation mechanism changes from photophysical to thermal.

30.3 Shock Waves, Plume Expansion

In this section we summarize investigations on laser-induced shock waves and the expansion of vapor/plasma plumes in gases and liquids.

30.3.1 Propagation in Gases

In a vacuum the plume undergoes free expansion and ultimately reaches some terminal velocity. The situation is different in the presence of an ambient medium. Here, the expanding products ejected from the substrate act like a piston which accelerates the ambient species to supersonic speeds causing a shock wave *ahead* of the contact surface (Fig. 30.3.1a–c). The position of the shock front can be described, within a certain range, by the model for strong explosions [*Zeldovich* and *Raizer* 1966; *Sedov* 1959]

$$r_{sw} \approx \xi \left(\frac{E}{\rho(\infty)} t^2 \right)^n, \qquad (30.3.1)$$

where $\xi \equiv \xi(\gamma) \approx 1$ (for air $\gamma = c_p/c_v = 7/5$). ξ and the exponent n depends on the symmetry of the shock front. For spherical expansion $n = 1/5$ ($\xi = 1.033$), for cylindrical symmetry $n = 1/4$, and for plane waves $n = 1/3$. $\rho(\infty)$ is the (undisturbed) density of the background atmosphere. E is essentially given by the sum of the kinetic energy of the shock wave and the thermal energy of the vapor plume; for spherical symmetry, it can be approximated by $E \approx A\phi F - E_{loss} - \rho_s F \Delta h \Delta H$ where A is the absorptivity of the material, F the irradiated area, E_{loss} the energy loss due to heat conduction into the solid and $\Delta H \approx \Delta H_m + \Delta H_v + \Delta H_i$ the enthalpy for melting, vaporization, and ionization, if relevant. The shock wave compresses the ambient medium and thereby increases its temperature. The thickness of the shell which contains most of the mass of the shocked gas is

$$\Delta r = r_{sw} \frac{\gamma - 1}{3(\gamma + 1)}. \qquad (30.3.2)$$

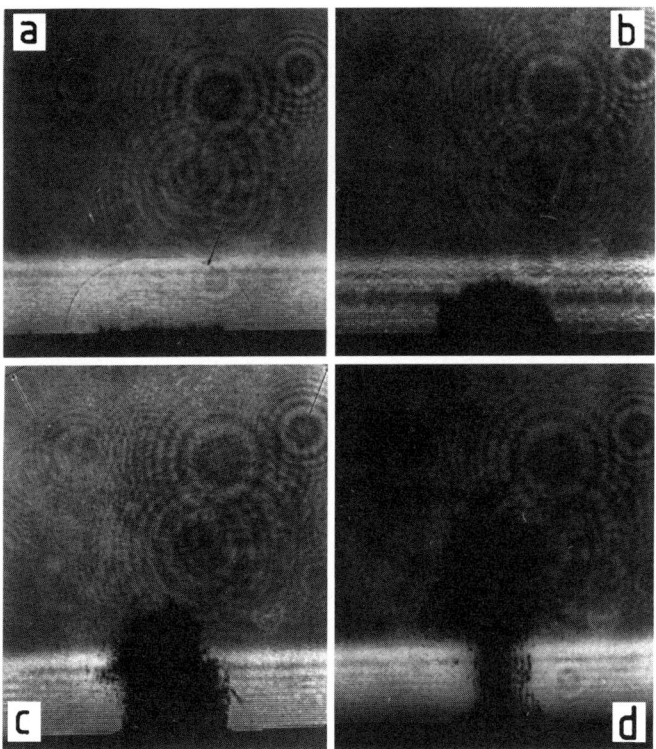

Fig. 30.3.1a–d. PMMA exposed to a single KrF-laser pulse in air [ϕ (248 nm) $= 2$ J/cm^2, $\tau_l \approx$ 20 ns, $2w = 970\,\mu$m]. The material ejected was photographed by firing above the target a second laser pulse ($\lambda = 596$ nm, $\tau'_l \approx 1$ ns) with a known delay t_d. This permits one to image both the *shock wave* (indicated by arrows) and the ejected material. The interface between the dense ejected material and the ambient medium (dark to light) is termed *contact surface*. (**a**) $t_d = 0.75\,\mu$s. Only the shock wave is seen. (**b**) $t_d = 1\,\mu$s. The column of ejecta has the width of the laser spot, a height $l = 0.42$ mm and an average velocity $v_p \approx 4.2 \times 10^4$ cm/s. The nearly hemispherical shock wave has an extension normal to the target of 1.24 mm and an average velocity of $v_{sw} \approx 1.2 \times 10^5$ cm/s. (**c**) $t_d = 6\,\mu$s, $l \approx 1$ mm, $v_p \approx 1.7 \times 10^4$ cm/s, $v_{sw} \approx 5.6 \times 10^4$ cm/s. (**d**) $t_d = 15\,\mu$s, $l = 1.8$ mm, $v_p \approx 1.2 \times 10^4$ cm/s [*Braren* et al. 1991]

For strong shock waves where $p_{sw} \gg p(\infty)$, the density and temperature at the shock front is [*Landau* and *Lifshitz* Fluid Mechanics]

$$\rho_{sw} = \rho(\infty)\frac{\gamma+1}{\gamma-1}; \quad T_{sw} = T(\infty)\frac{\gamma-1}{\gamma+1}\,\frac{p_{sw}}{p(\infty)}, \qquad (30.3.3)$$

where the (time-dependent) pressure is given by

$$p_{sw} = 2\left(\frac{2\xi}{5}\right)^2 \left(\frac{E^2\rho^3(\infty)}{t^6}\right)^{\frac{1}{5}} \gg p(\infty)\,. \qquad (30.3.4)$$

30.3 Shock Waves, Plume Expansion

γ and $T(\infty)$ refer to the background gas. The typical behavior of the density, temperature, pressure, and mass velocity *behind* the shock front are illustrated in Fig. 30.3.2a. During the propagation of this front, the pressure, density, and velocity drop off. Their behavior at such a later stage (weak shock) is shown in (b). Due to rarefaction within the volume just behind the shock front, p, ρ, and v decrease and can become even smaller than within the pre-shock ambient medium. After degeneration of the shock wave into a sound wave, the *central* region near the target has still a high temperature and low density.

With subsequent cooling, the pressure decreases and the gas moves towards the center. If cooling is fast enough, this mechanism can put a strong mechanical force onto the substrate which, besides thermal and recoil stresses, may play an important role during pulsed-laser ablation. The inward moving gas stream ($v < 0$) can also favor recondensation of debris at or near the ablated surface (Chap. 12).

With very high laser fluences, the shock wave ionizes the ambient medium and thereby gives rise to plasma formation. This process can result in the formation of a detonation wave (Sect. 11.2).

It is evident that a proper description of gas-phase processes requires *simultaneous* consideration of both the propagation of the shock wave and the ablated material. Nevertheless, the simple equations (30.3.1 to 4) permit a *qualitative* understanding of many experimental observations.

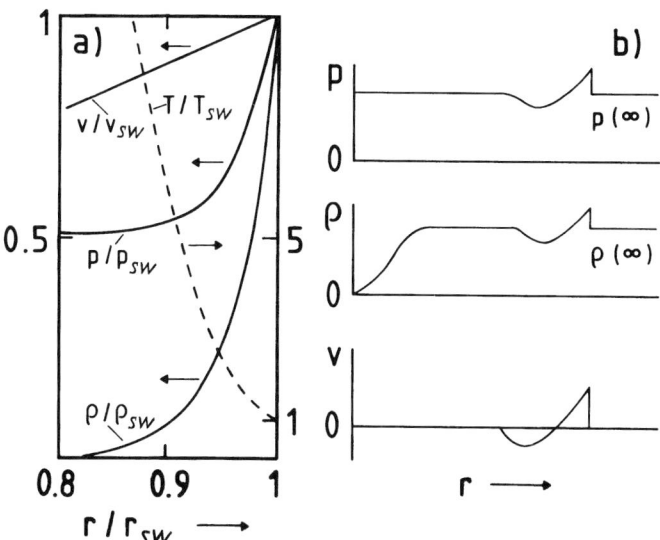

Fig. 30.3.2 (a) Typical behavior of density ρ/ρ_{sw}, pressure p/p_{sw}, velocity v/v_{sw}, and temperature T/T_{sw}, behind the shock front in a gas with $\gamma = 1.23$. **(b)** Characteristics of shock wave at a later stage. Within a certain range there is an inward motion where $v < 0$, adapted from [*Zeldovich* and *Raizer* 1966]

Experimentally, the *dynamics* of the shock wave and the expansion of the plume have been studied by fast photography using a conventional camera [*Grun* et al. 1991; *Srinivasan* et al. 1990a,b; *Simon* 1989] or a streak camera [*Geohegan* 1992; *Dyer* et al. 1992b] or Schlieren photography [*Ventzek* et al. 1990]. The *overall* expansion and shape of the plume has been investigated by time-integrated photography [*Proyer* and *Stangl* 1995; *Dyer* et al. 1992b]. The time scale and surface pressure associated with the *initial* ablation step have been studied by employing acoustic and pyroelectric monitors [*Grad* and *Mozina* 1993; *Dyer* et al. 1992a; *Leung* and *Tam* 1992; *Dienstbier* et al. 1990], time-resolved reflection [*Paraskevopoulos* et al. 1991; *Pang* and *Yeung* 1990], and laser-beam deflection [*Ediger* and *Pettit* 1992; *Ventzek* et al. 1991; *Sell* et al. 1991].

Figure 30.3.1 exhibits the time-resolved KrF-laser ablation of PMMA in air. In the initial stage (Fig. 30.3.1a) the most prominent feature is the shock wave. With the parameters employed, no significant ionization within the pressure front is observed. The *initial* thickness of the vapor plume is, typically, $l_p \approx v_0 \tau_l$ [*Anisimov* et al. 1993]. With $v_0 = 10^5 - 10^6$ cm/s and $\tau_l = 10^{-8}$ s we find $l_p \approx 10-100$ μm. With the (*high*) ambient pressure employed in these experiments (about 1 bar), the dense vapor plume remains far behind the shock wave (Fig. 30.3.1a,b,c). Later on, the ejected stream narrows, slows down, and the large fragments even seem to be melted (Fig. 30.3.1d). The material recondensed on the surface forms the debris. The situation is different if ablation is performed at *low* pressures.

Chemical reactions

With certain systems, the ablated material reacts with the ambient medium. This reaction is particularly strong near the *contact surface*. Here, the reaction rate may be strongly enhanced by turbulence which causes fast mixing. For diffusional mixing the thickness of the reaction zone around the contact surface is $\Delta r_r \propto t^{1/2}$; with the development of instabilities, we find $\Delta r_r \propto t^2$. Furthermore, the high temperatures generated by the shock wave can thermally enhance reactions or radical formation from gas- or liquid-phase molecules which, in turn, may strongly react with each other or with ablation products. Such reactions are of great importance for the suppression of debris (Chap. 12) and in many cases of PLD (Chap. 22). In other words, reactions between ablated species and the ambient medium can take place *far away* from the target at a distance l_r. This has been investigated for excimer-laser ablation of YBCO in O_2. Here, the spatial and temporal behavior of the luminosity of the plume, which is related to emissions from BaO, YO, and CuO, can be interpreted by reactions between the metal species ablated from the target and shock-heated oxygen. This interpretation is supported by streak and "species-resolved" measurements of l_r (Fig. 30.3.3). The data are in reasonable agreement with each other. With the pressures employed, the temporal

30.3 Shock Waves, Plume Expansion

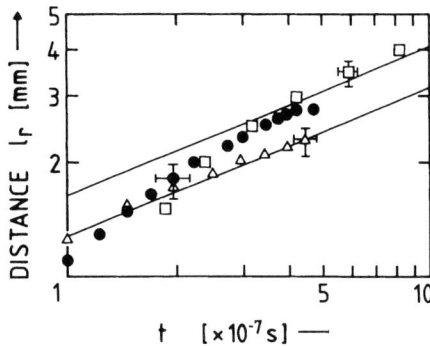

Fig. 30.3.3. Temporal dependence of the distance of the luminous (reaction) front l_r, from the target during KrF-laser ablation of YBCO in O_2. Data have been derived from: • streak photography ($\phi = 5.5$ J/cm^2, $p(O_2) = 6.5$ mbar); □ time-resolved recording of the Y^+ emission (6 J/cm^2, 6.5 mbar); Δ streak photography ($\phi = 5$ J/cm^2, $p(O_2) = 20$ mbar). The full lines have the form $l_r \propto r_{sw} \propto t^{0.4}$ [*Dyer* et al. 1992*b*]

dependence of l_r can be described, within a certain range, by $l_r \approx r_{sw} \propto t^{0.4}$. This is in agreement with (30.3.1). It should be noted, however, that the *same* dependence is expected at later times, i.e., for weak shocks [*Zeldovich* and *Raizer* 1966]. The thickness of the "shell-type" reaction zone, Δr, can be approximated by (30.3.2) [*Dyer* et al. 1990*a*]. In contrast to the situation observed at higher pressures the shock wave and the expanding plume seem to propagate together [*Dyer* et al. 1992*b*; *Geohegan* 1992]. The interpretation of these investigations, however, is still under discussion [*Gupta* et al. 1991].

Another important parameter is the maximum *length* of the plume, l_p. This determines the target–substrate distance in PLD. If we assume that the products expand adiabatically until their pressure is equal to the pressure of the ambient gas, $p(\infty)$, this length can be estimated, for spherical expansion, from

$$l_p = l_0 f(\gamma) \left[\frac{E_p}{p(\infty) l_0^3} \right]^{1/3\gamma}, \tag{30.3.5}$$

where l_0 is the initial radius of the plume and $f(\gamma) \approx 1$. Note that $\gamma = c_p/c_v$ refers now to the *ablated* material. E_p is the energy within the plume, $E_p \approx E - E_{sw}$ where E is defined together with (30.3.1).

Figure 30.3.4 shows the plume length as a function of pressure calculated from the gas dynamic equations for adiabatic expansion. For an elliptic plume and a wide range of ambient pressures, we obtain $l_p = l_{0z} f'(\gamma, \zeta_0, \eta_0) p^{-\beta}(\infty)$ where $\beta = \beta(\gamma, \zeta_0, \eta_0) \approx 1/3\gamma \approx 0.3$. $\zeta_0 \approx l_{0z}/w_x$ and $\eta_0 \approx w_y/w_x$ characterize the "initial" geometry of the plume. For $p^* < 10^{-7}$ the length of the visible plume was found to become almost independent of pressure, $p(\infty)$.

30.3.2 Propagation in Liquids

Laser-induced shock waves in liquids play an important role in medical applications, in particular in stone fragmentation (lithotripsy). *Quantitative*

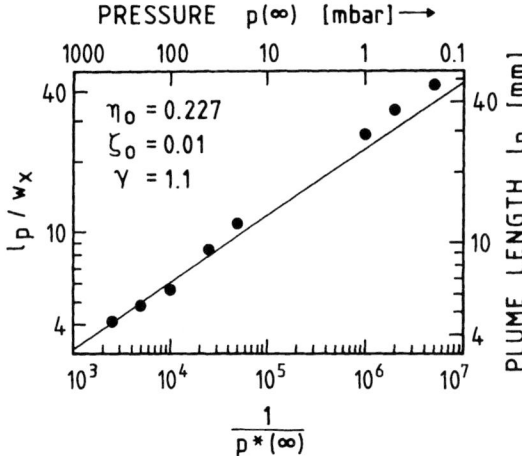

Fig. 30.3.4. Data show the (visible) plume length (right-hand scale) measured for different gas pressures (upper scale). The full curve represents a theoretical fit in dimensionless quantities $p^*(\infty) = p(\infty)w_x^3/E_p$ where E_p was assumed to be equal to the absorbed laser-light energy, adapted from [Stangl et al. 1994]

investigations have been performed for water and various materials. The fragmentation mechanism depends on the laser parameters.

With μs *laser pulses* and energies of, typically, some 100 mJ the (primary) shock wave induced by the expansion of the vapor plume plays a *minor* role in the fragmentation process. More important is the (secondary) shock which is related to the collapse of the cavitation bubble [*Rink* et al. 1992; *Ihler* 1992]. The formation of large cavitation bubbles observed with these laser parameters is probably related to the large amount of material ablated from the surface.

With ns *laser pulses* and energies of a few mJ, the primary shock may significantly exceed the secondary shock. Figure 30.3.5 shows the peak pressures of (primary) shock waves induced during XeCl-laser ablation of PI [*Zweig* and *Deutsch* 1992]. Here, the velocities of shock waves, v_{sw}, have been determined by HeNe-laser probe beams parallel to the PI target. The pressure is then estimated from Newton's law

$$p_{sw} = \rho_l(\infty)v_{sw}v , \qquad (30.3.6)$$

and the relation

$$v_{sw} \approx c_1 + c_2 v , \qquad (30.3.7)$$

where v is the mass velocity (of the water) behind the shock wave. Equation (30.3.7) is a good approximation for "weak" shock waves. For water, the (undisturbed) density is $\rho_l(\infty, 20°C) \approx 0.987$ g/cm³, $c_1 = 1.483 \times 10^5$ cm/s, and $c_2 = 2.07$.

The dependence of the shock pressure on laser fluence and pulse length can be calculated from

$$p_{sw} \approx \frac{2}{1+Z_1/Z_2}\left(\frac{Z}{1+f/2}\frac{\phi - \phi_{th}}{\tau_l}\right)^{1/2} . \qquad (30.3.8)$$

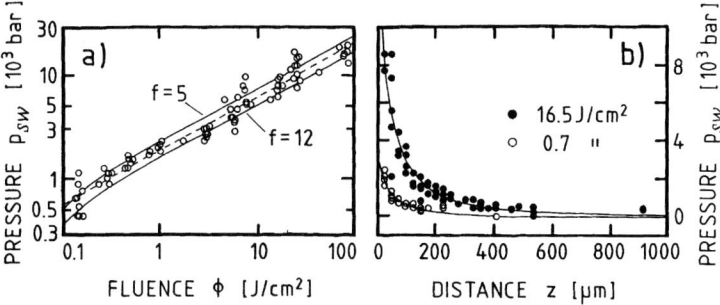

Fig. 30.3.5 a,b. Pressures of (primary) shock waves, p_{sw}, transmitted into water at the *rear* surface of a 25 μm-thick PI foil irradiated with XeCl–laser light. (**a**) p_{sw} (measured at a distance $z = 25$ μm away from the rear surface) as a function of fluence, ϕ. The dashed line is a least-squares fit to data with a slope of 0.52. The full curves have been calculated from (30.3.6). (**b**) p_{sw} versus distance z for two laser fluences. Full curves are fits to (30.3.9) [*Zweig* and *Deutsch* 1992]

Here it has been assumed that the products ablated from the substrate can be treated as an ideal gas expanding within two (plane) media with shock impedances $Z_i = \rho_i v_{0i}$. Here v_{0i} is the sound velocity within medium i and $Z = Z_1 Z_2 / (Z_1 + Z_2)$ the effective impedance of the two materials [Z_1(PI) = 30.7 bar s/m; $Z_2(H_2O) = 14.6$ bar s/m]. The number f is given by $f = 2c_v/R_G$ where the (molar) specific heat, c_v, depends on the number of (molecular) degrees of freedom.

At large enough distances from the irradiated zone, the shock wave propagates spherically. Because of momentum conservation, the pressure, p_{sw}, varies inversely with the square of the distance from the shock center, i.e.,

$$p_{sw} \propto \frac{1}{(r_{sw} - r_0)^2}, \qquad (30.3.9)$$

where r_0 is a fit parameter. This simple relation describes the data in Fig. 30.3.5b quite well.

The dependence $p_{sw} \propto (\phi - \phi_{th})^{1/2}$ as given by (30.3.8) is in good agreement with the data in Fig. 30.3.5a. The equation may also explain why primary shock waves play a more important role with short laser pulses τ_l. In fact, with the energies employed in the experiments of *Zweig* and *Deutsch*, any shocks due to bubble collapses were below the detection limit.

30.4 Processed Surfaces and Thin Films

The tools most commonly employed for the diagnostics of laser-processed surfaces and thin films include various optical techniques, electrical measurements, X-ray diffraction (XRD), and many of the techniques employed in surface science. Subsequently, we describe some applications of these techniques with special emphasis on their use for in situ diagnostics.

30.4.1 Optical Techniques

Raman microprobe spectroscopy and LIF have been applied for *in situ* measurements of laser-induced heating and melting of Si and Ge [*Tang* and *Herman* 1991] during laser-CVD of Si from SiH_4 [*Magnotta* and *Herman* 1986], etching of Cu and Si in Cl_2 [*Herman* et al. 1991; *Tang* and *Herman* 1990], and oxidation of Cu [*Herman* et al. 1991].

Coherent anti-Stokes Raman spectroscopy (CARS) has been used to investigate on a ps time scale the (surface) dynamics during laser ablation of PMMA [*Hare* and *Dlott* 1994].

Time-resolved reflection, absorption, and transmission measurements have been employed to study changes in surface morphology during laser-induced solid-phase epitaxy [*Olson* et al. 1981] and LCVD of Au [*Comita* et al. 1992]. These techniques have also been used to investigate transient phenomena during UV-laser irradiation of organic polymers [*Bor* et al. 1995; *Srinivasan* 1994; *Ediger* and *Pettit* 1992; *Zweig* and *Deutsch* 1992; *Frisoli* et al. 1991; *Pettit* and *Sauerbrey* 1991].

30.4.2 Other Techniques

Techniques typically employed in surface science such as X-ray photoemission spectrospocy (XPS), Auger electron spectroscopy (AES), Rutherford backscattering (RBS), secondary ion mass spectroscopy (SIMS), etc., [*Lüth* 1995; *Ertl* and *Küppers* 1985] have been applied also to laser-processed surfaces. Some of the related investigations have been mentioned in various chapters. There are, however, only a few experiments where these techniques have been used for *in situ* analysis. Among those are XPS and LEED investigations on ArF-laser irradiated (100) Si/SiO_2 interfaces [*Kubatova* et al. 1989], and XPS studies on Al deposited from $Al(C_4H_9)_3$ [*Mantell* and *Orlowski* 1988].

Laser annealing has been studied by ns time-resolved XRD [*Larson* et al. 1982] and by high-speed ellipsometry [*Moritani* and *Hamaguchi* 1985]. These techniques permit one to derive lattice strains, annealing rates, and temperatures from in situ measurements.

30.4.3 Transport Measurements

The resistance of a rectangularly shaped film of length l, width d, and height h is given by

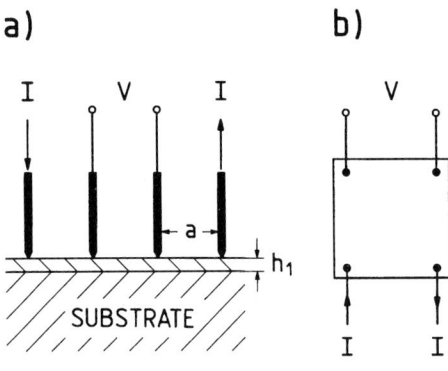

Fig. 30.4.1. Four-point-probe methods for sheet resistance measurements. I and V denote the current and voltage, respectively

$$R = \frac{\rho}{h}\frac{l}{d} = \frac{\rho}{h} \equiv \mathcal{R}\,[\Omega/\square]\,.$$

The latter equality holds for one *square* of the film, i.e., for $l = d$. $\mathcal{R}$ is termed the *sheet resistance* and depends only on the resistivity and thickness of the film but not on its size. Sheet resistances are most commonly measured by means of *four-point-probe* methods.

In the configuration shown in Fig. 30.4.1a the resistivity of a semi-infinite substrate ($h_1 = 0$) is given by

$$\rho_s = 2\pi \frac{V}{I} a\,, \tag{30.4.1}$$

where a is the distance between the *equidistant* electrodes. Probe spacings typically employed lie between $a = 200$ μm and several mm. If a film of resistivity ρ is placed on an insulating substrate so that $\rho \ll \rho_s$ the sheet resistance is

$$\mathcal{R} = \frac{\rho}{h_1} = \frac{\pi}{\ln 2}\frac{V}{I} = 4.532\,\frac{V}{I}\,. \tag{30.4.2}$$

Higher spatial resolution can be achieved with the configuration depicted in Fig. 30.4.1b. The sheet resistance is then

$$\mathcal{R} = \frac{2\pi}{\ln 2}\frac{V}{I}\,. \tag{30.4.3}$$

For further details see, e.g., [*Maissel* and *Glang* 1970] and references therein.

APPENDIX A: Definitions and Formulas

A.1: Symbols and Conversion Factors

A absorptivity
a distance
     aperture
b net increase in number of molecules per formula unit; $b = \mu - 1$
C constant
C Euler's constant; $C = 0.577$
c speed of light; $c = 2.998 \times 10^{10}$ cm/s
c_p specific heat at constant pressure [J/gK, J/mol K]
c_v specific heat at constant volume [J/gK, J/mol K]
D_i molecular diffusion coefficient of species i [cm^2/s]
D heat diffusivity [cm^2/s]
     transmittivity
d lateral width of laser-processed features [μm, cm]
     diameter
E electric field [V/cm]
     energy [J]
     $k_B T (T = 273.15 \text{ K}) = 2.354 \times 10^{-2}$ eV
     1 kcal/mol $\triangleq$ 0.043 eV $\triangleq$ 5.035 × 10^2 K
     1 eV $\triangleq$ 1.1604 × 10^4 K $\triangleq$ 1.602 × 10^{-19} J
     1 kcal $\triangleq$ 4.187 × 10^3 J
     1 cm^{-1} $\triangleq$ 1.24 × 10^{-4} eV $\triangleq$ 1.439 K
     1 J $\triangleq$ 2.39 × 10^{-4} kcal
$\mathscr{E}$ activation temperature [K]; $\mathscr{E} = \Delta E/k_B$
$\mathscr{E}^*$ normalized activation temperature; $\mathscr{E}^* = \mathscr{E}/T(\infty)$
ΔE activation energy [eV; kcal/mol]
ΔE_m activation energy for melting
ΔE_v activation energy for vaporization
E_g gap energy = energy distance between (lowest) conduction and (highest) valence band
e elementary charge; $e = 1.602 \times 10^{-19}$ C
e $e \approx 2.718$
eV   electron Volt
     1 eV/particle = 23.04 kcal/mol

A.1: Symbols and Conversion Factors

F	area
	Faraday Constant; $F = 96\,485$ C/mol
f	focal length [cm]
$\mathcal{G}r$	Grashof number
ΔG	Gibbs free energy
g	acceleration due to gravity
g_T	temperature discontinuity coefficient
ΔH	total enthalpy [J/cm^3, J/g, J/mol] reaction enthalpy
	ΔH^a [J/atom] $= \Delta H$ [J/cm^3]$\cdot M/\rho L = \Delta H$ [J/g]$\cdot M/L = \Delta H$ [J/mol]$/L$
ΔH_v	heat of vaporization
ΔH_m	heat of melting
ΔH_t	total latent heat $\Delta H_t = \Delta H_m + \Delta H_v$
h	Planck's constant; $h = 6.626 \times 10^{-34}$ Js
	height, thickness or depth of laser-processed patterns [Å, μm]
h_1	thickness of single evaporated or sputtered layer on a substrate
h_i	thickness of layer i on a substrate
h_l	thickness of a liquid layer, or an adsorbate
h_s	thickness of slab or substrate
Δh	change in layer thickness
	ablated layer thickness per pulse [Å/pulse]
$h\nu$	photon energy
	$h\nu$ [eV] $\approx 1240/\lambda$ [nm]
I	intensity [W/cm^2]
I_a	absorbed laser-light intensity
I_{th}	threshold intensity
I_v	evaporation intensity
J	flux
J_i	flux of species i [species/cm^{2}s]
j	current density
k	kinetic (rate) constant
k_0	preexponential factor
k_B	Boltzmann constant; $k_B = 1.381 \times 10^{-23}$ Ws/K
k_i^{rec}	recombination constant for species i
k_l	wave vector of laser radiation
k_T	thermal diffusion ratio
L	Avogadro number (Loschmidt number); $L = 6.022 \times 10^{23}$/mol
L	Langmuir [1 L $= 10^{-6}$ Torr s]
l	characteristic length, depth [μm]
l_T	heat diffusion length [μm]
l_α	optical penetration depth [μm]; $l_\alpha = \alpha^{-1}$
M	molar mass [g/mol]
m	mass
	exponent, e.g., in $\kappa(T)$

N	total number of species (atoms, molecules, electrons, holes, etc.) per volume [cm^{-3}] or per area [cm^{-2}]
	number of laser pulses
N_i	number of species i per volume [cm^{-3}] or per area [cm^{-2}]
n	refractive index (real part)
	exponent, e.g., in $D_i(T)$
$\boldsymbol{n}$	normal vector
$\hat{\boldsymbol{n}}$	unit vector
$\tilde{n}$	complex refractive index; $\tilde{n} = \sqrt{\varepsilon} = n + i\kappa_a \equiv n(1 + i\kappa_0)$
P	laser power [W]
P_a	absorbed laser power [W]
p	total gas pressure [mbar]
	1 mbar $\triangleq 10^2$ N/m^2 $\triangleq 10^2$ Pa ≈ 0.750 Torr $\triangleq$
	1.02×10^{-3} at [kp/cm^2] $\triangleq 9.87 \times 10^{-4}$ atm
p_i	partial pressure of species i [mbar]
Q	source term
q	exponent, e.g. in equation of state
$\boldsymbol{q}$	wave vector
R	optical (power) reflectivity
	electrical resistance [Ω]
$\mathscr{R}$	sheet resistance [$\Omega/\square$]
R_D	optical reflection coefficient of deposited material
R_G	gas constant; $R_G = 8.314$ J/K mol $\triangleq 1.987$ cal/K mol
$\mathscr{R}a$	Rayleigh number
r	radial distance
r_D	radius of deposit
S	stress
	oversaturation
$\boldsymbol{S}$	Poynting vector
s	sticking coefficient
T	temperature [K]
T_c	center temperature
T_G	gas-phase temperature
T_l	temperature within liquid
T_M	temperature within medium
T_m	melting temperature
T_s	substrate temperature
	surface temperature
T_{th}	threshold temperature
T_v	vaporization temperature
$T(\infty)$	temperature far away from irradiated zone
ΔT	temperature rise
T^*	normalized temperature, e.g., $T/T(\infty)$
t	time
t_v	time of vaporization

A.1: Symbols and Conversion Factors

Δt_m	time of existence of melt on surface
V	volume [cm^3]
V_n	volume per molecule/atom
v	velocity [cm/s]
	mass average velocity
v_{ls}	velocity of liquid–solid interface
v_{vl}	velocity of vapor–liquid interface
v_0	sound velocity
v_s	scanning velocity of laser beam or substrate [µm/s]
W	reaction rate
	heterogeneous reactions [number of species/s cm^2]
	homogeneous reactions [number of species/s cm^3]
W_A	ablation rate [µm/s; Å/pulse]
W_D	deposition rate [µm/s; Å/pulse]
W_E	etch rate [µm/s; Å/pulse]
W_{ex}	excitation rate
w	radius of laser focus with constant intensity distribution [µm]
	radius of laser focus at FWHM
w_e	radius of laser focus (1/e^2 intensity); $w_e = \sqrt{2}\, w_0$
w_0	radius of laser focus of Gaussian beam (1/e intensity) [µm]
$\wp$	probability
	width of reaction zone
x_i	molar ratio of species i; $x_i = N_i/N$
$\boldsymbol{x}, x_\alpha$	set of space coordinates with $\alpha = 1, 2, 3$, e.g., x, y, z
Y	Young's modulus
Z	number of condensed atoms per molecule
z	charge of ions in units of e
z_R	Rayleigh length of laser focus [µm]
α	optical absorption coefficient [cm^{-1}]
α_T	thermal diffusion constant
β	exchange coefficient
	exponent
	parameter
	symmetry factor
	factor
β_T	coefficient of thermal expansion
Γ	increment
	parameter
	ratio
	aspect ratio [ratio of depth or height to width of pattern]; $\Gamma = h/d$
γ	exponent
	total reaction order
	adiabatic index; $\gamma = c_p/c_v$
	real part of increment
γ_i	reaction order with respect to species i

Δ	difference
δ	delta function
	parameter
ε	dielectric constant
	permittivity
	spectral emissivity
ε_a	apparent emissivity
ε_0	dielectric constant in vacuum; $\varepsilon_0 = 8.854 \times 10^{-12}$ As/Vm
ε_t	total emissivity
ζ	parameter
	integer
	factor
ζ_i	stoichiometric coefficient of species i
η	dissociation yield
	dynamic viscosity [g/cm s]; $\eta = \rho v_k$
	reaction probability
	surface conductance [coefficient of surface heat transfer] [W/cm²K]
Θ	angle
θ	linearized temperature
θ_c	center temperature rise for Gaussian beam; $\theta_c = \sqrt{\pi} I_a w_0 / 2\kappa$, see (7.1.4)
Θ_i	coverage by species i
ϑ	angle
κ	thermal conductivity [W/cmK]; 1 W/mK $\hat{=}$ 2.39×10^{-3} cal/cmKs
κ_a	absorption index $\kappa_a = n\kappa_0$
κ_D	thermal conductivity of deposit
κ_L, κ_1	thermal conductivity of thin layer
κ_M	thermal conductivity of medium
κ_s	thermal conductivity of substrate
κ_0	attenuation index
Λ	parameter
	spacing
Λ	function
λ	wavelength of electromagnetic radiation [nm, μm] λ [nm] $\approx 1240/h\nu$ [eV]
λ_m	mean free path of molecules [cm]
μ	factor
	index
	integer
	chemical potential
	Poisson ratio
	$\mu = b + 1$
μ_e, μ_h	mobility of electrons and holes [cm²/Vs]
ν	frequency [s^{-1}]
	index

v_k kinematic viscosity [cm^2/s]
ξ overpotential
     parameter
Π product
     parameter
π 3.14159
ρ electrical resistivity [Ωcm]
     mass density [g/cm^3]
Σ summation sign
$\sum_{\pm}$ e.g., $a \pm b \mp c \equiv (a+b-c)+(a-b+c)$
σ electrical conductivity [Ωcm]$^{-1}$
     surface tension [J/cm^2]
     excitation cross section of species [cm^2]
σ_r Stefan–Boltzmann constant; $\sigma_r = 5.67 \times 10^{-12}$ W/cm^{2}K^4
τ relaxation time [s]
τ_l laser pulse duration [s]
     laser beam dwell time [s]; $\tau_l = 2w/v_s$
τ_m time for surface melting
τ_T thermal relaxation time [s]
Φ electrical potential
ϕ laser fluence [J/cm^2]
     angle
ϕ_{th} threshold fluence
φ angle
χ magnetic susceptibility
     parameter
Ψ function
ψ wave function
Ω total solid angle; $\Omega = 4\pi$
     Ohm
$d\Omega$ solid angle [sr]
ω angular frequency [s^{-1}]; $\omega = 2\pi v$
$\perp$ Index indicating laser-light incidence normal to substrate
$\parallel$ Index indicating laser-light incidence parallel to substrate
∇^2 Laplace operator
∇ Nabla operator

A.2: Abbreviations, Acronyms

acac   [CH$_3$COCHCOCH$_3$]$^-$ = acetylacetonate anion
AdGC   allyl-diglycol-carbonate
AES   Auger electron spectroscopy
ALE   atomic layer epitaxy

ALE	atomic layer epitaxy
AM1	sunlight illumination
APD	ablative photodecomposition
CAD	computer aided design
CAM	computer aided manufacturing
CARS	coherent anti-Stokes Raman scattering
CBE	chemical beam epitaxy
CCD	charge coupled device
CVD	chemical vapor deposition
DLC	diamond-like carbon
EAL	etching of atomic layers
EB	electron beam
EBCVD	electron beam-induced chemical vapor deposition
EDX	energy-dispersive X-ray analysis
EELS	electron energy loss spectroscopy
EMF	electromotive force
ESCA	electron spectroscopy for chemical analysis
ESR	electron spin resonance
FH	fourth harmonic
FWHM	full width at half maximum
HAZ	heat affected zone
hfacac	$[CF_3COCHCOCF_3]^-$ = hexafluoroacetylacetonate anion
HTS	high temperature superconductors
HV	high vacuum ($10^{-7} < p < 10^{-3}$ mbar)
IC	integrated circuit
IR	infrared radiation
ITO	indium tin oxide
Kapton	polyimide (Du Pont)
LA	laser annealing
LCP	laser-induced chemical processing
LCVD	laser-induced CVD
LEC	laser-enhanced electrochemistry
LEE	laser-enhanced electrochemical etching
LEED	low-energy electron diffraction
LEP	laser-enhanced electrochemical plating
LID	laser-induced desorption
LIF	laser-induced fluorescence
LIFT	laser-induced forward transfer
LIS	laser isotope separation
LMBE	laser molecular beam epitaxy
LPCVD	laser-enhanced PCVD
LPE	laser-enhanced plasma etching
LPPC	laser-pulsed plasma chemistry
LSA	laser-surface alloying
LSAW	laser supported absorption wave
LSD	laser-sputter deposition

A.2: Abbrevations, Acronyms

LSCW	laser supported combustion wave
LSDW	laser supported detonation wave
MBE	molecular beam epitaxy
ME	metal
ML	multiline operation of laser
	monolayer
MMA	methylmethacrylate
MOCVD	metal-organic CVD
MOMBE	metalorganic molecular beam epitaxy
MP	multiphoton
MPD	multiphoton dissociation
MPI	multiphoton ionization
Mylar	same as PET
NC	nitrocellulose
NEP	noise equivalent power
NIR	near IR
OMA	optical multichannel analyzer
PC	polycarbonate
PCVD	plasma CVD
PE	plasma etching
	polyethylene
PEEK	polyetheretherketone
PES	polyethersulfone
PET	polyethylene-terephthalate (same as mylar)
PI	polyimide [Kapton, Upilex]
PL	photoluminescence
PLA	pulsed-laser annealing
PLD	pulsed-laser deposition
PLE	pulsed-laser evaporation
PLPC	pulsed-laser plasma chemistry
PLZT	lanthanum doped PZT, i.e. $Pb_{1-3y/2}La_yTi_{1-x}Zr_xO_3$
PMMA	polymethyl-methacrylate (Plexiglas)
PP	polypropylene
pps	pulses per second
PS	polystyrene
PSUL	polysulfone
PTFE	polytetrafluoroethylene (Teflon)
PVAC	polyvinylacetate
PVC	polyvinyl chloride
PVDF	polyvinylidene fluoride
PXE	same as PZT ($PbTi_{1-x}Zr_xO_3$)
Pyrex	borosilicate glass (80% SiO_2, 12% B_2O_3, 3% Al_2O_3, 4% Na_2O)
PZT	lead titanate zirconate $PbTi_{1-x}Zr_xO_3$
QCM	quartz crystal microbalance
QMS	quadrupole mass spectrometer
RBS	Rutherford backscattering spectroscopy

RF	radio frequency
RHEED	reflection high-energy electron diffraction
RIE	reactive ion etching
rms	root mean square
RTA	rapid thermal annealing
SAW	surface acoustic wave
SEM	scanning electron microscopy
SERS	surface enhanced Raman scattering
SEW	surface electromagnetic wave
SH	second harmonic
SI	semi-insulating
SIMS	secondary ion mass spectroscopy
SOI	silicon on insulator
SOS	silicon on sapphire
SQUID	superconducting quantum interference device
STE	self-trapped exciton
TEM	transmission electron microscopy
TEOS	tetraethylorthosilicate
TFT	thin-film transistor
TG	thermogravimetry
TH	third harmonic
TiBAl	$Al(C_4H_9)_3$
TM	trade mark
TMVS	trimethylvinylsilane
TOF	time-of-flight
UHV	ultrahigh-vacuum ($p < 10^{-7}$ mbar)
ULSI	ultra-large-scale integrated systems
UPS	ultraviolet photo-spectroscopy
UV	ultraviolet radiation
VIS	visible radiation
VLSI	very-large-scale integrated systems
VUV	vacuum UV
XAFS	X-ray absorption fine structure spectroscopy
XPS	X-ray photoemission spectroscopy
XRD	X-ray diffraction
YBCO	$YBa_2Cu_3O_{7-\delta}$
YSZ	8 mol % Y_2O_3 stabilized ZrO_2

A.3: Mathematical Functions and Relations

A.3.1 Bessel Function

$$J_n(x) = \frac{1}{\pi}\int_0^\pi \cos(x\cos\zeta - n\zeta)\,d\zeta = \frac{i^{-n}}{\pi}\int_0^\pi \exp(ix\cos\zeta)\cos n\zeta\,d\zeta$$

A.3: Mathematical Functions and Relations

$$J_n(x \ll 1) \approx x^n\left(\frac{1}{2^n n!} - \frac{x^2}{2^{n+2}(n+1)!} + \frac{x^4}{2^{n+4}(n+2)!} + \cdots\right)$$

$$J_n(x \gg 1) \approx \left(\frac{2}{\pi x}\right)^{1/2}\left[\cos\left[\left(\frac{n}{2}+\frac{1}{4}\right)\pi - x\right] + \cdots\right]$$

Modified Bessel function $I_n(x)$ of order n

$$I_n(x) = (-1)^n I_n(-x) = \frac{1}{\pi}\int_0^\pi \exp(x\cos\zeta)\cos n\zeta\, d\zeta$$

$$I_0(x \ll 1) \approx 1 + \frac{x^2}{4} + \frac{x^4}{64} + \cdots$$

$$I_0(x \gg 1) \approx \frac{1}{(2\pi x)^{1/2}}\left(1 + \frac{1}{8x} + \frac{9}{128\,x^2} + \cdots\right)\exp x$$

Modified Bessel function $K_n(x)$ of order n

$$K_n(x > 0) = \int_0^\infty \exp(-x\cosh\zeta)\cosh n\zeta\, d\zeta$$

$$K_0(x \ll 1) = -\ln\left(\frac{x}{2}\right) - C - \frac{x^2}{4}\ln x + \cdots$$

$$K_1(x \ll 1) = \frac{1}{x} + \frac{x}{2}\ln\left(\frac{x}{2}\right) + \frac{x}{2}\left(C - \frac{1}{2}\right) + \cdots$$

$$K_n(x \gg 1) = \left(\frac{\pi}{2x}\right)^{1/2}\left(1 + \frac{4n^2-1}{8x} + \frac{(4n^2-1)(4n^2-9)}{128\,x^2} + \cdots\right)\exp(-x)$$

$C = 0.577$ is Euler's constant

A.3.2 Error Function

$$\operatorname{erf} x = -\operatorname{erf}(-x) = \frac{2}{\sqrt{\pi}}\int_0^x \exp(-\zeta^2)\,d\zeta$$

$$\operatorname{erf}(x \ll 1) \approx \frac{2x}{\pi^{1/2}} - \frac{2x^3}{3\pi^{1/2}} + \frac{x^5}{5\pi^{1/2}} + \cdots$$

$$\operatorname{erf}(x \gg 1) \approx 1 + \left(-\frac{1}{\pi^{1/2}x} + \frac{1}{2\pi^{1/2}x^3} + \cdots\right)\exp(-x^2)$$

Complementary error function

$$\operatorname{erfc} x = 1 - \operatorname{erf} x = \frac{2}{\pi^{1/2}}\int_x^\infty \exp(-\zeta^2)\,d\zeta$$

$$\operatorname{erfc}(x \ll 1) \approx 1 - \frac{2x}{\pi^{1/2}} - \frac{2x^3}{3\pi^{1/2}} - \cdots$$

$$\operatorname{erfc}(x \gg 1) \approx \left(\frac{1}{\pi^{1/2} x} - \frac{1}{2\pi^{1/2} x^3} + \frac{3}{4\pi^{1/2} x^5} - \cdots \right) \exp(-x^2)$$

$$i^{-1} \operatorname{erfc} x = \frac{2}{\pi^{1/2}} \exp(-x^2)$$

$$i^0 \operatorname{erfc} x = \operatorname{erfc} x$$

$$i \operatorname{erfc} x = \frac{\exp(-x^2)}{\pi^{1/2}} - x \operatorname{erfc} x$$

$$i^2 \operatorname{erfc} x = \frac{1}{4}[\operatorname{erfc} x - 2x i \operatorname{erfc} x]$$

$$i^n \operatorname{erfc} x = \int_x^\infty i^{n-1} \operatorname{erfc} \zeta \, d\zeta = -\frac{x}{n} i^{n-1} \operatorname{erfc} x + \frac{1}{2n} i^{n-2} \operatorname{erfc} x$$

A.3.3 Exponential Integral Function

$$\operatorname{Ei}(x < 0) = -\int_{-x}^\infty \zeta^{-1} \exp(-\zeta) \, d\zeta$$

$$\operatorname{Ei}(x > 0) = -\mathscr{P} \int_{-x}^\infty \zeta^{-1} \exp(-\zeta) \, d\zeta \qquad (\mathscr{P} \text{ stands for principal value})$$

$$\operatorname{Ei}(x \ll 1) \approx C + \ln|x| + x + \cdots$$

$$\operatorname{Ei}(x \gg 1) \approx \left(\frac{1}{x} + \frac{1}{x^2} + \frac{2}{x^3} + \cdots \right) \exp(x)$$

A.3.4 Gamma Function

$$\Gamma(x) = \int_0^\infty \zeta^{x-1} \exp(-\zeta) \, d\zeta$$

$$\Gamma(n) = (n-1)!, \quad \Gamma(x+1) = x\Gamma(x)$$

$$\Gamma(x \ll 1) \approx \frac{1}{x} - C + \left(\frac{C^2}{2} + \frac{\pi^2}{12} \right) x + \cdots$$

$$\Gamma(x \gg 1) \approx \left(\frac{x}{e} \right)^x \left(\frac{2\pi}{x} \right)^{1/2} \left(1 + \frac{1}{12x} + \cdots \right)$$

A.3.5 Heaviside Function

$$\mathscr{H}(x) = \begin{cases} 0 & \text{if } x \leq 0 \\ 1 & \text{if } x > 0 \end{cases}$$

A.3.6 Jacobian Theta Function

$$\theta_3^1[u|\exp(-\beta)] = \theta_3^1(u) = 1 + 2\sum_{n=1}^{\infty} \cos(2nu)\exp(-\beta n^2)$$

$$= \left(\frac{\pi}{\beta}\right)^{1/2} \sum_{n=-\infty}^{+\infty} \exp\left(-\frac{(u-n\pi)^2}{\beta}\right)$$

$$\theta_3^1(u|\beta \gg 1) \approx 1 + 2\exp(-\beta)\cos(2u) + \cdots$$

A.4: The Density of Dissociated Species

The average number density of species i in the photochemical decomposition process (2.3.2) shall be described by N_i where $i \equiv AB_\mu$, AB_μ^*, A, B. The total number density of species is $N = \Sigma N_i$. For an estimate of processing rates, the density of species A or AB_μ^*, depending on the particular system, must be known. The initial number density of species AB_μ shall be N_0. Because the number of atoms A remains unchanged

$$N_0 = N_{AB} + N_{AB^*} + N_A \quad \text{and} \quad N_B = \mu N_A.$$

The total number of species is

$$N = N_{AB} + N_{AB^*} + (1+\mu)N_A = N_0 + \mu N_A.$$

Light absorption by photoproducts, and mass transport limitations shall be ignored. We introduce the dimensionless variables

$$x_i = \frac{N_i}{N_0}; \quad t^* = \frac{t}{\tau_{em}}; \quad k_d^* = \frac{\tau_{em}}{\tau_d}; \quad k_{rec}^* = \frac{\tau_{em}}{\tau_{rec}}; \quad k_{rec}^{*\prime} = \frac{\tau_{em}}{\tau_{rec}'}; \quad (A.4.1)$$

$$I^* = \sigma_0 \tau_{em} \frac{I}{h\nu}; \quad \sigma_1^* = \frac{p_0}{\sigma_0} \frac{\partial \sigma}{\partial p}\bigg|_{p=p_0}.$$

The pressure dependence of the cross section shall be described by

$$\sigma \equiv \sigma(p) = \sigma_0 + \frac{\partial \sigma}{\partial p}\bigg|_{p=p_0}(p - p_0) \qquad (A.4.2)$$

with $\sigma_0 \equiv \sigma(p_0)$ and $\partial \sigma/\partial p \gtrless 0$ depending on the laser wavelength. p_0 is the pressure at $t = 0$. For an ideal gas and constant temperature, $p \equiv p(t)$ is given by $p = p_0 N/N_0$.

The kinetic equations for A and AB_μ^* can be written in the form

$$\frac{dx_A}{dt^*} = k_d^* x_{AB^*} - \left(k_{rec}^* + k_{rec}^{*\prime}\right) \mu^\mu x_A^{\mu+1} \qquad (A.4.3a)$$

$$\frac{dx_{AB^*}}{dt^*} = I^* \left(1 + \sigma_1^* \mu x_A\right)\left(1 - 2x_{AB^*} - x_A\right)$$

$$- \left(1 + k_d^*\right) x_{AB^*} + k_{rec}^{*\prime} \mu^\mu x_A^{\mu+1} . \qquad (A.4.3b)$$

The initial condition is

$$x_{AB^*}(t^* = 0) = x_A(t^* = 0) = 0 . \qquad (A.4.4)$$

Non-stationary analytic solutions can be obtained only in the linear case $\mu = 0$, i.e., when species B are immediately removed from the gas so that the total pressure remains constant.

Stationary solutions can be found even with $\mu \neq 0$ and they are given by

$$x_{AB^*}(\infty) = \frac{k_{rec}^* + k_{rec}^{*\prime}}{k_d^*} \mu^\mu x_A^{\mu+1}(\infty) \qquad (A.4.5a)$$

and

$$I^*[1 + \sigma_1^* \mu x_A(\infty)]\left[1 - \frac{2(k_{rec}^* + k_{rec}^{*\prime})}{k_d^*} \mu^\mu x_A^{\mu+1}(\infty) - x_A(\infty)\right]$$

$$- \frac{k_d^* k_{rec}^* + k_{rec}^* + k_{rec}^{*\prime}}{k_d^*} \mu^\mu x_A^{\mu+1}(\infty) = 0 . \qquad (A.4.5b)$$

$x_A(\infty)$ can be determined from this transcendental equation. If $I^* \gg 1$ and if σ_1^*, k_d^*, k_{rec}^*, $k_{rec}^{*\prime}$ and μ are of the order of unity, the last term in (A.4.5b) can be ignored and $x_A(\infty)$ can be derived from

$$x_A(\infty) + \frac{2(k_{rec}^* + k_{rec}^{*\prime})}{k_d^*} \mu^\mu x_A^{\mu+1}(\infty) - 1 = 0 . \qquad (A.4.6)$$

In this limit, the reaction is saturated and becomes independent of intensity I^*. If $I^* \ll 1$ we find

$$x_A(\infty) \approx \left(\frac{k_d^* I^*}{\mu^\mu (k_d^* k_{rec}^* + k_{rec}^* + k_{rec}^{*\prime})}\right)^{1/(\mu+1)} \ll 1 . \qquad (A.4.7)$$

A.5: The $\mathscr{F}$-Function

The temperature distribution along the axis of laser-beam propagation (z-direction) is determined by the $\mathscr{F}$-function only. This function depends on the absorption coefficient α^*, the heat loss described by η^*, and the thickness of the substrate (Fig. 6.1.1). Interference effects are ignored (Chap. 8). The

A.5: The $\mathscr{F}$-Function

$\mathscr{F}$-function can be written in the form

$$\mathscr{F}(z^*, t_1^*) = \alpha^* \sum_{n=-\infty}^{\infty} F_n(z^*, t_1^*) =$$

$$\alpha^* \left[F_0(z^*, t_1^*) + 2 \sum_{n=1}^{\infty} F_n(z^*, t_1^*) \right] \tag{A.5.1}$$

with

$$F_n(z^*, t_1^*) = A_n f_n(z^*, t_1^*), \tag{A.5.2}$$

$$f_n(z^*, t_1^*) = B_n Z_n(z^*) \exp(-v_n^2 t_1^*),$$

$$= B_n \left[\cos(v_n z^*) + \frac{\eta^*}{v_n} \sin(v_n z^*) \right] \exp(-v_n^2 t_1^*) \tag{A.5.3}$$

$$B_n = \frac{v_n^2}{h_s^*(v_n^{*2} + \eta^{*2}) + 2\eta^*}.$$

v_n are the roots of

$$\tan(h_s^* v_n) = \frac{2\eta^* v_n}{v_n^2 - \eta^{*2}}. \tag{A.5.4}$$

The coefficients A_n in (A.5.2) are given by

$$A_n = \int_0^{h_s^*} Z_n(z_1^*) \exp(-\alpha^* z_1^*) dz_1^*$$

$$= \frac{1}{\alpha^{*2} + v_n^2} \left\{ (\alpha^* + \eta^*) \left[1 - \cos(v_n h_s^*) \exp(-\alpha^* h_s^*) \right] \right.$$

$$\left. + \frac{v_n^2 - \alpha^* \eta^*}{v_n} \sin(v_n h_s^*) \exp(-\alpha^* h_s^*) \right\}. \tag{A.5.5}$$

Subsequently, we discuss some limiting cases for infinite slabs and semi-infinite substrates.

A.5.1 Axial Temperature Distribution for Infinite Slabs

Case 1: $\alpha^* \neq \infty$, $\eta^* = 0$

For finite absorption we obtain from (A.5.4) in the absence of heat losses for a slab of thickness h_s

$$\tan(v_n h_s^*) = 0 \quad \text{or} \quad v_n h_s^* = n\pi \quad \text{with } n = 0, \pm 1, \pm 2, \ldots$$

Thus, (A.5.3 and 5) yield

$$f_n(z^*, t_1^*) = \frac{1}{h_s^*} \cos\left(n\pi \frac{z^*}{h_s^*}\right) \exp\left(-\frac{n^2 \pi^2}{h_s^{*2}} t_1^*\right),$$

$$A_n = \frac{\alpha^* h_s^{*2}}{\alpha^{*2} h_s^{*2} + n^2 \pi^2} \left[1 - (-1)^n \exp(-\alpha^* h_s^*)\right].$$

The $\mathcal{F}$-function can then be written as

$$\mathcal{F}(z^*, t_1^*) = \frac{1}{h_s^*} \left\{ \left[1 - \exp(-\alpha^* h_s^*)\right] + \right.$$

$$+ 2 \sum_{n=1}^{\infty} \frac{\alpha^{*2} h_s^{*2}}{\alpha^{*2} h_s^{*2} + n^2 \pi^2} \left[1 - (-1)^n \exp(-\alpha^* h_s^*)\right]$$

$$\left. \times \cos\left(n\pi \frac{z^*}{h_s^*}\right) \exp\left(-\frac{n^2 \pi^2}{h_s^{*2}} t_1^*\right) \right\}. \quad (A.5.6)$$

In the limit $t_1^* \to \infty$ $(t_1^* \gg h_s^{*2}/\pi^2)$ we obtain

$$\mathcal{F}(z^*, t_1^*) \to \mathcal{F} = \frac{1}{h_s^*}[1 - \exp(-\alpha^* h_s^*)].$$

Case 2: $\alpha^* \to \infty$, $\eta^* \neq 0$

For surface absorption and finite heat losses we obtain from (A.5.5) $\lim_{\alpha^* \to \infty}[\alpha^* A_n] = 1$ and thus

$$\mathcal{F}(z^*, t_1^*) = \sum_{n=-\infty}^{\infty} f_n(z^*, t_1^*) \quad (A.5.7)$$

where f_n is given by (A.5.3).

Case 3: $\alpha^* \to \infty$, $\eta^* = 0$

With surface absorption and no heat losses, the $\mathcal{F}$-function becomes

$$\mathcal{F}(z^*, t_1^*) = \frac{1}{h_s^*} \theta_3^J \left[\frac{\pi z^*}{2 h_s^*} | \exp\left(-\frac{\pi^2 t_1^*}{h_s^{*2}}\right)\right] \quad (A.5.8)$$

where θ_3^J is the Jacobian theta function (Appendix A.3).

A.5.2 Axial Temperature Distribution for Semi-infinite Substrates

To obtain the temperature distribution along the z-axis for semi-infinite substrates we have to consider the $\mathcal{F}$-function (A.5.1) in the limit $h_s^* \to \infty$. This yields

$$\mathcal{F}(z^*, t_1^*) = \frac{1}{2} \alpha^* \exp(\alpha^{*2} t_1^*) \left[\exp(\alpha^* z^*) \operatorname{erfc}\left(\alpha^* t_1^{*1/2} + \frac{z^*}{2 t_1^{*1/2}}\right)\right.$$

A.5: The $\mathscr{F}$-Function

$$+ \exp(-\alpha^* z^*) \operatorname{erfc}\left(\alpha^* t_1^{*1/2} - \frac{z^*}{2t_1^{*1/2}}\right)\Bigg]$$

$$-\frac{\alpha^* \eta^*}{\sqrt{\pi}(\eta^* - \alpha^*)} \int_0^{t_1^*} \frac{dt_2^*}{(t_1^* - t_2^*)^{1/2}} \exp\left(-\frac{z^{*2}}{4(t_1^* - t_2^*)}\right)$$

$$\times \Bigg[\eta^* \exp(\eta^{*2} t_2^*) \operatorname{erfc}(\eta^* t_2^{*1/2})$$

$$- \alpha^* \exp(\alpha^{*2} t_2^*) \operatorname{erfc}\left(\alpha^* t_2^{*1/2}\right)\Bigg]. \tag{A.5.9}$$

We now discuss some special cases of (A.5.9).

Case 1: $\eta^* = 0$

In the absence of heat losses (A.5.9) yields

$$\mathscr{F}(z^*, t_1^*) = \frac{1}{2}\alpha^* \exp(\alpha^{*2} t_1^*) \Bigg[\exp(\alpha^* z^*) \operatorname{erfc}\left(\alpha^* t_1^{*1/2} + \frac{z^*}{2t_1^{*1/2}}\right)$$

$$+ \exp(-\alpha^* z^*) \operatorname{erfc}\left(\alpha^* t_1^{*1/2} - \frac{z^*}{2t_1^{*1/2}}\right)\Bigg]. \tag{A.5.10}$$

Case 2: $\alpha^* \to \infty$, $\eta^* \neq 0$

With surface absorption and finite heat losses (A.5.9) yields

$$\mathscr{F}(z^*, t_1^*) = \frac{1}{(\pi t_1^*)^{1/2}} \exp\left(-\frac{z^{*2}}{4t_1^*}\right)$$

$$-\frac{\eta^*}{\sqrt{\pi}} \int_0^{t_1^*} dt_2^* \frac{1}{(t_1^* - t_2^*)^{1/2}} \exp\left(-\frac{z^{*2}}{4(t_1^* - t_2^*)}\right)$$

$$\times \Bigg[\frac{1}{(\pi t_2^*)^{1/2}} - \eta^* \exp(\eta^{*2} t_2^*) \operatorname{erfc}(\eta^* t_2^{*1/2})\Bigg] \tag{A.5.11}$$

where we have used the approximation

$$\operatorname{erfc}\left(\alpha^* t_1^{*1/2} + \frac{z^*}{2t_1^{*1/2}}\right) \approx \frac{1}{\sqrt{\pi}} \frac{1}{\alpha^* t_1^{*1/2} + \frac{z^*}{2t_1^{*1/2}}}$$

$$\times \exp\left(-\alpha^{*2} t_1^* - \alpha^* z^* - \frac{z^{*2}}{4t_1^*}\right)$$

Case 3: $\alpha^* \to \infty$, $\eta^* = 0$

With surface absorption one obtains from (A.5.11) in the absence of heat losses

$$\mathscr{F}(z^*, t_1^*) = \frac{1}{(\pi t_1^*)^{1/2}} \exp\left(-\frac{z^{*2}}{4t_1^*}\right). \tag{A.5.12}$$

This equation can also be obtained from (A.5.8) with $h_s^* \to \infty$. All terms in the Jacobian theta function vanish, except that for $n = 0$.

Appendix B: Tabular Presentation of the Materials Investigated

The tabular presentation summarizes the materials and chemical systems investigated, mainly after 1985. Systems without reference were investigated before 1986. The corresponding references can be found in *D. Bäuerle: Chemical Processing with Lasers, Springer Ser. Mater. Sci., Vol. 1* (Springer, Berlin, Heidelberg 1986), or in *D. J. Ehrlich* and *J.Y. Tsao: Laser Microfabrication – Thin Film Processes and Lithography* (Academic Press, London 1989).

The presentation is subdivided into fields of practical importance. Precursors, and atmospheres employed in processing applications are placed in square brackets.

B.1: Ablation of Inorganic Materials

Al_2O_3	*Rothenberg* and *Kelly* (1984)
$BaTiO_3$	*Bäuerle* et al. (1988)
Bi-Ca-Sr-Cu-O	*Bäuerle* et al. (1990b); *Kullmer* and *Bäuerle* (1988b); *Scheuermann* et al. (1987)
CdTe	*Brewer* et al. (1991)
Glasses (*borosilicate, Pyrex, SiO_2, sodiumtrisilicate, etc.*)	*Braren* and *Srinivasan* (1988); *Eschbach* et al. (1989); *Hogan* and *Lunney* (1988); *Ihlemann* et al. (1992); *Sugioka* et al. (1993, 1995)
$LiNbO_3$	*Beuermann* et al. (1990); *Eyett* and *Bäuerle* (1987); *Omori* and *Inoue* (1992)
Metals (Au, Cu, In, Mo, Ni, W)	*Preuss* et al. (1995)
MgO	*Dirnberger* et al. (1993)
Mn-Zn, Ni-Zn (*ceramic*)	*Tam* et al. (1991)
$PbTi_{1-x}Zr_xO_3$ (PZT)	*Eyett* et al. (1987, 1986)
Si	*Horiike* et al. (1987)
Y-Ba-Cu-O	*Heitz* et al. (1990); *Inam* et al. (1987); *Pandey* et al. (1988); *Proyer* et al. (1994); *Schwab* et al. (1991)

B.2: Ablation of Organic Polymers and Biological Materials

AdGC	
NC	
PC	*Srinivasan* (1994)
PEEK	*Dyer* et al. (1990b); *Sumiyoshi* et al. (1994)
PES	*Niino* and *Yabe* (1992a); *Sumiyoshi* et al. (1994)
PET	*Dyer* et al. (1989); *Srinivasan* (1994)
PI	*Brannon* (1989); *Heszler* et al. (1989); *Küper* et al. (1993); *Phillips* et al. (1992); *Srinivasan* (1994); *Zhang* et al. (1993)
PMMA	*Chuang* et al. (1988); *Costela* et al. (1995); *Goodall* et al. (1986); *Horiike* et al. (1987); *Küper* and *Stuke* (1987); *Preuss* and *Stuke* (1993); *Srinivasan* (1994)
PTFE	*Basting* et al. (1991); *Brannon* et al. (1991); *Brannon* (1989); *Egitto* and *Davis* (1992); *Haba* et al. (1995); *Küper* and *Stuke* (1989); *Preuss* et al. (1993); *Wada* et al. (1993)
BIOLOGICAL MATERIALS	*Asshauer* et al. (1994); *Cross* et al. (1988); *Deutsch* (1991); *Esenaliev* et al. (1989); *Fischer* et al. (1994); *Furzikov* (1987); *Husinsky* et al. (1989); *Kautek* et al. (1994); *Kitai* et al. (1991); *Lambda Highlights* (1990); *Oraevsky* et al. (1991, 1992); *Rink* et al. (1994a, b); *Schomacker* et al. (1991); *Srinivasan* et al. (1987); *Srinivasan* (1986); *Verdaasdonk* and *Borst* (1991); *Vogel* et al. (1996)

B.3: Materials Etching

If not otherwise indicated, wet etching is performed in *aqueous* solutions.

Ag [Cl_2]	*Mogyórosi* (1989); *Sesselmann* and *Chuang* (1987)
Al [Cl_2]	*Koren* et al. (1986)
Al[$HNO_3 + H_3PO_4 + K_2Cr_2O_7$]	
Al$_2$O$_3$ [H_3PO_4]	*Sugioka* et al. (1991)

Al₂O₃/TiC(*ceramic*) [CF₄, CF₃Cl, CCl₄, SF₆; KOH]
Au [Cl₂]
BaTiO₃ [H₂]
C [H₂] — Rothschild et al. (1986)
CdS [KCl, KBr, KI, HCl + HNO₃, H₂SO₄ + H₂O₂]
Cu [Cl₂] — Brannon and Brannon (1989); Chuang (1987); van Veen et al. (1988)

Cu [HCl, Br₂ + KBr]
CuCl [Cl₂] — Sesselmann et al. (1986b)
Fe [Cl₂]; **Fe$_x$Ni$_y$** [SF₆, CF₄, CCl₄]
Fe (*Stainless steel*) — Datta et al. (1987)
[NiCl₂; NaCl, NaNO₃, K₂SO₄]
Ferrite (MnO: ZnO: Fe₂O₃) [CCl₄, CCl₂F₂, CF₄, SF₆, CF₃Cl] — Y. F. Lu et al. (1991a); Takai et al. (1988a)
Ferrite (MnO: ZnO: Fe₂O₃; Fe:Al:Si) [KOH, H₃PO₄] — Takai et al. (1994)
GaAs [Cl₂, Br₂, HCl, HBr, O₃, CCl₄, CH₃Cl, CH₃Br, CF₃Br, CF₃I] — Berman (1991); Brewer et al. (1986); Foulon and Green (1995); Haase et al. (1992); Heydel et al. (1993); Koren and Hurst (1988); Matz et al. (1990); Meiler et al. (1989); Takai et al. (1988b)

GaAs [HF/H₂O, HCl, HCl + HNO₃, HI, HNO₃, H₂SO₄; H₂SO₄ + H₂O₂; H₃PO₄] — Matz and Zirrgiebel (1988); Podlesnik et al. (1986); Ruberto et al. (1991); Svorcik et al. (1988); Willner et al. (1990)

GaAs [KOH; NaOH] — Lee et al. (1993)
GaAs [NaBr, NaI, KBr, KI, CsBr, CsI + Br₂, I₂/H₂O, H₂SO₄ + NaSCN]
GaAs$_{1-x}$P$_x$ [HCl/He]
Ga$_x$Al$_{1-x}$As [H₃PO₄ + H₂O₂ + CH₃OH]
Ga$_{0.47}$In$_{0.53}$As [KOH, KOH/C₂H₅OH] — Moutonnet (1987)
Ge [Br₂, CF₃I]
Glass (*Pyrex, Corning,* BK-7) [H₂; HF; CF₂Br₂]
InP [Cl₂, HCl, HBr, CCl₄; CH₃Cl; CH₃Br; CF₃I] — Donnelly and Hayes (1990); Matz et al. (1990); Takai et al. (1988b)
InP [HCl + HNO₃, H₃PO₄, HNO₃ + HCl; FeCl₃; KOH/H₂O, C₂H₅OH] — Moutonnet (1986)
InSb [CCl₄] — Takai et al. (1988b)
LiNbO₃ [Cl₂, KF] — Ashby and Brannon (1986);

Mo [Cl_2, NF_3] Beeson et al. (1988)
Mogyorósi et al. (1989a);
Rothschild et al. (1987)

Ni [Cl_2]
Ni [$NiSO_4$]
Ni$_x$Fe$_{1-x}$ [Cl_2, CF_4, CCl_4, SF_6]
PbTi$_{1-x}$Zr$_x$O$_3$ (PZT, PXE)   Eyett et al. (1986)
[H_2, KOH]
Si [Cl_2, Br_2, HCl, NF_3, XeF_2,   Brannon (1988); Ehrlich (1993); Horiike
  COF_2/He, $CF_4 + O_2$, SF_6]   et al. (1987); Houle (1989); Konuma
et al. (1989); Kullmer and Bäuerle
(1987, 1988a); Mogyorósi et al. (1988);
Reksten et al. (1986); Treyz et al. (1988);
Watanabe et al. (1986)

Si [HF, HF + HNO_3, NaOH, KOH]   Grebel and Fang (1995)
SiC [HF/H_2O]   Zhang et al. (1990)
Si$_x$N$_y$ [H_2O]   Morita et al. (1988)
SiO$_2$ [H_2, Cl_2, NF_3/N_2, C_2F_4,   Agrawalla et al. (1987); Pan and Chen
  CF_2Cl_2, CF_2Br_2, CF_3Br, CDF_3]   (1988)
SiO$_2$ [H_2SO_4]
SrTiO$_3$ [H_2]
Steel [$NiCl_2$]
Ta [SF_6, XeF_2]
Ti [Br_2, NF_3, CCl_3Br]   Tyndall and Moylan (1990)
W [Cl_2, I_2, COF_2/He, air]   Koren (1986); Rothschild et al. (1987)
ZnSe [HCl + HNO_3]

B.4: LCVD of Microstructures

Elements and compounds deposited by laser-CVD, mainly as microstructures. The carrier gases mainly employed are H_2, He, and Ar.

Al [$Al_2(CH_3)_6$; $Al(C_4H_9)_3$;   Foulon and Stuke (1993);
  $AlH_3(CH_3)_3N$; $AlH_3(C_2H_5)N$; AlI_3]   Frugier et al. (1993);
Han et al. (1994);
Tonneau et al. (1994)

Au [$Au(CH_3)_2$(acac); $Au(CH_3)_2$ Baum et al. (1991); Baum (1987);
  (tfacac); $Au(CH_3)_2$(hfacac);   Ganz (1988); Jubber et al. (1989);
  $Au(CH_3)[(C_2H_5)_3P]$]   Metzger and Reichl (1993);
Morishige and Kishida (1994)

B [$BCl_3 + H_2$]   Boman and Bäuerle (1995);
Johansson et al. (1992);
Wallenberger et al. (1994)

C [C_2H_2/H_2, He, Ar; C_2H_4; CH_4]	*Doppelbauer* and *Bäuerle* (1986)
Cd [$Cd(CH_3)_2$]	
CoO [$Co(acac)_3/N_2$]	
Cr [$Cr(CO)_6$]	
Cu [$Cu(hfacac)_2$; Cu(hfac) TMVS]	*Han* and *Jensen* (1994); *Markwalder* et al. (1989); *Moylan* et al. (1986); *Preuss* and *Stafast* (1992); *Widmer* and *v.d. Bergh* (1995)
Fe [$Fe(CO)_5$]	*Jackman* et al. (1986); *Swanson* et al. (1987)
GaAs [$Ga(CH_3)_3 + AsH_3$]	*Aoyagi* et al. (1987); *Bedair* et al. (1986); *Doi* et al. (1986); *Karam* et al. (1986)
GaP [$Ga(CH_3)_3 + P(t\text{-}C_4H_9)_3 + H_2$]	*Solanki* et al. (1988)
Ge [GeH_4]	
In [InI]	
Mn [$Mn_2(CO)_{10}$]	
Mo [$Mo(CO)_6$]	*Gilgen* et al. (1987)
Ni [$Ni(CO)_4$]	*Bezuk* et al. (1987); *Boughaba* and *Auvert* (1993); *Tonneau* et al. (1989)
Os [$OsH_2(PF_3)_4$]	*Ganz* (1988)
Pd [Pd(hfacac); Pd-Allyl]	*Ganz* (1988)
Pt [$Pt(hfacac)_2$; $Pt(PF_3)_4$]	*Braichotte* et al. (1990); *Braichotte* and *v.d. Bergh* (1989); *Gilgen* et al. (1987)
Si [SiH_4; $SiCl_4$]	*Bäuerle* (1985); *Nagahori* and *Matsumoto* (1989); *Nordine* et al. (1993); *Tonneau* et al. (1987); *Westberg* et al. (1993)
Si-doped [$SiH_4 + BCl_3$, $B(CH_3)_3$, B_2H_6; $SiH_4 + Al_2(CH_3)_6$, PH_3]	*Herman* et al. (1986)
SiO_2, SiO_x [SiH_4, $Si_2H_6 + N_2O$]	*Hiura* et al. (1991)
Sn [$Sn(CH_3)_4$; $Sn(C_2H_5)_4$]	
SnO_2 [$(CH_3)_2SnCl_2 + O_2$]	
Ti [$TiCl_4$]	
TiC, TiO_2, $TiSi_2$ [$TiCl_4 + CH_4$; $TiCl_4 + H_2 + CO_2$; $TiCl_4 + SiH_4$]	*Reisse* et al. (1993)
TiN [$TiCl_4 + N_2/H_2$]	*Silvestre* et al. (1994)
W [$WF_6 + H_2$; $WF_6 + SiH_4$; $WF_6 + Si_2H_6$; $WCl_6 + H_2$/He, Ar, Kr, Xe]	*Bäuerle* (1990); *Black* et al. (1990); *Grossman* and *Karnezos* (1987); *Kullmer* et al. (1992); *Lecours* et al. (1993); *Meunier* et al. (1994); *Szörenyi* et al. (1988); *Toth* et al. (1992);

W [$W(CO)_6$]   *Westberg* et al. (1993); *Zhang* et al. (1987)
Gilgen et al. (1987)
WSi$_2$, WSi$_x$ [$WF_6 + SiH_4$]   *Desjardins* et al. (1993)
Zn [$Zn(CH_3)_2$; $Zn(C_2H_5)_2$]

B.5: Thin-Film Formation by LCVD

For laser-CVD of microstructures see Appendix B.4.

Al [$Al_2(CH_3)_6/H_2$; $AlH(CH_3)_2$]   *Hanabusa* and *Ikeda* (1991); *Hanabusa* et al. (1989)

Al$_2$O$_3$ [$Al_2(CH_3)_6 + N_2O/Ar$, He]
Au [$(CH_3)_3Au - P(CH_3)_3$]
C (*amorphous, graphite, diamond-like*)   *Kitahama* et al. (1986);
[C_2H_2/H_2; C_2H_5Cl/Ar, N_2]   *Tachibana* et al. (1988)
CdTe [$Cd(CH_3)_2 + Te(C_2H_5)_2$,   *Bicknell* et al. (1986);
$Te(CH_3)_2$]   *Irvine* et al. (1989); *Zinck* et al. (1988)
Co [$Co_2(CO)_8$]   *Schulmeister* et al. (1992)
Cr [$Cr(CO)_6/Ar$, He]   *Houle* and *Yeh* (1992); *Konstantinov* et al. (1988); *Okada* et al. (1992)

Cr + Mo [$Cr(CO)_6 + Mo(CO)_6$]   *Okada* et al. (1994)
Cu [$Cu(hfacac)_2$]   *Moylan* et al. (1986)
Fe [$Fe(CO)_5$]
GaAs [$Ga(CH_3)_3 + AsH_3/H_2$]   *Aoyagi* et al. (1990); *Balk* et al. (1987); *Chu* et al. (1988); *Kukimoto* et al. (1986); *Nishizawa* et al. (1986)

GaP [$Ga(CH_3)_3 [(C_4H_9)_3P]$]   *Sudarsan* et al. (1990)
Ge [GeH_4/He]   *Kiely* et al. (1989a, b); *Tavitian* et al. (1988)

a-Ge:H [$GeH_4 + SF_6$]   *Barth* et al. (1994)
Ge$_x$Si$_y$ [$GeH_4 + Si_2H_6/He$]   *Burke* et al. (1989)
HgTe [$Te(C_2H_5)_2/H_2$ + Hg vapor]   *Fujita* et al. (1989)
Hg$_{1-x}$Cd$_x$Te, HgTe/CdTe   *Ahlgren* et al. (1988);
[$Hg(CH_3)_2 + Cd(CH_3)_2$ *Morris* (1986)
$+ Te(CH_3)_2/He$; $Te(C_2H_5)_2$]
In [$In(CH_3)_3$, $In(C_5H_5)_2$]
InP [$(CH_3)_3InP(CH_3)_3$ *Donnelly* et al. (1986)
$+ P(CH_3)_3/He + H_2$]
InSb [$In(CH_3)_3 + Sb(CH_3)_3$]   *Zuhoski* et al. (1988)
In$_2$O$_3$ [$(CH_3)_3InP(CH_3)_3 + P(CH_3)_3$
$+ O_2$ or H_2O vapor/H_2, He]

Mo [$Mo(CO)_6$/He] — Flynn et al. (1986)
Pb [$Pb(C_2H_5)_4$]
Se [$Se(CH_3)_2$/Ar]
Si [SiH_4; Si_2H_6/H_2; SiH_2Cl_2/H_2] — Meguro et al. (1988b); Yamada et al. (1989)

a-Si:H (deposited by excimer lasers or UV lamps) [SiH_4/He; Si_2H_6/He, Ar, H_2; Si_3H_8/He, Ar] — Dietrich et al. (1989); Eres et al. (1988); Kawasaki et al. (1988); Kim et al. (1989); Kumata et al. (1986); Mizukawa et al. (1989)

a-Si:H (deposited by CO_2 lasers) [SiH_4/Ar, He, H_2, N_2; Si_2H_6; $SiH_4 + B_2H_6$, PH_3; SiH_2Cl_2/N_2; $RSiH_3(R=C_2H_5, C_4H_9, C_6H_5, SiH_3)$] — Golusda et al. (1993, 1992); Metzger et al. (1988); Meunier et al. (1987a,b); Tonneau et al. (1986)

a-Si/a-Ge; a-Si/a-Si_3N_4 [Si_2H_6, GeH_4; $Si_2H_6 + NH_3$] — Lowndes et al. (1988)

a-Si:H/a-$Al_{1-x}O_x$ [Si_2H_6/H_2, $Al_2(CH_3)_6 + O_2$] — Uwasawa et al. (1991)

SiO_2 [$SiH_4 + N_2O/N_2$, He; $SiH_4 + O_2 + N_2/H_2$; $SiH_6 + N_2O$; TEOS] — Fernandez et al. (1994); Inushima et al. (1988); Klumpp and Sigmund (1989); Marks and Robertson (1988); Scoles et al. (1988); Shirafuji et al. (1988); Szörényi et al. (1994)

Si_3N_4 [$SiH_4 + NH_3$/Ar, N_2; $Si_2H_6 + NH_3$/He] — Inushima et al. (1988); Petitjean et al. (1992); Sugii et al. (1988)
SnO_2 [$Sn(CH_3)_4$] — Larciprete et al. (1993)
Ta_2O_5 [$Ta(OC_2H_5)_5 + N_2O$] — Nishimura et al. (1993)
Ti [$TiCl_4$; $TiBr_4$/Ar] — Chou et al. (1989); Kubát and Engst (1993); Lavoie et al. (1991b)

TiB_2 [$TiCl_4 + BCl_3$] — Elders and v.Voorst (1994)
TiC, TiN [$TiCl_4 + H_2 + C_2H_4$, NH_3] — Cao et al. (1995); Conde et al. (1992)
TiO_2 [$TiCl_4 + H_2 + CO_2$]
$TiSi_2$ [$TiCl_4 + SiH_4$]
Tl [TlI, TlBr]
W [$WF_6 + H_2$/Ar] — van Maaren et al. (1991)
W [$W(CO)_6$/He] — X. Lu et al. (1991)
Zn [$Zn(CH_3)_2$/Ar, He]
ZnO [$Zn(CH_3)_2 + NO_2$, N_2O/He]
Zn_xSe_y [$Zn(CH_3)_2 + Se(CH_3)_2$/Ar] — Shinn et al. (1989)
ZnTe [$Zn(CH_3)_2$, $Zn(C_2H_5)_2$] — Ikejiri et al. (1993)

B.6: Deposition From Adsorbed Layers, Laser-MBE, Laser-ALE

Al [Al$_2$(CH$_3$)$_6$; Al(C$_4$H$_9$)$_3$]   *Foulon* and *Stuke* (1993); *Hanabusa* et al. (1989); *Higashi* (1989)
Al$_2$O$_3$ [Al$_2$(CH$_3$)$_6$ + N$_2$O]   *Ishida* et al. (1989)
Al$_x$Ti$_y$ [Al$_2$(CH$_3$)$_6$ + TiCl$_4$]
Cd [Cd(CH$_3$)$_2$]
CdTe [Cd(CH$_3$)$_2$ + Te(C$_2$H$_5$)$_2$]   *Irvine* et al. (1989)
Fe [Fe(CO)$_5$]   *Swanson* et al. (1987)
GaAs [Ga(C$_2$H$_5$)$_3$ + As$_4$, AsH$_3$]   *Donnelly* and *McCaulley* (1989); *Donnelly* et al. (1988); *Isshiki* et al. (1993); *McCaulley* et al. (1989)
HgTe/CdTe   *Cheung* and *Madden* (1987)
Mo [Mo(CO)$_6$]   *Radloff* et al. (1990)
Ni [Ni(CO)$_4$]
Pb [Pb(C$_2$H$_5$)$_4$]
PMMA [MMA]
Pt [Pt(hfacac)$_2$]   *Braichotte* and *v.d. Bergh* (1988)
Si [Si$_2$H$_6$]   *Tanaka* et al. (1987)
Ti [TiCl$_4$]   *Lavoie* et al. (1991b)
W [W(CO)$_6$]   *Radloff* et al. (1990)
Zn [Zn(CH$_3$)$_2$; Zn(C$_2$H$_5$)$_2$]   *Krchnavek* et al. (1987)

B.7: Deposition from Liquids

If not otherwise indicated, aqueous solutions have been employed.

Ag [AgNO$_3$/N-methyl-2-pyrrolidinone sol.; AgNO$_3$ + NH$_3$ + KNaC$_4$H$_4$O$_6$; AgCF$_3$SO$_3$, AgPF$_6$, AgBF$_4$/ C$_6$H$_5$CH$_3$, CH$_3$CN]   *Krabe* and *Radloff* (1988); *Montgomery* and *Mantei* (1986)
Al [AlH$_3$·N(C$_2$H$_5$)$_3$ (TEAA)]   *Lehmann* and *Stuke* (1992)
Au [Triphenyl-phosphine complexes; Gold-cyanide sulfite sol.; K [Au(CN)$_2$]; HAuCl$_4$/CH$_3$OH]   *Brook* et al. (1991); *Gelchinski* et. al. (1987); *Jacobs* and *Nillesen* (1990); *Nanai* et al. (1989a); *Sugioka* and *Toyoda* (1992)
C   *Singh* et al. (1993)
Cd [CdSO$_4$/H$_2$SO$_4$]
Cr [Cr(C$_6$H$_6$)$_2$]
Cu [CuSO$_4$/H$_2$SO$_4$; CuSO$_4$/H$_2$SO$_4$, HCl; CuSO$_4$/HF; CuSO$_4$/NaOH;   *Mini* et al. (1994); *Nanai* et al. (1989a); *Niino* and *Yabe* (1993b);

$Cu_2P_2O_7$; $CuClO_4/CH_3CN$]   Niino and Yabe (1992b)
Mo [$Mo(C_6H_6)_2/C_6H_6$;   Geretovszky et al. (1994)
$(NH_4)_6Mo_7O_{24}$]
Ni [$NiCl_2$; $NiSO_4$; $Ni(NH_2SO_3)_2$]   Y.-F. Lu et al. (1991b); Niino and Yabe
(1993a,b); Niino and Yabe (1992b)
Pd [Palladium cyanide; H_2PtCl_6]
Zn [$ZnSO_4/H_2SO_4$]

B.8: Formation of Thin Films and Heterostructures by PLD

Classification according to metals, semiconductors and insulators, and high-temperature superconductors

Metals

Ag, Bi, Cd, Cr, Dy, Fe, Gd, Hf, In, Mo, Nb, Sn, Ta, Ti, Te, W, Zr	Bykovskii et al. (1978)
Ag-Ni	van Ingen et al. (1994)
Al, Sb, Sn, Ti	Gaponov et al. (1979a)
Cr	Klimer (1973)
Cu-Ni	van Ingen (1994)
Er	Osterreicher et al. (1978)
Fe	Krebs and Bremert (1993)
In, Ta	Gaponov et al. (1979a)
Ir	Samson et al. (1967)
Mo	Gaponov et al. (1977)
Ni_3Mn	Desserre and Eloy (1975)
Os	Maier-Komor (1979)
Pt	Cillessen et al. (1993a); Hess and Milkosky (1972)
$ReBe_{22}$	Desserre and Eloy (1975)
Ru	Kliwer (1973)
Ti/TiC	Scheibe et al. (1990)
W	Schwartz and Tourtellotte (1969)

Semiconductors and Insulators

Al_xN_y	Norton et al. (1991)
Al_2O_3	Ban and Kramer (1970); Bykovskii et al. (1978)

As$_2$S$_3$	*Smith* and *Turner* (1965)
BaF$_2$	*Sankur* (1984)
Ba$_{1-x}$K$_x$BiO$_3$	*Moon* et al. (1991)
Ba$_x$Sr$_{1-x}$TiO$_3$	*Kobayashi* and *Kobayashi* (1994); *Quadri* et al. (1995)
BaTiO$_3$	*Davis* and *Gower* (1989); *Koinkar* and *Ogale* (1991); *Nashimoto* et al. (1992); *Yeh* et al. (1993)
BeO	*Mitrikov* et al. (1984)
Bi$_2$Te$_3$	*Bykovskii* et al. (1974)
Bi/CdTe	*Gaponov* et al. (1980)
BN	*Ballal* et al. (1993); *Friedmann* et al. (1994)
C (including diamond)	*Fabisiak* et al. (1991); *Müller* et al. (1993); *Müller* and *Mann* (1992); *Murray* and *Peeler* (1993); *Rengan* and *Narayan* (1992); *Sato* et al. (1988); *Scheibe* et al. (1992)
CaF$_2$	*Sankur* et al. (1987)
Ca$_{10}$(PO$_4$)$_6$(OH)$_2$	*Cotell* (1993)
CaTiO$_3$	*Ban* and *Kramer* (1970)
CeF$_3$	*Sankur* (1984)
Cd$_3$As$_2$	*Dubowski* and *Williams* (1984)
CdCr$_2$S$_4$, CdCr$_2$Se$_4$	*Ban* and *Kramer* (1970)
CdS	*Pashmakov* et al. (1985)
CdS/Y-Ba-Cu-O	*Shi* et al. (1991)
CdSe	*Ban* and *Kramer* (1970)
CdS$_x$Se$_{1-x}$	*Kwok* et al. (1988)
CdSnAs$_2$, CdTe, InAs, PbTe	*Gaponov* et al. (1981)
CdTe	*Cheung* et al. (1986); *Dubowski* et al. (1985)
Cu$_x$O$_y$	*Ogale* et al. (1992)
Fe$_x$O$_y$	*Ogale* et al. (1988)
GaAs	*Baleva* et al. (1986); *Gaponov* et al. (1981)
GaP	*Gaponov* et al. (1981)
GaSb	*Ban* and *Kramer* (1970)
Ge	*Cheung* and *Sankur* (1992); *Lubben* et al. (1985)
HfC	*Desserre* and *Eloy* (1975)
HfO$_2$	*Sankur* (1984)
Hg$_{1-x}$Cd$_x$Te	*Cheung* and *Sankur* (1992)
HgTe	*Cheung* et al. (1986)
HgTe/CdTe	*Cheung* et al. (1986)
HgCdTe$_{x1}$/HgCdTe$_{x2}$	*Cheung* et al. (1988)

InSb	*Dimitrov* et al. (1982); *Gaponov* et al. (1981); *Sheftal* and *Cherbakov* (1981)
InSb/CdTe, InSb/PbTe	*Gaponov* et al. (1980)
InSnO (ITO)	*Zheng* and *Kwok* (1993)
$KTa_{1-x}Nb_xO_3$	*Yilmaz* et al. (1991)
$LaAlO_3$	*Groh* (1968)
LaF_3	*Sankur* et al. (1987)
La_2O_3	*Sankur* and *Hall* (1985)
$La_{1-x}Sr_xCoO_3$	*Cillessen* et al. (1993b)
$LiNbO_3$	*Afonso* et al. (1993)
$LiCoO_2$	*Antaya* et al. (1993)
$MgAl_2O_4$	*Ban* and *Kramer* (1970); *Hass* and *Ramsey* (1969)
MgF_2	*Sankur* et al. (1987)
MgO	*Sankur* and *Hall* (1985)
MoO_3	*Smith* and *Turner* (1965)
MoS_2	*Donley* et al. (1991)
Na_3AlF_6, NaF	*Sankur* (1984)
Nb_2O_5	*Sheftal* and *Cherbakov* (1981)
NdCeCuO	*Kussmaul* et al. (1992)
NdF_3	*Sankur* (1984)
PbC_{12}	*Smith* and *Turner* (1965)
$Pb_{1-x}Cd_xSe$	*Baleva* et al. (1986)
PbF_2	*Sankur* (1986)
$Pb_5Ge_3O_{11}$	*Peng* et al. (1992)
PbO_2	*Bykovskii* et al. (1978)
PbSe	*Gaponov* et al. (1977)
PbTe	*Volodin* et al. (1981)
PbTe/CdTe	*Gaponov* et al. (1979b)
$PbTiO_3$	*Tabata* et al. (1994); *Tanaka* et al. (1994)
$PbTi_{1-x}Zr_xO_3$ (PZT)	*Otsubo* et al. (1990); *Roy* et al. (1992, 1991); *Saenger* et al. (1990)
Sb_2S_3	*Schwartz* and *Tourtellotte* (1969)
Sc_2O_3	*Sankur* (1984)
Se	*Hansen* and *Robitaille* (1987)
Si	*Hanabusa* et al. (1983); *Lubben* et al. (1985)
SiO	*Hass* and *Ramsey* (1969)
SiO_2	*Slaoui* et al. (1992)
Si_3N_4	*Sankur* (1984)
SnO_2	*Koinkar* and *Ogale* (1991)
SrF_2	*Sankur* et al. (1987)
$SrTiO_3$	*Hiratani* et al. (1993); *Rao* and *Krupanidhi* (1994)
Ta_2O_5	*Sankur* and *Hall* (1985)

TiO_2	*Cheung* et al. (1986)
Ti_2O_3	*Sankur* (1984)
V_2O_5	*Sankur* and *Hall* (1985)
Y_2O_3	*Sankur* (1984)
ZnO	*Sankur* and *Cheung* (1983); *Craciun* et al. (1994)
ZnS	*Bykovskii* et al. (1978); *Myl'nikov* et al. (1978)
$ZnSe_xTe_{1-x}$	*Aydinli* et al. (1991); *Misiewicz* et al. (1993)
ZnSe/MnSe	*Misiewicz* et al. (1993)
ZrC	*Desserre* and *Eloy* (1975)
ZrO_2	*Murray* et al. (1987)

High-Temperature Superconductors

Bi-Pb-Sr-Ca-Cu-O	*Razavi* and *Habermeier* (1991)
Bi-Sr-Ca-Cu-O	*Arnold* et al. (1993); *Kanai* et al. (1989); *Kumar* et al. (1990); *Kung* and *Muenchausen* (1993); *Li* et al. (1996); *Lin* and *Wu* (1992); *Ludorf* et al. (1989); *Tabata* et al. (1989); *Viret* et al. (1993)
$Bi_4Ti_3O_{12}$/Y-Ba-Cu-O	*Buhay* et al. (1991); *Maffei* and *Krupanidhi* (1992); *Ramesh* et al. (1991, 1990)
Hg-Ba-Ca-Cu-O	*Krusin-Elbaum* et al. (1995)
Lu-Ba-Sr-Cu-O	*Schwab* et al. (1992)
RE-Ba-Sr-Cu-O	*Stangl* et al. (1996)
Tl-Ba-Ca-Cu-O	*Eddy* et al. (1991); *Piehler* et al. (1994); *R. S. Liu* et al. (1992); *Michael* et al. (1994)
Tm-Ba-Sr-Cu-O	*Stangl* et al. (1995)
Y-Ba-Cu-O	*Auciello* et al. (1988); *Bäuerle* et al. (1990b); *Basovich* et al. (1993); *von d. Burg* et al. (1992); *Chang* et al. (1990); *Char* et al. (1990); *Chiba* et al. (1991); *Eibl* and *Roas* (1990); *Foltyn* et al. (1991); *Gervais* and *Keller* (1995); *Habermeier* (1992); *Heitz* et al. (1990); *Inam* et al. (1990); *Kautek* et al. (1990); *Kilibarda* and *Ng* (1990); *Kumar* et al. (1993); *Lowndes* et al. (1992); *G. Lu* et al. (1992); *Norton* and *Lowndes* (1993); *Pinto* et al. (1992); *Roas* et al. (1989, 1988);

	Schwab et al. (1993, 1992); Schwab and Bäuerle (1991); Singh and Narayan (1990); Vase et al. (1990); Venkatesan et al. (1988); Watanabe et al. (1994); Watanabe (1994); Wiener-Avnear et al. (1990); Witanachchi et al. (1988); X. D. Wu et al. (1990a); Ying et al. (1989); S. Zhu et al. (1993)
Y-Ba-Cu-O with buffer layers	Fork et al. (1991a, b); Kumar et al. (1991); Saitoh et al. (1991)
Y-Ba-Cu-O/Y-Pr-Ba-Cu-O; Y-Ba-Cu-O/Nd-Ga-O	Venkatesan et al. (1990); X. D. Wu et al. (1990b); Yoshida et al. (1992)
Polymers	Blanchet et al. (1993); Chaudhari and Rao (1993); Hansen and Robitaille (1988)

B.9: Surface Oxidation and Nitridation

Cd [air]	Wautelet et al. (1990)
Cr [air]	Birjega et al. (1986)
Cu [air; CS_2]	Baufay et al. (1987); Wautelet and Hanus (1991)
Fe [liquid N_2; NH_3; H_2O]	Illgner et al. (1995); Ogale et al. (1987)
GaAs [O_2; N_2O; NH_3]	Z. Lu et al. (1990); Schmidt et al. (1991); Seo et al. (1990); X.-Y. Zhu et al. (1993)
Ge	Craciun et al. (1990)
(Hg, Cd)Te [N_2O]	
In [air]	Wautelet et al. (1990)
InP [air, N_2O]	
Nb [O_2; liquid N_2]	
Ni [air]	Mesarwi and Ignatiev (1989)
Si [air; O_2; $O_2 + H_2O$; N_2O; CO_2; O^+ implanted; $O_2 + Cl_2$; $H_2 + O_2$; $O_2 + NF_3$; N_2; NH_3]	Boyd et al. (1993); Cohen et al. (1986); Craciun et al. (1994); Fogarassy et al. (1988); Orlowski and Mantell (1988); Slaoui et al. (1988); Young (1988)
Sn [air]	Wautelet et al. (1990)
Te [air]	Wautelet (1989)
Ti [air; N_2; NH_3; O_2; and liquid N_2, NH_3, H_2O]	D'Anna et al. (1991); Gasser et al. (1989); Nanai et al. (1989b); Ogale et al. (1987); Ursu et al. (1986a, b)
V [air]	Nanai et al. (1993)

Y-Ba-Cu-O [air; O_2]   *Liberts* et al. (1988a); *Shen* et al. (1991); *Sobolewski* et al. (1994)
Zn [air]   *Wautelet* et al. (1990)
Zr [air; N_2]   *Ursu* et al. (1986c)
Zr [liquid N_2]   *Ursu* et al. (1986d)

B.10: Surface Modifications, Transformation of Solid Films

Ag [$AgNO_3$/polymer, $AgOOCCH_3$]   *Y.-F. Lu* et al. (1992)
Au [Au-metallo-polymer]   *Comita* et al. (1994); *Gross* et al. (1986)
$BaTiO_3$ *Bäuerle* et al. (1988)
a-C: H   *Spousta* et al. (1993)
Cu [$Cu(HCOO)_2 \cdot 2H_2O$/aq.]   *Gupta* and *Jagannathan* (1987); *Hoffmann* et al. (1989)
Ir [[$Ir_4(CO)_{11}Br$] [$N(C_2H_5)_4$]; [$Ir_6(CO)_{15}$] [$N(CH_3)_3C_8H_{17}$]]   *Hoffmann* et al. (1989)
$LiNbO_3$ *Eyett* and *Bäuerle* (1988); *Lavoie* et al. (1991a)
$Pb_{1-3y/2}La_yTi_{1-x}Zr_xO_3$ (PLZT)   *Kapenieks* et al. (1986a,b)
Pd [$Pd(OOCCH_3)_2$]   *Esrom* and *Wahl* (1989); *Gross* et al. (1987); *Zhang* and *Stuke* (1989)
PET   *Arenholz* et al. (1993); *Bahners* et al. (1990); *Heitz* et al. (1993); *Lazare* et al. (1988); *Srinivasan* (1994); *Silvain* et al. (1991)
PI   *Arenholz* et al. (1993); *Bäuerle* et al. (1996); *Kokai* et al. (1989); *Phillips* et al. (1993b); *Schumann* et al. (1991); *Srinivasan* et al. (1993)
PMMA   *Fukumura* et al. (1991)
Pt [Pt-metallo-polymer (Engelhard Bright Pt-0.5X, 7.5 wt% Pt)]   *Sausa* et al. (1987)
PTFE (Teflon)   *Murahara* (1994); *Niino* et al. (1995)
SiO_2 [$Si(OR)_x(OH)_{4-x}$]   *Kurosawa* et al. (1993)
Si_3N_4   *Donnelly* et al. (1990)
$SrTiO_3$ *Otto* et al. (1984)
V_xO_y   *Okabe* et al. (1987)
Y-Ba-Cu-O   *Aizaki* et al. (1988); *Dye* et al. (1990); *Koinuma* et al. (1988); *Liberts* et al. (1988a, b); *Minamikawa* et al. (1988); *Murakami* et al. (1987); *Rothschild* et al. (1988); *Shen* et al. (1991)
ZrO_2   *Hontzopoulos* and *Damigos* (1991)

Table I. Commercial lasers most commonly used in materials processing. Only strongest lines are listed. The wavelengths are given in nanometers, if not otherwise indicated. Within the text, laser wavelengths are sometimes rounded. Wavelengths and pulse energies (cw powers) of higher harmonics are written in parentheses.

Laser	Wavelength λ[nm]	Pulse length τ_l	Pulse energy [J]; cw Power [W]
F_2	157	ns	0.060 J; $\leq$ 3 W
ArF	193	ns	0.8 J; 75 W
KrCl	222	ns	0.2 J; $\leq$ 10 W
KrF	248	ns	2 J; 150 W
XeCl	308	ns	3 J; 200 W
XeF	351	ns	0.65 J; 70 W
N_2	337	ns	0.01 J; 0.5 W
HeCd	441.6	cw	0.2 W
Ar^+	275–306	cw	ML 1.6 W
	334–364	cw	ML 7 W
	458–515	cw	ML 25 W
	457.9 (229)	cw	1.5 W (0.035 W)
	476.5	cw	3 W
	488.0 (244)	cw	8 W (0.35 W)
	496.5	cw	3 W
	501.7	cw	1.9 W
	514.5 (257)	cw	10 W (0.5 W)
	528.7	cw	1.8 W
Kr^+	337–356	cw	ML 2 W
	413.1	cw	1.8 W
	476.2	cw	0.5 W
	520.8	cw	1 W
	530.9	cw	1.5 W
	568.2	cw	1.1 W
	647.1	cw	4 W
	676.4	cw	1 W
	752.5	cw	1.5 W
HeNe	632.8	cw	0.120 W
Ruby	694.3	ms (ns)	50 J (1 J)
Alexandrite	701–818	cw (μs, ns)	60 W (100 W, 20 W)
Ti:sapphire	720–1080	ps (fs)	1.5 W (2 W)
Nd:glass	1062.3	ms (ns)	70 J (35 J)
Nd:YAG	1064.1	cw	2 kW
		ms (ns)	100 J (2 J)
	532/355/266	ns	1 J/0.75 J/0.2 J
CO	5–7 μm	cw	25 W
CO_2	9–11 μm	cw	25 kW
		μs	100 J

Table II. Thermophysical properties of materials. Mass density ρ, glass temperature T_g, melting temperature T_m, vaporization temperature T_v (at 1013 mbar), specific heat c_p, thermal conductivity κ, and thermal diffusivity D. If not otherwise indicated in parentheses, values of ρ, c_p, κ, and D refer to $T \approx 300$ K

Material	ρ [g/cm³]	T_m [K]	T_v [K]	c_p [J/gK]	κ(T[K]) [W/cm K]	D(T[K]) [cm²/s]
Ag	10.5	1234	2483	0.23	4.28	1.72
					4.12 (500)	1.61 (500)
					3.75 (1000)	1.3 (1000)
					1.97 (2000)	
					1.91 (3000)	
Al	2.7	933	2720	0.90	2.4	1.03
					2.38 (500)	0.88 (500)
					1.07 (1500)	
AlAs	4.22	1323				
AlN	3.1	2673		0.78	2.0	
Al₂O₃	4.0	2324		0.75	0.30	
Al₂O₃ (ceramic)	3.89	2340	3800	0.9	0.30	
p-Al₂O₃					0.40	1.0
					0.20 (500)	0.048 (500)
					0.078 (1000)	0.016 (1000)
					0.06 (2000)	0.012 (2000)
AlP	3.81	> 1873			1.3	
AlSb	6.1				0.57	
As₂S₃	3.43	585		0.50		
Au	19.3	1338	2980	0.13	3.18 (273)	
					3.15	1.22
					3.09 (500)	1.19 (500)
					2.78 (1000)	0.93 (1000)
					2.44 (1500)	
					1.20 (2000)	
					1.25 (3000)	
BaTiO₃	6.02	1891		0.49	0.062	0.023
Be	1.85	1556	2753	1.8	2.2 (273)	0.42
					0.96 (1000)	
BeO	3.05	2853		1	3.0	
Bi	9.8	544	1833	0.12	0.11	
Bi-Sr-Ca-Cu-O		1160				≈ 0.5
BN	2.25	3273		0.81	1.80	
C (graph.)	2.24	3923	4623	0.71	20	
					22.3 ∥	
					0.11 ⊥	
					11.3 (500) ∥	
					0.05 ⊥	
					5.3 (1000) ∥	
					0.03 ⊥	
					2.5 (2000) ∥	
					0.01 ⊥	
C (diam.)	3.52	> 3822		0.50	20	
Ca	1.55	1112	1713			
CaF₂	3.18	1691		0.85		
Cd	8.65	594	1038	0.23	0.98 (273)	
CdS	4.82	1653		0.35	0.16	
CdSe	5.81	1623		0.26		

Table II. *Contd.*

Material	ρ [g/cm^3]	T_m [K]	T_v [K]	c_p [J/gK]	κ(T[K]) [W/cm K]	D(T[K]) [cm^2/s]
CdTe	6.0	1314		0.21	0.06	
Co	8.9	1768	3200	0.43	1.04	
					0.74 (500)	
CoSi$_2$	4.9	1550		0.65		
Cr	7.2	2130	2945	0.46	0.95	0.29
					0.85 (500)	0.23 (500)
					0.65 (1000)	0.16 (1000)
					0.67 (1300)	
CrSi$_2$	5.0	1823		0.7		
Cu	8.95	1357	2840	0.39	4.0 (273)	1.14
					3.95	
					3.88 (500)	1.04 (500)
					3.56 (1000)	0.85 (1000)
					1.82 (2000)	
					1.80 (3000)	
Brass (70% Cu; 30% Zn)	8.5			0.38	1.05	0.33
Fe	7.86	1808	3135	0.46	0.83	0.23
					0.62 (500)	0.15 (500)
					0.33 (1000)	
					0.43 (2000)	
					0.46 (3000)	
Cast Iron	7.4			0.57	0.56	0.12
Mild steel (0.1%C)	7.85			0.49	0.46	0.12
Stainless steel (304)	8.03	1723	3273	0.5	0.15	
					0.16 (500)	
					0.25 (1000)	
FeSi$_2$	4.9	1485		0.67		
Ga	5.91	303	2510	0.36		
GaAs	5.32	1511		0.35	0.47	0.24
GaN	4.09	1738		0.88	1.7	
GaP	4.13	1739			1.0	
GaSb	4.79	1328			0.39	
Ge	5.33	1210	3105	0.32	0.6	0.36
GeO$_2$	4.7	1389		0.72		
Glass						
Crown	2.4			0.89	0.01	0.0058
BK7	2.51			0.86	0.011	0.0052
H$_2$O (water)	1	273	373	4.18	0.06	0.014
HgSe	8.25	1073		0.18		
HgTe	8.42	943		0.15		
In	7.31	429	2270	0.22	0.87	
InAs	5.78	796			0.27	
InP	4.79	1330			0.68	
InSb	5.77	798			0.17	
KCl	1.99	1045		0.65		
LiF	2.60	1121		1.9		
LiNbO$_3$	4.46	1526		0.64	0.042	0.015
Mg	1.74	923	1380	1.03		
MgO	3.65	3090		1.0	0.36	0.11
					0.27 (500)	0.065 (500)
					0.10 (1000)	0.022 (1000)
						0.159 (1500)
					0.09 (2000)	0.02 (2000)

Table II. Contd.

Material	ρ [g/cm³]	T_m [K]	T_v [K]	c_p [J/gK]	κ(T [K]) [W/cm K]	D(T [K]) [cm²/s]
Mn	7.3	1517		0.48	0.077	
Mo	10.2	2891	5050	0.26	1.35	0.52
					1.30 (500)	0.49 (500)
					1.12 (1000)	0.38 (1000)
					0.98 (1500)	0.29 (1500)
					0.88 (2000)	0.22 (2000)
						0.17 (2500)
						0.13 (3000)
Mo₂C	9.2	2843		0.62	0.07	
NaCl	2.16	1074		0.83		
Nb	8.4	2741	5017	0.27	0.51 (273)	
					0.60 (1000)	
NbC	7.85	4033		0.47	0.14	
NbN	7.28	2573		0.47	0.03	
Ni	8.90	1726	3187	0.44	0.91 (273)	
					0.87	0.24
					0.72 (500)	0.17 (500)
					0.72 (1000)	0.14 (1000)
NiSi₂	4.8	1263		0.65		
Os	22.6	3273	5773	0.13	0.88	
Pb	11.3	601	2023	0.13	0.35	0.24
					0.33 (500)	0.21 (500)
					0.22 (1000)	
PbS	7.5	1392		0.21	0.024	
PbSe	8.1	1349		0.17	0.017	
PbTe	8.16	1193		0.15	0.022	
Pd	12.1	1825	3413	0.24	0.76 (273)	0.24
					0.71	
a-PE	0.852	($T_g \approx 252$)		1.55	0.0039	
c-PE	1.004	415		2.2		
PET	1.4	5 3 0				
		($T_g \approx 340$)		2.1	0.0015	0.0013
PI (Kapton)	1.42	subl.		0.96	0.0016	
		($T_g \approx 508$)		1.28 (400)	0.0017 (400)	
				1.55 (500)	0.0018 (500)	
PMMA	1.18	($T_g \approx 378$)		1.42	0.0019	
Pt	21.5	2045	4100	0.13	0.71 (273)	
					0.71	0.25
					0.73 (500)	0.24 (500)
					0.79 (1000)	0.24 (1000)
						0.25 (1500)
					0.90 (1500)	0.27 (2000)
a-PTFE	2.15	($T_g \approx 240$)		0.9	0.004	
c-PTFE	2.89	605		1.03		
PtSi	12.4	2046		0.23		
PVC	1.39	546		0.95	0.0016	
		($T_g \approx 354$)				
PZT	7.6	1660		0.38	0.012	0.0042
Re	20.5	3453	5873	0.14	0.48 (1500)	
Rh	12.4				1.51	
					1.40 (500)	
					1.21 (1000)	
Sb	6.69	904	2000	0.21	0.26	
Se	4.82	490	958			

Table II. Contd.

Material	ρ [g/cm^3]	T_m [K]	T_v [K]	c_p [J/gK]	κ(T[K]) [W/cm K]	D(T[K]) [cm^2/s]
a-Si	2.32	1420		0.8	0.018	
				1.15	0.010 (1000)	
c-Si	2.32	1690	2680	0.71	1.5	0.85
				0.99 (1500)	0.23 (1500)	0.10
l-Si	2.52			0.91	0.53 (1800)	0.29
					0.6 (2000)	
a-Si$_3$N$_4$	3.1					
SiO	2.13			0.94	0.015	
a-SiO$_2$	2.2	1873		0.72	0.014	0.009
				1.1 (500)	0.021 (500)	0.009 (500)
				1.22 (1000)	0.033 (1000)	0.013 (1000)
					0.076 (1500)	
SiO$_2$	2.2	1940		0.74	0.14	
α-SiC	3.21	3102				
SiC	3.22	≈ 3073		1.24	4.9	
Si$_3$N$_4$	3.0	2173		1.1		
Sn	7.30	505	2876	0.23	0.67 (273)	
					0.63	0.38
					0.60 (500)	
					0.41 (1000)	
Sr	2.6	1041	1653			
SrTiO$_3$	5.11	2183		0.69		
Ta	16.6	3270	5700	0.14	0.55 (273)	
					0.55	0.24
					0.58 (500)	0.24 (500)
					0.61 (1000)	0.235 (1000)
						0.23 (1500)
					0.64 (2000)	0.22 (2000)
						0.20 (2500)
					0.665 (3000)	0.17 (3000)
TaC	14.5	4153		0.26	0.22	
Ta$_2$N	14.1	3363		0.20	0.05	
TaSi$_2$	9.1	2473		0.32		
Te	6.25	722	1263	0.2	0.04	0.032
ThO$_2$	9.7			0.23	0.15	0.07
					0.06 (500)	0.02 (500)
					0.03 (1000)	0.01 (1000)
Ti	4.52	1943	3560	0.52	0.22	0.094
					0.20 (500)	0.075 (500)
					0.21 (1000)	0.062 (1000)
TiC	4.9	3423		0.84	0.24	
TiN	5.43	3478		0.81	0.2	
TiO$_2$	4.28	2143		0.93	0.089	0.031
					0.065	
					0.059 (500)	0.017 (500)
					0.035 (1000)	0.009 (1000)
TiSi$_2$	4.0	1813		0.73		
V	5.96	2170	3670	0.49	0.30 (273)	
					0.31	0.11
					0.33 (500)	0.11 (500)
					0.39 (1000)	0.10 (1000)
					0.51 (2000)	0.10 (2000)

Table II. Contd.

Material	ρ [g/cm³]	T_m [K]	T_v [K]	c_p [J/gK]	κ(T [K]) [W/cm K]	D(T [K]) [cm²/s]
VC	5.8	3103		0.79	0.25	
VN	6.1	2633		0.77	0.18	
VSi₂	4.6	1943		0.70		
W	19.3	3660	5830	0.13	1.70 (273)	
					1.80	0.65
					1.49 (500)	0.56 (500)
					1.20 (1000)	0.41 (1000)
					1.06 (1500)	0.35 (1500)
					1.0 (2000)	0.30 (2000)
						0.26 (2500)
					0.91 (3000)	0.23 (3000)
WC	15.8	3143		0.25	0.29	
WSi₂	9.9	2438		0.31		
Y		1782	3200			
Y-Ba-Cu-O	6.4	1400–1570		1	0.06	0.01
Zn	7.14	693	1180	0.39	1.2 (273)	
					1.17	0.43
					1.11 (500)	0.36 (500)
					0.67 (1000)	
ZnO	5.63	2247	2310	4.97	0.29	
				6.54 (1000)	0.13 (500)	
					0.04 (>1300)	
ZnS	4.1	1973		0.49		
ZnS (α)	3.98		1458			
ZnS (β)	4.09		1293			
ZnSe	5.42	1788		0.35		
ZnTe	5.72	1511		0.26		
Zr	6.49	2127	4680	0.28	0.22 (273)	
					0.23	0.12
					0.21 (500)	0.10 (500)
					0.24 (1000)	0.10 (1000)
					0.31 (2000)	
ZrC	6.4	3803		0.48	0.20	
ZrN	7.35	3253		0.48	0.17	
ZrO₂	5.82	2950		0.61	0.02	0.0074
					0.02 (500)	0.0063 (500)
					0.02 (1000)	0.0053 (1000)
ZrSi₂	4.9	1973		0.51		

Table III. Optical normal-incidence reflectivity R (mainly for polished surfaces) and optical absorption coefficient α [cm⁻¹] at $T \approx 300$ K

Material	R	α [cm⁻¹]	λ [μm]
Ag	0.25		0.2
	0.30	5 E5	0.25
	0.34		0.251
	0.09		0.305
	0.75		0.357
	0.91	7.14 E5	0.5

Table III. *Contd.*

Material	R	α [cm^{-1}]	λ [μm]
	0.98		0.5
	0.95		0.7
	0.97		1
	0.99	8.33 E5	1.06
	0.98		5
	0.99		9
	0.99	8.33 E5	10.6
Al	0.93		0.248
	0.92	1.25 E6	0.25
	0.86		0.305
	0.90	1.43 E6	0.5
	0.92		0.5
	0.87		0.7
	0.85		0.8
	0.91		1
	0.94	1 E6	1.06
	0.96		5
	0.97		9
	0.98	8.33 E5	10.6
Al$_2$O$_3$ (ceramic)	0.9	30–70	0.694
	0.3 (3800 K)		0.694
Au	0.22		0.193
	0.33	5.6 E5	0.25
	0.39		0.251
	0.28		0.357
	0.47	4.6 E5	0.5
	0.84		0.6
	0.92		0.7
	0.95		0.8
	0.98	7.7 E5	1.06
	0.97		5
	0.98		9
	0.98	7.1 E5	10.6
BaTiO$_3$	0.29	3.6 E5	0.308
	0.26	3 E4	0.35
	0.16	< 10	0.647
	0.27	1.6 E4	10.6
CdTe		0.002	10.6
Cornea		2.7 E3	0.193
Cr	0.55		0.6
	0.551		0.7
	0.555		0.8
	0.56		0.9
	0.56		1
	0.625		2
	0.698		3
	0.75		4
	0.8		5
Cu	0.1[a]		0.25
	0.255		0.3
	0.31		0.4

Table III. *Contd.*

Material	R	α [cm^{-1}]	λ [μm]
	0.43	7.14 E5	0.5
	0.725		0.6
	0.83		0.7
	0.86		0.8
	0.89		0.9
	0.908		1
	0.98	7.7 E5	1.06
	0.95		2
	0.96		3
	0.964		4
	0.97		5
	0.99	7.7 E5	10.6
Fe	0.57		0.6
	0.587		0.7
	0.609		0.8
	0.619		0.9
	0.639		1
	0.769		2
	0.835		3
	0.873		4
	0.907		5
c-GaAs	0.6	1.67 E6	0.25
	0.39	1 E5	0.5
	0.31	1.43 E2	1.06
	0.28	0.02	10.6
a-Ge	0.48	1 E6	0.25
	0.47	2 E5	0.5
	0.42	1 E4	1.06
	0.34	0.032	10.6
c-Ge	0.42	1.43 E6	0.25
	0.49	6.7 E5	0.5
	0.38	50	1.06
	0.36	0.032	10.6
KCl	0.05	<1	0.25
	0.04	<1	0.5
	0.04	<1	1.06
	0.03	0.001	10.6
LiNbO$_3$	0.20	280	0.308
	0.18	<1	0.35
	0.16	<1	0.647
	0.01	890	10.6
Mo	0.63		0.248
	0.58–0.66		0.633
	0.61–0.7		1.064
NaCl		0.002	10.6
Ni	0.44		0.2
	0.15[a]		0.25
	0.49		0.357
	0.62	8.33 E5	0.5
	0.65		0.6
	0.69		0.7
	0.70		0.8

Table III. Contd.

Material	R	α [cm^{-1}]	λ [μm]
	0.72		1
	0.67	6.7 E5	1.06
	0.94		5
	0.96		9
	0.97	2.7 E5	10.6
PC		5.5 E5	0.193
		1 E4	0.248
		22	0.308
		4	0.351
PE		630	0.193
		< 10	0.248
		< 10	0.308
		< 10	0.351
PET		3 E5	0.193
		1.6 E5	0.248
		1.3 E5	0.254
		4 E3	0.308
		1.1 E4	0.3
	0.1	3 E3	9
PI (Kapton)	0.05	3.4 E5	0.193
	0.12	3.0 E5	0.248
	0.11	0.86 E5	0.308
	0.09	0.36 E5	0.351
PMMA		2 E3	0.193
		130 ± 70	0.248
		< 20	0.308
		< 10	0.351
PP		5.3 E2	0.193
		< 10	0.248
		< 10	0.308
		< 10	0.351
PS		8 E5	0.193
		6.5 E3	0.248
		80	0.308
		≈ 10	0.351
PSUL		4 E5	0.193
		1.5 E5	0.248
		8.1 E2	0.308
		≈ 10	0.351
Pt	0.46		0.248
	0.34		0.251
	0.40		0.305
	0.43		0.357
	0.58		0.5
	0.64		0.6
	0.69		0.7
	0.70		0.8
	0.73		1
	0.94		5
	0.95		9
PTFE		2.6 E2	0.193
		< 160	0.248
		< 10	0.308

Table III. Contd.

Material	R	α [cm^{-1}]	λ [μm]
		< 10	0.351
PVAC		1 E3	0.193
		< 100	0.248
		< 10	0.308
		< 10	0.351
Re	0.56		0.248
a-Si	0.55	1.5 E6	0.193
	0.75	1 E6	0.25
	0.48	1 E5	0.5
	0.44	7 E4	0.694
	0.35	1 E4	1.06
	0.32	<1	10.6
c-Si	0.59	1 E6	0.193
	0.61	1.67 E6	0.25
	0.730	2.09 E6	0.276
	0.60	1.48 E6	0.308
	0.591	1.38 E6	0.31
	0.56	1.12 E6	0.337
	0.571	1 E6	0.354
	0.58	1.07 E6	0.355
	0.591		0.37
	0.48	9.01 E4	0.405
	0.467	7.11 E4	0.413
	0.42	2.64 E4	0.458
	0.39	1.71 E4	0.485
	0.39	1.56 E4	0.488
	0.392	1.39 E4	0.496
	0.36	2 E4	0.5
	0.38	1.12 E4	0.514
	0.37	9 E3	0.53
	0.348	3.8 E3	0.62
	0.35	3.6 E3	0.633
	0.34–0.44	2.5 E3–7 E4	0.694
	0.33	50	1.06
	0.30		1.064
	0.30	10	10.6
l-Si	0.68	1.67 E6	0.193
	0.69	1.46 E6	0.308
	0.72	1.25 E6	0.5
	0.72	7.69 E5	1.06
SiO$_2$	0.06	< 1	0.25
	0.04	< 1	0.5
	0.04	< 1	1.06
	0.2	2.5 E2	10.6
Si$_3$N$_4$		1.5 E5	0.193
W	0.51		0.248
	0.51	1.43 E6	0.25
	0.49	7.69 E5	0.5
	0.58	4.35 E5	1.06
	0.98	5 E5	10.6
Y-Ba-Cu-O	0.2	2.3 E5	0.248
ZnO	0.2	3.3 E5	0.248
ZnSe		0.005	10.6

[a] unpolished

Table IV. Melting and vaporization enthalpies

Material	ΔH_m 10^3 J/g (10^3 J/mol)	ΔH_v 10^3 J/g (10^3 J/mol)
Ag	0.10 (11)	2.32 (250)
Al	0.41 (11)	10.75 (290)
Al_2O_3(ceramic)	1.08 (110)	18.18 (1850)
AlSb	0.55 (82)	
As_2S_3	0.12 (29)	
Au	0.06 (12)	1.73 (340)
Be	1.3 (12)	32.2 (290)
BeO	3.24 (81)	
Bi	0.05 (11)	0.81 (170)
C (diamond)	8.3 (100)	
CaF_2	0.38 (30)	
Cd	0.06 (6.2)	0.89 (100)
Co	0.27 (16)	6.52 (380)
Cr	0.29 (15)	5.96 (310)
Cu	0.20 (13)	4.7 (300)
Fe	0.27 (15)	6.3 (350)
Stainless Steel (304)	0.3	6.5
Ge	0.51 (37)	4.13 (300)
GeO_2	0.42 (44)	
In	0.03 (3.3)	2.0 (230)
InAs	0.41 (77)	
InP	0.51 (75)	
KCl	0.36 (27)	1.61 (120)
LiF	1.04 (27)	
Mg	0.35 (8.5)	5.35 (130)
MgO	1.91 (77)	
Mn	0.27 (15)	4.19 (230)
Mo	0.29 (28)	6.15 (590)
NaCl	0.48 (28)	
Nb	0.28 (26)	7.32 (680)
Ni	0.31 (18)	6.3 (370)
Os	0.14 (27)	4.10 (780)
Pb	0.02 (4.8)	0.87 (180)
PbS	0.15 (36)	
PbSe	0.17 (49)	
PbTe	0.17 (57)	
Pd	0.16 (17)	3.4 (360)
PET	0.017 (3.32)	
PMMA		0.97 (98)
Pt	0.10 (20)	2.61 (510)
PTFE	0.068 (3.42)	2.64 (132)
Sb	0.16 (20)	1.6 (190)
a-Si	1.25 (35)	
c-Si	1.78 (50)	15.0 (420)
SiO_2	0.14 (8.5)	
Sn	0.06 (7.0)	1.94 (230)
Ta	0.17 (31)	3.87 (700)
Te	0.14 (18)	0.38 (49)
Ti	0.40 (19)	8.8 (420)

Table IV. Contd.

Material	ΔH_m 10^3 J/g (10^3 J/mol)	ΔH_v 10^3 J/g (10^3 J/mol)
TiO$_2$	0.84 (67)	
V	0.41 (21)	
W	0.19 (35)	3.86 (710)
Zn	0.11 (7.4)	1.8 (120)
ZnO	0.97	7.66
Zr	0.19 (17)	5.5 (500)
ZrO$_2$	0.71 (87)	

Table V. Absorption/dissociation cross sections, σ, for precursor molecules commonly used in LCP. The values listed in the table have been obtained by averaging "most reliable" results reported by different authors [see, e.g., *Rothschild* 1989]

Molecule	σ [10^{-18} cm^2]	λ [nm]
Al(CH$_3$)$_3$	20 ± 3	193
	0.011	248
	3 E-3	257
Al(C$_2$H$_5$)$_3$	4.7 ± 1.4	193
Al$_2$(CH$_3$)$_6$	20	193
	2 E-3	257
As(CH$_3$)$_3$	45	193
As(C$_2$H$_5$)$_3$	18	193
AsH$_3$	18	193
B(C$_2$H$_5$)$_3$	0.44	193
BCl$_3$	0.045	193
B$_2$H$_6$	0.13 ± 0.09	193
Br$_2$	< E-4	193
	< E-4	248
	< E-4	266
	3.8 E-4	308
	0.034	351
Cd(CH$_3$)$_2$	7 ± 3	193
	3 ± 1	248
	1.3 ± 0.4	257
	0.45	266
CCl$_4$	0.7 ± 0.3	193
	2.3 E-3	248
CF$_2$Br$_2$	1.1	193
	0.63 ± 0.03	248
	0.24	257
CF$_3$Br	0.073 ± 0.01	193
	(5 ± 3) E-3	248
	4.5 E-4	257
C$_2$F$_5$Br	0.14 ± 0.04	193
	4.2 E-3	248
	E-3	257
1,2-C$_2$F$_4$Br$_2$	1.1	193
	0.051	248
	0.016	257

Table V. *Contd.*

Molecule	$\sigma\,[10^{-18}\mathrm{cm}^2]$	λ [nm]
CF_2Cl_2	0.32	193
CF_3Cl	(1.4 ± 0.6) E-3	193
C_2F_5Cl	1.9 E-3	193
$1,2\text{-}C_2F_4Cl_2$	0.044	193
CF_3I	3.8 E-3	193
	0.27 ± 0.01	248
	0.034	308
	< E-3	351
C_2F_5I	< E-3	193
	0.33	248
	0.64	268
	0.058	308
$1,2\text{-}C_2F_4I_2$	0.97	248
	1.5	257
	2.1	266
	0.10	308
CH_2Br_2	0.36	248
	0.11	257
CH_3Br	0.55 ± 0.03	193
	0.016 ± 0.003	248
	(3.7 ± 0.6) E-3	257
CH_2Cl_2	0.37 ± 0.02	193
$CHCl_3$	0.89 ± 0.02	193
CH_3Cl	0.062 ± 0.008	193
CH_4	< E-3	193
C_2H_4	0.015	193
C_2H_3Cl	6	193
C_2H_5Cl	0.035	193
CH_2FCl	0.010	193
CHF_2Cl	1.2 E-3	193
$CHFCl_2$	0.20	193
CH_2I_2	29	193
	1.6	248
	1.3	257
	1.7 ± 0.3	266
	3.3	308
	0.25	351
CH_3I	0.82	248
	1.1	257
	0.95	266
	8.4 E-3	308
Cl_2	2.5 E-3	193
	< E-3	248
	0.010	266
	0.19	308
	0.14 ± 0.03	351
	0.25 E-2	458
	6 E-4	488
	1.5 E-4	515
$Cr(CO)_6$	19 ± 7	193
	47 ± 10	248
	25	257

Table V. Contd.

Molecule	σ [10^{-18}cm^2]	λ [nm]
	25	266
	5.3 ± 0.2	308
CrO$_2$Cl$_2$	3.0	248
	3.1	257
	5.5	266
	5.1	308
F$_2$	< E3	193
	0.013	248
	0.015	266
	0.019	308
	5.7 E-3	351
Fe(CO)$_5$	57	193
	27	248
	14 ± 5	257
	13	266
	2.4	308
	1.3	355
Ga(CH$_3$)$_3$	20 ± 6	193
	2.1	248
	1.8	257
Ga(C$_2$H$_5$)$_3$	7.4 ± 1.6	193
GeH$_4$	0.025	193
HBr	1.7	193
HCl	0.89	193
HF	< E-3	193
HNO$_3$	12 ± 1	193
	0.020	248
	0.019	257
	0.017 ± 1	266
	(1.1 ± 0.1) E-3	308
H$_2$O$_2$	0.60	193
	0.08 ± 0.003	248
	0.061 ± 0.004	257
	0.04 ± 0.003	266
	4.2 E-3	308
H$_2$S	6.4 ± 0.6	193
	0.048 ± 0.023	248
Hg(CH$_3$)$_2$	26 ± 1	193
	0.2 ± 0.12	248
	0.05	257
HI	0.56 ± 0.02	193
	0.51	248
	0.18	266
I$_2$	6.1	193
	0.027	248
	0.054	266
In(CH$_3$)$_3$	12 ± 2	193
	2 ± 0.8	248
	1.5	257
	1.5	266
InI	< 7	193

Table V. *Contd.*

Molecule	σ [10^{-18}cm^2]	λ [nm]
Mo(CO)$_6$	56 ± 5	193
	51 ± 7	248
	20	257
	17	266
	13	308
	0.5	350–360
MoF$_6$	4.9	193
	0.20	248
	0.14	257
	0.090	266
	< E-3	308
NF$_3$	(5.7 ± 0.4) E-3	193
NH$_3$	9 ± 3	193
NO	0.020	193
NO$_2$	0.6 ± 0.19	193
	0.03 ± 0.01	248
	0.03 ± 0.012	257
	0.035 ± 0.01	266
	0.16	308
N$_2$O	0.09 ± 0.007	193
	2 E-6	248
Ni(CO)$_4$	30	248
	2.4	308
O$_2$	(7.5 ± 7) E-4	193
	4 E-6	248
O$_3$	0.4	193
	5	248
PET (Mylar)	20[a]	193
PH$_3$	27 ± 14	193
P(CH$_3$)$_3$	34	193
P(C$_2$H$_5$)$_3$	8.5	193
Pb(CH$_3$)$_4$	0.37	257
	0.062	266
Pt(PF$_3$)$_4$	0.19	248
Pt(hfacac)	100	
ReF$_6$	7.3	193
SO$_2$	9.3	193
	0.075	248
	0.19	257
	0.45	266
	0.63	308
SF$_6$	6.8 E-4	193
Si(CH$_3$)$_4$	< E-3	193
Si$_2$(CH$_3$)$_6$	46	193
SiH$_3$(C$_6$H$_5$)	170 ± 130	193
SiH$_4$	1.1 E-3	193
Si$_2$H$_6$	2	193
Sn(CH$_3$)$_4$	40	193
SnCl$_4$	38	193
	8.3	248
	5.2	257
	1.3	266

Table V. Contd.

Molecule	σ [10^{-18}cm^2]	λ [nm]
Te(CH$_3$)$_2$	5.2 ± 1.4	193
	7 ± 3	248
	1.7 ± 0.6	257
Te(C$_2$H$_5$)$_2$	15	193
	8 ± 2	248
	1.4 ± 0.5	257
	1	266
TiBr$_4$	20	248
	23	257
	30	266
	1.2	308
TiCl$_4$	30	193
	11	248
	5.5	257
	10	266
	2.3	308
TlBr	22	193
TlI	24	193
	2.6	248
VCl$_4$	10	248
	9.6	257
	8.9	266
	11	308
VOCl$_3$	18	248
	13	257
	8.9	266
	5.0	308
W(CO)$_6$	12	193
	4.5	248
	2.4	308
	0.5	350–360
WF$_6$	0.35	193
Zn(CH$_3$)$_2$	15	193

[a] per monomer unit

References

Abraham F.F.: *Homogeneous Nucleation Theory* (Academic, London 1978)

Adrian F.J., J. Bohandy, B.F. Kim, A.N. Jette, P. Thompson: A study of the mechanism of metal deposition by the laser-induced forward transfer process. J. Vac. Sci. Technol. B **5**, 1490 (1987)

Afonso C.N, F. Vega, J. Gonzalo, C. Zaldo: Lithium niobate films grown by excimer laser deposition. Appl. Surf. Sci. **69**, 149 (1993)

Agrawalla B.S., B.T. Dai, S.D. Allen: Laser ablative chemical etching of SiO_2. J. Vac. Sci. Technol. B **5**, 601 (1987)

Ahlgren W.L., E.J. Smith, J.B. James, T.W. James, R.P. Ruth, E.A. Patten: Photo-MOCVD growth of HgTe-CdTe superlattices. J. Cryst. Growth **86**, 198 (1988)

Aizaki N., K. Terashima, J. Fujita, S. Matsui: $YBa_2Cu_3O_y$ superconducting thin film obtained by laser annealing. Jpn. J. Appl. Phys. **27**, L231 (1988)

Akhmanov S.A, V.I. Emel'yanov, N.I. Koroteev, V.N. Seminogov: Interaction of powerful laser radiation with the surfaces of semiconductors and metals: Nonlinear optical effects and nonlinear optical diagnostics. Sov. Phys. – Usp. **28**, 1084 (1985)

Akulin V.M., N.V. Karlov: *Intense Resonant Interactions in Quantum Electronics* (Springer, Berlin, Heidelberg 1992)

Alimov D.T., V.A. Bobyrev, F.V. Bunkin, V.L. Zhuravskii, B.S. Luk'yanchuk, E.A. Morozova, S.A. Ubaidullaev, P.K. Khabibullaev: Non-equilibrium kinetics of oxide layer grains growth at laser heating of metals in air. Sov. Phys. – Dokl. **29**, No. 12 (1984)

Aliouchouche A., J. Boulmer, B. Bourguignon, J.P. Budin, D. Débarre, A. Desmur: Laser etching of silicon by chlorine: Effect of post-desorption collisions and chlorine in-diffusion on the laser desorption yield. Appl. Surf. Sci. **69**, 52 (1993)

Allen F.G.: Emissivity at 0.65 micron of silicon and germanium at high temperatures. J. Appl. Phys. **28**, 1510 (1957)

Allmen M. von, A. Blatter: *Laser-Beam Interactions with Materials*, 2nd edn., Springer Ser. Mater. Sci. Vol.2 (Springer, Berlin, Heidelberg 1995)

Anisimov S.I., B.S. Luk'yanchuk, A. Luches: Dynamics of 3D expansion of vapor at pulsed laser deposition. Appl. Surf. Sci. (1996), in print

Anisimov S.I., V.A. Khokhlov: *Instabilities in Laser-Matter Interaction* (CRC, Boca Raton 1995)

Anisimov S.I., D. Bäuerle, B.S. Luk'yanchuk: Gas dynamics and film profiles in pulsed-laser deposition of materials. Phys. Rev. B **48**, 12076 (1993)

Anisimov S.I., M.I. Tribel'skii, Ya.G. Epelbaum: Instability of plane evaporation front in interaction of laser radiation with a medium. Sov. Phys. – JETP **51**, 802 (1980)

Anisimov S.I., Y.A. Imas, G.S. Romanov, Y.V. Khodyko: *Action of High-Power Radiation on Metals* (Consult. Bureau, Springfield, VA 1971)

Antaya M., J.R. Dahn, J.S. Preston, E. Rossen, J.N. Reimers: Preparation and characterization of $LiCoO_2$ thin films by laser ablation deposition. J. Electrochem. Soc. **140**, 575 (1993)

Aoyagi Y., T. Meguro, S. Iwai: Beam assisted layer-by-layer processes and the mechanism in III-V compounds. 1st. Int'l Symp. on Atomic Layer Epitaxy, ed. by L. Niinistö, Acta Polytechnica Scandinavica, Chemical Technology and Metallurgy Series No. 195, Helsinki (1990) p.55

Aoyagi Y., A. Doi, S. Iwai, S. Namba: Atomic-layer growth of GaAs by modulated-continuous-wave laser metal-organic vapor-phase epitaxy. J. Vac. Sci. Technol. B **5**, 1460 (1987)

Arenholz E., J. Heitz, V. Svorcik, D. Bäuerle: Non-coherent structure formation on UV-laser irradiated polymers, in *Excimer Lasers*, ed. by L.D. Laude, NATO ASI Series (Kluwer, Dordrecht 1994) p.237

Arenholz E., J. Heitz, M. Wagner, D. Bäuerle, H. Hibst, A. Hagemeyer: Laser-induced surface modification and structure formation of polymers. Appl. Surf. Sci. **69**, 16 (1993)

Arenholz E., M. Wagner, J. Heitz, D. Bäuerle: Structure formation in UV-laser-ablated polyimide foils. Appl. Phys. A **55**, 119 (1992)

Arenholz E., V. Svorcik, T. Kefer, J. Heitz, D. Bäuerle: Structure formation in UV-laser ablated poly-ethylene-terephthalate (PET). Appl. Phys. A **53**, 330 (1991)

Arjavalingam G., G. Hougham, J.P. LaFemina: Emission mechanisms in polyimide. Polymer **31**, 840 (1990)

Armstrong D.A., J.L. Holmes: Decomposition of halides and derivatives, in *Decomposition of Inorganic and Organometallic Compounds*, ed. by C.H. Bamford, C.F.H. Tipper, Chem. Kinetics, Vol.4 (Elsevier, Amsterdam 1972) p.143

Arnold J., U. Dasbach, W. Ehrfeld, K. Hesch, H. Löwe: Combination of excimer laser micromachining and replication processes suited for large scale production. Appl. Surf. Sci. C **86**, 251 (1995)

Arnold J., A. Pfuch, J. Borck, K. Zach, P. Seidel: Preparation and characterization of $Bi_2 Sr_2 Ca_1 Cu_2 O_{8+x}$ thin films made by LPVD. Physica C **213**, 71 (1993)

Arnold N., E. Thor, N. Kirichenko, D. Bäuerle: Pyrolytic LCVD of fibers: A theoretical description. Appl. Phys. A **62** (1996), in print

Arnold N., P.B. Kargl, D. Bäuerle: Laser direct writing and instabilities: A one-dimensional approach. Appl. Surf. Sci. **86**, 457 (1995)

Arnold N., R. Kullmer, D. Bäuerle: Simulation of growth in pyrolytic laser-CVD of microstructures: I. One-dimensional approach. Microelectronic Eng. **20**, 31 (1993)

Arnold N., D. Bäuerle: Simulation of growth in pyrolytic laser-CVD of microstructures: II. Two-dimensional approach. Microelectronic Eng. **20**, 43 (1993)

Arnold N.D., N.A. Kirichenko: Phenomenological description of spiral waves arising under radiant heating of metals. Sov. Phys. – JETP **74**, 750 (1992)

Ashby C.I.H., D.R. Myers, G.A. Vawter, R.M. Biefeld, A.K. Datye: Selective suppression of photochemical dry etching using elevated surface impurity concentrations: A new technique for self-aligned etching. J. Appl. Phys. **68**, 2406 (1990)

Ashby C.I.H., P.J. Brannon: Laser-driven chemical reaction for etching $LiNbO_3$. Appl. Phys. Lett. **49**, 475 (1986)

Ashby M.F., K.E. Easterling, W.B. Li: Modelling the laser transformation hardening of steel, in *Laser Processing of Materials*, ed. by K. Mukherjee, J. Mazumder (Metallurgical Society of AIME, Warrendale, PA 1985)

Ashcroft N.W., N.D. Mermin: *Solid State Physics* (Holt-Saunders, New York 1976)

Asshauer T., K. Rink, G. Delacrétaz: Acoustic transient generation by holmium-laser-induced cavitation bubbles. J. Appl. Phys. **76**, 5007 (1994)

Auciello O., A.R. Krauss, J. Santiago-Aviles, A.F. Schreiner, D.M. Gruen: Surface compositional and topographical changes resulting from excimer laser impacting on $YBa_2Cu_3O_7$ single phase superconductors. Appl. Phys. Lett. **52**, 239 (1988)

Aussenegg F.R., A. Leitner, M.E. Lippitsch (eds.): *Surface Studies with Lasers*, Springer Ser. Chem. Phys., Vol.33 (Springer, Berlin, Heidelberg 1983)

Auston D.H., J.A. Golovchenko, A.L. Simons, R.E. Slusher, P.R. Smith, C.M. Surko, T.N.C. Venkatesan: Dynamics of laser annealing, in *Laser-Solid Interactions and Laser Processing*, ed. by S.D. Ferris, H.J. Leamy, J.M. Poate (AIP, New York 1979) p.11

Avouris P., W.M. Gelbart, M.A. El-Sayed: Nonradiative electronic relaxation under collision-free conditions. Chem. Rev. **77**, 793 (1977)

Aydinli A., G.C. Puente, A. Bhat, A. Compaan, A. Chan: $ZnSe_xTe_{1-x}$ films grown by pulsed laser deposition. J. Vac. Sci. Technol. A **9**, 3031 (1991)

Bachmann F.: Excimer lasers in a fabrication line for a highly integrated printed circuit board. Chemtronics **4**, 149 (1989)

Baeri P., S.U. Campisano, F. Priolo, E. Rimini: Transient temperature measurement by thin film thermocouple during nanosecond laser-induced mixing of Ni-Si layers, in *Energy Beam-Solid Interactions and Transient Thermal Processing*, ed. by V.T. Nguyen and A.G. Cullis (Physique, Les Ulis 1985) p.237

Bäuerle D., E. Arenholz, V. Svorcik, J. Heitz, B. Luk'yanchuk, N. Bityurin: Laser-induced surface modifications, structure formation, and ablation of organic polymers. SPIE Proc. **2403** (1996), in print

Bäuerle D., E. Arenholz, J. Heitz, S. Proyer, E. Stangl, B. Luk'yanchuk: Surface patterning and thin-film formation by pulsed-laser ablation, in *Semiconductor Processing and Characterization with Lasers – Applications in Photovoltaics*, ed. by M. Brieger, H. Dittrich, M. Klose, H.W. Schock, J. Werner. Mater. Sci. Forum **173 & 174**, 41 (1995)

Bäuerle D., B. Luk'yanchuk, P. Schwab, X.Z. Wang, E. Arenholz: Laser-ablation: Fundamentals and recent developments, in *Laser Ablation of Electronic Materials – Basic Mechanisms and Applications*, ed. by E. Fogarassy, S. Lazare (Elsevier, Amsterdam 1992) p.39

Bäuerle D., B. Luk'yanchuk, K. Piglmayer: On the reaction kinetics in laser-induced pyrolytic chemical processing. Appl. Phys. A **50**, 385 (1990a)

Bäuerle D., J. Heitz, W. Ludorf, P. Schwab, X.Z. Wang: Laser deposition and patterning of high-temperature superconductors, in *In-situ Patterning: Selective Area Deposition and Etching*, ed. by R. Rosenberg, A.F. Bernhardt, J.G. Black, MRS Proc. **158**, 451 (1990b)

Bäuerle D.: Chemical processing with lasers: An overview, in *Spatial Inhomogeneities and Transient Behaviour in Chemical Kinetics*, ed. by P. Gray, G. Nicolis, F. Baras, P. Borckmans, S.K. Scott (Manchester Univ. Press, Manchester 1990) p.541

Bäuerle D., M. Eyett, U. Kolzer, R. Kullmer, P. Mogyorosi, K. Piglmayer: Laser-induced surface modification and etching of materials. MRS Proc. **101**, 411 (1988)

Bäuerle D.: *Chemical Processing with Lasers*, Springer Ser. Mater. Sci., Vol.1 (Springer, Berlin, Heidelberg 1986)

Bäuerle D.: Materialbearbeitung mit Laserlicht. Laser und Optoelektronik **1**, 29 (1985)

Bäuerle D.: Laser induced chemical vapor deposition, in *Laser Processing and Diagnostics*, ed. by D. Bäuerle, Springer Ser. Chem. Phys., Vol.39 (Springer, Berlin, Heidelberg 1984) p.166

Bäuerle D.: Production of microstructures by laser pyrolysis, in *Laser Diagnostics and Photochemical Processing for Semiconductor Devices*, ed. by R.M. Osgood, S.R.J. Brueck, H.R. Schlossberg. MRS Proc. **17**, 19 (1983a)

Bäuerle D.: Laser induced chemical vapor deposition, in *Surface Studies with Lasers*, ed. by F.R Aussenegg, A. Leitner, M.E. Lippitsch, Springer Ser. Chem. Phys., Vol.33 (Springer, Berlin, Heidelberg 1983b) p.178

Bäuerle D., G. Leyendecker, D. Wagner, E. Bauser, Y.C. Lu: Laser grown single crystals of silicon. Appl. Phys. A **30**, 147 (1983)

Bäuerle D., P. Irsigler, G. Leyendecker, H. Noll, D. Wagner: Ar^+ laser-induced chemical vapor deposition of Si from SiH_4. Appl. Phys. Lett. **40**, 819 (1982)

Bäuerle D., D. Wagner, M. Wöhlecke, B. Dorner, H. Kraxenberger: Soft modes in semiconducting $SrTiO_3$: II. The ferroelectric mode. Z. Physik B **38**, 335 (1980)

Bagratashvili V.N., M.V. Kuz'min, S.V. Letokhov, A.N. Shibanov: Observation of proton and electron detachment from an anthracene molecule during pronounced IR many-photon superexcitation. JETP Lett. **37**, 112 (1983)

Bagratashvili V.N., V.N. Doljikov, V.S. Letokhov, E.A. Ryabov: Isotopically-selective dissociation of the CF_3I molecules at high pressure under the action of the pulse radiation of CO_2 laser. Zh. Tekh. Phys. Letters (Pis'ma Red.) **4**, 1181 (1978)

Bagratashvili V.N., I.N. Knyazev, V.S. Letokhov, V.V. Lobko: Optoacoustic detection of multiple photon molecular absorption in a strong IR field. Opt. Commun. **18**, 525 (1976)

Bahners T., D. Knittel, F. Hillenkamp, U. Bahr, C. Benndorf, E. Schollmeyer: Chemical and physical properties of laser-treated poly(ethyleneterephthalate). J. Appl. Phys. **68**, 1854 (1990)

Bahners T., A. Bossmann, E. Schollmeyer: Oberflächenstrukturierung polymerer Fasern durch UV-Laserbestrahlung, Teil 4. Angew. Makromol. Chem. **170**, 203 (1989)

Baleva M.I., M.H. Maksimov, S.M. Metev, M.S. Sendova: Laser-assisted sputtering of $Pb_{1-x}Cd_xSe$ films. J. Mater. Sci. Lett. **5**, 533 (1986)

Balk P., M. Fischer, D. Grundmann, R. Lückerath, H. Lüth, W. Richter: Ultraviolet-assisted growth of GaAs. J. Vac. Sci. Technol. B **5**, 1453 (1987)

Balkanski M., R.F. Wallis, E. Haro: Anharmonic effects in light scattering due to optical phonons in silicon. Phys. Rev. B **28**, 1928 (1983)

Ballal A.K., L. Salamanca-Riba, C.A. Taylor, G.L. Doll: Structural characterization of preferentially oriented cubic BN films grown on Si(001) substrates. Thin Solid Films **224**, 46 (1993)

Baller T.: Chemical etching induced by a pulsed laser beam. Dissertation, University of Twente, Holland (1990)

Ban V.S., D.A. Kramer: Thin films of semiconductors and dielectrics produced by laser evaporation. J. Mater. Sci. **5**, 978 (1970)

Barr W.P.: The production of low scattering dielectric mirrors using rotating vane particle filtration. J. Phys. E **2**, 2 (1969)

Barth M., P. Hess, G. Mollekopf, H. Stafast: SF_6 sensitized CO_2-laser chemical vapor deposition of a-Ge:H. Thin Solid Films **241**, 61 (1994)

Baseman R.J., A. Gupta, R.C. Sausa, C. Progler: Laser induced forward transfer. MRS Proc. **101**, 237 (1988)

Basovich A., S. Gaponov, L. Jastrabik, M. Jelinek, N. Kiselev, E. Kluenkov, O. Lebedev, L. Mazo, L. Soukup, M. Strikovskij, V. Talanov, A. Vasiliev: Laser deposition of YBaCuO on ZrO_2-coated sapphire substrates. Thin Solid Films **228**, 193 (1993)

Basting D., U. Sowada, F. Voß, P. Oesterlin: Processing of PTFE with high power VUV laser radiation. SPIE Proc. **1412**, 80 (1991)

Battuello M., T. Ricolfi: Effect of the emissivity of real bodies on pyrometer readings. High Temperatures – High Pressures **12**, 247 (1980)

Baufay L., F.A. Houle, R.J. Wilson: Optical self-regulation during laser-induced oxidation of copper. J. Appl. Phys. **61**, 4640 (1987)

Baum T.H., P.B. Comita, T.T. Kodas: Laser-induced gold deposition for thin-film circuit repair. SPIE Proc. **1598**, 122 (1991)

Baum T.H.: Laser chemical vapor deposition of gold. J. Electrochem. Soc. **134**, 2616 (1987)

Bedair S.M., J.K. Whisnant, N.H. Karam, D. Griffis, N.A. El-Masry, H.H. Stadlmaier: Laser selective deposition of III-V compounds on GaAs and Si substrates. J. Cryst. Growth **77**, 229 (1986)

Beeson K.W., V.H. Houlding, R. Beach, R.M. Osgood Jr.: Laser etching of $LiNbO_3$ in a Cl_2 atmosphere. J. Appl. Phys. **64**, 835 (1988)

Benedek G.: Molecule-surface interaction: Vibrational excitations, in *Interfaces under Laser Irradiation*, ed. by L.D. Laude, D. Bäuerle, M. Wautelet, Nato ASI Series (Nijhoff, Dordrecht 1987) p.27

Beneking H.: Laser deposition of single crystalline GaAs and stimulated sheet doping, in *Laser Processing and Diagnostics*, ed. by D. Bäuerle, Springer Ser. Chem. Phys., Vol.39 (Springer, Berlin, Heidelberg 1984) p.188

Benes E.: Improved quartz crystal microbalance technique. J. Appl. Phys. **56**, 608 (1984)

Bennett T.D., C.P. Grigoropoulos, D.J. Krajnovich: Near-threshold laser sputtering of gold. J. Appl. Phys. **77**, 849 (1995)

Ben-Shaul A., Y. Haas, K.L. Kompa, R.D. Levine: *Lasers and Chemical Change*, Springer Ser. Chem. Phys., Vol.10 (Springer, Berlin, Heidelberg 1981)

Bentini G.G., M. Bianconi, C. Summonte: Surface doping of semiconductors by pulsed-laser irradiation in reactive atmosphere. Appl. Phys. A **45**, 317 (1988)

Bentini G.G., M. Servidori, C. Cohen, R. Nipoti, A.V. Drigo: Titanium and nickel silicide formation after Q-switched laser and multiscanning electron beam irradiation. J. Appl. Phys. **53**, 1525 (1982)

Berman M.R.: Laser-assisted etching of gallium arsenide in chlorine at 308 nm. Appl. Phys. A **53**, 442 (1991)

Bertness K.A., T.T. Chiang, C.E. McCants, P.H. Mahowald, A.K. Wahi, T. Kendelewicz, I. Lindau, W.E. Spicer: Comparative uptake kinetics of N_2O and O_2 chemisorption on GaAs(110). Surf. Sci. **185**, 544 (1987)

Beuermann Th., H.J. Brinkmann, T. Damm, M. Stuke: Picosecond UV excimer laser ablation of $LiNbO_3$. MRS Proc. **191**, 37 (1990)

Bezuk S.J., R.J. Baseman, C. Kryzak, K. Warner, G. Thomes: Pyrolytic laser direct writing of nickel over polyimides. MRS Proc. **75**, 75 (1987)

Bhattacharyya A., B.G. Streetman: Theoretical considerations regarding pulsed CO_2 laser annealing of silicon. Solid State Commun. **36**, 671 (1980)

Bialkowski M.M., G.S. Hurst, J.E. Parks, D.H. Lowndes, G.E. Jellison: Charge emission from silicon and germanium surfaces irradiated with KrF excimer laser pulses, in *Laser Ablation – Mechanisms and Applications*, ed. by J.C. Miller, R.F. Haglund, Lecture Notes Phys., Vol.389 (Springer, Berlin, Heidelberg 1991) p.265

Bicknell R.N., N.C. Giles, J.F. Schetzina: p-type CdTe epilayers grown by photoassisted molecular beam epitaxy. Appl. Phys. Lett. **49**, 1735 (1986)

Bird R.B., W.E. Stewart, E.N. Lightfoot: *Transport Phenomena* (Wiley, New York 1960)

Birjega M.I., M. Dinescu, I.N. Mihailescu, L. Nanu, C.A. Constantin, I.T. Florescu, M. Popescu-Pogrion, C. Sarbu: On cw CO_2 laser oxidation and crystallization of thin amorphous sputtered Cr films. Phys. Status Solidi (a) **95**, 423 (1986)

Birnbaum M.: Semiconductor surface damage produced by ruby lasers . J. Appl. Phys. **36**, 3688 (1965)

Bixon M., J. Jortner: Intramolecular radiationless transitions. J. Chem. Phys. **48**, 715 (1968)

Black J.G., S.P. Doran, M. Rothschild, D.J. Ehrlich: Low-temperature laser deposition of tungsten by silane- and disilane-assisted reactions. Appl. Phys. Lett. **56**, 1072 (1990)

Blanchet G.B., C.R. Fincher, C.L. Jackson, S.I. Shah, K.H. Gardner: Laser ablation and the production of polymer films. Science **262**, 719 (1993)

Bockris J.O.M., A.K.N. Reddy: *Modern Electrochemistry* I and II (Plenum, New York 1977)

Bohac V., E. D'Anna, G. Leggieri, S. Luby, A. Luches, E. Majkova, M. Martino: Tungsten silicide formation by XeCl excimer-laser irradiation of W/Si samples. Appl. Phys. A **56**, 391 (1993)

Bolle M., S. Lazare: Large scale excimer laser production of submicron periodic structures on polymer surfaces. Appl. Surf. Sci. **69**, 31 (1993)

Bollmann D., G. Neumayer, R. Buchner, K. Haberger: Shallow p-n junctions produced by laser doping with boron silicate glass. Appl. Surf. Sci. **69**, 249 (1993)

Boman M., D. Bäuerle: Laser-assisted chemical vapor deposition of boron. J. Chinese Chem. Soc. **42**, 405 (1995)

Bor Z., B. Racz, G. Szabo, D. Xenakis, C. Kalpouzos, C. Fotakis: Femto-second transient reflection from polymer surfaces during femtosecond UV photoablation. Appl. Phys. A **60**, 365 (1995)

Born M., E. Wolf: *Principles of Optics* (Pergamon, Oxford 1980)

Borsella E., S. Botti, R. Giorgi, S. Martelli, S. Turtu, G. Zappa: Laser-driven synthesis of nanocrystalline alumina powders from gas-phase precursors. Appl. Phys. Lett. **63**, 1345 (1993*a*)

Borsella E., S. Botti, R. Alexandrescu, I. Morjan, T. Dikonimos-Makris, R. Giorgi, S. Martelli: Nanocomposite ceramic powder production by laser-induced gas-phase reactions. Mater. Sci. Eng. A **168**, 177 (1993*b*)

Boughaba S., G. Auvert: Growth kinetics of micron-size nickel lines produced by laser-assisted decomposition of nickel tetracarbonyl. J. Appl. Phys. **73**, 8590 (1993)

Bowers J.E., B.R. Hemenway, D.P. Wilt: Etching of deep grooves for the precise positioning of cleaves in semiconductor lasers. Appl. Phys. Lett. **46**, 453 (1985)

Boyd I.W.: ULSI dielectrics: Low-temperature silicon dioxides. Mater. Chem. and Phys. **41**, 266 (1995)

Boyd I.W., V. Craciun, A. Kazor: Vacuum-ultra-violet and ozone induced oxidation of silicon and silicon-germanium. Jpn. J. Appl. Phys. **32**, 6141 (1993)

Boyd I.W.: *Laser Processing of Thin Films and Microstructures*, Springer Ser. Mater. Sci., Vol.3 (Springer, Berlin, Heidelberg 1987)

Braichotte D., C. Garrido, H.v.d. Bergh: The photolytic laser chemical vapor deposition rate of platinum, its dependence on wavelength, precursor vapor pressure, light intensity, and laser beam diameter. Appl. Surf. Sci. **46**, 9 (1990)

Braichotte D., H.v.d. Bergh: Temperature effects in the photolytic LCVD of platinum. Appl. Phys. A **49**, 189 (1989)

Braichotte D., H.v.d. Bergh: Gas phase vs. surface contributions to photolytic LCVD rates. Appl. Phys. A **45**, 337 (1988)

Brailovsky A.B., S.V. Gaponov, V.I. Luchin: Mechanisms of melt droplets and solid-particle ejection from a target surface by a pulsed laser action. Appl. Phys. A **61**, 81 (1995)

Brannon J., C. Snyder: Pulsed 532 nm laser wirestripping: Removal of dye doped polyurethane insulation. Appl. Phys. A **59**, 73 (1994)

Brannon J.H., D. Scholl, E. Kay: Ultraviolet photoablation of a plasma-synthesized fluorocarbon polymer. Appl. Phys. A **52**, 160 (1991)

Brannon J.H.: Micropatterning of surfaces by excimer laser projection. J. Vac. Sci. Technol. B **7**, 1064 (1989)

Brannon J.H.: Chemical etching of silicon by CO_2-laser-induced dissociation of NF_3. Appl. Phys. A **46**, 39 (1988)

Brannon J.H.: Excimer laser induced photochemical etching of glass, in *Laser Chemical Processing of Semiconductor Devices*, ed. by F.A. Houle, T.F. Deutsch, R.M. Osgood (Materials Research Society, Boston 1984) Extended Abstracts, p.112

Branz H.M., S. Fan, J.H. Flint, B.T. Fiske, D. Adler, J.S. Haggerty: Doped hydrogenated amorphous silicon films by laser-induced chemical vapor deposition. Appl. Phys. Lett. **48**, 171 (1986)

Braren B., K.G. Casey, R. Kelly: On the gas dynamics of laser-pulse sputtering of polymethylmethacrylate. Nuclear Instrum. Methods B **58**, 463 (1991)

Braren B., R. Srinivasan: Controlled etching of silicate glasses by pulsed ultraviolet laser radiation. J. Vac. Sci. Technol. B **6**, 537 (1988)

Brekel C.H.J. van der: Mass transport and morphology in chemical vapour deposition processes, Dissertation, University of Nijmegen (1978)

Brewer P., J.J. Zinck, G.L. Olson: Excimer laser ablation of CdTe, in *Laser Ablation - Mechanisms and Applications*, ed. by J.C. Miller, R.F. Haglund. Lect. Notes Phys., Vol.389 (Springer, Berlin, Heidelberg 1991) p.96

Brewer P., D. McClure, R.M. Osgood Jr.: Excimer laser projection etching of GaAs. Appl. Phys. Lett. **49**, 803 (1986)

Brewer P., S. Halle, R.M. Osgood: Photon-assisted dry etching of GaAs. Appl. Phys. Lett. **45**, 475 (1984)

Brook M.R., K.I. Grandberg, G.A. Shafeev: Kinetics of laser-induced Au pyrolytic deposition from the liquid phase. Appl. Phys. A **52**, 78 (1991)

Brown W.L.: Laser processing of semiconductors. *Laser Materials Processing* **3**, 337 (North-Holland, Amsterdam 1983)

Bruines J.J.P., R.P.M.van Hal, H.M.J. Boots, W. Sinke, F.W. Saris: Direct observation of resolidification from the surface upon pulsed-laser melting of amorphous silicon. Appl. Phys. Lett. **48**, 1252 (1986)

Brunauer S., P.H. Emmett, E. Teller: Adsorption of gases in multimolecular layers. J. Am. Chem. Soc. **60**, 309 (1938)

Brunco D.P., M.O. Thompson, C.E. Otis, P.M. Goodwin: Temperature measurements of polyimide during KrF excimer laser ablation. J. Appl. Phys. **72**, 4344 (1992)

Buene L., D.C. Jacobson, S. Nakahara, J.M. Poate, C.W. Draper, J.K. Hirvonen: Laser irradiation of nickel: Defect structures and surface alloying, in *Laser and Electron-Beam Solid Interactions and Materials Processing*. MRS Proc. **1**, 583 (1981)

Buhay H., S. Sinharoy, W.H. Kasner, M.H. Francombe, D.R. Lampe, E. Stepke: Pulsed laser deposition and ferroelectric characterization of bismuth titanate films. Appl. Phys. Lett. **58**, 1470 (1991)

Bunkin F.V., A.K. Dmitriyev, B.S. Luk'yanchuk, G.A. Shafeev: Thermo-diffusional instability and potential distribution in laser-heated absorbing electrolytes. Appl. Phys. A **40**, 159 (1986)

Bunkin F.V., B.S. Lukyanchuk, G.A. Shafeev, E.K. Kozlova, A.I. Portniagin, A.A. Yeryomenko, P. Mogyorosi, J.G. Kiss: Si etching affected by IR laser irradiation. Appl. Phys. A **37**, 117 (1985)

Bunkin F.V., N.A. Kirichenko, B.S. Luk'yanchuk: Characteristics of deep-penetration melting by a moving laser beam. Sov. J. Quantum Electron. **11**, 277 (1981)

Bunkin F.V., N.A. Kirichenko, B.S. Luk'yanchuk: Optimal regimes of material heating by laser radiation. Preprint FIAN, No.146 (P.N. Lebedev Phys. Inst., Moscow 1978) (in Russian)

Burg E. von der, M. Diegel, H. Stafast, W. Grill: $Y_1 Ba_2 Cu_3 O_{7-\delta}$ thin films from KrF laser ablation with substrate heating and in situ patterning by CO_2 laser radiation. Appl. Phys. A **54**, 373 (1992)

Burgener M.L., R.E. Reedy: Temperature distributions produced in a two-layer structure by a scanning cw laser or electron beam. J. Appl. Phys. **53**, 4357 (1982)

Burke H.H., I.P. Herman, V. Tavitian, J.G. Eden: Laser photochemical deposition of germanium-silicon alloy thin films. Appl. Phys. Lett. **55**, 253 (1989)

Bykovskii Y.A., V.M. Boyakov, V.T. Galochkin, A.S. Molchanov, I.N. Nokolaev, A.N. Orevskii: Deposition of metal, semiconductor, and oxide films with a periodically pulsed CO_2 laser. Sov. Phys. – Tech. Phys. **23**, 578 (1978)

Bykovskii Y.A., A.G. Dudoladov, V.P. Kozlenkov, P.A. Leont'ev: Oriented crystallization of thin films that are obtained with the aid of a laser. JETP Lett. **20**, 135 (1974)

Cabrera N., N.F. Mott: Theory of the oxidation of metals. Rep. Prog. Phys. **12**, 163 (1949)

Calder I.D., R. Sue: Modeling of cw laser annealing of multilayer structures. J. Appl. Phys. **53**, 7545 (1982)

Calvert J.G., J.N. Pitts: *Photochemistry* (Wiley, New York 1966)

Campbell E.E.B., G. Ulmer, B. Hasselberger, H.-G. Busmann, I.V. Hertel: An intense, simple carbon cluster source. J. Chem.Phys. **93**, 6900 (1990)

Campisano S.U., D.C. Jacobson, J.M. Poate, A.G. Cullis, N.G. Chew: Impurity and interfacial effects on the formation of amorphous Si from the melt. Appl. Phys. Lett. **45**, 1216 (1984)

Cao L.X., Z.C. Feng, Y. Liang, W.L. Hou, B.C. Zhang, Y.Q. Wang, L. Li: LCVD of TiN and TiC films. Thin Solid Films **257**, 7 (1995)

Carslaw H.S., J.C. Jaeger: *Conduction of Heat in Solids* (Clarendon, Oxford 1988)

Celler G.K.: Modification of silicon properties with lasers, electron beams, and incoherent light. CRC Crit. Rev. Solid State and Mater. Sci. **12**, 193 (1984)

Chan C.L., J. Mazumder: One-dimensional steady-state model for damage by vaporization and liquid expulsion due to laser-material interaction. J. Appl. Phys. **62**, 4579 (1987)

Chandrasekhar S.: *Hydrodynamic and Hydromagnetic Stability* (Clarendon, Oxford 1961)

Chang C.C., X.D. Wu, R. Ramesh, X.X. Xi, T.S. Ravi, T. Venkatesan, D.M. Hwang, R.E. Muenchhausen, S. Foltyn, N.S. Nogar: Origin of surface roughness for c-axis oriented Y-Ba-Cu-O superconducting films. Appl. Phys. Lett. **57**, 1814 (1990)

Chang Y.C., J.T. Cheung, A. Chiou, M. Khoshnevisan: Optical properties of $Hg_{1-x}Cd_xTe$ sawtooth superlattices. J. Appl. Phys. **66**, 829 (1989)

Chapman R.L., J.C.C. Fan, H.J. Zeiger, R.P. Gale: Crystallization-front velocity during scanned laser crystallization of amorphous Ge films. Appl. Phys. Lett. **37**, 292 (1980)

Char K., D.K. Fork, T.H. Geballe, S.S. Laderman, R.C. Taber, R.D. Jacowitz, F. Bridges, G.A.N. Connell, J.B. Boyce: Properties of epitaxial $YBa_2Cu_3O_7$ Thin Films on Al_2O_3 {$\bar{1}012$}. Appl. Phys. Lett. **56**, 785 (1990)

Chaudhari G.N., V.J. Rao: Electrical properties of polyimide on n-GaAs (100) interfaces by a pulsed laser evaporation technique. Appl. Phys. A **56**, 353 (1993)

Chen C.J., R.M. Osgood: Direct observation of the local-field-enhanced surface photochemical reactions. Phys. Rev. Lett. **50**, 1705 (1983); Surface-catalyzed photochemical reactions of physisorbed molecules. Appl. Phys. A **31**, 171 (1983)

Chen X., J. Mazumder, A. Purohit: Optical emission diagnostics of laser-induced plasma for diamond-like film deposition. Appl. Phys. A **52**, 328 (1991)

Cheung J.T., H. Sankur: The unique applications of pulsed laser deposition to the epitaxial growth of semiconductor films, in *Laser Ablation of Electronic Materials – Basic Mechanisms and Applications*, ed. by E. Fogarassy, S. Lazare (Elsevier, Amsterdam 1992) p.325

Cheung J.T., E.H. Cirlin, N. Otsuka: Structure of nonrectangular HgCdTe superlattice grown by laser MBE. Appl. Phys. Lett. **53**, 310 (1988)

Cheung J.T., J. Madden: Growth of HgCdTe epilayers with any predesigned compositional profile by laser molecular beam epitaxy. J. Vac. Sci. Technol. B **5**, 705 (1987)

Cheung J.T., G. Niizawa, J. Moyle, N.P. Ong, B.M. Paine, T. Vreeland, Jr.: HgTe and CdTe epitaxial layers and HgTe-CdTe superlattices grown by laser molecular beam epitaxy. J. Vac. Sci. Technol. A **4**, 2086 (1986)

Chiba H., K. Murakami, O. Eryu, K. Shihoyama, T. Mochizuki, K. Masuda: Laser excitation effects on laser ablated particles in fabrication of high-T_c superconducting thin films. Jpn. J. Appl. Phys. **30**, L732 (1991)

Chou W.B., M.N. Azer, J. Mazumder: Laser chemical vapor deposition of Ti from $TiBr_4$. J. Appl. Phys. **66**, 191 (1989)

Chrisey D.B., J.S. Horwitz, R.E. Leuchtner: Excimer laser ablation of a $YBa_2Cu_3O_{7-\delta}$ target in a vacuum: Characterization of the mass and energy of ejected material. Thin Solid Films **206**, 111 (1991)

Chu S.S., T.L. Chu, C.L. Chang, H. Firouzi: Laser-induced homoepitaxial growth of gallium arsenide films. Appl. Phys. Lett. **52**, 1243 (1988)

Chuang T.J., H. Hiraoka, A. Mödl: Laser-photoetching characteristics of polymers with dopants. Appl. Phys. A **45**, 277 (1988)

Chuang T.J.: Gas-surface interactions stimulated by laser radiation: Bases and applications, in *Interfaces under Laser Irradiation*, ed. by L.D. Laude, D. Bäuerle, M. Wautelet, Nato ASI Series (Nijhoff, Dordrecht 1987) p.235

Chuang T.J., I. Hussla, W. Sesselmann: Laser-assisted chemical etching of inorganic materials: Mechanistic studies, in *Laser Processing and Diagnostics*, ed. by D. Bäuerle, Springer Ser. Chem. Phys., Vol.39 (Springer, Berlin, Heidelberg 1984) p.300

Chuang T.J.: Laser-enhanced chemical etching of solid surfaces. IBM J. Res. Dev. **26**, 145 (1982)

Chuang T.J.: Multiple photon excited SF_6 interaction with silicon surfaces, J. Chem. Phys. **74**, 1453 (1981)

Chvoj Z.: Kinetics of the non-equilibrium phase transformations. J. Non-Equilib.Thermodyn. **18**, 201 (1993)

Cillessen J.F.M., R.M. Wolf, D.M.de Leeuw: Pulsed laser deposition of hetero-epitaxial thin Pt films on MgO (100). Thin Solid Films **226**, 53 (1993a)

Cillessen J.F.M., R.M. Wolf, A.E.M. De Veirman: Hetero-epitaxial oxidic conductor $La_{1-x}Sr_xCoO_3$ prepared by pulsed laser deposition. Appl. Surf. Sci. **69**, 212 (1993b)

Clark J.H., R.G. Anderson: Silane purification via laser-induced chemistry. Appl. Phys. Lett. **32**, 46 (1978)

Cleland T.A., D.W. Hess: Diagnostics and modeling of N_2O RF glow discharges. J. Electrochem. Soc. **136**, 3103 (1989)

Clyne M.A.A., D.H. Stedman: Recombination of ground-state halogen atoms I. Radiative recombination of chlorine atoms. Trans. Faraday Soc. **64**, 1816 (1968); ibid. II. Kinetics of the overall recombination of chlorine atoms. Trans. Faraday Soc. **64**, 2698 (1968)

Cohen C., J. Siejka, G.G. Bentini, M. Berti, L.F. Dona Dalle Rose, A.V. Drigo: Loss and incorporation mechanisms during pulsed laser irradiation of silicon in gaseous atmosphere, in *Laser Processing and Diagnostics II*, ed. by D. Bäuerle, K.L. Kompa, L.D. Laude (Physique, Les Ulis 1986) p.249

Comita P.B., E. Kay, R. Zhang, W. Jacob: Laser-induced coalescence of gold clusters in fluorocarbon composite thin films. Appl. Surf. Sci. **79/80**, 196 (1994)

Comita P.B., P.E. Price, T.T. Kodas: Time-resolved reflectance studies of surface melting during laser-assisted deposition with a modulated laser source. J. Appl. Phys. **71**, 221 (1992)

Compaan A.: Phonon populations during pulsed laser annealing. J. Luminesc. **30**, 425 (1985)

Conde O., A. Kar, J. Mazumder: Laser chemical vapor deposition of TiN dots: A comparison of theoretical and experimental results. J. Appl. Phys. **72**, 754 (1992)

Costela A., J.M. Figuera, F. Florido, I. Garcia-Moreno, E.P. Collar, R. Sastre: Ablation of poly(methyl methacrylate) and poly(2-hydroxyethyl methacrylate) by 308, 222 and 193 nm excimer-laser radiation. Appl. Phys. A **60**, 261 (1995)

Cotell C.M.: Pulsed laser deposition and processing of biocompatible hydroxylapatite thin films. Appl. Surf. Sci. **69**, 140 (1993)

Coufal H.: Photothermal spectroscopy using a pyroelectric thin-film detector. Appl. Phys. Lett. **44**, 59 (1984)

Craciun V., S. Amirhaghi, D. Craciun, J. Elders, J.G.E. Gardeniers, I.W. Boyd: Effects of laser wavelength and fluence on the growth of ZnO thin films by pulsed laser deposition. Appl. Surf. Sci. **86**, 99 (1995)

Craciun V., I.W. Boyd, A.H. Reader, W.J. Kersten, F.J.G. Hakkens, P.H. Oosting, D.E.W. Vandenhoudt: Microstructure of oxidized layers formed by the low-temperature ultraviolet-assisted dry oxidation of strained $Si_{0.8}Ge_{0.2}$ layers on Si. J. Appl. Phys. **75**, 1972 (1994)

Craciun V., I.N. Mihailescu, A. Luches, S.G. Kiyak, G.N. Mikhailova: Laser processing of germanium. SPIE Proc. **1392**, 629 (1990)

Crank J.: *Free and Moving Boundary Problems* (Clarendon, Oxford 1988)

Cronberg H., M. Reichling, E. Broberg, H.B. Nielsen, E. Matthias, N. Tolk: Effects of inverse Bremsstrahlung in laser-induced plasmas from a graphite surface. Appl. Phys. B **52**, 155 (1991)

Cross F.W., R.K. Al-Dhahir, P.E. Dyer: Ablative and acoustic response of pulsed UV laser-irradiated vascular tissue in a liquid environment. J. Appl. Phys. **64**, 2194 (1988)

Cullis A.G.: Transient annealing of semiconductors by laser, electron beam and radiant heating techniques. Rep. Prog. Phys. **48**, 1155 (1985)

Cullis A.G., H.C. Webber, P. Bailey: A device for laser beam diffusion and homogenisation. J. Phys. E **12**, 688 (1979)

Curcio F., I. Gianinoni, M. Musci: CO_2 laser-assisted deposition of amorphous semiconductors, in *Laser Processing and Diagnostics II*, ed. by D. Bäuerle, K.L. Kompa, L.D. Laude (Les Editions de Physique, Les Ulis 1986) p.117

D'Anna E., G. Leggieri, A. Luches, M. Martino, A.V. Drigo, I.N. Mihailescu, S. Ganatsios: Synthesis of pure titanium nitride layers by multipulse excimer laser irradiation of titanium foils in a nitrogen-containing atmosphere. J. Appl. Phys. **69**, 1687 (1991)

D'Anna E., G. Leggieri, A. Luches: Laser synthesis of metal silicides. Appl. Phys. A **45**, 325 (1988)

Datta M., L.T. Romankiw, D.R. Vigliotti, R.J. von Gutfeld: Laser etching of metals in neutral salt solutions. Appl. Phys. Lett. **51**, 2040 (1987)

Davies A.J.: *The Finite Element Method* (Clarendon, Oxford 1980)

Davis G.M., M.C. Gower: Epitaxial growth of thin films of $BaTiO_3$ using excimer laser ablation. Appl. Phys. Lett. **55**, 112 (1989)

Davis G.M., M.C. Gower, C. Fotakis, T. Efthimiopoulos, P. Argyrakis: Spectroscopic studies of ArF laser photoablation of PMMA. Appl. Phys. A **36**, 27 (1985)

Davis M., P. Kapadia, J. Dowden, W.M. Steen, C.H.G. Courtney: Heat hardening of metal surfaces with a scanning laser beam. J. Phys. D **19**, 1981 (1986)

Deal B.E., A.S. Grove: General relationship for the thermal oxidation of silicon. J. Appl. Phys. **36**, 3770 (1965)

Demtröder W.: *Laser Spectroscopy*, 2nd edn. (Springer, Berlin, Heidelberg 1996)

DePuydt Y., P. Bertrand, P. Lutgen: Study of the Al/PET interface in relation with adhesion. Surf. Interface Anal. **12**, 486 (1988)

Desjardins P., R. Izquierdo, M. Meunier: Diode laser induced CVD of WSi_x on TiN from WF_6 and SiH_4. J. Appl. Phys. **73**, 5216 (1993)

Desserre J., J.F. Eloy: Interaction d'un Faisceau de Lumiere Coherente Pulse avec une Cible Complexe: Application a d'Elaboration de Composes en Couches Minces. Thin Solid Films **29**, 29 (1975)

Deutsch T.F.: IR-laser ablation in medicine: Mechanisms and applications, in *Laser Ablation – Mechanisms and Applications*, ed. by J.C. Miller, R.F. Haglund, Lecture Notes Phys., Vol.389 (Springer, Berlin, Heidelberg 1991) p.109

Deutsch T.F.: Applications of excimer lasers to semiconductor processing, in *Laser Processing and Diagnostics*, ed. by D. Bäuerle, Springer Ser. Chem. Phys., Vol.39 (Springer, Berlin, Heidelberg 1984) p.239

Dickey F.M., B.D. O'Neil: Multifaceted laser beam integraters: General formulation and design concepts. Opt. Eng. **27**, 999 (1988)

Dickinson J.T., S.C. Langford, J.J. Shin, D.L. Doering: Positive ion emission from excimer laser excited MgO surfaces. Phys. Rev. Lett. **73**, 2630 (1994)

Dickinson J.T., S.C. Langford, L.C. Jensen: Simultaneous bombardment of wide bandgap materials with UV excimer irradiation and keV electrons, in *Laser Ablation – Mechanisms and Applications*, ed. by J.C. Miller, R.F. Haglund. Lecture Notes Phys., Vol.389 (Springer, Berlin, Heidelberg 1991) p.301

Dienstbier M., R. Benes, P. Rejfir, P. Sladky: Study of UV laser ablation by optothermal methods. Appl. Phys. B **51**, 137 (1990)

Dietrich T.R., S. Chiussi, H. Stafast, F.J. Comes: ArF laser CVD of hydrogenated amorphous silicon: The role of buffer gases. Appl. Phys. A **48**, 405 (1989)

Dimitrov D., S. Metev, I. Gugov, V. Kozhukharov: Properties of laser beam deposited thin polycomponent films. J. Mater. Sci. Lett. **1**, 334 (1982)

Dirnberger L., P.E. Dyer, S. Farrar, P.H. Key, P. Monk: Laser ablation studies of magnesium oxide. Appl. Surf. Sci. **69**, 216 (1993)

Doi A., Y. Aoyagi, S. Namba: Stepwise monolayer growth of GaAs by switched laser metalorganic vapor phase epitaxy. Appl. Phys. Lett. **49**, 785 (1986)

Donley M.S., J.S. Zabinski, V.J. Dyhouse, P.J. John, P.T. Murray, N.T. McDevitt: Pulsed laser deposition of tribological materials, in *Laser Ablation – Mechanisms and Applications*, ed. by J.C. Miller, R.F. Haglund. Lecture Notes Phys., Vol. 389 (Springer, Berlin, Heidelberg 1991) p.271

Donnelly V.M.: Products of pulsed laser induced thermal decomposition of triethylgallium and trimethylgallium adsorbed on GaAs (100). J.Vac. Sci. Technol. A **9**, 2887 (1991)

Donnelly V.M., T.R. Hayes: Excimer laser-induced etching of InP. Appl. Phys. Lett. **57**, 701 (1990)

Donnelly V.M., J.A. Mucha, V.R. McCrary: Mechanisms of dehydrogenation during ArF excimer laser patterning of plasma-deposited silicon nitride films. J. Appl. Phys. **67**, 3337 (1990)

Donnelly V.M., J.A. Mucha: Direct pattern replication in plasma deposited silicon nitride films by 193 nm ArF excimer laser-induced suppression of etching. Appl. Phys. Lett. **54**, 1567 (1989)

Donnelly V.M., J.A. McCaulley: Selected area growth of GaAs by laser-induced pyrolysis of adsorbed triethylgallium. Appl. Phys. Lett. **54**, 2458 (1989)

Donnelly V.M., C.W. Tu, J.C. Beggy, V.R. McCrary, M.G. Lamont, T.D. Harris, F.A. Baiocchi, R.C. Farrow: Laser-assisted metalorganic molecular beam epitaxy of GaAs. Appl. Phys. Lett. **52**, 1065 (1988)

Donnelly V.M., D. Brasen, A. Appelbaum, M. Geva: Excimer laser induced deposition of InP. J. Vac. Sci. Technol. A **4**, 716 (1986)

Donnelly V.M., M. Geva, J. Long, R.F. Karlicek: Excimer laser-induced deposition of InP and indium-oxide films. Appl. Phys. Lett. **44**, 951 (1984)

Donohue T.: Laser surface modification below a liquid layer, in *Laser Processing and Diagnostics*, ed. by D. Bäuerle, Springer Ser. Chem. Phys., Vol.39 (Springer, Berlin, Heidelberg 1984) p.332

Doppelbauer J., D. Bäuerle: Kinetics of laser-induced pyrolytic chemical processes and the problem of temperature measurements, in *Interfaces under Laser Irradiation*, ed. by L.D. Laude, D. Bäuerle, M. Wautelet, NATO ASI Series (Nijhoff, Dordrecht 1987) p.277

Doppelbauer J.: Kinetische Untersuchungen an laserinduzierten pyrolytischen Prozessen, Dissertation, Universität Linz, Austria (1987)

Doppelbauer J., D. Bäuerle: Kinetic studies of pyrolytic laser-induced chemical processes, in *Laser Processing and Diagnostics II*, ed. by D. Bäuerle, K.L. Kompa, L.D. Laude (Les Editions de Physique, Les Ulis 1986) p.53

Driel H.M. van, J.E. Sipe, J.F. Young: Laser-induced coherent modulation of solid and liquid surfaces. J. Luminesc. **30**, 446 (1985)

Dubowski J.J., D.F. Williams, P.B. Sewell, P. Norman: Epitaxial growth of (100) CdTe on (100) GaAs induced by pulsed laser evaporation. Appl. Phys. Lett. **46**, 1081 (1985)

Dubowski J.J., D.F. Williams: Pulsed laser evaporation of Cd_3As_2. Appl. Phys. Lett. **44**, 339 (1984)

Duley W.W.: *Laser Processing and Analysis of Materials* (Plenum, New York 1983)

Duley W.W.: CO_2 *Lasers – Effects and Applications* (Academic, London 1976)

Dye R.C., R.E. Muenchausen, N.S. Nogar, A. Mukherjee, S.R.J. Brueck: Laser writing of superconducting patterns on $YBa_2Cu_3O_x$ Films. Appl. Phys. Lett. **57**, 1149 (1990)

Dyer P.E., S.R. Farrar, P.H. Key: Fast time-response photoacoustic studies and modelling of KrF laser ablated $YBa_2Cu_3O_7$. Appl. Surf. Sci. **54**, 255 (1992a)

Dyer P.E., S.R. Farrar, A. Issa, P.H. Key: Laser ablation dynamics of superconductors: Photoacoustic and spectroscopic studies, in *Laser Ablation of Electronic Materials*, ed. by E. Fogarassy and S. Lazare (Elsevier, Amsterdam 1992b) p.101

Dyer P.E., G.A. Oldershaw, J. Sidhu: Ultraviolet-laser-induced ablation of poly (ethylene terephthalate). J. Phys. Chem. **95**, 10004 (1991)

Dyer P.E., A. Issa, P.H. Key: Dynamics of excimer laser ablation of superconductors in an oxygen environment. Appl. Phys. Lett. **57**, 186 (1990a)

Dyer P.E, G.A. Oldershaw, D. Schudel: XeCl laser ablation of poly-etheretherketone. Appl. Phys. B **51**, 314 (1990b)

Dyer P.E., R.J. Farley: Periodic surface structures in the excimer laser ablative etching of polymers. Appl. Phys. Lett. **57**, 765 (1990)

Dyer P.E., G.A. Oldershaw, J. Sidhu: CO_2 laser ablative etching of polyethylene terephthalate. Appl. Phys. B **48**, 489 (1989)

Dyer P.E., R.D. Greenough, A. Issa, P.H. Key: Spectroscopic and ion probe measurements of KrF laser ablated Y-Ba-Cu-O bulk samples. Appl. Phys. Lett. **53**, 534 (1988)

Dyer P.E., S.D. Jenkins, J. Sidhu: Development and origin of conical structures on XeCl laser ablated polyimide. Appl. Phys. Lett. **49**, 453 (1986)

Dyer P.E., S.A. Ramsden, J.A. Sayers, M.A. Skipper: The interaction of CO_2 laser radiation with various solid targets. J. Phys. D **9**, 373 (1976)

Dykhne A.M., B.P. Rysev: On the simultaneous excitation of surface acoustic and electromagnetic waves under the thermoelastic action of laser radiation. Bull. USSR Acad. Sci., Ser. Phys, **50** (3), 609 (1986)

Eddy M.M., J.Z. Sun, R.D. Hammond, L. Drabeck, I.B. Ferreira, K. Holczer, G. Grüner: Surface resistance studies of laser-deposited superconducting $Tl_2Ba_2CaCu_2O_8$ films. J. Appl. Phys. **70**, 496 (1991)

Eden J.G.: Photochemical vapor deposition, in *Thin Film Processes II* (Academic, London 1991)

Ediger M.N., G.H. Pettit: Time-resolved reflectivity of ArF laser-irradiated polyimide. J. Appl. Phys. **71**, 3510 (1992)

Egitto F.D., C.R. Davis: Dopant-induced excimer laser ablation of poly (tetrafluoroethylene) II. Effect of dopant concentration. Appl. Phys. B **55**, 488 (1992)

Ehrlich D.J.: Critical issues for single-chamber manufacturing: The role of laser technology. Appl. Surf. Sci. **69**, 115 (1993)

Ehrlich D.J., J.Y. Tsao (eds.): *Laser Microfabrication – Thin Film Processes and Lithography* (Academic, London 1989)

Ehrlich D.J., J.Y. Tsao, C.O. Bozler: Submicrometer patterning by projected excimer-laser-beam induced chemistry. J. Vac. Sci. Technol. B **3**, 1 (1985)

Ehrlich D.J., R.M. Osgood, Jr., T.F. Deutsch: Photodeposition of metal films with UV laser light. J. Vac. Sci. Technol. **21**, 23 (1982)

Ehrlich D.J., J.Y. Tsao: Submicrometer-linewidth doping and relief definition in silicon by laser-controlled diffusion. Appl. Phys. Lett **41**, 297 (1982)

Ehrlich D.J., R.M. Osgood, T.F. Deutsch: Direct writing of regions of high doping on semiconductors by UV-laser photodeposition. Appl. Phys. Lett. **36**, 916 (1980)

Eibl O., B. Roas: Microstructure of $YBa_2Cu_3O_{7-x}$ thin films deposited by laser evaporation. J. Mater. Res. **5**, 2620 (1990)

Elders J., J.D.W.v.Voorst: Laser-induced chemical vapor deposition of titanium diboride. J. Appl. Phys. **75**, 553 (1994)

Eliasson B., U. Kogelschatz: Electron impact dissociation in oxygen. J. Phys. B **19**, 1241 (1986)

Emery K., P.K. Boyer, L.R. Thompson, R. Solanki, H. Zarnani, G.J. Collins: Thin film deposition by UV laser photolysis, in *Laser Assisted Deposition, Etching and Doping*, ed. by S.D. Allen, SPIE Proc. **459**, 9 (1984)

Emmerich R., S. Bauer, B. Ploss: Temperature distribution in a film heated with a laser spot: Theory and measurement. Appl. Phys. A **54**, 334 (1992)

Eres D., D.H. Lowndes, D.B. Geohegan, D.M. Mashburn: Laser photochemical growth of amorphous silicon at low temperatures and comparison with thermal chemical vapor deposition. MRS Proc. **101**, 355 (1988)

Ertl G., J. Küppers: *Low Energy Electrons and Surface Chemistry* (VCH, Weinheim 1985)

Eschbach P.A., J.T. Dickinson, S.C. Langford, L.R. Pederson: The interaction of ultraviolet excimer laser light with sodium trisilicate. J. Vac. Sci. Technol. A **7**, 2943 (1989)

Esenaliev R.O., A.A. Oraevsky, V.S. Letokhov: Laser ablation of atherosclerotic blood vessel tissue under various irradiation conditions. IEEE Trans. BE-**36**, 1188 (1989)

Esrom H., G. Wahl: UV excimer laser-induced pre-nucleation of surfaces followed by electroless metallization. Chemtronics **4**, 216 (1989)

Eyett M., D. Bäuerle: Laser-induced surface reduction of $LiNbO_3$. Ferroelectrics Lett. **8**, 93 (1988)

Eyett M., D. Bäuerle: Influence of the beam spot size on ablation rates in pulsed-laser processing. Appl. Phys. Lett. **51**, 2054 (1987)

Eyett M., D. Bäuerle, W. Wersing, H. Thomann: Excimer laser-induced etching of ceramic $PbTi_{1-x}Zr_xO_3$. J. Appl. Phys. **62**, 1511 (1987)

Eyett M., D. Bäuerle, W. Wersing, K. Lubitz, H. Thomann: Laser-induced chemical etching of ceramic $PbTi_{1-x}Zr_xO_3$. Appl. Phys. A **40**, 235 (1986)

Eyring H., S.H. Lin, S.M. Lin: *Basic Chemical Kinetics* (Wiley, New York 1980)

Fabisiak K., P. Hoffmann, J.-M. Philippoz, H. van den Bergh: A laser-induced chemical transport reaction method for diamond deposition. Surface and Coatings Technology **47**, 528 (1991)

Fairand B.P., A.H. Clauer, R.G. Jung, B.A. Wilcox: Quantitative assessment of laser-induced stress waves generated at confined surfaces. Appl. Phys. Lett. **25**, 431 (1974)

Fairand B.P., B.A. Wilcox, W.J. Gallagher, D.N. Williams: Laser shock-induced microstructural and mechanical property changes in 7075 aluminum. J. Appl. Phys. **43**, 3893 (1972)

Fan J.C.C., H.J. Zeiger: Crystallization of amorphous silicon films by Nd:YAG laser heating. Appl. Phys. Lett. **27**, 224 (1975)

Feigelson R.S.: Opportunities for research on single-crystal fibers. Mater. Sci. Eng. B **1**, 67 (1988)

Feldmann D., J. Kutzner, J. Laukemper, S. MacRobert, K.H. Welge: Mass spectroscopic studies of the ArF-laser photoablation of polystyrene. Appl. Phys. B **44**, 81 (1987)

Fennell M.D., G.J. Fisanick, D.K. Atwood: Electrical characterization of Au lines produced by laser-direct-writing of metallopolymer films, in *Beam Induced Chemical Processes*, ed. by R.J. von Gutfeld, J.E. Greene, H. Schlossberg (Materials Research Society, Boston 1985) Extended Abstracts, p.39

Fernandez D., P. Gonzalez, J. Pou, E. Garcia, J. Serra, B. Leon, M. Perez-Amor, C. Garrido: CO_2 laser chemical vapor deposition of silica films in a parallel configuration: A study of gas phase phenomena. J. Vac. Sci. Technol. A **12**, 484 (1994)

Ferziger J.H., H.G. Kaper: *Mathematical Theory of Transport Processes in Gases* (North Holland, Amsterdam 1972)

Feurer T., R. Sauerbrey, M.C. Smayling, B.J. Story: UV-laser-induced permanent electrical conductivity in polyimide. Appl. Phys. A **56**, 275 (1993)

Fisanick G.J., M.E. Gross, J.B. Hopkins, M.D. Fennell, K.J. Schnoes, A. Katzir: Laser-initiated microchemistry in thin films: Development of new types of periodic structures. J. Appl. Phys. **57**, 1139 (1985)

Fischer J.P., J. Dams, M.H. Götz, E. Kerker, F.H. Loesel, C.J. Messer, M.H. Niemz, N. Suhm, J.F. Bille: Plasma-mediated ablation of brain tissue with picosecond laser pulses. Appl. Phys. B **58**, 493 (1994)

Fisher V.I., V.M. Kharash: Fast gas-ionization wave in a laser beam. Sov. Phys. – JETP, **56** (5), 1004 (1982)

Flynn D.K., J.I. Steinfeld, D.S. Sethi: Deposition of refractory metal films by rare-gas halide laser photodissociation of metal carbonyls. J. Appl. Phys. **59**, 3914 (1986)

Fogarassy E., B. Prevot, S. De Unamuno, M. Elliq, H. Pattyn, E.L. Mathe, A. Naudon: Pulsed laser crystallization of hydrogen-free a-Si thin films for high-mobility poly-Si TFT fabrication. Appl. Phys. A **56**, 365 (1993)

Fogarassy E.: Basic mechanisms and applications of the laser-induced forward transfer for high T_c superconducting thin film deposition. SPIE Proc. **1394**, 169 (1990)

Fogarassy E., C.W. White, A. Slaoui, C. Fuchs, P. Siffert, S.J. Pennycook: Excimer laser induced oxidation of ion-implanted silicon. Appl. Phys. Lett. **53**, 1720 (1988)

Fogarassy E.P., D.H. Lowndes, R. Narayan, C.W. White: UV laser incorporation of dopants into silicon: Comparison of two processes. J. Appl. Phys. **58**, 2167 (1985)

Foltyn S.R., R.E. Muenchausen, R.C. Dye, X.D. Wu, L. Luo, D.W. Cooke, R.C. Taber: Large-area, two-sided superconducting $YBa_2Cu_3O_{7-x}$ films deposited by pulsed laser deposition. Appl. Phys. Lett. **59**, 1374 (1991)

Fork D.K., F.A. Ponce, J.C. Tramontana, T.H. Geballe: Epitaxial MgO on Si(001) for Y-Ba-Cu-O thin-film growth by pulsed laser deposition. Appl. Phys. Lett. **58**, 2294 (1991*a*)

Fork D.K., F.A. Ponce, J.C. Tramontana, N. Newman, J.M. Phillips, T.H. Geballe: High critical current densities in epitaxial $YBa_2Cu_3O_{7-\delta}$ thin films on silicon-on-sapphire. Appl. Phys. Lett. **58**, 2432 (1991*b*)

Foulon F., M. Green: Laser projection-patterned etching of (100) GaAs by gaseous HCl and CH_3Cl. Appl. Phys. A **60**, 377 (1995)

Foulon F., M. Stuke: Argon-ion laser direct-write Al deposition from trialkylamine alane precursors. Appl. Phys. A **56**, 283 (1993)

Foulon F., M. Stuke: Excimer laser projection-patterned deposition of Al via photolytically driven decomposition of trialkylamine alane as adsorbate precursor. Appl. Phys. A **56**, 267 (1993)

Freeman D.L., J.D. Doll: The influence of diffusion on surface reaction kinetics. J. Chem. Phys. **78**, 6002 (1983)

Friedmann T.A., K.F. McCarty, E.J. Klaus, J.C. Barbour, W.M. Clift, H.A. Johnsen, D.L. Medlin, M.J. Mills, D.K. Ottesen: Pulsed laser deposition of BN onto silicon (100) substrates at 600°C. Thin Solid Films **237**, 48 (1994)

Frisoli J.K., Y. Hefetz, T.F. Deutsch: Time-resolved UV absorption of polyimide – Implications for laser ablation. Appl. Phys. B **52**, 168 (1991)

Fritzsche H.: Noncrystalline semiconductors. Physics Today **37**, 34 (October 1984)

Frugier T., A. Boulahia, A. Sayah, D. Tonneau, J.E. Bourée, J.M. Siffre, D. Mencaraglia: Laser-induced deposition of aluminium on gallium arsenide and silicon nitride from trimethylamine alane. Appl. Surf. Sci. **69**, 305 (1993)

Fujii T., I. Morioka, H. Uehara: Buoyant plume above a horizontal line heat source. J. Heat Mass Transfer **16**, 755 (1973)

Fujita Y., S. Fujii, T. Iuchi: Ultraviolet spectra of II-VI organometallic compounds and their application to in situ measurements of the photolysis in a metalorganic chemical vapor deposition reactor. J. Vac. Sci. Technol. A **7**, 276 (1989)

Fukumura H., N. Mibuka, S. Eura, H. Masuhara: Porphyrin-sensitized laser swelling and ablation of polymer films. Appl. Phys. A **53**, 255 (1991)

Furzikov N.P.: Different lasers for angioplasty: Thermooptical comparison. IEEE J. **QE-23**, 1751 (1987)

Fuss W., T.P. Cotter: Energy and pressure dependence of the CO_2 laser-induced dissociation of sulfur hexafluoride. Appl. Phys. **12**, 265 (1977)

Ganz J.: Einsatzmöglichkeiten des Laser-CVD Verfahrens bei der Erzeugung feiner Strukturen. Opto Elektronik **5** (1988)

Gaponov S.V., A.A. Gudkov, B.M. Luskin, V.I. Luchin, N.N. Salashchenko: Formation of semiconductor films from a laser erosion plasma scattered by a heated screen. Sov. Phys. – Tech. Phys. **26**, 598 (1981)

Gaponov S.V., B.M. Luskin, N.N. Salashchenko: Superlattices based on InSb-CdTe, InSb-PbTe, Bi-CdTe pairs. Sov. Phys. – Semicond. **14**, 873 (1980)

Gaponov S.V., A.A. Gudkov, B.M. Luskin, V.I. Luchin, N.N. Salashenko: Reflection of a laser plasma from a hot shield. Sov. Tech. – Phys. Lett. **5**, 195 (1979*a*)

Gaponov S.V., B.M. Luskin, N.N. Salashenko: Synthesis of a superlattice structure by laser deposition. Sov. Phys. – Tech. Phys. **5**, 210 (1979*b*)

Gaponov S.V., B.M. Luskin, B.A. Nesterov, N.N. Salashchenko: Morphological features and structure of films condensed from a laser plasma. Sov. Phys. – Solid State **19**, 1736 (1977)

Gaponov-Grekhov A.V., A.S. Lomov, G.V. Osipov, M.I. Rabinovich: Pattern formation and dynamics of two-dimensional structures in non-equilibrium dissipative media, in *Nonlinear Waves 1 – Dynamics and Evolution*, ed. by A.V. Gaponov-Grekhov, M.I. Rabinovich, J. Engelbrecht (Springer, Berlin, Heidelberg 1989)

Garcia B.J., J. Martinez, J. Piqueras: Ge diffusion into GaAs by pulsed laser irradiation. Appl. Phys. A **46**, 191 (1988)

Garrison B.J., R. Srinivasan: Laser ablation of organic polymers: Microscopic models for photochemical and thermal processes. J. Appl. Phys. **57**, 2909 (1985)

Gasser A., E.W. Kreutz, K. Wissenbach: Beschichten mit CO_2-Laserstrahlung (Cladding with CO_2-laser radiation), in *Oberflächentechnik*, SURTEC Berlin'89 (Hanser, München 1989) p.545

Gauthier R., C. Guittard: Mechanism investigations of a pulsed laser light induced desorption. Phys. Status Solidi (a) **38**, 477 (1976)

Gelchinski M.H., R.J. von Gutfeld, L.T. Romankiw: Laser enhanced electroplating, in *Interfaces under Laser Irradiation*, ed. by L.D. Laude, D. Bäuerle, M. Wautelet, Nato ASI Series (Nijhoff, Dordrecht 1987) p.349

Geohegan D.B.: Physics and diagnostics of laser ablation plume propagation for high-T_c superconductor film growth. Thin Solid Films **220**, 138 (1992)

Geretovszky Zs., T. Szörenyi, K. Bali, A. Toth: Dependence of deposition kinetics on precursor concentration and writing speed in pyrolytic laser deposition from solution. Thin Solid Films **241**, 67 (1994)

Gerischer H.: Electrochemical photo and solar cells – Principles and some experiments. Electroanal. Chem. Interfacial Electrochem. **58**, 263 (1975)

Gervais A., D. Keller: Nucleation of $YBa_2Cu_3O_{7-\delta}$ films on MgO substrates by pulsed laser deposition. Physica C **246**, 29 (1995)

Gilgen H.H., T. Cacouris, P.S. Shaw, R.R. Krchnavek, R.M. Osgood: Direct writing of metal conductors with near UV-light. Appl. Phys. B **42**, 55 (1987)

Girault C., D. Damiani, C. Champeaux, P. Marchet, J.P. Mercurio, J. Aubreton, A. Catherinot: Characterization of the KrF laser-induced plasma plume created above a BiSrCaCuO target. Appl. Phys. Lett. **56**, 1472 (1990)

Girault C., D. Damiani, J. Aubreton, A. Catherinot: Time-resolved spectroscopic study of the KrF laser-induced plasma plume created above an YBaCuO superconducting target. Appl. Phys. Lett. **55**, 182 (1989)

Golusda E., P. Hessenthaler, G. Mollekopf, H. Stafast: SF_6 sensitized CO_2 laser CVD of amorphous silicon. Appl. Surf. Sci. **69**, 258 (1993)

Golusda E., R. Lange, K.D. Lühmann, G. Mollekopf, M. Wacker, H. Stafast: CW CO_2 laser CVD of amorphous hydrogenated silicon (a-Si:H): Influence of the deposition geometry. Appl. Surf. Sci. **54**, 30 (1992)

Golusda E., K.D. Lühmann, G. Mollekopf, H. Stafast, M. Wacker: CO_2 laser chemical vapor deposition of amorphous silicon: Gas phase processes and thin film properties. Ber. Bunsenges. Phys. Chem. **95**, 1414 (1991)

Goodall F.N., R.A. Moody, W.T. Welford: Reduction photolithography by ablation at wavelength 193 nm. Opt. Commun. **57**, 227 (1986)

Grad L., J. Mozina: Acoustic in situ monitoring of excimer laser ablation of different ceramics. Appl. Surf. Sci. **69**, 370 (1993)

Grebel H., K.J. Fang: Photoablation: Schottky barriers on patterned Si surfaces. J. Appl. Phys. **77**, 367 (1995)

Gregson V.G.: Laser heat treatment. *Materials Processing – Theory and Practices* **3**, 201 (North-Holland, Amsterdam 1983)

Griffith A.A.: The phenomena of rupture and flow in solids. Phil. Trans. Roy. Soc. (London), A **221**, 163 (1920)

Grimaldi M.G., P. Baeri, E. Rimini, G. Celotti: Epitaxial $NiSi_2$ formation by pulsed laser irradiation of thin Ni layers deposited on Si substrates. Appl. Phys. Lett. **43**, 244 (1983)

Groh G.: Vacuum deposition of thin films by means of a CO_2 laser. J. Appl. Phys. **39**, 5804 (1968)

Gross M.E., A. Appelbaum, P.K. Gallagher: Laser direct-write metallization in thin palladium acetate films. J. Appl. Phys. **61**, 1628 (1987)

Gross M.E., A. Appelbaum, K.J. Schnoes: A chemical and mechanistic view of reaction profiles in laser direct-write metallization in metallo-organic films: Gold. J. Appl. Phys. **60**, 529 (1986)

Grossman W.M., M. Karnezos: Localized laser chemical processing of tungsten films. J. Vac. Sci. Technol. B **5**, 843 (1987)

Grun J., J. Stamper, C. Manka, J. Resnick, R. Burris, B.H. Ripin: Observation of high-pressure blast-wave decursors. Appl. Phys. Lett. **59**, 246 (1991)

Gunton J.D., M. San Miguel, P.S. Sahni: The dynamics of first-order phase transitions, in *Phase Transitions and Critical Phenomena*, ed. by C. Domb, J.L. Lebowitz (Academic, London 1983)

Guosheng Z., P.M. Fauchet, A.E. Siegmann: Growth of spontaneous periodic surface structures on solids during laser illumination. Phys. Rev. B **26**, 5366 (1982)

Gupta A., B. Braren, K.G. Casey, B.W. Hussey, R. Kelly: Direct imaging of the fragments produced during excimer laser ablation of $YBa_2Cu_3O_{7-\delta}$. Appl. Phys Lett. **59**, 1303 (1991)

Gupta A., R. Jagannathan: Laser writing of copper lines from metalorganic films. Appl. Phys. Lett. **51**, 2254 (1987)

Gutfeld R.J. von, F.A. McDonald, R.W. Dreyfus: Surface deformation measurements following excimer laser irradiation of insulators. Appl. Phys. Lett. **49**, 1059 (1986)

Gutfeld R.J. von: Laser enhanced plating and etching: A review, in *Laser Processing and Diagnostics*, ed. by D. Bäuerle, Springer Ser. Chem. Phys., Vol.39 (Springer, Berlin, Heidelberg 1984) p.323

Gutfeld R.J. von, R.T. Hodgson: Laser enhanced etching in KOH. Appl. Phys. Lett. **40**, 352 (1982)

Haase G., V. Liberman, R.M. Osgood: Ultraviolet laser-induced interaction of Cl_2 with GaAs (110). J. Vac. Sci. Technol. B **10**, 206 (1992)

Haba B., Y. Morishige, S. Kishida: Laser through-hole drilling and laser cutting in teflon. Appl. Phys. A **60**, 27 (1995)

Habermeier H.-U.: In-situ preparation of high T_c YBCO thin films by pulsed laser deposition, in *Laser Ablation of Electronic Materials*, ed. by E. Fogarassy, S. Lazare, E-MRS Monographs, Vol.4 (North-Holland, Amsterdam 1992) p.281

Hänggi P., P. Talkner, M. Borkovec: Reaction-rate theory: Fifty years after Kramers. Rev. Mod. Phys. **62**, 251 (1990)

Hagemeyer A., H. Hibst, J. Heitz, D. Bäuerle: Improvements of the peel test for adhesion evaluation of thin metallic films on polymeric substrates. J. Adhesion Sci. Technol. **8**, 29 (1994)

Haglund R.F., M. Affatigato, J. Arps, K. Tang: UV laser ablation from ionic solids, in *Laser Ablation – Mechanisms and Applications*, ed. by J.C. Miller, R.F. Haglund. Lecture Notes Phys. **389**, 246 (Springer, Berlin, Heidelberg 1991)

Haken H.: *Synergetics, An Introduction*, Springer Ser. Syn., Vol.1 (Springer, Berlin, Heidelberg 1978)

Han C.H., Y. Ishii, K. Murata: Reshaping collimated laser beams with Gaussian profile to uniform profiles. Appl. Opt. **22**, 3644 (1983)

Han J., K.F. Jensen, Y. Senzaki, W.L. Gladfelter: Pyrolytic laser assisted chemical vapor deposition of Al from dimethylethylamine-alane: Characterization and a new two-step writing process. Appl. Phys. Lett. **64**, 425 (1994)

Han J., K.F. Jensen: Combined experimental and modeling studies of laser-assisted chemical vapor deposition of copper from copper(I)-hexafluoro-acetylacetonate trimethylvinylsilane. J. Appl. Phys. **75**, 2240 (1994)

Hanabusa M., M. Ikeda: Wavelength dependence in photochemical vapor deposition of aluminum film using dimethylaluminum hydride. Appl. Organometallic Chemistry **5**, 289 (1991)

Hanabusa M., A. Oikawa, P.Y. Cai: Deposition of aluminum thin films by photochemical surface reaction. J. Appl. Phys. **66**, 3268 (1989)

Hanabusa M., H. Kikuchi, T. Iwanaga, K. Sugai: IR laser photo-assisted deposition of silicon films, in *Laser Processing and Diagnostics*, ed. by D. Bäuerle, Springer Ser. Chem. Phys., Vol.39 (Springer, Berlin, Heidelberg 1984) p.197

Hanabusa M., S. Moriyama, H. Kikuchi: Laser-induced deposition of silicon films. Thin Solid Films **107**, 227 (1983)

Hansen S.G.: Velocity profiles of species ejected in UV laser ablation of several polymers examined by time-of-flight mass spectroscopy. J. Appl. Phys. **66**, 3329 (1989)

Hansen S.G., T.E. Robitaille: Formation of polymer films by pulsed laser evaporation. Appl. Phys. Lett. **52**, 81 (1988)

Hansen S.G., T.E. Robitaille: Characterization of the pulsed laser evaporation process: Selenium thin-film formation. Appl. Phys. Lett. **50**, 359 (1987)

Hare D.E., D.D. Dlott: Picosecond coherent Raman study of solid-state chemical reactions during laser polymer ablation. Appl. Phys. Lett. **64**, 715 (1994)

Harradine D., F.R. McFeely, B. Roop, J.I. Steinfeld, D. Denison, L. Hartsough, J.R. Hollahan: Reactive etching of semiconductor surfaces by laser-generated free radicals. SPIE Proc. **270**, 52 (1981)

Hass G., J.B. Ramsey: Vacuum deposition of dielectric and semiconductor films by a CO_2 laser. Appl. Opt. **8**, 1115 (1969)

Hattori K., A. Okano, Y. Nakai, N. Itoh: Laser-induced electronic processes on GaP (110) surfaces: Particle emission and ablation initiated by defects. Phys. Rev. B **45**, 8424 (1992)

Heaven M.C., M.A.A. Clyne: Interpretation of the spontaneous predissociation of $Cl_2[B^3II(O_u^+)]$. J. Chem. Soc., Faraday. Trans. 2, **78**, 1339 (1982)

Heitz J., E. Arenholz, D. Bäuerle, K. Schilcher: Growth of excimer-laser-induced dendritic surface structures on polyethylene-terephthalate. Appl. Surf. Sci. **81**, 103 (1994)

Heitz J., E. Arenholz, D. Bäuerle, H. Hibst, A. Hagemeyer, G. Cox: Dendritic surface structures on excimer-laser irradiated PET foils. Appl. Phys. A **56**, 329 (1993)

Heitz J., X.Z. Wang, P. Schwab, D. Bäuerle, L. Schultz: KrF laser-induced ablation and patterning of Y-Ba-Cu-O films. J. Appl. Phys. **68**, 2512 (1990)

Heraeus Inc., Germany: Data sheet Q-A 1/112 (1979)

Herman I.P., H. Tang, P.P. Leong: Real time optical diagnostics in laser etching and deposition. MRS Proc. **201**, 563 (1991)

Herman I.P., F. Magnotta, D.E. Kotecki: Direct-laser writing of silicon microstructures: Raman microprobe diagnostics and modeling of the nucleation phase of deposition. J. Vac. Sci. Technol. A **4**, 659 (1986)

Herziger G., E.W. Kreutz: Fundamentals of laser microprocessing of metals. Physica Scripta T **13**, 139 (1986)

Herziger G., E.W. Kreutz: Fundamentals of laser micromachining of metals, in *Laser Processing and Diagnostics*, ed. by D. Bäuerle, Springer Ser. Chem. Phys., Vol. 39 (Springer, Berlin, Heidelberg 1984) p. 90

Hess M.S., J.F. Milkosky: Vapor deposition of platinum using cw laser energy. J. Appl. Phys. **43**, 4680 (1972)

Heszler P., P. Mogyorosi, J.O. Carlsson: Time resolved spectroscopy of the emission of fluorescent light upon UV laser assisted CVD of W from WF_6. Appl. Surf. Sci. **69**, 376 (1993)

Heszler P., Zs. Bor, G. Hajos: Incubation process in polyimide upon UV photoablation. Appl. Phys. A **49**, 739 (1989)

Heydel R., R. Matz, W. Göpel: Maskless excimer laser induced projection patterning of InP in Cl_2 etch gas. Appl. Surf. Sci. **69**, 38 (1993)

Higashi G.S.: Excimer laser projection-patterned deposition of aluminium using alkyl-aluminium precursor molecules. Chemtronics **4**, 123 (1989)

Higashi G.S., L.J. Rothberg: Surface photochemical phenomena in laser chemical vapor deposition. J. Vac. Sci. Technol. B **3**, 1460 (1985)

Hill C., A.L. Butler, J.A. Daly: Shaping of dopang concentration profiles in silicon by multiple-pulse laser processing, in *Laser and Electron-Beam Interactions with Solids*, ed. by B.R. Appleton, G.K. Celler. MRS Proc. **4**, 579 (1982)

Hiratani M., Y. Tarutani, T. Fukazawa, M. Okamoto, K. Takagi: Growth of $SrTiO_3$ thin films by pulsed-laser deposition. Thin Solid Films **227**, 100 (1993)

Hirose M., S. Yokoyama, Y. Yamakage: Characterization of photochemical processing. J. Vac. Sci. Technol. B **3**, 1445 (1985)

Hirose M.: Glow discharge, in *Hydrogenated Amorphous Silicon*, ed. by J.I. Pankove (Academic, London 1984), Vol. 21A, Chap. 2, p. 9

Hirschfelder J.O., C.F. Curtiss, R.B. Bird: *Molecular Theory of Gases and Liquids* (Wiley, New York 1964)

Hirvonen J.K., J.M. Poate, A. Greenwald, R. Little: Pulsed electron beam irradiation of ion-implanted copper single crystals. Appl. Phys. Lett. **36**, 564 (1980)

Hitchman M.L., A.D. Jobson, L.F.T. Kwakman: Some considerations of the thermodynamics and kinetics of the chemical vapour deposition of tungsten. Appl. Surf. Sci. **38**, 312 (1989)

Hiura Y., Y. Morishige, S. Kishida: Laser chemical vapor deposition direct patterning of insulating film. J. Appl. Phys. **69**, 1744 (1991)

Ho C.Y., R.W. Powell, P.E. Liley: Thermal conductivity of the elements: A comprehensive review. J. Phys. Chem. Ref. Data **3**, Suppl. 1, 588 (1974)

Hoffmann P., B. Lecohier, S. Goldoni, H. van den Bergh: Fast laser writing of copper and iridium lines from thin solid surface layers on metalorganic compounds. Appl. Surf. Sci. **43**, 54 (1989)

Hogan M., J.G. Lunney: Laser photoablation of spin-on-glass. Appl. Phys. Lett. **53**, 831 (1988)

Holber W., G. Reksten, R.M. Osgood: Laser-enhanced plasma etching of silicon. Appl. Phys. Lett. **46**, 201 (1985)

Holzapfel B., B. Roas, L. Schultz, P. Bauer, G. Saemann-Ischenko: Off-axis laser deposition of $YBa_2Cu_3O_{7-\delta}$ thin films. Appl. Phys. Lett. **61**, 3178 (1992)

Honig R.E.: Laser-induced emission of electrons and positive ions from metals and semiconductors. Appl. Phys. Lett. **3**, 8 (1963)

Hontzopoulos E., E. Damigos: Excimer laser surface treatment of bulk ceramics. Appl. Phys. A **52**, 421 (1991)

Horiike Y., N. Hayasaka, M. Sekine, T. Arikado, M. Nakase, H. Okano: Excimerlaser etching on silicon. Appl. Phys. A **44**, 313 (1987)

Houle F.A., L.I. Yeh: Continuous wave visible laser assisted decomposition of $Cr(CO)_6$ on a growing film: In situ observations. J. Phys. Chem. **96**, 2691 (1992)

Houle F.A.: Doping effects on the etching chemistry of GaAs and Si. MRS Proc. **204**, 25 (1991)

Houle F.A.: Photochemical etching of silicon: The influence of photogenerated charge carriers. Phys. Rev. B **39**, 10120 (1989)

Houle F.A., R.J. Wilson, T.H. Baum: Surface processes leading to carbon contamination of photochemically deposited copper films. J. Vac. Sci. Technol. A **4**, 2452 (1986)

Hunger E., H. Pietsch, S. Petzoldt, E. Matthias: Multishot ablation of polymer and metal films at 248 nm. Appl. Surf. Sci. **54**, 227 (1992)

Husinsky W., S. Mitterer, G. Grabner, I. Baumgartner: Photoablation by UV and visible laser radiation of native and doped biological tissue. Appl. Phys. B **49**, 463 (1989)

Ihlemann J., B. Wolff, P. Simon: Nanosecond and femtosecond excimer laser ablation of fused silica. Appl. Phys. A **54**, 363 (1992)

Ihler B.C.: Laser-Lithotripsie – System und Fragmentierungsprozesse unter der Lupe (Laser lithotripsy – system and fragmentation processes closely examined). Laser und Optoelektronik **24**, 76 (1992)

Ikejiri M., H. Nakayama, M. Nishio, H. Ogawa, A. Yoshida: ZnTe growth by photoassisted metalorganic vapor phase epitaxy at atmospheric pressure. Appl. Surf. Sci. **70/71**, 755 (1993)

Illgner C., P. Schaaf, K.P. Lieb, E. Schubert, R. Queitsch, H.W. Bergmann: Laser nitriding of iron: Nitrogen profiles and phases. Appl. Phys. A **61**, 1 (1995)

Inam A., X.D. Wu, L. Nazar, M.S. Hedge, C.T. Rogers, T. Venkatesan, R.W. Simon, K. Daly, H. Padamsee, J. Kirchgessner, D. Moffat, D. Rubin, Q.S. Shu, D. Kalokitis, A. Fathy, V. Pendrick, R. Brown, B. Brycki, E. Belohoubek, L. Drabeck, G. Gruner, R. Hammond, F. Gamble, B.M. Lairson, J.C. Bravman: Microwave properties of highly oriented $YBa_2Cu_3O_{7-\delta}$ thin films. Appl. Phys. Lett. **56**, 1178 (1990)

Inam A., X.D. Wu, T. Venkatesan, S.B. Ogale, C.C. Chang, D. Dijkkamp: Pulsed laser etching of high-T_c superconducting films. Appl. Phys. Lett. **51**, 1112 (1987)

Industrial Laser Annual Handbook (Penwell Books, Tulsa 1990)

Ingen R.P. van, R.H.J. Fastenau, E.J. Mittemeijer: Laser ablation deposition of Cu-Ni and Ag-Ni films: Nonconservation of alloy composition and film microstructure. J. Appl. Phys. **76**, 1871 (1994)

Inui S., T. Nii, S. Matsumoto: Precise control of sheet resistance in boron doping of silicon by excimer laser irradiation. IEEE EDL-12, 702 (1991)

Inushima T., N. Hirose, K. Urata, K. Ito, S. Yamazaki: Film growth mechanism of photo-chemical vapor deposition. Appl. Phys. A 47, 229 (1988)

Irvine S.J.C., H. Hill, G.T. Brown, S.J. Barnett, J.E. Hails, O.D. Dosser, J.B. Mullin: Selected area epitaxy in II-VI compounds by laser-induced photo-metalorganic vapor phase epitaxy. J. Vac. Sci. Technol. B 7, 1191 (1989)

Irvine S.J.C, J.B. Mullin, J. Tunnicliffe: Low temperature growth of HgTe by a UV photosensitisation method, in *Laser Processing and Diagnostics*, ed. by D. Bäuerle, Springer Ser. Chem. Phys., Vol.39 (Springer, Berlin, Heidelberg 1984) p.234

Isenor N.R.: CO_2 laser-produced ripple patterns on Ni_xP_{1-x} surfaces. Appl. Phys. Lett. 31, 148 (1977)

Ishida M., A. Eto, T. Nakamura, T. Suzaki: Decomposition of trimethylaluminum and N_2O on Si surfaces using UV laser photolysis to produce Al_2O_3 films. J. Vac. Sci. Technol. A7, 2931 (1989)

Ishikawa Y., C.E. Brown, P.A. Hackett, D.M. Rayner: Excimer laser photolysis of group 6 metal carbonyls in the gas phase. J. Phys. Chem. 94, 2404 (1990)

Isshiki H., Y. Aoyagi, T. Sugano, S. Iwai, T. Meguro: Formation of low-dimensional structures by atomic layer epitaxy. Optoelectron. – Devices and Technologies 8, 509 (1993)

Itoh N., K. Hattori, Y. Nakai, J. Kanasaki, A. Okano, R.F. Haglund: Laser-induced particle emission from surfaces of non-metallic solids: A search for primary processes of laser ablation, in *Laser Ablation – Mechanisms and Applications*, ed. by J.C. Miller, R.F. Haglund. Lecture Notes Phys. 389, 213 (Springer, Berlin, Heidelberg 1991)

Iwabuchi M., K. Kinoshita, H. Ishibashi, T. Kobayashi: Reduction of pinhole leakage current of $SrTiO_3$ films by ArF excimer laser deposition with shadow mask ("eclipse method"). Jpn. J. Appl. Phys. 33, L610 (1994)

Jackman R.B., J.S. Foord, A.E. Adams, M.L. Lloyd: LCVD of patterned Fe on silica glass: Observation and origins of periodic ripple structures. J. Appl. Phys. 59, 2031 (1986)

Jackson R.L., G.W. Tyndall: Quartz crystal microbalance determination of laser photochemical deposition rates: Mechanism of laser photochemical deposition from the group 6 hexacarbonyls. MRS Proc. 101, 207 (1988)

Jacobs J.W.M., C.J.C.M.Nillesen: Repair of transparent defects on photomasks by laser-induced metal deposition from an aqueous solution. J. Vac. Sci. Technol. B 8, 635 (1990)

Jain A.K., V.N. Kulkarni, D.K. Sood, J.S. Uppal: Periodic surface ripples in laser-treated aluminum and their use to determine absorbed power. J. Appl. Phys. 52, 4882 (1981)

Jain K., M.R. Latta, G.T. Sincerbox: Holographic method and apparatus for transforming of a light beam into a line source of required curvature and finite numerical aperture. U.S. Patent Nos. 4,444,456 and 4,516,832 (1984)

JANAF Thermochemical Tables (National Bureau of Standards, Washington, DC 1970)

Jellison G.E., F.A. Modine: Optical functions of silicon between 1.7 and 4.7 eV at elevated temperatures. Phys. Rev. B 27, 7466 (1983)

Johansson S., J. Schweitz, H. Westberg, M. Boman: Microfabrication of three-dimensional boron structures by laser chemical processing. J. Appl. Phys. **72**, 5956 (1992)

Jones S.C., P. Braunlich, R.T. Casper, X.-A. Shen, P. Kelly: Recent progress on laser-induced modifications and intrinsic bulk damage of wide-gap optical materials. Opt. Eng. **28**, 1039 (1989)

Jubber M., J.I.B. Wilson, J.L. Davidson, P.A. Fernie, P. John: Laser writing of high-purity gold lines. Appl. Phys. Lett. **55**, 1477 (1989)

Juch K., W. Löschau: Herstellung titankarbidhaltiger Verschleiß-Schutzschichten durch Laserbestrahlung vorbeschichteter Stahloberflächen (Production of TiC containing layers with a high wear resistance on steel surfaces by laser melting). Laser und Optoelektronik **26**, 60 (1994)

Jyumonji M., K. Sugioka, H. Takai, H. Tashiro, K. Toyoda: Mechanism of silicon implant-deposition for surface modification of stainless steel 304 using KrF-excimer laser. Appl. Phys. A **60**, 41 (1995)

Kahlert H.J., U. Sarbach, B. Burghardt, B. Klimt: Excimer laser illumination and imaging optics for controlled microstructure generation. SPIE Proc. **1835**, 110 (1993)

Kamimura T., M. Hirose: Effect of hydrogen dilution of silane in hydrogenated amorphous silicon films prepared by photochemical vapor deposition. Jpn. J. Appl. Phys. **25**, 1778 (1986)

Kanai M., T. Kawai, S. Kawai, H. Tabata: Low-temperature formation of multilayered Bi(Pb)-Sr-Ca-Cu-O thin films by successive deposition using laser ablation. Appl. Phys. Lett. **54**, 1802 (1989)

Kantor Z., Z. Toth, T. Szörenyi, A.L. Toth: Deposition of micrometer-sized tungsten patterns by laser transfer technique. Appl. Phys. Lett. **64**, 3506 (1994)

Kapenieks A., M. Eyett, D. Bäuerle: Laser-induced surface metallization of ceramic PLZT. Appl. Phys. A **41**, 331 (1986a)

Kapenieks A., M. Eyett, R. Stumpe, D. Bäuerle: Laser direct writing of electrodes on PLZT ceramics, in *Laser Processing and Diagnostics II*, ed. by D. Bäuerle, K.L. Kompa, L.D. Laude (Editions de Physique, Les Ulis 1986b) p.165

Kar A., M.N. Azer, J. Mazumder: Three-dimensional transient mass transfer model for laser chemical vapor deposition of titanium on stationary finite slabs. J. Appl. Phys. **69**, 757 (1991)

Kar A., J. Mazumder: Two-dimensional model for material damage due to melting and vaporization during laser irradiation. J. Appl. Phys. **68**, 3884 (1990)

Karam N.H., N.A. El-Masry, S.M. Bedair: Laser direct writing of single-crystal III-V compounds on GaAs. Appl.Phys.Lett. **49**, 880 (1986)

Kargl P.B., R. Kullmer, D. Bäuerle: Periodic structures in pyrolytic laser-CVD of W from WCl_6. Appl. Phys. A **57**, 175 (1993a)

Kargl P.B., R. Kullmer, D. Bäuerle: Bistable growth in laser chemical vapor deposition. Appl. Phys. A **57**, 577 (1993b)

Karlicek R.F., V.M. Donnelly, G.J. Collins: Laser-induced metal deposition on InP. J. Appl. Phys. **53**, 1084 (1982)

Karlov N.V., N.A. Kirichenko, B.S. Luk'yanchuk: Microscopic kinetics of thermochemcial processes on laser heating: Current state and prospects. Russian Chem. Rev. **62**, 203 (1993)

Karlov N.V., N. Kirichenko, B. Luk'yanchuk: *Laser Thermochemistry* (Nauka, Moscow 1992) (in Russian)

Karlov N.V., B.S. Luk'yanchuk, E.V. Sissakyan, G.A. Shafeev: Etching of semiconductors by photothermal dissociation of gases. Sov. J. Quantum Electr. **15**, 803 (1985)

Karner C., A. Mandel, F. Träger: Pulsed laser photothermal displacement spectroscopy for surface studies. Appl. Phys. A **38**, 19 (1985)

Kautek W., S. Mitterer, J. Krüger, W. Husinsky, G. Grabner: Femtosecond-pulse laser ablation of human corneas. Appl. Phys. A **58**, 513 (1994)

Kautek W., B. Roas, L. Schultz: Formation of Y-Ba-Cu-Oxide thin films by pulsed laser deposition: A comparative study in the UV, visible and IR range. Thin Solid Films **191**, 317 (1990)

Kawasaki M., Y. Tsukiyama, H. Hada: Study on the early stage of photochemical vapor deposition of amorphous silicon from disilane on a SiO_2 substrate. J. Appl. Phys. **64**, 3254 (1988)

Keilmann F.: Laser-driven corrugation instability of liquid metal surfaces. Phys. Rev. Lett. **51**, 2097 (1983)

Kelly R., A. Miotello, B. Braren, C.E. Otis: On the debris phenomenon with laser-sputtered polymers. Appl. Phys. Lett. **60**, 2980 (1992)

Kiely C.J., V. Tavitian, J.G. Eden: Microstructural studies of epitaxial Ge films grown on [100] GaAs by laser photochemical vapor deposition. J. Appl. Phys. **65**, 3883 (1989a)

Kiely C.J., V. Tavitian, C. Jones, J.G. Eden: NH_3 as a photosensitizer in the epitaxial growth of Ge on GaAs by laser photochemical vapor deposition. Appl. Phys. Lett. **55**, 65 (1989b)

Kilibarda S., H.K. Ng: Superconducting free-standing thin films of $YBa_2Cu_3O_x$. Appl. Phys. Lett. **57**, 201 (1990)

Kim D.Y, S.K. Tripathy, L. Li, J. Kumar: Laser-induced holographic surface relief gratings on nonlinear optical polymer films. Appl. Phys. Lett. **66**, 1166 (1995)

Kim W.-Y., A. Shibata, Y. Kazama, M. Konagai, K. Takahashi: Optimum cell design for high-performance a-Si:H solar cells prepared by photo-CVD. Jpn. J. Appl. Phys. **28**, 311 (1989)

Kimerling L.C., J.L. Benton: Defects in laser processed semiconductors, in *Laser and Electron Beam Processing of Materials*, ed. by C.W. White, P.S. Peercy (Academic, London 1980) p.385

Kirichenko N., D. Bäuerle: The influence of heterogeneous and homogeneous reactions in laser-chemical processing. Thin Solid Films **218**, 1 (1992)

Kirichenko N., K. Piglmayer, D. Bäuerle: On the kinetics of non-equimolecular reactions in laser chemical processing. Appl. Phys. A **51**, 498 (1990)

Kitahama K., K. Hirata, H. Nakamatsu, S. Kawai, N. Fujimori, T. Inai, H. Yoshino, A. Doi: Synthesis of diamond by laser-induced CVD. Appl. Phys. Lett. **49**, 634 (1986)

Kitai M.S., V.L. Popkov, V.A. Semchishen, A.A. Kharizov: The physics of UV laser cornea ablation. IEEE J. QE-27, 302 (1991)

Kitai M.S., V.L. Popkov, V.A. Semchishen: Dynamics of UV-laser ablation of PMMA, caused by mechanical stresses. Theory and experiment. Macromol. Chem., Macromol. Symp. **37**, 257 (1990)

Kline L.E., W.D. Partlow, R.M. Young, R.R. Mitchell, T.V. Congedo: Diagnostics and modeling of RF discharge dissociation in N_2O. IEEE Trans. PS.-19, 278 (1991)

Kliwer J.: Laser evaporation and elemental analysis. J. Appl. Phys. 44, 490 (1973)

Klumpp A., H. Sigmund: Deposition of High Quality SiO_2 layers from TEOS by excimer laser. Appl. Surf. Sci. 36, 141 (1989)

Kobayashi H., T. Kobayashi: Heteroepitaxial growth of quaternary $Ba_xSr_{1-x}TiO_3$ thin films by ArF excimer laser ablation. Jpn. J. Appl. Phys. 33, L533 (1994)

Kodas T.T., P.B. Comita: A diffusive transport relaxation technique for studying laser-induced chemical vapor deposition reactions at high pressures. J. Appl. Phys. 65, 2513 (1989)

Kogelschatz U., H. Esrom: Neue inkohärente Ultraviolett-Excimerstrahler zur photolytischen Materialabscheidung (New incoherent UV excimer sources for photolytic material deposition). Laser und Optoelektronik 22, 55 (1990)

Koinkar V.N., S.B. Ogale: Pulsed excimer laser processing of optical thin films. Thin Solid Films 206, 259 (1991)

Koinuma H., Y. Takemura, T. Hashimoto, K. Takeuchi, K. Fueki: Reversible resistivity control of $Ba_2YCu_3O_{7-\delta}$ thin films by laser annealing. Jpn. J. Appl. Phys. 27, L652 (1988)

Kokai F., H. Saito, T. Fujioka: X-ray photoelectron spectroscopy studies on modified polyimide surface after ablation with a KrF excimer laser. J. Appl. Phys. 66, 3252 (1989)

Konstantinov L., R. Novak, P. Hess: Film growth and mechanism of LICVD of chromium films from $Cr(CO)_6$ at 248 nm. Appl. Phys. A 47, 171 (1988)

Konuma M., H. Stützler, J. Kuhl, E. Bauser: Laser-induced chemical etching of silicon in NF_3 atmosphere. Appl. Phys. 48, 465 (1989)

Kools J.C.S., T.S. Baller, S.T. De Zwart, J. Dieleman: Gas flow dynamics in laser ablation deposition. J. Appl. Phys. 71, 4547 (1992)

Koren G., J.J. Donelon: CO_2 laser cleaning of black deposits formed during the excimer laser etching of polyimide in air. Appl. Phys. B 45, 45 (1988)

Koren G., J.E. Hurst Jr.: 248 nm laser etching of GaAs in chlorine and ozone gas environments. Appl. Phys. A 45, 301 (1988)

Koren G., F. Ho, J.J. Ritsko: XeCl laser controlled chemical etching of aluminum in chlorine gas. Appl. Phys. A 40, 13 (1986)

Koren G.: Ar ion laser assisted chemical etching of via holes in tungsten sheets in air. Appl. Phys. A 40, 215 (1986)

Koren G., J.T.C. Yeh: Emission spectra and etching of polymers and graphite irradiated by excimer lasers. J. Appl. Phys. 56, 2120 (1984)

Krabe D., W. Radloff: Laser-aided structured silver deposition from the liquid phase (in German). Experimentelle Technik der Physik 36, 501 (1988)

Kräuter W., D. Bäuerle, F. Fimberger: Laser-induced chemical vapor deposition of Ni by decomposition of $Ni(CO)_4$. Appl. Phys. A 31, 13 (1983)

Kräutle H., P. Roentgen, M. Maier, H. Beneking: Laser-induced doping of GaAs. Appl. Phys. A 38, 49 (1985)

Kräutle H., D. Wachenschwanz: Ohmic contacts on n- and p-layers of GaAs using laser-induced diffusion. Solid-State Electron. 28, 601 (1985)

Krajnovich D.J.: Laser sputtering of highly oriented pyrolytic graphite at 248 nm. J. Chem. Phys. 102, 726 (1995)

Krajnovich D.J., J.E. Vázquez: Formation of "intrinsic" surface defects during 248 nm photoablation of polyimide. J. Appl. Phys. **73**, 3001 (1993)

Kramer K.J., S. Talwar, P.G. Carey, E. Ishida, D. Ashkenas, K.H. Weiner, T.W. Sigmon: Impurity distribution and electrical characteristics of boron-doped $Si_{1-x}Ge_x/Si$ p^+/N heterojunction diodes produced using pulsed UV-laser-induced epitaxy and gas-immersion laser doping. Appl. Phys. A **57**, 91 (1993)

Krchnavek R.R., G.R. Lalk, D.H. Hartman: Laser direct writing of channel waveguides using spin-on polymers. J. Appl. Phys. **66**, 5156 (1989)

Krchnavek R.R., H.H. Gilgen, J.C. Chen, P.S. Shaw, T.J. Licata, R.M. Osgood: Photodeposition rates of metal from metal alkyls. J. Vac. Sci. Technol. B **5**, 20 (1987)

Krchnavek R.R., H.H. Gilgen, R.M. Osgood Jr.: Maskless laser writing of silicon dioxide. J. Vac. Sci. Technol. B **2**, 641 (1984)

Krebs H.-U., O. Bremert: Pulsed laser deposition of thin metallic alloys. Appl. Phys. Lett. **62**, 2341 (1993)

Kreitschitz O., W. Husinsky, G. Betz, N.H. Tolk: Time-of-flight investigation of the intensity dependence of laser-desorbed positive ions from SrF_2. Appl. Phys. A **58**, 563 (1994)

Kreuzer H.J., Z.W. Gortel: *Physisorption Kinetics*, Springer Ser. Surf. Sci., Vol.1 (Springer, Berlin, Heidelberg 1986)

Krusin-Elbaum L., C.C. Tsuei, A. Gupta: High current densities above 100 K in the high-temperature superconductor $HgBa_2CaCu_2O_{6+\delta}$. Nature **373**, 679 (1995)

Kubát P., P. Engst: Study of the formation of titanium deposit in the excimer-laser decomposition of titanium tetrachloride by time-resolved UV spectroscopy. Appl. Surf. Sci. **64**, 97 (1993)

Kubatova J., V. Chab, I. Lukes, P. Jiricek, F. Fendrych: ArF excimer laser induced changes in the $Si(100)/SiO_2$ interface studied in situ by ESCA and LEED. Appl. Surf. Sci. **43**, 297 (1989)

Kubo M., M. Hanabusa: Fabrication of microlenses by laser chemical vapor deposition. Appl. Opt. **29**, 2755 (1990)

Kuehn T.H., R.J. Goldstein: Numerical solution to the Navier-Stokes equations for laminar natural convection about a horizontal isothermal circular cylinder. Int'l J. Heat Mass Transfer **23**, 971 (1980)

Küper S., J. Brannon, K. Brannon: Threshold behavior in polyimide photo-ablation single-shot rate measurements and surface-temperature modeling. Appl. Phys. A **56**, 43 (1993)

Küper S., J. Brannon: Ambient gas effects on debris formed during KrF laser ablation of polyimide. Appl. Phys. Lett. **60**, 1633 (1992)

Küper S., M. Stuke: Ablation of UV-transparent materials with femtosecond UV excimer laser pulses. Microelectron. Eng. **9**, 475 (1989)

Küper S., M. Stuke: Femtosecond UV excimer laser ablation. Appl. Phys. B **44**, 199 (1987)

Kukimoto H., Y. Ban, H. Komatsu, M. Takechi, M. Ishizaki: Selective area control of material properties in laser-assisted MOVPE of GaAs and AlGaAs. J. Cryst. Growth **77**, 223 (1986)

Kullmer R., B. Kargl, D. Bäuerle: Laser-induced deposition of tungsten from tungsten hexachloride. Thin Solid Films **218**, 122 (1992)

Kullmer R., D. Bäuerle: Laser-induced chemical etching of silicon in chlorine atmosphere: III. Combined cw- and pulsed-irradiation. Appl. Phys. A **47**, 377 (1988*a*)

Kullmer R., D. Bäuerle: Excimer-laser-induced ablation of the high-T_c superconductor Bi-Ca-Sr-Cu-O. Appl. Phys. A **47**, 103 (1988*b*)

Kullmer R., D. Bäuerle: Laser-induced chemical etching of silicon in chlorine atmosphere: I. Pulsed-irradiation. Appl. Phys. A **43**, 227 (1987)

Kumagai H., M. Ezaki, K. Toyoda, M. Obara: Fabrication of periodic submicron dot structures of N-InP by laser-induced surface electromagnetic wave etching. Jpn. J. Appl. Phys. **31**, L928 (1992)

Kumar A., L. Ganapathi, S.M. Kanetkar, J. Narayan: Single-chamber, in situ processing of superconducting $YBa_2Cu_3O_{7-\delta}$ thin films on stainless steel with yttria-stabilized zirconia buffer layer. J. Appl. Phys. **69**, 2410 (1991)

Kumar A., L. Ganapathi, J. Narayan: In situ processing of textured superconducting thin film of Bi(-Pb)-Ca-Sr-Cu-O by excimer laser ablation. Appl. Phys. Lett. **56**, 2034 (1990)

Kumar D., M. Sharon, R. Pinto, P.R. Apte, S.P. Pai, S.C. Purandare, L.C. Gupta, R. Vijayaraghavan: Large critical currents and improved epitaxy of laser ablated Ag-doped $YBa_2Cu_3O_{7-\delta}$ thin films. Appl. Phys. Lett. **62**, 3522 (1993)

Kumata K., U. Itoh, Y. Toyoshima, N. Tanaka, H. Anzai, A. Matsuda: Photochemical vapor deposition of hydrogenated amorphous silicon films from disilane and trisilane using a low pressure mercury lamp. Appl. Phys. Lett. **48**, 1380 (1986)

Kung P.J., R.E. Muenchausen: Growth and characterization of laser-deposited superconducting Bi-Sr-Ca-Cu-O thin films. J. Vac. Sci. Technol. A **11**, 1354 (1993)

Kurosawa K., W. Sasaki, Y. Takigawa, M. Ohmukai, M. Katto, M. Okuda: Growth of silicon microcrystals in thin surface layers of quartz glass with vacuum ultraviolet laser processing. Appl. Surf. Sci. **70/71**, 712 (1993)

Kurtze D.A., W.v. Saarloos, J.D. Weeks: Front propagation in self-sustained and laser-driven explosive crystal growth: Stability analysis and morphological aspects. Phys. Rev. B **30**, 1398 (1984)

Kussmaul A., J.S. Moodera, P.M. Tedrow, A. Gupta: Improved laser-ablated thin films of NdCeCuO by use of N_2O. Appl. Phys. Lett. **61**, 2715 (1992)

Kwok H.S., J.P. Zheng, S. Witanachchi, L. Shi, D.T. Shaw: Growth of CdS_xSe_{1-x} thin films by laser evaporation deposition. Appl. Phys. Lett. **52**, 1815 (1988)

Kwong D.L., D.M. Kim: Pulsed laser heating of silicon: The coupling of optical absorption and thermal conduction during irradiation. J. Appl. Phys. **54**, 366 (1983)

Lambda Highlights: Ablation of bone and cartilage with 308 nm radiation guided by a new type of fiber optics. Lambda Highlights Nr.25 (Lambda Physik, Göttingen, October 1990) p.6

Lampert M.O., J.M. Koebel, P. Siffert: Temperature dependence of the reflectance of solid and liquid silicon. J. Appl. Phys. **52**, 4975 (1981)

Landau L.D., E.M. Lifshitz: *Statistical Physics V*, Pt.1 (Pergamon, Oxford 1980)

Landau L.D., E.M. Lifshitz: *Fluid Mechanics VI* (Pergamon, Oxford 1974)

Landau L.D., E.M. Lifshitz: *Theory of Elasticity VII* (Pergamon, Oxford 1976)

Landau L.D., E.M. Lifshitz: *Electrodynamic of Continuous Media VIII* (Pergamon, Oxford 1976)

Landau L.D., E.M. Lifshitz: *Physical Kinetics X* (Pergamon, Oxford 1974)

Lankard J.R., G. Wolbold: Excimer laser ablation of polyimide in a manufacturing facility. Appl. Phys. A **54**, 355 (1992)

Larciprete R., E. Borsella, P.DePadova, M. Mangiantini, P. Perfetti, M. Fanfoni: Synchrotron radiation photoemission analysis of ArF laser deposited tin oxide. J.Vac. Sci. Technol. A **11**, 336 (1993)

Larson B.C., C.W. White, T.S. Noggle, D. Mills: Synchrotron X-ray diffraction study of silicon during pulsed-laser annealing. Phys. Rev. Lett. **48**, 337 (1982)

Laude L.D. (ed.): *Excimer Lasers*, NATO ASI Series – Applied Sciences, Vol.265 (Kluwer, Dordrecht 1994)

Laude L.D., M. Wautelet, R. Andrew: Laser-induced synthesis of compound semiconducting films. Appl. Phys. A **40**, 133 (1986)

Lavoie C., M. Meunier, S. Boivin, R. Izquierdo, P. Desjardins: Profile of titanium lines produced by excimer laser direct writing on lithium niobate. J. Appl. Phys. **70**, 2343 (1991*a*)

Lavoie C,, M. Meunier, R. Izquierdo, S. Boivin, P. Desjardins: Large area excimer laser induced deposition of titanium from titanium tetrachloride. Appl. Phys. A **53**, 339 (1991*b*)

Lazare S., V. Granier: Mechanism of polymer photoablation explored with a quartz crystal microbalance, in *Polymers in Microlithography: Materials and Processes*, ed. by E.Reichmanis, S.A.MacDonald, T. Iwayanagi. ACS Symp. Ser. **412**, 411 (Am. Chem. Soc., Washington, DC 1989)

Lazare S., V. Granier, P. Lutgen, G. Feyder: Controlled roughening of poly(ethyleneterephthalate) by photoablation: Study of wetting and contact angle hysteresis. Revue Phys. Appl. **23**, 1065 (1988)

LeComber P.G., A.J. Snell, K.D. Mackenzie, W.E. Spear: Applications of a-Si field effect transistors in liquid crystal displays and in integrated logic circuits. J. Physique (Paris) C **4**, 423 (1981)

Lecours A., R. Izquierdo, M. Tabbal, M. Meunier, A. Yelon: Laser-induced deposition of tungsten on GaAs from WF_6. J. Vac. Sci. Technol. B **11**, 51 (1993)

Lee C., H. Sayama, M. Takai: Comparison of laser-induced etching behavior of III-V compound semiconductors. Appl. Phys. A **56**, 343 (1993)

Lee S., X. Wen, W.A. Tolbert, D.D. Dlott, M. Doxtader, D.R. Arnold: Direct measurement of polymer temperature during laser ablation using a molecular thermometer. J. Appl. Phys. **72**, 2440 (1992)

Lehmann O., M. Stuke: Liquid precursor two-step aluminum thin-film deposition on KrF-laser patterned palladium. Appl. Phys. Lett. **61**, 2027 (1992)

Lehmann O., M. Stuke: Generation of three-dimensional free-standing metal microobjects by laser chemical processing. Appl. Phys. A **53**, 343 (1991)

Letokhov V.S.: Laser-induced chemistry – Basic nonlinear processes and applications. Appl. Phys. B **46**, 237 (1988)

Letokhov V.S.: *Nonlinear Laser Chemistry, Multiple Photon Excitation*, Springer Ser. Chem. Phys., Vol.22 (Springer, Berlin, Heidelberg 1983)

Leung W.P., A.C. Tam: Noncontact monitoring of laser ablation using a miniature piezoelectric probe to detect photoacoustic pulses in air. Appl. Phys. Lett. **60**, 23 (1992)

Levich V.G.: *Physicochemical Hydrodynamics* (Prentice Hall, Englewood Cliffs, NJ 1962)

Leyendecker G., J. Doppelbauer, D. Bäuerle, P. Geittner, H. Lydtin: Raman diagnostics of CVD systems: Determination of local temperatures. Appl. Phys. A **30**, 237 (1983*a*)

Leyendecker G., H. Noll, D. Bäuerle, P. Geittner, H. Lydtin: Rapid determination of apparent activation energies in chemical vapor deposition. J. Electrochem. Soc. **130**, 157 (1983*b*)

Li L.J., J. Mazumder: A study of the mechanism of laser cladding processes, in *Laser Processing of Materials*, ed. by K. Mukherjee, J. Mazumder (The Metallurgical Society of AIME, Warrendale, PA 1985) p.35

Li S.T., A. Ritzer, S. Proyer, E. Stangl, D. Bäuerle, N. Reschauer: Anisotropic resistivity in pulsed-laser deposited $Bi_2Sr_2CaCu_2O_{8+\delta}$ films. Appl. Surf. Sci. (1996)

Libenson M.N., M.N. Nikitin: On the diffusion of film atoms into the substrate under the action of laser radiation. Fiz. i Khim. Obrabotki Materialov (Physics and chemistry of material processing), **1**, 9 (1973) (In Russian)

Liberts G., M. Eyett, D. Bäuerle: Direct laser writing of superconducting patterns into semiconducting ceramic Y-Ba-Cu-O. Appl. Phys. A **46**, 331 (1988*a*)

Liberts G., M. Eyett, D. Bäuerle: Laser-induced surface reduction of the high T_c superconductor $YBa_2Cu_3O_{7-x}$. Appl. Phys. A **45**, 313 (1988*b*)

Licata T.J., M.T. Schmidt, R.M. Osgood, W.K. Chan, R. Bhat: Application of photodeposited Cd to Schottky barrier diode and transistor fabrication on InP and $In_{0.53}Ga_{0.47}As$ substrates. Appl. Phys. Lett. **58**, 845 (1991)

Licata T.J., D.V. Podlesnik, H. Tang, I.P. Herman, R.M. Osgood, S.A. Schwarz: Continuous-wave laser doping of micrometer-sized features in gallium arsenide using a dimethylzinc ambient. J. Vac. Sci. Technol. A **8**, 1618 (1990)

Lifshitz I.M., V.V. Slyozov: The kinetics of precipitation from supersaturated solid solutions. J. Phys. Chem. Solids **19**, 35 (1961)

Lin W.T., K.C. Wu: Effects of oxygen pressure and substrate temperature on the in situ grown BiSrCaCuO films by laser ablation. J. Appl. Phys. **72**, 4812 (1992)

Liu B., R.F. Hicks, J.J. Zinck, J.E. Jensen, G.L. Olson: Photoassisted organometallic vapor-phase epitaxy of CdTe. J. Vac. Sci. Technol. B **10**, 1384 (1992)

Liu R.S., W.C. Shih, K. Scott, P.P. Edwards, W.A. Phillips, A.L. Greer: Preparation and characterization of $Tl_2Ba_2CaCu_2O_8$ thin films by laser ablation and thallium diffusion. J. Appl. Phys. **71**, 4085 (1992)

Liu Y.S.: Laser direct writing of tungsten lines for VLSI applications, in *Tungsten and Other Refractory Metals for VLSI Applications I*, ed. by R.S. Blewer (Materials Research Society, Pittsburgh, PA 1986) p.43

Lompré L.A., J.M. Liu, H. Kurz, N. Bloembergen: Time-resolved temperature measurement of picosecond laser irradiated silicon. Appl. Phys. Lett. **43**, 168 (1983)

Loper G.L., M.D. Tabat: UV laser-induced radical-etching for micro-electronic processing, in *Laser Assisted Deposition, Etching, and Doping*, ed. by S.D. Allen. SPIE Proc. **459**, 121 (1984)

Lowndes D.H., D.P. Norton, S. Zhu, X.Y. Zheng: Laser ablation synthesis and properties of epitaxial $YBa_2Cu_3O_{7-\delta}/PrBa_2Cu_3O_{7-\delta}$ superconducting superlattices, in *Laser Ablation of Electronic Materials*, ed. by E. Fogarassy, S. Lazare (Elsevier, Amsterdam 1992) p.265

Lowndes D.H., D.B. Geohegan, D. Eres, S.J. Pennycook, D.N. Mashburn, G.E. Jellison: Photon-controlled fabrication of amorphous superlattice structures using ArF (193nm) excimer laser photolysis. Appl. Phys. Lett. **52**, 1868 (1988)

Lu G., F. Phillipp, B. Leibold, A. Lourenco, H.-U.Habermeier: Interface structures of YBCO thin films on different substrates, in *High-T_c Superconductor Thin Films*, ed. by L. Correra (Elsevier, Amsterdam 1992) p.689

Lu X., J. Zhang, M. Qui: Laser-induced thermochemical vapor deposition of tungsten films and its applications. Thin Solid Films **196**, 95 (1991)

Lu Y.F., M. Takai, S. Komuro, T. Shiokawa, Y. Aoyagi: Surface cleaning of metals by pulsed-laser irradiation in air. Appl. Phys. A **59**, 281 (1994)

Lu Y.F., M. Takai, S. Nagatomo, K. Kato, S. Namba: Direct writing of Ag-lines on Mn-Zn ferrite by laser-induced thermal decomposition of CH_3COOAg. Appl. Phys. A **54**, 51 (1992)

Lu Y.F., M. Takai, S. Nagatomo, S. Namba: Laser-induced dry chemical etching of Mn-Zn ferrite in CCl_2F_2 atmosphere. Appl. Phys. B **53**, 39 (1991a)

Lu Y.F., M. Takai, T. Nakata, S. Nagatomo, S. Namba: Laser-induced deposition of Ni lines on ferrite in $NiSO_4$ aqueous solution. Appl. Phys. A **52**, 129 (1991b)

Lu Z., M.T. Schmidt, D.V. Podlesnik, C.F. Yu, R.M. Osgood: Ultraviolet-light-induced oxide formation on GaAs surfaces. J. Chem. Phys. **93**, 7951 (1990)

Lubben D., S.A. Barnett, K. Suzuki, S. Gorbatkin, J.E. Greene: Laser-induced plasmas for primary ion deposition of epitaxial Ge and Si films. J. Vac. Sci. Techn. B **3**, 968 (1985)

Ludorf W., X.Z. Wang, D. Bäuerle: In situ preparation of Bi-Sr-Ca-Cu-O superconducting films by laser sputtering. Appl. Phys. A **49**, 221 (1989)

Lüth H.: *Surface and Interfaces of Solid Materials*, 3rd edn. (Springer, Berlin, Heidelberg 1995)

Luk'yanchuk B., N. Bityurin, S. Anisimov, N. Arnold, D. Bäuerle: The role of excited species in UV-laser materials ablation III. Non-stationary ablation of organic polymers. Appl. Phys. A **62** (1996a)

Luk'yanchuk B., N. Bityurin, S. Anisimov, D. Bäuerle: The role of excited species in UV-laser materials ablation IV. The influence of stresses. Appl. Phys. A **62** (1996b)

Luk'yanchuk B., N. Bityurin, S. Anisimov, D. Bäuerle: Photophysical ablation of organic polymers, in *Excimer Lasers*, ed. by L.D. Laude, NATO ASI Ser. (Kluwer, Dordrecht 1994) p.59

Luk'yanchuk B., N. Bityurin, S. Anisimov, D. Bäuerle: The role of excited species in UV-laser materials ablation I. Photo-physical ablation of organic polymers. Appl. Phys. A **57**, 367 (1993a)

Luk'yanchuk B., N. Bityurin, S. Anisimov, D. Bäuerle: The role of excited species in UV-laser materials ablation II. The stability of the ablation front. Appl. Phys. A **57**, 449 (1993b)

Luk'yanchuk B., K. Piglmayer, N. Kirichenko, D. Bäuerle: Inversion effects in the kinetics of laser-chemical processing. Physica A **180**, 285 (1992)

Maaren A.J.P. van, R.L. Krans, E. de Haas, W.C. Sinke: A high vacuum system for laser-induced deposition of tungsten. J. Vac. Sci. Technol. B **9**, 89 (1991); Excimer laser induced deposition of tungsten on silicon. Appl. Surf. Sci. **38**, 386 (1989)

Maaren A.J.P. van, R.L. Krans, W.C. Sinke: HF formation in laser-induced CVD of tungsten. Appl. Surf. Sci. **46**, 138 (1990)

Maffei N., S.B. Krupanidhi: Excimer laser-ablated bismuth titanate thin films. Appl. Phys. Lett. **60**, 781 (1992)

Magnotta F., I.P. Herman: Raman microprobe analysis during the direct laser writing of silicon microstructures. Appl. Phys. Lett. **48**, 195 (1986)

Maier-Komor P.: Preparation of isotope targets using laser beam evaporation techniques. Nucl. Instrum. Methods **167**, 73 (1979)

Maissel L.I., R. Glang: *Handbook of Thin Film Technology* (McGraw Hill, New York 1970)

Malvezzi A.M., H. Kurz, N. Bloembergen: Nonlinear photoemission from picosecond irradiated silicon. Appl. Phys. A **36**, 143 (1985)

Mantell D.A., T.E. Orlowski: XPS surface composition analysis during UV laser photodeposition of Al from TIBA on Si (100) substrates: Direct observation of the prenucleation regime. MRS Proc. **101**, 171 (1988)

Marks J., R.E. Robertson: Silicon dioxide deposition at 100°C using vacuum ultraviolet light. Appl. Phys. Lett. **52**, 810 (1988)

R.F. Marks, R.A. Pollak, P. Avouris, C.T. Lin, Y.J. Théfaine: Laser-pulsed plasma chemistry: Laser-initiated plasma oxidation of niobium. J. Chem. Phys. **78**, 4270 (1983)

Markwalder B., M. Widmer, D. Braichotte, H.v.d. Bergh: High-speed laser chemical vapor deposition of copper: A search for optimum conditions. J. Appl. Phys. **65**, 2470 (1989)

Marsal D.: *Die Numerische Lösung Partieller Differentialgleichungen* (Bibliographisches Institut, Mannheim 1976)

Massoud H.Z., J.D. Plummer: Analytical relationship for the oxidation of silicon in dry oxygen in the thin-film regime. J. Appl. Phys. **62**, 3416 (1987)

Matsumoto S., S. Yoshioka, J. Wada, S. Inui, K. Uwasawa: Boron doping of silicon by ArF excimer laser irradiation in B_2H_6. J. Appl. Phys. **67**, 7204 (1990)

Matthias E., M. Reichling, J. Siegel, O.W. Käding, S. Petzoldt, H. Skurk, P. Bizenberger, E. Neske: The influence of thermal diffusion on laser ablation of metal films. Appl. Phys. A **58**, 129 (1994)

Matthias E., T.A. Green: Laser-induced desorption, in *Desorption Induced by Electronic Transitions, DIET IV*, ed. by G. Betz, P. Varga, Springer Ser. Surf. Sci., Vol. 19 (Springer, Berlin, Heidelberg 1990) p. 112

Matz R., J. Meiler, D. Haarer: X-ray photoemission investigation of excimer laser induced etching of InP. MRS Proc. **158**, 307 (1990)

Matz R., J. Zirrgiebel: Fast photoelectrochemical etching of quarter-micrometer diffraction gratings in n-InP. J. Appl. Phys. **64**, 3402 (1988)

McCaulley J.A., V.R. McCrary, V.M. Donnelly: Laser-induced decomposition of triethylgallium and trimethylgallium adsorbed on GaAs(100). J. Phys. Chem. **93**, 1148 (1989)

McFeely F.R., J.F. Morar, F.J. Himpsel: Soft X-ray photoemission study of the silicon-fluorine etching reaction. Surf. Sci. **165**, 277 (1986)

Meguro T., T. Suzuki, K. Ozaki, Y. Okano, A. Hirata, Y. Yamamoto, S. Iwai, Y. Aoyagi, S. Namba: Surface processes in laser-atomic layer epitaxy (laser-ALE) of GaAs. J. Cryst. Growth **93**, 190 (1988a)

Meguro T., N. Ikeda, T. Itoh: Analytical study on interface of epitaxial Si on Si substrate grown by CO_2 laser CVD. J. Electrochem. Soc. **135**, 2046 (1988b)

Meiler J., R. Matz, D. Haarer: Laser-induced photochemical etching of InP by HBr and HCl. Appl. Surf. Sci. **43**, 416 (1989)

Melcher R.L.: Thermal and acoustic techniques for monitoring pulsed laser processing, in *Laser Processing and Diagnostics*, ed. by D. Bäuerle, Springer Ser. Chem. Phys., Vol. 39 (Springer, Berlin, Heidelberg 1984) p. 418

Mesarwi A., A. Ignatiev: Laser-enhanced oxidation of nickel. J. Vac. Sci. Technol. A **7**, 1754 (1989)

Metzger D., H. Reichl: Laser direct writing of gold to repair defective lines in thin-film metallizations. Appl. Surf. Sci. **69**, 69 (1993)

Metzger D., K. Hesch, P. Hess: Process characterization and mechanism for laser-induced chemical vapor deposition of a-Si:H from SiH_4. Appl. Phys. A **45**, 345 (1988)

Meunier M., M. Suys, M. Tabbal, R. Izquierdo, A. Yelon, E. Sacher: Excimer laser-induced metallization for in situ processing on Si and GaAs. Appl. Surf. Sci. **79/80**, 208 (1994)

Meunier M., J.H. Flint, J.S. Haggerty, D. Adler: Laser-induced chemical vapor deposition of hydrogenated amorphous silicon I. Gas-phase process model, J.Appl. Phys. **62**, 2812 (1987*a*)

Meunier M., J.H. Flint, J.S. Haggerty, D. Adler: Laser-induced chemical vapor deposition of hydrogenated amorphous silicon II. Film properties. J.Appl. Phys. **62**, 2822 (1987*b*)

Michael P.C., L.-G. Johansson, L. Bengtsson, T. Claeson, Z.G. Ivanov, E. Olsson, P. Berastegui, E. Stepantsov: Optimisation of growth of epitaxial $Tl_2Ba_2Ca_1Cu_2O_8$ superconducting thin films for electronic device applications. Physica C **235-240**, 717 (1994)

Migoni R., H. Bilz, D. Bäuerle: Origin of Raman scattering and ferroelectricity in oxidic perovskites. Phys. Rev. Lett. **37**, 1155 (1976)

Miller J.C. (ed.): *Laser Ablation – Principles and Applications*, Springer Ser. Mater. Sci., Vol.28 (Springer, Berlin, Heidelberg 1994)

Minamikawa T., Y. Yonezawa, S. Otsubo, T. Maeda, A. Moto, A. Morimoto, T. Shimizu: Preparation of $Ba_2YCu_3O_x$ superconducting films by laser evaporation and rapid laser annealing. Jpn. J. Appl. Phys. **27**, L619 (1988)

Mini L., C. Giaconia, C. Arnone: Copper patterning on dielectrics by laser writing in liquid solution. Appl. Phys. Lett. **64**, 3404 (1994)

Miotello A., R. Kelly, B. Braren, C.E. Otis: Novel geometrical effects observed in debris when polymers are laser sputtered. Appl. Phys. Lett. **61**, 2784 (1992)

Misiewicz J., C. Huber, D. Heiman, T.Q. Vu: Pulsed laser deposition growth of ZnSe, MnSe, and ZnSe/MnSe epilayers. Appl. Surf. Sci. **69**, 156 (1993)

Mitrikov M., D. Dimitrov, S. Metev: Thermally stimulated exoelectron emission and optically stimulated exoelectron emission of laser evaporated beryllium oxide thin films. J. Phys. D **17**, 1563 (1984)

Mitzner R., A. Rosenfeld, R. König: Time-resolved absorption studies of excimer laser ablation of CaF_2. Appl. Surf. Sci. **69**, 180 (1993)

Mizukawa S., K. Sato, K. Yasuhiro, M. Isawa, K. Kuroiwa, Y. Tarui: Optical studies of doping-characteristics in boron-doped a-Si:H films prepared by the photo-CVD technique. Jpn. J. Appl. Phys. **28**, 961 (1989)

Mönch W.: *Semiconductor Surfaces and Interfaces*, 2nd edn., Springer Ser. Surf. Sci., Vol.26 (Springer, Berlin, Heidelberg 1995)

Mönch W.: On the oxidation of III-V compound semiconductors. Surf. Sci. **168**, 577 (1986)

Mogyorosi P., K. Piglmayer, D. Bäuerle: Ar^+ laser-induced chemical eching of molybdenum in chlorine atmosphere. Surf. Sci. **208**, 232 (1989*a*)

Mogyorosi P., T. Szörenyi, K. Bali, Z. Toth, I. Hevesi: Pulsed laser ablative deposition of thin metal films. Appl. Surf. Sci. **36**, 157 (1989*b*)

Mogyorosi P.: CW-laser induced chemical etching of thin silver and molybdenum films. Appl. Surf. Sci. **36**, 332 (1989)

Mogyorosi P., K. Piglmayer, R. Kullmer, D. Bäuerle: Laser-induced chemical etching of silicon in chlorine atmosphere: II. Continuous irradiation. Appl. Phys. A **45**, 293 (1988)

Monreal R., S.P. Apell: Electromagnetic-field-enhanced desorption of atoms. Phys. Rev. B **41**, 7852 (1990)

Montgomery R.K., T.D. Mantei: UV laser deposition of metal films by photogenerated free radicals. Appl. Phys. Lett. **48**, 493 (1986)

Moon B.M., C.E. Platt, R.A. Schweinfurth, D.J.Van Harlingen: In situ pulsed laser deposition of superconducting $Ba_{1-x}K_xBiO_3$ thin films. Appl. Phys. Lett. **59**, 1905 (1991)

Morishige Y., S. Kishida: Thick gold-film deposition by high-repetition visible pulsed-laser CVD. Appl. Phys. A **59**, 395 (1994)

Morita N., S. Ishida, Y. Fujimori, K. Ishikawa: Pulsed laser processing of ceramics in water. Appl. Phys. Lett. **52**, 1965 (1988)

Moritani A., C. Hamaguchi: High-speed ellipsometry of arsenic-implanted Si during cw laser annealing. Appl. Phys. Lett. **46**, 746 (1985)

Morris B.J.: Photochemical organometallic vapor phase epitaxy of mercury cadmium telluride. Appl. Phys. Lett. **48**, 867 (1986)

Morrison S.R.: *The Chemical Physics of Surfaces* (Plenum, New York 1977)

Motorin V.I: Vitrification kinetics of pure metals. Phys. Status Solidi (a) **80**, 447 (1983)

Mott N.F., E.A. Davis: *Electronic Processes in Non-Crystalline Materials*, 2nd edn. (Clarendon, Oxford 1979)

Moutonnet D.: Maskless photoassisted etching of n-type $Ga_{0.47}In_{0.53}As$ in basic solutions. Appl. Phys. B **42**, 221 (1987)

Moutonnet D.: Photochemical wet etching of MBE epitaxial $Ga_{0.47}In_{0.53}As$ on InP, in *Laser Processing and Diagnostics II*, ed. by D. Bäuerle, K.L. Kompa, L.D. Laude (Edition de Physique, Les Ulis 1986) p.173

Moylan C.R., T.H. Baum, C.R. Jones: LCVD of copper: Deposition rates and deposit shapes. Appl. Phys. A **40**, 1 (1986)

Müllenborn M., H. Dirac, J.W. Petersen: Three-dimensional nanostructures by direct laser etching of Si. Appl. Surf. Sci. **86**, 568 (1995)

Müller F., K. Mann, P. Simon, J.S. Bernstein, G.J. Zaal: A comparative study of deposition of thin film by laser induced PVD with femtosecond and nanosecond laser pulses. SPIE Proc. **1858**, 464 (1993)

Müller F., K. Mann: Laser-induced physical vapour deposition of diamond-like carbon films. Diamond Rel. Mater. **2**, 233 (1992)

Murahara M.: Photochemical modification of fluorocarbon resin to generate adhesive properties. Lamdba Physik Highlights **43**, 1 (1994)

Murahara M., K. Toyoda: Linear-focused ArF excimer laser beam for depositing hydrogenated silicon films, in *Laser Processing and Diagnostics*, ed. by D. Bäuerle, Springer Ser. Chem. Phys., Vol.39 (Springer, Berlin, Heidelberg 1984) p.252

Murakami K.: Dynamics of laser ablation of high-T_c superconductors and semiconductors, and a new method for growth of films, in *Laser Ablation of Electronic Materials – Basic Mechanisms and Applications*, ed. by E. Fogarassy, S. Lazare (Elsevier, Amsterdam 1992) p.125

Murakami K., O. Eryu, K. Takita, K. Masuda: Energy beam irradiation of high-T_c superconductors $Y_1Ba_2Cu_3O_{7-y}$ and $Ho_1Ba_2Cu_3O_{7-y}$. Jpn. J. Appl. Phys. **26**, L1731 (1987)

Murray P.T., D.T. Peeler: Dynamics of graphite photoablation: Kinetic energy of the precursors to diamond-like carbon. Appl. Surf. Sci. **69**, 225 (1993)

Murray T., J.D. Wolf, J.A. Mescher, J.T. Grant, N.T. McDevitt: Growth of yttria stabilized cubic zirconia on GaAs(100) by pulsed laser evaporation. Mater. Lett. **5**, 250 (1987)

Myl'nikov V.S., T.S. Turoskaya, N.I. Pozdnyak, G.S. Zhdanov, G.P. Tikhomirov: Structure of zinc sulfide film produced by laser evaporation. Sov. Phys. – Tech. Phys. **23**, 1389 (1978)

Nagahori T., S. Matsumoto: Growth mechanism of direct writing of silicon in Ar^+ laser CVD. MRS Proc. **129**, 189 (1989)

Nakai Y., K. Hattori, A. Okano, N. Itoh, R.F. Haglund: Nonthermal laser sputtering from solid surfaces. Nucl. Instrum. Methods B **58**, 452 (1991)

Nanai L., R. Vajtai, I. Hevesi, D.A. Jelski, T.F. George: Metal oxide layer growth under laser irradiation. Thin Solid Films **227**, 13 (1993)

Nanai L., I. Hevesi, F.V. Bunkin, B.S. Luk'yanchuk, M.R. Brook, G.A. Shafeev, D.A. Jelski, Z.C. Wu, T.F. George: Laser-induced metal deposition on semiconductors from liquid electrolytes. Appl. Phys. Lett. **54**, 736 (1989*a*)

Nanai L., I. Hevesi, B.S. Luk'yanchuk, E.A. Morozova, A.S. Rogachev, A.V. Simakin, N.V. Sukonkina, G.A. Shafeev: Characteristics of laser-heated titanium in nitrogen atmosphere. Acta Physica Hungarica **65**, 405 (1989*b*)

Narayan J., R.T. Young, R.F. Wood, W.H. Christie: p-n junction formation in boron-deposited silicon by laser-induced diffusion. Appl. Phys. Lett. **33**, 338 (1978)

Nashimoto K., D.K. Fork, T.H. Geballe: Epitaxial growth of MgO on GaAs(001) for growing epitaxial $BaTiO_3$ thin films by pulsed laser deposition. Appl. Phys. Lett. **60**, 1199 (1992)

Nassuphis N., R.H. Mathews, S.T. Palmacci, D.J. Ehrlich: Three-dimensional laser direct writing: Applications to multichip modules. J. Vac. Sci. Technol. B **12**, 3294 (1994)

Newman C.G., H.E. O'Neal, M.A. Ring, F. Leska, N. Skipley: Kinetics and mechanism of the silane decomposition. Int'l J. Chem. Kinet. **11**, 1167 (1979)

Nguyen V.T., SOI Group: Laser processing in silicon on insulator (SOI) technologies, in *Laser Processing and Diagnostics*, ed. by D. Bäuerle, Springer Ser. Chem. Phys., Vol.39 (Springer, Berlin, Heidelberg 1984) p.73

Nicolis G., I. Prigogine: *Self-Organization in Non-Equilibrium Systems* (Wiley, New York 1977)

Niino H., T. Murao, S. Matsumura, A. Yabe: Hydrophilic surface modification and metallization of poly(tetrafluoroethylene) film by excimer laser processing. Mol. Cryst. Liq. Cryst. **267**, 365 (1995)

Niino H., A. Yabe: Surface modification and metallization of fluorocarbon polymers by excimer laser processing. Appl. Phys. Lett. **63**, 3527 (1993*a*)

Niino H., A. Yabe: Excimer laser polymer ablation: Formation of positively charged surfaces and its application into the metallization of polymer films. Appl. Surf. Sci. **69**, 1 (1993*b*)

Niino H., A. Yabe: Excimer laser ablation of polyethersulfone derivatives: Periodic morphological micro-modification on ablated surface. J. Photochem. Photobiol. A **65**, 303 (1992*a*)

Niino H., A. Yabe: Positively charged surface potential of polymer films after excimer laser ablation: Application to selective-area electroless plating on the ablated films. Appl. Phys. Lett. **60**, 2697 (1992*b*)

Nishida S., T. Shiimoto, A. Yamada, S. Karasawa, M. Konagai, K. Takahashi: Epitaxial growth of silicon by photochemical vapor deposition at a very low temperature of 200°C. Appl. Phys. Lett. **49**, 79 (1986)

Nishimura Y., K. Tokunaga, M. Tsuji: Deposition of tantalum oxide films by ArF excimer laser chemical vapour deposition. Thin Solid Films **226**, 144 (1993)

Nishizawa J., H. Shimawaki, Y. Sakuma: Reaction mechanism of GaAs vapor-phase epitaxy. J. Electrochem. Soc. **133**, 2567 (1986)

Nitzan A., L.E. Brus: Theoretical model for enhanced photochemistry on rough surfaces. J. Chem. Phys. **75**, 2205 (1981)

Nordine P.C., S.C.de la Veaux, F.T. Wallenberger: Silicon fibers produced by high-pressure LCVD. Appl. Phys. A **57**, 97 (1993)

Northrop D.A.: Vaporization of lead zirconate-lead titanate materials II. Hot-pressed compositions at near theoretical density. J. Am. Ceram. Soc. **51**, 357 (1968)

Norton D.P., D.H. Lowndes: Enhanced superconducting properties in ultrathin $YBa_2Cu_3O_{7-\delta}$ layers. Appl. Phys. Lett. **63**, 1432 (1993)

Norton M.G., P.G. Kotula, C.B. Carter: Oriented aluminum nitride thin films deposited by pulsed-laser ablation. J. Appl. Phys. **70**, 2871 (1991)

Novicki R.S.: Properties of rf-sputtered Al_2O_3 films deposited by planar magnetron. J. Vac. Sci. Technol. **14**, 127 (1977)

Novis Y., J.J.Pireaux, A. Brezini, E. Petit, R. Caudano, P. Lutgen, G. Feyder, S. Lazare: Structural origin of surface morphological modifications developed on poly(ethylene terephthalate) by excimer laser photoablation. J. Appl. Phys. **64**, 365 (1988)

Ogale S.B., P.G. Bilurkar, N. Mate, S.M. Kanetkar, N. Parikh, B. Patnaik: Deposition of copper oxide thin films on different substrates by pulsed excimer laser ablation. J. Appl. Phys. **72**, 3765 (1992)

Ogale S.B., V.N. Koinkar, S. Joshi, V.P. Godbole, S.K. Date, A. Mitra, T. Venkatesan, X.D. Wu: Deposition of iron oxide thin films by pulsed laser evaporation. Appl. Phys. Lett. **53**, 1320 (1988)

Ogale S.B., A. Polman, F.O.P. Quentin, S. Roorda, F.W. Saris: Pulsed laser oxidation and nitridation of metal surfaces immersed in liquid media. Appl. Phys. Lett. **50**, 138 (1987)

Oguz S., W. Paul, T.F. Deutsch, B.Y. Tsaur, D.V. Murphy: Synthesis of metastable, semiconducting Ge-Sn alloys by pulsed UV-laser crystallization. Appl. Phys. Lett. **43**, 848 (1983)

Okabe K., T. Mitsuishi, Y. Sasaki: Reduction and sintering of vanadium oxide films by laser-beam irradiation. Jpn. J. Appl. Phys. **26**, 1802 (1987)

Okada N., Y. Katsumura, K. Ishigure: Improvement of corrosion resistance of carbon steel using chemical vapor deposition from the mixture of $Mo(CO)_6$ and $Cr(CO)_6$ with an ArF-excimer laser. Appl. Phys. A **58**, 99 (1994)

Okada N., Y. Katsumura, K. Ishigure: Improvement of corrosion resistance of carbon steel using chemical vapor deposition from $Cr(CO)_6$ with an ArF excimer laser. Appl. Phys. A **55**, 207 (1992)

Okano A., A.Y. Matsuura, K. Hattori, N. Itoh, J. Singh: A model of laser ablation in nonmetallic inorganic solids. J. Appl. Phys. **73**, 3158 (1993)

Ollier B., N. Pirch, E.W. Kreutz: Ein numerisches Modell zum einstufigen Laserstrahlbeschichten (A numerical model of the one-step laser cladding process). Laser und Optoelektronik **27**, 63 (1995)

Olson G.L., S.A. Kokorowski, J.A. Roth, L.D. Hess: Direct measurement of cw laser-induced crystal growth dynamics by time-resolved optical reflectivity, in *Laser and Electron-Beam Solid Interactions and Materials Processing*, ed. by J.F. Gibbons, L.D. Hess, T.W. Sigmon. MRS Proc. **1**, 125 (1981)

Omori N., M. Inoue: Excimer laser ablation of inorganic materials. Appl. Surf. Sci. **54**, 232 (1992)

Oraevsky A.A., S.L. Jacques, G.H. Pettit, I. Saidi, F.K. Tittel, R.A. Sauerbrey, P.D. Henry: XeCl ablation of aorta tissue: Optical properties and energy pathways. Lasers in Surgery and Medicine **12**, 585 (1992)

Oraevsky A.A., R.O. Esenaliev, V.S. Letokhov: Pulsed laser ablation of biological tissue: Review of the mechanisms, in *Laser Ablation – Mechanisms and Applications*, ed. by J.C. Miller, R.F. Haglund, Lecture Notes Phys. **389**, 112 (Springer, Berlin, Heidelberg 1991)

Orlowski T.E., D. A. Mantell: Aluminum deposition by ultraviolet laser photofragmentation of trimethylaluminum on Al: Identification of photoproducts and desorption dynamics. J. Vac. Sci. Technol. A **7**, 2598 (1989)

Orlowski T.E., D.A. Mantell: Ultraviolet laser-induced oxidation of silicon: The effect of oxygen photodissociation upon oxide growth kinetics. J. Appl. Phys. **64**, 4410 (1988)

Ostermayer F.W., P.A. Kohl, R.M. Lum: Hole transport equation analysis of photoelectrochemical resolution. J. Appl. Phys. **58**, 4390 (1985)

Ostermayer F.W., P.A. Kohl: Photoelectrochemical etching of p-GaAs. Appl. Phys. Lett. **39**, 76 (1981)

Osterreicher H., H. Bittner, B. Kothari: Laser evaporation and condensation of Er in hydrogen and inert atmosphere. J. Solid State Chem. **26**, 97 (1978)

Otis C.E., P.M. Goodwin: Internal energy distributions of laser ablated species from $YBa_2Cu_3O_{7-\delta}$. J. Appl. Phys. **73**, 1957 (1993)

Otsubo S., T. Maeda, T. Minamikawa, Y. Yonezawa, A. Morimoto, T. Shimizu: Preparation of $Pb(Zn_{0.52}Ti_{0.48})O_3$ films by laser ablation. Jpn. J. Appl. Phys. **29**, L133 (1990)

Otto J., R. Stumpe, D. Bäuerle: Laser induced reduction and etching of oxidic perovskites, in *Laser Processing and Diagnostics*, ed. by D. Bäuerle, Springer Ser. Chem. Phys., Vol.39 (Springer, Berlin, Heidelberg 1984) p.320

Ozaki Y., K. Takamato: Cylindrical fly's eye lens for intensity redistribution of an excimer laser beam. Appl. Opt. **28**, 106 (1989)

Pan D., J. Chen: Etching enhancement of SiO_2 in a two-laser field. Chem. Phys. Lett. **143**, 599 (1988)

Pandey H.C., Y.K. Jain, S.K. Bhatnagar, B.R. Singh, W.S. Khokle: Direct laser beam writing on YBaCuO film for superconducting microelectronic devices. Jpn. J. Appl. Phys. **27**, L1517 (1988)

Pang H.M., E.S. Yeung: Absorption spectroscopy in laser-generated plumes by surface reflection. Appl. Spectrosc. **44**, 1218 (1990)

Paraskevopoulos G., D.L. Singleton, R.S. Irwin, R.S. Taylor: Time-resolved reflectivity as a probe of the dynamics of laser ablation of organic polymers. J. Appl. Phys. **70**, 1938 (1991)

Pashmakov B., D. Nesheva, E. Vateva, D. Dimitrov, S. Metev: Properties of laserbeam sputtered CdS thin film. J. Mater. Res. Sci. Lett. **4**, 442 (1985)

Pauleau Y., D. Tonneau, G. Auvert: Deposition of silicon films by photodissociation of silane under IR laser irradiation, in *Laser Processing and Diagnostics*, ed. by D. Bäuerle, Springer Ser. Chem. Phys., Vol. 39 (Springer, Berlin, Heidelberg 1984) p. 215

Paulsen-Boaz C., W.L. O'Brien, T. Rhodin: Pulsed ultraviolet laser stimulated chlorination mechanisms for Si(111). J. Vac. Sci. Technol. B **10**, 216 (1992)

Pazionis G.D., H. Tang, I.P. Herman: Raman microprobe analysis of temperature profiles in cw laser heated silicon microstructures. IEEE J. QE-**25**, 976 (1989)

Pechen E.V., A.V. Varlashkin, S.I. Krasnosvobodtsev, B. Brunner, K.F. Renk: Pulsed-laser deposition of smooth high-T_c superconducting films using a synchronous velocity filter. Appl. Phys. Lett. **66**, 2292 (1995)

Peng C.J., D. Roy, S.B. Krupanidhi: Oriented lead germanate thin films by excimer laser ablation. Appl. Phys. Lett. **60**, 827 (1992)

Perkins G.G.A., E.R. Austin, F.W. Lampe: The 147-nm photolysis of monosilane. J. Am. Chem. Soc. **101**, 1109 (1979)

Petitjean M., N. Proust, J.-F. Chapeaublanc, J. Perrin: UV-CVD deposited SiNH films on InP: Optimization of the physico-chemical and interface properties. Appl. Phys. A **54**, 95 (1992)

Pettit G.H., R. Sauerbrey: Fluence-dependent transmission of polyimide at 248 nm under laser ablation conditions. Appl. Phys. Lett. **58**, 793 (1991)

Petzoldt F., K. Piglmayer, W. Kräuter, D. Bäuerle: Lateral growth rates in laser CVD of microstructures. Appl. Phys. A **35**, 155 (1984)

Pflügl W., U.M. Titulaer: The size distribution of liquid droplets during their growth from a vapor. Physica A **198**, 410 (1993)

Phillips H.M., D. Sarkar, N.J. Halas, R.H. Hauge, R. Sauerbrey: Excimer-laser-induced electric conductivity in thin-film C_{60}. Appl. Phys. A **57**, 105 (1993*a*)

Phillips H.M., S. Wahl, R. Sauerbrey: Submicron electrically conducting wires produced in polyimide by ultraviolet laser irradiation. Appl. Phys. Lett. **62**, 2572 (1993*b*)

Phillips H.M., D.L. Callahan, R. Sauerbrey, G. Szabo, Z. Bor: Direct laser ablation of Sub-100 nm line structures into polyimide. Appl. Phys. A **54**, 158 (1992)

Picraux S.T., D.M. Follstaedt: Surface modification and alloying: Aluminium, in *Surface Modification and Alloying by Laser, Ion, and Electron Beams*, ed. by J.M. Poate, G. Foti, D.C. Jacobson (Plenum, New York 1983) p. 287

Piehler A., N. Reschauer, U. Spreitzer, J.P. Ströbel, R. Schönberger, K.F. Renk, G. Saemann-Ischenko: Preparation of epitaxial $TlBa_2Ca_2Cu_3O_9$ high T_c thin films on $LaAlO_3(100)$ substrates. Appl. Phys. Lett. **65**, 1451 (1994)

Piglmayer K., D. Bäuerle: On the reaction kinetics in laser-induced photochemical gas-phase processing. Appl. Phys. B **48**, 453 (1989)

Piglmayer K., D. Bäuerle: Temperature distributions in pyrolytic laser-induced chemical processing, in *Laser Processing and Diagnostics II*, ed. by D. Bäuerle, K.L. Kompa, L.D. Laude (Les Editions de Physique, Les Ulis 1986) p. 79

Piglmayer K., J. Doppelbauer, D. Bäuerle: Temperature distributions in CW-laser pyrolysis, in *Laser Controlled Chemical Processing of Surfaces*, ed. by A.W. Johnson, D.J. Ehrlich, H.R. Schlossberg. MRS Proc. **29**, 47 (1984)

Pinto R., S.P. Pai, C.P. D'Souza, L.C. Gupta, R. Vijayaraghavan, D. Kumar, M. Sharon: Optimization of KrF laser ablation parameters for in-situ growth of $Y_1Ba_2Cu_3O_{7-\delta}$ thin films. Physica C **196**, 264 (1992)

Podlesnik D.V., H.H. Gilgen, R.M. Osgood: Waveguiding effects in laser-induced aqueous etching of semiconductors. Appl. Phys. Lett. **48**, 496 (1986)

Podlesnik D.V., H.H. Gilgen, R.M. Osgood: Deep-ultraviolet induced wet etching of GaAs. Appl. Phys. Lett. **45**, 563 (1984)

Podlesnik D.V., H.H. Gilgen, R.M. Osgood, A. Sanchez: Maskless chemical etching of submicrometer gratings in single-crystalline GaAs. Appl. Phys. Lett. **43**, 1083 (1983)

Possin G.E., H.G. Parks, S.W. Chiang, Y.S. Liu: The effects of selectively absorbing dielectric layers and beam shaping on recrystallization and FET characteristics in laser recrystallized silicon on amorphous substrates. MRS Proc. **13**, 549 (1983)

Prawer S., R. Kalish, M. Adel: Pulsed laser treatment of diamondlike carbon films. Appl. Phys. Lett. **48**, 1585 (1986)

Press W.H., S.A. Teukolsky, W.T. Vetterling, B.P. Flannery: *Numerical Recipes in FORTRAN – The Art of Scientific Computing* (Cambridge Univ. Press, Cambridge 1992)

Preuss S., A. Demchuk, M. Stuke: Sub-picosecond UV laser ablation of metals. Appl. Phys. A **61**, 33 (1995)

Preuss S., M. Späth, Y. Zhang, M. Stuke: Time resolved dynamics of subpicosecond laser ablation. Appl. Phys. Lett. **62**, 3049 (1993)

Preuss S., H. Stafast: CO_2 laser CVD of copper lines with twofold periodic structures. Appl. Phys. A **54**, 152 (1992)

Price S.J.W.: The decomposition of metal alkyls, aryls, carbonyls and nitrosyls, in *Decomposition of Inorganic and Organometallic Compounds*, ed. by C.H. Bamford, C.F.H. Tipper, Chem. Kinetics, Vol.4 (Elsevier, Amsterdam 1972) p.197

Prokhorov A.M., V.I. Konov, I. Ursu, I.N. Mihailescu: *Laser Heating of Metals* (Hilger, Bristol 1990)

Proyer S., E. Stangl, M. Borz, B. Hellebrand, D. Bäuerle: Particulates on pulsed-laser deposited Y-Ba-Cu-O films. Physica C **257**, 1 (1996)

Proyer S., E. Stangl: Time-integrated photography of laser-induced plasma plumes. Appl. Phys. A **60**, 573 (1995)

Proyer S., E. Stangl, P. Schwab, D. Bäuerle, P. Simon, C. Jordan: Patterning of YBCO films by excimer-laser ablation. Appl. Phys. A **58**, 471 (1994)

Puippe J.C., R.E. Acosta, R.J. von Gutfeld: Investigation of laser-enhanced electroplating mechanisms. J. Electrochem. Soc. **128**, 2539 (1981)

Qadri S.B., J.S. Horwitz, D.B. Chrisey, R.C.Y. Auyeung, K.S. Grabowski: X-ray characterization of extremely high quality $(Sr,Ba)TiO_3$ films grown by pulsed laser deposition, Appl. Phys. Lett. **66**, 1605 (1995)

Radloff W., E. Below, H. Dürr, V. Stert: Laser photolytic deposition of molybdenum and tungsten thin film microstructures. Appl. Phys. A **50**, 233 (1990)

Raizer Y.P.: Optical discharges. Sov. Phys. – Usp. **23**, 789 (1980)

Ramesh R., A. Inam, B. Wilkens, W.K. Chan, T. Sands, D.K. Fork, T.H. Geballe, J. Evans, J. Bullinton: Ferroelectric bismuth titanate/superconductor (Y-Ba-Cu-O) thin-film heterostructures on silicon. Appl. Phys. Lett. **59**, 1782 (1991)

Ramesh R., K. Luther, B. Wilkens, D.L. Hart, E. Wang, J.M. Tarascon, A. Inam, X.D. Wu, T. Venkatesan: Epitaxial growth of ferroelectric bismuth titanate thin films by pulsed laser deposition. Appl. Phys. Lett. **57**, 1505 (1990)

Rao G.M., S.B. Krupanidhi: Study of electrical properties of pulsed excimer laser deposited strontium titanate films. J. Appl. Phys. **75**, 2604 (1994)

Razavi F.S., H.-U. Habermeier: Preparation of thin films of BiPbSrCaCuO using the laser deposition technique. Physica C **180**, 81 (1991)

Ready J.F.: *Industrial Applications of Lasers* (Academic, London 1978)

Ready J.F: *Effects of High-Power Laser Radiation* (Academic, London 1971)

Reisse G., F. Gänsicke, A. Fischer, H. Johansen: Laser direct writing of titanium silicide thin films. Appl. Surf. Sci. **69**, 412 (1993)

Reksten G.M., W. Holber, R.M. Osgood: Wavelength dependence of laser enhanced plasma etching of semiconductors. Appl. Phys. Lett. **48**, 551 (1986)

Rengan A., J. Narayan: Optical, structural and mechanical properties of diamondlike carbon films deposited by laser ablation and laser-plasma ablation techniqes, in *Laser Ablation of Electronic Materials*, ed. by E. Fogarassy, S. Lazare (Elsevier, Amsterdam 1992) p.363

Richardson T., E. Swenson: Generating holograms with a computer. Photon. Spectra **23**, 133 (1989)

Rink K., G. Delacrétaz, G. Pittomvils, R. Boving, J.P. Lafaut: Incidence of cavitation in the fragmentation process of extracorporeal shock wave lithotriptors. Appl. Phys. Lett. **64**, 2596 (1994a)

Rink K., G. Delacrétaz, R.P. Salathé: Influence of the pulse duration on laser-induced mechanical effects. SPIE Proc. **2077**, 181 (1994b)

Rink K., G. Delacrétaz, R.P. Salathé: Fragmentation process induced by microsecond laser pulses during lithotripsy. Appl. Phys. Lett. **61**, 258 (1992)

Roas B., L. Schultz, G. Endres: Superconducting properties of laser evaporated epitaxial Y-Ba-Cu-O thin films. J. Less-Common Met. **151**, 413 (1989)

Roas B., L. Schultz, G. Endres: Epitaxial growth of $YBa_2Cu_3O_{7-x}$ thin films by a laser evaporation process. Appl. Phys. Lett. **53**, 1557 (1988)

Roberts M.W., C.S. McKee: *Chemistry of the Metal-Gas Interface* (Clarendon, Oxford 1978)

Rothenberg J.E., R. Kelly: Laser sputtering II: The mechanism of sputtering of Al_2O_3. Nucl. Inst. Meth. Phys. Res. B **1**, 291 (1984)

Rothschild M.: Spectroscopy and photochemistry of gases, adsorbates, and liquids, in *Laser Microfabrication – Thin Film Processes and Lithography*, ed. by D.J Ehrlich, J.Y Tsao (Academic, London 1989) p.163

Rothschild M., J.H.C. Sedlacek, J.G. Black, D.J. Ehrlich: Reversible laser chemically induced phase transformations in thin-film $Ba_2YCu_3O_x$ superconductors. Appl. Phys. Lett. **52**, 404 (1988)

Rothschild M., J.H.C. Sedlacek, J.G. Black, D. J. Ehrlich: Visible-laser photochemical etching of Cr, Mo, and W. J. Vac. Sci. Technol. B **5**, 414 (1987)

Rothschild M., C. Arnone, D.J. Ehrlich: Excimer-laser etching of diamond and hard carbon films by direct writing and optical projection. J. Vac. Sci. Technol. B **4**, 310 (1986)

Roy D., S.B. Krupanidhi, J.P. Dougherty: Excimer laser ablation of ferroelectric $Pb(Zr,Ti)O_3$ thin films with low pressure direct-current glow discharge. J. Vac. Sci. Technol. A **10**, 1827 (1992)

Roy D., S.B. Krupanidhi, J.P. Dougherty: Excimer laser ablated lead zirconate titanate thin films. J. Appl. Phys. **69**, 7930 (1991)

Ruberto M.N., X. Zhang, R. Scarmozzino, A.E. Willner, D.V. Podlesnik, R.M. Osgood: The laser-controlled micrometer-scale photoelectrochemical etching of III-V semiconductors. J. Electrochem. Soc. **138**, 1174 (1991)

Saenger K.L., R.A. Roy, K.F. Etzold, J.J. Cuomo: Lead zirconate titanate films produced by pulsed laser deposition. MRS Proc. **200**, 115 (1990)

Saenger K.L.: Time-resolved optical emission during laser ablation of Cu, CuO, and high-T_c superconductors: $Bi_{1.7}Sr_{1.3}Ca_2Cu_3O_x$ and $Y_1Ba_{1.7}Cu_{2.7}O_y$. J. Appl. Phys. **66**, 4435 (1989)

Saitoh J., M. Fukutomi, K. Komori, Y. Tanaka, T. Asano, H. Maeda, H. Takahara: Preparation of $YBa_2Cu_3O_y$ superconducting thin films on metallic substrates by excimer laser ablation. Jpn. J. Appl. Phys. **30**, L898 (1991)

Samadi Hosseinali G., R.M. Schalk, H.W. Weber, A. Pönninger, S. Proyer, P. Schwab: Angular dependence of J_c in laser ablated YBCO thin films. Physica B **194-196**, 2357 (1994)

Sameshima T., S. Usui: Analysis of dopant diffusion in molten silicon induced by a pulsed excimer laser. Jpn. J. Appl. Phys. **26**, L1208 (1987)

Samsonov G.V. (ed.): *The Oxide Handbook* (Plenum, New York 1973)

Sankur H., F. Woodberry, R. Hall, W.J. Gunning: Properties of metal fluoride films deposited by laser and e-beam evaporation. MRS Proc. **77**, 727 (1987)

Sankur H.: Properties of thin PbF_2 films deposited by cw and pulsed laser-assisted evaporation. Appl. Opt. **25**, 1962 (1986)

Sankur H., R. Hall: Thin-film deposition by laser-assisted evaporation. Appl. Opt. **24**, 3343 (1985)

Sankur H.: Properties of dielectric thin films formed by laser evaporation. MRS Proc. **29**, 373 (1984)

Sankur H., J.T Cheung: Highly oriented ZnO films grown by laser evaporation. J. Vac. Sci. Technol. A **1**, 1806 (1983)

Sato T., S. Furuno, S. Iguchi, M. Hanabusa: Diamond-like carbon films prepared by pulsed-laser evaporation. Appl. Phys. A **45**, 355 (1988)

Sato T.: Spectral emissivity of silicon. Jpn. J. Appl. Phys. **6**, 339 (1967)

Sausa R.C., A. Gupta, J.R. White: Laser decomposition of platinum metallo-organic films for electroless copper plating. J. Electrochem. Soc. **134**, 2707 (1987)

Scheibe H.J., P. Siemroth, B. Schöneich, A. Mucha: Diamond-like carbon film preparation by laser arc. Surface and Coatings Technology **52**, 129 (1992)

Scheibe H.J., W. Pompe, P. Siemroth, B. Buecken, D. Schulze, R. Wilberg: Preparation of multilayered film structures by laser arcs. Thin Solid Films **193/194**, 788 (1990)

Scheuermann M., C.C. Chi, C.C. Tsuei, D.S. Yee, J.J. Cuomo, R.B. Laibowitz, R.H. Koch, B. Braren, R. Srinivasan, M.M. Plechaty: Magnetron sputtering and laser patterning of high transition temperature Cu oxide films. Appl. Phys. Lett. **51**, 1951 (1987)

Schmidt M.T., Z. Wu, R.M. Osgood: A marker technique to identify diffusing elements during initial reactions using ion scattering spectroscopy. Surface and Interface Analysis **17**, 43 (1991)

Scholz M., W. Fuß, K. Kompa: CVD of silicon carbide powders using pulsed CO_2 lasers. Adv. Mater. **5**, 38 (1993)

Schomacker K.R., Y. Domankevitz, T.J. Flotte, T.F. Deutsch: Co:MgF$_2$ laser ablation of tissue: Effect of wavelength on ablation threshold and thermal damage. Lasers in Surgery and Medicine **11**, 141 (1991)

Schröder H., B. Rager, S. Metev, N. Rösch, H. Jörg: Photochemistry of transition metal complexes, in *Interfaces under Laser Irradiation*, ed. by L.D. Laude, D. Bäuerle, M. Wautelet, NATO ASI Ser. (Nijhoff, Dordrecht 1987) p.255

Schuegraf K.K.: Low temperature photochemical vapor deposition of SiO$_2$ and Si$_3$N$_4$. Microelectronic Manufacturing and Testing (March 1983) p.23

Schulmeister K., J.G. Lunney, B. Buckley: Laser chemical vapor deposition of cobalt thin films. J. Appl. Phys. **72**, 3480 (1992)

Schumann M., R. Sauerbrey, M.C. Smayling: Permanent increase of the electrical conductivity of polymers induced by ultraviolet laser radiation, Appl. Phys. Lett. **58**, 428 (1991)

Schuöcker D., A. Kaplan: Overview over modelling for laser applications. SPIE Proc. **2207**, 236 (1994)

Schwab P., X.Z. Wang, S. Proyer, A. Kochemasov, D. Bäuerle: Synthesis of YBa$_{2-x}$Sr$_x$Cu$_3$O$_{7-\delta}$ by pulsed-laser deposition. Physica C **214**, 257 (1993)

Schwab P., A. Kochemasov, R. Kullmer, D. Bäuerle: The influence of photodissociated N$_2$O in pulsed-laser deposition of Y-Ba-Cu-O Films. Appl. Phys. A **54**, 166 (1992)

Schwab P., X.Z. Wang, D. Bäuerle: In situ fabrication of superconducting Lu-Ba-Sr-Cu-O films by pulsed-laser deposition. Appl. Phys. Lett. **60**, 2023 (1992)

Schwab P., J. Heitz, S. Proyer, D. Bäuerle: Femtosecond-excimer-laser patterning of YBa$_2$Cu$_3$O$_7$ films. Appl. Phys. A **53**, 282 (1991)

Schwab P., D. Bäuerle: Pulsed-laser deposition of Y-Ba-Cu-O in O$_2$ and N$_2$O atmospheres. Physica C **182**, 103 (1991)

Schwartz B.: *GaAs Surface Chemistry – A Review* (CRC Press, Boca Raton, FL 1975) p.609

Schwartz H., H.A. Tourtellotte: Vacuum deposition by high-energy laser with emphasis on barium titanate films. J. Vac. Sci. Technol. **6**, 3763 (1969)

Scoles K.J., A.H. Kim, M.-H. Jiang, B.C. Lee: Deposition and characteristics of silicon dioxide thin films deposited by mercury-arc-source driven photon-activated chemical-vapor deposition. J. Vac. Sci. Technol. B **6**, 470 (1988)

Scott B.A.: Homogeneous chemical vapor deposition, in *Hydrogenated Amorphous Silicon* A **21**, 123 (Academic, London 1984)

Seder T.A., S.P. Church, A.J. Ouderkirk, E. Weitz: Gas-phase photofragmentation of Cr(CO)$_6$: Time-resolved infrared spectrum and decay kinetics of "naked" Cr(CO)$_5$. J. Am. Chem. Soc. **107**, 1432 (1985)

Sedov L.I.: *Similarity and Dimensional Methods in Mechanics* (Academic, London 1959)

Seel M., P.S. Bagus: Ab initio cluster study of the interaction of fluorine and chlorine with the Si(111) surface. Phys. Rev. B **28**, 2023 (1983)

Sell J.A., D.M. Heffelfinger, P.L.G. Ventzek, R.M. Gilgenbach: Photoacoustic and photothermal beam deflection as a probe of laser ablation of materials. J. Appl. Phys. **69**, 1330 (1991)

Sell J.A., D.M. Heffelfinger, P. Ventzek, R.M. Gilgenbach: Laser beam deflection as a probe of laser ablation of materials. Appl. Phys. Lett. **55**, 2435 (1989)

Seo J.M., Y.Z. Li, S.G. Anderson, D.J.W. Aastuen, U.S. Ayyala, G.H. Kroll, J.H. Weaver: X-ray-induced low-temperature oxidation: $N_2O/GaAs(110)$. Phys. Rev. B **42**, 9080 (1990)

Sesselmann W., T.J. Chuang: Reaction of chlorine with Ag surfaces and radiation effects by X-ray photons and Ar^+-ions. Surf. Sci. **184**, 374 (1987)

Sesselmann W., E.E. Marinero, T.J. Chuang: Laser stimulated desorption from noble metal surfaces reacted with chlorine. Surf. Sci. **178**, 787 (1986*a*)

Sesselmann W., E.E. Marinero, T.J. Chuang: Laser-induced desorption and etching processes on chlorinated Cu and solid CuCl surfaces. Appl. Phys. A **41**, 209 (1986*b*)

Sesselmann W., T.J. Chuang: Chlorine surface interaction and laser-induced surface etching reactions. J. Vac. Sci. Technol. B **3**, 1507 (1985)

Sheftal R.N., I.V. Cherbakov: Mechanism of condensation of hetero-epitaxial III-V layers deposited by a laser pulse of moderate power. Cryst. Res. Technol. **16**, 887 (1981)

Shen Y.Q., T. Freltoft, P. Vase: Laser writing and rewriting on $YBa_2Cu_3O_7$ films, Appl. Phys. Lett. **59**, 1365 (1991)

Shewmon P.G.: *Diffusion in Solids* (McGraw-Hill, New York 1963)

Shi L., Y. Hashishin, S.Y. Dong, J.P. Zheng, H.S. Kwok: Laser deposition of CdS/ Y-Ba-Cu-O heterostructures, Appl. Phys. Lett. **59**, 1377 (1991)

Shibata T., J.F. Gibbons, T.W. Sigmon: Silicide formation using a scanning cw laser beam. Appl. Phys. Lett. **36**, 566 (1980)

Shinn G.B., P.M. Gillespie, W.L. Wilson, Jr., W.M. Duncan: Laser-assisted metalorganic chemical vapor deposition of zinc selenide epitaxial films, Appl. Phys. Lett. **54**, 2440 (1989)

Shiosaki T., M. Tanizawa, H. Kamei, A. Kawabata: Laser micromachining of a modified $PbTiO_3$ ceramics in KOH water solution. Jpn. J. Appl. Phys. **22**, Suppl. 22-2, 109 (1983)

Shirafuji J., S. Miyoshi, H. Aoki: Laser-induced chemical vapour deposition and characterization of amorphous silicon oxide films. Thin Solid Films **157**, 105 (1988)

Shklovskii V.A., V.M. Kuz'menko: Explosive crystallization of amorphous substances. Sov. Phys. – Usp. **32**, 163 (1989)

Siegman A.E.: *Lasers* (University Science Books, Mill Valley, CA 1986)

Silvain J.F., J.J. Ehrhardt, P. Lutgen: Interfacial analysis of aluminum and copper thin films evaporated on polyetheneterephthalate. Thin Solid Films **195**, L5 (1991)

Silvestre A.J., M.L.G.F. Parames, O. Conde: Investigation of the microstructure, chemical composition and lateral growth kinetics on TiN films deposited by laser-induced chemical vapor deposition. Thin Solid Films **241**, 57 (1994)

Simon P.: Time-resolved ablation-site photography of XeCl-laser irradiated polyimide. Appl. Phys. B **48**, 253 (1989)

Singh J., M. Vellaikal, J. Narayan: Laser-enhanced synthesis and processing of diamond films from liquid hydrocarbons. Appl. Phys. **73**, 4351 (1993)

Singh R.K., J. Narayan: Nature of epitaxial growth of high-T_c laser-deposited Y-Ba-Cu-O films on (100) strontium titanate substrates. J. Appl. Phys. **67**, 3785 (1990)

Singleton D.L., G. Paraskevopoulos, R.S. Irwin: XeCl laser ablation of polyimide: Influence of ambient atmosphere on particulate and gaseous products. J. Appl. Phys. **66**, 3324 (1989)

Singmaster K.A., F.A. Houle: Effect of laser heating on compositions of films deposited from the metal hexacarbonyls. MRS Proc. **201**, 159 (1991)

Singmaster K.A., F.A. Houle, R.J. Wilson: Photochemical deposition of thin films from the metal hexacarbonyls. J. Phys. Chem. **94**, 6864 (1990)

Sladek K.J.: The role of homogeneous reactions in chemical vapor deposition. J. Electrochem. Soc. **118**, 654 (1971)

Slaoui A., E. Fogarassy, C. Fuchs, J.P. Stoquert: Influence of ambient gas and substrate temperature in preparation of silicon dioxide films by laser ablation, in *Laser Ablation of Electronic Materials – Basic Mechanisms and Applications*, ed. by E. Fogarassy, S. Lazare (Elsevier, Amsterdam 1992) p.351

Slaoui A., F. Foulon, C. Fuchs, E. Fogarassy, P. Siffert: Photoabsorption of BCl_3 gas under pulsed ArF excimer laser irradiation. Appl. Phys. A **50**, 317 (1990)

Slaoui A., E. Fogarassy, C.W. White, P. Siffert: Infrared characterization of UV laser-induced silicon oxide films. Appl. Phys. Lett. **53**, 1832 (1988)

Smith H.M., A.F. Turner: Vacuum deposited thin films using a ruby laser. Appl. Opt. **4**, 147 (1965)

Smoluchowski M.: Zur Theorie der Wärmeleitung in verdünnten Gasen und der dabei auftretenden Druckkräfte. Annal. Physik **35**, 983 (1911)

Sobolewski R., W. Xiong, W. Kula, J.R. Gavaler: Laser patterning of Y-Ba-Cu-O thin-film devices and circuits. Appl. Phys. Lett. **64**, 643 (1994)

Solanki R., U. Sudarsan, J.C. Johnson: Laser-induced homoepitaxy of GaP. Appl. Phys. Lett. **52**, 919 (1988)

Solanki R., W.H. Ritchie, G.J. Collins: Photodeposition of aluminum oxide and aluminum thin films. Appl. Phys. Lett. **43**, 454 (1983)

Solanki R., G.J. Collins: Laser-induced deposition of zinc oxide. Appl. Phys. Lett. **42**, 662 (1983)

Sona A.: Metallic materials processing: Cutting and drilling, in *Applied Laser Tooling*, ed. by O.D.D. Soares, M. Perez-Amor (Nijhoff, Dordrecht 1987) p.105

Spalding I.J.: Non-metallic materials processing: An introduction, in *Applied Laser Tooling*, ed. by O.D.D. Soares, M. Perez-Amor (Nijhoff, Dordrecht 1987) p.81

Spousta J., J. Perrière, A. Laurent, E. Fogarassy, B. Prevot, S. de Unamuno: Laser-induced modifications in a-C:H thin films. Appl. Surf. Sci. **69**, 242 (1993)

Srinivasan R.: Interaction of laser radiation with organic polymers, in *Laser Ablation*, ed. by J.C. Miller, Springer Ser. Mater. Sci., Vol.28 (Springer, Berlin, Heidelberg 1994) p.107

Srinivasan R.: Ablation of polyimide (kapton) films by pulsed (ns) ultraviolet and infrared (9.17 μm) lasers: A comparative study. Appl. Phys. A **56**, 417 (1993)

Srinivasan R., R.R. Hall, D.C. Allbee: Generation of electrically conducting features in polyimide (kapton) films with continuous wave, ultraviolet laser radiation. Appl. Phys. Lett. **63**, 3382 (1993)

Srinivasan R., B. Braren, K.G. Casey: Nature of "incubation pulses" in the ultraviolet laser ablation of polymethyl methacrylate. J. Appl. Phys. **68**, 1842 (1990*a*)

Srinivasan R., K.G. Casey, B. Braren, M. Yeh: The significance of a fluence threshold for UV laser ablation and etching of polymers. J. Appl. Phys. **67**, 1604 (1990*b*)

Srinivasan R., B. Braren: Ultraviolet laser ablation of organic polymers. Chem. Rev. **89**, 1303 (1989)

Srinivasan R., P.E. Dyer, B. Braren: Far-UV laser ablation of the cornea: Photoacoustic studies. Lasers in Surgery and Medicine **6**, 514 (1987)

Srinivasan R.: Ablation of polymers and biological tissue by UV lasers. Science **234**, 559 (1986)

Srinivasan R., B. Braren, R.W. Dreyfus, L. Hadel, D.E. Seeger: Mechanism of the UV laser ablation of polymethyl methacrylate at 193 and 248 nm: Laser-induced fluorescence analysis, chemical analysis, and doping studies. J. Opt. Soc. Am. B **3**, 785 (1986)

Stafast H.: Initial steps in the photochemical vapour deposition of amorphous silicon. Appl. Phys. A **45**, 93 (1988)

Stangl E., S. Proyer, B. Hellebrand, D. Bäuerle: Synthesis of RE-Ba-Sr-Cu-O by pulsed-laser deposition. Appl. Surf. Sci. (1996), to be published

Stangl E., S. Proyer, B. Hellebrand: Properties of laser-deposited $TmBaSrCu_3O_{7-\delta}$ thin films. Physica C **243**, 69 (1995)

Stangl E., B. Luk'yanchuk, H. Schieche, K. Piglmayer, S. Anisimov, D. Bäuerle: Dynamics of the vapor plume in laser materials ablation, in *Excimer Lasers*, ed. by L.D. Laude (Kluwer, Dordrecht 1994) p.79

Steen W.M.: *Laser Material Processing* (Springer, Berlin, Heidelberg 1991)

Steinfeld J.I. (ed.): *Laser-Induced Chemical Processes* (Plenum, New York 1981)

Stock D., H.-D. Geiler, K. Hehl: Theoretical evidence for opposite moving phase fronts during ultrafast solidification processes. Phys. Status Solidi (a) **87**, K115 (1985)

Strikovsky M.D., E.B. Klyuenkov, S.V. Gaponov, J. Schubert, C.A. Copetti: Crossed fluxes technique for pulsed laser deposition of smooth $YBa_2Cu_3O_{7-x}$ films and multilayers. Appl. Phys. Lett. **63**, 1146 (1993)

Stritzker B., A. Pospieszczyk, J.A. Tagle: Measurement of lattice temperature of silicon during pulsed laser annealing. Phys. Rev. Lett. **47**, 356 (1981)

Stuck R., E. Fogarassy, J.C. Muller, M. Hodeau, A. Wattieux, P. Siffert: Laser-induced diffusion by irradiation of silicon dipped into an organic solution of the dopant. Appl. Phys. Lett. **38**, 715 (1981)

Sudarsan U., N.W. Cody, T. Dosluoglu, R. Solanki: Excimer laser assisted selective epitaxy of GaP. Appl. Phys. A **50**, 325 (1990)

Sugii T., T. Ito, H. Ishikawa: Low-temperature fabrication of silicon nitride films by ArF excimer laser irradiation. Appl. Phys. A **46**, 249 (1988)

Sugioka K., S. Wada, H. Tashiro, K. Toyoda, Y. Ohnuma, A. Nakamura: Multiwavelength excitation by vacuum-UV beams coupled with fourth harmonics of a Q-switched Nd:YAG laser for high-quality ablation of fused quartz. Appl. Phys. Lett. **67**, 2789 (1995)

Sugioka K., S. Wada, H. Tashiro, K. Toyoda, A. Nakamura: Novel ablation of fused quartz by preirradiation of vacuum-UV laser beams followed by fourth harmonics irradiation of Nd:YAG laser. Appl. Phys. Lett. **65**, 1510 (1994)

Sugioka K., K. Toyoda: Generation mechanism and thermal stability of high carrier concentrations by KrF-excimer-laser doping of Si into GaAs. Appl. Phys. A **59**, 233 (1994)

Sugioka K., S. Wada, A. Tsunemi, T. Sakai, H. Takai, H. Moriwaki, A. Nakamura, H. Tashiro, K. Toyoda: Micropatterning of quartz substrates by multi-wavelength vacuum-UV laser ablation. Jpn. J. Appl. Phys. **32**, 6185 (1993)

Sugioka K., K. Toyoda: Selective deposition of Au films on GaAs by projection-patterned excimer laser doping combined with electroless plating. Appl. Phys. A **54**, 380 (1992)

Sugioka K., J.F. Fan, K. Kita, S. Tanaka, K. Toyoda: Selective etching of Al_2O_3 on GaAs using excimer lasers. Jpn.J. Appl. Phys. **30**, No.11B, 3182 (1991)

Sugioka K., K. Toyoda: Self-aligned microfabrication of metal-semiconductor contacts by projection-patterned excimer laser doping. Jpn.J. Appl. Phys. **29**, 2255 (1990)
Sugioka K., K. Toyoda, K. Tachi, M. Otsuka: Formation of p-type layer by KrF excimer laser doping of carbon into GaAs in CH_4 gas ambient. Appl. Phys. A **49**, 723 (1989)
Suits J.C., R.H. Geiss, C.J. Lin, D. Rugar, A.E. Bell: Lorentz microscopy of micronsized laser-written magnetic domains in TbFe. Appl. Phys. Lett. **49**, 419 (1986)
Sumiyoshi T., Y. Ninomiya, H. Ogasawara, M. Obara, H. Tanaka: Efficient ablation of organic polymers polyether sulphone and polyether ether ketone by a TEA CO_2 laser with high perforrnation ability. Appl. Phys. A **58**, 475 (1994)
Sutcliffe E., R. Srinivasan: Dynamics of UV laser ablation of organic polymer surfaces. J. Appl. Phys. **60**, 3315 (1986)
Suzaki Y., A. Tachibana: Measurement of the μm sized radius of Gaussian laser beam using the scanning knife-edge. Appl. Opt. **14**, 2809 (1975)
Suzuki N., C. Anayama, K. Masu, K. Tsubouchi, N. Mikoshiba: Pyrolysis and photolysis of trimethylaluminum. Jpn. J. Appl. Phys. **25**, 1236 (1986)
Svantesson K.G., N.G. Nilsson: Determination of the absorption and the free carrier distribution in silicon at high level photogeneration at 1.06 μm and 294 K. Phys. Scr. **18**, 405 (1978)
Svorcik V., V. Rybka, V. Myslik: Laser-stimulated etching of n-type semiconductors Chem. Phys. Lett. **144**, 548 (1988)
Swanson J.R., C.M. Friend, Y.J. Chabal: Laser-assisted deposition of iron o Si(111)-(7×7): The mechanism and energetics of $Fe(CO)_5$ decomposition. J. Chem Phys. **87**, 5028 (1987)
Sytov I.P.: Model of laser-induced chemical etching of silicon in chlorine atmospher Appl. Phys. A **53**, 372 (1992); Estimation of the capabilities of maskless micropa terning by laser-induced chemical etching. Appl. Phys. A **61**, 75 (1995)
Szörényi T., P. González, E. Garcia, J. Pou, D. Fernández, J. Serra, B. Léon, N Pérez-Amor: Tailoring silicon oxide film properties by tuning the laser beam-t substrate distance in ArF laser-induced chemical vapor deposition. Thin Solid Filr **241**, 80 (1994)
Szörényi T., P. González, M.D. Fernández, J. Pou, B. Léon, M. Pérez-Amor: Silic oxide films deposited by excimer laser chemical vapor deposition. Thin Solid Fili **193/194**, 619 (1990)
Szörényi T., K. Piglmayer, G.Q. Zhang, D. Bäuerle: Lateral growth rates in laser CV of tungsten microstructures. Surf. Sci. **202**, 442 (1988)

Tabata H., O. Murata, T. Kawai, S. Kawai, M. Okuyama: c-axis preferred orientat of laser ablated epitaxial $PbTiO_3$ films and their electrical properties. Appl. Ph Lett. **64**, 428 (1994)
Tabata H., T. Kawai, M. Kanai, O. Murata, S. Kawai: Formation of the high phase of the superconducting Bi-Pb-Sr-Ca-Cu-O thin film by the laser ablat method. Jpn. J. Appl. Phys. **28**, L430 (1989)
Tachibana H., A. Nakaue, Y. Kawate: Deposition of amorphous carbon films by la induced CVD. MRS Proc. **101**, 367 (1988)
Takahashi K., M. Konagai: *Amorphous Silicon Solar Cells* (Academic, London 19;
Takai M., S. Nagatomo, H. Kohda, C. Yada, H. Sandaiji, F. Takeya: Laser chem processing of magnetic materials for recording-head application. Appl. Phys. A 359 (1994)

Takai M., Y.F. Lu, T. Koizumi, S. Namba, S. Nagatomo: Thermo-chemical dry etching of single-crystal ferrite by laser irradiation in CCl_4 gas atmosphere. Appl. Phys. A **46**, 197 (1988*a*)

Takai M., J. Tsuchimoto, J. Tokuda, H. Nakai, K. Gamo, S. Namba: Laser-induced thermochemical maskless-etching of III-V compound semiconductors in chloride gas atmosphere. Appl. Phys. A **45**, 305 (1988*b*)

Tam A.C., W.P. Leung, W. Zapka: Laser cleaning techniques for the removal of small surface particulates, in *Particles on Surfaces*, ed. by K.L. Mittal (Dekker, New York 1995) p.405

Tam A.C., W.P. Leung, D. Krajnovich: Excimer laser ablation of ferrite ceramics, in *Laser Ablation – Mechanisms and Applications*, ed. by J.C. Miller, R.F. Haglund. *Lecture Notes Phys.* **389**, 260 (Springer, Berlin, Heidelberg 1991)

Tanaka H., H. Tabata, T. Kawai, S. Kawai: Dominant factors for formation of perovskite $PbTiO_3$ films using excimer laser ablation. Jpn. J. Appl. Phys. **33**, L451 (1994)

Tanaka T., K. Deguchi, M. Hirose: Initial stage of laser-induced selective chemical vapor deposition of silicon. Jpn. J. Appl. Phys. **26**, 2057 (1987)

Tang H., I.P. Herman: Raman microprobe scattering of solid silicon and germanium at the melting temperature. Phys. Rev. B **43**, 2299 (1991)

Tang H., I.P. Herman: Laser-induced and room temperature etching of copper films by chlorine with analysis by Raman spectroscopy. J. Vac. Sci. Technol. A **8**, 1608 (1990)

Tavitian V., C.J. Kiely, D.B. Geohegan, J.G. Eden: Epitaxial growth of Ge films on GaAs (285-415°C) by laser photochemical vapor deposition. Appl. Phys. Lett. **52**, 1710 (1988)

Tokuyama M., Y. Enomoto: Dynamics of crossover phenomenon in phase-separating systems, Phys. Rev. Lett. **69**, 312 (1992)

Tonneau D., M. Cuniot, J.M. Laroche, H. Orain, J.F. Rommeluere, J.E. Bouree: Electrical characterization of laser-induced deposits of aluminium on gallium arsenide. Thin Solid Films **241**, 47 (1994)

Tonneau D., G. Auvert, Y. Pauleau: Deposition of nickel microstructures by CO_2 laser-assisted decomposition of nickel tetracarbonyl. J. Appl. Phys. **66**, 165 (1989)

Tonneau D., G. Auvert, Y. Pauleau: CO_2-laser-induced chemical vapour deposition of polycrystalline silicon from silane. Thin Solid Films **155**, 75 (1987)

Tonneau D., G. Auvert, Y. Pauleau: Absorption and decomposition of silane under infrared laser irradiation. J. Chem. Phys. **103**, 353 (1986)

Toth Z., T. Szörenyi, A.L. Toth: Ar^+ laser-induced forward transfer (LIFT): A novel method for micrometer-size surface patterning. Appl. Surf. Sci. **69**, 317 (1993)

Toth Z., P. Kargl, C. Grivas, K. Piglmayer, T. Szörenyi, D. Bäuerle: LCVD of tungsten microstructures on quartz. Appl. Phys. **54**, 189 (1992)

Toulemonde M., S. Unamuno, R. Heddache, M.O. Lampert, M. Hage-Ali, P. Siffert: Time-resolved reflectivity and melting depth measurements using pulsed ruby laser on silicon. Appl. Phys. A **36**, 31 (1985)

Trajanovic Z., L. Senapati, R.P. Sharma, T. Venkatesan: Stoichiometry and thickness variation of $YBa_2Cu_3O_{7-x}$ in off-axis pulsed laser deposition. Appl. Phys. Lett. **66**, 2418 (1995)

Treyz G.V., R. Beach, R.M. Osgood Jr.: Rapid direct writing of high-aspect ratio trenches in silicon: Process physics. J. Vac. Sci. Technol. B **6**, 37 (1988)

Tsao J.Y., D.J. Ehrlich: Patterned photonucleation of chemical vapor deposition of Al by UV-laser photodeposition. Appl. Phys. Lett. **45**, 617 (1984)

Tyndall G.W., C.R. Moylan: Laser-induced etching of titanium by Br_2 and CCl_3Br at 248 nm. Appl. Phys. A **50**, 609 (1990)

Ujihara K.: Reflectivity of metals at high temperatures. J. Appl. Phys. **43**, 2376 (1972)

Ulmer G., B. Hasselberger, H.-G. Busmann, E.E.B. Campbell: Excimer laser ablation of polyimide. Appl. Surf. Sci. **46**, 272 (1990)

Ursu I., I.N. Mihailescu, L.C. Nistor, V.S. Teodorescu, A.M. Prokhorov, V.I. Konov, S.A. Uglov: Zirconium and titanium nitridation by repeated action of a breakdown plasma induced in nitrogen as a result of microsecond-pulsed TEA CO_2 laser irradiation. Appl. Opt. **25**, 2725 (1986a)

Ursu I., I.N. Mihailescu, L. Nanu, L.C. Nistor, M.Popescu, V.S. Teodorescu, A.M. Prokhorov, V.I. Konov, S.A. Uglov, V.G. Ralchenko: Nitridation of Ti and Zr by multi-pulse TEA CO_2 laser irradiation in liquid nitrogen. J. Phys. D **19**, 1183 (1986b)

Ursu I., I.N. Mihailescu, L.C. Nistor, M. Popescu, V.S. Teodorescu, A.M. Prokhorov, V.G. Ralchenko, V.I. Konov: Mechanism of surface compound formations by cw CO_2 laser irradiation of zirconium samples in air. J. Appl. Phys. **59**, 668 (1986c)

Ursu I., I.N. Mihailescu, I. Gutu, A. Hening, T. Julea, L.C. Nistor, M. Popescu, V.S. Teodorescu, A.M. Prokhorov, V.I. Konov, V.G. Ralchenko: Surface nitridation of zirconium and hafnium by powerful CW CO_2 laser irradiation in air. Appl. Opt. **25**, 2720 (1986d)

Uwasawa K., F. Ishihara, J. Wada, S. Matsumoto: Fabrication of a Si:H/a-$Al_{1-x}O_x$ superlattice by excimer laser MOCVD and its properties, MRS Proc. **201**, 147 (1991)

Vase P., S. Yueqiang, T. Freltoft: Deposition, characterization, and laser ablation patterning of YBCO thin films. Appl. Surf. Sci. **46**, 61 (1990)

Vedavarz A., K. Mitra, S. Kumar: Hyperbolic temperature profiles for laser surface interactions. J. Appl. Phys. **76**, 5014 (1994)

Veen G.N.A. van, T.S. Baller, J. Dieleman: A time-of-flight study on the nanosecond laser induced etching of Cu with Cl_2 at 308 nm. Appl. Phys. A **47**, 183 (1988)

Veldkamp W.B.: Laser beam profile shaping with interlaced binary diffraction gratings. Appl. Opt. **21**, 3209 (1982)

Venkatesan T., A. Inam, B. Dutta, R. Ramesh, M.S. Hedge, X.D. Wu, L. Nazar, C.C. Chang, J.B. Barner, D.M. Hwang, C.T. Rogers: Epitaxial $Y_1Ba_2Cu_3O_{7-y}$/$Y_{1-x}Pr_xBa_2Cu_3O_{7-y}$ heterostructures. Appl. Phys. Lett. **56**, 391 (1990)

Venkatesan T., C.C. Chang, D. Dijkkamp, S.B. Ogale, E.W. Chase, L.A. Farrow, D.M. Hwang, P.F. Miceli, S.A. Schwarz, J.M. Tarascon, X.D. Wu, A. Inam: Substrate effects on the properties of Y-Ba-Cu-O superconducting films prepared by laser deposition. J. Appl. Phys. **63**, 4591 (1988)

Ventzek P.L.G., R.M. Gilgenbach, D.M. Heffelfinger, J.A. Sell: Laser-beam deflection measurements and modeling of pulsed laser ablation rate and near-surface plume densities in vacuum. J. Appl. Phys. **70**, 587 (1991)

Ventzek P.L.G., R.M. Gilgenbach, J.A. Sell, D.M. Heffelfinger: Schlieren measurements of the hydrodynamics of excimer laser ablation of polymers in atmospheric pressure gas. J. Appl. Phys. **68**, 965 (1990)

Verdaasdonk R.M., C. Borst: Ray tracing of optically modified fiber tips II. Laser scalpels. Appl. Opt. **30**, 2172 (1991)

Vicanek M., A. Rosch, F. Piron, G. Simon: Thermal deformation of a solid surface under laser irradiation. Appl. Phys. A **59**, 407 (1994)

Viret M., J.F. Lawler, J.G. Lunney: Synthesis of BiSrCaCuO thin films by in situ and ex situ pulsed laser deposition. Supercond. Sci. Technol. **6**, 490 (1993)

Vogel A., R. Engelhardt, U. Behnle, U. Parlitz: Minimization of cavitation effects in pulsed laser ablation – illustrated on laser angioplasty. Appl. Phys. B **62**, 173 (1996)

Volodin B.A., S.V. Gaponov, E.B. Klyeukov, B.A. Nesterov, N.N. Salashchenko: Influence of pulsed heating of substrates on oriented growth of films by laser evaporation. Sov. Phys. – Semicond. **15**, 688 (1981)

Vorobyev A.Y., V.A. Petrov, V.E. Titov, A.P. Chernyshov: Experimental investigation of the refractory oxides vapor condensation during the surface heating of the target in air by laser radiation. Teplophys. High Temperat. **29**, 981 (1991)

Wada S., H. Tashiro, K. Toyoda: Direct photoetching of teflon films using a VUV anti-Stokes Raman laser. RIKEN Rev. No.1 (April 1993) p.37

Wagner D., D. Bäuerle, F. Schwabl, B. Dorner, H. Kraxenberger: Soft modes in semiconducting $SrTiO_3$ I. Zone boundary mode. Z. Physik B **37**, 317 (1980)

Walkup R.E., J.M. Jasinski, R.W. Dreyfus: Studies of excimer laser ablation of solids using a Michelson interferometer. Appl. Phys. Lett. **48**, 1690 (1986)

Wallenberger F.T., P.C. Nordine, M. Boman: Inorganic fibers and microstructures directly from the vapor phase. Composites Science and Technology **51**, 193 (1994)

Wallraff G.M., R.D. Allen, W.D. Hinsberg, C.F. Larson, R.D. Johnson, R. DiPietro, G. Breyta, N. Hacker, R.R. Kunz: Single-layer chemically amplified photoresists for 193-nm lithography. J. Vac.Sci. Technol. B **11**, 2783 (1993)

Watanabe S., S. Ueda, N. Nakazato, M. Takai: Photolytic etching of polycrystalline silicon in SF_6 atmosphere. Jpn. J. Appl. Phys. **25**, L881 (1986)

Watanabe Y., M. Tanamura, H. Asami, Y. Matsumoto, Y. Seki, S. Matsumoto: Macroscopically and atomically smooth $YBa_2Cu_3O_7$ film grown by reactive rapid sequential pulse laser deposition. Physica C **235-240**, 579 (1994)

Watanabe Y.: Reactive rapid sequential pulse laser deposition of $YBa_2Cu_3O_7$: A candidate to eliminate particulate formation. Appl. Phys. Lett. **64**, 1295 (1994)

Wautelet M., F. Hanus: Thickness-dependent kinetics of laser-induced oxidation of thin copper films. Appl. Phys. Lett. **58**, 1355 (1991)

Wautelet M.: Laser-assisted reaction of metals with oxygen. Appl. Phys. A **50**, 131 (1990)

Wautelet M., V. Miroir, Z.H. Huang: CW laser-assisted oxidation of thin Cd, In, Sn, and Zn films in air. Appl. Phys. A **50**, 311 (1990)

Wautelet M.: Maximum lateral dimensions of laser-induced oxidation of thin tellurium films: A negative feedback effect. J. Appl. Phys. **65**, 4033 (1989)

Wautelet M., P. Quenon, A. Jadin: Origin of laser-assisted and doping-assisted phenomena in semiconductors. Semicond. Sci. Technol. **3**, 54 (1988)

Weerasinghe V.M., W.M. Steen: Laser cladding with pneumatic powder delivery, in *Proc. 4th Int'l Conf. on Lasers in Materials Processing*, Los Angeles 1983, ed. by E.A. Metzbower (Publ. ASM, Materials Park, OH 1984) p.166

Wel H. van der, J. Lub: Surface modification of polymethylmethacrylate by UV light as studied by TOF-SIMS. Surface and Interface Analysis **20**, 373 (1993)

Westberg H., M. Boman, S. Johansson, J. Schweitz: Free-standing silicon microstructures fabricated by laser chemical processing. J. Appl. Phys. **73**, 7864 (1993)
White C.W., S.R. Wilson, B.R. Appleton, F.W. Young: Supersaturated substitutional alloys formed by ion implantation and pulsed laser annealing of group-III and group-V dopants in silicon. J. Appl. Phys. **51**, 738 (1980)
Widmer M., H. v.d. Bergh: Laser-induced pyrolytic deposition of Cu from (hexafluoroacetylacetonate) (trimethylvinylsilane) copper. J. Appl. Phys. **77**, 5464 (1995)
Wiener-Avnear E., G.L. Kerber, J.E. McFall, J.W. Spargo, A.G. Toth: In situ laser deposition of $Y_1Ba_2Cu_3O_{7-x}$ high T_c superconducting thin films with $SrTiO_3$ underlayers. Appl. Phys. Lett. **56**, 1802 (1990)
Williams J.G.: *Stress Analysis of Polymers*, 2nd edn. (Ellis Horwood, Chichester 1980)
Willner A.E., D.V. Podlesnik, R.M. Osgood: 600 µm/min laser-induced nonthermal etching of GaAs in an HF solution. Electron. Lett. **26**, 568 (1990)
Winters H.F., I.C. Plumb: Etching reactions for silicon with F atoms: Product distributions and ion enhancement mechanisms. J. Vac. Sci. Technol. B **9**, 197 (1991)
Winters H.F., D. Haarer: Influence of doping on the etching of Si(111). Phys. Rev. B **36**, 6613 (1987)
Wissenbach K., A. Gillner, F. Dausinger: Umwandlungshärten mit CO_2-Laserstrahlung. Laser und Optolelektronik **3**, 291 (1985)
Witanachchi S., K. Ahmed, P. Sakthivel, P. Mukherjee: Dual-laser ablation for particulate-free film growth. Appl. Phys. Lett. **66**, 1469 (1995)
Witanachchi S., H.S. Kwok, X.W. Wang, D.T. Shaw: Deposition of superconducting Y-Ba-Cu-O Films at 400°C without post-annealing. Appl. Phys. Lett. **53**, 234 (1988)
Wolff-Rottke B., J. Ihlemann, H. Schmidt, A. Scholl: Influence of the laser-spot diameter on photo-ablation rates. Appl. Phys. A **60**, 13 (1995)
Wood R.F., G.A. Geist: Modeling of nonequilibrium melting and solidification in laser-irradiated materials. Phys. Rev. B **34**, 2606 (1986)
Wright W.W.: Application of thermal methods to the study of the degradation of polyimides, in *Developments in Polymer Degradation III*, ed. by N. Grassie (Applied Science, London 1981) p.1
Wu X.D., R.E. Muenchausen, S. Foltyn, R.C. Estler, R.C. Dye, C. Flamme, N.S. Nogar, A.R. Garcia, J. Martin, J. Tesmer: Effect of deposition rate on properties of $YBa_2Cu_3O_{7-\delta}$ superconducting thin films. Appl. Phys. Lett. **56**, 1481 (1990a)
Wu X.D., X.X. Xi, Q. Li, A. Inam, B. Dutta, L. DiDomenico, C. Weiss, J.A. Martinez, B.J. Wilkens, S.A. Schwarz, J.B. Barner, C.C. Chang, L. Nazar, T. Venkatesan: Superlattices of Y-Ba-Cu-O/Y_y-Pr_{1-y}-Ba-Cu-O grown by pulsed laser deposition. Appl. Phys. Lett. **56**, 400 (1990b)
Wu X.D., B. Dutta, M.S. Hedge, A. Inam, T. Venkatesan, E.W. Chase, C.C. Chang, R. Howard: Optical spectroscopy: An in situ diagnostic for pulsed laser deposition of high T_c superconducting thin films. Appl. Phys. Lett. **54**, 179 (1989)
Wu Z.: Laser-induced ion emission from Si and Ge surfaces, in *Desorption Induced by Electronic Transitions, DIET IV*, ed. by G. Betz, P. Varga, Springer Ser. Surf. Sci., Vol.19 (Springer, Berlin, Heidelberg 1990) p.163

Yabe A., H. Niino: Surface modification of polymers with excimer lasers and its applications, in *Laser Ablation of Electronic Materials – Basic Mechanisms and Applications*, ed. by E. Fogarassy, S. Lazare (Elsevier, Amsterdam 1992) p.199

Yamada A., A. Satoh, M. Konagai, K. Takahashi: Low-temperature (600-650 °C) silicon epitaxy by excimer laser-assisted chemical vapor deposition. J. Appl. Phys. **65**, 4268 (1989)

Yavas O., P. Leiderer, H.K. Park, C.P. Grigoropoulos, C.C. Poon, W.P. Leung, N. Do, A.C. Tam: Optical and acoustic study of nucleation and growth of bubbles at a liquid-solid interface induced by nanosecond-pulsed-laser heating. Appl. Phys. A **58**, 407 (1994)

Yeh M.H., Y.C. Liu, K.S. Liu, I.N. Lin, J.Y.M. Lee, H.F. Cheng: Electrical characteristics of barium titanate films prepared by laser ablation. J. Appl. Phys. **74**, 2143 (1993)

Yilmaz S., T. Venkatesan, R. Gerhard-Multhaupt: Pulsed laser deposition of stoichiometric potassium-tantalate-niobate films from segmented evaporation targets. Appl. Phys. Lett. **58**, 2479 (1991)

Ying Q.Y., H.S. Kim, D.T. Shaw, H.S. Kwok: Nature of in situ superconducting film formation. Appl. Phys. Lett. **55**, 1041 (1989)

Yoffa E.J.: Dynamics of dense laser-induced plasmas. Phys. Rev. B **21**, 2415 (1980)

Yoshida A., H. Tamura, H. Takauchi, T. Imamura, S. Hasuo: Dielectric-base transistor using $YBa_2Cu_3O_{7-x}/NdGaO_3/SrTiO_3$ heterostructures, J. Appl. Phys. **71**, 5284 (1992)

Young E.M.: Electron-active silicon oxidation. Appl. Phys. A **47**, 259 (1988)

Young E.M., W.A. Tiller: Photon enhanced oxidation of silicon. Appl. Phys. Lett. **42**, 63 (1983)

Zarnani H., H. Demiryont, G.J. Collins: Optical properties of UV laser photolytic deposition of hydrogenated amorphous silicon. J. Appl. Phys. **60**, 2523 (1986)

Zeiger H.J., D.J. Ehrlich, J.Y. Tsao: Transport and kinetics, in *Laser Microfabrication – Thin Film Processes and Lithography*, ed. by D.J. Ehrlich, J.Y. Tsao (Academic, London 1989) p.285

Zeldovich Y.B., Yu.P. Raizer: In *Physics of Shock Waves and High-Temperature Hydrodynamic Phenomena*, ed. by W.D. Hayes, R.F. Probstein (Academic, London 1966) Vol.1, p.93

Zhang G.Q., T. Szörenyi, D. Bäuerle: Kr^+ laser-induced chemical vapor deposition of W. J. Appl. Phys. **62**, 673 (1987)

Zhang J.Y., H. Esrom, U. Kogelschatz, G. Emig: Large area photochemical dry etching of polyimide with excimer UV lamps. Appl. Surf. Sci. **69**, 299 (1993)

Zhang S.K., K. Sugioka, J. Fan, K. Toyoda, S.C. Zou: Studies on excimer laser doping of GaAs using sulphur adsorbate as dopant source. Appl. Phys. A **58**, 191 (1994)

Zhang X.C., J.S. Shor, M.N. Ruberto, M.T. Schmidt, R.M Osgood: In Laser Electrochemical Etching of SiC, Proc. 12th State of the Arted Program on Compound Semiconductors (SOTAPOCS XII), Montreal, Quebec (1990) Vol.90-15, p.271

Zhang Y., M. Stuke: Synchrotron processing of thin palladium acetate spin-on films for metallic surface patterning. Chemtronics **4**, 212 (1989)

Zheng J.P., H.S. Kwok: Preparation of indium tin oxide films at room temperature by pulsed laser deposition. Thin Solid Films **232**, 99 (1993)

Zhu S., D.H. Lowndes, B.C. Chakoumakos, S.J. Pennycook, X.Y. Zheng, R.J. Warmack: Growth mechanisms and superconductivity of ultrathin $Y_1Ba_2Cu_3O_{7-x}$ epitaxial films on (001) MgO substrates. Appl. Phys. Lett. **62**, 3363 (1993)

Zhu X.-Y., M. Wolf, J.M. White: Laser-assisted nitridation of GaAs: Mechanisms. J. Vac. Sci. Technol. A **11**, 838 (1993)

Ziman J.M.: *Principles of the Theory of Solids* (Cambridge Univ. Press, London 1972)

Zinck J.J., P.D. Brewer, J.E. Jensen, G.L. Olson, L.W. Tutt: Excimer laser-assisted metalorganic vapor phase epitaxy of CdTe on GaAs. Appl. Phys. Lett. **52**, 1434 (1988)

Zuhoski S.P., K.P. Killeen, R.M. Biefeld: Photolytic deposition of InSb films. MRS Proc. **101**, 313 (1988)

Zweig A.D., T.F. Deutsch: Shock waves generated by confined XeCl excimer laser ablation of polyimide. Appl. Phys. B **54**, 76 (1992)

Zweig A.D.: A thermo-mechanical model for laser ablation. J. Appl. Phys. **70**, 1684 (1991)

Subject Index

Page numbers printed in italics refer to tables or listings of data or materials.

Ablation   191, 209, *558*
– biological materials   193, 226, *559*
– congruent   399
– debris   207, 506
– defect-related   224
– fragments   532
– incongruent   400
– incubation   197, 224
– influence of ambient medium   207, 533
– influence of laser spot size   202
– inorganic materials   *558*
– interpolation formula   222
– liquid-phase expulsion   169
– material damage   204
– models   209
– non-stationary   221
– organic polymers   193, 201, 207, 221, *559*
– photochemical   209, 223
– photomechanical   226
– photophysical   209, 217
– pulsed-laser   191
– rate   198
– stationary   219
– surface patterning   192
– thermal   209, 213
– thermomechanical   226
– threshold   196
– time-resolved   203
– velocity   176, 212
Absorption, see Optical
Accomodation coefficient   66
Acoustic monitors   520
Activation energy   41
Adhesion   472, 477
Adsorbates   370
– adsorbed-layer photolysis   381
– BET   379
– chemical (chemisorption)   371
– coverage   371
– deposition from   375
– desorption   373
– dissociative   372
– doping from   436
– influence of laser light   374
– Langmuir equation   376
– molecular   372
– multilayer coverages   379
– physical (physisorption)   371
– vibrations   37
Affinity level   253
Alloying   443
Analysis
– plasma/vapor plumes   529
– processed surfaces   539
– species   526
– thin films   539
Angioplasty   194, *559*
Annealing   427
Applications
– circuit repair   335
– contacts   334
– gratings   270
– junctions   439
– LIGA   *194*
– link cutting   *194*
– lithography   *194*, 478
– mask repair   335
– medical   *194*, 537, *559*
– micropackaging   *194*
– non-planar fabrication   336
– optical waveguides   133, 275, 449
– scribing   5
– solar cells   354
– surface cleaning   *194*
– thin-film transistors (TFT)   429
– 3-D structures   336
– trimming   *194*
– tunnel barriers   459
– via holes   270
– wirestripping   *194*

Arrhenius law 41
Aspect ratio 174
Auger process 35
Autonomous systems 481
Avalanche ionization 36, 182

Ballistic approximation 235
Band bending 253, 271
Beam
– deflection measurements 204, 520
– shapes 101, see also Laser beam
Beer's law 20, 223
Bessel function *550*
BET 379
Biological tissues 193
Bistabilities 130, 500
Blackbody radiance 523
Bleaching 220
Bond-breaking 223
Bouguer-Lambert-Beer law 20
Brewster effect 426
Bubble
– collapse 226, 538
– formation 181
Buffer layers 418
Butler-Volmer equation 394

Cabrera-Mott theory 252, 451
Capillary waves 409, 479
Carbonization 475
Catalytic effects 37
Cavitation 181, 226, 538
Center temperature rise 102, 106, 124
Ceramics
– ablation 192
– etching 247
– metallization 467
– oxidation 465
– reduction 467
Charge transfer 252
Chemical
– activation energy 41
– reactor 491
– relaxation 18
Chemisorption 241, 371
Chemophoretic forces 62, 69
Cladding 6, 442
Clausius-Clapeyron relation 63
Clusters 65, 532, see also Nucleation
Coalescence 71
Cold melting 435
Collisions 31
Color centers 197, 225
Compound formation 444

Concentration battery 392
Conductivity
– electrical, see Electrical
– thermal 19, 50, 110, 113, 161, *573*
Confinement of excitations 82, 260, 274
Conical structures 506
Contact front 402, 536
Convection
– chemical 55
– forced 94
– free 94, 147
– laminar 165
– Marangoni 165
– velocity 148
Coupling of fluxes 309
Coverage 371, 376
Crank transform 22
Crystallization
– amorphous films 427
– explosive 494
– fibers 304, 447
Cutting 187
CVD 10, 338, 348, 355, *358*, 368, 382

Debris 207, 506
Debye length 453
Degress in freedom 481, 491
Dember effect 273, 392
Dendrites 72, 472
Deposition of microstructures *561*, *565*, *571*
– adsorbed layers 378
– applications 334
– bistabilities 500
– direct writing 77, 317, 381, 476
– electrical properties 320
– electrochemical 387
– electroless 389
– fibers 304
– gas-phase (LCVD) 281, 304, 317
– growth rates 43
– hybrid 332
– influence of thermal diffusion 309
– kinetics 39, 307
– liquid-phase 387, *565*, 587
– modelling pyrolytic LCVD 287, 315, 321
– morphology, microstructure 282, 306, 317
– periodic structures 495
– photolytic (photochemical) 41, 298
– photophysical 332
– precursor molecules 281
– process limitations 303
– projection 382
– pyrolytic (photothermal) 43, 282
– rates 43, 517

Subject Index 643

- resolution 77, 82, 303, 319, 383
- rods (fibers) 304
- single crystals 306
- solid-phase 442, 476
- spots 282
- techniques of measurement 305, 517
- temperature distributions 291, 305, 311, 321
- temperature measurements 305, 523
- transport of species 45, 309
Deposition of thin films 563, 565, 566
- adsorbed layers 375, 565
- buffer layers 418
- cladding 442
- electrical properties 350, 416, 470
- electrochemical 393
- electroless 389
- epitaxial 82, 354, 359, 410
- evaporation 397
- heterostructures 366, 412, 419
- high temperature superconductors 414, 569
- insulators 362, 363, 365, 413, 566
- kinetics 343
- laser-ALE 385
- laser-CVD (LCVD) 338, 563
- laser-MBE (LMBE) 383, 402
- liquid-phase 387, 565
- metals 348, 411, 563, 565, 566
- metastable compounds 419
- modelling LCVD 339
- photosensitization 28, 355
- PLD 397, 566, see also Pulsed-laser deposition
- rates 517
- semiconductors 352, 358, 411, 566
- sputtering 397
- temperature distributions 339
- thickness profiles 343, 399
Dermatology 194
Desorption 351, 377
Diagnostic techniques 515
Diamond like carbon 412
Dielectric permittivity 20
Diffusion
- atomic/molecular 49
- binary 46
- equations 46
- flux 46
- interstitial 453
- length thermal 21
- liquid-phase 435
- solid-phase 374, 432
- substitutional 453
- surface 373
- thermal 50

Diffusivity 19, 573
Dimensionality of heat flow 21
Dimensionless variables 96
Direct writing 77
- deposition 317, 381, 476
- doping 440
- grooves 247, 254
- metallization 467
- oxidation 465
Dissociation 25, see also Excitation, Photochemistry
Doping 432
- applications 439
- diffusion length 433
- ion implantation 427
- liquid-phase 435
- local 440
- profiles 428, 433
- semiconductors 436
- sheet 436
- solid-phase 432
- stainless steel 444
Drift velocity 53
Drilling 187
Droplets 65, 406
Dry etching, see also Etching
- insulators 244
- metals 241
- semiconductors 251
Dufour effect 45
Dulong-Petit law 158
Dynamic viscosity 165

Electrical properties
- conductivity of deposits 320
- doped surfaces 438
- pn junctions 439
- sheet resistance 438, 540
- surface metallized structures 468
- thin films 350, 416, 470
Electrochemical plating 393
Electrochemical (Nernst) potential 390
Electroless, see Deposition, Etching
Electromagnetic field enhancement 37
Electromotive force (EMF) 387
Electron-beam (EB) processing 10, 365, 367, 447
Electron-hole pairs 35, 252, 258, 455
Electronic excitation, see Excitation
Emissivity, see Optical
Energy
- activation 41
- balance 155, 175
- collisional transfer 31

Enthalpy 22, 41
– melting 152, *582*
– vaporization 173, *582*
Epitaxial growth 82, 354, 359, 410
Error function *551*
Etching 229, *559*
– atomic layers (EAL) 267
– concentration of radicals 234
– dark 251
– diffusive 242
– direct writing 254, 275
– dry 241, 251
– electrochemical 393
– electroless 271, 390
– gratings 270
– influence of crystal orientation 260
– influence of doping 260
– influence of reaction chamber 237
– inorganic insulators 244, 269
– metals 241, 249
– passivating 243
– photochemical 255
– precursor molecules 230
– projection 255, 267
– rates 241, 249, 255, 268, 517
– resolution 254, 260, 274
– semiconductors 251, 265
– spontaneous 241, 251
– transport of species 234
– via holes 270
– wet 249, 268
Excitation
– coherent 28
– confinement 82, 260, 274
– electronic
– – molecules 14, 26, 231, 299, 349
– – predissociation 27
– – surfaces 34, 255, 271
– mechanisms 13
– selective 25, 32
– sequential 28
– vibrational
– – molecules 29, 233, 375
– – quasi-continuum 30
– – relaxation times 32
Experimental aspects 73, 396, 398, 515, 526
Explosive crystallization 494
Exponential integral *552*
Extinction coefficient 20

Feedback 479, 486
Fermi level 253
Fibers 304, 447
– LCVD 304

– pedestal growth 447, *448*
– tensile strength 307
Fluxes
– coupling of 309
– diffusion 46
Focus 74
Franck-Condon principle 26
Frank-Kamenetsky expansion 41
Frenkel-Wilson law 157, 394

Gamma function *552*
Gas-phase
– condensation 408
– heating 339
– nucleation 62, 315
– recombination 239, 346
– transport 45, 309, 343
Gibbs free energy 63
Glassy alloys 444
Glazing 430
Grashof number 148
Green's function 96
Growth rate 43, see also Deposition

Hardening 425, 430
Heat
– affected zone (HAZ) 426
– flow dimensionality 21
– transfer coefficient 94
Heat equation 19, 23, 95
– attenuation function 92
– general solutions 91
– point source 21
– source term 19, 92
Heaviside function *553*
Heterogeneous reactions 39
Heterostructures
– laser-CVD 366
– pulsed-laser deposition 412, 419
Hydrophilic 476
Hysteresis 500

Increment 481, 503
Incubation 197, 224
Infinite slabs
– cutting speed 188
– interferences 128
– temperature distributions 124
Instabilities 479, see also Structure formation
– ablation 502
– description 480
– direct writing 496, 500
– evaporation 505

Subject Index

- hydrodynamic 507
- Kelvin-Helmholtz 166, 408, 507
- oxidation 493
- Rayleigh-Taylor 166, 409, 507
- spiral waves 507
- stress-related 507
- thermochemical 493
Interconnects 335, see also Applications
Interference 78, 128, 135, 270, 518
Internal conversion 27
Ion implantation 427
Ion probe measurements 404
Isotope separation 33

Jacobian Theta function *553*
Jahn-Teller distortions 225
Jet-plating 395

Kelvin-Helmholtz instabilities 166, 408, 507
Kerf width 188
Kinetics
- ablation of organic polymers 209
- adsorbate-adsorbent interfaces 370
- gas-solid interfaces 39, 51, 234, 252, 287, *308*
- liquid-solid interfaces 271, 387
- mass transport limited 40
- oxidation 450
- solid-solid interfaces 432, 450
Kirchhoff transform 21
Knudsen layer 173
Kune segment 227

Landau-Teller relation 34
Landau-Zener transition 27
Langmuir equation 376
Laplace formula 64
Laser
- ALE 385
- annealing 427
- commercial types 73, *572*
- CVD of microstructures 281, *561*
- CVD of thin films 338, *563*
- direct writing, see Direct writing
- effective power 73
- enhanced plating 387
- high-power 75
- induced etching 229
- induced flash evaporation (LIFE) 397
- induced forward transfer (LIFT) 420
- induced oxidation 450
- induced structural transformations 425
- induced vaporization 173
- machining 187

- MBE 383
- pulsed plasma chemistry 456
- recrystallization 427
- sputter deposition (LSD) 397
- supported combustion waves 184
- supported detonation waves 185
- surface hardening 425, 430
Laser beam
- characterization 515
- deflection 204, 520
- divergence 75
- dwell time 5
- focus length 75
- focus width 74
- Gaussian 73
- homogenization 515
- intensity 73
- power 73, 515, *572*
- profile 515
- pulse lengths *572*
- pulse shapes 98
- Rayleigh length 75
Latent time 204
Law of mass action 42
LCVD
- applications 334
- microstructures 281
- photolytic 41, 298
- photophysical (hybrid) 332
- pyrolytic 43, 282
- thin films 338
LIF 526
LIFT 420
LIGA 194
Liquid-phase
- deposition 387
- expulsion 169
- transport 435
Liquid-solid interface 152, 271, 387
Lithography 194, 478
Lithotripsy 537, *559*
LPPC 456
LSCW 184

Marangoni convection 165
Mass spectroscopy 527
Mass transport
- gases 45, 234, 309, 343
- liquids 45, 164, 394, 435
- solids 432
Mean free path of molecules 150
Medical applications 194, 537, *559*
Melting 152
- depth 154

Melting (contd.)
– energy balance   155
– enthalpy   152, *582*
– Frenkel-Wilson law   157
– solidification   160
– Stefan problem   156
– surface   159
– temperature   *573*
– time   154
Metal-liquid interfaces   390
Metallization, see Deposition, Surface modification
Metastable materials   419, 444
Microbalance   518
Micromechanics   11, 191, 255
Microprocessing   77
Microsurgery   194, *559*
Mie theory   65
Monolayer coverage   372
Multilayer structures
– absorptivity   133
– LCVD   366
– PLD   419
Multiphoton excitation   28, see also Excitation

Nernst–Planck equation   394
Nernst potential   390
Neumann solutions   156
Nitridation   459, *570*
Nonlinear phenomena   15, 36, 84, 182, 479
Non-radiative waves   482
Normalized quantities   96, 212
Nucleation   62
– droplets within a laser beam   65
– laser-CVD   69
Nusselt number   149

Oleophilic   476
Ophthalmology   194, *559*
Optical
– absorption coefficient   20, 112, 121, *122*, *577*
– absorption cross section   27, 31, *583*
– absorptivity   128
– breakdown   36, 183
– deflection   526
– emissivity   95, 305, 521
– finite absorption   92
– penetration depth   92
– reflectivity   92, 114, 134, *577*
– refractive index   20
– spectroscopy   524, 526, 540
– surface absorption   93
– total reflection   275

– transmittivity   128
– waveguides   133, 449
Order parameters   481, 491
Ostwald ripening regime   65
Overheating   160
Overpotential   393
Oxidation   450, *570*
– instabilities   493
– mechanisms   451
– metals   452, 456
– native oxide   451
– pulsed-laser plasma chemistry   459
– semiconductors   460
– spontaneous   450
Oxide formation   362, see also Deposition
Oxide transformation   465

Parabolic approximation   103
Partial reaction orders   41
Particulates   207, 405
Passivation   274
PCVD   10, 338, 355, *358*, *365*, *367*, 368
Pearlite   426
Pedestal growth   447
Peel strength   473
Penetration depth   21, 92
Periodic structures   479
Permittivity   20
Persistent length   227
Photochemical processes   16
Photochemistry
– adsorbates   375
– alkyls   281, 300, 350
– carbonyls   281, 301, 350
– concentration of species   234
– gas-solid interfaces   255
– halides   231
– halogen compounds   232, 349
– heterogeneous   271, 392
– homogeneous   263, 348
– hydrides   281
– liquid-solid interfaces   271, 392
– NH_3 365
– N_2O 362, 417
– SF_6 233, 262
– silanes   352
Photodeposition, see Deposition
Photodesorption   374
Photoeffect   34, 390, 532
Photoelectric methods   520
Photoenhanced growth, see Deposition
Photolytic (photochemical) processes   16
Photophysical processes   17
Photopolymerization   449

Subject Index

Photosensitization 28, 355
Photothermal deflection 525
Physisorption 242, 371
Plasma
– CVD (PCVD) 10, 338, 355, *358*, *365*, *367*, 368
– optical properties 182
– oxidation 450
– plume 174, 403
– plume analysis 529
– plume expansion 533
– shielding 182, 185, 202
Plasmatron 185
Plating
– electrochemical 393
– electroless 389
– jet 395
– thermobattery 391
Point source 21
Poisson-Boltzmann equation 254
Poisson equation 452
Polymerization 449
Poynting vector 19
Prandtl number 148
Predissociation 27
Prenucleation 376
Processing
– large-area 80
– micro 77
– non-planar 12, 270, 336
– optimization 164
– planar 12, 334
Projection patterning 78, 192, 255, 381
Pulsed-laser ablation 191, see also Ablation
Pulsed-laser deposition (PLD) 397, *566*
– buffer layers 418
– energetic species 405
– experimental requirements 398
– heterostructures 412, 419
– high-temperature superconductors 414
– insulators 413
– metals 411
– metastable compounds 419
– particulates 405
– reactive/non-reactive 398
– semiconductors 411
– surface processes 402
– targets 401
– volume processes 402
Pulsed-laser evaporation (PLE) 397
Pulsed-laser plasma chemistry (PLPC) 455
Pulse shapes 98
Pyroelectric monitors 520
Pyrolytic (photothermal) processes 15
Pyrometry 305, 521

QCM 518
Quasi-continuum 30

Radiation pressure 179
Raman spectroscopy 524, 540
Rayleigh
– length 75
– number 148
– waves 485
Rayleigh-Taylor instabilities 166, 409, 507
Reaction
– adsorbed layers 375
– anodic 273
– cathodic 273
– chamber 79, 237
– enthalpy 41
– equimolecular 53
– heterogeneous 39, 51
– homogeneous 39, 58
– influence of gas-phase heating 56
– influence of scanning 56
– kinetics 39
– non-equimolecular 55
– order 41
– photochemical 43, 61
– photothermal 41
– rate 41, 54
Recoil pressure 178, 409
Recombination 239, 346
Recrystallization 427, 494
Redox equations 390
Reduction, see Surface modification
Reflectivity, see Optical
Refractive index 20
Relaxation
– chemical 18
– stresses 509
– times 15, 210
Reoxidation 450, 465
Resolution 77, 254, 274
– confinement of excitations 82, 260, 274
– deposition 85, 303, 319, 383
– etching 254, 260, 274
Reststrahl oscillator 112
RF/DC-sputtering *363*, 398
Ripples 482, *488*
Rods 304
Runaway thermal 121

Scanning knife-edge technique 515
Schlieren photography 536
Self-focusing 133
Self-organization 479
Self-trapped excitons (STE) 225

Semiconductor-liquid interfaces 268, 392
Sensitizer molecules 28, 355, 449
Shaping materials 187
Sharpening 114
Sheet resistance 474, 541
Shielding, see Plasma
Shock
– hardening 430
– waves 186, 226, 533
Silicides 444
Skin depth 93
Smoluchowski equation 54
Solidification 160
Soret effect 312
Space charge layer 259
Spatial confinement 82, see also Resolution
Specific heat 19, 161, *573*
Speckles 517
Spectroscopy
– mass 527
– optical 524, 526, 540
Standard electrode potential 390
Stefan-Maxwell equations 46
Stefan problem 156
Steric factor 41
Sticking coefficient 71, 375, 464
Stoichiometric coefficient 41
Streak photography 526, 536
Stresses 307, 509
Strong explosion model 533
Structural transformations 425
Structure formation 479, see also Instabilities
– autowaves 493
– circular waves 493
– coherent 479
– conical 506
– degrees of freedom 491
– description 480
– direct writing 496
– exothermal reactions 495
– explosive crystallization 494
– feedback 486
– nap-type 509
– non-coherent 479
– order parameters 481
– oscillations 496
– ripples 482, *488*
– spatio-temporal oscillations 490
– spiral waves 493
– star-like 508
– stress related 507
– technological aspects 511
– wall-type 509
– zero isoclines 492

Substrates
– cleaning 82, 194
– infinite slabs 124
– non-uniform 132
– pretreatment 81
– semi-infinite 106
Superdetonation 186
Superexcitation 31
Superheating 408
Surface
– absorption 93
– acoustic waves (SAW) 479
– adsorbed layers, see Adsorbates
– cleaning 82
– deformations 166
– electric field 252
– electromagnetic waves (SEW) 482
– energy 509
– excitation 34
– melting 152
– polaritons 486
– tension 165
Surface modification 467, *571*
– changes in crystallinity 473
– chemical 474
– doping 432, see also Doping
– metallization of oxides 467
– nitridation 459
– oxidation 450, see also Oxidation
– photochemical exchange of species 476
– polymers 472
– qualitative description 470
– reduction 467
– superconductors 469
Synthesis 445

Targets 401
Temperature
– jump 94, 150
– measurements 305, 521
– melting 152, *573*
– vaporization *573*
Temperature distributions
– ambient medium 144, 339
– analytical solutions 96
– characteristics 102
– deposits 293, 315, 321
– infinite slabs 124
– melting 152
– multilayer structures 143
– non-uniform materials 132
– pulsed irradiation 117
– scanned cw laser 115, 125
– semi-infinite substrates 106

- solidification   160
- thin films   137
- width   102
Thermal
- activation energy   41
- conductivity   573, see Conductivity
- desorption   373
- diffusion   46, 312
- diffusivity   19, 573
- processes   15
- runaway   121
Thermobattery   391
Thermocapillary effect   165
Thermophoretic forces   62, 69
Thin film formation, see Deposition
Thin-film transistors (TFT)   429
Threshold
- fluence for ablation   196
- optical breakdown   15, 183
Time of flight (TOF)   525, 528
Transformation hardening   425
Transmission measurements   518
Transmittivity   128, see also Optical
Turbulence   149

Undercooling   153

Vaporization   173, 573
- enthalpy   173, 582

- temperature   573
- velocity   176
Variable range hopping   475
Velocity
- ablation   176, 215
- arithmetic mean   373
- average   46
- convection   148
- drift   53
- drilling   189
- solidification   158, 161
- sound   182
- species   46
Vibrational excitation   29
Viscosity   148, 165

Wagner law   453
Wagner-Valency theory   454
Waveguides   133, 275, 449
Welding   167
Wet etching, see also Etching
- ferrites   248
- metals   249
- semiconductors   268
Wien's approximation   522

X-rays   187, 532

Druck: Saladruck, Berlin
Verarbeitung: Buchbinderei Lüderitz & Bauer, Berlin